# Mechanisms and Machine Science

Volume 108

This book series establishes a well-defined forum for monographs, edited Books, and proceedings on mechanical engineering with particular emphasis on MMS (Mechanism and Machine Science). The final goal is the publication of research that shows the development of mechanical engineering and particularly MMS in all technical aspects, even in very recent assessments. Published works share an approach by which technical details and formulation are discussed, and discuss modern formalisms with the aim to circulate research and technical achievements for use in professional, research, academic, and teaching activities.

This technical approach is an essential characteristic of the series. By discussing technical details and formulations in terms of modern formalisms, the possibility is created not only to show technical developments but also to explain achievements for technical teaching and research activity today and for the future.

The book series is intended to collect technical views on developments of the broad field of MMS in a unique frame that can be seen in its totality as an Encyclopaedia of MMS but with the additional purpose of archiving and teaching MMS achievements. Therefore, the book series will be of use not only for researchers and teachers in Mechanical Engineering but also for professionals and students for their formation and future work.

The series is promoted under the auspices of International Federation for the Promotion of Mechanism and Machine Science (IFToMM).

Prospective authors and editors can contact Mr. Pierpaolo Riva (publishing editor, Springer) at: pierpaolo.riva@springer.com

Indexed by SCOPUS and Google Scholar.

More information about this series at http://www.springer.com/series/8779

Giuseppe Quaglia · Alessandro Gasparetto ·
Victor Petuya · Giuseppe Carbone
Editors

# Proceedings of I4SDG Workshop 2021

## IFToMM for Sustainable Development Goals

Springer

*Editors*
Giuseppe Quaglia
DIMEAS
Politecnico di Torino
Torino, Italy

Victor Petuya
Department of Mechanical Engineering
University of the Basque Country
Bilbao, Spain

Alessandro Gasparetto
Università degli Studi di Udine
Udine, Italy

Giuseppe Carbone
DIMEG
University of Calabria
Rende, Italy

ISSN 2211-0984 ISSN 2211-0992 (electronic)
Mechanisms and Machine Science
ISBN 978-3-030-87382-0 ISBN 978-3-030-87383-7 (eBook)
https://doi.org/10.1007/978-3-030-87383-7



This Springer imprint is published by the registered company Springer Nature Switzerland AG
The registered company address is: Gewerbestrasse 11, 6330 Cham, Switzerland

# Preface

The 1st IFToMM for Sustainable Development Goals workshop, I4SDG 2021, is a new initiative addressed to present and highlight the research activities realized in IFToMM world community that can contribute to reach some targets within SDGs, Sustainable Development Goals.

Founded in the Post-Second World War era, IFToMM's mission has always been to bring the people from different nationalities and systems together for the exchange of science and the strengthening of international bonds in the general area of mechanism and machine science. Such strong weaving of peoples is more important than ever, and IFToMM is continuing this mission with a strongly growing network of its constituting member organizations.

Every day is clearer that sustainability, in the wide sense expressed by Sustainable Development Goals (SDGs), is the center of tomorrow's technical challenges. The 17 goals, adopted by all United Nations Member States in 2015, as part of the 2030 Agenda for Sustainable Development, represent a road map which can also address and inspire the research within typical areas of mechanism and machine science. Some, non-exhaustive, examples of connection between IFToMM's technical areas and SDGs can be summarized as follows:

- Research on biomechanical engineering is correlated to SDG 3, Good Health and Well-being.
- Studies on sustainable energy systems are addressed to SDG 7, Affordable and Clean Energy.
- All the topics covered by IFToMM Technical Committees can be related to SDG 8, Decent Work and Economic Growth; SDG 9, Industry, Innovation and Infrastructure; and SDG 12, Responsible Consumption and Production.
- Humanitarian engineering and appropriate technologies are linked to SDG 10, Reduced Inequalities, and SDG 11, Sustainable Cities and Communities.

Starting from this analysis is possible to highlight the main aims of the conference:

- Present IFToMM as an active world organization that can contribute to United Nation 2030 Development Agenda.
- Increase the awareness related to SDGs in IFToMM scientific community.
- Generate a network of active researchers, with specific interest on effects of new technologies on sustainable world.
- Promote and increase the dissemination of research.
- Involve new generations of researchers, with an active and proactive role.
- Use "COVID time" that inhibits face-to-face meeting, to start new method for networking and dissemination activities.

I4SDG 2021, organized by IFToMM Italy, has been strongly supported by the following IFToMM Technical Committees and Permanent Commissions: Sustainable Energy Systems; Robotics and Mechatronics; Tribology; Linkages and Mechanical Control; Biomechanical Engineering; Rotordynamics; Multibody Dynamics; Transportation Machinery; Micromachines; Engines and Powertrains; PC for the History of Mechanism and Machine Science; and PC for Education. This large endorsement expresses the significance of the topic covered and is the key element that enable the born of a successful and participated new event.

Moreover, in 2021 IFToMM created the new organizational entity of "Cross Disciplinary Groups" (CDG) and sustainability falls within the aims of CDG 1: Securing our Future Environment, Air, Water, Energy and represents a clear example of theme that will benefit from this cross-disciplinary approach.

I4SDG workshop is an example of collaboration and participation of the whole IFToMM world community, and without the help of IFToMM constitutional bodies, of all the organizers, and of all the authors, it would have been impossible to get so many significant contributions.

A special acknowledgment is due to IFToMM president, to all the TC/PC chairs, to IFToMM Italy, to the members of scientific and organizing committee, and to all the reviewers which gave their valuable collaboration to realize this initiative.

This book collects 69 scientific papers and is articulated in nine chapters, which reflect the technical sessions of the conference. All papers have been selected through an accurate peer review process, which considered their relevance, novelty, and clarity, and which guaranteed the high-quality level of this work.

Each paper is related to one or more SDGs and to one TC/PC/topic. To clearly identify the connection between SDGs and the presented research, it was requested to include as first keyword the most relevant SDG.

The topics are presented in a sequence of chapters that includes Sustainable Energy Systems; Appropriate Technologies and Technical Systems for Sustainable and Inclusive Development; Biomechanical Engineering; Robotics and Mechatronics; Green Tribology; Micromachines; Linkage and Mechanical Control; Transportation Machinery, Engines and Powertrains; Rotordynamics, Multibody Dynamics and Vibrations; and History of Mechanism and Machine Science, and Education.

We are confident that any researcher who is interested to shape his research activity considering its impact on building an inclusive, sustainable, and resilient future for people and planet will find this book an extraordinary and up-to-date window on the last findings in this area.

November 2021

Giuseppe Quaglia
Alessandro Gasparetto
Victor Petuya
Giuseppe Carbone

# Organization

## Honorary Chair

| | |
|---|---|
| Andres Kecskemethy | University of Duisburg-Essen, Germany |

## Conference Chair

| | |
|---|---|
| Giuseppe Quaglia | Politecnico di Torino, Italy |

## Conference Co-Chair

| | |
|---|---|
| Alessandro Gasparetto | Università degli Studi di Udine, Italy |
| Victor Petuya | University of the Basque Country (UPV/EHU), Spain |

## Scientific Committee

| | |
|---|---|
| Giuseppe Quaglia | Politecnico di Torino, Italy |
| Alessandro Gasparetto | Università degli Studi di Udine, Italy |
| Giuseppe Carbone | University of Calabria, Italy |
| Vincenzo Niola | University of Naples, Italy |
| Macedon Moldovan | Transilvania University of Brasov, Romania |
| Victor Petuya | University of the Basque Country (UPV/EHU), Spain |
| Carlo Ferraresi | Politecnico di Torino, Italy |
| Enrico Ciulli | University of Pisa, Italy |
| Erwin-Chr. Lovasz | Politehnica University of Timisoara, Romania |
| Paulo Flores | University of Minho, Portugal |
| Cristina Castejon | University Carlos III of Madrid, Spain |
| Giovanni Boschetti | University of Padova, Italy |
| Marco Ceccarelli | 'Tor Vergata' University of Rome, Italy |

| | |
|---|---|
| Mauro Velardocchia | Politecnico di Torino, Italy |
| Lena Zentner | Technical University of Ilmenau, Germany |
| José Serrano | Politechnica University of Valencia, Spain |

## Organizing Committee

| | |
|---|---|
| Giovanni Boschetti | University of Padova, Italy |
| Giuseppe Carbone | University of Calabria, Italy |
| Carmen Visconte | Politecnico di Torino, Italy |
| Luca Carbonari | Politecnico di Torino, Italy |
| Andrea Botta | Politecnico di Torino, Italy |
| Walter Franco | Politecnico di Torino, Italy |
| Dario Padovan | University of Torino, Italy |

## Workshop Secretariat

| | |
|---|---|
| Carmen Visconte | Politecnico di Torino, Italy |

# Contents

# Sustainable Energy Systems

# Design and Numerical Study of a 2-DoF Mechanism for Efficient Wave Energy Extraction in Onshore Application

Luigi Solazzi and Nicola Pellegrini(✉)

Department of Mechanical and Industrial Engineering, Università degli Studi di Brescia, Brescia, Italy
{luigi.solazzi,nicola.pellegrini}@unibs.it

**Abstract.** The electricity generation from wave energy represents an important step towards meeting renewable energy targets. The challenge of efficiently capturing the wave power is still an open topic for academies, scientific communities and companies' commercial competitiveness. This paper presents a study on the design and numerical simulation of a two degrees of freedom mechanism able to maximize the mechanical energy transformation in onshore application. The effects of the geometric parameters, leverage lengths, heavy-buoy and paddles shapes, relative position between hinges, auxiliary masses position for semi-submerged conditions were systematically studied. The numerical results demonstrated that the resulting configuration setting is significantly influenced by hydrodynamic input performance, nevertheless the proposed device is able to guarantee a stable performance and it has an efficiency that could reach up to 64.9%.

**Keywords:** SDG7 · Wave energy extraction · 2-DoF system · Onshore application

## 1 Introduction

Advances in mechanism design offers numerous opportunities to progress renewable efficient energy extraction in offshore to coastal applications range [1–3]. The demand for clean energy to meet sustainable development goals has become one of the most challenging topics of the global research. Wave power is the science studying the extraction of energy from sea and ocean waves. Literature presents many devices and some of them have reached a pre-commercial stage [4–8]. Three clusters are predominant: attenuator – the system is aligned to the wave direction; point absorber – the device is a floating structure that heave on the surface of the water and the terminator – the principal axis is perpendicular to the wave direction. The design driver in a wave energy converter is a tradeoff between the wave power and site accessibility. The advantage in offshore application is that the wave energy is larger than in coastal depths. Many wave energy converter devices are currently designed for offshore deployment (70–150 m depth) [9–11]. On the other side, the construction costs, the cabling costs, the installation and maintenance activities can be reduced selecting coastal application, i.e. 10–30


G. Quaglia et al. (Eds.): I4SDG 2021, MMS 108, pp. 3–11, 2022.
https://doi.org/10.1007/978-3-030-87383-7_1

m depths. Energy losses in coastal application are mainly due to bottom terrain friction and there are relatively small compared to the energy losses in the breaker zone, as 5–15 m depth [12, 13]. The power loss is approximately 25% at the depth range of 20–40 m. The OBREC (Overtopping BReakwater for Energy Conversion) efficiency is closed to 20%, estimated with the prototype installed by University of Campania "Luigi Vanvitelli". Meanwhile, the efficiency in Backward Bent Duct Buoy (BBDB) wave energy conversion technology in installed in China for offshore application reaches 42%. In this work the water depth ranges from 5 to 30 m, in addition the influence of wind on the wave has been neglected. The optimal solution for wave energy efficient extraction is to take advantage of a novel enabling technologies that involve several domains of knowledge, as: concurrent mechanism design, energy storage, data acquisition, system perception and adaptation, communication and maintenance paradigms. In this way, this study aims at proposing an effective mechanism designed to convert the waves in onshore zone. The innovative contribution of proposed device compared with the systems existing in literature is the ability to transform simultaneously and efficiently the vertical and the horizontal force vectors by a single 2-DoF mechanism, using a heaving buoy and two paddles. The proof-of-concept and the mathematical formulation, are described in Sect. 2. In Sect. 3, the numerical analyses and the sensitivity evaluation are presented. Discussion and conclusions about findings, limitations and possible future works are provided in Sect. 4, 5.

## 2 2-DoF Mechanism: Concept and Formulation

In this section, the concept is described and a schematic representation is presented in Fig. 1. The mechanism is composed by a buoy (1), two paddles (2), a principal arm (3), able to rotate around the hinge (5) and it is connected at the distal end to a power take-off (6) that exert reaction on the ground frame at (7). The secondary arm is connected to the principal one by the hinge (4) which supports the two paddles; the spring-damper system (8) is rigidly coupled to the principal arm, while the buoy and the paddles are immersed in the water. Aluminum and composite materials have been studied and compared as alternative of traditional material in order to evaluate the reduction of the structure weight and to minimize the system inertia [14].

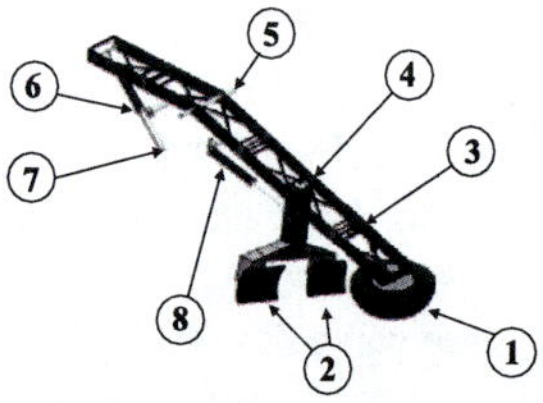


**Fig. 1.** Proposed structure of wave energy extraction system.

The proposed system is functionally defined as two lumped-mass systems, as shown in Fig. 2. The analyzed variables that are able to determine the system configuration based on the hinges $O_1$ are: the position of the buoy ($l_1$), the position of the hinge of the principal arm ($l_2$), the position of secondary hinge $O_2$ ($l_3$). Instead, based on $O_2$ position: the distance between $O_2$ and the secondary spring ($b_2$), the distance between $O_2$ and the paddle ($b_1$). The complete formulation considers simultaneously the heaving-buoy and the two paddles in order to maximize the energy extraction provided by the incident wave. The buoy is subjected mainly to the vertical action and so this element extracts the vertical component of the wave. The paddles, positioned rear of the buoy, are submitted essentially to the horizontal action and so they transform the horizontal fraction of the pressure provided by the single frequency wave. Consequently, two spring elements, $k_1$ and $k_2$, connected to the principal arm need to be characterized. The mechanism has two centers of rotation: the angle $\theta$, rotation on $O_1$ and the angle $\alpha$ that represents the angle of rotation of the paddles around $O_2$.

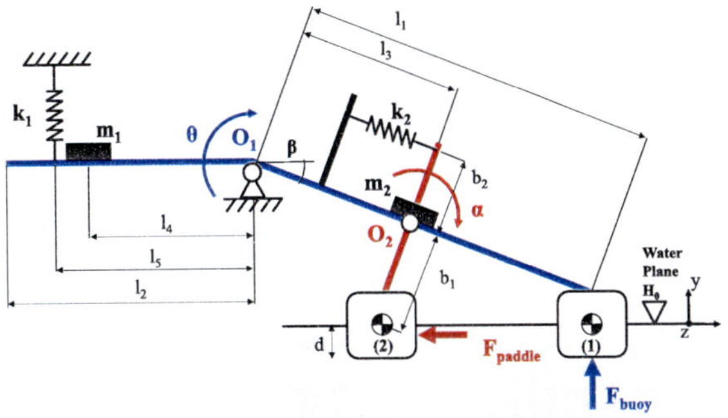


**Fig. 2.** 2-DoF mechanism representation.

In this research, the wave conditions are limited to the oscillation in the heaving-buoy and paddles modes. The equilibrium in semi-submerged condition can be studied. The resultant loads acting on system based on center of rotation $O_1$ is described as:

$$\sum M_{O_1} = J \cdot \theta \tag{1}$$

where $\mathbf{M_{O1}}$ is the moment around $O_1$, $\mathbf{J}$ is the moment of inertia, $\boldsymbol{\theta}$ is the rotation in $O_1$ on z-axis. For the minimum and maximum wave fluctuation instants, the forces magnitude acting on principal arm (1) are derived in Eqs. (2), (3). During the heaving ascending phase of the float can be obtained as in Eq. (2) and the dual descending phase in Eq. (3), as

$$F^{y+}{}_{k_1} = \frac{F_{buoy} \cdot l_1 \cdot cos\beta + (m_1 \cdot l_4 - m_2 \cdot l_3 \cdot cos\beta) \cdot g}{l_5} + \frac{J \cdot \ddot{\theta}}{l_5 \cdot cos\theta} \tag{2}$$

$$F^{y-}{}_{k_1} = \frac{F_{buoy} \cdot l_1 \cdot cos\beta - (m_1 \cdot l_4 - m_2 \cdot l_3 \cdot cos\beta) \cdot g}{l_5} + \frac{J \cdot \ddot{\theta}}{l_5 \cdot cos\theta} \tag{3}$$

where $F^{y+}{}_{k_1}$ comprises the effect of auxiliar masses $m_1$ and $m_2$ (positive in ascending and negative in descending phase, respectively). The $F_{buoy}$ is the excitation force on heaving buoy that covers the dynamic effects and it is related to the buoy geometry. Additionally, $m_1$ and $m_2$ are auxiliary masses that guarantee the semi-submerged condition in dynamic mode. Wave fluctuation caused by a single frequency excitation in vertical and horizontal directions can be presented as:

$$y(x;t) = \frac{H}{2} \cdot \cos(k \cdot x - \omega \cdot t) \tag{4}$$

where H is the wave height [m], x is the horizontal position in propagation x-direction [m], t is the time period [s], k is wave number [$m^{-1}$], ω angular frequency [$s^{-1}$].

The excitation force resultant on buoy is calculated as follows:

$$F_{buoy} = -m_{buoy} \cdot (\frac{H}{2} \cdot \omega^2 \cdot \cos(k \cdot x - \omega \cdot t) + g) \tag{5}$$

where $g$ represent the gravity acceleration. The influence of wave height to the time-dependent $\ddot{\theta}$ is formulated at minimum and maximum wave amplitude as:

$$\ddot{\theta}_{Min,Max} = \pm \frac{H}{2} \cdot \frac{\omega^2}{l_1 \cdot cos\theta} \tag{6}$$

In Lagrangian formulation, the Lagrangian of the entire system is defined by the total kinetic energy minus the total potential energy. The equation of motion of the entire system, composed by the heavy-buoy and the paddles system, is considered in terms of θ and α, as follows:

$$\begin{aligned} \frac{d}{dt}\frac{\partial L}{\partial \dot{\theta}} - \frac{\partial L}{\partial \theta} = 0 \\ \frac{d}{dt}\frac{\partial L}{\partial \dot{\alpha}} - \frac{\partial L}{\partial \alpha} = 0 \end{aligned} \tag{7}$$

Considering the center of rotation $O_1$ the kinetic and potential energy is given by:

$$\begin{aligned} L = & -\frac{1}{2} \cdot J_1^{O_1} \cdot \dot{\theta}^2 + \frac{1}{2} \cdot J_2^{O_2} \cdot \dot{\alpha}^2 + F_{Paddle} \cdot (l_3 \cdot cos\theta - b_1 \cdot \sin\alpha) + \frac{1}{2} \cdot k_1 \\ & \cdot (l_5 \cdot \sin\theta)^2 - \frac{1}{2} \cdot k_2 \cdot (b_2 \cdot \sin\alpha)^2 - g \cdot \sin\theta \\ & \cdot (m_1 \cdot l_3 - m_2 \cdot l_4) + F_{buoy} \cdot (l_1 \cdot \sin\theta) \end{aligned} \tag{8}$$

Based on Eq. (6), the term θ provides:

$$\frac{d}{dt}\frac{\partial L}{\partial \dot{\theta}} - \frac{\partial L}{\partial \theta} = 0 \tag{9}$$

Then using the Eqs. (8) and (9), it can be re-written as:

$$\begin{aligned} & J_1^{O_1} \cdot \ddot{\theta} - F_{Paddle} \cdot (l_3 \cdot \sin\theta) + k_1 \cdot l_5^2 \cdot \sin\theta \cdot \cos\theta \\ & - g \cdot \cos\theta \cdot (m_1 \cdot l_3 - m_2 \cdot l_4) + F_{buoy} \cdot (l_1 \cdot \cos\theta) = 0 \end{aligned} \tag{10}$$

The force acting on the spring $k_1$, connected to the main lever including the paddle contribution is formulated as:

$$F^{y+}{}_{k_1} = \frac{F_{buoy} \cdot l_1 \cdot cos\beta + (m_1 \cdot l_4 - m_2 \cdot l_3 \cdot cos\beta) \cdot g}{l_5} + \frac{J_1^{O_1} \cdot \ddot{\theta} - F_{paddle} \cdot l_3 \cdot \sin\theta}{l_5 \cdot cos\theta} \tag{11}$$

$$F^{y-}{}_{k_1} = \frac{F_{buoy} \cdot l_1 \cdot cos\beta - (m_1 \cdot l_4 - m_2 \cdot l_3 \cdot cos\beta) \cdot g}{l_5} + \frac{J_1^{O_1} \cdot \ddot{\theta} + F_{paddle} \cdot l_3 \cdot \sin\theta}{l_5 \cdot cos\theta} \tag{12}$$

From Eq. (7) for α term can be expressed as in (13):

$$J_2^{O_2} \cdot \ddot{\alpha} - F_{Paddle} \cdot (b_1 \cdot \cos\alpha) + k_2 \cdot {b_2}^2 \cdot \sin\alpha \cdot \cos\alpha = 0 \tag{13}$$

The force acting on the spring $k_2$, connected to the main lever, during the wave horizontal propagation can be obtained as in Eq. (14):

$$F^{x+}{}_{k_2} = \frac{F_{paddle} \cdot b_1}{b_2} + \frac{J_2^{O_2} \cdot \ddot{\alpha}}{b_2 \cdot cos\alpha} \tag{14}$$

Thus, the external forces acting on the two springs during the wave period are formulated. The force profile is influenced by the vertical fraction on the heaving-buoy and the horizontal fraction on the two paddles. The adopted power-take-off technique is based on the leverage motion conversion to hydraulic power system, where the pump stiffness influences the pressured oil flow. Then, the generator converts the mechanical pump to electric energy. The energy extracted is synthesized by the spring (1) at the maximum and minimum range are derived as in (15), (16), respectively:

$$E_{k_1;Max} = \frac{1}{2} \cdot k_1 \cdot {\Delta l_{k_1}}^2 \tag{15}$$

$$E_{k_1;Min} = \frac{1}{2} \cdot \frac{{F_{Min}}^2}{k_1} \tag{16}$$

Then, the energy converted by the 2-DoF system in $k_1$ considering a single wave period, T, is defined as:

$$E_{k_1} = \frac{1}{T/2} \cdot \left(E_{k_1;Max} + E_{k_1;Min}\right) \tag{17}$$

The energy extracted by the power-take-off mechanism in $k_2$ can be evaluated as:

$$E_{k_2} = \frac{1}{2} \cdot k_2 \cdot {\Delta l_{k_2}}^2 \tag{18}$$

The total wave energy transformed can be expressed as the sum of the energies stored within the two spring elements. Using Eqs. (15), (16) and (18) the energy provided by the complete system in a wave period can be computed:

$$E_{k_1;k_2} = \frac{1}{T/2} \cdot \left(E_{k_1} + E_{k_2}\right) \tag{19}$$

## 3 Numerical Analyses and Sensitivity Evaluation

In order to investigate the ability of proposed wave energy extraction mechanism, the synthesis of system model and numerical simulation have been performed. Consequently, sensitivity evaluations on geometrical features and boundary conditions are executed to validate the system performance. In this research, for each input as well as output, five levels are used to tuning the system model. The inputs considered are the number of paddles, the relative position between the two paddles and the relative immersion percentage between buoy and paddle. In addition, the boundary conditions such as the velocity of the fluid has been studied. Computational fluid dynamics (CFD) allows the investigation of the physics of the blade profile interaction with the water wave flow. The CFD supports the design of relative position between buoy and paddles, the buoy and paddles geometry and if the configuration generates turbulence near the system with the consequent efficiency degradation. Table 1 shows some results of the numerical simulations of resultant fluid velocity and pressure, based on FLUENT SW, considering the input velocity of water as 4 m/s. Furthermore, the table highlights the influence of relative position between buoy and paddles, as immersion percentage parameter in 20%–80% range.

**Table 1.** Hydrodynamic simulation wave velocity input as 4.0 m/s.

| Immersion % | Wave Velocity [m/s] / Pressure [Pa] |
|---|---|
| 20% | 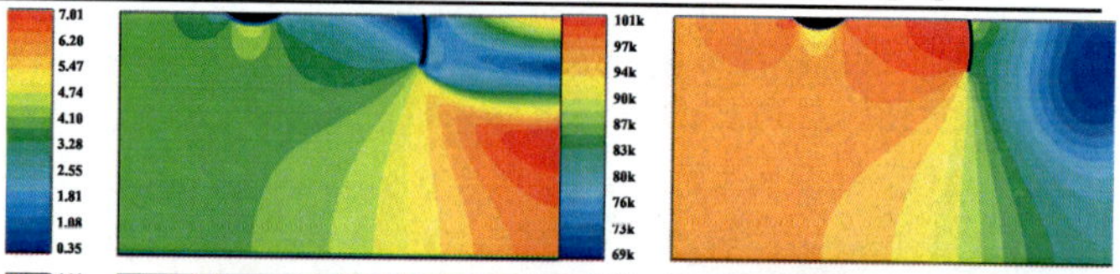 |
| 50% | 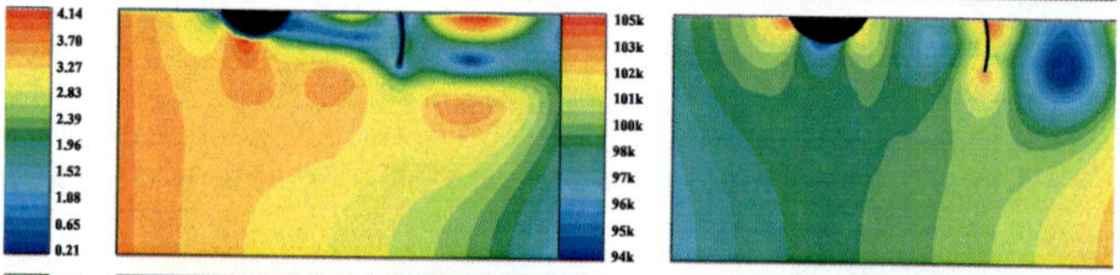 |
| 80% | 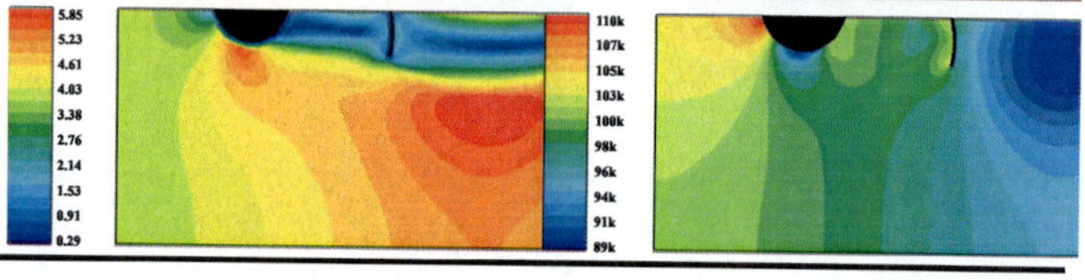 |

The magnitude of the forces resulting in the horizontal and vertical directions on the heaving-buoy and paddles are shown in the following Figs. 3a,b. The analysis remarks the influence between the buoy-paddle immersion percentage and loads profile on x-, y-directions. The study considers the incident wave velocity feature perturbation that an increment of wave velocity by 25% affects the vertical force increased by 45% in top-level immersion percentage setting, and a reduction of 32% of horizontal force in low-level of immersion configuration.

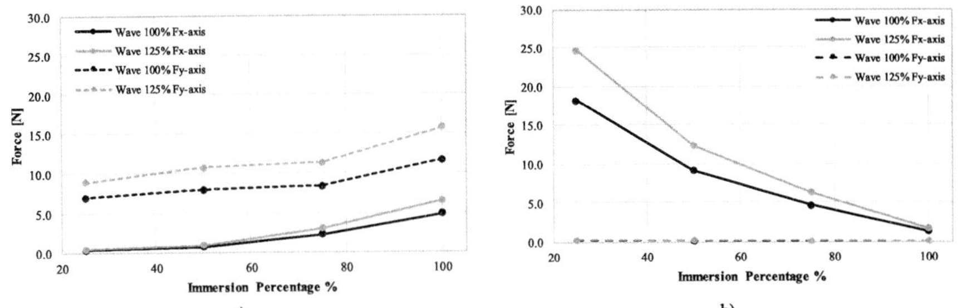


**Fig. 3.** Buoy loads influence to buoy/paddle immersion percentage, a), Paddle loads influence to buoy/paddle immersion percentage, b).

# 4 Discussion

All wave energy extraction parameter combinations have the same heavy-buoy and paddles shape. The wave interactions among selected features are studied, hence the suitable configuration of the 2-DoF system is defined and summarized in Table 2. The investigated scale factor is in 3.5–4.5 range to represent the Mediterranean Sea application. The parameters setting is obtained evaluating the optimal trade-off between the device compactness and the energy extraction for a set of wave inputs seeing the future work extension with experimental test.

**Table 2.** Parameters of energy extraction mechanism.

| $l_1$[m] | $l_2$[m] | $l_5$[m] | $b_1$[m] | $b_2$[m] | $l_3$[m] | $k_1$[N/m] | $k_2$[N/m] |
|---|---|---|---|---|---|---|---|
| 0.75 | 0.75 | 0.40 | 0.25 | 0.25 | 0.25 | 490 | 196 |

The auxiliary masses, $m_1$ and $m_2$, that guarantee the semi-submerged condition in dynamic mode are determined as 3.0 kg. The wave reference features are: amplitude as 0.12 m, period as 1.35 s and wavelength $\lambda$ equal to 1.10 m. Based on the study above, the energy extracted by the system is estimated as 3.94 J from the heaving buoy and 1.48J from the paddles leverage for each wave period. Considering the buoy width equal to 0.40 m, the average total energy provided by the wave to the buoy is estimated equal to 7.95 J. The efficiency loss ratio is the percentage between the energy absorbed by the spring $k_1$ in a period and the energy provided by the wave to the buoy, is obtained as:

$$\eta_{k_1} = \frac{E_{k_1}}{E_{Wave}} = \frac{3.94J}{7.95J} = 49.5\% \tag{20}$$

It is evident an efficiency reduction of the buoy with the presence of the paddles due to the hydrodynamic disturbance. Similarly, it is possible to calculate the total average energy supplied by the wave to the paddles $k_2$ imposing the paddle width of 0.24 m. The resultant energy is closed to 4.77 J. The percentage of energy absorbed by the $k_2$ system in a period is stated as:

$$\eta_{k_2} = \frac{E_{k_2}}{E_{Wave}} = \frac{1.48J}{4.77J} = 31.2\% \tag{21}$$

In addition, the paddles are perturbed by the buoy presence that partially damps the horizontal component of the incident wave. The overall performance of the 2-DoF system can be formulated by the sum of the efficiency obtained respectively by the buoy in Eq. (21) and the paddle assembly in Eq. (22), as follows:

$$\eta_{k_1} = \eta_{k_1} + f \cdot \eta_{k_{2\ldots n}} = 64.9\% \; n = 2, \ldots, i \tag{22}$$

where $f$ is the loss-fraction due to multiple paddles interactions in parallel configuration. The Eqs. (21) and (22) comparison shows an increased efficiency of the complete system of about 15% due to the contribution of paddles that recovers part of the energy available from the wave. In general, it can be concluded that using a system in combined form, a higher energy efficiency is obtained compared to mechanism that optimize the contributions in a single direction.

## 5 Conclusions

This paper has presented a 2-DOF mechanism for energy extraction from waves using a combined heaving-buoy and multiple paddles system. The description of main mechanical design concepts and mathematical model are discussed. Simulations have been performed to investigate the working performance of the suggested device. The simulation results indicated an optimal configuration setting that is an effective solution and possible to fabricate in a near future for wave energy generation. A further works will consider the similitude effects on the model structure optimization, the material in challenging corrosive environment, the effect of nonlinearity on power-take-off element and establish a robust model that considers also different irregular sea conditions.

## References

1. Bracco, G., Giorcelli, E., Mattiazzo, G.: ISWEC: a gyroscopic mechanism for wave power exploitation. Mech. Mach. Theory **46**(10), 1411–1424 (2011)
2. Chen, F., Duan, D., Han, Q., Yang, X., Zhao, F.: Study on force and wave energy conversion efficiency of buoys in low wave energy density seas. Energy Conv. Manage. **182**, 191–200 (2019)
3. Cruz, J.: Ocean Wave Energy: Current Status and Future Perspectives. Springer, Heidelberg (2008)
4. Falcão, A.: Wave energy utilization: a review of the technologies. Renew. Sust. Energy Rev. **14**, 899–918 (2010)

5. Negri, M., Clerici, F., Malavasi, S.; A breaker-zone wave energy converter. In: International Conference on Renewable Energies and Power Quality (ICREPQ'13) Bilbao (Spain), 20th to 22th March (2013)
6. Pizer, D., et al.: PELAMIS WEC – recent advances in the numerical and experimental modelling programme. In: Proceedings of 6th European Wave Energy Conference Glasgow, UK, pp. 373–378 (2005)
7. Wu, B., Li, M., Wu, R., Chen, T., Zhang, Y., Ye, Y.: BBDB wave energy conversion technology and perspective in China. Ocean Eng. **169**, 281–291 (2018)
8. Berenjkoob, M., Ghiasi, M., Guedes Soares, C.; Influence of the shape of a buoy on the efficiency of its dual-motion wave energy conversion. Energy **214**, 118998 (2021)
9. Clemente, D., Rosa-Santos, P., Taveira-Pinto, F., Martins, P., Paulo-Moreira, A.: Proof-of-concept study on a wave energy converter based on the roll oscillations of multipurpose offshore floating platforms. Energy Conv. Manage. **224**, 113363 (2020)
10. Giannini, G., López, M., Ramos, V., Rodríguez, C.A., Rosa-Santos, P., Taveira-Pinto, F.: Geometry assessment of a sloped type wave energy converter. Renew. Energy **171**, 672–686 (2021)
11. Ma, Y., Zhang, A., Yang, L., Li, H., Zhai, Z., Zhou, H.: Motion simulation and performance analysis of two-body floating point absorber wave energy converter. Renew. Energy **157**, 353–367 (2020)
12. Bergillos, R.J., Rodriguez-Delgado, C., Iglesias, G.: Ocean Energy and Coastal Protection: A Novel Strategy for Coastal Management Under Climate Change. Springer International Publishing, New York (2020)
13. Contestabile, P., Crispino, G., Di Lauro, E., Ferrante, V., Gisonni, C., Vicinanza, D.: Overtopping breakwater for wave Energy Conversion: Review of state of art, recent advancements and what lies ahead. Renew. Energy **147**(1), 705–718 (2020)
14. Solazzi, L., Buffoli, A.: Telescopic hydraulic cylinder made of composite material. Appl. Compos. Mater. **26**, 1189–1206 (2019)

# Scaled Prototyping of an Onshore Oscillating-Arm Wave Energy Converter

Luca Bruzzone(✉), Giovanni Berselli, and Pietro Fanghella

University of Genoa, 16147 Genoa, GE, Italy
{luca.bruzzone,giovanni.berselli,pietro.fanghella}@unige.it

**Abstract.** The paper presents the development of the scaled prototype of an oscillating-arm Wave Energy Converter (WEC), for onshore installation, suitable to be placed above breakwaters. The mechanical architecture is based on a four-bar linkage moved by a floating buoy with alternate heave motion. The mechanical energy is transferred to the electrical rotary generator with the interposition of a mechanical rectifier, to obtain unidirectional motion at higher speed. Since the power take-off system is fully electromechanical, without pneumatic or hydraulic components, the proposed WEC is characterized by high efficiency and reliability. The functional design of the proposed WEC and its control layout are summarized. Finally, the methodology for designing the 1:15 scale model, destined for the preliminary experimental validation, is discussed, and the embodiment design of the model is outlined.

**Keywords:** SDG7 · Wave energy converter · Oscillating arm · Scaled prototype

## 1 Introduction

Within the scenario of renewable energies, one of the most promising research areas is represented by Wave Energy Converters (WECs). As a matter of fact, WECs are generally at an early stage if compared to other renewable energy plants, as pointed out in several reports [1, 2], with few examples of systems at operative Technology Readiness Level (TRL).

Nevertheless, the huge amount of available sea energy greatly encourages the efforts to achieve sufficient TRL and to fulfil the requirements of levelized cost of energy, especially by introducing simple, reliable and cost-effective WEC designs.

WECs can be classified in many different ways, on the basis of their distance from the shore (onshore, nearshore, offshore, [3]), of their size and orientation with respect to the incoming wave (attenuators, terminators, point absorbers, [4]), of the working principle (pressure differential WECs, floating WECs, overtopping WECs, impact WECs, [5]), of the power take-off (PTO) system (air turbines [6], hydraulic systems [7], linear generators [8]).

A general tendency in WEC design is the attention to the aspects of reliability and easiness of installation and maintenance. As a consequence, WECs based on oscillating arms carrying floating buoys with heave motion and hinged to shore have been recently


G. Quaglia et al. (Eds.): I4SDG 2021, MMS 108, pp. 12–19, 2022.
https://doi.org/10.1007/978-3-030-87383-7_2

proposed: the WaveStar [9], the WECs of the Israeli Eco Wave Power [10], the plant installed at the port of Pecem by the COPPE Institute of the Federal University of Rio de Janeiro [11]. With this layout, the power take-off systems are placed onshore, with great benefits for installation and maintenance.

In [12] a low-cost oscillating-arm WEC is proposed, characterized by a fully electromechanical power take-off system, without pneumatic or hydraulic transmission lines, differently from the previously cited oscillating-arm WECs. Its architecture is based on an oscillating arm carrying a floating buoy with heave motion, connected to a rotating electrical generator by means of a mechanical rectifier, equipped with two one-way clutches and a speed multiplier gearbox. This WEC transforms directly the wave energy into electrical energy by means of a mechanical transmission without fluid interposition, potentially leading to high efficiency and reliability.

In this paper, the development of the 1:15 model, preliminary to the full-scale prototype, is outlined: Sect. 2 introduces the functional design of the proposed WEC, Sect. 3 presents the control layout, Sect. 4 discusses the scaling criteria for the 1:15 model, Sect. 5 summarizes conclusions and future developments.

## 2 Functional Design of the Oscillating-Arm WEC

The architecture of the proposed WEC, installed on a breakwater wall, is represented in Fig. 1 (left). A floating buoy, *d*, with hemisphere-cylinder shape, is rigidly connected to a rocker arm, *c*. The motion of the floating buoy, caused by the waves, is transferred to the upper rocker arm *a* by the rod *b*. The motion of the rocker arm *a* is low speed and alternate, therefore it is convenient to transform it into a unidirectional motion with higher speed, more suitable for a rotary electrical generator, by means of the mechanical rectifier represented in Fig. 1 (right).

This transmission comprises two one-way clutches (*OC+* and *OC-*) which are interposed between the shaft connected to the rocker arm *a* and the two gear wheels *r+* and *r-*. The one-way clutches transfer torque in opposite directions, therefore when the buoy *d* moves down, the input shaft of the speed multiplier MG is actuated by the gear wheel *o-*, coupled with *r-*; on the contrary, when the buoy *d* moves up, the input shaft of MG is actuated by the gear wheel *o+*, in the same direction, since an idle wheel *id* is interposed between *r+* and *o+*. With this transmission scheme the oscillating motion of the upper rocker arm *a* is transformed into unidirectional motion of the multiplier gearbox MG. In order to store kinetic energy, thus regularizing the generator speed, a flywheel FW is fixed to the output shaft of *MG*, which drives the electrical generator *G*. Since the ratio between the number of teeth of *r+* and *o+* is equal to the ratio between the number of teeth of *r-* and *o-*, the rectifying mechanism has the same transmission ratio in both directions.

The geometry of the WEC is represented in Fig. 2. This preliminary design has been developed considering a possible installation site (latitude: 44.400 N, longitude: 8.871 E) characterized by the sea-state probabilities represented in the scatter diagram of Fig. 3 in terms of mean wave period $T_0$ and significant wave height $H_s$ (i.e. the mean wave height of the one third highest waves), based on the historical time series from 1979 to 2017 [12]. The operative range of motion of the WEC, adequate to this location, corresponds

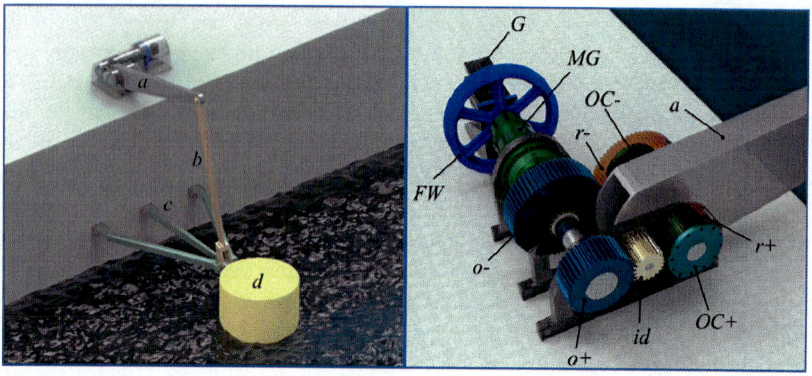


**Fig. 1.** The proposed oscillating-arm WEC with mechanical rectifier: complete view (left) and detail of the mechanical rectifier (right).

to 4.75 m of vertical motion of the centre of the buoy *d*, and to ±25° of inclination of the lower rocker arm *c*. In case of storm with extreme sea conditions, the WEC can be placed in vertical recovery position (Fig. 2, red), leaning against the breakwater wall. The main WEC inertial and mechanical data, tuned by means of the model discussed in Sect. 3, are collected in Table 1.

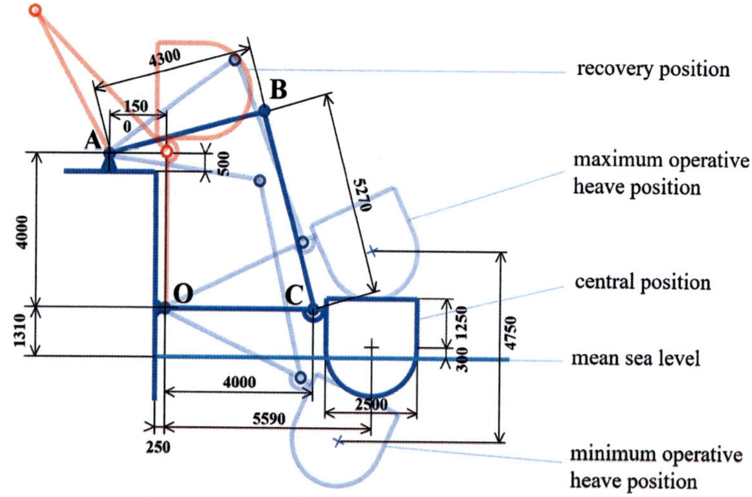


**Fig. 2.** Geometry of the WEC four-bar mechanism.

| $H_s$[m] ↓ $T_0$[s] → | 2.5 | 3 | 3.5 | 4 | 4.5 | 5 | 5.5 | 6 | 6.5 | 7 | 7.5 | 8 | 8.5 |
|---|---|---|---|---|---|---|---|---|---|---|---|---|---|
| 0.2 | 280.0 | 391.3 | 502.5 | 414.8 | 211.5 | 122.2 | 71.1 | 36.0 | 20.2 | 10.0 | 5.0 | 1.9 | 0.9 |
| 0.4 | 272.8 | 324.9 | 275.7 | 283.6 | 270.7 | 150.9 | 86.0 | 52.3 | 28.0 | 14.7 | 6.6 | 3.4 | 2.3 |
| 0.6 | 45.4 | 377.5 | 244.6 | 181.3 | 180.8 | 181.3 | 104.8 | 55.2 | 34.3 | 20.4 | 10.5 | 3.6 | 3.0 |
| 0.8 | 0.1 | 150.0 | 301.2 | 145.2 | 129.4 | 128.1 | 116.6 | 66.3 | 41.3 | 27.8 | 14.5 | 6.5 | 4.8 |
| 1.0 | 0.0 | 2.5 | 282.2 | 129.4 | 88.2 | 87.8 | 88.7 | 81.4 | 54.3 | 34.4 | 19.8 | 9.5 | 7.1 |
| 1.2 | 0.0 | 0.0 | 71.6 | 185.3 | 63.1 | 55.3 | 61.1 | 61.8 | 51.5 | 32.5 | 21.4 | 12.9 | 10.3 |
| 1.4 | 0.0 | 0.0 | 0.3 | 113.0 | 46.5 | 32.4 | 32.4 | 38.3 | 35.8 | 31.7 | 22.7 | 12.4 | 11.8 |
| 1.6 | 0.0 | 0.0 | 0.0 | 26.0 | 48.7 | 18.6 | 18.7 | 22.3 | 23.1 | 23.5 | 19.3 | 11.0 | 10.9 |
| 1.8 | 0.0 | 0.0 | 0.0 | 0.2 | 29.4 | 11.0 | 11.0 | 10.8 | 13.6 | 15.7 | 14.1 | 9.5 | 10.1 |
| 2.0 | 0.0 | 0.0 | 0.0 | 0.0 | 9.4 | 7.2 | 5.5 | 8.0 | 7.8 | 8.8 | 9.0 | 8.1 | 7.9 |
| 2.2 | 0.0 | 0.0 | 0.0 | 0.0 | 1.7 | 3.9 | 4.7 | 4.9 | 5.0 | 5.4 | 5.2 | 3.6 | 3.5 |

**Fig. 3.** Number of occurrences per sea-state (hours/year); $T_0$: wave period; $H_s$: significant wave height (latitude: 44.400, longitude: 8.871).

**Table 1.** Main WEC inertial and mechanical features.

| Parameter | Symbol | Value |
|---|---|---|
| Mass of the rocker arm *a* (estimated) | $m_a$ | 200 kg |
| Mass of the rod *b* (estimated) | $m_b$ | 250 kg |
| Mass of body *c* (estimated) | $m_c$ | 500 kg |
| Mass of body *d* (estimated) | $m_d$ | 2700 kg |
| Speed multiplier transmission ratio | $\tau_{MG}$ | 200 |
| Mechanical rectifier efficiency (estimated) | $\eta_{MR}$ | 0.96 |
| Speed multiplier efficiency (estimated) | $\eta_{MG}$ | 0.85 |
| Overall rotary inertia of generator, flywheel, speed multiplier gearbox (referred to the generator shaft, estimated) | $J^*_{gen}$ | 10.5 kgm$^2$ |

## 3 WEC Control Layout

In [13], the dynamic model of the WEC is discussed. The performance of the system has been assessed by means of a multibody model in which the hydrostatic and hydrodynamic interactions between sea and buoy are calculated using the Linear Wave Theory implemented in the time domain through the Cummins approach [14]. The generator is controlled adopting the widely used resistive model, in which the PTO applies to the system a torque/force proportional to its speed [15, 16], therefore:

$$M_{gen} = c_{PTO}\omega_{gen} \tag{1}$$

where $M_{gen}$ and $\omega_{gen}$ are the torque and the speed of the electrical generator and $c_{PTO}$ is the speed-torque ratio. The response surfaces of Fig. 4 show the generated power as a function of the sea state (polychromatic wave with Jonswap spectrum characterized by period $T_0$ and significant wave height $H_s$) for $c_{PTO}$ ranging from 4 to 36 Nms/rad, obtained by simulation.

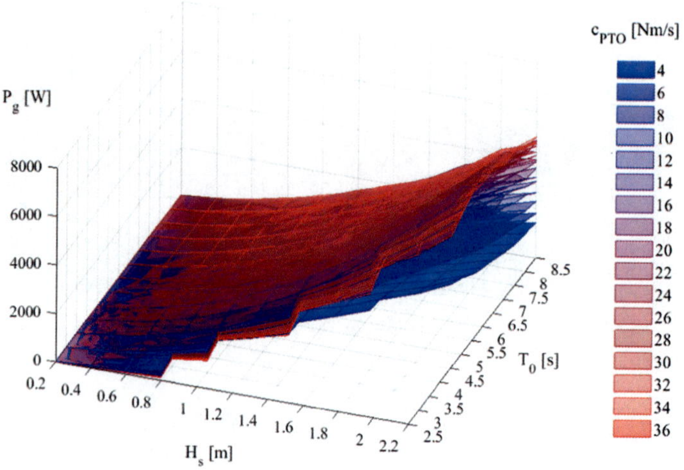


**Fig. 4.** Generated power $P_g$ [W] as a function of the sea state for different PTO speed-torque ratios $c_{PTO}$ [13].

Adopting a model-based approach, for any sea state the optimal value of the speed-torque ratio can be calculated and then applied in the real system, provided that the actual values of wave period $T_0$ and significant wave height $H_s$ are obtained in real-time by measuring the recent past of the sea level [17].

This approach has a great drawback: its effectiveness relies on the WEC model accuracy in the evaluation of the generated power. While the multibody model of the WEC can be considered accurate, the assessment of the sea-buoy interaction forces is much more critical: the Linear Wave Theory does not consider viscous losses, and it is accurate only for small motion amplitudes. Moreover, the real WEC behaviour is influenced by the sea-bed configuration, which can change over time. For all of these reasons, a model-free control system has been proposed in [13], based on Reinforcement Learning [18], capable of tuning the speed-torque ratio during working conditions on the basis of the past experience, avoiding modelling approximations, and updating its knowledge by alternating exploitation of the past experience and exploration of new tunings, to compensate for long-term modifications of the environment.

## 4 Scaled Development of the WEC Prototype

The costs of construction and installation of a WEC prototype are in general high, therefore it is necessary to validate simulation results by means of experimental test on small-scale models before the realization of the full-scale WEC system.

In order to obtain a correct scaling between the model and the full-scale prototype, both the geometric and the hydrodynamic similitudes must be satisfied [19]. The geometric similitude is obtained by scaling all the model linear dimensions by the same scale factor λ:

$$l_p = \lambda l_m \tag{2}$$

where the subscript $p$ and $m$ stand for prototype and model. The hydrodynamic similitude is imposed by means of the Froude, Strouhal and Reynolds numbers. The Froude number corresponds to:

$$Fr = \frac{u^2}{gD} \tag{3}$$

where $u$ is the fluid velocity, $g$ is the gravitational acceleration and $D$ is a characteristic dimension of the body in contact with the fluid. Imposing equal Froude numbers for prototype and model, we obtain scaling factors respectively for velocity, acceleration, time, frequency, mass and force [19]:

$$u_p = \sqrt{\lambda} u_m \tag{4}$$

$$\dot{u}_p = \dot{u}_m \tag{5}$$

$$t_p = \sqrt{\lambda} t_m \tag{6}$$

$$f_p = f_m \Big/ \sqrt{\lambda} \tag{7}$$

$$m_p = \lambda^3 m_m \tag{8}$$

$$F_p = \lambda^3 F_m \Big/ 0.975 \tag{9}$$

In Eq. (9) the quantity 0.975 derives from the ratio between the densities of fresh water (for the model) and sea-water (for the prototype).

While the Strouhal similitude is coherent with the Froude similitude, the Reynolds similitude is not respected in small scale models. As a matter of fact, the Reynolds number for a model scaled according to the Froude similitude is [19]:

$$Re_p = \lambda^{3/2} Re_m \tag{10}$$

Therefore, for example, for $\lambda = 50$ $Re_p \cong 354 \cdot Re_m$; therefore, even if the prototype flow regime is turbulent, the model flow regime is probably laminar, with higher drag coefficients. To overcome this model deficiency, a possible solution for ship models is to stimulate the turbulence of the fluid approaching the model surface by means of grids located in front of the model. In the discussed application this solution is not feasible, since the buoy heave motion is almost vertical and alternate. Fortunately, since the motion of the buoy is alternate and characterized by low relative velocity between buoy surface and fluid, and consequently low drag force, the differences in drag scaling between prototype and model are not critical.

Figure 5 represents some details of the embodiment design of the 1:15 WEC model, scaled considering Eqs. (4) to (9), in order to perform preliminary experimental tests in the basin of the University of Genoa.

The beams composing the four-bar mechanism are realized by means of aluminum tubes ($a_{m1}$, $b_{m1}$, $c_{m1}$, Fig. 5) and 3D printed ABS end connectors ($a_{m2}$, $a_{m3}$, $b_{m2}$, $b_{m3}$,

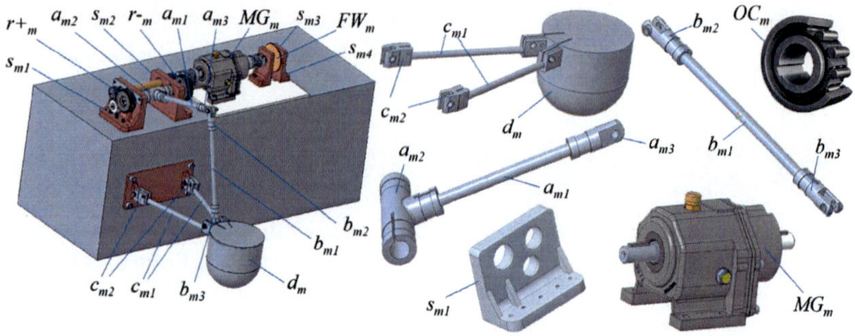


**Fig. 5.** Embodiment design of the 1:15 WEC model.

$c_{m2}$, Fig. 5). Also the floating buoy ($d_m$, Fig. 5) and the transmission supports ($s_{m1}$ … $s_{m4}$, Fig. 5) are 3D printed in ABS, while all the gears wheels are realized in steel for structural requirements. In particular, the two gear wheels ($r+_m$ and $r\text{-}_m$, Fig. 5) host two commercial one-way clutches (NBS KKPP, $OC_m$, Fig. 5). The multiplier gearbox is a commercial gearbox (SITI NHL 20/2, $MG_m$, Fig. 5) with transmission ratio $\tau_{\mathrm{MG,m}} = 24.1$, different from the prototype transmission ratio ($\tau_{\mathrm{MG,p}} = 200$, Table 1). Consequently, the rotary inertia of the flywheel, which is an important parameter in the WEC design [12], has been calculated for the model by imposing equivalent kinetic energy and imposing the scaling criteria discussed in this section. For the prototype, the rotary inertia reduced to the slow shaft ($J_{\mathrm{fw,eq,p}}$) can be calculated from the actual flywheel rotary inertia ($J_{\mathrm{fw,\,p}}$):

$$J_{\mathrm{fw,eq,p}} = \tau^2_{\mathrm{MG,p}} J_{\mathrm{fw,p}} \tag{11}$$

Then, considering Eqs. (2) and (8) and the dimensions of the rotary inertia, we obtain the actual flywheel rotary inertia for the model:

$$J_{\mathrm{fw,m}} = \frac{J_{\mathrm{fw,eq,m}}}{\tau^2_{\mathrm{MG,m}}} = \frac{J_{\mathrm{fw,eq,p}}}{\lambda^5 \tau^2_{\mathrm{MG,m}}} = \frac{\tau^2_{\mathrm{MG,p}} J_{\mathrm{fw,p}}}{\lambda^5 \tau^2_{\mathrm{MG,m}}} \tag{12}$$

## 5 Conclusions and Future Developments

The functional and control architecture of a novel, fully electromechanical oscillating-arm WEC have been outlined. The scaling criteria and the embodiment design of the 1:15 model have been discussed. In the following of the research, the experimental tests on the model shall be performed, to validate the feasibility of the project before full-scale prototyping.

## References

1. Pecher, A., Kofoed, J.P. (eds.): Handbook of Ocean Wave Energy. OEO, vol. 7. Springer, Cham (2017). https://doi.org/10.1007/978-3-319-39889-1

2. Mwasilu, F., Jung, J.W.: Potential for power generation from ocean wave renewable energy source: a comprehensive review on state-of-the-art technology and future prospects. IET Renew. Power Gener. **13**, 2019363–2019375 (2019)
3. López, I., Andreu, J., Ceballos, S., Martínez de Alegría, I., Kortabarria, I.: Review of wave energy technologies and the necessary power-equipment. Renew. Sustain. Energy Rev. **27**, 413–434 (2013)
4. Drew, B., Plummer, A., Sahinkaya, M.: A review of wave energy converter technology. Proc. Inst. Mech. Eng. Part A J. Power Energy **223**, 887–902 (2009)
5. Lagoun, M., Benalia, A., Benbouzid, M.: Ocean wave converters: state of the art and current status. In: Proceedings of 2010 IEEE International Energy Conference, Manama, Bahrain, pp. 636–641 (2010)
6. Gato, L.M.C., Falcao, F.: On the theory of the Wells turbine. J. Eng. Gas Turb. Power **106**(3), 628–633 (1984)
7. Henderson, R.: Design, simulation, and testing of a novel hydraulic power takeoff system for the Pelamis wave energy converter. Renew. Energy **31**(2), 271–283 (2006)
8. Mueller, M.A.: Electrical generators for direct drive wave energy converters. IEEE Gener. Trans. Distrib. **149**(4), 446–456 (2002)
9. Marquis, L., Kramer, M.M., Kringelum, J., Chozas, J.F., Helstrup, N.E.: Introduction of Wavestar Wave Energy Converters at the Danish offshore wind power plant Horns Rev 2. In: Proceedings of 4th International Conference on Ocean Energy ICOE 2012 (2012)
10. Eco Wave Power (2019). http://www.ecowavepower.com. Accessed 13 May 2021
11. Estefen, S., Castello, X., Lourenço, M.I., Rossetto, R.: Design analyses applied to a hyperbaric wave energy converter. In: Proceedings of 11th International Symposium on Practical Design of Ships and Other Floating Structures, PRADS, vol. 2 (2010)
12. Albert, A., Berselli, G., Bruzzone, L., Fanghella, P.: Mechanical design and simulation of an onshore four-bar wave energy converter. Renew. Energy **114**(part B), 766–774 (2017)
13. Bruzzone, L., Fanghella, P., Berselli, G.: Reinforcement learning control of an onshore oscillating arm wave energy converter. Ocean Eng. **206**, Article no. 107346 (2020)
14. Folley, M.: Numerical modelling of Wave Energy Converters. Elsevier, New York (2016)
15. Shadman, M., Estefen, S.F., Rodriguez, C.A., Nogueira, I.C.M.: A geometrical optimization method applied to a heaving point absorber wave energy converter. Renew. Energy **115**, 533–546 (2018)
16. Zang, Z., Zhang, Q., Qi, Y., Fu, X.: Hydrodynamic responses and efficiency analyses of a heaving-buoy wave energy converter with PTO damping in regular and irregular waves. Renew. Energy **116**(part A), 527–542 (2018)
17. Holthuijsen, L.H.: Waves Oceanic Coastal Waters. Cambridge University Press, Cambridge (2007)
18. Sutton, R.S., Barto, A.G.: Reinforcement Learning. 2nd ed. The MIT Press, Cambridge (2018)
19. Chakrabarti, S.K.: Offshore Structure Modeling. World Scientific Publishing, Singapore (1994)

# Heat Transfer Behaviour of Hybrid Solar Collector Module for Liquid-Based Type

Ahssan M. A. Alshibil[1,4], Piroska Víg[2], and Istvan Farkas[3](✉)

[1] Doctoral School of Mechanical Engineering, Hungarian University of Agriculture and Life Sciences, Godollo, Hungary
ihssanm.ali@uokufa.edu.iq

[2] Institute of Mathematics and Basic Science, Hungarian University of Agriculture and Life Sciences, Godollo, Hungary
vig.piroska@uni-mate.hu

[3] Institute of Technology, Hungarian University of Agriculture and Life Sciences, Godollo, Hungary
farkas.istvan@uni-mate.hu

[4] Department of Mechanical Engineering, College of Engineering, University of Kufa, Kufa, Iraq

**Abstract.** The solar thermal collector and photovoltaic systems are combined to form what is known as the hybrid solar collectors (PV/T). This method takes advantage of solar light and heat to generate electricity and thermal energy. This study conducts the heat transfer directions into the PV/T materials sections. The general objective of this paper is to know how heat is transferred through these sections via building a thermal model. The thermal behavior of PV/T collectors has a significant impact on how the PV/T outputs appear concerning each other. The common tool TRNSYS in the simulation of solar systems was selected to describe the thermal behavior of the PV/T. Using a standard pattern of meteorologicaldata for the middle days of three months in February, June, and October as reference weather data for TRNSYS components. The variations of the thermal and electrical efficiencies which aid the working fluid temperature due to the typical months have been investigated.

**Keywords:** Affordable and clean energy · Sustainable energy system · Hybrid solar collector · Thermal model · Heat transfer

## 1 Introduction

Solar systems are considered one of the energy production alternatives that do not make the contaminations or serious wastes and, in this way, reduce the hazards of environmental pollution. Thus, it produces clean energy and is considered a sustainable energy system. Utilizing solar energy as an alternative energy source has high reliability because there is no cost of fuel or any moving mechanical parts, thus, less maintenance. Tthe sun's energy is converted into electrical energy by solar photovoltaic cells and thermal energy by solar collectors [1].


G. Quaglia et al. (Eds.): I4SDG 2021, MMS 108, pp. 20–29, 2022.
https://doi.org/10.1007/978-3-030-87383-7_3

The solar photovoltaic cell that produces electricity from the photons of solar radiation. The other is the solar collector, generates thermal energy from solar radiation. Between 7% to 16% of solar radiation that falls on photovoltaic cells is converted to electricity. Part of the residual radiation is reflected and a significant amount is turned into heat. This heat increases the solar cell temperature and reduces their efficiency and lifetime [2].The combination that includes the PV module and a heat exchanger at its back is called a photovoltaic thermal collector (PV/T). The heat absorbed by the solar collector from therear of the solar modules can decrease the surface PV module's temperature. Thus, electricity yield is increased, and the gained heat energy can be used in different applications [3, 4].

In addition to energy generation, this system contributes to reducing global warming to enhance the world's sustainability. According to the Technology Roadmap reported by the International Energy Agency [5], power from PV modules deployed worldwide may exceed over 4.6 TW by 2050 and create 6,300 TWh of electrical. Considering a 15% PV efficiency, the remaining 85% of the radiation from the sun is partly reflected and partly converted to thermal energy by the PV modules, equal to a yearly temperature rise of 0.024 °C of the Earth's atmosphere. It has heated up four times more than the average worldwide temperature rise in the last 100 years. This is the significant background for the exploitation of the that makes PV/T combination.

In terms of the nominal temperature of the PV module and solar radiation, Ross [6] predicted the PV cells operating temperature. The prediction was unsuitable due to its difference with wind velocity. King et al. [7] have offered another definition for the temperature of PV cells depending on the module's rear temperature. Considering the thermal effect generated by the PV cells, Jones and Underwood [8] have created a thermal model for a photovoltaic cell as a non-steady state.

In the ability to forecast thermal and electrical performances, Raghuraman [9] has developed a 1-D analysis for fluid flat-plate PV/T collectors. The study asserted the enhancement in the heat transfer between the PV module and absorber plate, and the collector performance greatly improved. According to an open thermodynamic system, Chow [10] tried to investigate the PV/T module performance, of which a dynamic model was created for a glazed flat-plate PV/T water heating device. Diego and Seppo [11] considered specific radiation features, including view variables, to provide a thermal model assumes an average temperature in the module to transmit radiation through the layers. These measurements established a thermal model using a mean temperature as a constant and accounted for all radiation losses with that information. A thermal model suggestion from Armstrong and Hurley [12] matches the PV cells' electrical characteristics with the thermal output. They studied the effects of changing wind speed on the panel PV.

This study investigates the heat transfer dynamics of the PV/T materials. The general objective of this paper is to know how heat is transferred through these sections via building a thermal model as the thermal behavior of the PV/T collectors. The common tool TRNSYS for the simulation of solar systems was selected to describe the thermal behavior of the PV/T. A standard pattern of meteorologicaldata for three middle days in February, June, and October was used reference weather data for the TRNSYS components. Finally, it is shown how the research could positively enhance affordable and clean energy use.

## 2 Thermodynamic Analysis of Layers of PV/T Collectors

The PVT collectors can be classified into two essential categories. Based on which media is used for heat removal, water-based and air-based. Besides that glass cover, PV cells, Poly Vinyl Fluoride polymer (Tedlar), ethylene-vinyl acetate layers (EVA), absorber plate, and thermal insulation are the common layers for the PV/T collectors [13]. The heat transfer goes through these layers. This section will describe the mathematical model of the thermal behavior for the liquid-based PV/T collector layers used in this study. Figure 1 illustrates the directions of heat transfer into these layers.

Generally, solar energy systems received energy as input energy from the sun $E_{in}$ as has shown in the Eq. 1:

$$E_{in} = I_G \cdot A_m \tag{1}$$

where $I_G$ is the global irradiance, $A_m$ is the area of the collector surface.

The solar energy falling on the PV/T module is distributed into the output energies and losses, as shown in Fig. 2. Thus, using the Hottel model configured by Florschuetz [14], the energy balance equation of the flat-plate based PV/T collector as a steady-states is as follows:

$$I_G \cdot A_m = E_u + E_e + E_l + E_{rl} \tag{2}$$

$I_G \bullet A_m$ is the incoming solar energy on the PV/T module area $A_m$ ($m^2$), $E_u$ is the useful energy gain to the heat removal fluid, $E_e$ is the electrical energy obtained, $E_l$ the thermal energy losses, $E_{rl}$ the reflection losses.

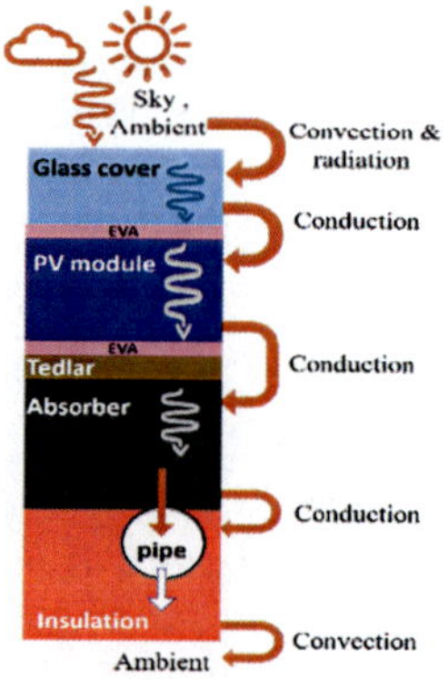


**Fig. 1.** Heat transfer directions of the PV/T

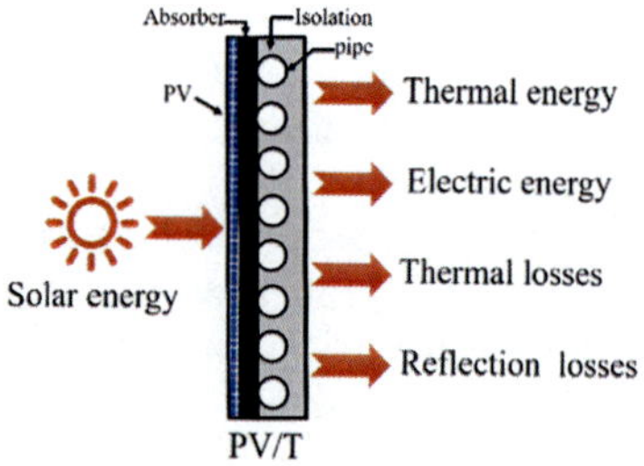


**Fig. 2.** Distribution of solar energy falling on PV/T

## 3 Heat Transfer Direction of the PV/T

Most of the radiation that hits the transparent glass surface has its way through the PV cells and the Tedlar adhesive, and the remaining is transformed into heat. The absorber plate conducts heat energy to the pipes. Once the heat is absorbed, the heat pipes deliver it to the liquid.

For the model to optimize estimation, certain simplifying assumptions have been made since some variables have been neglected, like Tedlar heat capacity, EVA heat capacity, and the longitudinal heat transfer of the absorber plate. In the same direction, the fluid temperature is assumed to be the same [15].

### 3.1 Transparent Glass Surface

At any position of the glass cover, the energy balance indicates the solar energy received by that surface, the losses described as convected heat, the radiation heat transmitted to the surroundings through the surface of the glass, and the conducted heat transferred to the solar cells through the glass surface. Equation 3 describes the heat transfer direction for this layer:

$$\rho_g\, A_g\, cp_g\, \delta_g\, \frac{dT_g}{dt} = E_g - h_{rd,g\sim s}\, A_g\, \Delta T_{g\sim s} - h_{cv,g\sim a}\, A_g\, \Delta T_{g\sim a} - h_{cd,g\sim PV}\, A_g\, \Delta T_{g\sim PV} \tag{3}$$

The solar radiation received by the transparent glass surface $E_g$ as energy is [16]:

$$E_g = \alpha_g A_g I_G \tag{4}$$

where $\alpha_g$, $\rho_g$, $A_g$, $cp_g$, $\delta_g$ are the absorption coefficient, density, area, the specific heat of glass, and glass surface thickness, respectively. $h_{rd,g\sim s}$, $h_{cv,g\sim a}$, $h_{cd,g\sim PV}$ are the radiation, convection, and conduction coefficients of heat transfer between the transparent glass surface and sky, ambient, and PV cells, respectively. $\Delta T_{g\sim s}$, $\Delta T_{g\sim a}$, $\Delta T_{g\sim PV}$ are the temperature differences between the transparent glass surface and sky, ambient, and PV cells, respectively. $I_G$ is the global solar irradiance.

Equation 5 describes the heat transfer energy as radiation between the sky and the transparent glass surface [17].

$$h_{rd,g\sim s} = \varepsilon_g \sigma (T_g^2 + T_s^2)(T_g + T_s) \tag{5}$$

where $\varepsilon_g$ and $\sigma$ are the glass emissivity and constant of Stefan–Boltzmann.

Using Eq. 6, Swinbank determined the sky's temperature as [18, 19]:

$$T_s = 0.0552 T_a^{1.5}. \tag{6}$$

Besides, Eq. 7 describes transferred heat in the glass surface along the outer sides, which is by convection coefficients of heat energy transfer [20, 21].

$$h_{cv,g\sim a} = 2.8 + 3V_w, \tag{7}$$

where $V_w$ is the wind velocity.

The conduction coefficient of heat transfer between the transparent glass surface and the PV cells is calculated by Eq. 8 [1].

$$h_{cd,g\sim PV} = \left[\frac{thk_g}{cod_g} + \frac{thk_{PV}}{cod_{PV}}\right]^{-1}, \quad (8)$$

where $thk_g$ and $thk_{PV}$ are the thickness of the transparent glass surface and PV module, $cod_g$ and $cod_{pv}$ are the conductivity of the transparent glass surface and PV module.

## 3.2 Photovoltaic Cells Section

The incoming solar radiation transferred through the glass surface is absorbed by the PV cells, and the remaining is transmitted to the absorber layer through conduction. Equation 9 describes this heat direction.

$$\rho_{PV} A_{PV} cp_{PV} \delta_{PV} \frac{dT_{PV}}{dt} = E_{PV} + h_{cd,g\sim PV} A_{PV} \Delta T_{g\sim PV} - h_{cd,PV\sim abp} A_{PV} \Delta T_{PV\sim abp} - E_e, \quad (9)$$

where $\rho_{PV}$, $A_{PV}$, $cp_{PV}$, $\delta_{PV}$ are the density, area, specific heat of PV, and thickness of the PVmoduel, respectively. $h_{cd,PV\sim abp}$ is the conduction coefficient of heat transfer between the absorber and PV module. $\Delta T_{PV\sim abp}$ is the temperature difference between the PV module and absorber. $E_{PV}$ is determined by the expression:

$$E_{PV} = \alpha_{PV} \tau_g A_{PV} I_G, \quad (10)$$

where $\alpha_{PV}$ is the PV absorption coefficient of solar cells, $E_{PV}$ is the sun radiation per unit area received as energy by solar cells, and $\tau_g$ is the transmittance of the glass surface.

The conduction coefficient of heat transfer between the absorber plate and the PV solar module is calculated from:

$$h_{cd,PV\sim abp} = \left[\frac{thk_{abp}}{cod_{abp}} + \frac{thk_{PV}}{cod_{PV}}\right]^{-1}, \quad (11)$$

where $cod_{abp}$ and $thk_{abp}$ are the conductivity and the thickness of the absorber.

## 3.3 Absorber Plate Section

Through the heat transfer between the PV module and the absorber plate, pipe and insulation, the energy consumed by the absorber plate can be measured by Eq. 12 as follows:

$$\rho_{abp} A_{abp} cp_{abp} \delta_{abp} \frac{dT_{abp}}{dt} = h_{cd,PV\sim abp} A_{abp} \Delta T_{PV\sim abp} - h_{cd,abp\sim pipe} A_{abp,pipe} \Delta T_{abp\sim pipe} - h_{cd,abp\sim is} A_{abp,is} \Delta T_{abp\sim is} \quad (12)$$

where $\rho_{abp}$, $A_{abp}$, $cp_{abp}$, $\delta_{abp}$ are the density of absorber, area, the specific heat of absorber, and thickness of the absorber plate, respectively. $h_{cd,abp\sim is}$, $h_{cd,abp\sim pipe}$ are conduction coefficients of heat transfer between the absorber plate, insulation and pipe. $\Delta T_{abp\sim pipe}$, $\Delta T_{abp\sim is}$ are the temperature differences between the absorber with insulation and pipe.

The heat transfer conduction coefficients between the absorber and insulation, absorber and pipe, respectively, are as follow:

$$h_{cd,abp\sim is} = \left[\frac{thk_{abp}}{cod_{abp}} + \frac{thk_{is}}{cod_{is}}\right]^{-1}, \tag{13}$$

$$h_{cd,abp\sim pipe} = \left[\frac{thk_{abp}}{cod_{abp}} + \frac{thk_{pipe}}{cod_{pipe}}\right]^{-1}, \tag{14}$$

where $thk_{pipe}$ and $thk_{is}$ are the thickness of the pipe and insulation, $cod_{pipe}$ and $cod_{is}$ are the conductivity of the pipe and insulation.

## 3.4 Pipe Section

The total amount of energy being absorbed by the pipe, minus the amount expended to the layers of liquid and insulation, is the energy balance equation in Eq. 15, as follows:

$$\rho_{pipe} A_{pipe} cp_{pipe} \delta_{pipe} \frac{dT_{pipe}}{dt} = h_{cd,abp\sim pipe} A_{abp,pipe} \Delta T_{abp\sim pipe} - h_{cv,pipe\sim l} A_{abp,pipe} \Delta T_{pipe\sim l} - h_{cd,pipe\sim is} A_{pipe,is} \Delta T_{pipe\sim is} \tag{15}$$

where $\rho_{pipe}$, $A_{pipe}$, $cp_{pipe}$, $\delta_{pipe}$ are the density of pipe, area, the specific heat of pipe, and pipe thickness, respectively. $h_{cv,pipe\sim l}$, $h_{cd,pipe\sim is}$ are convection and conduction coefficients of heat transfer between the pipe with liquid and insulation. $\Delta T_{pipe\sim l}$, $\Delta T_{pipe\sim is}$ are the temperature differences between the pipe with liquid and insulation.

The following equation articulates the conduction coefficient of heat transfer presented by this layer:

$$h_{cd,pipe\sim is} = \left[\frac{thk_{pipe}}{cod_{pipe}} + \frac{thk_{is}}{cod_{is}}\right]^{-1}. \tag{16}$$

Flow in the pipe relies on the liquid Nusselt number. The equation of the convection coefficient of heat transfer is expressed by Eq. 17 [22, 23]:

$$h_{cv,pipe\sim is} = 0.5 Nu_l \frac{cod_l}{R_i}, \tag{17}$$

where $R_i$ is the tube inner radius and $cod_l$ is the thermal conductivity.

## 3.5 Liquid Section

The equation of the thermal balance Eq. 18 assumes that the amount of heat conveyed to the liquid is equal to the convection heat received by the pipe, subtracting the liquid thermal energy [24].

$$\rho_l A_l cp_l \frac{dT_l}{dt} = h_{cv,pipe\sim l} A_{pipe,l} \Delta T_{pipe\sim l} - \dot{m}_l cp_l \Delta T_{PV\sim abp}, \tag{18}$$

where $\rho_l$, $A_l$, $cp_l$, $\delta_l$ are the density of the liquid, area, the specific heat of the liquid, and liquid thickness, respectively. $\dot{m}_l$ is the liquid mass flow rate.

### 3.6 Insulation Section

Equation 19 describes the heat transfer of this section, where the energy is transferred from the pipe and absorber plate to the insulation layer, and the convection heat lost transferred from this layer to the surrounding.

$$\rho_{is}A_{is}cp_{is}\delta_{is}\frac{dT_{is}}{dt} = h_{cd,abp\sim is}A_{abp,is}\Delta T_{abp\sim is} + h_{cd,pipe\sim is}A_{pipe,is}\Delta T_{pipe\sim is} - h_{cv,is\sim a}A_{is}\Delta T_{is\sim a}, \tag{19}$$

where $\rho_{is}$, $A_{is}$, $cp_{is}$, $\delta_{is}$ are the density of insulation, area, the specific heat of insulation, and insulation thickness, respectively. $h_{cv,is\sim a}$ is the convection coefficient of the heat transfer between the insulation and ambient. $\Delta T_{is\sim a}$, is the temperature differences between the insulation and ambient.

## 4 Thermal Behaviours Model Simulation of PV/T

The thermal behavior of PV/T collectors has a significant impact on how the PV/T outputs appear concerning each other. The common tool TRNSYS for the simulation of solar systems was selected to describe the thermal behavior of the PV/T. Using a standard pattern of meteorological data for the middle days of the months of February, June and October was used as a reference weather data for TRNSYS components for a specific design and weather conditions [25].

The thermal and electrical behavior for these three months is shown in Fig. 3 in parallel with the cell's temperature. Based on the results, the thermal efficiency during

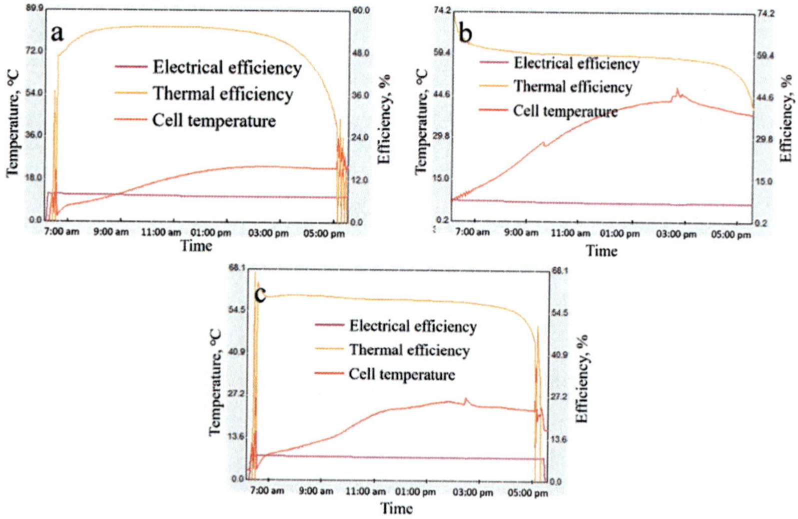


**Fig. 3.** PV/T efficiencies with cell temperature for three months, (a) February, (b) June, (c) October.

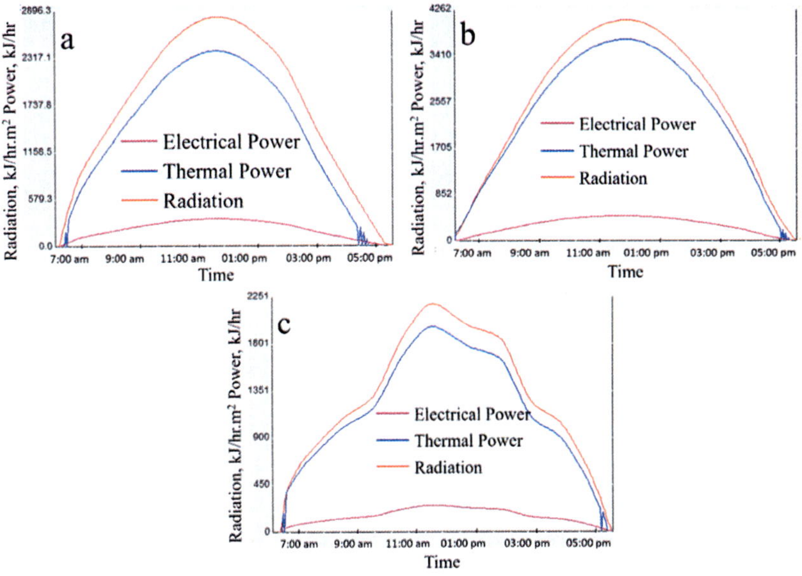


**Fig. 4.** PV/T energies with solar radiation for three months, (a) February, (b) June, (c) October

in June was higher than in February and in October by 12.7% and 5.1%. Besides, the electrical efficiency was higher by 5.2% and 2.5%.

Furthermore, Fig. 4 describes the relation of radiation with thermal and electrical energies of the PV/T. In June, the electrical and thermal energy system was 24.4% and 35% higher than in February, 48.8% and 43.3% higher than in October.

The most significant factor in the PV/T system is the outlet temperature of the fluid. Its value in June was 40.2% and 44.11% higher than in February and October.

## 5 Conclusion

Hybrid solar thermal collectors are a promising technology because they have a dual-energy source, thermal and electrical. The thermal model of the PV/T has been presented in detail for its most common sections. The thermal behavior was examined using the TRNSYS tools. According to what has been presented in the thermal model for the PV/T layers, the most significant findings were that the best thermal efficiency in June was higher by 12.7% than in the worst case in February, and the electrical efficiency increased by 5.2%, respectively. The relationship between solar radiation and thermal and electrical energies of PV/T is also defined.

On the contrary, June's electricity and thermal productivity were higher by 48.8%, and 43.3% in the worst case in October. The outlet temperature of the fluid is the crucial ingredient in the PV/T system. Its value in June was higher by 40.2% and 44.11% than in February and October. For the future, the materials used as a coolant with high thermal conductivity can improve the hybrid solar collector efficiency and change the general configuration of PV/T collectors, which enhances heat transfer between the absorber and coolant. It has been shown how the research can positively advance the use of affordable and clean energy.

Besides, the heat transfer enhancement for the PV/T sections lead to a decrease in heat losses, thus, this contributes to curbing global warming for an excellent sustainable system.

**Acknowledgment.** This work was supported by the Stipendium Hungaricum Programme and the Doctoral School of Mechanical Engineering, Hungarian University of Agriculture and Life Sciences, Gödöllő, Hungary.

## References

1. Duffie, J.A.: Beckman: Solar Engineering of Thermal Processes, 4th edn. Wiley, Canada (2013)
2. Rodrigues, E.M.G., Melício, R., Mendes, V.M.F., Catalão, J.P.S.: Simulation of a solar cell considering single-diode equivalent circuit model. Renew. Energy Power Qual. J. **1**, 369–373 (2011)
3. Bődi, S., Víg, P., Farkas, I.: Possibilities of improving PV/T system efficiency. Hung. Agri. Eng. **33**, 55–58 (2018)
4. Kumar, A., Baredar, P., Qureshi, U.: Historical and recent development of photovoltaic thermal (PVT) technologies. Renew. Sustain. Energy Rev. **42**, 1428–1436 (2015)
5. IEA: Technology Roadmap, Solar photovoltaic energy, International Energy Agency, pp. 20–21 (2014)
6. Ross, R.G.: Design techniques for flat-plate photovoltaic arrays. In: Proceedings of the 15th Photovoltaic Specialists Conference, Orlando, FL, USA. pp. 12–15 (1981).
7. Lawrence, M.G.: The relationship between relative humidity and the dewpoint temperature in moist air: A simple conversion and applications. Bull. Am. Meteorol. Soc. **86**, 225–233 (2005)
8. Jones, A.D., Underwood, C.P.: A thermal model for photovoltaic systems. Fuel Energy Abs. **70**, 349–359 (2001)
9. Raghuraman, P.: Analytical predictions of liquid and air photovoltaic/thermal, flat-plate collector performance. J. Solar Energy Eng. **103**, 291–298 (1981)
10. Chow, T.T.: Performance analysis of photovoltaic-thermal collector by explicit dynamic model. Sol. Energy **75**, 143–152 (2003)
11. Torres-Lobera, D., Valkealahti, S.: Inclusive dynamic thermal and electric simulation model of solar PV systems under varying atmospheric conditions. Sol. Energy **105**, 632–647 (2014)
12. Armstrong, S., Hurley, W.G.: A thermal model for photovoltaic panels under varying atmospheric conditions. Appl. Thermal Eng. **30**(11–12), 1488–1495 (2010)
13. Michael, J.J., Iniyan, S., Goic, R.: Flat plate solar photovoltaic-thermal (PV/T) systems: a reference guide. Renew. Sustain. Energy Rev. **51**, 62–88 (2015)
14. Florschuetz, L.W.: Extension of the Hottel-Whillier model to the analysis of combined photovoltaic/thermal flat plate collectors. Sol. Energy **22**, 361–366 (1979)
15. Rejeb, O., Dhaou, H., Jemni, A.: Parameters effect analysis of a photovoltaic thermal collector: case study for climatic conditions of Monastir, Tunisia. Energy Conv. Manag. **89**, 409–419 (2015)
16. Boumaaraf, B., Touafek, K., Ait-cheikh, M.S.: Comparison of electrical and thermal performance evaluation of a classical PV generator and a water glazed hybrid photovoltaic – thermal collector. Math. Comput. Simul. **167**, 176–193 (2020)

17. Sarhaddi, F., Farahat, S., Ajam, H., Behzadmehr, A.: Exergetic performance assessment of a solar photovoltaic thermal (PV/T ) air collector. Energ Build. **42**, 2184–2199 (2010)
18. Aste, N., Del Pero, C., Leonforte, F., Manfren, M.: Performance monitoring and modeling of an uncovered photovoltaic-thermal (PVT) water collector. Sol. Energy **135**, 551–568 (2016)
19. Barroso, J.C.S., Barth, N., Correia, J.P.M., Ahzi, S., Khaleel, M.A.: A computational analysis of coupled thermal and electrical behavior of PV panels. Solar Energy Mater. Solar Cells **148**, 73–86 (2016)
20. Aste, N., Leonforte, F., Del Pero, C.: Design, modeling and performance monitoring of a photovoltaic-thermal (PVT) water collector. Sol. Energy **112**, 85–99 (2015)
21. Fudholi, A., Sopian, K., Yazdi, M.H., Ruslan, M.H., Ibrahim, A., Kazem, H.A.: Performance analysis of photovoltaic thermal (PVT) water collectors. Energy Conv. Manage. **78**, 641–651 (2014)
22. Antonanzas, J., del Amo, A., Martinez-Gracia, A., Bayod-Rujula, A.A., Antonanzas-Torres, F.: Towards the optimization of convective losses in photovoltaic-thermal panels. Sol. Energy **116**, 323–336 (2015)
23. Maadi, S.R., Kolahan, A., Passandideh-Fard, M., Sardarabadi, M., Moloudi, R.: Characterization of PVT systems equipped with nanofluids-based collector from entropy generation. Energy Conv. Manage. **150**, 515–531 (2017)
24. Das, D., Kalita, P., Dewan, A., Tanweer, S.: Development of a novel thermal model for a PV/T collector and its experimental analysis. Solar Energy **188**, 631–643 (2019)
25. Alshibil, A.M.A., Víg, P., Farkas, I.: Transient simulation of a hybrid solar collector system. Mech. Eng. Lett. **20**, 101–106 (2020)

# Thermal Improvement in Parabolic Trough Solar Collector Using Receiver Tube Design and Nanofluid

Asaad Yasseen Al-Rabeeah[1,4], Istvan Seres[2], and Istvan Farkas[3](✉)

[1] Doctoral School of Mechanical Engineering, Hungarian University of Agriculture and Life Sciences, Gödöllő, Hungary
Al-Rabeeah.Asaad.Yasseen.Ali@phd.uni-mate.hu
[2] Institute of Mathematics and Basic Science, Hungarian University of Agriculture and Life Sciences, Gödöllő, Hungary
seres.istvan@uni-mate.hu
[3] Institute of Technology, Hungarian University of Agriculture and Life Sciences, Pater K. u. 1., Gödöllő, Hungary
farkas.istvan@uni-mate.hu
[4] Department of Mechanical Engineering, Faculty of Engineering, University of Kufa, Kufa, Iraq

**Abstract.** Improving the thermal performance of parabolic trough solar collector (PTSC) has been recently a topic of intensive research due to its high advantages, such as a reduction in system size and expense, as well as an improvement in thermal efficiency. Such studies strive for enhancing heat transfer in the receiver portion to reduce heat loss and improve heat transfer to the working fluid. The focus of the recent analysis is an experimental research relevant to the thermal performance of the receiver part. Different enhancement techniques, including nanofluids, the surface modifications and inserts or the two categories mixed, are discussed. The thermophysical properties of nanoparticles play an essential role in the absorption, the extinguishing coefficient, and penetration depth of the nanofluid's solar radiation. Finally, the key findings were presented for each work, and this review may open new horizons for more research in this area.

**Keywords:** Affordable and clean energy · PTSC · Thermal performance improvement · Receiver tube geometry · Nanofluid

## 1 Introduction

Energy demand is increasing every day, and non-renewable sources are decreasing at an alarming rate. Conventional energy sources, such as fossil fuels (oil, natural gas and coal), would be exhausted in a few decades [1]. Use of the renewable energy sources appears to be one of the most promising options for producing clean energy [2]. Therefore, renewable energy production is one of the key instruments to reduce greenhouse gas emission. PTSC is one of the most advanced large-scale solar power technologies available today. PTSC is an active technology used in the field of solar thermal applications.


G. Quaglia et al. (Eds.): I4SDG 2021, MMS 108, pp. 30–40, 2022.
https://doi.org/10.1007/978-3-030-87383-7_4

These collectors have many benefits, including high power density, high thermal performance, versatility and modularity [3]. It consists of a reflecting surface in a parabola shape, an absorber tube and the working fluid passing through the tube. The absorber tube is usually coated by a selective surface for greater absorption of solar radiation and reducing the losses [4, 5].

The design should be accurate to increase the thermal efficiency and the selection of material with less weight, greater mechanical strength and higher thermal conductivity is preferable. The material of the absorber tube effects the heat transfer between the metal and the working fluid which impact the performance of PTSC [6].

In addition, recent developments in nanotechnology have led to nanofluid application in a wide range of fields, such as solar collectors [7]. The addition of nanofluids to the base fluid causes a change in the physical properties of thermal conductivity and the optical properties of the base fluid.

The nanoparticles can be from metallic (Au, Fe, Cu, Al) and non-metallic ($Al_2O_3$, CuO, $TiO_2$, $SiO_2$) [8]. The nanoparticles volumetric concentration is usually low (up to 8%), and the diameters of nanoparticle are usually up to 100 nm. Nanofluids are promising options for thermal applications because the suspension of nanoparticles in the base fluid results in superior thermal properties compared to base fluids [9]. Nanofluids have higher thermal conductivity than base fluids, which raises the heat transfer coefficient in the flow considerably. So, thermal losses are reduced, and performance is improved [10].

This study reviews innovations and technologies on the PTSC and different modifications and assessment techniques to improve their effectiveness. The literature review has categorized into the following themes of heat transfer element, governing equations, and performance enhancement techniques. Finally, it is shown how the research can positively interact with the use of affordable and clean energy.

## 2 Heat Transfer Element for PTSC

The heat transfer element of PTSC is the major component and contains an absorber tube (Fig. 1). The solar radiation has focused the heat on the absorber tube, and a heat transfer fluid is moving through the tube, such as thermal oil, water, nanoparticle laden fluid, gaseous and molten salt [11].

The schematic diagram of the PTSC receiver illustrates (Fig. 2) the losses from the cross-section of the tube. The absorber tube is covered by an evacuated glass envelope to reduce the heat losses. The fluid flow inside the absorber tube by forced convection may be single-phase or two-phase [13].

The total power by the sun to the collector (absorbed radiation):

$$Q_{abs} = I_B a \eta_o, \tag{1}$$

where: $I_B$- beam radiation, $\alpha$ - the absorptivity, $\eta_o$- optical efficiency.

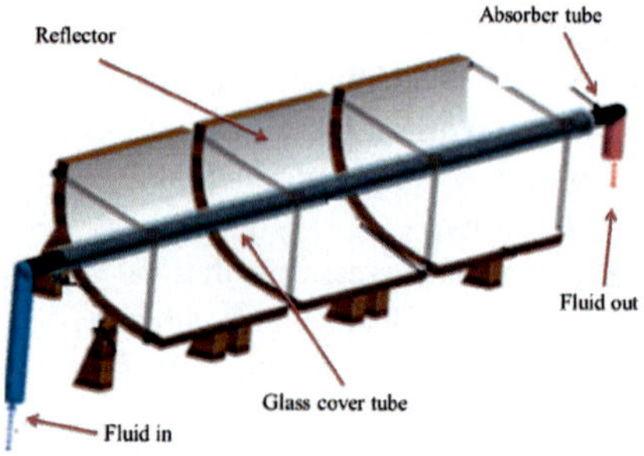


**Fig. 1.** Schematic of a typical PTSC [12]

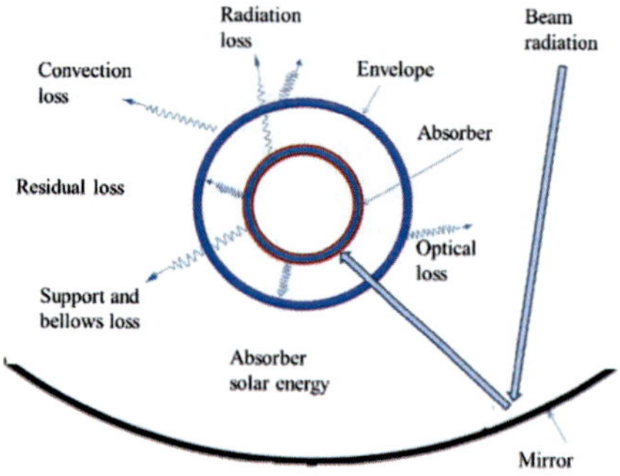


**Fig. 2.** The schematic diagram of PTSC receiver

Convection inside the absorber tube between the fluid and the absorber

$$Q_{conv} = \pi D_{ia} h_{htf} (T_{ia} - T_{htf}), \tag{2}$$

$$h_{htf} = Nu_{htf} \frac{k_{htf}}{D_{ia}}, \tag{3}$$

where: $h_{htf}$ - convection heat transfer coefficient *(W/m$^2$.K)*, $D_{ia}$ - inside diameter of the absorber tube (*m*), $T_{htf}$ - mean (bulk) temperature of the heat transfer fluid ($^oC$), $T_{ia}$ - inside surface temperature (°C), $Nu_{htf}$ - Nusselt number based on $D_{ia}$, $k_{htf}$ - thermal conductance of the heat transfer fluid at $T_{htf}$*(W/mK)*.

The Nusselt number is constant when the laminar option is selected, and the Reynolds number is less than 2300. The value for pipe flow would be 4.36 [14].
Laminar:

$$Nu_{htf} = 4.36, \tag{4}$$

The following Nusselt number correlation developed by Gnielinskiis, to model the convective heat transfer from the absorber to the heat transfer fluid for turbulent and transitional cases (Reynolds number > 2300) is used [15].

Turbulence:

$$Nu_{htf} = \frac{\frac{f}{8}(Re_{htf} - 1000)Pr_{htf}}{1 + 12.7\sqrt{\frac{f}{8}}(Pr_{htf}^{\frac{2}{3}} - 1)}\left(\frac{Pr_{htf}}{Pr_{ia}}\right)^{0.11}, \quad (5)$$

$$f = (0.79 lnRe_{htf} - 1.64)^{-2}, \quad (6)$$

where: $f$ - friction factor for the fluid, $Re_{htf}$ - Reynolds number of the fluid, $Pr_{htf}$ - Prandtl number evaluated at the temperature of the heat transfer fluid, $Pr_{ia}$ - Prandtl number evaluated at the absorber inner surface temperature.

The convective heat transfer between the absorber and the glass cover can be written in the annular space per meter as:

$$Q_{conv} = \frac{2\pi k_{eff}}{\ln(D_{ie} - D_{oa})}(T_{oa} - T_{ie}), \quad (7)$$

$$\frac{k_{eff}}{k} = \left(\frac{Pr}{0.861 + Pr}\right)^{\frac{1}{4}}(FcRa)^{\frac{1}{4}}, \quad (8)$$

$$Fc = \frac{(\ln\left(\frac{D_{ie}}{D_{oa}}\right))^4}{(\frac{D_{ie}-D_{oa}}{2})^3(D_{ie}^{-\frac{3}{5}} - D_{oa}^{-\frac{3}{5}})^5}, \quad (9)$$

where: $K_{eff}$ - effective thermal conductivity, $Fc$ - geometric factor for concentric cylinders, $Ra$ - the Rayleigh number.

Free molecular convection in the annulus

$$Q_{conv} = \frac{2\pi k_{eff}}{\ln(D_{ie} - D_{oa})}(T_{oa} - T_{ie}), \quad (10)$$

$$\frac{k_{eff}}{k} = \left[1 + \frac{(2-a)(9\gamma - 5)\lambda}{a(\gamma + 1)\ln\left(\frac{D_{ie}}{D_{oa}}\right)}\left(\frac{1}{D_{ie}} + \frac{1}{D_{oa}}\right)\right]^{-1}, \quad (11)$$

$$\lambda = 2.331 \times 10^{-20}\frac{T_{ae}}{Pa^{\delta 2}} \quad (12)$$

where: $\lambda$ - mean-free-path between collisions of a molecule (*cm*), $\gamma$ - ratio of specific heats for the annulus gas, $a$ - accommodation coefficient, $T_{ae}$ - average temperature ($^oC$), P - annulus gas pressure (*mmHg*), $\delta$ - molecular diameter of annulus gas (*cm*).

The radiation heat transfer from the absorber outside wall to the glass cover inner wall is calculated as [16]

$$Q_{rad} = \frac{\sigma\pi D_{oa}(T_{oa}^4 - T_{ia}^4)}{\frac{1}{\varepsilon_a} + \frac{(1-\varepsilon_c)D_{oa}}{\varepsilon_c D_{ia}}}, \quad (13)$$

where, $\sigma$ - the Stefan-Boltzmann constant ($\sigma = 5.67 \times 10^{-8}\ W/m^2K^4$), $\varepsilon_a$ = selective coating emissivity, and $\varepsilon_c$ - glass cove emissivity.

We can calculate the conduction through supports as

$$Q_{cond} = \sqrt{h_s P_s k_b A_s}(T_a - T_b)/L_{HSC} \tag{14}$$

where $h_s$- the convection heat transfer coefficient, $A_s$– surface area, $k_b$-conduction coefficient (W/m-K), $L_{HSC}$- length (m).

The greatest cause of heat loss is convection from the glass envelope to the atmosphere, particularly if there is wind [17]:

$$Q_{conv} = \pi D_{oe} h_a (T_{oe} - T_a) \tag{15}$$

natural:

$$h_a = Nu \frac{k_a}{D_{oe}}, \tag{16}$$

$$Nu = [0.6 + \frac{0.387 RA_{D_{oe}}^{1/6}}{[1 + (0.599\mathrm{Pr})^{9/16}]^{8/27}}]^2, \tag{17}$$

where: $T_{oe}$ - outer surface temperature of glass envelope (°C), $T_a$ - ambient temperature (°C), $h_a$ - convection heat transfer coefficient *(Toe – Ta)/2 (W/m$^2$-K)*, $k_a$ - thermal conductance *(Toe – Ta)/2 (W/m-K)*, $D_{oe}$ - outer diameter of glass envelope (*m*), *Nu* - average Nusselt number.

## 3 Thermal Enhancement Techniques

Improving thermal performance inside the receiver is essential to enhance PTSC efficiency. This aims to increase heat transfer flow from inside face of the receiver to the thermal fluid, resulting in lower heat losses and higher thermal performance [18]. There are many ways to improve heat transfer: passive, active or their combination.

The passive ways can be summarized to improve heat transfer without using any external force and they include: inserting different versions or altering the receiver tube's inner and outer surfaces, adding metallic and non-metallic nanoparticles to a variety of base fluids to create nanofluid.

All categories aim to increase heat transfer of PTSC, reduce heat loss by reducing thermal boundary conditions and increasing turbulent fluid flow, and enhance thermophysical properties such as thermal conductivity of the resulting fluids [19].

### 3.1 Novel Designs

Various new designs were implemented to improve the efficiency of PTSC, as outlined in Table 1. The novel design experiments have centred on either increasing the thermal and optical efficiency by absorbing a large amount of solar radiation or reducing the amount of heat lost from the heat collector element [20]. In order to reduce the losses of solar radiation, various designs suggest various shapes, such as triangular, V-shaped, and cylindrical [21].

**Table 1.** Novel design studies in PTSC

| Ref. | Enhancement technique | Solution method | Finding |
|---|---|---|---|
| [22] | V-shape cavity, insulated envelope with copper rows pipes | Finite difference method | The optimal collector performance shifts toward a higher flow rate value at high inlet temperatures |
| [23] | Porous heat collector element with an envelope of partially insulated | Finite difference method | Insulation conductivity affects a significant impact on thermal efficiency. As the envelope diameter increases, less solar radiation strikes on the mirror, lowering thermal efficiency |
| [24] | Absorber tube has two long fins at externally attached | Iteration | Thermal efficiency improves with fin height/absorber diameter ratio and determines the intercept factor |
| [25] | vacuum shell of double wall | Thermal resistance network | The evacuated tube performance is superior in vacuum shell to the non-evacuated tubes |
| [26] | Cylindrical air-based cavity heat collector element | Finite difference method | Spillage loses a third of an incident radiation on the heat collector element, as well as one-sixth due to reflection |
| [27] | Twin glass tube | Thermal resistance network | The proposed collector element is optimal for low heat transfer flow of inlet temperatures due to efficiency losses |

## 3.2 Passive Heat Transfer Enhancement

This section is concerned with improving the performance measures of the receiver tube by modifying its geometry by inserts or fins [28]. Several researchers examined the use of above techniques to improve passive heat transfer in the PTSC absorber, as shown in Table 2.

**Table 2.** Passive studies in PTSC

| Ref. | Enhancement technique | Solution method | Finding |
|---|---|---|---|
| [29] | Porous fins | Finite volume method | Porous fins increase the overall performance of heat collector element in comparison with solid longitudinal fins |
| [30] | Porous triangular, square, and circular inserts trapezoidal | Finite volume method | In all porous configurations, the heat loss due to the heat collector element was roughly the same |
| [31] | Helical fin | Finite volume method | It leads to a 40% increase in parasitic losses, which corresponds to a 3% increase in collector efficiency |
| [32] | Twisted wall tape | Finite volume method | The heat transfer efficiency improves, with twist ratio reduced and the Reynolds number and width ratio increase |
| [33] | Helical fins, protrusions, and dimples | Finite volume method | The heat collector element with dimples performs better as a heat transfer compared to collector elements with protrusions or helical fins |
| [34] | Internal longitudinal fin | Finite difference method | Fins with a length of 10 mm are more effective in terms of exergetic. Helium is the most effective operating fluid as compared to air and carbon dioxide |

### 3.3 Nanofluids

Oil and water are common industrial fluids. However, in some implementations, these fluids are insufficiently thermally conductive.

Nanofluids were discovered to get through this barrier as they are made up of nanoscale components that are dispersed inside a fluid. In general, nanofluids are composed of two phases: nanoparticles in the solid phase and base fluid in the liquid phase [35].

In recent time, the field of nanofluids is very active, with several researchers contributing experimentally and theoretically to improving the thermal and physical properties of nanofluids [36]. The disadvantage of nano technology is its high cost which applied in PTSC. The nanofluids increase heat transfer efficiency in the heat collector element by capturing solar radiant energy more effectively.

Table 3 summarizes the research on nanofluid use in traditional PTSC and shows how this technique improves heat transfer and thermal efficiency.

**Table 3.** Nanofluid and prominent findings in PTSC

| Ref. | Nanofluid nanoparticle/base fluid | Volume concentration (%) | Main findings |
|---|---|---|---|
| [37] | $Al_2O_3$/water | 0.2<br>0.25<br>0.3 | At 2 l/min, the maximum thermal efficiencies obtained with $Al_2O_3$ and $Fe_2O_3$ nanofluids are 13% and 11% higher, respectively |
| [38] | $Al_2O_3$/water | 0.1<br>0.3 | At 1% and 3%volume concentration, the maximum efficiency obtained 57.7% and 57.7%, respectively. The thermal performance depends on the incident angle |
| [39] | $Fe_2O_3$/water | 0.05 | The PTSC efficiency reaches a maximum, which is 25% higher than the traditional |
| [40] | Fe3O4 CuO/Therminol 66 | 4 | The convective heat transfer coefficient is increased by increasing the Reynolds number |
| [41] | MWCNT/ Mineral oil | 0.2<br>0.3 | When 0.2% and 0.3% MCNT/mineral oil nanofluid is used instead of pure oil, the global efficiency is increased by 4–5% and 5–7%, respectively |
| [42] | CuO/water SiO2/water | 0.01 | When 0.01% concentration of CuO/water and $SiO_2$/water of the parabolic solar collector, performance is increased by around 7.64% and 6.68% per cent, respectively |

## 4 Conclusion

The current analysis concentrated on the energy crisis in the world, which improved ground-breaking research, especially in renewable energy. Solar energy is an excellent strategy for meeting present energy demand, which is expected to rise even further in the future. The PTSC can produce up to 500 °C temperature and contribute to the advancement of thermal applications. The overall performance of the PTSC improves by enhancing the thermophysical properties. According to the thermal analysis of PTSC, the thermal efficiency can be improved by modifying major determining parameters such as absorber materials, working fluid, and absorber coatings. The use of nanofluid is a good approach for enhancing the thermal efficiency of the PTSC was successful and demonstrated a variable enhancement ratio. Nanofluids have higher thermal conductivity than base fluids, which raises the heat transfer coefficient in the flow considerably. So, thermal losses are reduced, and performance is improved. In future work it will also be shown how the research can positively interact with the use of affordable and clean energy.

**Acknowledgement.** This work was supported by the Stipendium Hungaricum Programme and the Doctoral School of Mechanical Engineering, Hungarian University of Agriculture and Life Sciences, Gödöllő, Hungary.

## Referencesss

1. Bellos, E., Korres, D., Tzivanidis, C., Antonopoulos, K.A.: Design, simulation and optimization of a compound parabolic collector. Sustain. Energy Technol. Assess. **16**, 53–63 (2016)
2. Wang, K., He, Y.-L., Qiu, Y., Zhang, Y.: A novel integrated simulation approach couples MCRT and Gebhart methods to simulate solar radiation transfer in a solar power tower system with a cavity receiver. Renew. Energy **89**, 93–107 (2016)
3. Al-Rabeeah, A.Y., Seres, I., Farkas, I.: Recent improvements of the optical and thermal performance of the parabolic trough solar collector systems. Facta Univ. Ser. Mech. Eng. vol. 19 (2021). https://doi.org/10.22190/FUME201106030A.
4. Patil, R.G., Panse, S.V., Joshi, J.B.: Optimization of non-evacuated receiver of solar collector having non-uniform temperature distribution for minimum heat loss. Energy Convers. Manage. **85**, 70–84 (2014)
5. Reddy, K.S., Ananthsornaraj, C.: Design, development and performance investigation of solar Parabolic Trough Collector for large-scale solar power plants. Renew. Energy. **146**, 1943–1957 (2020)
6. Fuqiang, W., Jianyu, T., Lanxin, M., Chengchao, W.: Effects of glass cover on heat flux distribution for tube receiver with parabolic trough collector system. Energy Convers. Manage. **90**, 47–52 (2015)
7. Marefati, M., Mehrpooya, M., Shafii, M.B.: Optical and thermal analysis of a parabolic trough solar collector for production of thermal energy in different climates in Iran with comparison between the conventional nanofluids. J. Clean. Prod. **175**, 294–313 (2018)
8. Sahin, A.Z., Uddin, M.A., Yilbas, B.S., Al-Sharafi, A.: Performance enhancement of solar energy systems using nanofluids: an updated review. Renew. Energy. **145**, 1126–1148 (2020)
9. Bhalla, V., Beejawat, S., Doshi, J., Khullar, V., Singh, H., Tyagi, H.: Silicone oil envelope for enhancing the performance of nanofluid-based direct absorption solar collectors. Renew. Energy **145**, 2733–2740 (2020)
10. Sharafeldin, M.A., Grof, G.: Evacuated tube solar collector performance using CeO2/water nanofluid. J. Clean. Prod. **185**, 347–356 (2018)

11. Benoit, H., Spreafico, L., Gauthier, D., Flamant, G.: Review of heat transfer fluids in tube-receivers used in concentrating solar thermal systems: properties and heat transfer coefficients. Renew. Sustain. Energy Rev. **55**, 298–315 (2016)
12. Forristall, R.: Heat transfer analysis and modeling of a parabolic trough solar receiver implemented in engineering equation solver. Natl. Renew. Energy Lab. Golden, Co. (US) (2003)
13. Cengel, Y.A., Ghajar, A.J.: Heat and mass transfer (a practical approach, SI version). McGraw-700 Hill Educ., 701, 52 (2011)
14. Incropera, F.P., Dewitt, D.P.: Fundamentals of Heat and Mass Transfer, p. 17. Wiley, New york (1990)
15. Gnielinski, V.: New equations for heat and mass transfer in turbulent pipe and channel flow. Int. Chem. Eng. **16**, 359–368 (1976)
16. Bergman, T.L., Incropera, F.P., DeWitt, D.P., Lavine, A.S.: Fundamentals of Heat and Mass Transfer. Wiley, New York (2011)
17. Ratzel, A.C., Hickox, C.E., Gartling, D.K.: Techniques for reducing thermal conduction and natural convection heat losses in annular receiver geometries (1979)
18. Padilla, R.V., Demirkaya, G., Goswami, D.Y., Stefanakos, E., Rahman, M.M.: Heat transfer analysis of parabolic trough solar receiver. Appl. Energy **88**, 5097–5110 (2011)
19. Akbarzadeh, S., Valipour, M.S.: Heat transfer enhancement in parabolic trough collectors: A comprehensive review. Renew. Sustain. Energy Rev. **92**, 198–218 (2018)
20. Kajavali, A., Sivaraman, B., Kulasekharan, N.: Investigation of heat transfer enhancement in a parabolic trough collector with a modified absorber. Int. Energy J. **14** (2015)
21. Chandra, Y.P., Singh, A., Mohapatra, S.K., Kesari, J.P., Rana, L.: Numerical optimization and convective thermal loss analysis of improved solar parabolic trough collector receiver system with one sided thermal insulation. Sol. Energy **148**, 36–48 (2017)
22. Sandeep, H.M., Arunachala, U.C.: Solar parabolic trough collectors: a review on heat transfer augmentation techniques. Renew. Sustain. Energy Rev. **69**, 1218–1231 (2017)
23. Grald, E.W., Kuehn, T.H.: Performance analysis of a parabolic trough solar collector with a porous absorber receiver. Sol. energy. **42**, 281–292 (1989)
24. Hegazy, A.S.: Thermal performance of a parabolic trough collector with a longitudinal externally finned absorber. Heat mass Transf. **31**, 95–103 (1995)
25. Daniel, P., Joshi, Y., Das, A.K.: Numerical investigation of parabolic trough receiver performance with outer vacuum shell. Sol. Energy. **85**, 1910–1914 (2011)
26. Bader, R., Pedretti, A., Steinfeld, A.: Experimental and numerical heat transfer analysis of an air-based cavity-receiver for solar trough concentrators. J. Sol. energy Eng., 134, (2012)
27. Fan, M., Liang, H., You, S., Zhang, H., Zheng, W., Xia, J.: Heat transfer analysis of a new volumetric based receiver for parabolic trough solar collector. Energy **142**, 920–931 (2018)
28. Huang, Z.F., Nakayama, A., Yang, K., Yang, C., Liu, W.: Enhancing heat transfer in the core flow by using porous medium insert in a tube. Int. J. Heat Mass Transf. **53**, 1164–1174 (2010)
29. Reddy, K.S., Kumar, K.R., Satyanarayana, G.: V: Numerical investigation of energy-efficient receiver for solar parabolic trough concentrator. Heat Transf. Eng. **29**, 961–972 (2008)
30. Reddy, K.S., Satyanarayana, G.: V: Numerical study of porous finned receiver for solar parabolic trough concentrator. Eng. Appl. Comput. fluid Mech. **2**, 172–184 (2008)
31. Muñoz, J., Abánades, A.: Analysis of internal helically finned tubes for parabolic trough design by CFD tools. Appl. Energy. **88**, 4139–4149 (2011)
32. Mwesigye, A., Bello-Ochende, T., Meyer, J.P.: Heat transfer enhancement in a parabolic trough receiver using wall detached twisted tape inserts. In: ASME International Mechanical Engineering Congress and Exposition. p. V06BT07A031. American Society of Mechanical Engineers (2013)
33. Jebasingh, V.K., Herbert, G.M.J.: Numerical simulation of elliptical absorber tube in parabolic trough collector for better heat transfer. Indian J. Sci. Technol. **8**, 1–7 (2015)

34. Bellos, E., Tzivanidis, C., Daniil, I., Antonopoulos, K.A.: The impact of internal longitudinal fins in parabolic trough collectors operating with gases. Energy Convers. Manag. **135**, 35–54 (2017)
35. Mahian, O., Kianifar, A., Heris, S.Z., Wen, D., Sahin, A.Z., Wongwises, S.: Nanofluids effects on the evaporation rate in a solar still equipped with a heat exchanger. Nano Energy **36**, 134–155 (2017)
36. Mahian, O., Kianifar, A., Sahin, A.Z., Wongwises, S.: Performance analysis of a minichannel-based solar collector using different nanofluids. Energy Convers. Manag. **88**, 129–138 (2014)
37. Rehan, M.A., et al.: Experimental performance analysis of low concentration ratio solar parabolic trough collectors with nanofluids in winter conditions. Ren. Energy., 118, 742–751 (2018)
38. De los Rios, M.S.B., Rivera-Solorio, C.I., García-Cuéllar, A.J.: Thermal performance of a parabolic trough linear collector using Al2O3/H2O nanofluids. Renew. Energy., 122, 665–673 (2018)
39. Alsaady, M., Fu, R., Yan, Y., Liu, Z., Wu, S., Boukhanouf, R.: An experimental investigation on the effect of ferrofluids on the efficiency of novel parabolic trough solar collector under laminar flow conditions. Heat Transf. Eng. **40**, 753–761 (2019)
40. Malekan, M., Khosravi, A., Syri, S.: Heat transfer modeling of a parabolic trough solar collector with working fluid of Fe3O4 and CuO/Therminol 66 nanofluids under magnetic field. Appl. Therm. Eng., 163, 114435 (2019)
41. Kasaeian, A., Daviran, S., Azarian, R.D., Rashidi, A.: Performance evaluation and nanofluid using capability study of a solar parabolic trough collector. Energy Convers. Manag. **89**, 368–375 (2015)
42. Ajay, K., Kundan, L.: Combined experimental and CFD investigation of the parabolic shaped solar collector utilizing nanofluid (CuO-H2O and SiO2-H2O) as a working fluid. J. Eng., 2016, (2016)

# Robotics Applied to Precision Agriculture: The Sustainable Agri.q Rover Case Study

Andrea Botta(✉) and Paride Cavallone

Politecnico di Torino, 10129 Turin, Italy
{andrea.botta,paride.cavallone}@polito.it

**Abstract.** The Agri.q rover for precision agriculture is the main focus of the present paper. After a description of the prototype, a set of experimental results is presented to validate the energetic sustainability of the robot. Characterized by a modular articulated mechanical structure, and provided with specific sensors and tools, the Agri.q is able to operate in unstructured agricultural environments in order to fulfill several tasks as mapping, monitoring, and manipulating or collecting small soil and crop samples. In addition, the rover is equipped with a platform covered with solar panels, whose orientation can be exploited to maximize the efficiency of the energy harvesting, but also to function as a self-leveling landing platform for drones. The design of such system was mainly driven by a particular attention to energy consumption, bearing in mind the main objective of improving the profitability while decreasing the impact of the agricultural processes. In this paper, all these characteristics are described and analyzed in detail, with particular attention to the energy balance of the whole machine deriving from locomotion on different slopes and soils, panels orienting operations, and of course collection of solar energy.

**Keywords:** SDG2 · SDG12 · SDG15 · Precision agriculture · Robotics

## 1 Introduction

Among the many scientific and technical challenges deriving from the pursuit of the *Sustainable Development Goals* (SDGs) agenda, the application of robotics in agricultural processes particularly caught the attention of the engineering research community [1]. Such technologies, in fact, perfectly fit the *precision agriculture* paradigm, which aims at improving the sustainability of agricultural production by taking advantage of specialized equipment. As a side effect, also profitability and efficiency turn conveniently enhanced [2]. In the main scenario of the targets proposed by the SDG 2 (*Zero Hunger*), SDG 12 (*Responsible Consumption and Production*), and SDG 15 (*Life on Land*), the smart implementation of robotics to agriculture promotes a sustainable use of the terrestrial ecosystem, by enhancing the food production efficiency while maintaining a low


G. Quaglia et al. (Eds.): I4SDG 2021, MMS 108, pp. 41–50, 2022.
https://doi.org/10.1007/978-3-030-87383-7_5

impact both reducing the use of pesticides and fertilizers, and restraining the use of non-renewable energy.

**Fig. 1.** Prototype of the Agri.q mobile robot.

Many relevant studies demonstrated in the recent past the beneficial effect of precision agriculture on the environment and economics in many developed countries [3,4]. The interest demonstrated by private producers and governments is then well justified. The leading role played by robotics and automation can profit of the significant advancements already obtained in decades of industrial research, driving the modernization of the agricultural production. Many available examples in literature are re-adaptation of agricultural machines: Wang et al. [5], and later Zaman et al. [6], focused on the use of advanced navigation systems; in [7] a cooperative system of aerial and ground vehicles is used for monitoring tasks; Khaliq et al. [8] approached the problem under the point of view of trajectory planning using multispectral imagery. Aside such interesting advancements in the field of control, few examples of machines designed to encounter some specific requirements are available. Among others [9] presents an innovative solution for automated harvesting, while in [10] the cutting of a specific tree is considered.

In this scenario, the researchers of the Politecnico di Torino developed a novel wheeled UGV (shown in Fig. 1) tailored on precision agriculture tasks with a specific attention towards monitoring and sampling of crops and soil [11,12]. The variety of tasks that the robot is able to fulfill has been widened by providing it with a collaborative robotic arm [13] for samples and low-force operations. At present, the research is focusing on the benefits carried on by the Agri.q locomotion strategies [14] and on the strategies adoptable for an effective and efficient use of the 8-wheeled locomotion layout.

## 2 Rover Description

The rover object of this case study, named Agri.q, is a small sized electric vehicle built to operate in unstructured workspaces, to move through vineyards, and to cooperate with aerial drones. To improve the battery run-time, the robot was equipped with a set of solar panels (Fig. 2a). Such feature permits the robot to operate long mission without intervention of human operators but also to function as a self-leveling landing platform for drones. At the time of this work, this is the only implemented drone cooperation feature. Future plans include the use of drones as an aid to mapping and autonomous operation in general.

The main guideline which drove the functional design of the Agri.q locomotion system has been efficiency on uneven terrains. It is conceived to reach a convenient balance between the efficiency of wheels and the effectiveness of tracks on uneven terrains, from both energetic and mobility points of view. To this aim, the rover is provided with four driving units, each one composed of two wheels housed on a rocker arm (Fig. 2b). For further details on the robot structure, the interested reader is addressed to [12,15]. The connection of each rocker with the robot main body takes place by means of a passive revolute joint, whose free rotation allows the wheels to follow the ground slopes independently from the configurations of the other rockers. The high number of contact points with the ground ensures an improved weight distribution, similarly to a tracked robot, that reduces the single wheel contact force magnitude, and therefore the soil compaction, while maintaining an efficiency comparable to that of a wheeled machine.

The actuators of the modular locomotion systems have been carefully chosen to allow it to approach severe slopes (about 15°) at a maximum speed of about 5 km/h, a reasonable speed for the robot tasks. To achieve such goal, the transmission of each unit is composed by an electric DC motor, connected to a planetary gearbox. A further motion reduction is then provided by a chain system which synchronously moves the two wheels. Some details on the adopted mechanical components are provided in Table 1.

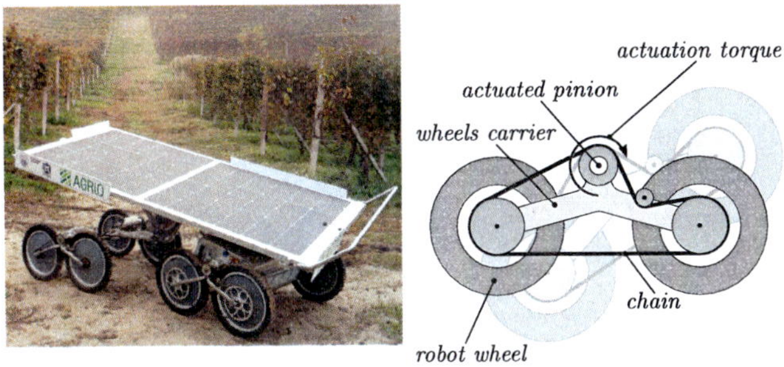


(a) (b)

**Fig. 2.** Usage of the orientable upper panel during solar recharge (a); efficient soil assessment mechanism (b).

In order to guarantee the locomotion of the system even in unfavorable conditions for sunrays collection, the energetic layout was completed by a battery able to feed the adopted locomotion systems (nominal voltage 29.7 V, capacity 56 Ah, total energy stored 4.84 MJ). The choice of the battery was also made considering

**Table 1.** Front and rear driving modules functional parameters.

| Description | Symbol | Front driving units | Rear driving units |
|---|---|---|---|
| Motor nominal power | $P_{M,Nom}$ | 120 W | 120 W |
| Motor nominal torque | $C_{M,Nom}$ | 0,38 Nm | 0,38 Nm |
| Motor nominal speed | $\omega_{M,Nom}$ | 3000 rpm | 3000 rpm |
| Motor nominal current | $I_{M,Nom}$ | 10 A | 10 A |
| Motor maximum current | $I_{M,Max}$ | 20 A | 20 A |
| Motor torque constant | $K_I$ | 0.055 Nm/A | 0.055 Nm/A |
| Gearbox reduction | $\tau_1$ | 1:15.88 | 1:28.93 |
| Chain system reduction | $\tau_2$ | 1:3 | 1:1 |
| Total reduction | $\tau$ | 1:47.64 | 1:28.93 |
| Wheels radius | $r_w$ | 0,200 m | 0,200 m |

that the adopted solar panels must meet in dimension and weight the requirements deriving from a mobile application. In particular, two flexible light-weighted panels were installed to guarantee an adequate recharging capability without affecting the robot dynamics with an excessive load. Some technical details are shown in Fig. 3. The data refers to a perpendicular solar irradiation of $G_0 = 1000\,\mathrm{W/m^2}$ at the working temperature of 25 °C. Given the panel rated peak power of $W_p = 154\,\mathrm{W}$, the single panel surface $S = 1.0418\,\mathrm{m^2}$, and the reference irradiation $G_0$, it is possible to estimate the panels efficiency as follow:

$$\eta_p = \frac{2W_pG_0}{2\,S} = 0.1478 \tag{1}$$

then, the panel efficiency enables to evaluate the nominal *photo-voltaic* (PV) power from the *solar global horizontal irradiance* (GHI) data:

$$P_{PV,nominal} = S\eta_p\mathrm{GHI} \tag{2}$$

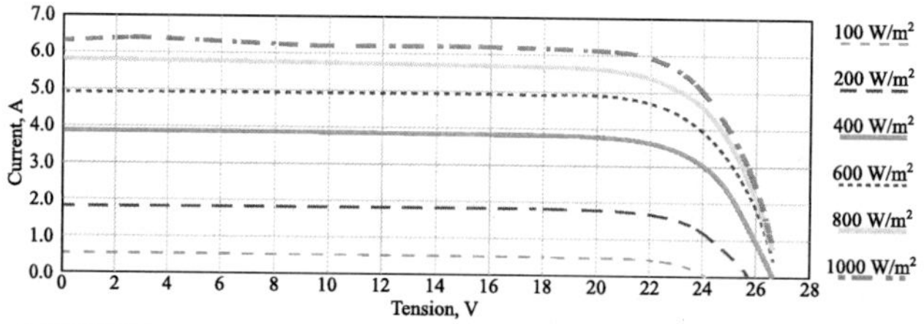


| | | | |
|---|---|---|---|
| Peak power | $154\,W$ | Maximum tension | $18.2\,V$ |
| Dimensions | $1046 \times 996\,mm$ | Maximum current | $4.84\,A$ |
| Mass | $2.4\,kg$ | Working temperature | $-45/+48°\,C$ |

**Fig. 3.** Current/voltage characteristic of the photo-voltaic panels at different levels of solar incident power.

## 3 Methodology

The main focus of this case study about sustainable mobile robot for precision agriculture is to evaluate Agri.q power expenditures and generation while it performs its typical tasks. In order to understand the robot power flows, the robot own sensors and additional power meters had been employed (Fig. 4). A custom power sensor was used to measure the robot battery voltage and the battery discharge current powering the whole robot. A similar custom sensor measured the power coming from the solar panels to the battery. While these two main sensors could be sufficient to estimate the robot power balance, during the experimental runs, all the traction motors speeds and currents were collected by the robot internal sensors. By means of these data and the motors torque constant $K_T$, it was possible to asses the required traction power. All these sensors, and other ones useful to keep track of Agri.q state, were logged at $10\,Hz$ during the test runs simulating typical activities.

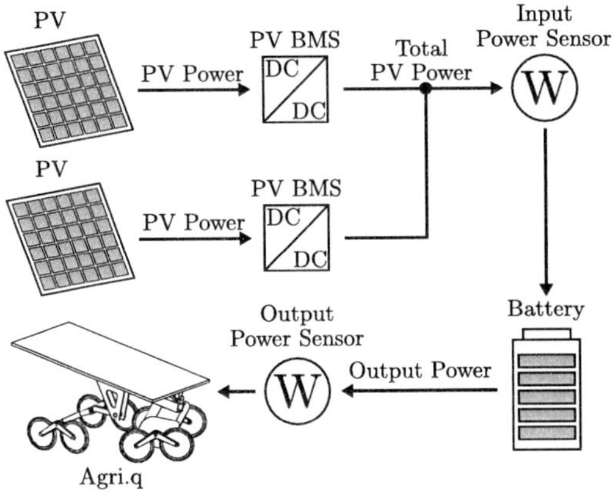


**Fig. 4.** Agri.q power flows schematic representation

A total of 8 test scenarios were adopted. In the first one, the robot was left idle in an open field with its solar panels horizontal recharging itself for about 5 min. In the second test scenario, the solar panels were oriented to maximize the solar energy collection while idle. During both these tests simulating two different approach to the robot recharge strategy, the sky was clear and no shadow was casting on the solar panel. In the following tests, Agri.q drove in a straight line on a paved road for about 300 m with just its front locomotion units (run 1) and with all its locomotion units (run 2). After that, the robot drove following a short straight path on a grass field, first activating only its front traction motors (run 3) and then using all the motors (run 4). To conclude the test scenarios, the robot followed a circular path on paved road (run 5) and on the grass (run 6), in both cases, all the traction motors were active. During all the test runs from 1 to 6 (navigation tests), the robot panels were held horizontal and the Sun exposure was not optimised.

The reference Sun radiation data was provided by Solcast [16] with a period of 30 min for the city of Turin on May 12th 2021.

## 4 Results

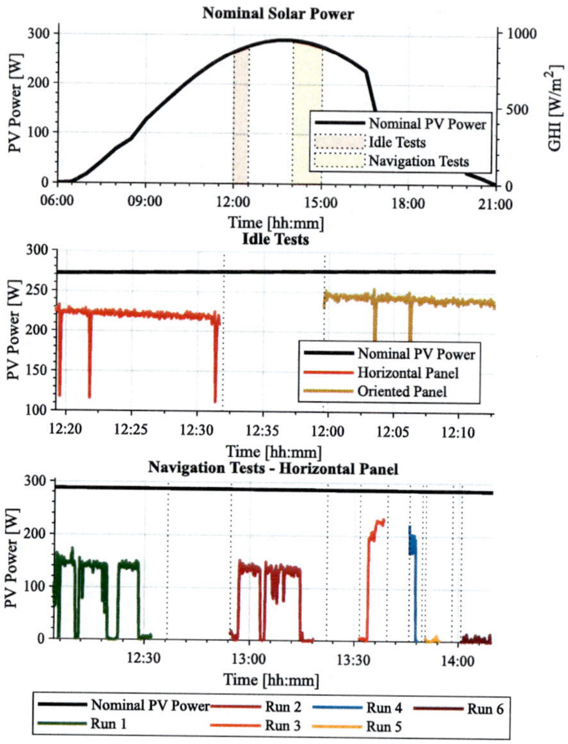


**Fig. 5.** Top: Reference Solar power incident on the panels surface on May 12th 2021 in Turin, Italy. Data provided by Solcast [16]. Centre: Solar power generated by Agri.q panels while idling. Bottom: Solar power generated by Agri.q during its navigation tests

Figure 5 collects some insight on how well Agri.q is able to harvest the solar energy in various scenario. The top figure shows the nominal solar power that the panels could ideally collect during the day and the relative GHI during the day. The centre figure depicts the detail about the two idle test where Agri.q were stationary and recharging its battery. As expected, when the panels are oriented to maximize the incident solar irradiance (Zenith = 29°, Azimuth = 151°), Agri.q generates about 9.8% more solar power than when the panels are kept horizontal. In both cases, the generated power is below the estimated ideal value, however the values are quite close, respectively about 19% less for the horizontal panels and 13% less for the oriented panels.The occasional peaks that appear in the measured data are partly due to noise and brief changes in lighting conditions. At last, the bottom figure shows how the panels behave while the robot is navigating without actively trying to recharge. The first two runs were done in partial or complete shadow, hence it can be clearly see that the generated power is much lower than the two previous scenario with complete Sun exposure. Runs 3 and 4 were instead in an open grass field with direct sun exposure, therefore the collected power is closer to the one while idling with the horizontal panel. The last two runs were done in areas shaded by buildings close by, therefore the harvested solar power is almost zero.

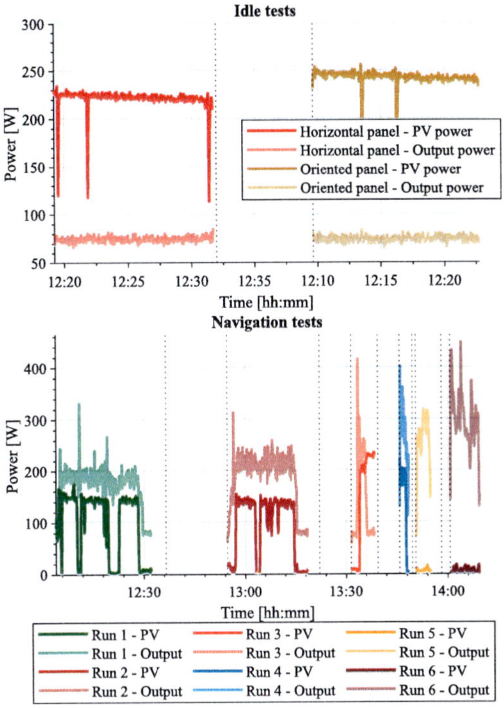


**Fig. 6.** Comparison between the robot required (output) power and the collected solar power (input).

To better evaluate the robot autonomy, it is useful to compare the power required by the robot to perform its task and the collected solar one. Figure 6 depicts this comparison for the different test scenarios. While idle, Agri.q consumed about 74 W of power to run its electronics, whose most energetically relevant elements are the *single board computer* (SBC), the micro-controller, and the motor drivers. The collected PV power is noticeable higher than the idle one both with the horizontal panels and with the oriented panels, hence the robot can easily recharge its battery with a good Sun exposure. When the robot is moving (runs 1–6), the solar exposition is not optimised, hence the solar power is never enough to recharge the robot, however it can largely contributes to reduce the battery discharge. For example, the required power to drive along a straight path on a paved road with only the front motors (run 1 in Fig. 6) is almost comparable to the power generated by the panels during the test run, even if they were partially or completely in shadow (when the PV power is zero) while driving along the straight path. The results of run 2 simply show that by activating all the locomotion units, the required power increases. Runs 3 and 4 were done with a better solar exposition, however, the power required to navigate the grass field is also larger. The last two tests (runs 5 and 6) provide irrelevant data about the power balance comparison, since during these tests the panels were shaded. Though, it can be seen that the power required to steer the robot on a paved surface is comparable to drive on a straight path on grass, and that the power required to follow a circular path on a grass field is even larger. Generally speaking, it is possible to observe a well known behaviour of this kind of robot architecture that are subject to significant lateral wheels slip while following curved paths [14,17].

Figure 7 represents how the required power is distributed in its 3 main components. The idle power is about 74 $W$, and, as said before, the main contributions to this value are the SBC, the micro-controller, and the motor drivers. The curve in yellow depicts the power contribution of the locomotion units (both the front ones and the rear ones in this case). The locomotion units power is an estimate of the actual mechanical power, since the motor speeds are directly measured but the motor torques are derived from the measured motor current. At last, the red curve is the total electric power required by the robot. The difference between the total power and the other two contributes is the unmeasured power required to run the rest of the robot. Similar results are found in the other runs.

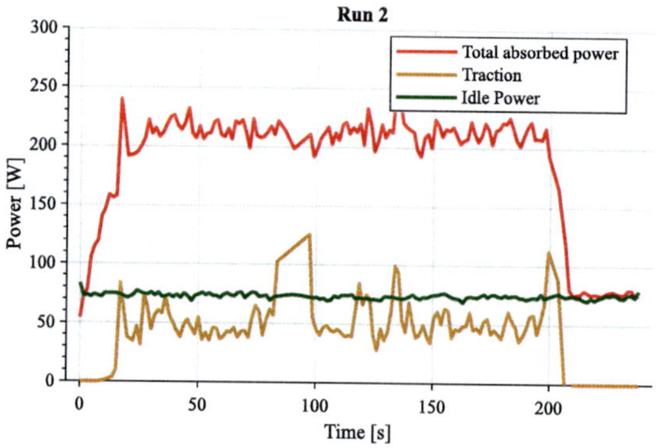


**Fig. 7.** Agri.q required power components during run 2: idle power, traction power, and total power.

## 5 Conclusions

This paper presented the case study of Agri.q, a mobile service robot designed for sustainable precision agriculture. The practical application of this modular rover is also a potential technological solution to address the SDGs 2, 12, and 15.

While precision agriculture could be a more sustainable approach to farming compared to traditional methods, the main focus of this study is to experimentally assess the energy balance of the rover in order to evaluate how much it can sustain itself by means of its solar panels. The main power flows of the robot have been monitored while it was doing its typical activities, such as navigating through different terrains, employing different traction control strategies or idling to recharge its battery.

The collected data demonstrates that in a sunny day with clear sky, Agri.q could sustain itself with a proper plan of its activities. If the robot is left idle with good Sun exposure the panels are able to recharge the battery. By optimising the solar panels orientation to maximize the collection of the direct solar irradiation, the recharge rate is further increased. While the robot drives, the solar panels are not enough to completely balance the required power, however, even if the Sun is not actively tracked, the panels provide a substantial contribute to reduce the battery discharge. Therefore by an accurate plan of Agri.q task in its typical day it is possible to largely extend the robot endurance. For example, the panels orientation could be controlled continuously to maximize the solar energy collection in accordance to its activities. In addition to that, the period of time around the solar irradiance peak could be exploited to perform an optimal recharge phase while idling. Further improvements could be reached with an optimisation of the robot power consumption with a partial redesign of its electronic architecture. For example, some components could be switched to low-power equivalent parts, components could be set in power-saving mode while idling, or the power distribution system could be redesigned to reduce inefficiency and thermal losses.

## References

1. Gil, J.D.B., Reidsma, P., Giller, K., Todman, L., Whitmore, A., van Ittersum, M.: Sustainable development goal 2: improved targets and indicators for agriculture and food security. Ambio **48**(7), 685–698 (2018). https://doi.org/10.1007/s13280-018-1101-4
2. Vecchio, Y., De Rosa, M., Adinolfi, F., Bartoli, L., Masi, M.: Adoption of precision farming tools: a context-related analysis. Land Use Policy **94**, 104481 (2020)
3. Lowenberg-DeBoer, J., Huang, I.Y., Grigoriadis, V., Blackmore, S.: Economics of robots and automation in field crop production. Precision Agric. **21**(2), 278–299 (2019). https://doi.org/10.1007/s11119-019-09667-5
4. Vecchio, Y., Agnusdei, G.P., Miglietta, P.P., Capitanio, F.: Adoption of precision farming tools: the case of Italian farmers. Int. J. Environ. Res. Public Health **17**(3), 869 (2020)

5. Wang, Y., Lan, Y., Zheng, Y., Lee, K., Cui, S., Lian, J.A.: A UGV-based laser scanner system for measuring tree geometric characteristics. In: International Symposium on Photoelectronic Detection and Imaging 2013: Laser Sensing and Imaging and Applications, vol. 8905, p. 890532. International Society for Optics and Photonics (2013)
6. Zaman, S., Comba, L., Biglia, A., Aimonino, D.R., Barge, P., Gay, P.: Cost-effective visual odometry system for vehicle motion control in agricultural environments. Comput. Electron. Agric. **162**, 82–94 (2019)
7. Tokekar, P., Vander Hook, J., Mulla, D., Isler, V.: Sensor planning for a symbiotic UAV and UGV system for precision agriculture. IEEE Trans. Robot. **32**(6), 1498–1511 (2016)
8. Khaliq, A., Comba, L., Biglia, A., Ricauda Aimonino, D., Chiaberge, M., Gay, P.: Comparison of satellite and UAV-based multispectral imagery for vineyard variability assessment. Remote Sensing **11**(4), 436 (2019)
9. Chatzimichali, A.P., Georgilas, I.P., Tourassis, V.D.: Design of an advanced prototype robot for white asparagus harvesting. In: 2009 IEEE/ASME International Conference on Advanced Intelligent Mechatronics, pp. 887–892. IEEE (2009)
10. Konam, S.: Agricultural aid for mango cutting (AAM). In: 2014 International Conference on Advances in Computing, Communications and Informatics (ICACCI), pp. 1520–1524. IEEE (2014)
11. Quaglia, G., Visconte, C., Carbonari, L., Botta, A., Cavallone, P.: Agri.q: a sustainable rover for precision agriculture. In: Visa, I., Duta, A. (eds.) Solar Energy Conversion in Communities, pp. 81–91. Springer, Cham (2020). https://doi.org/10.1007/978-3-030-55757-7_6
12. Cavallone, P., Botta, A., Carbonari, L., Visconte, C., Quaglia, G.: The agri.q mobile robot: preliminary experimental tests. In: Niola, V., Gasparetto, A. (eds.) The International Conference of IFToMM Italy, pp. 524–532. Springer, Cham (2020). https://doi.org/10.1007/978-3-030-55807-9_59
13. Quaglia, G., Visconte, C., Scimmi, L.S., Melchiorre, M., Cavallone, P., Pastorelli, S.: Robot arm and control architecture integration on a UGV for precision agriculture. In: IFToMM World Congress on Mechanism and Machine Science, pp. 2339–2348. Springer, Cham (2019). https://doi.org/10.1007/978-3-030-20131-9_231
14. Carbonari, L., Botta, A., Cavallone, P., Tagliavini, L., Quaglia, G.: Data driven analysis of locomotion for a class of articulated mobile robots. J. Mech. Robot. 1–15 (2021)
15. Visconte, C., Cavallone, P., Carbonari, L., Botta, A., Quaglia, G.: Design of a mechanism with embedded suspension to reconfigure the agri_q locomotion layout. Robotics **10**(1), 1–14 (2021)
16. Solar irradiance data (2021). https://solcast.com/
17. Botta, A., Cavallone, P., Tagliavini, L., Carbonari, L., Visconte, C., Quaglia, G.: An estimator for the kinematic behaviour of a mobile robot subject to large lateral slip. Appl. Sci. **11**(4), 1594 (2021)

# Design and Preliminary Testing of Cleanator a Mechatronic System for Solar Panels Cleaning

Marco Cristodaro, Santina Fortuna, Alessia Greco, Antonio Carmelo Petrolino, and Giuseppe Carbone(✉)

Department of Mechanics, Energy and Management Engineering (DIMEG), University of Calabria, Rende, Italy
giuseppe.carbone@unical.it

**Abstract.** The presence of dirt or dust causes losses of solar panels efficiency up to 40%. To ensure the maximum absorption rate of solar radiation, careful periodic cleaning of the panels surface is necessary. This paper proposes a mechatronic system for a safe and cost-effective cleaning of solar panels. The proposed mechatronic device Cleanator is an automated system that guarantees the washing of the photovoltaic modules by means of rotating brushes and demineralized water jets, which has been submitted for patenting. Specifically, the paper outlines the main design features of Cleanator with some insight on its modelling and design. Later a laboratory prototype is built and preliminarily tested to demonstrate the soundness and engineering feasibility of the proposed concept.

**Keywords:** SDG17 · Solar panels · Soiling impact · Cleaning robot · Mechanical design

## 1 Introduction

Solar energy can be used as an ideal and more conscious energy source for the direct production of electricity in order to reduce the use of fossil fuels to ensure high energy savings. According to the statistical report drawn up by national energy networks such as reported in [1] for Italy, the national solar energy market is constantly growing. this is also reported in statistical evidence such as shown in Fig. 1. The photovoltaic system in the entire period of activity must be able to generate electricity in a stable, safe way, and always guarantee maximum performance. The phenomenon of soiling impact is generated when, due to dirt or dust, the solar radiation is not captured at its best feasible efficiency rate by a solar panel module. This phenomenon, in addition to failing to generate the right amount of energy, also risks to damage or bring a solar panel out of order as reported for example in [2–5]. The problem of soiling power losses can be identified in all photovoltaic systems. However, this is a variable phenomenon.

It depends mainly on the environment in which the system is operating. Its effect goes from 5% per year in mild climates with moderate rainfall, up to 40% per year in particularly arid and desert climates [6]. In some areas, moreover, where there is the phenomenon of night frost and dust in the early hours of the morning, further dirt


G. Quaglia et al. (Eds.): I4SDG 2021, MMS 108, pp. 51–58, 2022.
https://doi.org/10.1007/978-3-030-87383-7_6

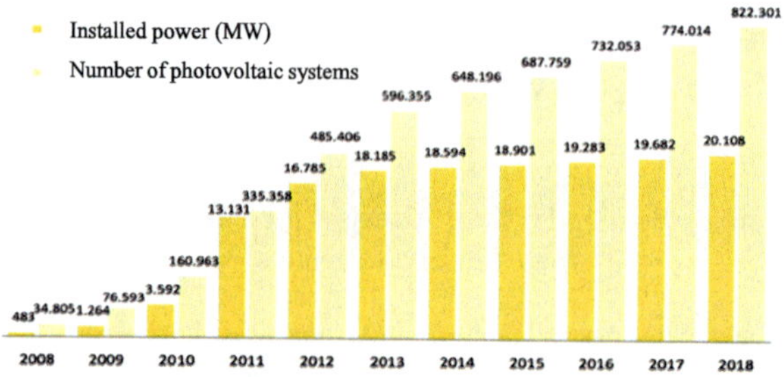


**Fig. 1.** Evolution of photovoltaic systems power and number, [1].

sedimentation can be generated on the panels surface. To this must be added the reduction in performance due to the deposit on the photovoltaic surface such as dry leaves guano, water stains due to rain or other materials [7]. To date, the traditional method to ensure maximum efficiency of the system is the periodic manual washing of the panels: a procedure that is particularly wasteful of economic resources, manpower and water. There are also robotic devices, which are controlled by expert personnel or completely autonomous as reported for example in [8–11].

The Cleanator project originated from the idea of designing a mechatronic system for the automated cleaning of photovoltaic panels for home uses. The choice arises from the need to ensure that the user does not find himself in dangerous conditions by manually washing the modules, especially if installed on very pitched roofs. In addition, the effectiveness of the device must be such as to remove not only dust but also limestone stains, foliage, and debris of various kinds, given the climatic and ecological conditions of the city environment, and not to damage the delicate surface of the module.

## 2 Proposed Solution and CAD Model

The requirements and constraints have been identified according to the solar panel standards and to the specific expected retail customers. For the purpose, the proposed system should be cheap, effective, easily constructed, and adaptable to almost all configurations of photovoltaic domestic facilities. Accordingly, Fig. 2a shows the proposed concept solution of the system. The whole device can slide longitudinally on the installation guides of photovoltaic modules through rolling bearings. The cleaning unit, mounted on the central frame, moves vertically on the roof using cables and pulleys. Furthermore, the cleaning unit slides on the central frame thanks to a network of rolling bearings, showed in Fig. 2b. It is possible to choose a suitable length of the frame to adapt the device on an existing photovoltaic system. The cleaning apparatus is rigidly linked with the cleaning unit and is composed of a rotating brush, kept at a correct distance from the panels surface thanks to the central frame, and a small pipe for liquid delivery. Figure 3a shows the device in its conceptual CAD scheme while Fig. 3b presents a detail of the cleaning unit that washes the photovoltaic modules using a rotating brush turned by a servomotor and jets of demineralized water driven by a cable chain. The cleaning unit moves longitudinally along the panel array by means of a friction wheel connected to a stepper motor to ensure the precise positioning of device. The proposed design of the proposed device has been submitted for patenting.

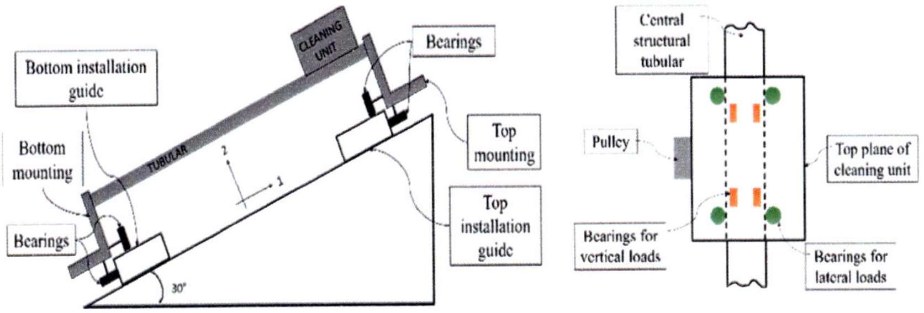


a) b)

**Fig. 2.** The proposed concept solution: a) scheme of the main components; b) Scheme of the cleaning unit.

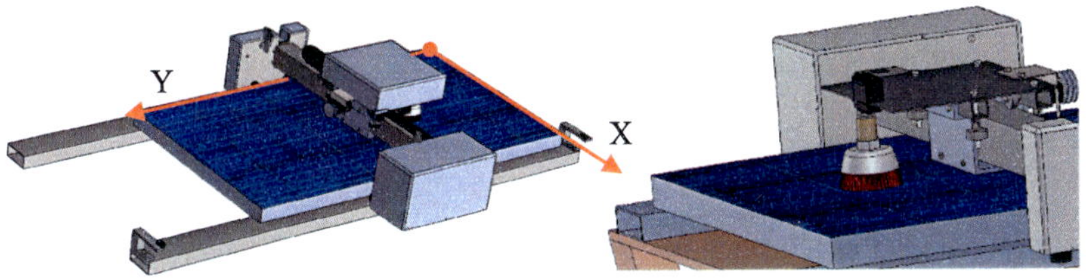


a) b)

**Fig. 3.** A *Cleanator* 3D CAD model: a) overall view with reference frame (X-horizontal; Y-vertical); b) a detail of the cleaning unit.

## 3 Analitycal and Numerical Calculations for Component Sizing

A dimensional synthesis has been made to demonstrate the effectiveness of the proposed concept. For the longitudinal motion, the drive torque required to move the device is estimated by analyzing the free body diagram of the friction wheel. The adhesion condition of the wheel has been verified to ensure the wheel non-slipping. To size the motor for the vertical movement, the cleaning unit is represented as a body on an inclined plane. The climbing phase is analyzed because it is the most critical phase. The driving torque for the vertical movement is then estimated through the free body diagram of the pulley. Friction effects are included in the whole analysis as proposed for example in [12] while inertia forces and moments are neglected since the mass of the system is small and the speed of the device remains almost constant.

Additional numerical analysis has been conducted to determine the tension force on the cable during the whole climbing phase modelling a pulley-cable mechanism on the commercial software ADAMS View. After the creation of the anchor points of the cable and the pulley, a linear guide simulates the sliding of the cleaning unit over the frame. Sliding friction is included in the simulation. Figure 4 presents the variation of the tension stress during mechanism climbing phase. Analytical calculations agree with the obtained numerical values. The step at 5 s in Fig. 4 refers to the change of direction by the brush.

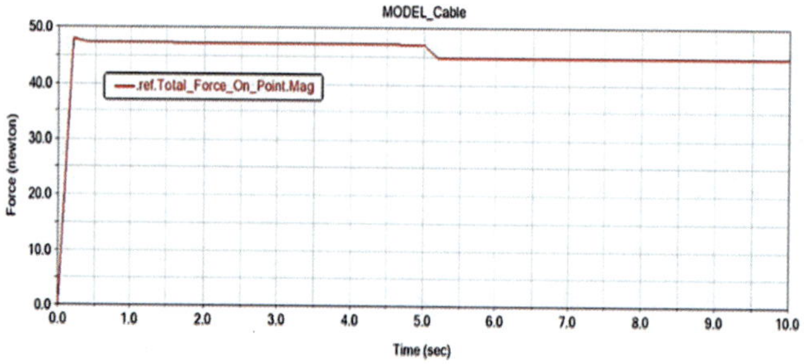


**Fig. 4.** Tension force on the cable of the cleaning unit during climbing phase.

## 4 A Built Prototype

For validating analytical calculations, a complete prototype of *Cleanator* has been realized. Little variations respect to the CAD model occurred to adapt the system to the available materials and production techniques. In fact, easy-to-find materials and rapid manufacturing techniques has been used to the fast realization of the prototype. In detail, the majority of *Cleanator* components are made of iron, steel, and wood. By contrast, parts characterized by complex shape and subjected to small loads are manufactured by 3D printing and are made of Polylactic Acid (PLA). In addition, all electric cables have been placed in a cable-chain to keep the system ordered. Figure 5 shows the built prototype of *Cleanator*.

**Fig. 5.** A *Cleanator* prototype at DIMEG.

## 4.1 Control Hardware and Programming

To ensure the correct operation and automated management of the mechatronic system, the device has been equipped with a control hardware structure that adopts an Arduino Mega board; the basis of the choice was precisely the high number of pins since there are numerous connections for the motors and sensors used [13].

The diagram in Fig. 6 shows the interactions between the user interface, the control unit and the apparatus of sensors and actuators installed on the device.

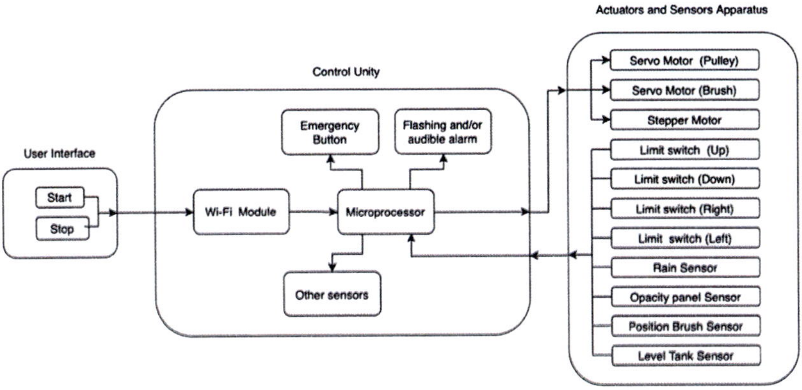


**Fig. 6.** Control system interactions scheme.

The sensors choice was influenced by the need to control the movement according to the climatic and structural conditions. The control system has been programmed in such a way that the device is able to clean the entire surface of the photovoltaic system and does it in the most accurate way; for this purpose, it must be able to issue instructions based on the data received from the sensors. In fact, the start and end of cleaning operations are regulated both by the values read by a button, manual management, and by the data received from the Wi-Fi network, managed by a Smartphone App. Also, in the presence of adverse weather conditions and/or low level of fluid in the tank, all operations will be suspended, and the device will receive instructions to return to its initial condition (HOME).

The flow chart in Fig. 7 illustrates how all the actions necessary to ensure the correct functionality of the prototype are coordinated.

To make it easier to control the cleaning system even remotely, a user interface was created using a smartphone app already developed: Blynk [14]. The App has been customized using widgets, each of which has a particular function. Two widgets were used to graphically represent the sensor readings in real time: an LED to indicate the presence of rain and a Level Indicator that monitors the fluid level in the tank. The user can start or end the cleaning process via a button.

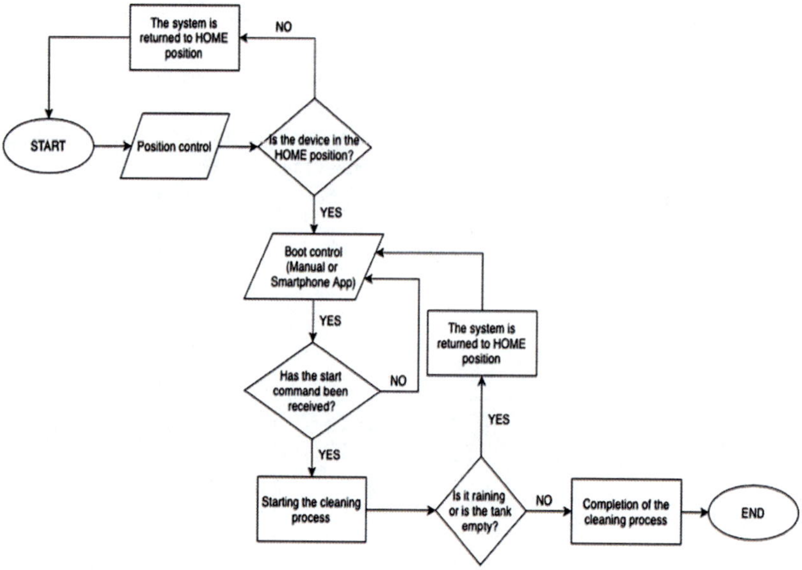


**Fig. 7.** Process flow diagram.

## 4.2 Preliminary Tests

In order to complete the design phase, experimental tests were carried out to verify the prototype correct functioning. Initially, the checks focused on calibration, reading data coming from the ultrasonic sensors and the rain sensor, and their interaction with the smartphone app. The experimental phase continued with tests relating to the system handling. The checks on horizontal direction were carried out to analyze the execution of commands given remotely and the reversal motion of the device when the limit switches are activated. To make the vertical movement of the cleaning unit more fluid, software corrections were made, and an additional tensioning element was added for the pulley cable. Figure 8 shows the horizontal and vertical movement sequences with the most significant frames. Sequence Fig. 8 a) The system starts from the home position and moves longitudinally over the photovoltaic panels; b) the device starts cleaning the panels by moving vertically with respect to the panel array; c) the device reverses motion in a vertical direction; d) the system repeats the cleaning operation in the return stroke; e) the system, moving longitudinally, finishes cleaning the array of panels; f) and g) the system repeats all the operations and movements in reverse to return to the starting position; i) the cleaning process is complete.

It is worth noting that the operation vertical and horizontal directions are arbitrary and can be modified according to the specific mounting of the solar panel. Also, the height that the rotating brush above the solar panel is adjustable and it can be tuned according to the wear of the brush. It is also worth noting the set up in Fig. 8 shows a laboratory demonstrator as based on a standard solar panel. Sizes and shapes can be modified according to the specific application. Further experimental tests will be carried out in future also by using 3D printing techniques as proposed for example in [15, 16]. An Italian patent application has been made accordingly, as reported in [17].

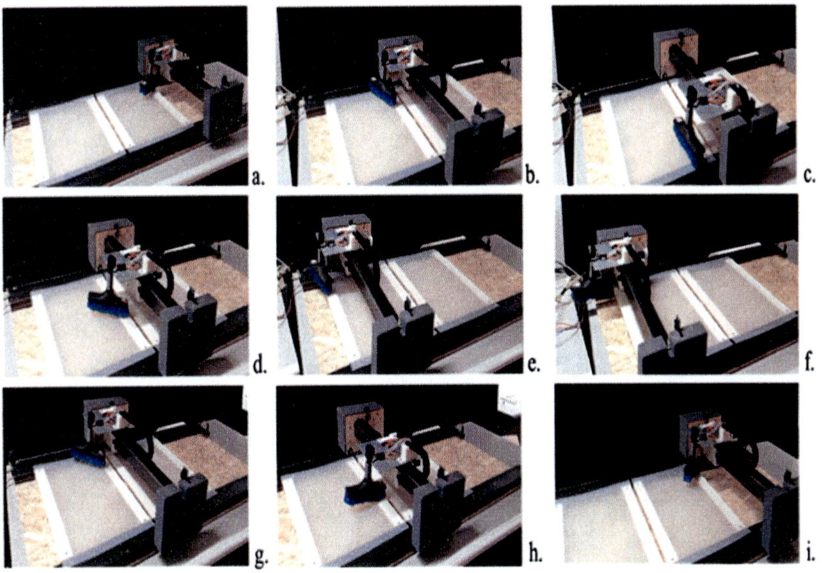


**Fig. 8.** Main phases of the movement sequence of *Cleanator* operation on a solar panel.

## 5 Conclusions

This paper proposes an automatic cleaning system for domestic photovoltaic panels, whose name is Cleanator. The system moves longitudinally, through a friction wheel, and vertically, through a cable-pulley mechanism, over domestic pre-existing photovoltaic roofs. Cleanator cleans the photovoltaic modules by means of demineralized water and a rotating brush. The proposed solution for the automatic cleaning of photovoltaic modules has been rapid prototyped to demonstrate its engineering feasibility and effectiveness. Preliminary experimental tests have shown the correct operating of the whole system. As a result, Cleanator comes as a light, flexible, reliable, and long-lasting system for automatic cleaning of photovoltaic modules. Further experiments of the system will be planned in future on real photovoltaic installations for verifying its reliability and energy efficiency.

## References

1. GSE, GDSES: Rapporto statistico. energia da fonti rinnovabili in Italia. anno 2018, giugno (2019)

2. Kazem, H.A., Chaichan, M.T., Al-Waeli, A.H.A., Sopian, K.: A review of dust accumulation and cleaning methods for solar photovoltaic systems. J. Cleaner Prod. **276**, 123187 (2020)
3. Xiaoyuan, L., Mauzerall, D.L., Bergin, M.H.: Global reduction of solar power generation efficiency due to aerosols and panel soiling. Nat. Sustain. **3**, 720–727 (2020)
4. Chen, Y., et al.: Experimental study on the effect of dust deposition on photovoltaic panels. Energy Procedia **158**, 483–489 (2019)
5. Sarver, T., Al-Qaraghuli, A., Kazmerski, L.L.: A comprehensive review of the impact of dust on the use of solar energy: history, investigations, results, literature, and mitigation approaches. Renew. Sustain. Energy Rev. **22**, 698–733 (2013)
6. Maghami, M.R., Hizam, H., Gomes, C., Radzi, M.A., Rezadad, M.I., Hajighorbani, S.: Power loss due to soiling on solar panel: a review. Renew. Sustain. Energy Rev. **59**, 1307–1316 (2016)
7. Klugmann-Radziemska, E., Rudnicka, M.: Decrease in photovoltaic module efficiency because of the deposition of pollutants. IEEE J. Photovoltaics **10**, 1772–1779 (2020)
8. Zatsarinnaya, Y.N., Amirov, D.I.: New equipment for cleaning solar panels. In: IOP Conference Series: Materials Science and Engineering, vol. 971 (2020)
9. Noh, F.H.M., Yaakub, M.F., et al.: Development of solar panel cleaning robot using Arduino. Indonesian J. Electr. Eng. Comput. Sci. **19**, 1245 (2020)
10. Mohammed, H., Al-Hilli, B.A.M., Al-Mejibli, I.S.: Smart system for dust detecting and removing from solar cells. J. Phys. Conf. Ser. **1032**, 012055 (2018)
11. Anilkumar, G., et al.: Design and development of wireless networking for solar PV panel cleaning robots. Conf. Ser. Mater. Sci. Eng. **937**, 012024 (2020)
12. Calligaris, L., Fava, S., Tomasello, C.: Manuale di Meccanica, 2nd edn. Hoepli, Milan (2016)
13. Arduino Mega 2560 Datasheet. https://store.arduino.cc/arduino-mega-2560-rev3. Accessed 23 May 2021
14. Blynk Inc., web page, 'Blynk IoT App'. https://blynk.io. Accessed 03 Feb 2021
15. Cafolla, D., Ceccarelli, M., Wang, M.F., Carbone, G.: 3D printing for feasibility check of mechanism design. Int. J. Mech. Control **17**(1), 3–12 (2016)
16. Ceccarelli, M., Carbone, G., Cafolla, D., Wang, M.: How to use 3D printing for feasibility check of mechanism design. Adv. Intell. Syst. Comput. **371**, 307–315 (2016)
17. Carbone, G., Cristodaro, M., Fortuna, S., Greco, A., Petrolino, A.C.: Mechatronic system for the automatic cleaning of solar panels (in Italian). Italian patent application no. 102021000006869, 22 March 2021

# Numerical Approaches for Loads and Motions Assessment of Floating WECs Moored by Means of Catenary Mooring Systems

Imanol Touzon[1(✉)], Victor Petuya[2], Vincenzo Nava[1,3], Maria Alonso-Reig[1], and Iñigo Mendikoa[1]

[1] Tecnalia, Basque Research and Technology Alliance (BRTA), Edificio 700, 48160 Derio, Bizkaia, Spain
imanol.touzon@tecnalia.com

[2] Department of Mechanical Engineering, University of the Basque Country - UPV/EHU, Leioa, Spain

[3] Basque Centre for Applied Mathematics, BCAM, Bilbao, Spain

**Abstract.** Technologies for harvesting offshore renewable energy based on floating platforms, such as offshore wind, wave and tidal energies, are currently being developed with the purpose of achieving a competitive cost of energy. The economic impact of the mooring system is significant within the total cost of such deployments, and large efforts are being carried out to optimize designs. Analysis of mooring systems at early stages generally require a trade-off between quick analysis methods and accuracy to carry out multi-variate sensitivity analyses. Even though the most accurate approaches are based on the non-linear finite element method in the time domain, these can result in being very time consuming.

The most widely used numerical approaches for mooring line load estimates are introduced and discussed in this paper. It is verified that accurate line tension estimates require lines drag and inertia forces to be accounted for. A mooring and floating structure coupled model based on the lumped mass finite element approach is also discussed, and it is confirmed that the differences found in the coupled numerical model are mainly produced by the uncertainty on hydrodynamic force estimates on the floating structure rather than by the lumped mass method. In order to enable quick line tension estimates, a linearization of the structure and mooring coupled model is discussed. It shows accurate results in operational conditions and enables modal analysis of the coupled system.

**Keywords:** Affordable and clean energy (SDG7) · Multibody dynamics · Catenary mooring · Offshore renewable energy · Oscillating water column

## 1 Introduction

Offshore renewable energies have already become relevant in the energy generation mix, especially bottom-fixed offshore wind, that has been commercially deployed for years. In addition, technologies based on floating structures for harvesting wind, tidal


G. Quaglia et al. (Eds.): I4SDG 2021, MMS 108, pp. 59–69, 2022.
https://doi.org/10.1007/978-3-030-87383-7_7

currents and wave energy, can be deployed in a wider variety of locations. They represent the largest untapped potential and exploiting a reasonable part of it aids in developing affordable and clean energy (SDG7).

Wave Energy Conversion technologies are being developed [1–3] with the purpose to be deployed in groups of several devices and the cost of the mooring system is a relevant item within the total array cost [4]. Therefore, it is very interesting that the estimates of the mooring cost are monitored along the path to commercialization, well estimated for single device and afterwards extrapolated to arrays of wave energy converters (*WECs*).

The sizing of mooring systems tends generally to be very influenced by extreme environmental conditions [5] but are also subject to fatigue loading under operational conditions. Unlike other mechanical systems, mooring lines are generally described by non-linear models and, given the required low natural frequencies to horizontal motions compared to wave frequencies (WF), mooring analysis needs long time domain (*TD*) simulations to define their statistics. Therefore, a design process in which multi-variate sensitivity analyses are required, can be difficult to be carried out with the most common numerical methods, based on the non-linear finite element method (*FEM*) solved in the time domain, as introduced by [6–9]. There are alternative methods to reduce the computation time, that consider catenary lines as quasistatic systems, solved both in the frequency [10] and in the time domains. Several standardization bodies provide guidelines on how to apply such methods for traditional offshore structures based on quasistatic mooring lines [11]. However, wave energy conversion structures are very dynamically excited and further assumptions may be necessary in order to use the mentioned quasistatic methods, as suggested in [12].

In order to ease the design process at different development stages, several models are discussed, combining floating structure and mooring models. The corresponding numerical approaches are introduced herein to consider floater and mooring lines coupled, of increasing complexity. The coupling scheme introduced herein is based on multibody dynamics, offering a systematic and feasible coupling between mechanical systems.

## 2 Numerical Approaches for Floating Offshore Structures

During the design of any moored floating structure, it is generally recommended, as a first step, to assess the influence of the mooring system on the structure through a single linearized stiffness term and, afterwards, assess mooring line loads imposing the resulting structure motions to a dynamic mooring system [13]. Once suitable designs have been found a fully coupled dynamic analysis is recommended. This process may lead to endless design loops, especially in early stages, when the number of variables is significant and the initial levelized cost of energy (*LCOE*) estimates of the whole system is to be reduced as much as possible.

In order to ease the design process at different development stages, the most relevant models are discussed, combining floating structure and mooring numerical models.

All the introduced approaches consider the floater as a dynamic mechanical system, whilst the main difference lies in considering the mooring system as a static or a dynamic system.

## 2.1 Quasistatic Linearized Frequency Domain

It consists in modelling mooring lines as a static mechanical system and the structure as a dynamic one, which is solved in the frequency domain (*QSFD*). It is introduced in [11] for traditional offshore structures, provided it is demonstrated that effects from mooring line dynamics are negligible. The catenary properties are computed based on the catenary equations, as specified in [10] and described in Fig. 1. The horizontal restoring force of the mooring system is linearized at the estimated mean horizontal offset, based on steady mean forces. The obtained horizontal stiffness is included in surge/sway motion and the equation of motion (1), accounting for the mooring horizontal stiffness, is solved in the frequency domain. Since drag forces are also considered, the system is solved iteratively through either harmonic or statistical linearization, for regular or irregular waves respectively. The maximum obtained motion, processed assuming specific statistical distributions, is added to the mean value so that the design offset is obtained, as shown in Fig. 1.

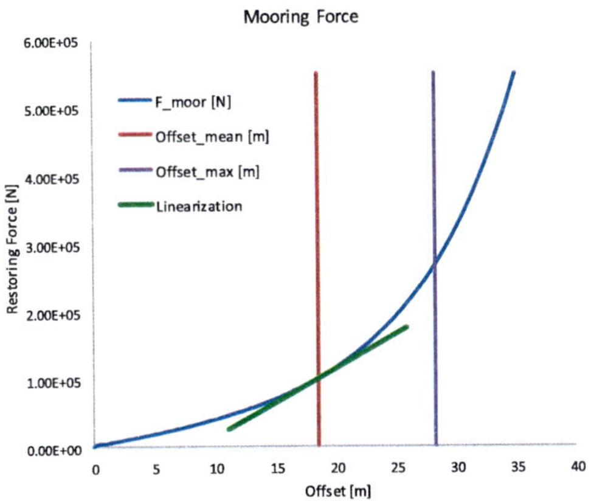


**Fig. 1.** Mooring static restoring force (blue) and the corresponding linearization (green) at the mean horizontal offset (red), along with the computed maximum offset (purple)

$$-\omega^2(\boldsymbol{M}+\boldsymbol{A}(\omega)+\boldsymbol{M_{dkin}})+i\omega\Big(\boldsymbol{B}(\omega)+\boldsymbol{B_d}\Big(\hat{\boldsymbol{\delta}}_{\boldsymbol{a}}(\omega)\Big)+\boldsymbol{C_{dkin}}+\boldsymbol{C_{pto}}\Big) \\ +(\boldsymbol{K_h}+\boldsymbol{K_m}+\boldsymbol{K_{dkin}})\cdot\hat{\boldsymbol{\delta}}_{\boldsymbol{a}}(\omega)=\hat{\boldsymbol{f}}_{\boldsymbol{w}}(\omega)\cdot\eta_a(\omega) \tag{1}$$

Where all bold variables represent vectors and matrices, indicated through lower- and uppercase respectively. $\boldsymbol{M}$ is the body mass matrix, $\boldsymbol{K_m}$ and $\boldsymbol{K_h}$ indicate linearized mooring and hydrostatic stiffness and $\boldsymbol{A}(\omega)$ and $\boldsymbol{B}(\omega)$ denote the added mass and radiation damping coefficients. Wave excitation force is represented through $\boldsymbol{f_w}(\omega)$, which accounts for Froude-Krylov and Diffraction and a linearized drag force is introduced through $\mathbf{B_d}\Big(\hat{\delta}_a(\omega)\Big)$. $\mathbf{M_{dkin}}$, $\mathbf{C_{dkin}}$, $\mathbf{K_{dkin}}$ indicate mass, damping and stiffness matrices from imposing kinematic relations between diffracting bodies, applying (1) with linear restrictions. The resulting motions are represented through the complex amplitude $\hat{\boldsymbol{\delta}}(\omega)$.

Equation (1) is introduced assuming the modelled floating *WEC* is made up of several diffracting bodies. In order to build up the corresponding numerical model mentioned bodies move rigidly in specific degrees of freedom whilst the power can be extracted from the relative motions of others. A floating spar type oscillating water column can be modelled trough two diffracting bodies [12], one representing the floating structure and one representing the surface water level of the internal water column. Therefore, restriction forces [14] arise in order to impose surge-sway-yaw motions to be rigidly joined, as per Eq. (2) that, as long as linear relations are set, can be reduced to the set of $\mathbf{M_{dkin}}$, $\mathbf{C_{dkin}}$ and $\mathbf{K_{dkin}}$ matrices. In addition, the power is extracted with the power take off mechanism, i.e. an air turbine, which is commonly modelled through a damping term, $C_{pto}$.

$$\left(\boldsymbol{M} + \alpha \cdot \boldsymbol{\Phi}_{\delta}^{t} \cdot \boldsymbol{\Phi}_{\delta}\right) \cdot \ddot{\boldsymbol{\delta}}(t) + \alpha \cdot \boldsymbol{\Phi}_{\delta}^{t} \cdot \left(\dot{\boldsymbol{\Phi}}_{\delta} \cdot \dot{\boldsymbol{\delta}}(t) + 2\xi\omega\boldsymbol{\Phi}_{\delta} \cdot \dot{\boldsymbol{\delta}}(t) + \omega^{2} \cdot \boldsymbol{\Phi}(t)\right) = \boldsymbol{f}(t) \tag{2}$$

The resulting system is solved separately for wave frequency and low frequency motions. First order *WF* excitation forces are computed with wave amplitudes derived from the spectrum of the corresponding sea state ($S_{\eta}(\omega)$). However, the low frequency forces (*LF*) are computed through the corresponding force amplitude of the slowly varying wave drift force spectrum, showed in Eq. (3), in the frequency domain, as introduced by Pinkster [15].

$$S_{SV}(\mu) = 8 \cdot \int_{0}^{\infty} S_{\tau}(\omega + \mu) \cdot S_{\tau}(\omega) \cdot |\boldsymbol{T}(\omega + \mu, \omega)|^{2} \cdot d\omega \tag{3}$$

Where $S_{SV}(\mu)$ is the slowly varying wave drift force spectrum and $\boldsymbol{T}(\omega, \omega)$ the drift force quadratic transfer function.

In order to ease the computation of the static properties of the mooring system on the floating structure, each mooring configuration, namely number of lines and radial distribution, can be characterized in advance, independently of the floating structure [10, 12].

### 2.2 Quasistatic Non-linear Time Domain

Solving the *Cummins equation* [16], showed in (3), accounting for all *DOF*s in the 3 dimensional space, coupled with the catenary mooring force $\mathbf{f_m}(t)$ is proposed in [17], and has been assessed in many works, such as [7, 8]. It assumes the mooring system as a static mechanical system and the floating structure as a dynamic one, which is solved in the time domain (*QSTD*). However, the static (non-linear) properties can be included not only in surge but also in all other degrees of freedom. In addition, since the proposed model is solved in the time domain, a nonlinear viscous drag force vector $\boldsymbol{f_d}(t)$ in all degrees of freedom can be included, as showed in the last term of the right hand side of (3). The convolution term for the radiation damping can be solved through several methods such as the prony approximation [18], frequency domain identification [19] or direct integration, as carried out in [9]. This model is advantageous since it considers the quadratic drag force as well as non-linear geometric stiffness of the mooring system and the influence of all degrees of freedom on mooring lines. However, it requires the catenary equations to be solved at every time step with its implicit iterative loop. The term $\mathbf{f_d}(t)$ in (4) represents the viscous drag force vector on the structure.

$$\begin{aligned}(\boldsymbol{M} + \boldsymbol{A}_{\infty} + \boldsymbol{M}_{\boldsymbol{dkin}}) \cdot \ddot{\boldsymbol{\delta}}(t) + \int_{-\infty}^{t} \boldsymbol{B}(t-\tau) \cdot \dot{\boldsymbol{\delta}}(\tau) \cdot d\tau + \left(\boldsymbol{C}_{\boldsymbol{dkin}} + \boldsymbol{C}_{\boldsymbol{pto}}\right) \cdot \dot{\boldsymbol{\delta}}(t) \\ +(\boldsymbol{K}_{\boldsymbol{dkin}} + \boldsymbol{K}_{\boldsymbol{h}}) \cdot \boldsymbol{\delta}(t) = \boldsymbol{f}_{\boldsymbol{w}}(t) + \boldsymbol{f}_{\boldsymbol{m}}(t) + \boldsymbol{f}_{\boldsymbol{d}}(t)\end{aligned} \tag{4}$$

## 2.3 Dynamic Non-linear Time Domain

In addition to the non-linear geometric stiffness of catenary lines, the most widely used approach to account for drag and inertia forces on lines is the *FEM* or lumped mass model coupled with the wave structure interaction model, resulting in a completely dynamic model solved in the time domain (*DynTD*). This model considers the wave structure interaction model based on the linear potential flow complemented with a Morison force term and dynamic moorings based on the lumped mass model. Both can be fully coupled through the use of kinematic relations, as introduced in [9], so that kinematic restrictions are imposed on all fairleads and anchors of each mooring line. Even though this approach is the most widely used due to its accuracy and availability in commercial codes, it can be too time consuming for sensitivity analyses of designs with several parameters.

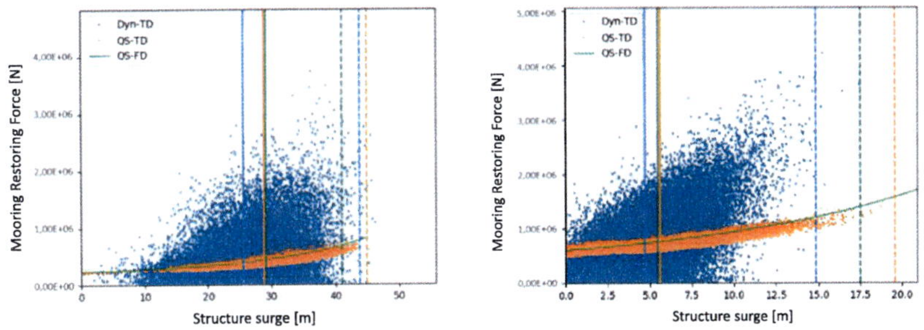


**Fig. 2.** Line tensions for the corresponding horizontal motions of the structure (surge) with the DynTD (blue), QSTD (orange) and QSFD (green) approaches. Vertical lines represent mean offsets (solid) and design offsets (dashed). Left: low mooring lines pretension; Right: high mooring lines pretension

Initial results from the case study showed in [12] and represented in Fig. 2, show that the *DynTD* approach provides higher uncertainty in terms of the correlation of the restoring force with the horizontal offset of the structure. However, given that *DynTD* approach accounts for lines drag and inertia forces, this model is considered the most accurate to compute line tensions. When compared with the *QSFD* approach, the design line tensions obtained with the *DynTD* model are two times higher with high line pretensions increasing up to seven times higher with low line pretensions, mostly due to increasing snap load events. Design offsets with low pretensions can be slightly larger with both time domain models due to the influence of the geometric stiffness, which can be covered in the *QSFD* approach through appropriate correction factors, see [12]. Nevertheless, with high pretensions, mooring induced damping becomes significant reducing the corresponding design offset, as shown in Fig. 2 right, which may require the use of mooring dynamic approaches in all stages of design.

Therefore, in order to obtain accurate line tensions and representation of the mooring induced damping with high pretensions, it is worth considering the non-linear *DynTD* approach. The resulting hydrodynamic and mooring coupled model can be summarized through a set of mass, damping and stiffness matrices together with a force vector as in Eq. (5):

$$\begin{bmatrix} (\boldsymbol{M}+\boldsymbol{A})_{str}+\boldsymbol{M}_{dkin} & \boldsymbol{M}_{f/a} \\ \boldsymbol{M}_{f/a} & \boldsymbol{M}_{moor} \end{bmatrix} \cdot \begin{Bmatrix} \ddot{\boldsymbol{\delta}}_{str}(t) \\ \ddot{\boldsymbol{\delta}}_{moor}(t) \end{Bmatrix} + \begin{bmatrix} \boldsymbol{C}_{pto}+\boldsymbol{C}_{dkin} & \boldsymbol{C}_{f/a} \\ \boldsymbol{C}_{f/a} & \boldsymbol{C}_{moor} \end{bmatrix} \cdot \begin{Bmatrix} \dot{\boldsymbol{\delta}}_{str}(t) \\ \dot{\boldsymbol{\delta}}_{moor}(t) \end{Bmatrix}$$
$$+ \begin{bmatrix} \boldsymbol{H}_{str}+\boldsymbol{K}_{pto}+\boldsymbol{K}_{dkin} & \boldsymbol{K}_{f/a} \\ \boldsymbol{K}_{f/a} & \boldsymbol{K}_{moor} \end{bmatrix} \cdot \begin{Bmatrix} \boldsymbol{\delta}_{str}(t) \\ \boldsymbol{\delta}_{moor}(t) \end{Bmatrix}$$
$$= \begin{Bmatrix} \boldsymbol{f}_{fk}(t)+\boldsymbol{f}^{sv}(t)+\boldsymbol{f}_{d}(t)-\boldsymbol{f}_{conv}(t)-\boldsymbol{f}_{f/a}(t) \\ \boldsymbol{f}_{z}(t)+\boldsymbol{f}_{f}(t)+\boldsymbol{f}_{g}+\boldsymbol{f}_{b}+\boldsymbol{f}_{morison}(t)+\boldsymbol{f}_{f/a}(t) \end{Bmatrix} \tag{5}$$

Subscripts *str* denotes structure, *moor* denotes mooring, *pto* denotes power take off and *dkin* and *f/a* denotes kinematic relations between diffracting bodies and fairleads/anchors respectively. In the force vector referred to the structure, a quadratic viscous drag term $\boldsymbol{f}_d(t)$ has been added, considering usual viscous drag coefficients as provided in offshore standards [20]. Due to the differences in natural frequencies of the floating structure and lines' elements, Eq. (5) represents a stiff system and implicit integration schemes are recommended such as the generalized-alpha methods or Newmark method [21]. It should be noted that the force $\boldsymbol{f}_{f/a}$ stands for the forces to maintain fairleads at a constant distance from the center of gravity of the structure and anchors fixed in the corresponding seabed positions.

The dynamic system described above is composed of three main parts, the floater, mooring lines and the lines fairleads and anchor. All of them have been included in the numerical model either through stiffness, damping and mass matrices or as time varying forces. Floater motions are linear since its wave structure interaction have been computed with a linear potential code and, hence, its matrices are time invariant as well as the power take off (*PTO*) forces, that are usually modelled through a set of linear stiffness and damping matrices.

On the other hand, kinematics of fairlead and anchor points have been defined by means of multibody restrictions. These restrictions are shown in Eq. (6).

$$\boldsymbol{\Phi}_{fair} = \begin{Bmatrix} x_{surge}+x_{roll}+x_{yaw}+\delta_{x-fair}-x_n \\ y_{sway}+y_{pitch}+y_{yaw}+\delta_{y-fair}-y_n \\ z_{yaw}+z_{roll}+z_{pitch}+\delta_{z-fair}-z_n \end{Bmatrix} \boldsymbol{\Phi}_{anchor} = \begin{Bmatrix} \delta_{x-anchor}-x_1 \\ \delta_{y-anchor}-y_1 \\ \delta_{z-anchor}-z_1 \end{Bmatrix} \tag{6}$$

In Eq. (6) $\delta_{x,y,z,-fair}$ indicates the position of the fairlead with respect to the center of gravity (*COG*) of the structure as well as $x_{surge,roll,yaw}$, $y_{sway,pitch,yaw}$ and $z_{heave,roll,pitch}$ denote the motions of the fairleads in the global X, Y and Z axis due to the corresponding motions of the structure. The anchor points are to be kept fixed on the seabed and to do

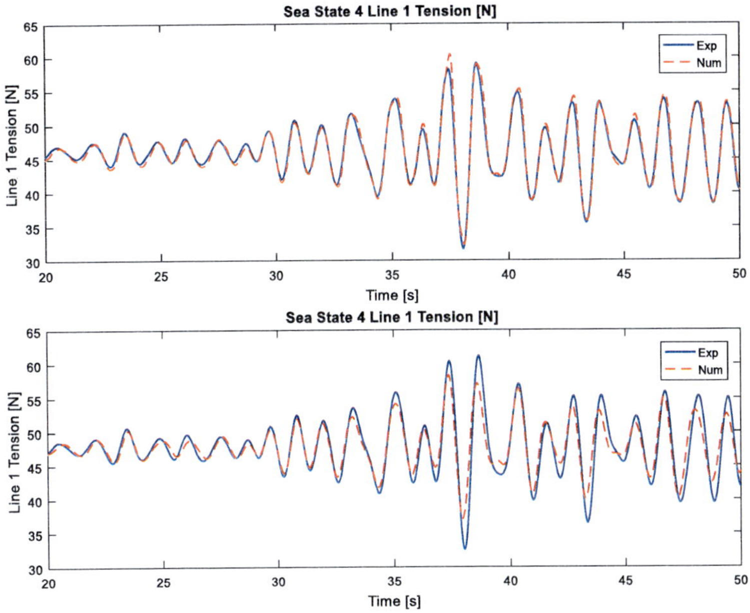


**Fig. 3.** Line tensions obtained with numerical model compared with the corresponding experimental results. Imposed motions on the fairlead (top) and fully coupled simulation (bottom)

so the position with respect to the *COG* of the structure $\delta_{x,y,z-anchor}$ are to be updated along the time.

Comparison between experimental and numerical results is introduced in Fig. 3. It represents and Catenary Anchor Leg Moored (CALM) buoy, moored by means of three catenary mooring lines (Lines 1 and 2 windward lines and Line 3 the leeward line). The experimental tests were carried out at a 1:13.6 scale obtaining good agreement between the numerical and experimental results. An in depth analysis of the comparison between experimental and numerical results of the *DynTD* approach has been introduced in [9]. The comparison is made in two main steps, imposed motions on the numerical model computing the corresponding line tensions, and simulating the fully coupled model of the floating structure and mooring system. It shows that the non-linear lumped mass method provides very accurate results when the motions of the floater are imposed. However, if the coupled system is solved, the accuracy of line tension estimates is reduced. It is attributed to not having modelled non-linearities of the floating structure.

## 2.4 Dynamic Linearized Frequency Domain

Considering mooring lines as dynamic mechanical systems, coupled to a dynamic floater in a model that can be solved in the frequency domain (*DynFD*), can be found just in a few references, such as in [22] applied in ultra-deep waters. It consists in solving the coupled system (5) in the frequency domain (*FD*), and in order to do so all forces arising from both the structure, the mooring system and line attachments need to be linearized. Main sources of non-linearity are the viscous forces on both the floater and mooring nodes, the geometric stiffness of catenary lines and the fairlead and anchor relations between the structure and mooring line ends. These effects are linearized in this approach in order to enable the *FD* solution of the whole coupled model, as specified in [23], as well as the modal analysis, represented in Fig. 4.

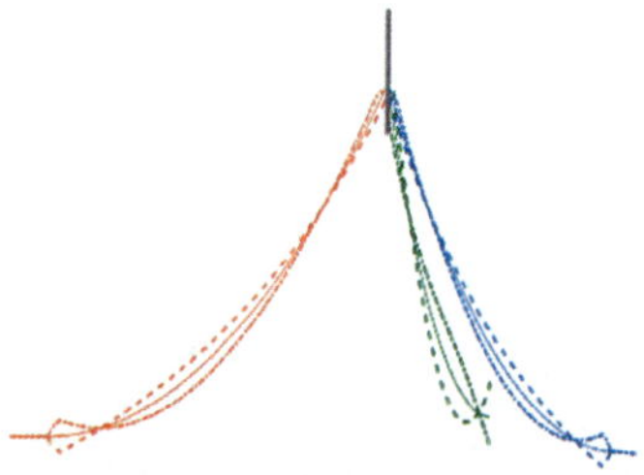

**Fig. 4.** Mode of the floating structure coupled with the mooring system, computed through modal analysis based in the DynFD approach

The linearization of the Morison viscous drag term ends up in a set of two linearized forces, proportional to the fluid and to the corresponding *DOF* velocities respectively. Consequently, the damping matrix and the velocity force depend on all *DOF* motions, implying the *FD* solution to be solved through a fixed-point iterative process, resulting in harmonic or statistical linearization process for regular or irregular waves respectively.

In contrast with the *DynTD* model, the *DynFD* model provides only the time-varying part of the solution and structural and stiffness matrix should be reviewed to work with the time varying part of the solution of motions.

The non-linear geometric stiffness contributes significantly to the system performance, especially in cases with significant mooring pretensions and in the *LF* range. Its influence on both the floater and mooring lines can be obtained through the secant stiffness force on the structure, as carried out in [23] accounting for two standard deviations of each degree of freedom of the floater about the mean position.

Kinematic relations, modeling fairlead and anchor points, are defined following Eq. (2) by means of Lagrange multipliers. Such relations can be broken down into two terms, those depending on model motions and constant forces. Constant forces are not included in the frequency domain solution since it is already in equilibrium at the mean position.

The slowly varying second order wave drift forces are included in the model through the spectrum defined in Eq. (3) and proposed in [15]. Seabed vertical reaction is modelled in the frequency domain through stiffness and damping matrices on the nodes on the seabed in the equilibrium position and the friction force has also been modelled through a damping matrix, applying the same damping coefficient as in the vertical motions.

Linearization of the coupled system provides very good results of both body motions and line tensions when compared with the corresponding non-linear time domain model, see Fig. 5. Accuracy is partially lost in surge and line tension estimates in very low and very high energy sea states, due to the influence of a coupled mode of motion and uncertainties in the mooring induced motion respectively.

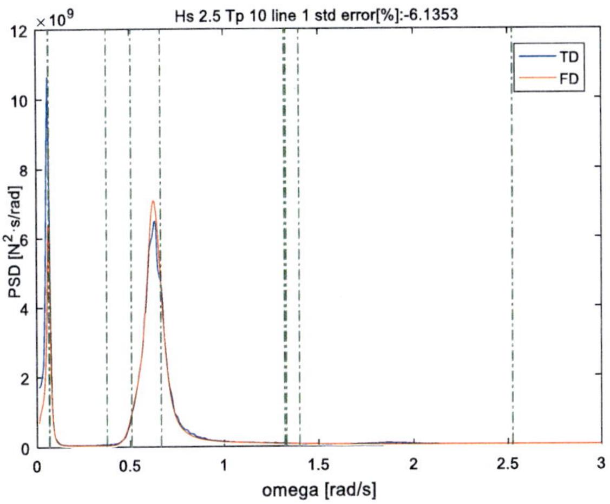


**Fig. 5.** Comparison of line tensions of the linearized frequency domain (red) with the non-linear time domain (blue) fully coupled models

## 3 Discussion

It is been verified that preliminary cost estimates, based on extreme environmental conditions, can be obtained with the *QSFD* approach for mid to low pretensions, provided total suspended lines length and lines tension are corrected with correction factors. In addition to the non-linear geometric stiffness, it is showed that drag and inertia effects are of utmost importance for line tension estimates. The non-linear lumped mass modelling approach provides very good results when the motion of the fairlead is imposed. Larger uncertainties are introduced in the fully coupled model due to non-linearities of floater body motions. Finally, since the mentioned *DynTD* model resolves the mooring system and floater motions numerical models in a fully coupled single mechanical system, it has been linearized, yielding the *DynFD* approach. Such a model is resolved in the frequency domain, coupling structure and line motions and accounting for lines drag and inertia as well as the linearized geometric stiffness on catenary mooring lines. The discussed *DynFD* model has been verified to provide good agreement in operational sea states.

## 4 Conclusions

In this paper a review of numerical modelling methods of catenary-moored floating structures is introduced, with an emphasis on *WEC* technologies, generally smaller and more dynamically excited than the traditional offshore structures. Summing up, preliminary cost estimates can be obtained with the *QSFD* approach for mid to low pretensions. The non-linear lumped mass modelling approach provides itself very good results when the motion of the fairlead is imposed. Larger uncertainties are introduced in floater body motions that are translated into line tension uncertainties. The fully coupled frequency domain approach discussed here has demonstrated good accuracy in operational conditions and it can efficiently be used to assess fatigue damage and control strategies accounting for the mooring influence.

The applicability of the fully coupled frequency domain approach to fatigue life estimates can be very beneficial in terms of computational time. It might, not only enable accounting for the fatigue damage earlier in the design process, but also to be integrated within different kinds of digital twins of floating platforms and its mooring systems. In addition, complex control algorithms of floating renewable energy structures can be implemented with the purpose of minimizing loads on mooring lines. Such control subroutines may require linearized systems of the floating structure and mooring lines to be solved systematically.

These numerically efficient approaches can facilitate further developing offshore renewable energy technologies at early stages, contributing in developing affordable and clean energy (SDG7). Authors are currently working in applying frequency domain approaches to floating offshore wind turbines, also aligned with the SDG7.

**Acknowledgments.** This work was funded by Project IT949-16 (Departamento de Educación, Política Lingüística y Cultura, Regional). Government of the Basque Country.

## References

1. MARMOK-A-5 Wave Energy Converter—Tethys. https://tethys.pnnl.gov/project-sites/marmok-5-wave-energy-converter
2. http://www.corpowerocean.com/technology/. http://www.corpowerocean.com/technology/
3. Ireland leads the way in ocean energy as SEAI funded wave energy device arrives in Hawaii for testing. Sustainable Energy Authority of Ireland. https://www.seai.ie/news-and-media/ocean-energy-test-device/
4. OES—News—International LCOE for Ocean Energy Technology. https://www.ocean-energy-systems.org/news/international-lcoe-for-ocean-energy-technology/
5. Muliawan, M.J., Karimirad, M., Gao, Z., Moan, T.: Extreme responses of a combined spar-type floating wind turbine and floating wave energy converter (STC) system with survival modes. Ocean Eng. **65**, 71–82 (2013)
6. Van den Boom, H.J.J.: Dynamic behaviour of mooring lines. Behav. Offsh. Struct. Elsevier Sci. Publ. B V 359–368 (1985)
7. Azcona, J., Munduate, X., González, L., Nygaard, T.A.: Experimental validation of a dynamic mooring lines code with tension and motion measurements of a submerged chain. Ocean Eng. **129**, 415–427 (2017)
8. Hall, M., Goupee, A.: Validation of a lumped-mass mooring line model with DeepCwind semisubmersible model test data. Ocean Eng. **104**, 590–603 (2015)
9. Touzon, I., Nava, V., Gao, Z., Mendikoa, I., Petuya, V.: Small scale experimental validation of a numerical model of the HarshLab2.0 floating platform coupled with a non-linear lumped mass catenary mooring system. Ocean Eng. **200**, 107036 (2020)
10. Touzon, I., et al.: Mooring system design approach: a case study for MARMOK-a floating OWC wave energy converter. In: American Society of Mechanical Engineers Digital Collection (2018). https://doi.org/10.1115/OMAE2018-77634
11. DNVGL-OS-E301: Position mooring (2018). http://rules.dnvgl.com/docs/pdf/dnvgl/os/2018-07/dnvgl-os-e301.pdf

12. Touzon, I., Nava, V., de Miguel, B., Petuya, V.: A comparison of numerical approaches for the design of mooring systems for wave energy converters. J. Mar. Sci. Eng. **8**, 523 (2020)
13. Offshore Renewable Energy, Committee v.4. in Proceedings of the 20th International Ship and Offshore Structures Congress (ISSC 2018) (Ed. by, International Ship & Offshore Structures Congress, Kaminski, M.L., Rigo, P., IOS Press) (2018). https://doi.org/10.3233/978-1-61499-864-8-193
14. Garcia De Jalon, J., Bayo, E.: Kinematic and Dynamic Simulation of Multibody Systems: The Realtime Challenge. Springer, New York (1994). https://doi.org/10.1007/978-1-4612-2600-0
15. Pinkster, J.A., Basin, N.S.M.: Low frequency second order wave exciting forces on floating structures, **220** (1980)
16. Cummins, W.E.: The impulse response function and ship motions (1962)
17. NI 461 DTO R00 E - Quasi-Dynamic Analysis of Mooring Systems using Ariane Software (1997)
18. Duclos, G., Clement, A.H., Chatry, G.: Absorption of outgoing waves in a numerical wave tank using a self-adaptive boundary condition. Int. J. Offsh. Polar Eng. **11**, 168–175 (2001)
19. Perez, T., Fossen, T.I.: Time-domain vs. frequency-domain identification of parametric radiation force models for marine structures at zero speed. Identification Control, 19 (2008)
20. DNV-RP-C205: Environmental Conditions and Environmental Loads, 124 (2010)
21. Naess, A., Moan, T.: Stochastic Dynamics of Marine Structures. Cambridge University Press, Cambridge (2012). https://doi.org/10.1017/CBO9781139021364
22. Low, Y.M., Langley, R.S.: Time and frequency domain coupled analysis of deepwater floating production systems. Appl. Ocean Res. **28**, 371–385 (2006)
23. Touzon, I., Nava, V., Gao, Z., Petuya, V.: Frequency domain modelling of a coupled system of floating structure and mooring Lines: an application to a wave energy converter. Ocean Eng. **220**, 108498 (2021)

# Design of a Reinforced Concrete Wave Energy Converter in Extreme Wave Conditions

Costanza Anerdi(✉), Bruno Paduano, Pietro Casalone, Giuliana Mattiazzo, and Luca Giordano

Politecnico di Torino, 10128 Turin, Italy
{costanza.anerdi,bruno.paduano,pietro.casalone, giuliana.mattiazzo,luca.giordano}@polito.it

**Abstract.** In the last decades, the growth of renewable energies showed to be more cost competitive, although traditional fossil fuels are still more affordable. Among the renewable energies, one of the most promising is wave energy, thanks to the high energy density stored in the waves. However, this precious resource requires further development, in particular to identify convenient and reliable production processes for conversion devices, known as Wave Energy Converters (WECs), and a view toward future changes and improvements of the existing prototypes. An interesting method to reduce the technological costs of energy and its environmental footprint could be found in the use of concrete structures, as opposed to traditional steel ones. This paper investigates the use of reinforced concrete for PeWEC, a floating wave energy converter, which converts wave energy into electrical energy thanks to its pitch motion. A preliminary design is carried out; pressure and mooring forces are evaluated and their structural effects are calculated by means of a finite element analysis. The design of reinforcements in a concrete shell is then reviewed. The general procedure is applicable to the case of a shell subjected to both bending and membrane stresses.

**Keywords:** SDG7 · Renewable energy · WEC · PeWEC · Concrete · FEM · LCOE

## 1 Introduction

Pollution, $CO_2$ emissions, and world temperature increase push the international community to intensify the application of renewable energy to satisfy human consumption: achieving the 60% of the energy from renewable sources by 2050 [1] is one of the most important goals in terms of sustainability.

Marine energy shows a great potential within renewable energies despite the fact that it is relatively new compared to the other more mature technologies. At the moment, technologies capable to extract energy from the sea are in full development. Some of the most common technologies use waves and sea current as driving forces or work through the exploitation of chemical or thermal potential of the environment. In a nutshell, the energy of the sea can be divided into the following sources: waves, tides, sea currents,


G. Quaglia et al. (Eds.): I4SDG 2021, MMS 108, pp. 70–77, 2022.
https://doi.org/10.1007/978-3-030-87383-7_8

temperature gradients, and salinity. In particular, the wave energy field has grown enormously [2] thanks to its power density [3] which is higher than other renewables sources. Nevertheless, these very young energy fields cannot be compared in terms of Levelised Cost Of Energy (LCOE) to traditional energy production methods [4, 6]. Hence, the necessity to minimize energy cost becomes every day more important. From the analysis of the different components of a technology, it seems clear that one of the main cost drivers is the capital expenditure for the external structure, so it is interesting to consider alternative solutions and materials. Moreover, concrete is more environment sustainable than steel [5], thanks to a lower carbon embodiment. For this reason, the use of concrete for wave energy converters [7] shows some advantages: it could decrease the LCOE by reducing the capital costs while also reducing the manufacturing carbon footprint, making the renewable energy technology more affordable and economically viable, and ultimately contributing to reaching global Sustainable Development Goals (SDGs), in particular SDG-7 "Affordable and Clean Energy". [8]. As previously mentioned, wave energy converters represent a new technology, compared to other marine structures or devices (naval, offshore structures, etc.) and dedicated standards or best practices are quite rare and incomplete. On the other side, adapting an offshore standard structure to the design process of Wave Energy Converter (WEC) could lead to oversizing, reducing the benefits in terms of costs optimization.

### 1.1 PeWEC

PeWEC is a floating offshore device that takes advantage from pitch motion produced by incoming waves and transforming it into electricity [11]. The hull is a reinforced concrete shell structure with a semi-cylindrical shape. The inertial properties able to guarantee the hydrodynamic performances are ensured by internal sand ballasts (located on the stern, bow, and keel). The internal pendulum is an oscillating steel mass that is supported by a trellis (attached to the hull) and connected to a PTO (Power Take-Off) (Fig. 1).

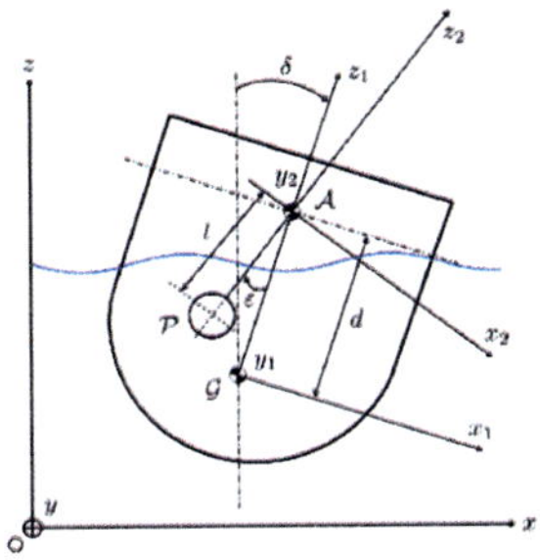

**Fig. 1.** PeWEC scheme.

The device is moored to the seabed through 4 catenary lines. Each line is formed by a chain with two attachments:

- a jumper that unloads the device from the vertical loads.
- a clump-weight that ensures a correct restoring force on the device.

This paper presents a preliminary analysis for the load assessment and structural design of the concrete structure; therefore, only major loads are considered (static and dynamic wave pressure field, viscosity, turbulence, mooring loads), while secondary or less impactful loads (splash, slam, spray, stress concentration) are neglected.

## 2 Advantages of Concrete Compared to Steel

In pursuing the SDGs, concrete industry has an important role [10]. The World Business Council for Sustainable Development (WBCSD) identifies a series of recommendations to keep the contribution of the concrete structures sustainable [8]. According to WBSCD [9], CO2 emissions per tonne of cementitious material correlated to cement and concrete industry has been decreased by 19.2% since 1990. This represents one of the most important goals achieved in terms of sustainability, considering the extensive use of the material all over the world. In addition, raw materials constituting the cementitious materials are widely available on the territory allowing a reduction on the carbon footprint from transport unlike steel. Concrete is a traditional and reliable construction material, and is well suited to the marine environment. It has numerous advantages both from an economic and performance points of view. In fact, reinforced concrete guarantees excellent durability (intended as maintenance of the performance and functional characteristics over time), especially when compared to steel, provided that the characteristics of the marine environment are appropriately considered. This feature significantly reduces the need for systematic maintenance with a relative reduction in operating costs. At the same time, the low unit cost of concrete ensure a significant drop in the capital cost. In addition, reinforced concrete has good fatigue resistance, which is essential in a dynamically stressed structure such as PeWEC. Concrete structures are generally heavier than steel ones, reducing the need for additional ballast. In this context, structural concrete can present an efficient solution by directly using the weight of the structure. A fundamental aspect to investigate is the mix design for offshore reinforced concrete stuctures which must be carefully considered following laboratory tests and developing durability models. Typically, marine concrete has a low water/cement ratio ($W/C < 0.45$) and involves the use of additives to facilitate its workability and protect the reinforcements from corrosion.

## 3 Load Definition

In the growth of renewable energies, international standards [11] are starting to take into account the related technologies in the certification process. Pressure field evaluation in extreme sea states could be achieved according to standards [12, 13] following semi-empirical formulations or simplified models or also developing hydrodynamic models.

For the hydrodynamic modeling of wave energy converters, there are different approaches with increasing accuracy and computational cost [15, 16]: starting from linear models that perform analysis in the frequency domain, through simplified time-domain analysis, to arrive, gradually removing all the simplifying hypotheses, to the application of Navier-Stokes equations in Computational Fluid Dynamic (CFD) codes. Low and medium-fidelity models, when applied appropriately, can provide reasonable results in terms of load assessment, at a fraction of the computational expense of CFDs. However, for extreme sea states and higher-order effects not included in the potential model [14, 15, 17, 18], linear results show significant errors compared to the CFD-based results and become an unsuitable tool for design. Considering that a statistical approach based on the governing response [19] of the device could not be used in the CFD environment for its computational time, several paths can be chosen to find short design waves [22, 23]. Such methodologies can be used for CFD simulations because they do not require a long time to reach regime condition. Detailed analysis on the wave choice lies outside this paper's purpose, but further details can be find in [20–23]. For the sake of clarity, CFD simulation has been performed considering an extreme regular wave [22] starting from an environmental contour (long-term probability distribution [24]) with 100 years of returning period with data from [25] for "Mazara del Vallo" site, in Southern Italy.

# 4 Numerical Simulation

## 4.1 CFD Model

A full 3D model has been created to consider frontal and not-frontal waves and use mesh techniques more efficiently, and the mooring effect has been included thanks to a coupling between STAR CCM+ [27] and Moordyn© software. The inclusion of mooring forces in CFD software is essential to have a realistic hull motion. The pressure field is extracted when a maximum of one of the kinematic degrees of freedom is reached and in correspondence with the maxima of local pressures recorded on the hull;to monitor the pressure trend over time in various points of the hull, point probes have been included in the model. The total fluid forces are instead obtained by integrating pressure and shear stresses on the whole hull. After the critical load conditions have been identified, these will be projected into the structural mesh that is different from the fluid dynamic mesh, since different physical phenomena require a different discretization of the continuum. In our case, the fluid dynamic mesh is a polyhedral mesh of smaller dimensions compared to the structural mesh, which is a quadrilateral mesh (Fig. 2). When pressure forces have been transferred from the fluid dynamic mesh to the structural mesh, it is necessary to ensure that the two load systems are equivalent from an energetic point of view, i.e. that no work has been created or eliminated by transferring the forces from one mesh to another. This is guaranteed by the mapping algorithm implemented in STAR CCM+ which, through a least-squares approximation, ensures a reliable mapping between the two meshes.

Thanks to the use of MoorDyn software, the mooring loads acting on the device can be considered and included on fairleads position.

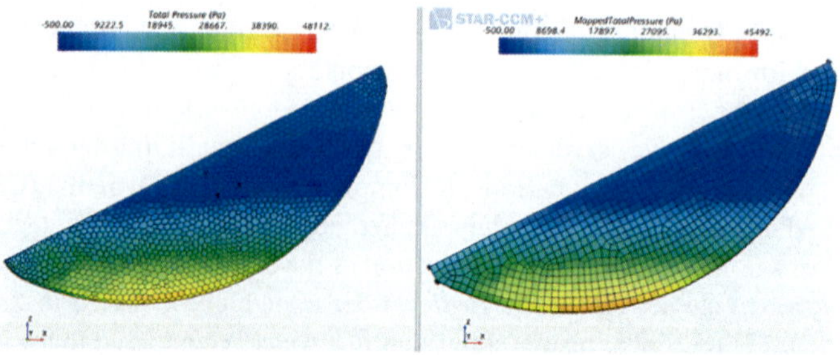


**Fig. 2.** Pressure mapping. On the left STAR-CCM+ mesh, on right the FEM mesh in ADINA.

### 4.2 Finite Element Model

ADINA Version 9.5, a general purpose finite element code has been selected as the basic platform for this study [29]. Structural analysis has been conducted based on the results obtained through a 3D numerical model of the device performing a linear elastic analysis. The numerical model is realized through bi-dimensional elements (Shell). A shell element is defined as a surface element with in-plane stiffness and out-of-plane bending stiffness with constant thickness. The reinforced concrete slabs have been modeled with quadrilateral 300 × 300 mm shell elements. The software can provide response information at the nodes and element stresses at the integration points within the element. Boundary conditions that represent structural supports specify values of displacement and rotation variables at appropriate nodes. The evaluation of the boundary conditions for PeWEC device, able to ensure a stable equilibrium of the structure, have been a challenging process that have led to the development of a two-step procedure:

- ***MODEL 1***: the node corresponding with the center of mass of the structure has been fully restrained in order to obtain the resultant forces and moments of the water pressure and mooring force. The center of mass has been temporarily connected with the structure, to transfer actions, by rigid frames. The material corresponding to those elements has been set to a null mass material, in order to not increase the mass of the hull.
- ***MODEL 2***: definition and application of an equivalent acceleration system (linear accelerations + angular accelerations) (Eq. 1, 2) that, togheter with water pressure and mooring forces, equilibrates the structure. The fixed boundary has been set to a generic node, and the rigid connections to the centroid have been removed.

$$\begin{Bmatrix} A_X \\ A_Y \\ A_Z \end{Bmatrix} = \begin{Bmatrix} R_X \\ R_Y \\ R_Z \end{Bmatrix} \cdot \{mass\}^{-1} \tag{1}$$

$$\begin{Bmatrix} \alpha_X \\ \alpha_Y \\ \alpha_Z \end{Bmatrix} = \begin{Bmatrix} M_X \\ M_Y \\ M_Z \end{Bmatrix} \cdot \begin{bmatrix} I_{XX} & -I_{XY} & -I_{XZ} \\ -I_{XY} & I_{YY} & -I_{YZ} \\ -I_{XY} & -I_{XY} & I_{ZZ} \end{bmatrix}^{-1} \tag{2}$$

*Ax, Ay, Az* are the linear accelerations along x, y and z; *Rx, Ry, Rz* are the constraint reactions on the center of mass. *Ax, Ay, Az*. Similarly, $\alpha x$, $\alpha y$, $\alpha z$ are the angular accelerations around the three axis divided by the inertial matrix [I]. This procedure is necessary to switch from a CFD dynamic model and the corresponding loads to a structural static model, avoiding unlikely reactions on arbitrary nodes. Once the internal actions are defined through the Finite Element Analysis (FEA) for each element, the design of reinforcement for concrete shells is based on the provisions in [28]. Generally, slab elements are subjected to eight stress resultants: three membrane force components ($n_x$, $n_y$ and $n_{xy}$), two flexural moment components ($m_x$ and $m_y$), the twisting moment ($m_{xy}$), and the two transverse shear force components ($v_x$ and $v_y$), as represented in Fig. 3. For the purpose of design, the slab is conceived as comprising two outer layers centred on the mid-planes of the outer reinforcement layers and an intermediate core. This is sometimes referred to as a "sandwich model". The covers of the sandwich model (i.e., the outer layers) are assumed to carry moments and membrane forces while internal layer carries the out-of-plane shear components and should be designed in accordance with the beam approach by using the principal shear $v_O$ (Eq. 3) [30].

$$v_O = \sqrt{v_x^2 + v_y^2} \tag{3}$$

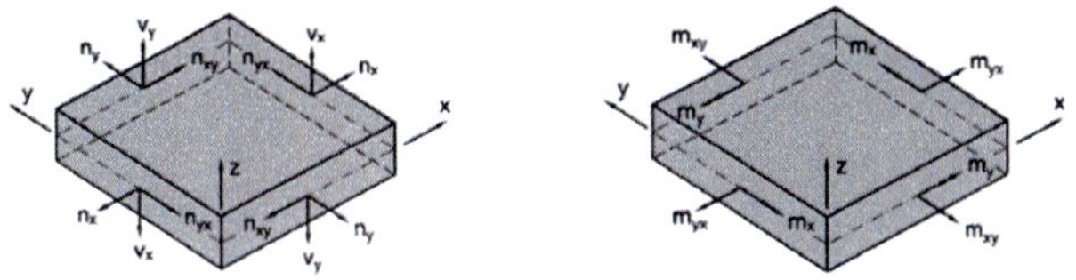

**Fig. 3.** Multiaxial state on shell elements

The design forces are calculated and are converted into reinforcement intensities (i.e., rebar area per unit width) using appropriate steel stress from the material properties assigned to the shell element. Finally, the concrete compressive forces are obtained and used to compute the compressive stresses to top and bottom layers. The amount of reinforcements in top, bottom and intermediate layers are calculatedelement by element. Then, an envelope representing the maximum amount of reinforcement for each extreme condition considered is used to assess the layout of steel reinforcements that have to be placed.

## 5 Results and Conclusions

This study was undertaken to assess the application of a well known technology, as reinforced concrete to a novel field. Given the lack of standards and the innovative nature of the project, the hull design has undergone a process of direct design through a Finite Element analysis and a previous definition of the loads by means of a CFD analysis. The results obtained, in terms of quantity of concrete and steel for the reinforcement, are typical for traditional reinforced concrete shell structures. Given the low unit cost of concrete and referring to a price-list for public works [31], a significant reduction of the capital cost is resumed.

$$\frac{Concrete\ hull\ cost*}{Steel\ hull\ cost} = 0.42$$

The cost reduction has been estimated considering also [32]. It is important to note that all the formworks considered are reusable and represent a cost that is amortised over time considering a large scale production. This suggest that the use of reinforced concrete for PeWEC hull could be a promising alternative, giving to a realistic reduction of LCoE.

*** average cost over 50 hulls, considering an initial expense for the formwork.**

## References


1. IRENA: Global Energy Transformation: A Roadmap to 2050, International Renewable Energy Agency, Abu Dhabi (2018)
2. Koca, K., et al.: Recent advances in the development of wave energy converters (2013)
3. Clément, A., et al.: Wave energy in Europe: current status and perspectives. Renew. Sustain. Energy Rev. **6**, 405–431 (2002). Author, F., Author, S.: Title of a proceedings paper. In: Editor, F., Editor, S. (eds.) CONFERENCE 2016, LNCS, vol. 9999, pp. 1–13. Springer, Heidelberg (2016)
4. Castro-Santos, L., García, G., Estanqueiro, A., Justino, P.: The levelized cost of energy (LCOE) of wave energy using GIS based analysis: the case study of Portugal. Int. J. Electr. Power Energy Syst. **65**, 21–25 (2015). https://doi.org/10.1016/j.ijepes.2014.09.022
5. Di Muro, A., et al.: Life cycle assessment for the ISWEC wave energy device. In: Niola, V., Gasparetto, A. (eds.) IFToMM ITALY 2020. MMS, vol. 91, pp. 515–523. Springer, Cham (2021). https://doi.org/10.1007/978-3-030-55807-9_58
6. Joint Research Centre – Low Carbon Energy Observatory "Ocean Energy, Technology development report". Publications Office of the European Union (2019)
7. Khosravi, N., Barker, L., O'Donoghue, V., Benzie, J., Newlands, M., Jones, M.: Use of concrete as the primary construction material for the pelamis wave energy converter (2014). https://doi.org/10.1201/b18973-44
8. WBCSD: Guidance for reducing and controlling emissions of mercury compounds in the cement industry, June 2016
9. WBSCD: Cement Industry Energy and CO2 Performance, Getting the Numbers Right (GNR)
10. Hajek, P.: IOP Conf. Ser.: Mater. Sci. Eng. **442** 012013 (2018)
11. Sirigu, S.A., et al.: Techno-economic optimisation for a wave energy converter via genetic algorithm. J. Mar. Sci. Eng. **8**, 482 (2020). https://doi.org/10.3390/jmse8070482Tidalturbines, DNVGL-ST-0164, October 2015
12. RINA - Regolamenti Per La Classificazione Delle Navi, Parte B
13. Offshore standard, DNVGL-OS-C101 – Edition July 2015 Design of offshore steel structures, general - LRFD method
14. Topper, M.: Guidance for Numerical Modelling in Wave and Tidal Energy (2010)
15. Penalba, M., Giorgi, G., Ringwood, J.V.: Mathematical modelling of wave energy converters: a review of nonlinear approaches. Renew. Sustain. Energy Rev. **78**, 1188–1207 (2017). ISSN 1364-0321
16. Giorgi, G., Rui, P.F., Gomes, J.C.C., Henriques, L.M.C., Gato, G.B., Mattiazzo, G.: Detecting parametric resonance in a floating oscillating water column device for wave energy conversion: numerical simulations and validation with physical model tests. Appl. Energy **276**, 115421 (2020), ISSN 0306-2619. https://doi.org/10.1016/j.apenergy.2020.115421

17. Fontana, M., Casalone, P., Sirigu, S.A., Giorgi, G., Bracco, G., Mattiazzo, G.: Viscous damping identification for a wave energy converter using CFD-URANS simulations. J. Mar. Sci. Eng. **8**, 355 (2020). https://doi.org/10.3390/jmse8050355
18. Van Rij, J., Yu, Y.-H., McCall, A., Coe, R.: Extreme Load Computational Fluid Dynamics Analysis and Verification for a Multibody Wave Energy Converter: Preprint. National Renewable Energy Laboratory, Golden (2019). NREL/CP5000-73474. https://www.nrel.gov/docs/fy19osti/73474.pdf
19. Offshore standards, DNVGL-OS-E301. Edition July 2018, Position mooring
20. Alford, L.: Estimating Extreme Responses Using a Non-uniform Phase Distribution (2008)
21. Kim, D.: Design Loads Generator: Estimation of Extreme Environmental Loadings for Ship and Offshore Applications (2012)
22. Coe, R., Yu, Y.-H., van Rij, J.: A survey of WEC reliability. Surv. Des. Pract. Energies **11**(1), 4 (2017). ISSN 1996-1073. https://doi.org/10.3390/en11010004. http://www.mdpi.com/1996-1073/11/1/4
23. Quon, E., Platt, A., Yu, Y.-H., Lawson, M.: Application of the most likely extreme response method for wave energy converters (2016). https://doi.org/10.1115/OMAE2016-54751
24. Recommended Practice, Det Norske Veritas, DNV-RP-C205 - Environmental Conditions and Environmental Loads, October 2010
25. http://dati.isprambiente.it/storia-della-ron/. Accessed 15 May 2021
26. Hall, M., Goupee, A.: Validation of a lumped-mass mooring line model with DeepCwind semisubmersible model test data. Ocean Eng. **104**, 590–603 (2015). https://doi.org/10.1016/j.oceaneng.2015.05.035
27. STAR CCM+ Users Manual. http://www.cd-adapco.com/products/star-ccm/documentation
28. CEN. EN 1992-1-1: Eurocode 2 – Design of concrete structures. Part 1–1: general rules and rules for buildings, CEN 2014, Brussels (2014)
29. ADINA R&D. Inc.: 71 Elton Avenue Watertown, MA 02472, USA (2014)
30. N°223: Ultimate Limit State Design Models - A State-of-Art Report: Ultimate Limit State Design of Structural Concrete Shell Elements (Fanti, Mancini) (1995)
31. ANAS: Listino prezzi 2019 – Nuove costruzioni manutenzione straordinaria
32. Costing Steelwork. Market and cost models update. Sponsored by AECOM, Steel for life, BCSA, May 2021

# Expanding Operational Bandwidth and Tuning a Wave Energy Converter Using U-shaped Sloshing Tanks

Fontana Marco(✉), Sergej A. Sirigu, Giuseppe Giorgi, and Giovanni Bracco

Marine Offshore Renewable Energy Laboratory (MORE) - Department of Mechanical and Aerospace Engineering, Politecnico Di Torino, Turin, Italy
marco_fontana@polito.it

**Abstract.** The ocean is the largest untapped energy site in the world and the recent decades have shown promising developments towards the wave energy converters (WEC). Among the problems related to the WECs are surely the optimal control strategy and the reliability in extreme waves conditions, and certainly also the tuning of the device, which due to the intrinsic nature of the sea will have to face wildly variable conditions compared to the nominal ones for which it was designed. This work aims to investigate the capability of tuning WECs through a U-shaped tuned liquid damper, expanding its operational bandwidth. To do this, a simply modified 1-D model of the U-tank coupled to the WEC is presented. The model has been validated by an experimental campaign concerning both kinematics, acquiring the water level via a motion tracking system, and dynamics, through load cells. The carried-out results show a good fit with the experiments both in regular and irregular imposed motion.

**Keywords:** SDG7 · Clean energy · SDG13 · Wave energy · WEC · ISWEC · U-tank

## 1 Introduction

Nowadays, climate change and the sustainable energy scenario represents ones the most relevant challenge to transform our world. The fulfilment of UN Sustainable Development Goals (SDG) relies upon the capability of the energy sector to provide clean and sustainable energy from renewable sources, in accordance the SDGs 7 that aims to ensure access to affordable, reliable, sustainable and modern energy for all, and the 13 focused on climate change. The theoretical resource potential of ocean energy ranges from 20,000 TWh to 80,000 TWh of electricity generation per year, representing 100% to 400% of the current global demand for electricity; within this framework, the European wave industry itself has set the ambition to provide 10% (or 188 GW) of European's electricity market by 2050 [1]. Despite its potential, the sector has to face the difficulties brought into play by the marine environment, which is particularly aggressive, both from a chemical point-of-view (for instance, salinity) and because of the extreme meteorological conditions. ISWEC (Inertial Sea WEC) [2] was born in 2011 in the European


G. Quaglia et al. (Eds.): I4SDG 2021, MMS 108, pp. 78–86, 2022.
https://doi.org/10.1007/978-3-030-87383-7_9

scenario, and more specifically in the Mediterranean Sea [3]. The device, developed by the Polytechnic of Turin and implemented by the company Wave for Energy Srl, is a device based on pitching motion caused by the frontal incoming wave. A gyroscope system and its power take-off allow producing energy in resonance conditions. So, the device, shown in Fig. 1 is designed on a specific marine site, characterized by its sea state.

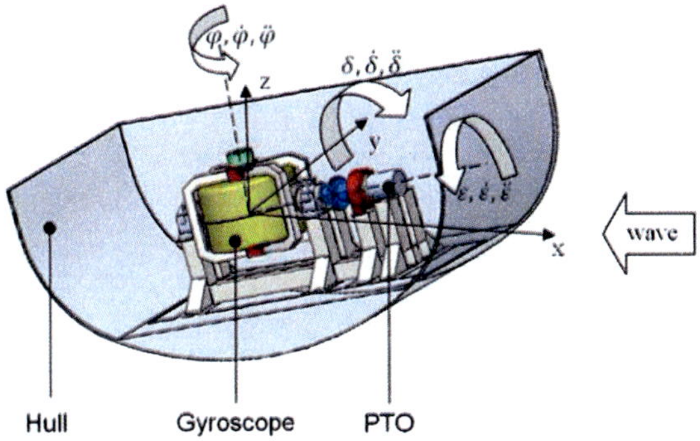


**Fig. 1.** ISWEC layout

The performances of the device depend on its frequency response and to match the resonance condition of the device with the more energetic condition of the particular sea, the mass and inertia of the device are calculated and fixed. The purpose of the designers is to increase the flexibility of the device making it effective for more than a single specific condition or sea state which is variable both in the short term (day/night cycle) and the long one (seasonality). There is more than one solution to module the bandwidth of a WEC depending on their technology, among which there is the pendulum-based PEWEC (PEndulum WEC) developed by Politecnico di Torino and ENEA [4, 5]. Beneath this consideration, this work aims to show a solution to improving the performances of the device, increasing the amplitude in resonance and the bandwidth. The idea is to use the so-called U-tank, a U-shaped Tuned Liquid Damper (TLD) in synergy with the ISWEC device to improve its performance. The original concept by Frahm [6] of the U-tank is to act as an anti-roll tank (ART) to reduce the roll motion of the ship, improving the seakeeping even at zero forward speed. U-tanks as ART are still in use in the naval field as shown in Fig. 2, and also in civil engineering to avoid resonance due to wind gusts in the skyscrapers. In the past few decades, the performance of anti-roll tanks has received considerable attention. Chadwick and Klotter [7] investigated the use of tanks for the US Navy. Van Den and Vugts [8] and Stigter [9] studied the performance of various types of anti-roll tanks for the Netherlands Ship Research Center. Webster et al. [10] studied the performance of free-flooding anti-roll tanks during the major upgrade of the USS Midway. Later, Lee and Vassalos [11] investigated the use of flow obstructions inside the tank.

The novelty here presented is to capsize the classical approach of using U-tank as ART, making it a stimulator of pitch motion in synergy with a gyroscope system. Other

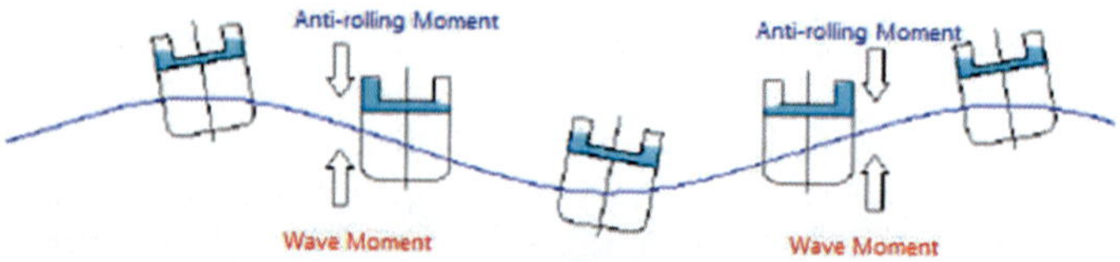


**Fig. 2.** U-shaped Anti-roll tanks in the naval field

systems use the U-tank in terms of direct exploitation of energy. Regarding models, the U-tank can be handled as Oscillating Water Column (OWC) [12].

After an introductory section, this paper is structured into three parts. In the first one, the model and dynamics of the coupling U-tank and WEC are shown. The second section describes the experimental setup. Finally, the results are reported.

## 2 Lumped Parameter Model

The numerical model presented in this section is derived from Lloyd's model [13] of a U-shaped anti-roll tank. The water motion inside a 1-DoF (degree of freedom) tank can be studied as a two-dimensional flow since the velocity gradients along the third axis are negligible. Regarding the fluid and its motion we can refer to the conservation and momentum equations, respectively Eqs. (1) and (2), written for a 2D incompressible flow in an inertial reference system.

$$\nabla \cdot v = 0 \tag{1}$$

$$\frac{\partial \mathbf{v}}{\partial t} + \nabla \cdot (vv) = -\frac{1}{\rho}\nabla p + g + \frac{1}{\rho}\nabla \cdot \overline{\tau} \tag{2}$$

Where ρ is the water density, **v** is the velocity field, p is the pressure field, g the gravitational field, $\mathbf{v}_b$ is the wall velocity, Ω is the pitch rotation imposed on the tank (due to the pitch of the WEC), r is the radius vector starting from the centre of rotation, τ is the shear stress tensor. The pitching motion imposed on the U-Tank generates a secondary wave inside the tank that propagates from one side to the other. The shape causes a water sloshing phenomenon inside the reservoirs (e.g., the vertical ducts). Figure 3 shows a simple U-tube passive tank made of two reservoirs and a connecting duct of constant rectangular cross-sections. The left reservoir is located on the bow of the WEC, and the right one on the stern. Along the port-starboard direction, the U-tank has a breadth $x_t$.

$n$ is the width of the tank perpendicular to the $y$ axis and it is assumed that there is no flow in the $n$ direction and the motions of a unit mass of fluid in the tank are governed by Euler's Eq. (3):

$$\frac{\partial v}{\partial t} + v\frac{\partial v}{\partial y} = Y - \frac{1}{\rho}\frac{\partial P}{\partial y} \tag{3}$$

Where $Y$ is the external force per unit mass. The assumptions of Lloyd's model are:

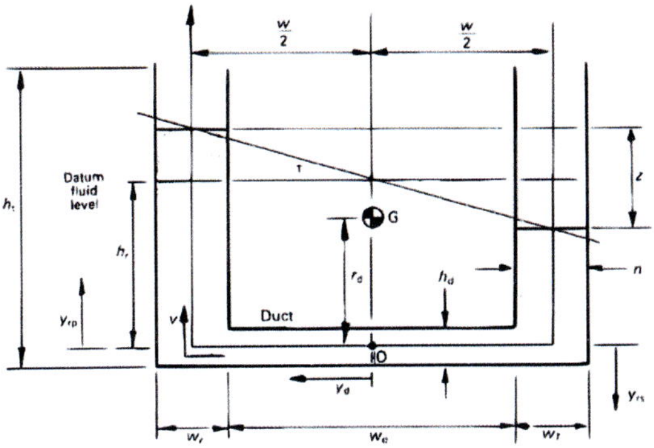


**Fig. 3.** U-tank scheme

- the motion inside the tank is one-dimensional
- the coupling U-tank – WEC only considers the pitch motion
- the system response is linear
- there are no velocity gradients perpendicular to the y-axis
- duct and reservoirs have a constant cross-section, hence $\frac{\partial v}{\partial y} = 0$
- the corner effects at the bows are neglected

Considering the external forces, Y is composed of the following contributions, for which the reader can refer to Fig. 4:

i. the component of the acceleration due to gravity along the y-direction
ii. the acceleration due to the pitch acceleration
iii. frictional and damping forces are linearized. Although these would be expected to be proportional to the square of the local velocity it is convenient to assume that the damping can be linearised [14].

The purpose is to obtain the motion of the fluid as a function of the moment applied to the tank (due to the pitch motion of the WEC). So, integrating along y and between the datum level of the reservoirs, after manipulations, we obtain:

$$a_{\tau\tau}\ddot{\tau} + b_{\tau\tau}\dot{\tau} + c_{\tau\tau}\tau = -\left[a_{\tau5}\ddot{\delta} + c_{\tau5}\delta\right] \tag{4}$$

Equation (4) represents a 1-DOF lumped parameter model of the second-order, dynamically coupled with the pitch acceleration and oscillation of the floater. The RHS of Eq. (4) is the torque due to the U-tank water motion acting on the floater pitch DOF, and the subscript 5 is equivalent to the pitch oscillation $\delta$ since the classical notation is indexed by a 6 components array composed by the three translation and the three rotations in "xyz" order. Once Eq. (7) is obtained it is possible to write the **transfer function** between the U-tank angle $\tau$ and the input oscillation $\delta$:

$$\frac{\tau(\omega)}{\delta(\omega)} = \frac{-a_{\tau5}\omega^2 + c_{\tau5}}{-a_{\tau\tau}\omega^2 + j\omega b_{\tau\tau} - c_{\tau\tau}} \tag{5}$$

Equation (5) presents two conjugate complexes poles and two coincident zeros. The resonance frequency of the sloshing water in motion inside the U-tank is a function of the geometrical properties and the datum level of the water.

$$\omega_{res} = \sqrt{\frac{c_{\tau\tau}}{a_{\tau\tau}}} = \sqrt{\frac{2gh_d}{w_r w + 2h_r h_d}}$$

## 3 U-tank and WEC Coupling

Since we refer to the only pitch motion, the motion equation in the frequency domain becomes:

$$[M_{55} + A_{55}(\omega)]\ddot{\delta} + B_{55}(\omega)\dot{\delta} + K_{55}\delta = F_{w,5} \tag{6}$$

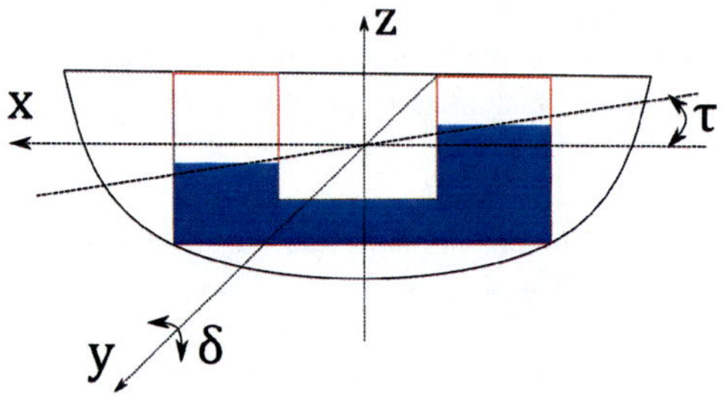


**Fig. 4.** Floater hull and U-tank coupling

The floater and the U-tank are coupled in acceleration and position, and the latter transfer its torque to the floater according to the following system of equations:

$$\begin{cases} I_{55}^{*}\ddot{\delta} + B_{55}(\omega)\dot{\delta} + K_{55}\delta = F_{w,5} + a_{5\tau}\ddot{\tau} + c_{5\tau}\tau \\ a_{\tau\tau}\ddot{\tau} + b_{\tau\tau}\dot{\tau} + c_{\tau\tau}\tau = a_{\tau 5}\ddot{\delta} + c_{\tau 5}\delta \end{cases} \tag{7}$$

The inertial coefficient $I_{55}^{*}$ embeds both the mass and added mass components. The subscript $w$ in the force term indicates the wave excitation. The last two terms on the right-hand side in the first Eq. (7) represent the torque applied on the hull by the U-Tank. Similarly on the second equation concerning the coupling hull-U-tank. Applying the Laplace transform to the equations the **transfer functions** can be obtained:

$$\frac{\overline{\delta}}{\overline{F_{w,5}}} = \frac{a_{\tau\tau}s^2 + b_{\tau\tau}s + c_{\tau\tau}}{\left(I_{55}^{*}s^2 + B_{55}s + K_{55}\right)\left(a_{\tau\tau}s^2 + b_{\tau\tau}s + c_{\tau\tau}\right) - \left(a_{5\tau}s^2 + c_{5\tau}\right)^2} \tag{8}$$

Showing that the system presents the typical dynamic of a $4^{th}$ order system, which implies two resonant peaks and an anti-resonance one as shown in Fig. 7. In blue is charted the Response Amplitude Operator of the WEC without the U-tank, while in red the coupling.

## 4 Experimental Setup

The experimental tests have involved the prototype in Fig. 5 put into by a 3 degrees of freedom HPR (Heave-Pitch-Roll) platform. The motion was imposed as regular sine varying amplitude and frequency. The water level, hence the angle $\tau$ has been tracked via a MoCap (Motion capture) acquiring system using floaters in which infrared-sensitive markers were applied. The other markers in Fig. 5 were used to capture the reference frames during the motion. The cameras that compose the acquisition system are not shown in Fig. 5.

**Fig. 5.** Light marker on the prototype

## 5 Results

By tracking the water levels it is possible to derive the water angle $\tau$ and the following Fig. 6 shows a time history of the latter one during a sinusoidal input in pitch DoF $\delta$.

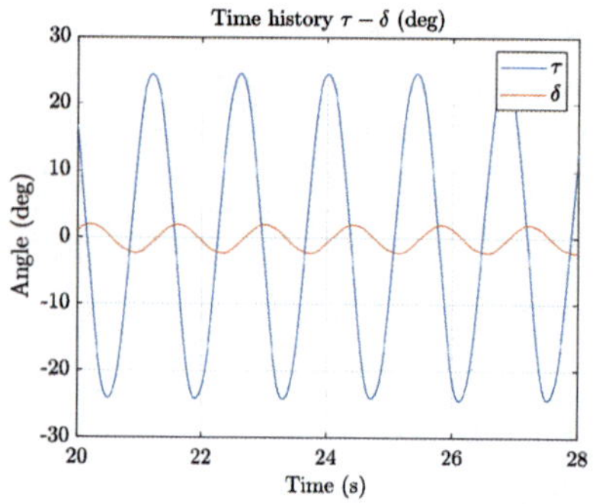


**Fig. 6. On the left** - Time history of the acquired water angle in regular wave

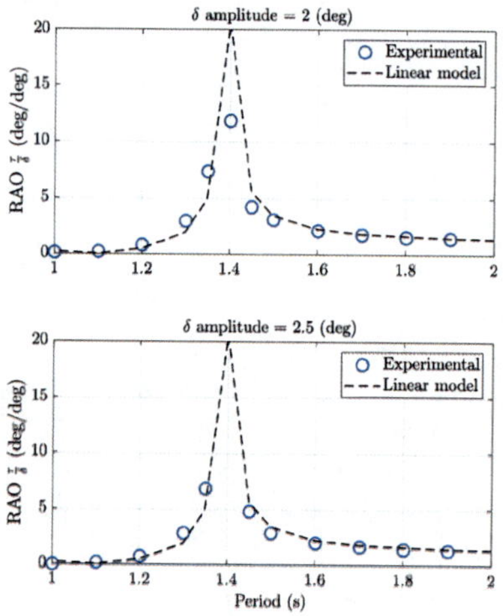


**Fig. 7. On the right** - RAO of the numerical model compared to the experimental data

# 6 Conclusions

In the frequency domain, the linear model presented in a previous section of this work seems to overestimate the response amplitude operator for periods close to the resonance condition (here set to 1.4 s) as shown in Fig. 7 where each point indicates the RAO amplitude for a giver frequency of the imposed sinusoidal motion, beyond this the fit is good and the linear model, much faster than experiments or eventually, fluid dynamics simulation, is enough accurate for preliminary design.

The device bodes well, by showing the capability to bring out a second peak in the RAO (Fig. 8) so increasing the productivity of the floating WEC. From the feasibility point of view, it is relatively easy to build and maintain. The dynamics are well-known for almost a century, but the capsize of the classical anti-roll tank into a stimulator tank constitute a new challenge and the opening of new possibilities in the wave energy sector.

It can be concluded that a passive U-tank can modify the dynamic properties of the WEC, tuning it for different sea state, so increasing its adaptability and therefore the performances.

The next steps will include the analysis of an Active U-tank, in which a proper control logic can tune the device in both the short and long term, increasing the magnitude of the RAO at the equal resonance frequency and/or expanding its bandwidth to match different sea states.

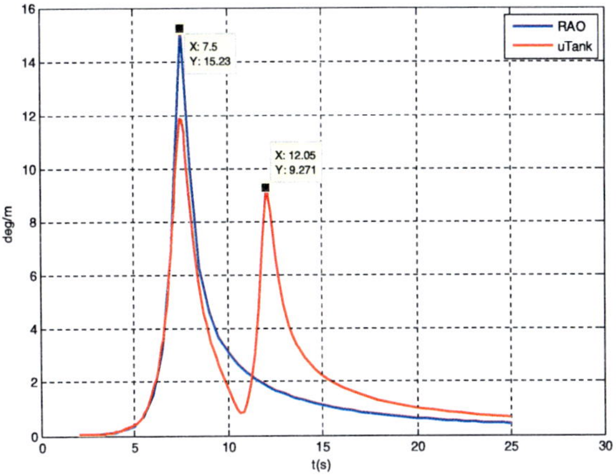


**Fig. 8.** Comparison of RAOs between WEC and the coupling with the U-Tank

## References

1. E. C. D.-G. for Energy, Energy: roadmap 2050. https://doi.org/10.2833/10759
2. Bracco, G., Giorcelli, E., Mattiazzo, G.: ISWEC: a gyroscopic mechanism for wave power exploitation. Mech. Mach. Theory **46**(10), 1411–1424 (2011). https://doi.org/10.1016/j.mechmachtheory.2011.05.012
3. Mattiazzo, G.: State of the art and perspectives of wave energy in the mediterranean sea: backstage of ISWEC. Front. Energy Res. **7**, 114 (2019). https://doi.org/10.3389/fenrg.2019.00114
4. Pozzi, N., Bracco, G., Passione, B., Sirigu, S.A., Mattiazzo, G.: PeWEC: experimental validation of wave to PTO numerical model. Ocean Eng. **167**, 114–129 (2018). https://doi.org/10.1016/j.oceaneng.2018.08.028
5. Sirigu, S.A., et al.: Techno-economic optimisation for a wave energy converter via genetic algorithm. J. Mar. Sci. Eng. **8**(8), 482 (2020). https://doi.org/10.3390/JMSE8070482
6. Frahm, H.H.: Results of trials of the anti-rolling tanks at sea. J. Am. Soc. Nav. Eng. **23**(2), 571–597 (1911). https://doi.org/10.1111/j.1559-3584.1911.tb04595.x
7. Chadwick, J.H., Klotter, K.: On the dynamics of anti-rolling tanks (1954)
8. Vugts, J.H.: On roll damping by free-surface tanks (1966)
9. Stigter, C.: The performance of U-tanks as a passive anti-rolling device 1. Int. Shipbuild. Prog. **13**(144), 249–275 (1966). https://doi.org/10.3233/isp-1966-1314401

10. Webster, W.C., Dalzell, J.F., Barr, R.A.: Prediction and measurement of the performance of free-flooding ship antirolling tanks. Soc. Nav. Archit. Mar. Eng. 96 (1988)
11. Lee, B.S., Vassaols, D.: An investigation into the stabilisation effects of anti-roll tanks with flow obstructions (1996)
12. Giorgi, G., Gomes, R.P.F., Henriques, J.C.C., Gato, L.M.C., Bracco, G., Mattiazzo, G.: Detecting parametric resonance in a floating oscillating water column device for wave energy conversion: numerical simulations and validation with physical model tests. Appl. Energy **276**, 115421 (2020). https://doi.org/10.1016/j.apenergy.2020.115421
13. Lloyd, A.R.: Seakeeping – ship behaviour in rough weather. J. Navig. **43**(1), 140 (1989). https://doi.org/10.1017/s0373463300013928
14. Fontana, M., Casalone, P., Sirigu, S.A., Giorgi, G., Bracco, G., Mattiazzo, G.: Viscous damping identification for a wave energy converter using CFD-URANS simulations. J. Mar. Sci. Eng. **8**(5) (2020). https://doi.org/10.3390/JMSE8050355

# Achieving the Sustainable Development Goals Through Education on Renewable Energy

Ion Visa and Macedon Moldovan(✉)

Renewable Energy Systems and Recycling Research Centre, Transilvania University of Brasov, Brasov, Romania
macedon.moldovan@unitbv.ro

**Abstract.** Education plays a decisive role in achieving the 17 United Nations Sustainable Development Goals, by developing human resources and raising awareness at community level in promoting and implementing the concept of Sustainable Development and Sustainable Energy. In this context, Higher Education Institutions are responsible for both the training of highly qualified human resources and the development together with various structures of Sustainable Development strategies. The paper describes the ways and tools of human resources training in universities through formal education and the involvement of universities in non-formal education. An essential role in promoting the concept of Sustainable Development is played by scientific research structures in the development of sustainable materials, products and systems. Ways and tools developed and implemented by the Transilvania University of Brasov, Romania to promote the concept of sustainable development at the institutional, local, regional, national and international levels are detailed in the paper.

**Keywords:** Quality education · Sustainable development · Sustainable energy

## 1 Introduction

### 1.1 The United Nations' Sustainable Development Goals

In 2015, nearly all the countries in the world have agreed to work together to create a better world by 2030 and better future for everyone ensuring that no one is left behind. The United Nations (UN) committed to 17 Global Goals, also known as the Sustainable Development Goals (SDGs). These life-changing goals include ending extreme poverty and hunger, giving people better healthcare and quality education, gender equality, clean water and energy, sustainable cities and communities, responsible consumption and production, climate action, life on land and below water [1].

The concept of sustainable energy is directly addressed by two SDGs: Goal 7 Affordable and Clean Energy and Goal 11 Sustainable Cities and Communities. Goal 7 has several targets by 2030: doubling the global rate of improvement in energy efficiency, substantially increasing the share of renewable energy in the global energy mix to ensure universal access to affordable, reliable and modern energy services.


G. Quaglia et al. (Eds.): I4SDG 2021, MMS 108, pp. 87–96, 2022.
https://doi.org/10.1007/978-3-030-87383-7_10

Goal 11 tackles the issues related to the rapid urbanization exerting a high pressure on this living environment where half of humanity lives today. Cities account for 60 to 80% of global energy consumption generating about 70% of anthropogenic induced greenhouse gas emissions. Considering that the world's cities occupy just 3% of the Earth's land, a high intensity of energy consumption results in these areas often translated in poor air quality and environmental pollution.

### 1.2 Education for Sustainable Development

Education is directly addressed in the fourth SDG (Quality Education). The aim of this SDG is to provide equitable and inclusive education as well as promoting learning opportunities. Avelar et al. [2] showed that the education has a direct impact on SDG8 - Decent work and economic growth, SDG9 – Industrial innovation and infrastructure, SDG12 – Responsible consumption and Production, and SDG17 – Partnerships for the goals, fostering innovation in the field of Clean water and sanitation (SDG 6), Affordable and clean energy (SDG 7), Life below water (SDG14) and Life on land (SDG15).

Sinakou et al. [3] conducted an on-line survey to investigate the academics in the field of Education for Sustainable Development. The participants decided which one of 12 pairs of statements better represented their interpretation of the Sustainable Development concept in the context of Education for Sustainable Development. The results show that there is a tendency to recognize the social and economic aspects of Sustainable Development more than the environmental ones. Another finding highlight that the three pillars of Sustainable Development (environment, society and economy) are seen as separated to each other and the concept of Sustainable Development is not conceived holistically.

Sinakou et al. [4] explore how three major factors (international policy documents, academic discourse, and teachers and students) view the concept of sustainable development within education for sustainable development. Through a holistic approach, the international policy documents are reviewed starting with the 1987 Brundtland report "Our Common Future" highlighting how the sustainable development related concerns are reflected in the Declarations issued over time. When analysing the discourse on SD concept, several perspectives were found upon the interests of different groups. Research also shows that neither teachers nor students hold a holistic view of the SD concept.

Similar findings are presented by Ocetkiewicz et al. [5] for Polish school teachers. It is shown that they are not well prepared to include the key issues of SD in school curricula such as responsible energy consumption, renewable energy use, or sustainable management of natural resources. It is concluded that the teachers are not aware of these issues, only a few taking part in specific workshops and training sessions. Some of the teachers argue that sustainable development should not be implemented as part of formal education while a large majority considers that extra-curricular classes should have as scope these aspects.

Nousheen et al. [6] showed that the teacher's education has an important role towards a sustainable future after conducting a study investigating the student-teachers' attitude towards sustainable development. The results indicate a positive effect on pre-service teachers who studied education for sustainable development subjects. It is highlighted the need of the education for sustainable development in different academic programs.

The perception about sustainability learning of 3364 engineering students from three Spanish universities was assessed by Sanchez-Carracedo et al. [7]. Four sustainability competencies were considered among which the application of ethical principles related to the values of sustainability in personal and professional behavior was found to be the least developed. The students declare less competence in the sustainable use of resources than in the critical contextualization of knowledge or in the participation in community processes. The paper also shows that the engineering curricula focus on economic aspects while neglecting environmental ones. It is concluded that the analyzed universities don't have a centralized policy on sustainable development education and the results may depend on the professors' motivation to include SD in their subjects.

Abad-Segura and Gonzales-Zamar [8] performed a global analysis within the SDGs framework on the sustainable economic development in higher education institutions. The global evolution of the research trends on sustainable development was assessed through bibliometric techniques applied on 1689 articles published between 1990 and 2018. The pillars of the SD are reflected by the most prominent subject areas: Social Sciences followed by Environmental Science, Business, Management and Accounting, and Engineering. Four main thematic axes of research were detected: sustainable development, environmental impact, economics, education and recycling. As future research direction, community wellbeing was found with the highest relevance score, followed by university students entrepreneurship, circular economy strategy and sustainable competitiveness.

### 1.3 Education for Renewable Energy

Considering that the energy is the main sector causing the environmental pollution due to unsustainable production and consumption, the development of new and clean technologies is necessary. This is possible through the use of renewable energy sources by increasing public awareness, knowledge and skills in this area. In this mission, the Universities have an important role in both formal and informal education by promoting strategies and the appropriate instruments in the field of renewable energy by education and training of students in all three Bologna cycles, professionals in companies, key players in public institutions, other members of communities.

A reduced number of Bachelor and Masteral Programs in Renewable Energy exists in European Universities in 2021 as shown in Table 1.

The objective of this paper is to identify the actions and the instruments to promote Sustainable Development with emphasis on Sustainable Energy by involving Higher Education Institutions in formal and non-formal education for different target groups. As case study, Transilvania University of Brasov, Romania is considered.

## 2 Sustainable Development in Formal Education

As we know, a Higher Education Institution prepares human resources for the labor market through study programs at bachelor's, master's, and doctoral level. The formulation of a strategy for implementing the concept of Sustainable Development at institutional level is the mission of the institution top management structure. This strategy aims to

**Table 1.** Bachelor and Masteral Programs in renewable energy in European Universities.

| University | Name of the programme study |
|---|---|
| Wels, AT | MSc on Sustainable Energy Systems |
| Freiburg, DE | MSc on Sustainable Systems Engineering |
| Rottenburg, DE | BSc on Renewable Energies |
| Stralsund, DE | BSc on Regenerative Energies |
| Reykjavik, IS | MSc on Sustainable Energy Engineering |
| Valletta, MT | BSc on Power Generation and Renewables |
| Eindhoven, NL | MSc on Sustainable Energy Technology |
| Castelo Branco, PT | BSc on Renewable Energy Engineering |
| Brasov, RO | BSc on Renewable Energy Systems Engineering<br>MSc on Sustainable Product Design and Environment Protection |
| Lund, SE | MSc on Sustainable Energy Engineering |
| Stockholm, SE | MSc on Sustainable Energy Engineering |
| Cardiff, UK | MSc on Sustainable Energy and Environment |
| Nottingham, UK | MSc on Electrical Engineering for Sustainable and Renewable Energy |
| Wrexham, UK | BSc on Renewable & Sustainable Engineering |

transform the main processes of education and scientific research for training graduates to understand and apply the concept of Sustainable Development.

The introduction in the existing courses of the concept of Sustainable Development, in the existing study programs of some course modules with this topic, the development of new study programs at BSc, MSc and PhD levels for the promotion and development of sustainable products are the main ways. Obviously it is mandatory to train the staff and create the necessary infrastructure.

As example, in the training of students from the Transilvania University of Brasov, the course module "Sustainable Development" was introduced for a series of study programs. New study programs for sustainable products were developed, such as Industrial Design and Renewable Energy Systems Engineering - at bachelor's degree and Product Design for Sustainable Development and Environment - at master's level. The entire activity of scientific research in the 18 faculties of the University (Fig. 1) has been completely restructured in Scientific Research Centers oriented towards Sustainable Development according to the specifics of each faculty. The scientific research centers from the nine technical faculties were oriented towards the development of products using the concept of Sustainable Energy (energy efficiency and saving, renewable energy systems, natural resources conservation and use), and the scientific research centers from the other nine faculties with a non-technical profile were oriented towards other elements of Sustainable Development, according to the each faculties specifics (economic development, education, communication, culture, sustainable legislation, health and life quality).

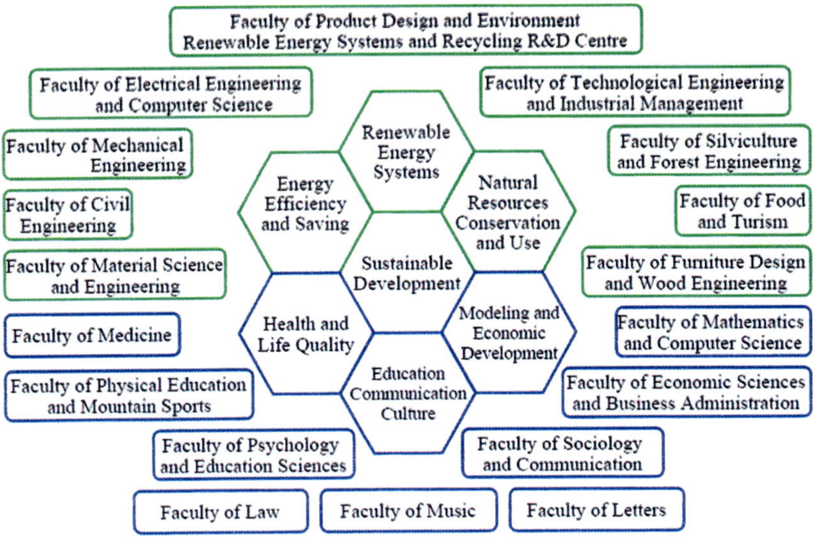


**Fig. 1.** Sustainable development in the Transilvania University of Brasov faculties

This development was possible due to the comprehensive structure of the University and all the Centers are located in the Research and Development Institute of Transilvania University (ICDT) developed through the major project "ICDT - High Tech Products for Sustainable Development" financed from European structural funds (Fig. 2). The location of the centers, developed through the project, is a sustainable community with a newly built and energy efficient environment (11 Low Energy Buildings) and the modern infrastructure for research used in developing sustainable products for the economic environment.

The Institute of the Transilvania University of Brasov was designed by in-house teams and developed between 2009 and 2013 through structural funds project [7]. The renewable based energy mix implemented here consists of: ground-water heat pumps used to extract the geothermal energy through horizontal (H-HX) or vertical (V-HX) heat exchangers for space heating, flat plate (FP) and evacuated tube (ET) solar thermal collectors (STC) to prepare domestic hot water, on-grid solar tracked photovoltaic platforms (PV) yearly balancing the electrical energy demand, off-grid wind turbines (WT) to supply the outdoor lighting system. An outdoor testing rig (OTR) is used to investigate the efficiency of novel trapeze [10] and triangle [11] solar thermal collectors with variously colored absorber plates. The fresh water is provided by a local water source (WS) and the waste water is cleaned by a local wastewater treatment plant (WW) before discharging on the geothermal heat exchangers (Fig. 2).

The activity of the Research Center Renewable Energy Systems and Recycling (RESREC) involves staffs and students from the Faculty of Product Design and Environment is focused on the Sustainable Energy in Energy Efficient Built Environment, mainly on solar energy conversion systems for thermal and electrical energy [12] and product development in the concept "from material to prototype" (Fig. 3).

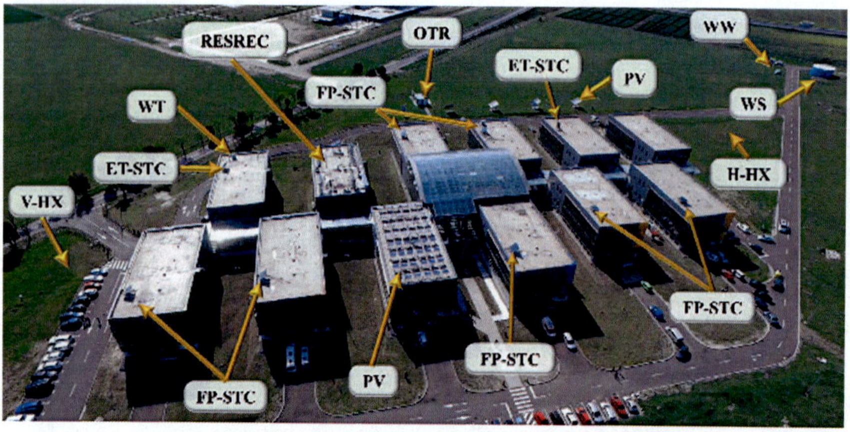


**Fig. 2.** The research and development institute of the Transilvania University of Brasov.

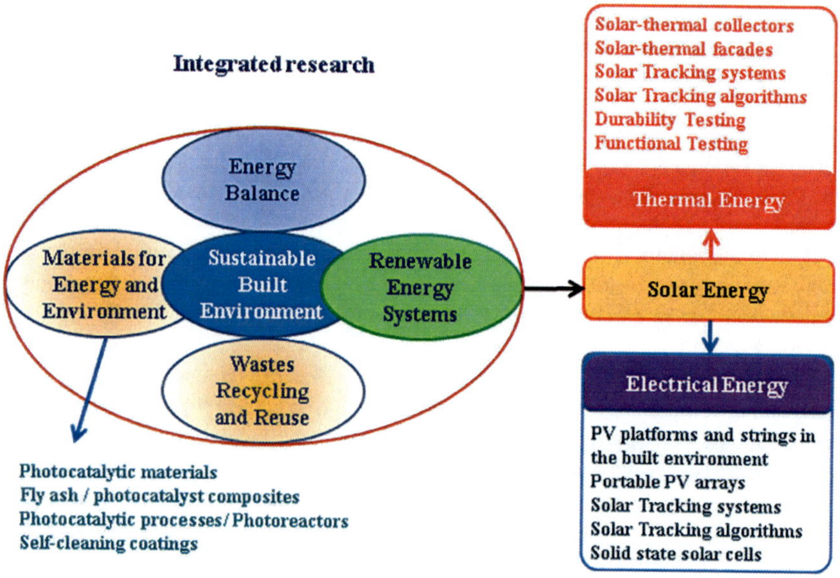


**Fig. 3.** Integrated education and research in the Renewable Energy Systems and Recycling Research Center of the Transilvania University of Brasov, Romania

The Renewable Energy Systems and Recycling Research Center has been organizing each three years the Conference on Sustainable Energy since 2005 (Table 2). The conference provides the floor for the dissemination of the new results, innovative concepts and solutions for the transition towards sustainable development that give use to the renewable energy sources. During these conferences, focus groups and workshops were organized with local authorities and other stakeholders.

**Table 2.** Topic of the conference for sustainable energy [13].

| Conference year | Conference name |
|---|---|
| CSE 2020 | Solar Energy Conversion in Communities |
| CSE 2017 | Nearly Zero Energy Communities |
| CSE 2014 | Nearly Zero Energy Buildings |
| CSE 2011 | Sustainable Communities |
| CSE 2008 | Solar Energy Conversion |
| CSE 2005 | Sustainable Energy |

Modern laboratories used in the training of an average number of 25 bachelor students per year were developed on the Colina Hill of the Transilvania University of Brasov (Fig. 4). A nearly Zero Energy Building, the Solar House [14], was developed for which almost the entire energy demand is covered by a renewable based energy mix consisting of a geothermal heat pump and a solar thermal system supplying thermal energy for space heating and domestic hot water preparation, and grid connected photovoltaic and wind turbine systems yearly balancing the electrical energy load. Approximatively 80% of the BSc graduates continue the study in MSc Sustainable Product Design and Environment Protection and around 70% are employed in companies in the field of renewable energy.

**Fig. 4.** The Colina Hill campus of the Transilvania University of Brasov.

## 3 Sustainable Development in Non-formal Education

In non-formal education, Higher Educations Institutions can play a major role in understanding, promoting and training on Sustainable Development by:

- in service-training for secondary schools teachers and pupils through certificated courses developed in the frame of European projects providing necessary instruments and materials resources (e.g. Comenius and Leonardo da Vinci projects);
- nationally certified long life learning training for professionals in companies developed based on a curriculum adapted to the needs of the target groups (e.g. sustainable built environment, energy efficiency, renewable energy systems, water treatment, waste recycling, sustainable transport etc.);
- communication with local authorities (Metropolitan Agency, Local Councils, Sustainable Development Agency) in developing the local Sustainable Development strategy;
- communication with the regional authorities in promoting the Sustainable Built Environment and Sustainable Energy through regional strategies and joint projects;
- promoting clean energy through training and joint projects with the Regional Clusters of Industrial Companies;
- collaboration at national level with representative structures such as Academies, in which there are departments and commissions with the objective of promoting sustainable development and renewable energy;
- promoting, as members of international structures, the sustainable energy through joint projects of education and scientific research, through the development of international curriculum and training through common curriculum, summer schools, joint conferences etc.

As example, one of the instruments developed in the frame of a Leonardo da Vinci project "*Renewable Energy Sources and Environment Management Friendly ICT Tools - RES & EM ICT Tools*" was designed as a result of an extended needs analysis for the awareness of key professional bodies, relevant authorities and energy users on the benefits of sustainable development at different levels (economic, environmental, quality of life) and in different sectors (urban, rural, agricultural, industrial). The objective of the project was to develop a dynamic and easy to use e-Learning tool, structured on 11 modules (Fig. 5) (Conversion of solar energy into thermal- and electrical- energy, Bioclimatic Architecture and Urbanism, Hydropower, Wind energy conversion systems, Hydrogen Technology, Hybrid Technologies, Energy Storage, Wastewater Treatment, Energy and Environment, Renewable Energy Sources Management) and 3 levels (fundamentals, design, advanced) according to the specific needs of a wide variety of interested target groups: program designers training, trainers, young graduates of technical schools or universities, companies, local, regional and national authorities. Developed by 9 partners from Romanian, German, Netherlands and Greek Universities, regional and national authorities (National Center for Sustainable Development), small and medium enterprises, the project essentially contributed to the development of continuous training through electronic resources and hard - soft material base.

Examples of RESREC staff cooperation with structures at different local, regional, national and international levels are presented in Fig. 6.

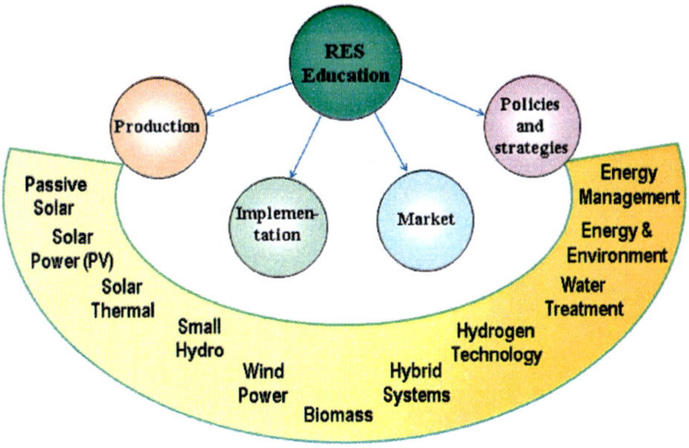


**Fig. 5.** Instrument of eLearning for in service-training

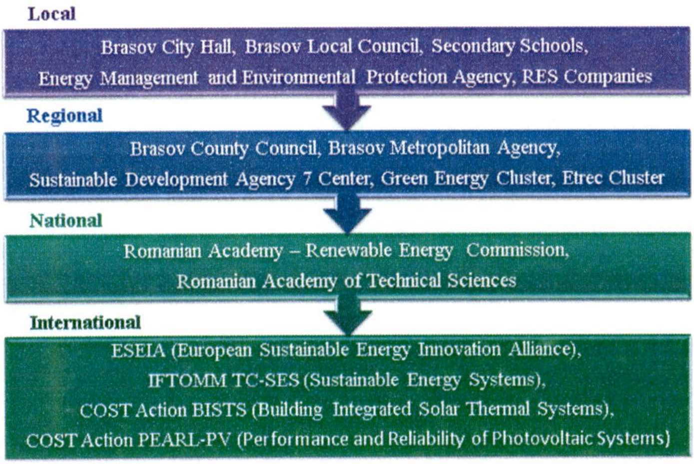


**Fig. 6.** Cooperation with different structures in the field of Sustainable Development

# 4 Conclusions

The implementation of the Sustainable Development concept requires qualified human resources, able to identify ways of implementation, to develop strategies and to promote concrete activities by collaborating with various structures at local, regional, national and international level.

In this sense, Higher Education Institutions have an important role in training these specialists by developing appropriate curriculum and tools for formal education at all three stages of the Bologna cycle.

Also, universities have the responsibility to develop curriculum and tools for non-formal education of various target groups through in service-training, long life learning, involvement in the development of Sustainable Development strategies at local, regional, national and international levels.

The paper describes the ways and exemplifies the involvement of the Transilvania University of Brasov in promoting the Sustainable Development concept, internally and in collaboration with various external structures.

## References

1. United Nations: Transforming our world: the 2030 agenda for sustainable development, New York, 12 August 2015
2. Avelar, A.B.A., da Silva-Oliveira, K.D., da Silva Pereira, R.: Education for advancing the implementation of the sustainable development goals: a systematic approach. Int. J. Manag. Educ. **17**(3), 100322 (2019)
3. Sinakou, E., Boeve-de Pauw, J., Goossens, M., Van Petegem, P.: Academics in the field of education for sustainable development: their conceptions of sustainable development. J. Clean. Prod. **184**, 321–332 (2018)
4. Sinakou, E., Boeve-de Pauw, J., Van Petegem, P.: Exploring the concept of sustainable development within education for sustainable development: implications for ESD research and practice. Environ. Dev. Sustain. **21**, 1–10 (2019)
5. Ocetkiewicz, I., Tomaszewska, B., Mróz, A.: Renewable energy in education for sustainable development. The Polish experience. Renew. Sustain. Energy Rev. (80) 92–97 (2017)
6. Nousheen, A., Zai, S.A.Y., Waseem, M., Khan, S.A.: Education for sustainable development (ESD): effects of sustainability education on pre-service teachers' attitude towards sustainable development (SD). J. Clean. Prod. **250**, 119537 (2020)
7. Sanchez-Carracedo, F., Sureda, B., Moreno-Pino, F.M., Romero-Portillo, D.: Education for sustainable development in Spanish engineering degrees. Case study. J. Clean. Prod. **294**, 126322 (2021)
8. Abad-Segura, E., Gonzales-Zamar, M.D.: Sustainable economic development in higher education institutions: a global analysis within the SDGs framework. J. Clean. Prod. **294**, 126133 (2021)
9. Visa, I., Duta, A., Moldovan, M., Burduhos, B.: implementing renewable energy systems in nearly zero energy communities. In: Visa, I., Duta, A. (eds.) CSE 2017. SPE, pp. 3–24. Springer, Cham (2018). https://doi.org/10.1007/978-3-319-63215-5_1
10. Visa, I., et al.: Design and experimental optimisation of a novel flat plate solar thermal collector with trapezoidal shape for facades integration. Appl. Therm. Eng. **90**, 432–443 (2015)
11. Visa, I., Moldovan, M., Duta, A.: Novel triangle flat plate solar thermal collector for facades integration. Renew. Energy **143**, 252–262 (2019)
12. Visa, I., Duta, A., Moldovan, M., Burduhos, B., Neagoe, M.: Solar Energy Conversion Systems in the Built Environment. Green Energy Technology Springer, Cham (2020). https://doi.org/10.1007/978-3-030-34829-8
13. CSE homepage. http://cse.unitbv.ro/. Accessed 12 May 2021
14. Visa, I., Moldovan, M., Comsit, M., Duta, A.: Improving the renewable energy mix in a building toward the nearly zero energy status. Energy Build. **68**, 72–78 (2014)

# Appropriate Technologies and Technical Systems for Sustainable and Inclusive Development

# Float-Ram: A Sustainable Machine for Buildings Made by Compressed Earth Blocks

Giuseppe Quaglia(✉), Carlo Ferraresi, and Walter Franco

Department of Mechanical and Aerospace Engineering, Politecnico di Torino, Torino, Italy
giuseppe.quaglia@polito.it

**Abstract.** This article describes a sequence of research, design, prototyping and utilization of a new manually operated press, which has the purpose of facilitating a construction technology that uses compressed earth blocks, CEB. Buildings and houses made of CEB have better quality than traditional in poor region. They are also economical, robust, aesthetically pleasing, and sustainable from an environmental point of view. The research is addressed to some targets of the SDG 11, Sustainable Cities and Communities: (i) "ensure access for all to adequate, safe and affordable housing and basic services and upgrade slums"; (ii) "support least developed countries, including through financial and technical assistance, in building sustainable and resilient buildings utilizing local materials" .

The paper presents the project path, the results and finally the impact in a specific context, the rural areas in Tanzania.

**Keywords:** SDG11 · Earth construction · Compressed earth block · Human powered press · Manual operated machine

## 1 Introduction

The right to adequate shelter is recognized as a basic and fundamental human right since 1948, with its introduction in article 25 of the Universal Declaration of Human Rights. Other international treaties have addressed the right to adequate housing with different approaches. The United Nations agency UN-Habitat promote socially and environmentally sustainable towns and cities to improve living conditions for all. Among the targets of the SDG 11, Sustainable Cities and Communities, we find: (i) "ensure access for all to adequate, safe and affordable housing and basic services and upgrade slums"; (ii) "support least developed countries, including through financial and technical assistance, in building sustainable and resilient buildings utilizing local materials" [1].

Nevertheless, the right to adequate shelter is far from its full implementation. The amount of slum dwellers reached more than 1 billion in 2018, the 24% of the urban population, even recording a slight increase compared to the 23% in 2014. In Eastern and South-Eastern Asia, 370 million people live in slums, 238 million in sub-Saharan Africa and 226 million in Central and Southern Asia [1]. In Latin America, 10 to 15 million households live in substandard housing and the deficit in adequate housing


G. Quaglia et al. (Eds.): I4SDG 2021, MMS 108, pp. 99–109, 2022.
https://doi.org/10.1007/978-3-030-87383-7_11

continues to grow. Faced with a demand of 2.5 million homes per year, only 1.5 million are adequately built [2].

This chronic lack of housing opens to practices of self-help housing in informal settlement, which house up to 60% of the urban population in some Latin American cities [2]. In fact, due to the limited planning and financing capacities of public bodies, the inefficiency of assistance mechanisms, the lack of interest of private individuals, and also the scarcity of resources on the part of single families, it seems that a contribution to the mitigation of the problem may come from low-cost construction technologies suitable for self-construction [3], that make use of local, widespread, easily available and low-cost materials, such straw [4–6], timber [7] or earth.

The history of earth construction technologies is as old as that of urban society. The first settlements in the Fertile Crescent, 10,000 years ago, were almost entirely in earth. Much of the building widespread in European cities until the industrial revolution was built with raw earth techniques [8]. Raw earth construction is widespread in a large part of the globe, in climatic regions very different from each other: in coastal areas, inland, in the plains and in areas mountainous or hilly. A common factor is the use of earth taken under the fertile soil layer and therefore devoid of the organic fraction, composed of appropriate quantities of clay, which acts as a binder, silt and sand. The earth building technique has been successful due to the almost unlimited availability of the building material, its cost-effectiveness, the possibility of use on site without the costs of transport and transformation. The numerous and different earth construction techniques, developed in innumerable regions of the world according to local conditions, can be grouped into two macro-categories: *rammed earth or pisé* in which the wall is built compacting the earth directly in a box; *raw brick* in which the wall is realized using earth blocks.

In the field of self-construction, in recent decades the technique that uses raw bricks has been the subject of renewed interest as the blocks can be produced off-site reducing drying and installation times, and allowing assembly also by unskilled labor [9, 10]. Blocks can be made manually pressing the earth into an open timber frame (adobe) or produced using proper presses (compressed earth block CEB). The CEBs technique permits to improve the mechanical characteristic of the block, and to avoid erosion during rains, in particular when the soil is stabilized with a small percentage of Portland cement (about 6–10%) [9]. For all these reasons, the CEBs construction technique is appropriate to improve the housing conditions of poor people [11, 12].

To produce CEBs, hundreds of block presses have been developed since the first invented by the French architect François Cointeraux in 1789 [13]. The Belgian engineer E. Gossiaux in 1904 designed a machine, La Madelon, improved in the Super Madelon model, produced by different manufacturers with different names (Stabibloc, Fib-SM, Terstaram, Ceraman) and used until the 1990s. These models were bulky, heavy and expensive, with consequent limitation of dissemination. In order to overcome these issues, the Chilean engineer Raul Ramirez developed in 1952 a low-cost model, called Cinva-Ram, one of the best-known and most widely used block presses in the worl. One of the limits of Cinva-Ram is to perform unidirectional compression. Bi-directional presses can compress two sides of the block at the same time, improving its homogeneity. Among the bidirectional presses we mention the Altech Geo50. In [14], the authors underlined some disadvantages of the bidirectional presses currently available on the market: constructive complexity, high cost, difficult to use, poor ergonomics, non-constant actuation force required to the operator, low compaction pressure, safety issues. In [15] they presented the concept and the general design of a new press, called Float-Ram. In this paper, first the local context of the application of the technology, the Njombe region of Tanzania, is described, then the detailed design of the machine is presented, finally the kindergarten of Mkiu made in self-construction wit CEBs produced using the Float-Ram is shown.

## 2 Local Needs and Design Requirements

### 2.1 Social Conditions and Traditional Buildings in Tanzania

To assess whether a technology can have a positive impact in a specific context, it is necessary to start from the needs and criticalities of that community, and define the scenario, considering the specific living conditions. An example can be that of the Njombe region, Tanzania, where the rural population lives mainly on small agricultural activities and the very humble houses are built using fired clay bricks with rudimentary techniques (Fig. 1a, b). The houses are not very durable, often unhealthy, and unsafe.

The workers, using their bare hands in the mud, fill moulds (Fig. 2a) which are overturned on an open ground (Fig. 2b) to allow drying in the sun. Subsequently, the bricks are fired creating small mounds under which the fire is kept burning for several days (Fig. 2c). This technology has numerous drawbacks: it generates deforestation and a significant environmental impact in the areas where it is used; the bricks produced are very inhomogeneous, brittle, fractured and irregular (Fig. 2d). Given the very irregular shape it is necessary to use a lot of mortar and therefore a lot of concrete between one brick course and the other, to build regular-looking walls, and a certain skill is required to make reasonably flat walls.

**Fig. 1.** Traditional buildings in fired clay bricks, Tanzania

**Fig. 2.** Production techniques of fired clay bricks in Tanzania

In this context, the local needs generate a request for appropriate technologies which is an indispensable condition for the implementation of successful projects. Appropriate technologies may have an impact not only on the aforementioned SDG11, but produce a social development, such as that desired in SDG1, and in particular in the targets 1.4, "ensure that all men and women, in particular the poor and the vulnerable, have access to appropriate new technology" and 1.5 "build the resilience of the poor and those in vulnerable situations and reduce their exposure and vulnerability to climate-related extreme events".

Construction techniques with CEB are a valid alternative, particularly suitable in poor countries with a rural economy, to provide a better quality of life and housing.

## 2.2 Design Requirements

*Requirements related to CEB and Production Process*

The authors chose to design a novel press, named Float-Ram, able to produce the CEB named "Blocco Mattone", whose geometry and dimensions are shown in Fig. 3. The geometry defined by prof. Roberto Mattone, of Politecnico di Torino, is effective for easy wall construction thanks to overhangs and holes, and guarantees high wall robustness [16, 17].

The earth must be compressed bi-directionally in order to increase the CEB quality, up to 60% of the initial volume, applying a compression pressure of $1 \div 2$ MPa [9].

Regardless of the type of press, the production process of a CEB must carry out seven phases summarized in Table 1.

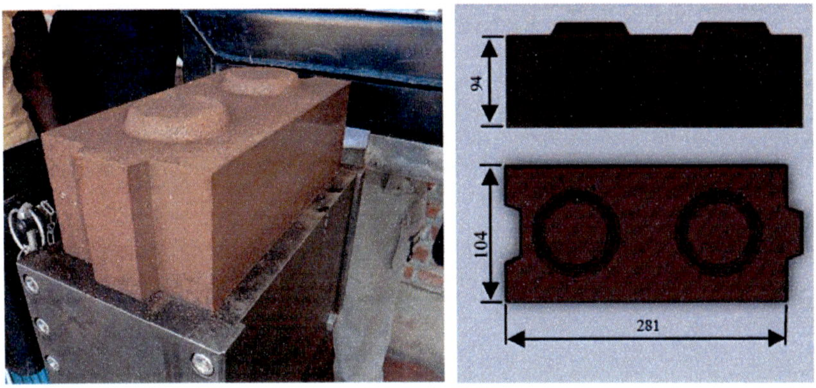


**Fig. 3.** "Mattone" compressed earth block

**Table 1.** CEB production cycle.

| Operation | Phase |
|---|---|
| 1. Mould filling | P1. press arrangement to earth filling |
| | P2. earth metering and mould filling |
| 2 Block forming | P3. press reconfiguration for block pressing |
| | P4. block pressing |
| 3. Block extraction | P5. press reconfiguration for block expulsion |
| | P6. block expulsion from mould |
| | P7. block removal |

*User Oriented Design*

The design process is focused on the users, their needs and requirements.

In the conception of the press, the following project objectives were therefore set: safety, simplicity of use, effectiveness and efficiency. sustainability, human wellbeing, ergonomy and performance. Furthermore, the environmental and social context makes manual operation preferable since the costs deriving from hydraulically operated presses would not be sustainable and in general the supply of energy in poor regions is often difficult or impossible.

*Operational Requirements*

Other requirements derive from the conditions of use: The press must be easily transportable on the construction site, loadable on a pick-up vehicle, must have a "closed" configuration very compact in a parallelepiped volume in order to be easily shipped to different countries, must be equipped with wheels for small mobility in the operational context, and must have the lowest possible construction cost, compatibly with high performance and reliability.

*Mechanical Synthesis Requirements*
A design challenge was to conceive a machine that could carry out the aforementioned seven phases basing all the movements on a single main kinematic element, as part of all the necessary kinematic pairs, with the general aim of reducing the press production costs and of making the press easily buildable and maintainable.

## 3 Functional Design

### 3.1 Mechanical Architecture

Starting from design requirements, the mechanical architecture of the machine was conceived (Fig. 4):

- a single lever, operated by the worker, is used to manage both the reconfigurations of the press and the compression and ejection phases of the block;
- bidirectional compression is obtained in an innovative way by moving only the lower plate and leaving free the translation of the mould walls;
- motion transmission occurs through a transformation of the rotation of the lever into a translation of the lower plate, thanks to a cam system; the synthesis of the variable transmission ratio mechanism allows the design of a system capable of exerting increasing pressure on the block, while requiring an almost constant force exerted from the operator;
- a single cylindrical kinematic element serves for the realization of all the kinematic pairs necessary for operations, reducing costs and complexity of the press and increasing its precision and repeatability.

The production cycle is presented in Fig. 4. First the mould must be positioned in the configuration of Fig. 4a, to start the production of the block (phase P1). The metering box must be completely filled by the operator with a mixture of earth, cement and water necessary for the creation of the block. Since each type of soil has specific properties, the volume of earth necessary for the creation of the single block varies and for this purpose a process set-up phase is necessary. By adjusting the position of the movable bottom of the metering box, the volume of soil that will be poured into the mould at each cycle is defined, allowing a regularity of the production process and a constancy of the mechanical properties of the bricks produced. To fill the mould (P2) the metering box is rotated quickly, pouring its contents into the mould.

The mould, together with the entire motion transmission system, is brought inside the pressing portal simply by dragging the lever in rotation, A, with respect to a vertical axis coinciding with the shaft (P3), Fig. 4b. During this operation, which takes place thanks to the presence of an axial support bearing on the shaft, the operator does not perceive significant forces since no part of the machine is raised or lowered.

The compression of the block (P4) occurs by rotating the lever downwards and thus activating the cam system, Fig. 4c. It is necessary to point out that there are two different cams, integral with each other and operated by the same lever. The pressing cam is designed in such a way as to optimize the compression phase; the expulsion cam to carry out the subsequent phase of ejection of the block. Contrary to usual schemes, the cam axis is free to translate and rotate, while the roller has a rotation axis in a fixed position with respect to the frame. The rotation of the cam is then transformed into a translation in the vertical direction of its axis, guided by the cylindrical kinematic pair consisting of the vertical shaft and the substructure that supports the lower plate of the press. In this way the soil compression process is accomplished. In the same phase it should be emphasized that the floating mould is free to translate in the vertical direction, and this expedient allows a bidirectional compaction of the earth.

By performing an inverse rotation B (Fig. 4b) with respect to phase P3, the mould is brought back out of the portal frame, thus carrying out phase P5, press reconfiguration for block expulsion. Finally, in Fig. 4d the phase P6, block expulsion from mould, is represented. In this phase the expulsion cam is active and the rotation of the cam is transformed into a wider translation of the lower plate, necessary for the complete expulsion of the block from the mould, and to allow the operator to take the block (P7), in ergonomic conditions (it is not necessary to bow down) simply by placing the hands on the side walls of the block.

Some further details are evident in Fig. 5, in which the press is seen in shipping conditions (closed, Fig. 5a) and in working conditions (open Fig. 5b).

The closed configuration allows for easy packaging and shipping anywhere in the world. The open condition is obtained with two simple actions: reassembling the folded operating lever in extended configuration, and rotating the lateral support foot 90°, which is used to prevent overturning during the expulsion phase.

Finally, it should be noted the presence of two wheels that facilitate the small movements on site and the loading and unloading operations. The press total mass is about 260 kg and for proper operation cleaning and lubrication of cylindrical pair are required.

Productivity can be close to 100 blocks/h, but it is strongly affected from the process of raw earth mixing and feeding.

### 3.2 Transmission System Design

The design of the cam system is a fundamental element to ensure ergonomic conditions for the operator. The cam profile was created starting from an experimental identification of the pressure trend during the compression of the block. Following an experimental campaign, example in Fig. 6a, the parameters $K_1$ and $K_2$ of an exponential model of the compression curve of the blocks, $p(y) = K_1 \cdot e^{K_2 \cdot y}$ were identified, where $p$ is the pressure exerted on the block surface $A$, and $y$ is the compression stroke.

The efficiency of the cam transmission system $\eta$ is equal to the ratio between the compression work $W_F$ of the block and the work $W_C$ provided by the operator, Eq. (1), indicating with $F$ the force on the compression plate, and $C$ the operating torque on the lever.

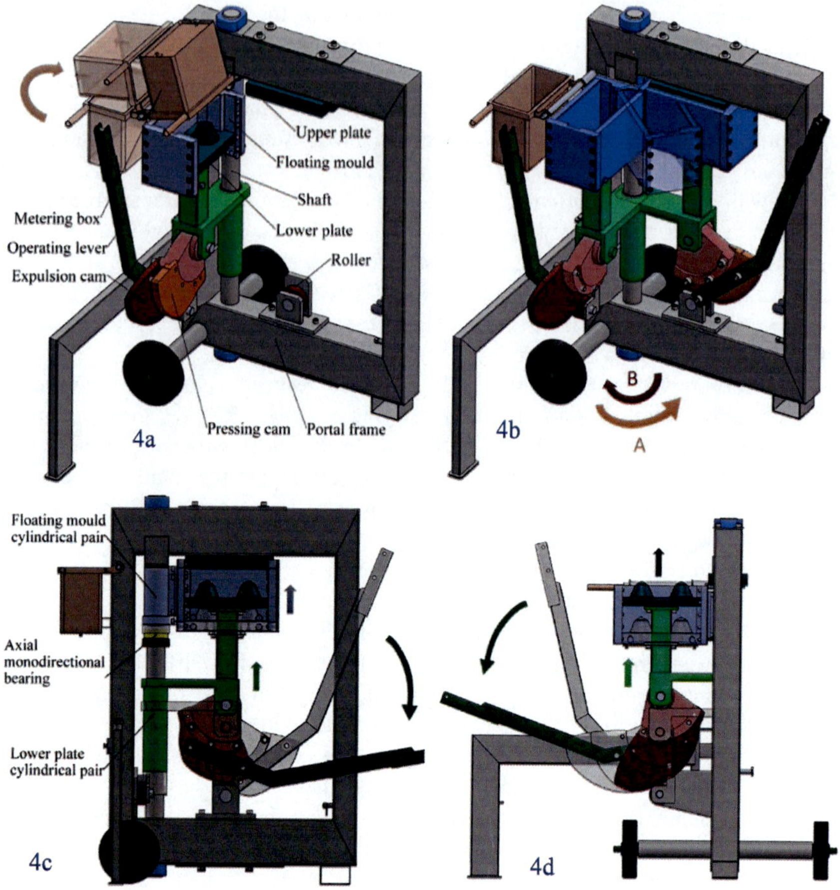


**Fig. 4.** Float-Ram mechanical architecture and operating phases

From Eq. 1 it is possible to express the instantaneous transmission ratio, $y'$, having the target to keep the requested torque as constant as possible (Eq. 2). After integration of Eq. 2, y(α) used in the typical synthesis algorithms of cam transmissions is obtained.

Figure 6b shows the profile of the compression cam. The synthesis of the profile also requires some further details regarding the initial clearance recovery phase and the final phase of resetting the transmission ratio, not described for sake of brevity.

$$\eta = \frac{dW_F}{dW_C} = \frac{F \cdot dy}{C \cdot d\alpha} = \frac{p \cdot A \cdot dy}{C \cdot d\alpha} \tag{1}$$

$$y' = \frac{dy}{d\alpha} = \frac{C \cdot \eta}{p \cdot A} = \frac{C \cdot \eta}{K_1 \cdot e^{K_2 \cdot y} \cdot A} \tag{2}$$

**Fig. 5.** Float-Ram transport configuration 800 × 830 × 1260 mm (a) and operative configuration (b)

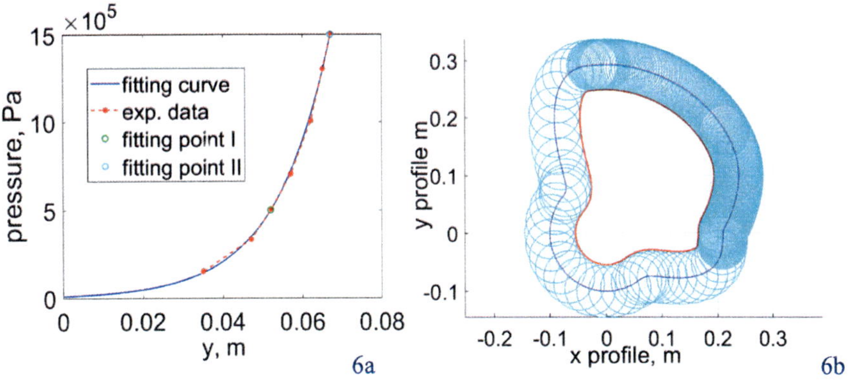


**Fig. 6.** Pressure experimental data and model (a) compression cam profile (b)

## 4 Conclusions

Several examples of Float-Ram press have been used in numerous projects in Africa and South America. In addition to fully satisfactory operational functionality and the quality of the blocks and houses built, numerous positive impacts have been observed. Better and safer working conditions have been created and new professions have been born. Above all, local communities have been able to create conditions for a better future. In Fig. 7 the kindergarten of Mkiu is presented, made in self-construction with CEB “Blocco Mattone”, using the Float-Ram press.

**Fig. 7.** Mkiu kindergarden, Tanzania

## References


1. United Nations. https://sdgs.un.org/goals/goal11. Accessed 12 May 2021
2. Habitat for Humanity. https://www.habitat.org/lc/lac/pdf/causes_of_inadequate_housing_in_lac.pdf. Accessed 12 May 2021
3. Foti, M.: Tecnologie per tutti. Soluzioni semplice a basso costo per l'habitat. Scuola di specializzazione in tecnologia architettura e città nei Paesi in via di sviluppo, Politecnico di Torino. Internal report (2003)
4. Kean, S.: Rebuilding: from the bottom up. Science **327**(5966), 638–639 (2010)
5. Franco, W., Iarussi, F., Quaglia, G.: Human powered press for producing straw bales for use in construction during post-emergency conditions. Biosys. Eng. **150**, 170–181 (2016)
6. Franco, W., Quaglia, G., Ferraresi, C.: Experimentally based design of a manually operated baler for straw bale construction. In: The 1st International Conference of IFToMM ITALY, Vicenza, Italy, 1–2 December 2016 in Mechanism and Machine Science, vol. 47, pp. 307–314 (2017)
7. ASF Piemonte: Haiti |1| Foyer d'accueil aux enfants démunis d'Haiti, in III Congress of the University Network for Development Cooperation (CUCS Torino 2013), Torino, 19–21 September 2013, p. 317 (2013)
8. Scudo, G., Narici, B., Talamo, C.: Costruire con la terra. Tecniche costruttive, campi di utilizzo e prestazioni, Sistemi Editoriali Gruppo Editoriale Esselibri, Napoli, IT (2001)
9. CRAterre: Compressed Earth Block, Production Equipments. http://craterre.org/. Accessed 12 May 2021
10. Association Mattone su Mattone. http://www.mattonesumattone.eu. Accessed 12 May 2021
11. Venkatarama Reddy, B.V.: Design of a manual press for the production of compacted stabilized soil blocks. Curr. Sci. **109**(9), 1651–1659 (2015)
12. Guillaud, H., Houben, H.: Earth Construction: A Comprehensive Guide. Intermediate Technology, London (1994)
13. Mukerji, K.: Soil Block Presses. German Appropriate Technology Exchange, GATE (1988)
14. Ferraresi, C., Franco, W., Quaglia, G.: Human powered press for raw earth blocks. In: Proceedings of ASME IMECE 2011 International Mechanical Engineering Congress and Exposition, pp. 579–588. ASME, Denver (2011)
15. Ferraresi, C., Franco, W., Quaglia, G.: Concept and design of Float-Ram, a new human powered press for compressed earth blocks. In: Proceedings of the ASME 2017 International Mechanical Engineering Congress and Exposition, IMECE2017, 3–9 November 2017, Tampa, Florida, USA (2017)

16. Melo, A.B., Barbosa, N.P., Lima, M.R.F., Silva, E.P.: Structural performance for a masonry prototype built with blocks of stabilized raw earth. Ambiente Construído **399**(11(2)), 111–124 (2011)
17. Sassu, M., et al.: Production procedures and mechanical behaviour of interlocking stabilized compressed earth blocks (ISCEBs) manufactured using float ram 1.0 press. Eng. Solid Mech. **6**(2), 89–104 (2018). https://doi.org/10.5267/j.esm.2018.3.004

# The SNAP: A Novel Four-Wheel Pedal-Assisted Electric Lightweight Vehicle

Francesco Passarella[1], Giacomo Mantriota[2], and Giulio Reina[2(✉)]

[1] SNAP s.r.l., 72100 Bari, BA, Italy
[2] Department of Mechanics, Mathematics and Management, Politecnico di Bari, Via Orabona 4, 70126 Bari, BA, Italy
{giacomo.mantriota,giulio.reina}@poliba.it

**Abstract.** Innovation in transportation and mobility is central to sustainable development. There is a widespread awareness that society would benefit if transportation became more sustainable, promoting economic growth while respecting the environment. This paper introduces SNAP, a four-wheel pedal-assisted electric vehicle that represents a new concept in sustainable mobility towards filling the gap between bicycle and automobile.

**Keywords:** SDG7 · SDG11 · Electric assisted vehicles · Quadricycle · Pedelec · Mobility · Last-mile delivery

## 1 Introduction

In the last few years, the interest for lightweight and efficient transportation systems, which are partially human powered has steadily increased. New concepts for individual, efficient and pollutant-free transportation of people and goods in local traffic and urban environments have been proposed. Some examples are the EE-Speed Bike [1], the Bike2 drivetrain [2], the Mando Footloose [3], the Podbike [4], the Flevobike Orca [5] and the RATH Racer [6]. Sales volume of about 30 million vehicles are forecasted by the market for China and 6.8 million for Europe for the year 2025 [7]. Although the idea of a bicycle including motor assistance is old and it started with small combustion engines, the demand has substantially increased with the introduction of electric motor assistance. The simple and clean management of the drivetrain is the main advantage. Typically, central motors are combined with conventional bicycle gearboxes. Recently the implementation of additional features to the motor control are used as unique selling propositions. Those features include smartphone functions, navigation, health monitoring and driving assistance systems. However, all the models currently available on the market are either fairing-less or lack a comfortable driving position or easy access to the vehicle. Both conditions can be hardly found in a model and have substantially limited the widespread adoption of this type of vehicle.

This paper presents a novel pedal-assisted electric quadricycle that can be used for sustainable personal or goods transportation in urban environments. It complies with the EN15194 [8] as an electrically power-assisted cycle. As such, SNAP does not require


G. Quaglia et al. (Eds.): I4SDG 2021, MMS 108, pp. 110–117, 2022.
https://doi.org/10.1007/978-3-030-87383-7_12

any insurance coverage or plate. It features four independently suspended wheels to improve both comfort and safety. A closed cockpit that represents a unique feature of SNAP provides protection from adverse weather conditions. It fulfills autonomy in the range of 60 km (expandable to 120 km) with a maximum velocity up to 25 km/h. It can host the driver plus one passenger or goods, for a total weight of 80 kg. The overall maximum payload is 160 kg.

## 2 Related Works and Design Requirements

The European standard EN15194 [8] applies to electrically power-assisted cycles. However, the proposed SNAP model represents a special case that is not explicitly considered in this standard. Based on EN15194, cycles have a maximum velocity of 25 km/h and maximum power of 250 W. When the power rate increases up to 1 kW, the vehicle falls in the L1e-A category, with mandatory insurance, that keeps the speed limit of 25 km/h but requires insurance coverage. Another possibility is to fall in the lightweight quadricycles, e.g., L6e category, with mandatory insurance that have a power limit of 4 kW, a maximum speed of 45 km/h, and a maximum net weight of 425 kg that is largely fulfilled by the SNAP model. Based on the existing standards, three steps of power are envisaged for the proposed vehicle in order to comply with all the above-mentioned categories based on the buyer requests. In order to comply with all those categories, the vehicle chassis has been designed to withstand the worst stress conditions, e.g., approval conditions for the L6e category.

Currently, the market for this type of hybrid vehicle is not fully developed. Most of the proposed models show limited accessibility for an average user and share many components with the "velomobiles" (Fig. 1a) that are vehicles with human propulsion featuring an aerodynamic fairing and that have been adopted since the beginning of the last century. Although velomobiles are commonly used for agonistic training, they are not well suited as an urban transportation system. Another typology is that of trikes (Fig. 1b), which are less extreme than velomobiles but they are open and not waterproof. Trikes have a height from the ground that makes them not practical and dangerous in the case of impact with a car.

Based on the previous considerations, the design requirements for the SNAP model have been obtained to cover the largest chunk of the market:

- ease of accessibility
- full waterproofness
- comfort
- removable battery
- net weight less than 200 kg
- one driver seat and one passenger
- trunk compartment
- low maintenance and easy spare part management
- cost effectiveness around 8,000 €.

The overall goal is to fill the gap between bicycles and automobiles. Many users find it important to drive under the rain or to be able to give a ride to their children to or back from school with a bag to be placed in the trunk compartment.

(a) (b)

**Fig. 1.** Examples of recumbent cycles: (a) velomobile [9], (b) trike [10].

## 3 The SNAP Vehicle

A rendering of the proposed urban vehicle SNAP is shown in Fig. 2. The main technical specifications are collected in Table 1.

**Fig. 2.** A rendering of the SNAP vehicle

The scheme of the power transmission is depicted in Fig. 3. One permanent magnet brushless motor delivers the required power to the rear wheels via a two-step chain transmission. A gearbox with 11 gear ratios of ebike derivation transforms the power parameters in input to the rear open differential (the commercial model available for trike is used). The electric motor is mounted in the front of the vehicle where also the pedals are present. This choice guarantees the weight balance between the front and rear axis.

**Table 1.** SNAP main technical specifications

| Design parameter | Value |
|---|---|
| Unladen mass (kg) | 185 |
| Length (mm) | 2550 |
| Width (mm) | 1250 |
| Height (mm) | 1610 |
| Tire | 24″ × 3″ |
| Battery | 2 × 1.1 kWh (48 V) Li-ion cells with integrated BMS (2 × 6 kg) |
| Range | Approx 60 km with 2 batteries<br>Expandable up to 120 km (4 batteries) |
| Accessories | Front electric wiper<br>Front electric defrost<br>GPS tracker and 2G/3G connectivity |

In order to evaluate the motion resistance encountered by SNAP during straight motion, its longitudinal dynamics is investigated.

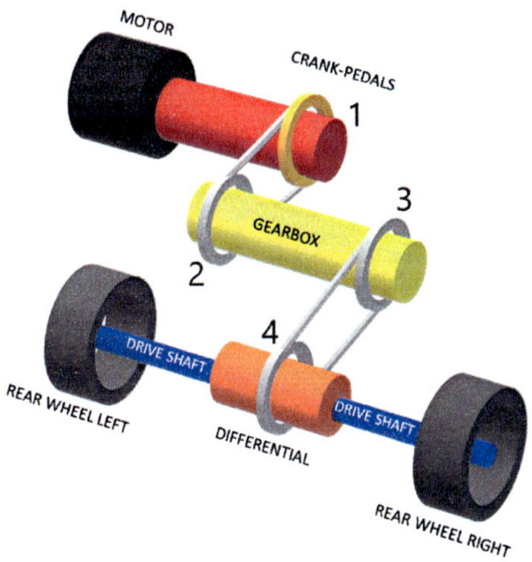


**Fig. 3.** Scheme of the power transmission

We recall that SNAP includes three power configurations: 250 W, 1000 W and 4000 W. Therefore, it is important to understand the behavior of the system in the lowest power configuration, especially in terms of gradeability and maximum speed. The power delivered by the drive motor can be expressed as

$$P_{\max}\eta_{tr} = \left(M_{vp}g\ \sin\ \alpha + M_{vp}g\ \cos\ \alpha f + \frac{1}{2}\rho C_x A V_{max}^2\right)V_{max}$$

where:

- $P_{max}$ is the maximum power (250 W);
- $\eta_{tr}$ is the efficiency of the transmission in the case of pure electric traction assumed equal to 78%. The loss in the two chain transmissions (motor to gearbox and gearbox to differential) is assumed approximately equal to 7% [11], while in the gearbox is assumed 8% [11], and finally 1% is considered in the semiaxes;
- $M_{vp}$ includes the vehicle mass (185 kg) and that of one driver (75 kg);
- $\alpha$ is the road slope;
- $f$ is the coefficient of motion resistance assumed constant and equal to 0.01, given an inflation pressure for the tires equal to 1 bar;
- $\rho$ is the air density (1.2 kg/m$^3$);
- $C_x$ is the aerodynamic resistance (0.64). This value is assumed equal to that of Renault "Twizy", which is very similar to SNAP from an aerodynamic standpoint;
- $A$ is the front area of the vehicle estimated as (total height - undercarriage clearance) $\times$ width $= 1.32$ m$^2$;
- $V_{max}$ is the maximum speed;
- the presence of wind is neglected.

Given the payload, the maximum achievable speed can be estimated varying the road slope for the 250 W configuration, as shown in Fig. 4. In the case of horizontal road, and with one driver only, the maximum velocity results in 17.5 km/h reducing to 14.7 km/h in the case of full payload consisting of a 190 kg payload (two 75 kg passengers + 40 kg luggage).

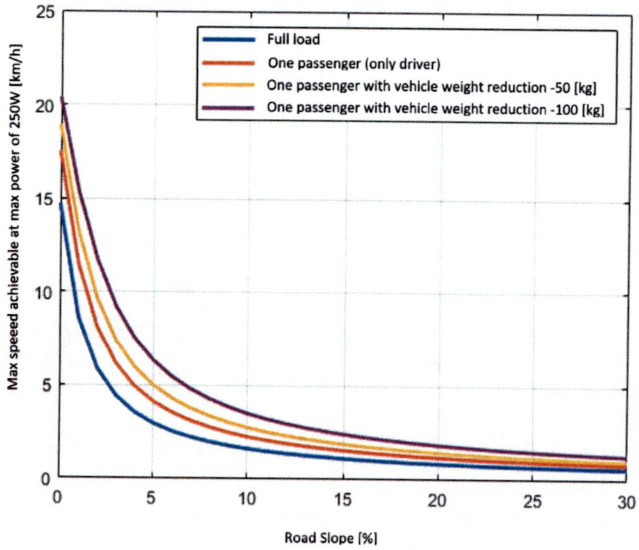


**Fig. 4.** Maximum achievable speed as a function of road slope

As expected, increasing the road slope results in a reduction in the maximum achievable speed. At the highest road slope of 30%, the maximum speed results in 1.5 and 1 km/h, respectively for the one driver and full load condition.

As for the SNAP chassis, the design choice has been made taking into account both the production costs and mechanical resistance. A trade-off solution is adopted with a tubular steel chassis that allows low investment costs and relatively small weight. Future developments may certainly include the adoption of composite materials although the costs may not be fully justified. The fairings are made of ABS and they are riveted on the chassis; a solution that allows easy replacement in the case of damage.

If we refer to the Renault Twizy again, both front and rear suspension adopt a MacPherson solution. However, this suspension type requires a dedicated strut (serving at the same time as steering axis, structural element, and spring-damper) that only large car manufacturers can afford and specialize for their model. Therefore, SNAP adopts a double wishbone solution with commercial spring-dampers. A tradeoff solution is pursued that limits the length of the suspension arms (about 200 mm) while ensuring a correct geometry that does not generate a large variation in the camber angle during the relative motion of tires with respect to the chassis [12].

### 3.1 The SNAP Design

Special care was paid to the process of developing the appearance, and to some extent the ergonomics, of the vehicle. This is a laborious task since the shape of each fairing panel has been optimized to reduce the cost of the molds and the successive machining operations, while preserving at the same time a captivating look as shown in Fig. 5.

The seats are adjustable and they ensure good lateral containment in the case of impact. However, their weight is about 10 kg each. Therefore, future developments will look at the opportunity to use seats made out of carbon or fiberglass with removable cover. In this case, the seat would be not reclinable, and a statistical study is required on a large user population to define the optimal size and geometry.

Vehicle safety is another important aspect in the SNAP design. Anti-intrusion bars are present in both the side doors and approved safety belts are provided for both the driver and the passenger of automotive derivation with a three-point configuration. A rollbar is also present that protects passengers within the vehicle in the case of roll-over.

In the case of front impact of rear collision, reinforcements distribute the load on the chassis and deform to absorb the impact energy.

(a)

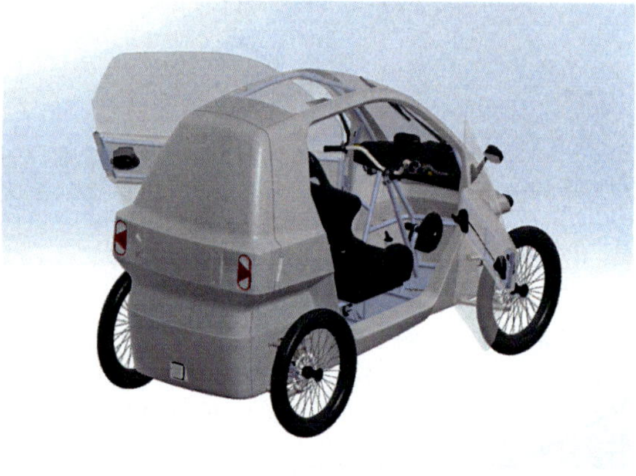

(b)

**Fig. 5.** Isometric view of SNAP: a) front, b) rear

## 4 Conclusions

The novel SNAP vehicle was illustrated as a new concept in sustainable mobility. The whole design process has been performed to pursue low maintenance and complexity. As a result, some components may appear oversized with respect to the load conditions expected for this type of vehicle. For example, the suspension arms could have been made out of composite materials ensuring lower weight but making repairments very expensive. In the case of SNAP, the choice of steel allows repairing with a standard continuous wire welding machine to restore both the chassis and the suspensions. Future developments will investigate the adoption of new materials as well as the adoption of novel drivetrain architectures that adopt two smaller electric motors [13]. The integration of driving assistance systems will be also investigated to monitor safety parameters, including tyre inflation pressure [14].

**Acknowledgements.** The research was financially supported by

Iniziativa finanziata con risorse del FSE - PO Puglia 2014/2020 Azione 8.4 e del Fondo per lo Sviluppo e la Coesione

Vincitore PIN/Winner of PIN – Iniziativa promossa dalle Politiche Giovanili della Regione Puglia e ARTI e finanziata con risorse del FSE – PO Puglia 2014/2020 Azione 8.4 e del Fondo per lo Sviluppo e la Coesione - Initiative promoted by the Youth Policies of the Puglia Region and ARTI and funded with resources from the ESF - OP Apulia 2014/2020, Action 8.4 and the Development and Cohesion Fund.

The partial financial support of the projects: Agricultural in TeroperabiLity and Analysis System (ATLAS), H2020 (Grant No. 857125), and multimodal sensing for individual plANT phenOtypiNg in agrIculture rObotics (ANTONIO), ICT-AGRI-FOOD COFUND (Grant No. 41946) is also gratefully acknowledged.

## References

1. Braune, S., Kramer, K., Meissner, K.: Electric drive system for a vehicle driven by muscle power. DE Patent 201310012208, 17 July 2013
2. Sveje, N., Hansan, J., Nielsen, O.: An electric vehicle with controllable generator. EP Patent 3110687, 01 April 2017
3. Son, H., Kawn, S.: Electric bicycle driving apparatus. EP Patent 2711239, 26 March 2014
4. About: PODBIKE. https://www.podbike.com/about/. Accessed 19 Apr 2018
5. HYBRID POWER: Hybrid Power. http://www.twike.com/enGB/vehicles/twike-3/hybrid-power/. Accessed 19 Apr 2018
6. Kronfeld, R., Smith, L., Bockin, R., Castellotti, S.: Human-rechargeable electric vehicle. US Patent 20130081892 A1, 04 April 2013
7. Gartner, J., Citron, R.: Electric bicycles Li-Ion and SLA E-bikes: drivetrain, motor, and battery technology trends, competitive landscape, and global market forecasts, Boulder, USA, 15 June 2016
8. EN 15194:2017, Cycles - Electrically power assisted cycles - EPAC Bicycles, Standards
9. https://www.velomobiles.co.uk/tag/bluevelo/. Accessed 15 May 2021
10. Ti-FLY X Homepage. https://azub.eu/recumbent-bikes-and-trikes/trikes/26-wheels/ti-fly-x/. Accessed 15 May 2021
11. Kyle, C.R., Caiozzo, V.: Experiments in human ergometry as applied to the design of human powered vehicles. J. Appl. Biomech. **2**(1), 6–19 (2001)
12. Reina, G., Foglia, M.: On the mobility of all-terrain rovers. Ind. Robot. **40**(2), 121–131 (2013)
13. Mantriota, G., Reina, G.: Dual-motor planetary transmission to improve efficiency in electric vehicles. Machines **9**(3), 58 (2021)
14. Reina, G., Gentile, A., Messina, A.: Tyre pressure monitoring using a dynamical model-based estimator. Veh. Syst. Dyn. **53**(4), 568–586 (2015)

# The Contribution of Ellul and Illich's Thought to the Design of Appropriate Machines for Communities in Socio-ecological Transition

Walter Franco[1](✉) and Osman Arrobbio[2]

[1] Department of Mechanical and Aerospace Engineering, Politecnico Di Torino, Turin, Italy
walter.franco@polito.it
[2] Department of Cultures, Politics and Society, Università Di Torino, Turin, Italy
osman.arrobbio@unito.it

**Abstract.** Ellul and Illich are two of the most prominent radical critics of technology of 20th century. Nevertheless, their fame in the academia is limited to small circles of technology sociologists and philosophers. Their thought could help technologists, on the one hand to become aware of the dynamics between technology and society and, on the other hand to develop new appropriate solutions that can improve the efficiency of production by limiting individual, collective and environmental negative effects in small communities in socio-ecological transition. In this contribution, a brief summary of Ellul and Illich's studies is presented, aimed in particular at technologists. Some features of the appropriate technologies that can be traced back to Illich and Ellul are discussed: low-tech, low-power, intermediate productivity, manual operated. Finally, by way of example, an intermediate machine designed at the Politecnico di Torino for threshing and cleaning wheat grown by farmers in mountain areas is presented.

**Keywords:** SDG12 · Technologies for social inclusion · Ecological transition · Grassroots innovation · Appropriate technology · Intermediate machines

## 1 Introduction

Ellul and Illich are considered two of the most important radical critics of technology of the last half of the 20th century [1]. They gave foundation and systematization to some principles of Appropriate Technologies, that may serve communities in socio-ecological transition.

An Appropriate Technology (AT), in a narrow sense, is a small scale, simple, energy efficient, environmentally sound, labor-intensive and community-driven technology [2]. AT movement, inspired by E.F. Schumacher [3], had remarkable success in the 1970s after the publication of the famous Schumacher's "Small is Beautiful" [4]. The AT movement is considered, together with others such as the 1980s' People's Science Movement in India, one of the historic grassroots innovation movements [5]. Grassroots innovation movements are informal phenomena which arise almost always in reaction to the perception of social injustices and environmental issues in contexts of conventional industrial


G. Quaglia et al. (Eds.): I4SDG 2021, MMS 108, pp. 118–126, 2022.
https://doi.org/10.1007/978-3-030-87383-7_13

production [6, 7], and that promote inclusive innovation processes to generate bottom-up solutions for sustainable development. In this sense, grassroots innovation movements can be considered as one of the possible catalysts for transition, in particular in marginal or underserved communities.

The first part of this contribution delivers a synthesis of the main suggestions coming from Ellul and Illich's thought that may support AT for small communities in socio-ecological transition. Finally, as examples of machine "compatible" with the suggestions of the two authors, a concept and a prototype of a machine for processing wheat will be described. This machine, suitable for mountain family farmers, was developed at the Appropriate Machines Laboratory of the Politecnico di Torino.

## 2 Critical Theory Foundations for AT: Jacques Ellul and Ivan Illich

### 2.1 Jacques Ellul

Jacques Ellul (1912–1994) was a French sociologist and theologist. In 1944, Ellul was appointed professor at the University of Bordeaux, position that he never left till he retired in 1980. He took part to several religious, environmentalist and civic groups, took position against communism, and self-defined-and/or accepted to be defined-as believer (protestant) and anarchist. These, and other reasons, all contributed to make Ellul a somehow marginal figure in the intellectual milieu of post-war France. He was indeed much more renown in the USA than in his home country. In this contribution we will pay exclusively attention to his works on technology. Or, better, on technique. This clarification is needed if we consider the meaning the term 'technology' (which, etymologically speaking, is a discourse on technique) is usually given nowadays, by both scholars and lay people who equally use it to refer to technical objects (e.g. smartphones, computers) and processes (e.g., nanotechnology, biotechnology) [8, 9].

On technique Ellul wrote many essays, among which: La Technique, ou l'enjeu du siècle (1954) (first English translation in 1964 with the title the technological society [10]); Le Système technicien (1977) (first English translation in 1980 with the title The technological system [11]); Le bluff technologique (1988) (first English translation in 1990 with the title The technological bluff [12]). Usually considered as a sort of trilogy, the two latter books could roughly be seen as a way for Ellul to refine and integrate the theory he first established in his 1954's book. However, how could an analysis written almost 70 years ago, when France (and Europe) had just recently left behind the destruction brought by the Second World War and had not entered yet the post-war economic boom-not to mention the informatisation of society as well as the appearance of other and more recent surprising technical advancements-be somehow useful today?

As said above, Ellul used the term "technique": "The term technique, as I use it, does not mean machines, technology, or this or that procedure for attaining an end. In our technological society, technique is the totality of methods rationally arrived at and having absolute efficiency (for a given stage of development) in every field of human activity. [...] Technique is not an isolated fact in society (as the term technology would lead us to believe) but is related to every factor in the life of modern man; it affects social facts as well as all others". [10], pp. xxv-xxvi.

He distinguishes between technical action and technical phenomenon. The latter being "[...] the quest of the one best means in every field. [...] It is the aggregate of these means that produces technical civilization". [10], p. 21.

Are specific techniques bad? Or are they good? For Ellul, it is impossible to draw up a balance. Experts may not even agree among them and, moreover, some aspects (be they positive or negative) may lay out of sight [10, 11]. The only sure thing is that the technicisation process continues relentlessly and extending to all domains of life as well as to ever more corners of the globe. In the words that Merton used to introduce Ellul's La Technique to the American reader [13], p. vi, progress "[...] consists in progressive de-humanization—a busy, pointless, and, in the end, suicidal submission to technique".

Technique creates problems that only technique can solve or, better, for which mostly technical solutions are looked for. What or who could stop this reinforcing loop? Technique is completely out of control, or at the point of becoming so. It is the determinant factor as it determines people lives as well as all social sub-systems, included the political one. Men and women surrendered to the logics of technique: "[...] man thinks in conformity with this environment. [...] If a person awakes to consciousness, he would no more dream of challenging or contesting the technological milieu in its perceptible aspects than a twelfth-century man would dream of objecting to trees, rain, a waterfall." [11], pp. 311–312.

Ellul deliberately refrained from proposing solutions: his work is a diagnosis and, moreover, he could see "[...] not even a beginning of a solution, no breach in the system of technical necessity" [10], pp. xxxi. He described the way of functioning of a system, and this has not changed since he wrote his book in 1954 [10]. The appearance of new rule-breaking technological gadgets or technical procedures can almost perfectly fit into his theoretical arguments.

The only way out of all that is to become aware. Should men completely abandon or reject technique? Of course not. Without technique mankind would have never appeared. First of all, man should be made aware of what determines him. Then, he may eventually put a break to the technicisation process, by deliberatively controlling it and steering it.

Usually not listed among the masters of sociology in university textbooks (or not named at all) his name mainly circulates among limited intellectual circles (e.g., technology sociologists and philosophers) and among thinkers and practitioners ascribable to what may be referred to as neoluddism. This could be defined as the scattered and unorganised movement that sees the negative effects of technique (and eventually also the ones deriving from specific-mostly modern-techniques) and that questions the inescapable character of technological progress.

## 2.2 Ivan Illich

The main character of the 2019 French movie Alice and the Mayor (orig. Alice et le maire) is Alice, a young philosophy and literature graduate who just got a post at the City Council of Lyon. As she was told during the first day, her task consists in stimulating the Mayor to think again. After having observed a City Council meeting, she goes to her office and takes notes. She writes down the names of three authors, likely to be the ones to take inspiration from on the theme of modesty: Rousseau, Orwell and Illich. To our knowledge this could be the first time the name of Illich, that we deem is the least known among the three, is cited in a movie targeted to the wider audience.

Ivan Illich (1926–2002) is a quite unclassifiable thinker, at times defined as free-thinker, others as philosopher, sociologist, pedagogist and theologian, among others. Author of provocative and influential books, whose production mainly concentrated in the 1970s, he was mainly renown (and still is, albeit to a minor extent) within alternative social movements.

We can see in Illich's suggestions a sort of operational translation of Ellul's insights. Indeed, Illich did offer solutions or, at least, directions to look to for solutions. It did so by concentrating its treatises on specific societal subsystems, mostly education [14], mobility [15], and health [16], of which he also depicted their perverse functioning. Perverse effects that start when the process he termed 'institutionalisation of values' occurs and that may bring-not just for the three above-mentioned subsystems-to the establishment of a 'radical monopoly' [17]. This describes those situations in which an industrial production process-not be intended as just related to the proper industry sector-exerts an exclusive control over the satisfaction of a need and excludes non-industrial activities from competition. When this happens, a new form of poverty makes its appearance (in both industrialised and non-industrialised countries) that makes people unable to satisfy, in a creative and autonomous way, their needs outside of market relations.

Illich asked for the recognition of the existence of thresholds, in the use of energy or in relying to industrially produced goods and services, whose crossing brings to environmental pollution, social polarisation, and psychological impotence. While the former of the three may even be avoided, the latter two will be not. Some tools are more suitable than others to prevent the critical thresholds to be crossed and, consequently, to prevent the three environmental, social and psychological bads to manifest in their most pernicious forms. Illich calls these "convivial tools". A more detailed description is delivered in Sect. 3.3.

## 3 Some Common Characteristics of Appropriate Machines

We define appropriate those machines that can be fully included in the category of AT. There are numerous examples of appropriate machines developed to serve small communities, including devices for self-building [18–20], housework [21], artisanal self-construction [22], peasant agriculture [23]. The great majority of them have some characteristics in common which are discussed and related to the thought of Ellul and Illich in the following of this section.

### 3.1 Low-Tech

Simplicity is one of the characteristics of appropriate machines [2], which arises from the need, on the part of small communities, to have the technology under full control, to be able to replicate it autonomously and to be able to maintain it over the years. Simplicity must be understood as resolved complexity. It consists in identifying a simple solution to a complex problem, for example through the hybridization of new knowledge on old practices, with the idea that technological innovation does not necessarily mean the new replacing the old, but also renewed use or redesign of traditional solutions [24, 25].

In our opinion, the propensity towards low-tech solutions in the context of small communities in socio-ecological transition also originates from the desire, more or less conscious, to want to bring technology back to the dimension of an instrument. In fact, as Ellul points out in [11], "even when technology is abstract, a procedure, an organization, it is far more of a mediation than an instrument". Technique takes the form of exclusive mediation between man and the natural environment, and in the relationships between individuals. "Thus, mediation by technology excludes any other, and this allows technology to entirely escape the desired or supposed values." Furthermore, mediation is: (i) "autonomous" since what is possible through Technique is necessary; (ii) "sterile and sterilizing", since Technique is univocal, stable and ordered, "is a truly efficient medium, and it has imposed itself in lieu of poetic mediations."; (iii) "non-mediated towards man", since "the relation to technology is immediate, which does not mean that consciousness has now become the simple reflection of the technological environment" [11].

Technique becomes the environment and removes the natural environment. In order to recover a conscious relationship with the natural environment, it is therefore necessary to reduce the complexity of technology, to identify and develop low-tech solutions.

### 3.2 Intermediate Productivity

According to Ellul, the fact that technique is an end and not a means is demonstrated by the low appreciation we have for the numerous objects we produce just because of the effectiveness of the technical means of production. These are often objects that are neither pleasant, nor truly useful, neither symbolic, nor affective. The focus is on the productivity of technology, not on the quality of the objects produced. "These objects have no value or importance, whatsoever, they exist only as products of the technological mechanism. What characterizes this society is not the object but the means." [11].

Numerous grassroots movements focus on product quality rather than quantity. For example, there is an orientation towards widespread and artisanal production [26], or towards peasant agriculture [27]. When the quality of the production is considered primary, other design parameters gain relevance, such as the quality of work, the positive interaction between worker and object of production, the possibility of realizing and expressing oneself through a working act.

It is necessary to try to understand what is really useful to us, to rediscover frugality, i.e. the individual and collective capacity to recognize what we need [28]. As Illich explains, productivity is opposed to conviviality, having to be, exasperation to self-limitation, oppression to equity [17].

### 3.3 Manual Operation

By conviviality, Illich means the opposite of industrial productivity [17]. "A convivial society would be the result of social arrangements that guarantee for each member the most ample and free access to the tools of the community and limit this freedom only in favour of another member's equal freedom".

Illich distinguishes two basic types of energy: a metabolic type, and an external one, and classifies the tools according to the type of energy used, distinguishing hand tools from power tools. As for the former, "mere transducers of the energy generated by man's extremities and fed by the intake of air and of nourishment", Illich believes that "they lend themselves to convivial use" being defined as convivial those tools that "can be easily used, by anybody, as often or as seldom as desired, for the accomplishment of a purpose chosen by the user. The use of such tools by one person does not restrain another from using them equally. They do not require previous certification of the user. Their existence does not impose any obligation to use them. They allow the user to express his meaning in action" [17, 29].

Conversely, tools that use external energy, called power tools, favour the centralization of control. Furthermore, they are often too expensive to be equally available worldwide and are therefore the prerogative of wealthy groups, or require a high degree of specialization, and therefore are unable to provide the freedom and versatility of simple hand tools.

### 3.4 Low-Power

The particular attention to hand tools also depends on the fact that, according to Illich, they are the tools that guarantee, par excellence, equity. Anyone who breathe and eat can use them. On the contrary, power tools become the prerogative of some and contributes, especially as power increases, to the creation or exacerbation of social inequalities.

Illich states that "What is generally overlooked is that equity and energy can grow concurrently only to a point. Below a threshold of per capita wattage, motors improve the conditions for social progress. Above this threshold, energy grows at the expense of equity. Further energy affluence then means decreased distribution of control over that energy [15]".

Illich, demonstrates his intuition by analyzing the theme of transportation, and highlighting how exceeding a certain threshold of individual power available for moving-corresponding to the muscular power for driving a bicycle-there is a systemic decrease in the efficiency of the transportation system. In societies whose transportation system is based on the automobile, most people spend an increasing chunk of their existence on unwanted travels, while only a small global elite can afford to travel enormous distances during their lifetime.

This thought of Illich has been confirmed in [30], in which the productivity and energy efficiency of different machines for threshing and cleaning grain are compared. Increasing the available power obviously increases productivity. However, the most evident increases manifest at low powers. In terms of energy efficiency, it is observed instead that this is maximum below a certain level of power, comparable to the metabolic one.

## 4 Re-thresher, an Example of Appropriate Machine

As an example, the prototype of an appropriate machine, designed within the Laboratory of Appropriate Machines of the Politecnico di Torino to meet the specific requests of peasants who produce small quantities of high-quality wheat grown in impervious mountains, is presented. A specific participatory design methodology was developed, which started from an accurate identification of users' needs and their involvement in the design phases [31]. Users need to have a machine that performs the threshing and cleaning operations of the grain where it is not possible to use a traditional combine harvester, either due to the impossibility of access, or due to the small size of the fields (as in the case of catalogue fields of historical varieties of cereals).

The solution designed (Fig. 1) has the characteristics discussed in Sect. 3, i.e. low-tech, intermediate productivity, manual operation and low-power. In effect, given the production of small quantities of wheat, for family use or for small local supply chains, the productivity of the machine is not central. This allows to move towards the design of a low-tech and low-power machine, possibly operated manually and electric-assisted. One or two people manually operate the machine through corresponding cranks. The mechanical transmission is of cycling derivation. Through several chain drives, the power is transmitted to the beater roller, to the ventilation system, and to the mechanism of the vibrating sieves. The grain, detached from the straw by the beater roller, is conveyed, by gravity, to the vibrating sieves, where it is hit by a flow of air which has the purpose of expelling impurities from the machine. The clean grain is collected in a special drawer.

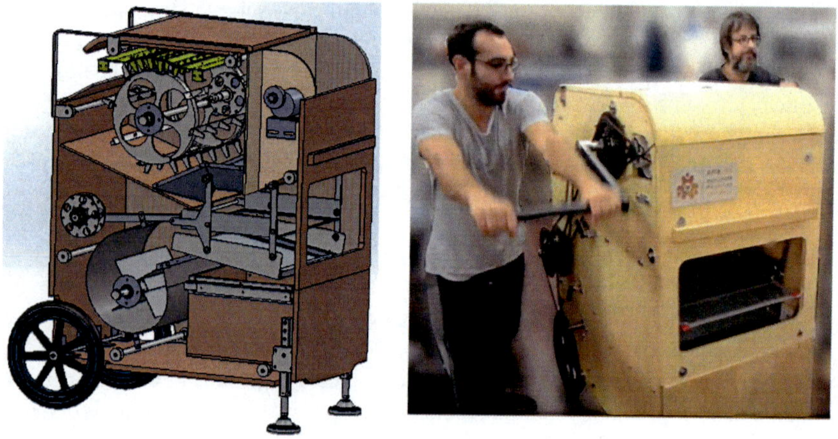

(a) (b)

**Fig. 1.** Re-Thresher, a manually operated power-assisted thresher-cleaner machine. a) The detailed design. b) The prototype

## 5 Conclusions

Ellul and Illich are two of the most important radical critics of technology of the last half of the 20th century. United by the idea that a systematic examination is necessary to identify the relation between technology and society, their analyses have contributed to exercise a unique intellectual influence on numerous scholars. They can be considered, in a sense, among the founding fathers of AT movement. In addition, the fact that Illich has complemented one of Ellul's last public presentations held at the end of the three days of the conference "Technique and Society in the Work of Jacques Ellul" at the University of Bordeaux in 1993 is read by some as confirmation of complementarity and affinity of the two intellectuals [1].

Ellul mainly studied the relations between technology and society, refraining from making judgments or proposing solutions. He highlighted how technique is the determining factor of society, more than economics and politics, and how, through a sort of automatism, it tends to develop into a universal and totalizing environment and system. Technique has become the only mediation between man and the environment and between men and men, who find themselves deprived of the mediation tools developed over millennia, such as narration, poetry, myth, tradition, symbolic and cultural codes.

Illich explained how the planetary crisis has its roots in the process of replacing the machine to man, the rational and unified processes to the occasional and punctual responses of the different communities. The relationship between man and tools of production is reversed into an opposite relationship. Tools make man a slave. Illich, unlike Ellul, proposes some solutions to the crisis, including the development of tools that improve efficiency while preserving the autonomy of individuals, extending their range of action without producing slaves or masters.

The analysis of Ellul and Illich, half a century later, still proves to be topical and possible source of inspiration for the design of technologies appropriate to small communities implementing socio-environmental transition processes. To limit the negative individual, social and environmental effects of a hypertrophic technology, it seems interesting to develop intermediate, simple, low-power, possibly manual solutions, capable of satisfying the limited productivity needs of frugal communities with improved energy efficiency.

In support of this reasoning, an assisted manually operated thresher-cleaner designed at the Politecnico di Torino for small scale mountain peasants is also presented. The machine is capable to produce, with high efficiency, small quantities of cereals for self-consumption or for local supply chains.

In conclusion, we believe that the thought of Ellul and Illich must come out of the narrow circle of sociologists and philosophers of technology, and also be the patrimony of technologists. There appears to be an untapped potential for cross-fertilization between humanists, social scientists and technologists that will be useful to fully exploit in the imminent phase of energy transition and resource scarcity.

## References

1. Ellul, J., Illich, I.: Statements by jacques ellul and ivan illich. Technol. Soc. **17**(2), 231–238 (1995)
2. Hazeltine, B., Bull, C.: Appropriate Technology. Choises, and Implications. Academic Press, San Diego, Tolls (1999)
3. Lissenden, J., Maley, S., Mehta, K.: An era of appropriate technology: evolutions, overights and opportunities. J. Human. Eng. **3**(1), 24–35 (2015)
4. Schumacher, E.F.: Small Is Beautiful. Economics as if People Mattered. Blond & Briggs, London (1973)

5. Smith, A., Fressoli, M., Thomas, H.: Grassroots innovation movements: challenges and contributions. J. Clean. Prod. **63**, 114–124 (2014)
6. Smith, A., Elisa, A., Fressoli, M., Thomas, H., Abrol, D.: Grassroots innovation for sustainable development: some enduring dilemmas, In: Proceedings of the Global Research Forum on Sustainable Consumption and Production Workshop 2012, Rio de Janeiro, Brazil (2012).
7. Smith, A., Fressoli, M., Abrol D., Elisa, A., Ely, A.: Grassroots Innovation Movements. Pathways to sustainability. Routledge, London & New York (2017).
8. Hickman, L.: Philosophical Tools for Technological Culture: Putting Pragmatism to Work. Indiana University Press, Bloomington (2001)
9. Schatzberg, E.: Technology: Critical History of a Concept. The University of Chicago Press, Chicago and London (2018)
10. Ellul, J.: The Technological Society. Knopf, New York (1964)
11. Ellul, J.: The Technological System. The Continuum Publishing Corporation, New York (1980)
12. Ellul, J.: The Technological Bluff. Eerdmans, Grand Rapids (1990).
13. Merton, R.K.: Foreword, in Ellul. J. Technol. Soc. Knopf, New York (1964)
14. Illich, I.: Deschooling Society. Harper & Row, New York (1971)
15. Illich, I.: Energy and Equity. Calder and Boyard, London (1974)
16. Illich, I.: Medical Nemesis. The Expropriation of Health. Calder and Boyars, London (1975)
17. Illich, I.: Tools for Conviviality. Harper & Row, New York (1973)
18. Mukerji, K.: Soil Block Presses. German Appropriate Technology Exchange, GATE, Eschborn (1988).
19. Ferraresi, C., Franco, W., Quaglia, G.: Concept and design of Float-Ram, a new human powered press for compressed earth blocks. In: ASME IMECE2017, International Mechanical Engineering Congress and Exposition, Tampa, Florida, USA (2017)
20. Franco, W., Quaglia, G., Ferraresi, C.: Experimentally based design of a manually operated baler for straw bale construction. In: Boschetti, G., Gasparetto, A. (eds.) Advances in Italian Mechanism Science. MMS, vol. 47, pp. 307–314. Springer, Cham (2017). https://doi.org/10.1007/978-3-319-48375-7_33
21. Dean, T.: The human powered home. Choosing muscles over motors. New Society Publishers, Gabriola Island (2008)
22. L'Atelier Paysan. Available from: https://www.latelierpaysan.org/ (Accessed 25 May 2021).
23. La Via Campesina. International Peasants' Movement. Available from: https://viacampesina.org/en/ (Accessed 26 May 2021).
24. De Decker, K.: Low Tech Magazine 2012–2018. Lulu.com, Milton Keynes (2019)
25. Edgerton, D.: The Shock of the Old. Technology and Global History since 1900. Oxford University Press, New York (2007).
26. Sennet, R.: The Craftsman. Yale University Press, New Haven-London (2008)
27. Slow Food. Available from: https://www.slowfood.it/ (Accessed 26 May 2021).
28. Berardi, F.: After the future. AK Press, Chico (2011)
29. Borremans, V. Guide to Convivial Tools. Library Journal, Special Report No.13. Bowker, New York (1979)
30. Franco, W., De Piccoli, M.: Intermediate agricultural machines energy efficiency: the example of harvesting and threshing. Mech Mach Sci **91**, 533–541 (2021)
31. Franco, W., Barbera, F., Bartolucci, L., Felizia, T., Focanti, F.: Developing intermediate machines for high-land agriculture. Dev Eng **5**, 100050 (2020). https://doi.org/10.1016/j.deveng.2020.100050

# Smaller is Often Better: A Revival of Interest in Research and Development in Cost Effective Smaller Scale Machines for Equitable Rural Economic Development

Stephen D. Biggs[1](✉) and Scott E. Justice[2]

[1] Department of Development Studies, SOAS University of London, London, UK
[2] Naypyidaw, Myanmar

**Abstract.** In the last few years there has been a revival of research and development (R&D) interest by the international agricultural and rural development community in the spread of smaller scale equipment and markets in customs services. This paper briefly reviews mainstream debates since the 1960s and evidence of a current revival of interest. It takes a case study of the recent rapid spread of mini tillers in the Nepal mid hills to show that this was the outcome of a long and complex R&D process. However, one of the central features of the history was the engagement of engineers with the reality of daily life of smaller farmers, small and medium entrepreneurs in local rural contexts. The case study illustrates that the R&D practice stretched over many years. The paper closes with five reflections, the importance of: 1) ongoing field work by engineers, economists, and anthropologists, 2) long term R&D interest in the complexities of local rural economies, 3) informal and artisanal R&D, 4) a policy/project/resources commitment to equitable economic development in rural areas, and 5) the individual.

**Keywords:** SDG2 · Small scale agricultural and rural mechanisation · Small farmers and rural entrepreneurs · Rural and agrarian economy · Markets in custom services · Rural capital goods

## 1 Introduction

There are many histories of rural mechanisation, but the one we want to concentrate on here is the history of research and development (R&D) practice associated with the spread of smaller engines in rural areas of low-income countries. We concentrate on South Asia as that reflects our experience and the activities of the International Agricultural Research Institutes of the Consultative Group on International Agricultural Research (CGIAR) system and as these institutions have often played a major role in directing the path and activities of agricultural and rural technology R&D.

Figure 1 shows a timeline of the history of international R&D and policy debates concerning the use of agricultural, irrigation, and other small-scale engines in rural areas from their start in late 1960s to the present.

Scott E. Justice—Independent consultant.


G. Quaglia et al. (Eds.): I4SDG 2021, MMS 108, pp. 127–137, 2022.
https://doi.org/10.1007/978-3-030-87383-7_14

| 1960 | 1970 | 1980 | 1990 | 2000 | 2910 | 2021 |
|---|---|---|---|---|---|---|

.................................Significant interest in field-based R&D on smaller engines

.........................Decreased interest and funding of R&D

...Renewed policy interest and field-based R&D

**Fig. 1.** Timeline of the history of international R&D and policy debates

### 1.1 Phase 1: Major Interest in Field-Based Research (1960 to Mid-1980s)

This period represented a time of great interest in field-based research in smaller scale rural engineering. One of the notable engineers who reflected this interest in field-based activity was Bill Chancellor. His work with students in India and Thailand was directed, not only at engineering considerations, but also at economic viability issues such as markets for engine services in the rural contexts in which the equipment was being developed (Singh and Chancellor 1975, Chancellor 1983).

During this period, the importance of small engines in rural areas was recognised by applied macro economists who pointed to the importance of small-scale engineering for economic development. Schumacher (1973) placed the emphasis on rural industrialisation based on smaller engines. Falcon (1967) demonstrated the importance of the linkages between industrial and agricultural sectors and the need to include the linkages in macro policy analyses. Timmer et al. (1975) highlighted the inappropriatness of larger scale irrigation technology in Bangladesh, then East Pakistan, and rice milling technology in Indonesia.

In this period, most of the international centres were promoting on-farm research of varietal/agronomy trials and economic surveys of farmers. Two exceptions were International Rice Research Institute (IRRI) and International Potato Center (CIP), which had village-level research with engineering R&D as a major concern. An illustration of IRRI's applied engineering R&D is recounted in the monitoring and evaluation research of Duff (1987) where he documented how the introduction of the IRRI axil flow thresher into Thailand and the Philippines gave rise to different socio-economic and equity outcomes in each country.

One of the most well known examples of applied engineering research in the CGIAR centres is the R&D work on potato storage technology. Although not small engine R&D, it is an important case study as it is an "engineering example" and gave rise to the widely known "farmer first" approaches to R&D promoted and developed by Chambers (Chambers et al. 1989). In their classic work, Rhoades and Booth recount how the engineering technology that subsequently spread was developed in an iterative process over time (Rhoades et al. 1986). Rhoades, an anthropologist, worked with CIP engineers and local farmers on their potato storage problems in an ever-changing socio-economic context (Rhoades et al. 1986). Rhoades went on to formalise some of the methods they had been developing in the 1970s into a guide for informal agricultural surveys (Rhoades 1982). The guide shows how quick, reliable, cost effective methods can give rise to timely information that can be used to guide engineering and other R&D.

Another engineering case study is the spread of bamboo tubewells and mobile pumpsets in the Kosi region of Bihar, India, in the early 1970s. What characterises this case study is the role detailed field work played in documenting and understanding the R&D processes at work (Clay 1980). The field work documents the government's introduction of a fixed package of a stationary pumpset and expensive well digging. This package, inappropriate for many reasons, set in motion an R&D process by local landowners, cultivators, artisans, and well diggers. This informal R&D resulted in the development of a cheap bamboo tubewell that was served by a mobile 5 HP locally made pumpset mounted in a bullock cart. Markets in engine services, well digging, and engine servicing also developed locally. The case study is also interesting because the senior Indian administrator in the region at the time, who was interested in broad-based equitable rural development, learned what was happening on the ground and changed local irrigation policy and projects (Appu 1974). In 1974, the Bangladesh Planning Commission sent an irrigation review team to the Kosi area to see if there were lessons that could be learnt for Bangladesh's policy and R&D. The Kosi region was only about 60 miles from the new country of Bangladesh. The myths that Clay (1980) was addressing were stories that described the invention of the bamboo tubewell in terms reminiscent of the deeds of great inventors, and that small farmers were the prime beneficiaries. Clay's detailed field-based research showed that this engineering, social, political, and technical change was far more complicated and led him to suggest that "The real significance of the bamboo tubewell phenomenon, when considered in the context of irrigation technology in the Indian sub-continent, is that it demonstrates the need to commit resources to research and development on agricultural engineering technologies comparable to those expanded on crop technology." (Clay 1980, p. 43).

There are other reasons why this period represented a time when field-based research and the involvement of engineers was high. Several large influential international NGOs were mainly staffed by highly respected engineers, for example, Appropriate Technology, Washington, and the Intermediate Technology Group, Rugby. Some engineering universities had units that looked at the engineering needs for smaller rural entrepreneurs and farmers, such as Silsoe College in the UK. The staff of these NGOs and specialised units were generally professionally trained engineers and often members of government advisory bodies.

A second group of professionals who were part of the international agricultural and rural mechanisation policy debates at the time was the staff of the Agricultural Development Council (ADC). The ADC was funded by the Rockefeller Foundation and concerned with social science institution building in Asia. Part of the strategy was to have ADC staff located in international agricultural research centres or in national agricultural institutions. While the ADC did not finance engineering R&D and training, it did promote field-based social science research, which often focussed on agricultural and rural mechanisation topics. Some of its staff included Clay, Gill, Binswanger, and Ruttan. These individuals played a major role in the agricultural and rural mechanisation policy debates in this period. ADC also published in this area (Stevenson & Locke, 1989). The ADC was closed in the late 1980s. Another group of people were those who had been in the US Peace Corps. They often came with grounded experiences of rural life, generally lived in village situations and some worked with local agricultural engineers as part of their experiences.

A large literature demonstrates a high interest in the formal and informal R&D behind the spread of small engines in rural areas during this time (Biggs and Farrington 1991).

However, most agricultural research interest was concentrated on technology associated with plant breeding and agronomy.

### 1.2 Phase 2: Late 1980s to the Late 2000s

The second phase was a period when field level research and policy informed by field data declined consistently (Biggs et al. 2015). There was little interest in the role of small engines used in rural areas, such as small pumpsets, two-wheel tractors (2WT), micro hydro, small mobile and station mills in the promotion of equitable rural development. Paradoxically, in the 1970s/80s there had been much written on the wealth, economi, and social inequities in many agrarian economies, and the nature of interlocking service markets in rural areas. One of the outcomes of this academic work led to the publication of "The Economics of Rural Organization: Theory, Practice and Policy" (Hoff et al. 1993). Sadly, the book was not followed up by field research that looked as these subjects with the spread of small engines and associated equipment in mind. There are exceptions in the more general academic literature, such as the work of Barbara Harriss-White (1996), but within the plant breeding based centres of CGIAR system this work did not take place.[1] The point is well-illustrated by a book edited by Cernea and Kassam (2006) entitled "Researching the Culture in Agri-culture: Social Research for International Development". The book brought together economists, anthropologists, and other social scientists, who recounted the experiences and the challenges for social scientists and natural scientists working together on R&D. From our perspective, it is noteworthy that there was virtually no mention of experiences of working with engineers on smaller engines used for irrigation, tillage, village transport, food processing, etc. Also, there were no papers by plant breeders, agronomists, or engineers, discussing their experiences and their problems of working with economists and other social scientists. One of the exceptions was a paper by Rhoades (2006), who documented why natural scientists find it difficult to work with social scientists, however, as an anthropologist he often felt that the engineers only wanted half his brain. A further issue that was not discussed, which is important for our concerns here, is the diversity of people in social and natural sciences. There is a whole spectrum of plant scientists, engineers, agronomists, social scientists mixing in the field, learning from colleagues from other disciplines, and from rural people. Unfortunately, siloes have been created by which economist are put in one box, plant breeders in another, etc., as if there were homogeneity within the boxes.

Another symptom of the lack of interest was the reduced funding of smaller scale engineering and the closure of departments at various international agricultural research institutes. ICRISAT's Engineering Group closed in 1994. Then IRRI's Agricultural Engineering Division closed in 1996 followed by the UK's Silsoe Research Institute's Agricultural Engineering International Development Group in 2003. Finally, not withstanding a growing recognition of the importance of engineering by economists to the growth of agricultural and nonfarm rural incomes (Haggblade et al. 2007) the lack of interest in rural mechanisation was reflected in the absence of coverage of agricultural and rural

[1] We do not discuss here the work of the International Water Management Institute and its relationship to other CGIAR centres as it is beyond the scope of this paper.

engineering in major assessments of global agriculture and food towards the end of the 2000s (World Bank 2008, McIntyre et al., 2008, Foresight 2011).

### 1.3 Phase 3: Revival of Policy and R&D Interest in Smaller Engines in Rural Economic Development

In the late 2000s and earlier 2010 attitudes began to change. Since then, there has been a general revival of interest in agricultural mechanisation. This is illustrated by the recently completed study by the International Food Policy Research Institute (IFPRI) (Diao et al. 2020), as well as by Daum (2020), Justice and Biggs (2020), Rigg (2016), and Hanlan and Kaplinsky (2016). We are not going to analyse the reasons for this revival, as that is the beyond the scope of this paper. Rather, we look at a case study of the recent rapid spread of small engines in a specific location and demonstrate how this relates to experiences of the past. Before getting into the case study, we want to note that the recent IFPRI study has, in many ways, legitimised a new research and policy interest in agricultural and rural engineering R&D in the international agricultural research community. In this, it has also shown the importance of and continuing spread of small-scale engines, markets in engine services, importance of local small and medium scale rural businesses and other rural entrepreneurs. Further, the study legitimizes the use of mixed methods in research and demonstrates that good empirical historical studies are as useful as, and compliment, econometric and bigger data analyses. The study demonstrates that empirical analyses of past patterns of rural mechanisation in different countries dismisses simple generalisations such as the need for land consolidation before rural mechanisation takes place, rural mechanisation takes on a generalizable path, and long-term local patterns of agricultural mechanisation can be separated from macro import/export policies for capital goods and from macro industrial policies.

The IFPRI study concentrated agricultural tractors and harvesters and the supply and demand for services of this type of machinery. The histories of individual countries clearly bring out how long there has been extensive small-scale engines in the field, and that they continue to spread rapidly. Many assume that these smaller engines are giving way to larger scale engines. However, in many countries if anything the opposite is happening. In Thailand, for example, in the mid 1970s there were about 90,000 2WTs, and by 1981 there were more than 284,000. Over 95% of the 2WTs were manufactured locally by small firms. Today, Cramb and Thepent (2020) estimated that there are more than 3 million 2WTs. They describe the extensive public sector support for field-based R&D over time and how the machinery manufacturing sector grew and is now a major exporter. Their work is an example of an agricultural economist writing with an experienced applied engineer, and in so doing make the material accessible to both disciplines, and to a wider audience.

## 2 Case Study: Recent Rapid Spread of Small Engines in the Nepal Mid-hills

In Nepal on April 25th, 2015, a magnitude 7.8 earthquake and its massive aftershocks not only killed over 9,000 people but killed tens of thousands of livestock, including draft animals. This led to mixed responses by the government and donors. One of the responses was to promote the use of the newly introduced mini-tillers (5–7 HP rotor-tillers) from China to replace draft animals. Nepal had a history of smaller engines being used in the mid hills that goes back to the 1970s when two-wheel tractors were introduced into the Kathmandu and Pokhara valleys. Many 2WTs lasted for many decades in agriculture and construction transport. Local repair and import industries for them were well established. The small scale rural capital goods such as MTs and the larger 2WTs have for decades mainly come from China. Figure 2 shows the timeline for the spread of small engines in Nepal.

**1960 1970 1980 1990 2000 2010 2020**

Micro hydro projects
Improved Water mills
2WTs & pumpsets from Japan
2WTs and pumpsets from Korea
2WTs from China
2WT imports stopped -road traffic ban KTM
Govt to stop subsidies for irr. pmps
2WT imports resumed by NTL
Market Experiments(MEs) with MTs
MEs with elect. thresher shellers
MEs with small elec. mills
Govt re-start subsidies
MTs spreading rapidly

**Fig. 2.** Timeline of the spread of small-scale engines and equipment in the Nepal mid-hills. Note: 2WT = two-wheel tractor. ADB = Asian Development Bank, AEPC = Alternative Energy Promotion Centre, MEs = market experiments, MT = mini tiller, NTL = National Trading Ltd. Source: Justice, Biggs & Heath (2020).

The International Centre for Maize and Wheat Improvement (CIMMYT), which had long term experience in Nepal working with smaller scale agri-machinery (Goodrich-Gurung et al. 2008), drew up an R&D programme in 2015 to work with government agencies on earthquake relief by promoting mini tillers (MTs). The relief programmes consisted of making equipment available in affected areas and provided grants and subsidies of up to 90%.

For decades prior to the 2015 earthquake, many researchers and development practitioners with experience of mechanisation in mountainous areas felt that the mechanisation of farming in these areas was limited and provided low benefit levels (Wymann et al. 2013). With high vertical relief and fragmented holdings, mountain farms were considered to be best managed by manual and animal drawn tools and equipment. Therefore, many national and international development specialists in Nepal thought that the small, cheap Chinese diesel, petrol and electric powered agricultural machinery, especially MTs, were not relevant for the mountains of Nepal. However, this had not stopped Nepal's private sector from experimenting with small scale agro-machinery in the market and sales prior to the earthquake (see Table 1). There had been some support for small equipment from the engineering unit in the Department of Agriculture. Later, more people became convinced that small equipment could be economically viable. In conjunction with government organisations, CIMMYT started a research programme on MTs.

**Table 1.** Spread over time of select agricultural machinery in the mountains of Nepal.

| | 1980 | 1990 | 2000 | 2010 | 2020 | % Change from 2010 to 2020 |
|---|---|---|---|---|---|---|
| 2WTs | <100 | <500 | <600 | >10,000 | >20,000 | 200 |
| Diesel-Petrol MTs | N/A | N/A | N/A | <500 | >40,000 | 7,900 |
| Peddle rice thresher | <300 | <800 | <1000 | 1500 | >3000 | 100 |
| Electric Pumps | 1000 | 2000 | 10,000 | 20,000 | 40,000 | 100 |
| Diesel-Petrol pump | <500 | <1000 | <2000 | <5000 | <10,000 | 100 |
| Powered-peddled rice-wheat threshers | N/A | N/A | N/A | 500 | 1000 | 100 |
| Powered maize shellers | N/A | N/A | N/A | 500 | 10,000 | 1,900 |
| Diesel-Petrol powered thresher | N/A | N/A | N/A | 100 | 1000 | 900 |
| Mini Rice Huller-Maize Flour Mill | N/A | N/A | N/A | 100 | 2000 | 1,500 |

Source: Justice, Biggs & Heath, 2020

In the months following the earthquake when the relief efforts came to an end, many people were surprised by the big rise in non-relief programme MT sales. By 2020, an estimated over 40,000 MTs had been sold in the mid hills—a dramatic increase from the estimated 500 sales in 2010 (see Table 1). In the relief effort and in the subsequent monitoring and evaluation (M&E), engineers were often part of research programme staff for the field level data collection. It was through this continued M&E of the relief programmes' market "push" of the MTs that the researchers learnt about who and how the introduced MTs were being used and maintained. The demand for the MTs came not only from the loss of draft bullocks and people during the earthquake, but also was as a result of the large-scale outmigration from rural areas.

After the earthquake, the MTs were being sold in the market and through government programmes. Local district government support programmes provided subsidies. Increased sales led to more intensive field studies that showed that this scale appropriate engineering technology was sound and reliable as regards the material reality of the rural economy (Paudel et al. 2019, Paudel et al. 2020).

The current spread of small machinery in the mountains is also an outcome of the smallholder-friendly Agricultural Mechanization Promotion Policy (AMPP) (AMPP 2014). AMPP was developed over several years with the involvement of engineers, social scientists, farmers, and the local agricultural engineering industry, and increased the focus on mechanisation in the Ministry of Agriculture. In 2016, an apex national extension programme was launched called the Prime Ministers Agricultural Modernization Project. This project has a substantial budget for agri-machinery extension and subsidies and is one of the largest employers of agricultural engineers in Nepal.

While it is hard to gauge its significance, but the boarder closure with India on fuel and other trade in 2015, very soon after the major earthquake probably gave additional political momentum to the government placing greater emphasis on national development.

Another cause of agricultural mechanization in the hills has been the recent introduction of a federal political system. This has led to increased government budgets going to promoting small scale machinery in the hills. Table 1 shows that, in addition to MTs and 2WTs other machinery, such as mini maize shellers, are spreading fast. Similar to MTs, they are often being bought without government loans or grants.

As a way to bring the current mountain mechanization processes unfolding in the Nepal mid hills to a larger international audience, the International Centre for Integrated Mountain Development and the UN's Food and Agricultural Organization began an ongoing webinar in March 2021. The objectives of the webinar, "Women farmers and sustainable mechanization: Improving lives and livelihoods in the Hindu Kush Himalaya", are to share experiences and lessons from various projects and programmes in the region, and to discuss examples of agricultural and other rural mechanisation that are evolving and that benefit mountain women.

# 3 Discussions and Conclusions

There are five conclusions we want to draw from the historical discussion and the current case study: ongoing field work, long-term R&D interest, importance of informal R&D, policy commitment, and importance of the individual.

## 3.1 The Importance of Ongoing Field Work by Engineers, Economists, and Anthropologists

The field work by engineers, economists and anthropologists illustrates the importance of the way ongoing data from the field effected and informed the R&D efforts. External events, such as the earthquake and Indian fuel and trade embargo soon after, led to unplanned experiments. This information was picked up in formal and informal ways and used to leverage a much larger programme of earthquake relief for farmers that included experiments with MTs and other smaller scale equipment. Literature from the early phase shows similar patterns of engineering R&D practice in those days. A feature of the current case study is the return of learning from what might be called "Monitoring and Evaluation" studies.

## 3.2 The Importance of a Long-Term R&D Interest in the Complexities of Local Rural Economies

A second conclusion is that rural mechanisation processes are long-term, complex processes that are location specific. The recent IFPRI study of several countries confirms this. The introduction and use of Korean and Japanese 2WTs into Nepal started in the 1970s, and some of them were in use for 30 odd years. Chinese 2WTs replaced these in the mid-1980s and then dominated the small engine industry since then. Experimentation with Chinese MTs was facilitated by the earlier growth of the local maintenance and machinery industries and the growing access to the wide range of cost-effective machinery from China. The rapid recent spread of MTs came out of a long history of building up a local engineering capability relevant to the Nepal economy.

### 3.3 The Importance of Informal R&D in Engineering Research

A common feature of detailed research on the spread of smaller engines and associated rural capital goods is the ongoing role of informal R&D. This is carried out by farmers, local artisans, workers in small and medium industries. While in the early period there were well known studies by engineers and social scientists on this subject, over the years the literature has declined. In the case of the recent spread of MTs and small engines for pumping and managing water in the hills, the activities are notable for technical and institutional innovations emerging out of informal R&D "much to the surprise" of some of the professional people involved. Part of the revival of general interest in rural mechanisation is illustrated by ethnographic field work that shows how important artisanal innovation is to engineering and by the widespread use of Chinese small engines on the waterways of the Congo (Lambertz 2021).

### 3.4 The Importance of Policy Commitments to Equitable Economic Development in Rural Areas

A feature of the earlier period was a national interest in economic development where equitable rural development in rural areas was a major goal of national economic planning. R&D on smaller engines was part of that planning. Interestingly, what we have seen in Nepal recently is a renewed interest in broad-based rural development, not only for national productivity goals, but also for broader social and environmental goals. In this, smaller engines are expected to continue to play a significant and growing role.

### 3.5 The Importance of the Individual

The final conclusion concerns the importance of the individual. While professional people are trained in the theories, methods, and skills of their profession, each one then uses those disciplinary skills in different ways. Not only this, but professional economic and engineering training often puts major emphasis on theory and models, rather than on applied field work engagement and training. Professional choices between these paths are always being made. Anthropologists, who place emphasis on field engagement, also have professional choices about how to collect and write up information. This is well illustrated by the contestation over the publishing of a book about an agricultural research project in India (Mosse 2006). Unfortunately, in some R&D project preparation documents there are calls for "an economist" as if all economists are trained in the same way and have the same professional experiences and abilities; and the same goes for other professionals. As is shown in the case study, there are engineers who are known as "engineers' engineers". They are known for a deep empirical field level engagement within their discipline. From history, it is clear that Bill Chancellor was a much-respected engineer who engaged for years in field level engineering problems that were facing poorer rural people. Finally, we suggest that in order to support such field-based engineering R&D, there should be greater funding for: 1) field work training for engineers, and 2) skills training for working in teams of engineers, economists and anthropologists in engineering R&D. These are not new suggestions, but we feel their revival needs supporting.

## References


Appu, P.: The bamboo tubewell: a low cost device for exploiting ground water. Econ. Pol. Wkly **9**(26), A63–A65 (1974)

Biggs, D.: Small machines in the garden: everyday technology and revolution in the Mekong Delta. Mod. Asian Stud. **46**(1), 47–70 (2012)

Biggs, S., Farrington, J.: Agricultural Research and the Rural Poor: A Review of Social Science Analysis. International Development Research Centre, Ottawa (1991)

Biggs, S., Justice, S., Lewis, D.: Patterns of rural mechanisation, energy and employment in South Asia: reopening the debate. Econ. Pol. Wkly **49**(9), 78–82 (2011)

Biggs, S., Justice, S.: Rural and agricultural mechanization: a history of the spread of small engines in selected Asian countries. In: IFPRI Discussion Paper 01443. Washington, DC (2015).

Cernea, M., Kasam, A. (eds.): Researching the Culture in Agri-Culture: Social Research for International Development. CABI Publishing, Oxford (2006)

Chambers, R., Pacey, A., Thrupp, L.A. (eds.): Farmer First: Farmer Innovation and Agricultural Research. Intermediate Technology Publications, London (1989)

Chancellor, W.J.: The sustainability of mechanization in Thailand. In: Consequences of small farm mechanization project. Asian Institute of Technology, Bangkok (1983).

Clay, E.: The economics of the bamboo tubewell: dispelling some myths about appropriate technology. CERES **13**(2), 43–47 (1980)

Cramb, R., Thepent, V.: Evolution of agricultural mechanisation in Thailand. In: Diao, X. Takeshima, H., Zhang, X. (eds.) An evolving paradigm of agricultural mechanization development: How much can Africa learn from Asia? International Food Policy. Research Institute. Washington, DC, pp. 165–202 (2020).

Daum, T., Birner, R.: Agricultural mechanization in Africa: Myths, realities and an emerging research agenda. Global Food Secur. **26**, 100393 (2020).

Diao, X., Takeshima, H., Zhang, X. (eds.): An Evolving Paradigm of Agricultural Mechanization Development: How Much Can Africa Learn From Asia? International Food Policy Research Institute, Washington, DC (2020)

Duff, B.: Changes in small farm paddy threshing technology in Thailand and the Philippines. In: Stewart, F. (ed.) Macro policies for appropriate technology in developing countries, pp. 95–139. Westview Press, Boulder, USA (1987)

Falcon, W.: Agricultural and industrial Interrelationships in West Pakistan. J. Farm Econ. **49**(5), 1139–1154 (1967)

Foresight: The future of food and farming. Final project report. The Government Office for Science, London: UK (2011).

Goodrich-Gurung, C.G., Justice, S., Biggs, S., Sah, G.: Participatory technology development in agricultural mechanisation in Nepal: how it happened and lessons learnt. Dev. Pract. **18**(4/5), 643–649 (2008)

Haggblade, S., Hazell, P.B.R., Readon, T. (eds.): Transforming the Rural Nonfarm Economy: Opportunities and Threats in the Developing World. World Bank and IFPRI, Washington (2007)
Hanlin, R., Kaplinsky, R.: South–south trade in capital goods –The market-driven diffusion of appropriate technology. Eur. J. Dev. Res. **28**, 361–378 (2016)
Harriss-White, B.: A Political Economy of Agricultural Markets in South India. Sage Publications, New Delhi (1996)
Hoff, K., Braverman, A., Stiglitz, J.: The economics of rural organization: Theory, practice, and policy. In: International Bank for Reconstruction and Development (1993).
Justice, S., Biggs, S.: The spread of smaller engines and markets in machinery services in rural areas of South Asia. J. Rural. Stud. **73**, 10–20 (2020)
Justice, S., Biggs, S., Heath, R.: Rural and agricultural mechanization in the Himalayan rural economy: The spread of small engines in the Nepal mid-Hills. Working Paper (2020).
Lambertz, P.: Longola marche arrière! chinese diesel engines on congo's inland waterways. Critic. African Stud. (2021). https://doi.org/10.1080/21681392.2021.1931385
McIntyre, B., Herren, H., Wakhungu, J., Watson, R. (eds.): International assessment of agricultural knowledge, science and technology for development (IAASTD). Island Press, Washington DC (2008)
Mosse, D.: Anti-social anthropology? Objectivity, objection, and the ethnography of public policy and professional communities. J. Roy. Anthropol. Inst. **12**, 935–956 (2006)
Nepal, Ministry of Agricultural Development: Agriculture mechanization promotion policy. Kathmandu (2014).
Paudel, G., Kc, D.B., Rahut, D., Justice, S., McDonald, A.: Scale-appropriate mechanization impacts on productivity among smallholders: evidence from rice systems in the mid-hills of Nepal. Land Use Policy **85**, 104–113 (2019)
Paudel, G., Gartaula, H., Rahut, D., Craufurd, P.: Gender differentiated small-scale farm mechanization in Nepal hills: an application of exogenous switching treatment regression. Technol. Soc. **61**, 101250 (2020).
Rhoades, R.: The art of the informal agricultural survey. In: Social Science Department Training Document 1982–2. International Potato Centre, Lima, Peru (1982).
Rhoades, R, Horton, D., Booth, R.: Anthropologists, biological scientists and economists: The three musketeers or three stooges of farming systems research? In Jones, G., Wallace, B. (eds.) Social sciences and farming systems research: Methodological perspectives on agricultural development. Westview Press, Boulder, USA (1986).
Rhoades, R.E.: Seeking half our brains: reflections on the social context of interdisciplinary research and development. In: Cernea, M., Kasam, A. (eds.) Researching the culture in agriculture: social research for international development, pp. 403–420. CABI Publishing, Oxford (2006).
Rigg, J., Salamanca, A., Thompson, E.: The puzzle of East and Southeast Asia's persistent smallholder. J. Rural. Stud. **43**, 118–133 (2016)
Schumacher, E.: Small Is beautiful: Economics as if people mattered. Harper & Row, New York (1973)
Singh, G., Chancellor, W.J.: Energy inputs and agricultural production under various regimes of mechanization in northern India. Trans. ASAE. **18**(2), 252–259 (1975).
Timmer, P., Thomas, J., Wells. L., Morawetz, D.: The choice of technology in developing countries: Some cautionary tales. In: Centre for International Affairs, Harvard University (1975).
Stevenson, R., Locke, V.O.: The Agricultural Development Council. A History. Winrock International Pubns Sales, Little Rock, AR (1989).
World Bank: Agriculture for development. The World Bank, Washington DC (2008)
Wymann von Dach, S., Romeo, R., Vita, A., Wurzinger, M., Kohler, T. (eds.): Mountain farming is family farming: a contribution from mountain areas to the International Year of Family Farming. FAO, CDE, BOKU, Rome (2013).

# Biomechanical Engineering

# Graph-Based Structural Analysis of Kinetic Art Mechanisms on the Example of a Moving Horse Mechanism

P. A. Andrienko, A. N. Evgrafov, D. P. Kozlikin(✉), Y. A. Semenov, and N. S. Semenova

Saint-Petersburg State Polytechnic University of Peter the Great, Saint Petersburg, Russia

**Abstract.** The authors consider the use of the graph method as structural analysis, using the mechanism of the famous artist Theo Jansen as an example. On the one hand, his work serves as an interesting piece of kinetic art using pure energy, wind power, and on the other hand, it has inventive solutions in the field of biomechanical engineering. This combination can refer simultaneously to several areas of the world's Sustainable Development Goals articulated in the United Nations 2030 Agenda. These include good health and well-being (SDG 3), affordable and clean energy (SDG 7). This intersection of topics has led authors to become more interested in the dissemination and use of the graph method as a simple and versatile way of carrying out structural analysis of mechanical systems, which is the starting point at the beginning of any study of mechanisms.

**Keywords:** Graph method · Structural analysis · Mechanisms · Assur groups · Kinematic pairs · Machine design · Walking mechanism · Theo Jansen's mechanism · Good health and well-being (SDG 3) · Affordable and clean energy (SDG 7)

## 1 Introduction

There is no doubt in the assertion that contemporary art in its various forms is an engine of technological progress. Modern art, on the one hand, rethinks reality, and on the other hand, it fantasizes about the future. What may it be? How can a man change? Some of these ideas could become reality if they inspire engineers, programmers, businessmen, etc.

It's not very common to see modern art that incorporates the concept of clean energy which is derived from natural sources: the sun, water, wind, etc. Such pieces of art include those of Dutch artist Theo Jansen. His exhibition of "Beach Animals" presented in Linz's town square in 2005 may serve as a good example. (see Fig. 1).

The direction of contemporary art Theo Jansen works in is called kinetic art. The main characteristic of a given artistic movement is the effect of playing with the real movement of the whole piece of art. Nowadays this term is used to describe three-dimensional sculptures that move on their own or use a motor. Works of Theo Jansen belong to the first category. The artist himself calls his kinetic sculptures "animals". They


G. Quaglia et al. (Eds.): I4SDG 2021, MMS 108, pp. 141–149, 2022.
https://doi.org/10.1007/978-3-030-87383-7_15

**Fig 1.** Beach animals. (Source: https://commons.wikimedia.org/wiki/File:Jansen-Strandbeest_crop.jpg).

can walk using only wind energy. These moving sculptures look especially impressive when they wander along the sandy beach, they can move in packs along the shore, and this view can remain one of the most unforgettable experiences.

The authors of this article are impressed by this phenomenon. But being scientific workers, mechanical engineers, the authors want to understand the basis of the structure of these "creatures" from a scientific point of view, and namely from the point of view of the theory of mechanisms and machines. The authors would like to consider the problem of structural analysis, using the method of graphs, which is quite widespread in the course "Theory of mechanisms and machines", taught to students in the Saint-Petersburg State Polytechnic University of Peter the Great.

## 2 Graph Theory in Structural Analysis of Mechanisms

As an example, let's consider the mechanism of Theo Jansen's horse, the kinematic diagram of which is shown in Fig. 2 [1–3, 4].

Before starting the study, let's define the physical models used in this analysis and the theory of the machines and mechanisms in general. The simplest model of a real mechanism is a mechanism with rigid links. By the link is meant a detail or a group of details representing from the kinematic point of view a single whole (i.e. a group of details rigidly connected with each other and moving as a single rigid body). All links are considered as non-deformable rigid bodies. Furthermore, speaking about physical model of connection of links among themselves, the concept of kinematic pair is introduced. A kinematic pair (KP) is a connection between two links that provides a certain relative motion. The KP is classified by the number $s$ of degrees of freedom in the relative motion of the links to be connected (mobility of the KP) and by the number $m$ of constraints imposed by the pair on this movement.

The number of constraints and their mobility can be defined as follows:

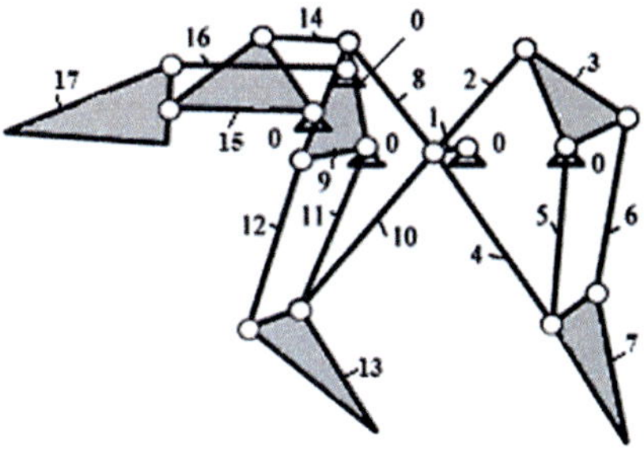


**Fig. 2.** Theo Jansen's horse mechanism

1) compose equations of the constraints imposed on the relative movement of links, their number $m$ indicates the number of constraints;

2) determine the mobility of a pair by the formula:

$$s = 6 - m. \tag{1}$$

Following (1) there are one-degree-of-freedom kinematic pair ($s = 1$), two-degree-of-freedom ($s = 2$) and so on up to $s = 5$ KP. According to another classification, kinematic pairs in which there is a contact area that belongs to conjugated links and coincide in any relative motion are called "lower kinematic pairs". KP in which there are only common lines or points that change their position during relative motion are called "higher kinematic pairs".

In this example of a mechanism (see Fig. 2), there are17 movable links in total, numbered with their respective numbers. All realized kinematic pairs are considered as inferior, and one-degree-of-freedom ($s = 1$) ones. A total of 25 kinematic pairs are realized. When counting the number of kinematic pairs it is necessary to make some remarks. In the abovementioned diagram the links 3, 7, 9, 13, 15 and 17 are rigid flat structure made like a triangle (darkened areas). There are also places in the diagram where the links are ambiguously connected. For example, the connection between links 2, 4, 8, 10 and 1. In the diagram, they are connected at one point. In the particular case, this connection must be treated as shown in Fig. 3, i.e., link 1 is connected to all others when links 2, 4, 8 and 10 are not connected. This ambiguity, for all other cases, will then be unambiguously determined by the constructed graph of the mechanism.

The next physical model, to be referred to is a kinematic chain. It is a set of links, which are connected to each other by kinematic pairs. The kinematic chain is characterized by its inputs and degrees of freedom. The inputs of a kinematic chain form connected links, the law of relative motion of which is set. These motions are set by motors. The number of motors determines the number of such inputs $n_c$of the kinematic chain. They may be internal, where driving forces are applied to the movable chain links, and external, where driving forces are applied to one of the movable chain links and to another link outside of the kinematic chain (in particular, to a frame, a fixed link).

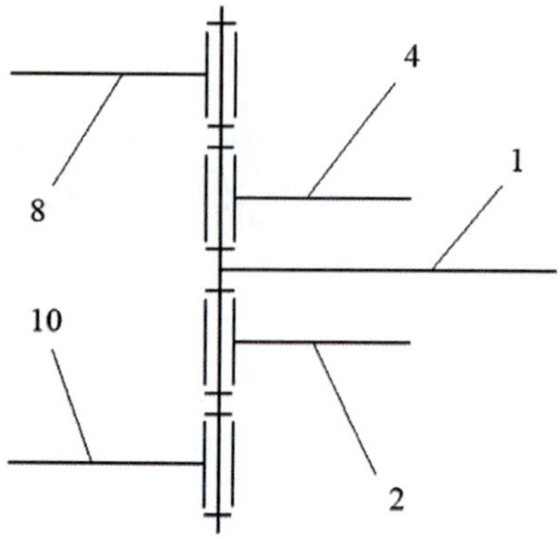


**Fig. 3.** Link connection 1, 2, 4, 8 and 10

If the kinematic chain contains $N_c$ links and $p_{sc}$ kinematic pairs of $s$ mobility, it has:

$$w_c = 6N_c - \sum_{s=1}^{s=5} (6 - s)p_{sc}. \quad (2)$$

degrees of freedom, since each $s$-movable pair imposes $(6-s)$ independent coupling conditions. The number of degrees of freedom $w_c$ of a kinematic chain with rigid links is usually called the number of its mobility degrees.

A kinematic chain, in which one of the links is taken to be stationary, is called a mechanism [4]. All the variety of existing mechanisms are divided into spatial and planar ones. Planar are those mechanisms in which all the links belonging to them move in parallel planes, other mechanisms are spatial.

Formula (2) is valid for spatial mechanisms. Since in the mechanism one link is taken as stationary, only the number of moving links $N$ should be taken into account in relation:

$$w = 6N - \sum_{s=1}^{s=5} (6 - s)p_s. \quad (3)$$

Formula (3) is called the Malyshev-Somov structural formula [4]. Using this formula, it is possible to determine the number of degrees of mobility of the mechanism $w$. In case of a planar mechanism, another formula is fair:

$$w = 3N - 2p_L - p_H, \quad (4)$$

which is called the Chebyshev formula [4], where $p_L$ is the number of the lower kinematic pairs, $p_H$ is the number of the higher kinematic pairs of the mechanism.

The mechanism under study is a planar mechanism, so we use formula (4) to determine the degree of mobility $w$. For this mechanism it has been shown above that $N = 17$, $p_L = 25$, $p_H = 0$, whence $w = 1$. The result shows that for the realization of movements of the mechanism one motor could be enough. In our case it is installed between links 0 and 1.

Strictly speaking, formulas (3) and (4) can lead to one general form:

$$w = \mu N - \sum_{\mu=1}^{s=\mu-1} (\mu - s)p_s, \tag{5}$$

where $\mu$ is the number of degrees of freedom of the space within which the mechanism operates (for spatial motion $\mu = 6$, for flat motion $\mu = 3$).

The main thrust of structural analysis is the division of the mechanism into simpler models-structural groups. A structural group is a kinematic chain in which the number of inputs is equal to the number of degrees of mobility $n_c = w_c$, such a structural group is also called a normal $n$-moving structural group [5]. This concept was introduced by M.Z. Kolovsky [6], to describe the structure of multi-engine machines and machines with internal inputs. A structural group is called a simple one if it cannot be divided into several groups with fewer links.

Elements of the graph theory have been proposed for identifying simple structural groups [4, 7, 8]. The graph is a schematic drawing, the elements of which are vertices (circles) and edges (lines). In the graph mechanism the vertices correspond to the links, so the number of the vertices coincides with the number of the link (this refers to the labelled graph). Edges of the graph correspond to kinematic pairs. The number of edges connecting the adjacent vertices coincides with $s$-mobility of the kinematic pair. If $s > 1$, the edges are called multiple edges. The root edges-the lines corresponding to the inputs of the mechanism, are marked on the graph of the mechanism with thickened lines.

Let's draw a graph for the example under consideration (see Fig. 4 a). Link 1 is taken as the input link and is highlighted with a thickened line.

After the graph is constructed, let us define the numerical characteristic that will be used to divide the graph of the mechanism into structural groups. For this purpose let us transform dependence (5). Let's assume that:

$$P = \sum_{s=1}^{s=\mu-1} p_s, \tag{6}$$

the number of generalized edges (the number of kinematic pairs of the mechanism, where $p_s$ is the number of kinematic pairs of $s$-mobility) and

$$S = \sum_{s=1}^{s=\mu-1} s p_s, \tag{7}$$

the number of edges of the graph (the total number of motions of all kinematic pairs).

Applying (6) and (7) in (5):

$$w = \mu N - \sum_{\mu=1}^{s=\mu-1} (\mu - s)p_s = \sum_{s=1}^{s=\mu-1} s p_s - \mu\left(\sum_{s=1}^{s=\mu-1} p_s - N\right) = S - \mu(P - N) = S - \mu K, \tag{8}$$

where $K = P - N$ is the number of independent loops of the graph. Independent contours of the graph are closed kinematic chains that differ from other closed kinematic chains by at least one generalized edge.

For division of the graph into structural groups it is necessary to separate it into subgraphs, closed kinematic chains, starting from the vicinity of vertex 0 (rack).

Let us get the numerical characteristic of such subgraph by going to the notation with the index c, indicating that we are hereafter talking about a kinematic chain. From the structural formula of the normal kinematic chain $n_c = w_c$ using (8):

$$n_c = w_c = S_c - \mu K_c. \tag{9}$$

From (9) it follows:

$$S_c - n_c = \mu K_c = \mu(P_c - N_c), \tag{10}$$

i.e. the difference between the total number of edges of subgraph $S_c$ and the number of root edges $n_c$ of the considered closed kinematic chain is multiple of six (for spatial mechanism), three (for planar mechanism) or equal to zero at $P_c = N_c$.

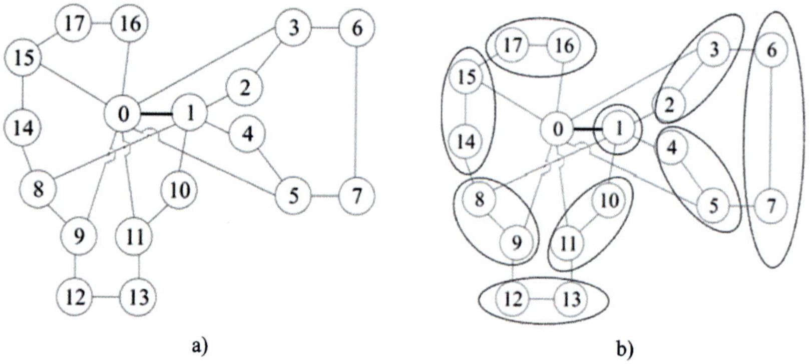


**Fig. 4.** Graph of an Theo Jansen's horse mechanism. (a - Graph mechanism, b - graph of the mechanism divided into structure groups)

According to the considered theory, first of all, it is necessary to search for subgraphs corresponding to the fulfilment of the condition $S_c = n_c$, such subgraphs correspond to the simple single-link single-moving structural groups. When all such options are pointed out, one can proceed to the definition of the subgraphs consisting of vertices and edges of independent loops.

If in any independent loop ($K_c = 1$) associated with vertex0 the number of thin edges is equal to three (or six in case of a spatial mechanism), the subgraph composed of vertices and edges of this loop, except for previously selected simplest structural groups ($S_c = n_c$), describes a simple structural group.

If there is no such a loop, then should be found two adjacent loops ($K_c = 2$) with the number of thin edges equal to six (or twelve)etc. This is how condition (10) is checked sequentially for all the loops, and thus simple structural groups of the mechanism are identified.

It is important that after defining the subgraph consisting of vertices and edges of independent loops, it is necessary to return to the search for the case of the subgraph when $S_c = n_c$. It may be so that the next vertex is attached to the subgraph just selected through the root edge, which means that this vertex and the attached edge area simple structural group, which requires their separate definition into a structural group.

At the final stage, it is possible to draw a structural graph of the mechanism, at the vertices of which the first digit indicates the number of links in the found simple groups and the second digit, separated by comma, indicates the number of inputs into them are specified. The order in which the structural groups are connected to the rack is sometimes indicated by arrows.

The graph drawn and divided in Fig. 4 b. has 8 independent loops ($K = 8$). Using the abovementioned rules of graph partitioning into structural groups we obtain a structural graph of Theo Jansen's horse mechanism. It is presented in Fig. 5.

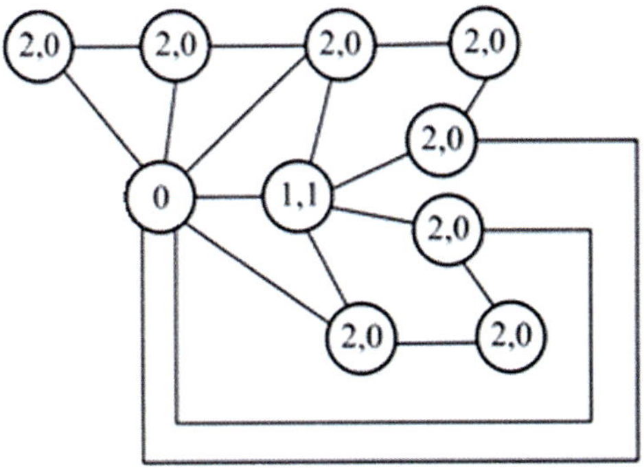


**Fig. 5.** Structural graph of an Theo Jansen's horse mechanism

It is possible to divide the graph into the simple structural groups from which one group of single-link single-moving (1,1) and 8 Assur groups two-link zero-moving are allocated (2,0). As can be seen, the resulting circuit, while seemingly cumbersome, is not a complex design. To realize the necessary movement of all the links it is enough to have one motor. In Theo Jansen's mechanisms, this motor is driven by wind power.

It should be noted that the use of graph theory in the study of mechanical systems is not a new scientific trend. There are quite a few works, which are devoted to the application of this mathematical theory to the needs of mechanics [9–16].

To summarize, below is a step-by-step algorithm for the method under consideration:

1. Construct a kinematic diagram of the mechanism with numbering all of the links, each link is assigned its own number, starting from the fixed links. All fixed links are numbered 0.
2. On the mechanism define inputs and locations where engines are installed.

3. Based on the numbered kinematic diagram of the mechanism construct a graph of the mechanism. Each vertex of the graph defines a link, all fixed links are combined in the vertex numbered 0.
4. Connect the numbered vertices of the graph by graph edges. The number of edges between two vertices depends on the degree of freedom of the kinematic pair which is formed between the considered links.
5. Mark the edges with a thickened line where the inputs (engines) are set. Make them root ones.
6. Divide the constructed graph into simple structural groups by their numerical characteristics. This process is iterative. Within each iteration, it is necessary to start distinguishing structural groups by searching for the simplest ones corresponding to the condition $S_c = n_c$. After selecting all of such structural groups at the first step of iteration go on to the next condition $S_c - n_c = \mu K_c$ where $K_c$ can be regarded as a singular counter the initial value of which is equal to one. If the condition is not met, the counter $K_c$ is increased by 1, and so on, until the condition is met. Then the iteration proceeds to the second step and starts all over again from the searching for the simplest structural groups from the condition $S_c = n_c$.

## 3 Conclusions and Remarks

1. If the constructed graph for the designed kinematic scheme cannot be divided into structural groups, it may mean that the scheme does not work. In this case, it is necessary to rebuild the kinematic diagram, change the number of links used or change the kinematic pairs. This statement does not always work, there are mechanisms with exceptions, for example, mechanisms with idle degrees of freedom and redundant connections.
2. The graph of a mechanism will always split into structural groups if it is assembled from normal structural groups.
3. The method under consideration can be used for mechanisms with internal inputs, which other ways of dividing mechanisms into structural groups often do not allow.
4. Given the method of dividing structural groups applies to both planar and spatial mechanisms. In this, the method is universal.

## References

1. Jansen, T.: The Great Pretender. Netherlands: 010 Uitgeverij. (2007).
2. Xu, K., Liu, H., Zhu, X., Song, Y.: Kinematic analysis of a novel planar six-bar bionic leg. Mech. Mach. Sci. **73**, 13–21 (2019)
3. Vujošević, V., Mumović, M., Tomović, A., Tomović, R.: Robot based on walking Jansen mechanism. IOP Conf. Mater. Sci. Eng. **393**(1), 012109 (2018)
4. Wang, C.-Y., Hou, J.-H.: Analysis and applications of theo jansen's linkage Mechanism: theo Jansen's linkage mechanism on kinetic architecture. In: CAADRIA 2018 - 23rd International Conference on Computer-Aided Architectural Design Research in Asia: Learning, Prototyping and Adapting, vol. 2, pp. 359–368 (2018)

5. Kolovsky, M.Z. (ed.) *Teorija mehanizmov i mashin: uchebnoe posobie dlja vuzov po mashinostroitel'nym special'nostjam* [The Theory of Machines and Mechanisms: Study Guide for Graduate Students of Mechanical EngineeringSpecialities], Moscow, p. 558 (2013)
6. Kolovsky, M.Z.: Teorija mehanizmov i mashin. Struktura i kinematika mehanizmov: Tekst lekcij [The theory of Mechanisms and Machines: Structure and Kinematics of Mechanisms: Lecture Texts]. Saint-Petersburg: Saint-Petersburg State Polytechnic University, p. 81 (1993)
7. Evgrafov, A.N., Karazin, V.I.: Vklad, M.Z.: Kolovskogo v teoriju mehanizmov i mashin [Contribution of M. Z. Kolovsky to the Theory of Mechanisms and Machines], https://doi.org/10.18721/JEST.25215, Nauchno-tehnicheskie vedomosti SPbPU. Ser.: Estestvennye i inzhenernye nauki [St. Petersburg polytechnic university journal of engineering science and technology]. Saint-Petersburg State Polytechnic University of Peter the Great. **25**(2), 183–191 (2019)
8. Semenov, Y.A., Semenova, N.S.: Strukturnyj i geometricheskij analiz mehanizmov: metodicheskie ukazanija k prakticheskim zanjatijam [Structural and GeometryAnalysis of Mechanisms: Methodology Guidlines for Practical Exercises], Saint-Petersburg: Saint-Petersburg State Polytechnic University of Peter the Great, p. 42 (1994)
9. Kolovsky, M.Z., Evgrafov, A.N., Semenov, Yu.A., Slousch, A.V.: Advanced Theory of Mechanisms and Machines. Springer - Verlag Berlin Heidelberg New York, p. 394 (2000)
10. Didenko, E.V.: Razrabotka i analiz mnogokonturnyh mehanizmov na osnove teorii grafov[Multi-loop Mechanisms Development and Analysis on the basis of the Graph Theory]. In: Moscow: Mechanical Engineering ResearchInstitute of the Russian Academy of Sciences. (2019)
11. Lung-Wen, T.: Mechanism Design: Enumeration of Kinematic Structures According to Function. CRC Press, p. 328 (2001)
12. Pavlova, L.A.: Metod grafov v strukturnom issledovanii prostranstvennyh mekhanizmov: special'nost' 05.00.00 "Tekhnicheskie nauki"[The method of graphs in the structural study of spatial mechanisms: specialty 05.00.00 "Technical Sciences"]. In: The sis for the degree of Candidate of Technical Sciences of Pavlova Lidiya Antonovna. Bauman State Technical University, Moscow, p. 187 (1976)
13. Zawislak, S.: Artificial intelligence aided design of gears based on graph-theoretical models. In: 12th IFToMM World Congress, Besancon (France) (2007)
14. Zawiślak, S., Rysiński, J. (eds.): Graph-Based Modelling in Engineering. MMS, vol. 42. Springer, Cham (2017). https://doi.org/10.1007/978-3-319-39020-8
15. Chen, I.M., Yang, T.H., Liu, T.: Function power graph a novel methodology for powertrain and hybrid system conceptual design and analysis. In: Proceedings of the 14th IFToMM World Congress, pp. 54452 (2015)
16. Wu, Z., Campbell, M.I., Fernandez, B.R.: Bond graph based automated modeling for computer-aided design of dynamic systems. J. Mech Des. **130**(4), 041102–1–041102–11 (2008)
17. Schmidt, L.C., Harshawardhan, S., Chase, S.C.: A graph grammar approach for structure synthesis of mechanisms. J. Mech. Des. **122**(4), 371–376 (2000)

# A New Sustainable Soft Body Protector for Increasing Human Safety in Combat Sports: The Case of Taekwondo

Denis Damiano[1], Angelo Davalli[2], Elvio Bonisoli[1], and Giovanni Gerardo Muscolo[3](✉)

[1] Department of Mechanical and Aerospace Engineering, Politecnico Di Torino, Turin, Italy
[2] FITA, Federazione Italiana Di Taekwondo, Roma, Italy
[3] Department of Computer Science, University of Verona, Verona, Italy
giovannigerardo.muscolo@univr.it

**Abstract.** Taekwondo is a martial art recognized as an Olympic sport since 2000. During competitions, athletes wear the typical protective garment composed by head, arms, hands, trunk, genitals, legs, foot protectors. The trunk protector has been developed with several modifications during history adding electronic devices to help referees in the points' assignment. In this paper, some recent technologies adopted in body protectors are analyzed with the purpose of modelling an innovative sustainable soft body protector specific for Taekwondo competitions. The proposed body protector is composed of internal embedded pneumatic devices to increase safety from injuries and reducing transmissibility of a kick or punch on the human's trunk. Making use of FEM analysis, several geometries of this device are examined to guarantee a high protection level. Some comparison on transmissibility of the impact on the human's trunk with our solutions and available results in literature of an electronic body protector on market underline the very good results obtained with our proposal. The transmissibility is reduced of 70% using the commercial product and of 30% using our engineering simulation. The excellent results open-up a new sustainable research field in the wearable protection systems.

**Keywords:** SDG03 · SDG09 · Taekwondo · Body protector · Pneumatic devices · Elastomers · Sustainability · Human's protection systems

## 1 Introduction

Taekwondo is a Korean fighting art born in the first century B.C. and recognized as a sport since the second half of the twentieth century. To date, there are 70 million athletes spread across 210 countries associated with the World Taekwondo [1]. In every contact sport, athletes go against injuries: among boxers the most common are concussions, cuts to the face, broken nose, wrist and finger sprains and hand fractures [2]. Roger E. Thomas, et al. highlighted that mixed martial art athletes present lacerations, abrasions, and contusions in most of the cases [3] and Del Vecchio et al. indicated that the most


G. Quaglia et al. (Eds.): I4SDG 2021, MMS 108, pp. 150–157, 2022.
https://doi.org/10.1007/978-3-030-87383-7_16

common injuries shared by martial arts and combat sports athletes are tendon injuries, sprain and dislocation/subluxation [4]. Focusing on Taekwondo, MinJoon Ji established that in this sport the most frequent injuries for athletes are contusions, strains and sprains and the most affected range is between 10 and 19 years old [5]. Daedo®, Adidas® and KSD® are some of the brands that produce forearm, shin, hand and foot protectors and Hee Seong Jeong et al. performed the comparison on these [6]. The chest protector of Taekwondo (called Hogu in Korean) was used taking inspiration from the one used in kendo martial art. However, the use of this protective system became mandatory only starting from 1968 [7]. Over the years, protection systems evolved, and cyber-physical systems and robotics have been introduced [8]. Nowadays wearable electronic textiles have found wide use in various fields such as energy storage materials, biosensor applications, sweat sensors, health systems and in sports, including Taekwondo, which allow to detect the intensity of a blow thanks to pressure sensors [9–12]. Already in 2004, the Stanford Taekwondo Program, the Palo Alto Research Center and Impact Measurement developed a chest protector with piezoelectric force sensors inside it to detect the intensity of the force and send the score wirelessly to a computer in order to help the referees in the points assignment [13]. Only in October 2009, during the nineteenth Taekwondo World Championship in Copenhagen, the electronic bodice made its first appearance in an official competition, leading important modifications in scoring techniques. From that moment there were several improvements, in fact during the 2012 Olympics in London the World Taekwondo Federation introduced a uniform that featured impact sensors in the chest area and socks coated with magnets. In 2016, during the Rio Olympics, sensors were also implemented in the headgear [14]. The 2009 was a crucial year for Taekwondo with the introduction of electronic scoring systems and the modification of scoring in favor of the most difficult movements, making the matches more spectacular [15]. The Electronic Scoring System became an excellent training tool to improve footwork and refine kicks, as well as increased competition [16]. As previously highlighted, various technological developments took place regarding the protection systems and various tests were carried out on them to evaluate their reliability. This is the case conducted by Niki Tasika who, using a 4 kg ball of iron released against an EBP (Electronic Body Protector) from a certain height, simulated the typical conditions of a kick and highlighted how the protection system suffered from a high instability in the estimation of the kinetic energy of impact [17]. The same problem was encountered by Rizal Razman and Ryan Wy Lunn Chong using a mechanical pendulum to analyze the phenomenon [18]. Although progress has been made in the last decades in the transparency of scoring, a study by Anna Perry and Juyoung Lee confirmed that the level of protection of uniforms worn during competitions is a feature that athletes would like to see improved. This result suggests that the protection offered by the current bodices is not yet fully satisfied [19].

The aim of this work is to model an innovative chest protector composed of elastic inflatable devices. Until now these soft devices have been used mainly in automation, for instance as grippers, in biomimetics and in medical applications [20–22].

The paper is structured as follows: Sect. 2 shows the design of the new sustainable body protector; Sect. 3 presents the FE modelling; in Sect. 4, comparisons and discussions are shown. Next steps are summarized in the conclusions.

## 2 Modelling and Design of the New Sustainable Body Protector

Figure 1 a, shows the classical Taekwondo trunk's protector. Two types of body protector solutions with embedded pneumatic devices of various geometries are proposed in this paper (see Fig. 1, c and d). The first solution provides a matrix arrangement of bubble soft devices in the internal part of the body protector (see Fig. 1 c). Four different geometrical shape are studied: spherical, cylindrical or truncated cones. The second solution presents cylindrical bubble soft devices positioned along the generatrix, as shown in Fig. 1 d.

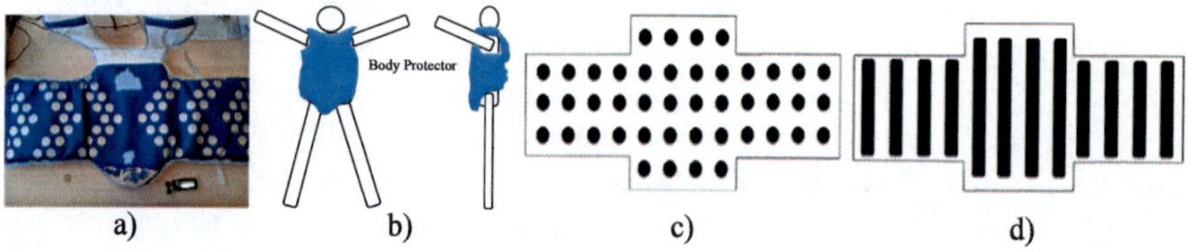


**Fig. 1.** From left to right: (a) real Taekwondo trunk's protector; (b) body protector worn by the user; (c) internal configuration of bubble soft devices for the solution 1; (d) internal configuration of cylindrical bubble soft devices for the solution 2.

Different device geometries are studied both from a static and a dynamic point of view in order to identify the optimal geometry of the devices and their arrangement inside the bodice. As shown in Fig. 2, the proposed solutions of the body protector are composed of the following layers (starting from the bottom): the surface in contact with the athlete's chest, the bodice composed of a first layer of fabric, a membrane of hyperelastic material that fix the device to the system, the device, a membrane and a final layer of fabric, covered by the same surface in contact with the athlete's chest.

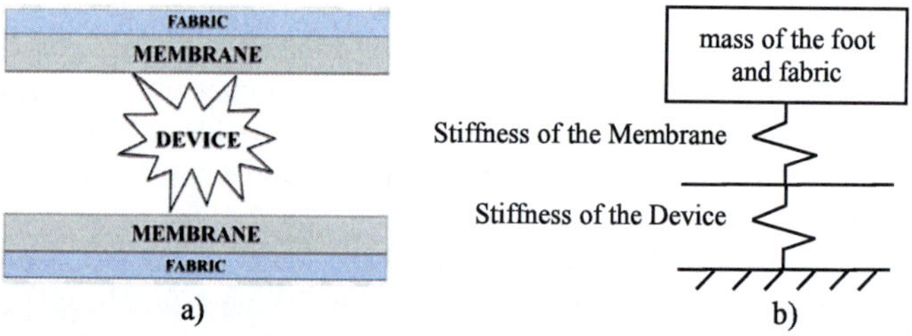


**Fig. 2.** Internal layers of the solution of the body protector (a) and the used model for the device and membrane (b).

Two different materials for the devices are tested: the Silicone and the EPDM (Ethylene Propylene Diene Monomer) hyperelastic materials. The material characteristics have been implemented in FE software by the Mooney-Rivlin mathematical model. In particular, for the Silicone is used the following formulation:

$$U = B_1(I_1 - 3) + B_2(I_2 - 3) \tag{1}$$

Where $U$ is the strain-energy function, $I_1$ and $I_2$ are the principal invariants of the tensor [12, 23], $B_1 = 0.25$ MPa and $B_2 = 0$ MPa [23] while for the EPDM the Mooney-Rivlin model has the following shape:

$$U = B_1(I_1 - 3) + B_2(I_2 - 3) + B_3(I_2 - 3)^2 \quad (2)$$

Where $B_1 = 20.7$ MPa, $B_2 = -4.4$ MPa, $B_3 = 77$ kPa [12]. Several geometries of the device are studied, in particular the hemispherical (S; external diameter: 40 mm; wall thickness: 2 mm and 4 mm), the cylindrical ($C_1$; external diameter: 20 mm; height: 20 mm; wall thickness: 1 mm and 2 mm) ($C_2$; external diameter: 40 mm; wall thickness: 2mm; length: 200 ÷ 400 mm), and the truncated cone in shape ($T_1$, $T_2$, $T_3$, $T_4$; larger-smaller diameter: 20–10 mm; height: 20 mm; wall thickness: 2 mm; throat radius: 0.5 ÷ 1 mm) (see Figs. 3 and 4). For the truncated cone type six different versions are modelled as shown in Fig. 4 also by taking inspiration from the work led by Mehmet A. Guler et al. [24] about the geometrical influence on the axial impact absorption of thin-walled structures. Any sharp edge of the models is rounded off with a radius of 1 mm.

**Fig. 3.** Half section of the devices studied with FE analysis: (a) From left to right: $C_{1,a}$, $C_{1,b}$, $S_a$, $S_b$, $C_2$ models; (b) examples of FE analysis with a plate on the upper part of the device used for transferring the resultant force.

**Fig. 4.** Half section of the devices studied with FE analysis. (a) From left to right: $T_1$, $T_{2,a}$, $T_{2,b}$, $T_3$, $T_{4,a}$, $T_{4,b}$ models; (b) examples of FE analysis with a steel plate on the upper part of the device used for transferring the resultant force.

## 3 FE Modelling of the Devices and Body Protector

In the static FE analysis, the whole bodice and the single devices have been studied. Every model was clamped and put in contact with a rigid plate; in our case the plate was made of steel, while a downward force was applied on the free surface of the plate in order to evaluate the vertical displacement of the device and, consequently, its stiffness. Every analysis has been repeated three times: in the first case the device presented an initial inner relative pressure of 0 bar, in the second case 1 bar and in last case 2 bar. After establishing the most suitable geometries thanks to the static analysis of the devices, that are respectively the $C_{1,b}$, $S_b$, $C_2$, $T_1$, $T_{2,b}$, $T_3$ and $T_{4,b}$ models (see Figs. 3 and 4), the dynamic FE analysis of the bodices has been performed. Several studies were made changing the pressure from 0 to 2 bar and the membrane thickness from 1 to 20 mm using EPDM material having a density of 1,430 kg/m$^3$ and a damping loss factor of 1 for both the membrane and the devices. The minimum distance among the geometrical limits of each device is assumed to be equal to 5 mm, the $C_2$ length of 300 mm, while the torso protective system is supposed to have a surface of 350 × 350 mm$^2$. This means that, taking in consideration the geometrical features of the devices, a matrix arrangement of 14 × 14 is adopted for the $C_1$, $T_1$, $T_{2b}$, $T_3$, $T_{4b}$ models, while the system with $S_b$ devices presents a matrix 8 × 8 and the one with $C_2$ devices is 1 × 8. In order to compare the results obtained by Nusret Ramazanoglu [25], who studied the kick impact transmission of a Daedo®'s body protector adopted in Taekwondo competitions, an impulse having the same characteristics of that applied by Nusret in his study has been applied on the steel plate, i.e. a maximum impact force $F_0$ of 264.17 N and an impact time of $24.7 \cdot 10^{-3}$ s. In order to evaluate the transmissibility of the proposed protection model, the following formulation has been used where $F_T$ is the maximum force transmitted to the torso and $F_0$ is the maximum impact force:

$$\mathrm{Tr} = \mathrm{F}_T/\mathrm{F}_0 \tag{3}$$

## 4 Comparisons and Discussions

Figure 5a, shows the transmissibility of the protector of the human body if only the cover membrane is used and without using the soft devices in the internal parts. In particular, the transmissibility is reduced only of 10% increasing the thickness of the membrane to a maximum of 20 mm. Silicone and EPDM contribute to a major reduction of transmissibility with respect to PMMA (PolyMethyl MethAcrylate), but the difference is not so high. In the following, the cover membrane of EPDM is used instead of Silicone. Figure 5b shows the transmissibility of the different used devices varying the stiffness of the membrane and using 0 bar as internal pressure. As can be deduced, the most performing models are those with $S_b$ and $C_2$ devices, especially with a membrane thickness that varies from 10 to 20 mm. The transmissibility of the reference Daedo®'s value of 68.09% is indicated as upper limit. Varying the internal pressure from 0 to 2 bar and using the EPDM membrane thickness of 1, 10 and 20 mm, the final results of the transmissibility of the protective system with spherical devices ($S_b$) is reported in Fig. 5c. If the spherical devices ($S_b$) are used, it is important to note that the inner pressure does not affect the transmissibility of the external force; however, the membrane thickness plays a crucial role for this purpose. Figure 5d shows how the increment of the transmissibility is proportional with the increment of the internal pressure from 0 to 2 bar if the cylindrical devices ($C_2$) are used. In order to further improve the presented versions of the protective system, the tests using a PMMA material for the membrane are performed. This material has a density of 1,190 kg/m$^3$, a Young's modulus of 3,000 MPa and a Poisson ratio equal to 0.4, while the damping loss factor is set equal to 0.1 [26]. Adopting the same impact force and impact time used in the previous tests, the final results are highlighted in Figs. 5e and 5f. From these graphs it is possible to underline how the use of PMMA material for the membrane improves the protection from kicks with a remarkable reduction of the transmissibility.

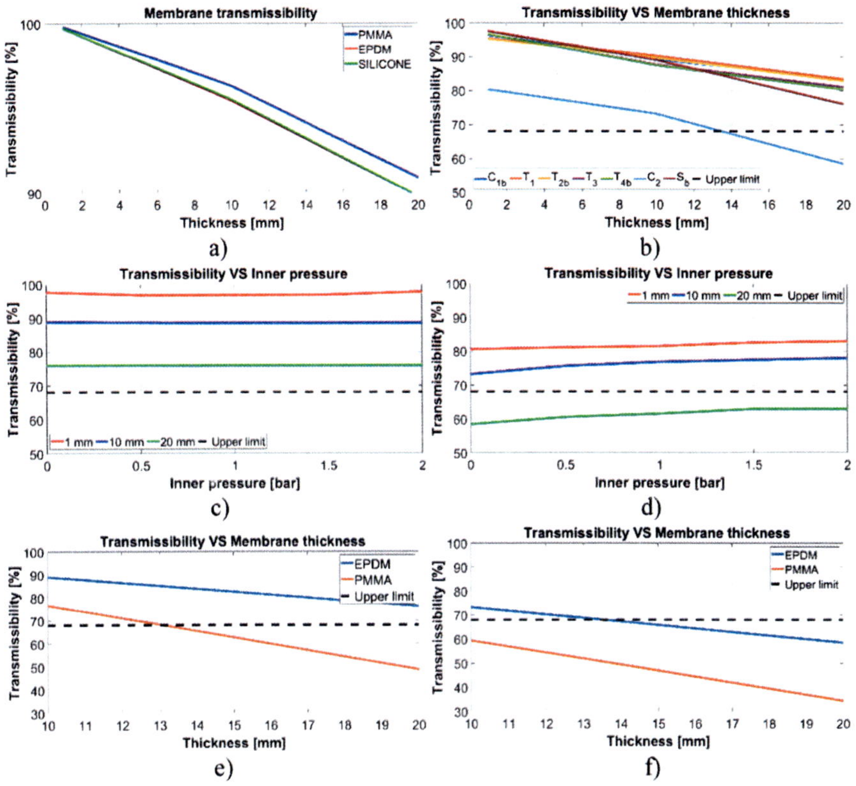


**Fig. 5.** (a) membrane transmissibility; (b) comparison of the transmissibility of the different devices; transmissibility of the protection system with embedded devices and with an EPDM membrane thickness of 1, 10 and 20 mm: $S_b$ devices (c) and $C_2$ devices (d); (e) transmissibility trend with embedded $S_b$ devices; (f) transmissibility trend with embedded $C_2$ devices.

## 5 Conclusion

The novel protection systems proposed in this paper are able to reduce the transmissibility of an impact on the human body of more than 70%. The commercial products are able to reduce at maximum of 30% of the external impact force. This is a first work in this direction, but the excellent numerical results underlined the research direction to be pursued to produce sustainable protection systems. Furthermore, in order to produce an ergonomic version of the new protection systems, it could be useful to choose the most appropriate arrangement of the devices taking in consideration the study conducted by Ki Jun Park et al. [27]. In fact, they highlighted that the thoracic spine, the lumbar spine, the sternum, the ribs, the pelvic, the sacrum and the buttock are the trunk areas more subjected to contusions, while the abdomen presents an injury frequency close to zero. These considerations could give more sustainability to the proposed excellent system in this work. The authors hope that further experimental validations will confirm these improvements. The important results obtained by the research presented in this paper underline that if the internal devices of the body protector are constituted by a cylindrical device connected with the user's body by the generatrix, the transmissibility of the impact force of the opponent transmitted to the user which wears the body protector is reduced considerably. In any case, the approach to include the devices in the internal part of the body protector, as proposed in this paper, gave very good results also for the spherical device, but minus with respect to the cylindrical one underlined before.

## References

1. "Il Taekwondo - La storia" https://www.taekwondoitalia.it/fita/il-taekwondo.html
2. "Do know the most common injuries in boxing?" http://fittoplay.org/sports/boxing1/the-most-common-injuries
3. Thomas, R.E., Thomas, B.C.: Systematic review of injuries in mixed martial arts. Phys. Sportsmed. **46**(2), 155–167 (2018). https://doi.org/10.1080/00913847.2018.1430451
4. Del Vecchio, F.B., Farias, C.B., de Leon, R.C., Rocha, A.C.C.A., Galliano, L.M., Coswig, V.S.: Injuries in martial arts and combat sports: Prevalence, characteristics and mechanisms. Sci. Sports **33**, 158–163 (2018). https://doi.org/10.1016/j.scispo.2018.02.003
5. Ji, M.: Analysis of injuries in taekwondo athletes. J. Phys. Ther. Sci. **28**, 231–234 (2016)
6. Jeong, H.S., O'Sullivan, D.M., Lee, S.-C., Lee, S.Y.: Safety evaluation of protective equipment for the forearm, shin, hand and foot in taekwondo . J. Sports. Sci. Med. **18**, 376–383 (2019)
7. Moenig, U., Cho, S., Song, H.: The modifications of protective gear, rules and regulations during taekwondo's evolution – From its obscure origins to the Olympics. Int J History Sport **29**(9), 1363–1381 (2012). https://doi.org/10.1080/09523367.2012.691474
8. Mencarini, E., Rapp, A., Tirabeni, L., Zancanaro, M.: Designing wearable systems for sports: a review of trends and opportunities in human–computer interaction. IEEE Trans. Hum. Mach. Syst. 49(4), 134–325 (2019). doi: https://doi.org/10.1109/THMS.2019.2919702.
9. Muscolo, G.G., Recchiuto, C.T.: T.P.T. a novel Taekwondo personal trainer robot. Robot. Autonom. Syst. 83, 150–157 (2016). https://doi.org/10.1016/j.robot.2016.05.009.
10. Molfino, R., Muscolo, G.G., Puig, D., Recchiuto, C.T., Solanas, A., Williams, A.M.: An embodied-simplexity approach to design humanoid robots bioinspired by taekwondo athletes. In: Natraj, A., Cameron, S., Melhuish, C., Witkowski, M. (eds.) TAROS 2013. LNCS (LNAI), vol. 8069, pp. 311–312. Springer, Heidelberg (2014). https://doi.org/10.1007/978-3-662-43645-5_34
11. Ismar, E., Bahadir, S.K., Kalaoglu, F., Koncar, V.: Futuristic clothes: electronic textiles and wearable technologies. Global Challenges. 4(7), 1900092 (2020). https://doi.org/10.1002/gch2.201900092.
12. Muscolo, G.G., Moretti, G., Cannata, G.: SUAS: a novel soft underwater artificial skin with capacitive transducers and Hyperelastic membrane. Robotica **37**(4), 756–777 (2019)
13. Chi, E.H.: Introducing wearable force sensors in martial arts. IEEE Pervasive Comput. 4(3), 47–53 (2005)
14. Cho, E.-H., Eom, H.-J., Jang, S.-Y.: Comparison of patterns of skill actions between analog and electronic protectors in taekwondo: a log-linear analysis. Int. J. Environ. Res. Public Health **17**, 3927 (2020). https://doi.org/10.3390/ijerph17113927

15. Pyciarz, T.: Analysis of sport fight structure in taekwondo during the Olympics in Beijing in 2008 and Senior World Championships in 2009 in terms of technical skills after regulation amendments and implementation of the electronic system of score recording. J. Combat Sports Martial Arts **2**(2), 109–115 (2011)
16. Sevinç, D., Çolak, M.: The effect of electronic body protector and gamification on the performance of taekwondo athletes. Int. J. Perform. Anal. Sport **19**(1), 110–120 (2019). https://doi.org/10.1080/24748668.2019.1570457
17. Tasika, N.: Reliability & linearity of an electronic body protector employed in taekwondo games: a preliminary study. J. Hum. Sport Exerc. **8**(3), S622–S632 (2013). https://doi.org/10.4100/jhse.2013.8.Proc3.08
18. Razman, R., Chong, R.W.L.: Reliability and validity of a taekwondo electronic body protector. Proc IMechE Part P: J. Sport. Eng. Technol. **233**(2), 202–209 (2019). https://doi.org/10.1177/1754337118815009
19. Perry, A., Lee, J.: Satisfaction with current martial arts' uniforms and purchase intention of new uniforms. Fash. Text. **4**(1), 1–18 (2017). https://doi.org/10.1186/s40691-016-0085-6
20. Gorissen, B., Reynaerts, D., Konishi, S., Yoshida, K., Kim, J.-W., De Volder, M.: Elastic inflatable actuators for soft robotic applications. Adv. Mater. **29**, 1604977 (2017). https://doi.org/10.1002/adma.201604977
21. Nguyen, P.H., Zhang, W.: Design and computational modeling of fabric soft pneumatic actuators for wearable assistive devices. Sci. Rep. **10**, 9638 (2020). https://doi.org/10.1038/s41598-020-65003-2
22. Moseley, P., Florez, J.M., Sonar, H.A., Agarwal, G., Curtin, W., Paik, J.: Modeling, design, and development of soft pneumatic actuators with finite element method. Adv. Eng. Mater. (2015). https://doi.org/10.1002/adem.201500503
23. Bottero, S., Muscolo, G.G., Ferraresi, C.: A new Soft RCC device with pneumatic regulation. Robotics. 98(9), 4 (2020). https://doi.org/10.3390/robotics9040098.
24. Guler, M.A., Cerit, M.E., Bayram, B., Gerçeker, B., Karakaya, E.: The effect of geometrical parameters on the energy absorption characteristics of thin-walled structures under axial impact loading. Int. J. Crashworthiness **15**(4), 377–390 (2010). https://doi.org/10.1080/13588260903488750
25. Ramazanoglu, N.: Transmission of impact through the electronic body protector in taekwondo. Int. J. Appl. Sci. Technol. 3(2), 1–7 (2013).
26. Das, R., Kumar, R., Banerjeea, S.L., Kundu, P.P.: Engineered elastomeric bio-nanocomposites from linseed oil/organoclay tailored for vibration damping. R. Soc. Chem. RSC Adv. **4**, 59265–59274 (2014)
27. Park, K.J., Song, B.B.: Injuries in female and male elite taekwondo athletes: a 10-year prospective, epidemiological study of 1466 injuries sustained during 250 000 training hours. Br. J. Sports Med. **52**, 735–740 (2018). https://doi.org/10.1136/bjsports-2017-097530

# New Sustainable Biped-Wheeled Exoskeleton Prototypes

Alessandro Loschi[1], Alessandro Smerchinich[1], and Giovanni Gerardo Muscolo[2(✉)]

[1] Politecnico di Torino, Turin, Italy
[2] Department of Computer Science, University of Verona, Verona, Italy
giovannigerardo.muscolo@univr.it

**Abstract.** This paper presents two versions of a new biped-wheeled prototype developed in a reduced scale using a 3D printer. The results of this paper allow to confirm the hypothesis of feasibility of a biped exoskeleton with wheeled feet. Recent works underlined the convenience to use wheels on feet for simplifying the control of biped locomotion. The new research field of including wheels in exoskeletons has been proposed in recent works and this paper confirms the feasibility of using wheels at feet for wearable human's supports. These results open new perspectives for the development of a real biped-wheeled platform for people without motion capabilities or with reduced mobility, increasing the sustainability in locomotion systems.

**Keywords:** SDG03 · SDG09 · SDG11 · Exoskeleton · Wearable personal vehicle · Wearable biped robot

## 1 Introduction

In recent years, the robotics world is focusing on developing robots which are in charge of perform different tasks such as obstacle avoidance, locomotion in unconventional scenarios and others [1]. Some examples of very advanced robots can be observed in platforms developed by Boston Dynamics® [2], producers of legged and biped robots. A very interesting field of the biped robotic research are the exoskeletons [3]. These types of robots may be "weared" by human. The purpose of using these machines is to help the user in many applications such us to pick and place loads in different working conditions [4, 5], to stand still and walk, even if the user is affected by some motor or cognitive disabilities, or to increase the strength of the human, helping him to perform tiring works [6, 7]. Some examples of these machines can be found in BLEEX, a lower extremity Exoskeleton developed by the University of Berkeley [8–10] and many others available in literature or on the market [11, 12]. In general, exoskeletons are biped machines like wearable humanoid robots, and not many differences in control and design could be underlined in complete exoskeletons with trunk, arms, and legs with respect to complete humanoid robots [13, 14]. With this preface, the authors of this paper would like to consider an exoskeleton like a humanoid robot weared by the user and employed by him as a personal vehicle [15, 16]. For the knowledge of the


G. Quaglia et al. (Eds.): I4SDG 2021, MMS 108, pp. 158–165, 2022.
https://doi.org/10.1007/978-3-030-87383-7_17

authors, not many biped wheeled humanoids were shown in literature five years ago and now some researchers started to use the wheels to bypass dynamic balance problems in biped robots. Muscolo and Recchiuto were some pioneers of this research field [17–19]. In the same direction, it is important to underline the novelty to include wheeled feet in the exoskeleton research field as presented by some authors of this paper in recent works [20–22]. In this paper, two versions of a biped-wheeled exoskeleton have been developed in a reduced scale and some simulations and experimentations to underline the novelty and feasibility of the biped-wheeled system are presented. The prototypes are controlled using Arduino®, which is becoming one of the most employed board used for testing prototypes in electronics, robotics, control, and similar engineering applications [23–25].

The paper is structured as follows: Sect. 2 shows the design of the biped-wheeled model; Sect. 3 shows the developed prototypes; Sect. 4 shows the implemented control system; Sect. 5 presents the performed simulation and tests. The paper ends with conclusion.

## 2 Biped-Wheeled Model

Figure 1 shows the sketch of the biped-wheeled exoskeleton model without wheels (left and center), and the schematic drawing of a developed prototype (on the right). The working capabilities of the system have been analyzed in other works [20–22], and the motion is performed actuating the driving wheels which permit to move the blocks 1 and 2 of the biped-wheeled exoskeleton model (see Fig. 1, center, case B). Two cases can be distinguished: case A in which the robot is in a straight position and case B in which the robot is walking. In the case A, the balancing condition is performed actuating all the seven actuators: two revolute joints in each leg which connect the links 1–4, 4–3, 2–5, 5–3, one linear joint which connects links 3–6 and two linear joints for links 1-ground and 2-ground. In the case B, the balancing condition may be performed without actuating all joints. Thanks to the design of the machine, the revolute joints which connect links 1–4 and 2–5 may be used in passive or active mode. The linear joints 1-ground and

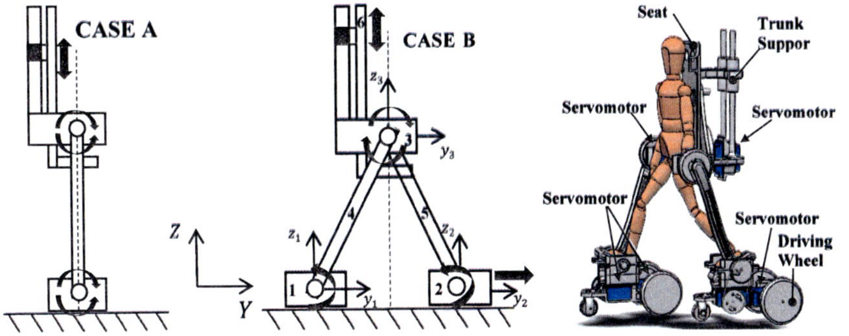


**Fig. 1.** Sketches of the biped-wheeled exoskeleton model without wheels [20–22] (left and center) and the schematic drawing of the second biped-wheeled exoskeleton prototype (right).

2-ground in the schemes of the two cases represent the linear motion of the centers of the right and left wheels in contact with the ground, as can be shown in the schematic drawing (Fig. 1, right) and in the two realized prototypes (see Figs. 2a and 2b).

## 3 Biped-Wheeled Exoskeleton Prototypes in a Reduced Scale

The prototypes have been manufactured using a 3D printer. They are conceived to support a reduced scale of a 200–300 mm tall dummy during walking [22]. Figure 2a shows a first version and Fig. 2b shows the second version of the biped-wheeled exoskeleton prototype.

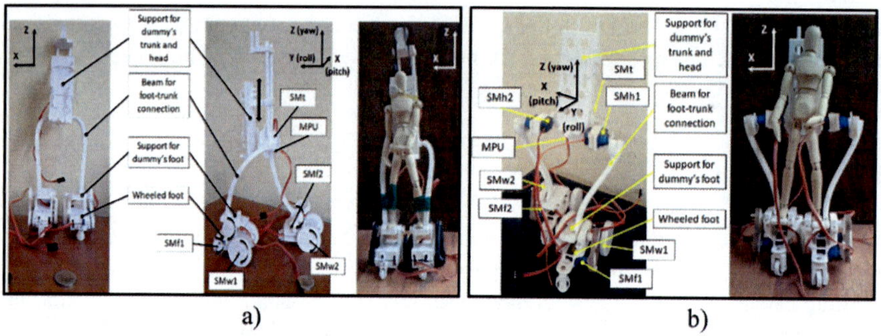


**Fig. 2.** Biped-Wheeled Exoskeleton prototype: a) first version with the servomotors (SMw1, SMw2, SMf1, SMf2, SMt) and the accelerometer (MPU); b) second version with the servomotors (SMw1, SMw2, SMf1, SMf2, SMh1, SMh2, SMt) and the accelerometer (MPU).

The prototypes are composed by wheeled feet attached to the dummy feet (as rollerblades in humans) and connected with the exoskeleton trunk (using beams) attached to the dummy's trunk. The first prototype is constituted by five servomotors (SMs) shown in Fig. 2a: two for the wheeled motion (SMw1 and SMw2), two for the foot revolution (SMf1 and SMf2), and one for the trunk motion (SMt) for adaptation of dummies with different tall and to allow motion perpendicular to the ground during walking. The second prototype has a total of seven servomotors (SMs), adding other two servomotors in the hip (SMh1 and SMh2) as shown in Fig. 2b. The robot may support the dummy to stay in a straight position and move it using only SMw1 and SMw2 and using the balancing performed thanks to the developed structure shown in the case B (see Fig. 1) or may allow a human-like motion actuating also feet. In the first version, the mechanical system which should control the height of the back, has too much friction to move as expected. In particular, during the preliminary design, we analyzed the causes of friction that could be avoided in future development of prototypes. In the first version prototype, the following aspects are underlined as causes of friction and solved in the second version prototype: 1) the cylindrical guide in the rear part of the seat, solved increasing the length of the cylinders and reducing friction on the shaft (see Figs. 1 and 2); 2) the length of the guide, solved reducing length; 3) the shaft of the pulley, solved eliminating the pulley; 4) the belt and the pulley, solved using another type of transmission. Other aspects could be solved increasing precision and modifying the material used for the prototypes. The problems noted in the first version have been taken into account for developing the second prototype.

## 4 Control Design

Figure 3 left shows the Arduino Mega® board used to actuate the control for the two prototypes taken into consideration. Since the structure of the prototypes is composed only by servomotors (SMs), the PWM controller integrated in the board is enough to control the actuators, and no external drivers are needed. To retrieve motion data, a generic accelerometer MPU (Motion Processing Unit) fits well for the purpose of this work. Since the exoskeletons must be controlled by some commands as input, an Android App has been developed, in order to establish a Bluetooth communication between the phone and the robot (see Fig. 3, right). This allows to create routines, depending on the received command, which can be employed to actuate the robot to perform different type of tasks. The App is in charge of sending control commands to the Bluetooth module, connected to Arduino®. It is composed of 8 soft-keys (u (up), d (down), w (walking), C (chair), l (left), f (forward), b (backward), r (right)) and two text fields (x and y) (see Fig. 3 right). The two text fields have been employed to set the steps period and the range of rotation of the servomotors at run-time: the desired number can be written in the field, and when the corresponding soft-key is pressed, the phone sends the string after the usual character. To test the app a simple code has been used, which is able to print data received by the Bluetooth module on the serial monitor. The used servomotors are the micro servo motor SG90 with the following characteristics: torque 1.6 kg-cm (0.1569 Nm), weight 9 g, voltage 3.0–7.2 V.

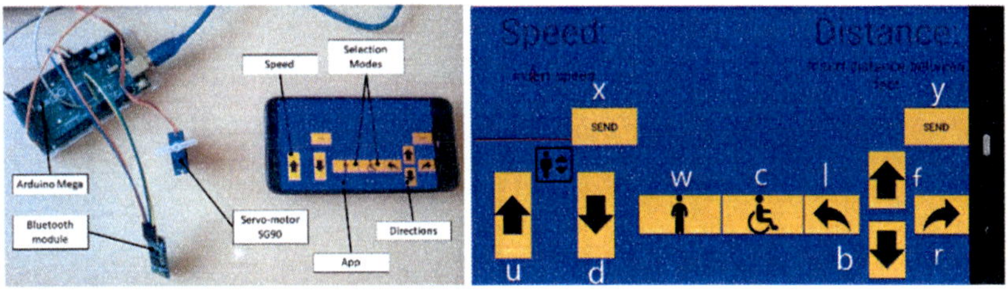


**Fig. 3.** Left: developed App, Arduino Mega®, Servomotor SG90, Bluetooth module. Right: Android Application.

## 5 Simulation, Experimentation and Discussion

Figure 4 shows the input angles of the gait cycle used for simulation using a multibody environment. In particular, the input of the wheel and the legs (left and right) are shown with two angular displacements of the seat actuator: fixed and mobile in order to evaluate the influence of the seat motion. In Fig. 5, the output of the simulation is shown. The higher value of torque is noted on the leg rod.

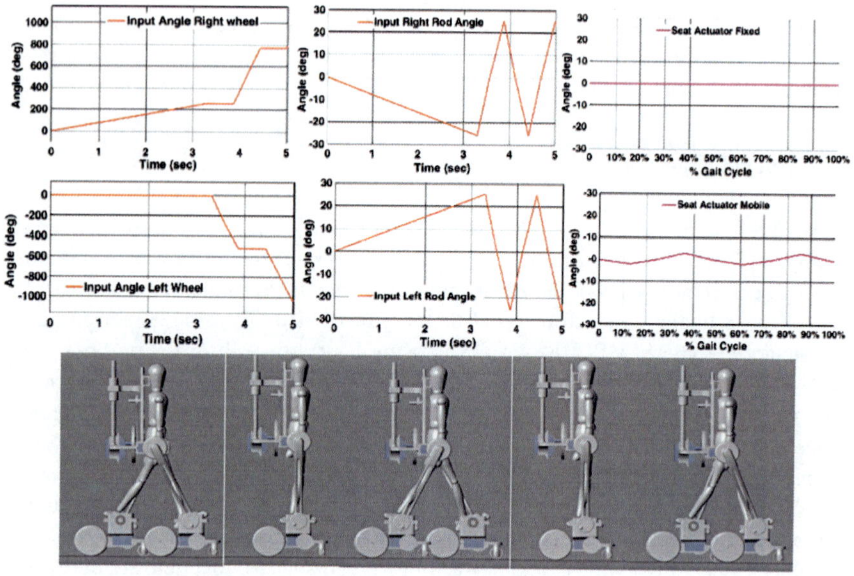


**Fig. 4.** Input angles used for the gait cycle. From the left to right: above) input angle of the right wheel, right leg rod and the actuator of the seat without perturbations; below): input angle of the left wheel, left leg rod and the actuator of the seat if a perturbation is performed.

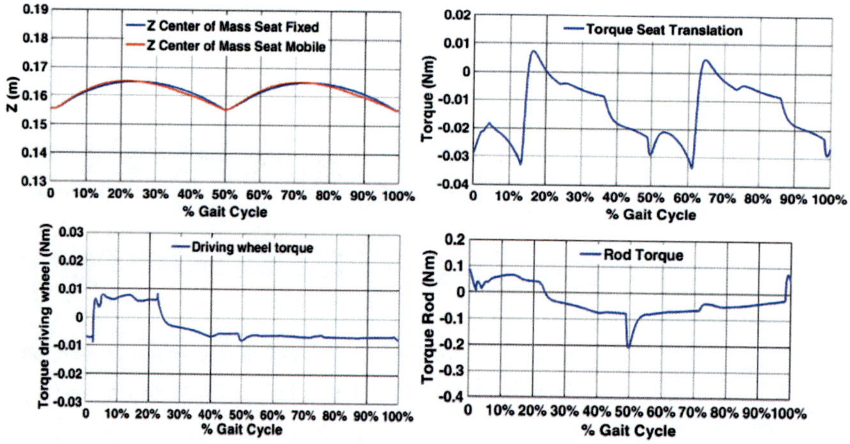


**Fig. 5.** Output of simulation. From left to right: above) center of mass position, torque of the seat; below) torque of the wheel and rod leg.

In order to test the exoskeletons, a control strategy has been developed. This strategy consists of using the command forward (f) and backward (b) of the application to apply one or more degrees of rotation to servomotors at every input (pushing the buttons of the application). The kinematic input is conceived to move the legs in opposite directions.

This operation allows to verify if the range of motion of the legs is wide enough to perform a complete walk, and to understand the range of motion to be applied to servomotors. The test consists of moving the legs in an alternate way at the same time. The algorithm allows to move forward the right leg while moving backward the left one when forward command (f) is received. When the backward command (b) is received, the legs are moved following the opposite behavior. The second prototype has two servos at the ankle, and this allows to have a higher range of motion for the legs. In Fig. 6 on the left, it can be clearly seen that the structure of the first prototype fails sometimes during the test. The problem can be observed when the pitch falls from 80° to 40. The robot cannot even survive for an entire step. This problem could be associated to some major factors: 1) the joint axis of the hip of the dummy and the biped-wheeled exoskeleton are not coincident including more resistance during motion and increasing unbalancing; 2) the support for dummy's foot is always perpendicular to the beam for the foot-trunk connection increasing unbalancing during motion; 3) the altitude of the total center of mass of the prototype increases unbalancing; 4) the friction of the seat produces unbalancing; 5) the passive joints of the hip of the biped-wheeled exoskeleton add friction. Applying the same control to the second prototype, where these and other problems are solved, good results have been obtained for this type of test. The robot never falls making some steps, as it can be observed in Fig. 6 on the right by the fact that the oscillations are all included in the range (−15, 10) Degrees. The graphs demonstrate that the second prototype is driven in a correct way, since the read data are equal to the expectations.

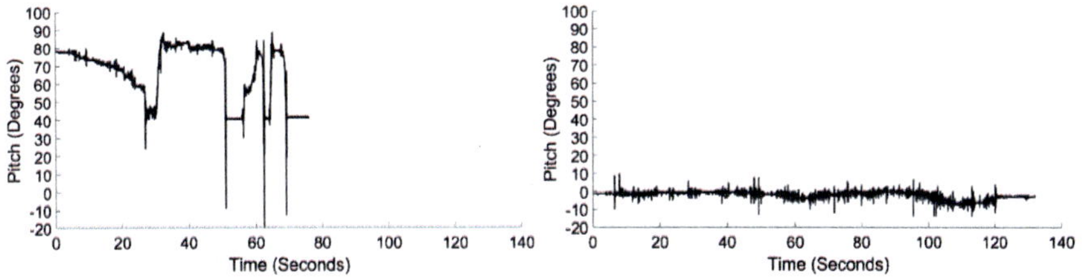


**Fig. 6.** Accelerometer (MPU) output for the test: first (on the left) and second (on the right) version prototype.

## 6 Conclusion

The two developed prototypes in reduced scale are respectively conceived with five and seven servomotors. The research work underlined the convenience to use seven servomotors for having the biped-wheeled exoskeleton most able to reproduce the human movements. Future works will be focused on the design and development of the full prototype in the real scale in order to be able to plan experiments on dummies with the same human's dimensions. The final objective is to develop a new biped-wheeled exoskeleton able to support people affected by complete paralysis and without motion capabilities. The project of the biped-wheeled exoskeleton named in other works as "HANDSHAKE" contributes to three sustainable development goals: 1) good health and well-being (SDG03), for allowing the biped locomotion to people without motion capabilities; 2) industry, innovation and infrastructure (SDG09), for the innovation proposed with the novel biped-wheeled exoskeleton concept and prototypes; 3) sustainable cities and communities (SDG11), for the novel conceptual system proposed which contributes to have cities and communities more sustainable allowing the use of electrical vehicles.

## References


1. Castano, A., Behar, A., Will, P.M.: The Conro modules for reconfigurable robots. IEEE/ASME Trans. Mechatron. **7**(4), 403–409 (2002)
2. Nelson, G., Saunders, A., Playter, R.: The PETMAN and atlas robots at Boston dynamics. Humanoid Robot. **169**, 186 (2019).
3. Walsh, C.J., Paluska, D., Pasch, K., Grand, W., Valiente, A., Herr, H.: Development of a lightweight, underactuated exoskeleton for load-carrying augmentation. In: Proceedings of the IEEE ICRA 2006, pp. 3485–3491 (2006).
4. Panero, E., Muscolo, G.G., Pastorelli, S., Gastaldi, L.: Influence of hinge positioning on human joint torque in industrial trunk exoskeleton. Mechan. Mach. Sci. **73**, 133–142 (2019)
5. Panero, E., Muscolo, G.G., Gastaldi, L., Pastorelli, S.: Multibody analysis of a 3D human model with trunk exoskeleton for industrial applications. Comput. Method Appl. Sci. **53**, 43–51 (2020)
6. Magnetti Gisolo, S., Muscolo, G.G., Paterna, M., De Benedictis, C., Ferraresi, C.: Feasibility study of a passive pneumatic exoskeleton for upper limbs based on a mckibben artificial muscle. In: Zeghloul, S., Laribi, M.A., Sandoval, J. (eds.) RAAD 2021. MMS, vol. 102, pp. 208–217. Springer, Cham (2021). https://doi.org/10.1007/978-3-030-75259-0_23
7. Fontana, M., Vertechy, R., Marcheschi, S., Salsedo, F., Bergamasco, M.: The body extender: a full-body exoskeleton for the transport and handling of heavy loads. IEEE Robot. Autom. Mag. **21**(4), 34–44 (2014)
8. Kazerooni, H., Steger, R., Huang, L.: Hybrid control of the berkeley lower extremity exoskeleton (BLEEX). Int. J. Robot. Res. **25**(5-6), 561–573 (2006)
9. Kazerooni, H., Racine, J.L., Huang, L., Steger, R.: On the control of the berkeley lower extremity exoskeleton (BLEEX). In: IEEE International Conference on Robotics and Automation, pp. 4353–4360 (2005).
10. Zoss, A.B., Kazerooni, H., Chu, A.: Biomechanical design of the berkeley lower extremity exoskeleton (BLEEX). IEEE/ASME Trans. Mechatron. **11**(2), 128–138 (2006). https://doi.org/10.1109/TMECH.2006.871087
11. Banala, S.K., Kim, S.H., Agrawal, S.K., Scholz, J.P.: Robot assisted gait training with active leg exoskeleton (ALEX). In: IEEE/RAS-EMBS International Conference on Biomedical Robotics and Biomechatronics, pp. 653–658 (2008).
12. Perry, J.C., Rosen, J., Burns, S.: Upper-limb powered exoskeleton design. IEEE/ASME Trans. Mechatron. **12**(4), 408–417 (2007). https://doi.org/10.1109/TMECH.2007.901934
13. Park, I.W., Kim, J.Y., Lee, J., Oh, J.H.: Mechanical design of humanoid robot platform KHR-3. In: IEEE/ASME International Conference on Advanced Intelligent Mechatronics (AIM), pp. 530–535 (2013).
14. Kaneko, K., Harada, K., Kanehiro, F., Miyamori, G., Akachi, K.: Humanoid Robot HRP-3. In: IEEE/RSJ International Conference on Intelligent Robots and Systems, pp. 2471–2478 (2008).
15. Michaud, F., et al.: Multi-modal locomotion robotic platform using leg-track-wheel articulations. Autonomous Robot. **18**(2), 137–156 (2005). https://doi.org/10.1007/s10514-005-0722-1

16. Kamen, D.L., Dastous, S.D., Duggan, R., Guay, G.M.: Fault tolerant architecture for a personal vehicle. In: US Patent Document. 6,223,104. pp. 1–3 (2001).
17. Muscolo, G.G., Recchiuto, C.T.: Flexible structure and wheeled feet to simplify biped locomotion of humanoid robots. Int. J. Humanoid Rob. **14**(1), 1–26 (2017)
18. Muscolo, G.G., Recchiuto, C.T.: T.P.T. a novel Taekwondo personal trainer robot. Robot. Autonom. Syst. **83**, 150–157 (2016).
19. Lisitano, D., Bonisoli, E., Recchiuto, C.T., Muscolo, G.G.: Dynamic balance of the head in a flexible legged robot for efficient biped locomotion. Appl. Sci. **11**, 2945 (2021). https://doi.org/10.3390/app11072945
20. Zoccali, A., Muscolo, G.G.: Comfort perception analysis of human models interfacing with novel biped-wheeled-exoskeletons. In: Rauter, G., Cattin, P.C., Zam, A., Riener, R., Carbone, G., Pisla, D. (eds.) MESROB 2020. MMS, vol. 93, pp. 21–28. Springer, Cham (2021). https://doi.org/10.1007/978-3-030-58104-6_3
21. Trono, G., Nicolì, A., Muscolo, G.G.: Sustainable compliant physical interaction in a biped-wheeled wearable machine. Front. Mech. Eng. **6**, 581626 (2020). https://doi.org/10.3389/fmech.2020.581626
22. Muscolo, G.G.: HANDSHAKE: handling system for human autonomous keeping. Int. J. Humanoid Rob. **18**(01), 2150003 (2021). https://doi.org/10.1142/S0219843621500031
23. Akbar, M.A.: Simulation of fuzzy logic control for dc servo motor using arduino based on Matlab/Simulink. In: International Conference on Intelligent Autonomous Agents, Networks and Systems, pp. 42–46 (2014).
24. Juang, H.S., Lum, K.Y.: Design and control of a two-wheel self balancing robot using the arduino microcontroller board. In: 10th IEEE International Conference on Control and Automation (ICCA), pp. 634–639 (2013).
25. Al-Busaidi, A.M.: Development of an educational environment for online control of a biped robot using MATLAB and arduino. In: 9th France-Japan and 7th Europe-Asia Congress on Mechatronics (MECATRONICS) / 13th Int'l Workshop on Research and Education in Mechatronics (REM), pp. 337–344 (2012).

# Preliminary Experimental Test of a Cable-Driven Wheelchair in Different Configurations

Paride Cavallone(✉), Andrea Botta, Elvio Bonisoli, and Giuseppe Quaglia

Politecnico Di Torino, 10129 Turin, Italy
{paride.cavallone,andrea.botta,elvio.bonisoli,
giuseppe.quaglia}@polito.it

**Abstract.** The paper presents the experimental tests of a manual wheelchair, named Handwheelchair.Q, that utilizes an innovative propulsion system composed of a system of cables and pulleys and a gesture inspired by the rowing motion. This prototype has been developed in order to increase the autonomy of the wheelchair users and to facilitate the practice of physical activity and sport. The aim of this paper is to evaluate the efficiency of this prototype in three different configurations. In the experimental trials, the wheel angular speeds and the left and right hand forces exerted by the user are monitored and presented. From the obtained results, the efficiency of each configuration and its specific characteristics are discussed.

**Keywords:** SDG3 · Manual wheelchair · Spinal cord injury

## 1 Introduction

Physical activity and exercise are important tools to be in good health physically [1] and psychologically [2]. They can significantly reduce the risk of several disease such as heart disease, diabetes, some type of cancer and so on [3]. In addition, physical activity and sports are in general important tools for social integration [4]. These considerations are particularly valuable for disabled people. In fact, the "Good Health and Well-being" SDG (SDG 3) aims "... to ensure healthy lives and promote well-being for all at all ages" [5]. In this scenario, a research group from Politecnico di Torino designed an innovative manual wheelchair, named Handwheelchair.Q [6–9] with the goal to increase the autonomy of the wheelchair users and to practice physical activity easily. The prototype is based on a commercial frame of a standard wheelchair on which is mounted an innovative propulsion system composed of a system of cables and pulleys and a gesture inspired by the rowing motion. The standard wheelchairs are provided with two rear wheels and one or two smaller pivoting wheels in the front. The small front wheels are a good solution in terms of lightness and firmness when the wheelchair is employed indoor on a flat and smooth surface. On the contrary, for outdoor use on roads, cycle paths and so on, the front wheels represent a significant limitation, due to their limited effectiveness in moving on uneven surfaces. In addition, the efficiency of the wheels with a small radius is low [10, 11].


G. Quaglia et al. (Eds.): I4SDG 2021, MMS 108, pp. 166–173, 2022.
https://doi.org/10.1007/978-3-030-87383-7_18

For this reason, there are different commercial solutions proposing a front pivoting wheel with a radius larger than the standard one, in order to solve this problem. The "Easywheel" (www.offcarr.com) and the "Freewheel" (www.gofreewheel.com) are two commercial devices that can be mounted on the standard wheelchairs.

The aim of this paper is to evaluate the efficiency of this prototype with or without the additional front wheel. In the experimental trials, the wheel angular speeds and the left and right hand forces exerted by the user are monitored and presented. From the obtained results, the efficiency of each configuration and its specific characteristics are discussed.

## 2 Handwheelchair.Q

Handwheelchair.Q is a prototype of manual wheelchair in which the propulsion motion is inspired by rowing motion. In the active phase the user pulls the cable using the handles to apply torque to the wheels. In the recovery phase the user returns to the initial configuration thanks to a rewinding elastic element. In Fig. 1 two frames during the active phase are shown.

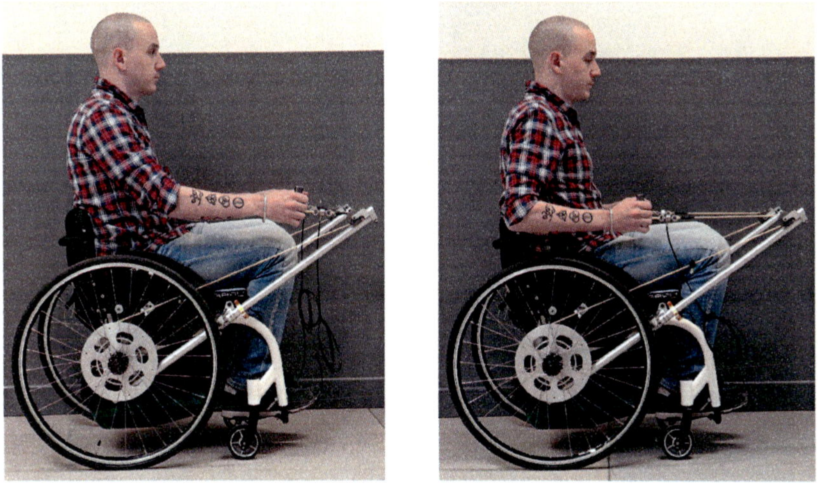

(a) (b)

**Fig. 1.** Handwheelchair.Q at the beginning (a) and at the end (b) of the active phase.

In these preliminary tests, the Handwheelchair.Q was tested in three different front wheels configurations. In the first configuration, Fig. 2, named "Standard", the wheelchair was provided with two pivoting wheels with a small radius. In the second and the third configuration, respectively Fig. 3 and 4, the wheelchair was provided with two different commercial pivoting wheels, named respectively "Easywheel" and "Free-wheel". Both commercial pivoting wheels have a radius higher than the pivoting wheels of the Standard configuration as reported in Table 1.

In Fig. 2b, 3b and 4b the functional design of the three configurations are reported.

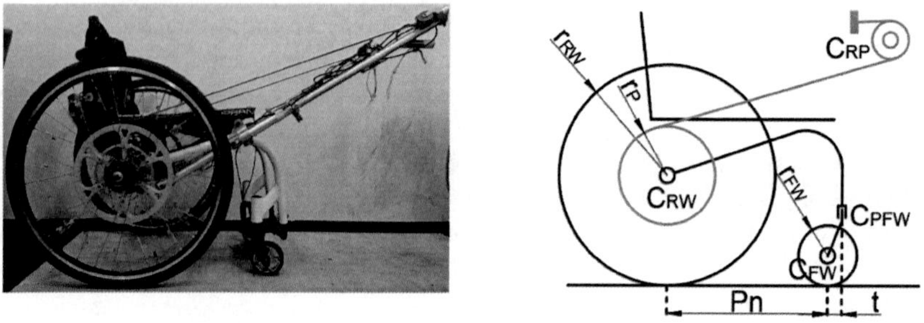


(a) (b)

**Fig. 2.** Handwheelchair.Q standard configuration, (a) photo and (b) functional design.

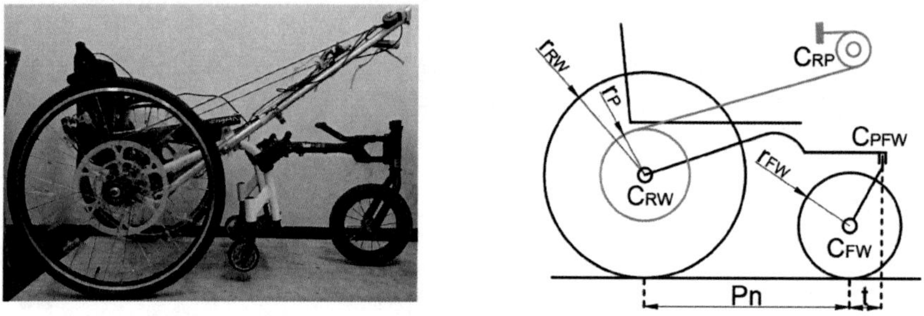


(a) (b)

**Fig. 3.** Handwheelchair.Q “Easywheel” configuration, (a) photo and (b) functional design.

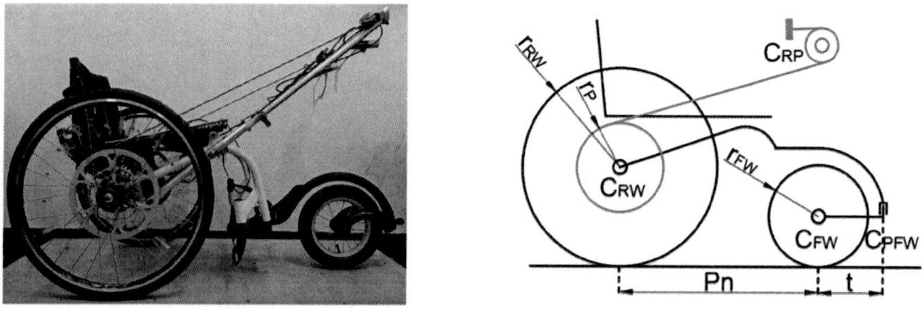


(a) (b)

**Fig. 4.** Handwheelchair.Q “Freewheel” configuration, (a) photo and (b) functional design.

**Table 1.** Main paramaters of Handhweelchair.Q in different configurations.

| Configuration | Mass [kg] | $r_{Fw}$ [m] | *Pn* [m] | *t* [m] |
|---|---|---|---|---|
| Standard | 15.5 | 0.045 | 0.4 | 0.035 |
| Easywheel | 19.0 | 0.145 | 0.84 | 0.08 |
| Freewheel | 17.5 | 0.155 | 0.72 | 0.18 |

# 3 Methods

## 3.1 Subject

One able-bodied non-wheelchairs user (29 years old, 170 cm, 65 kg) participated in the experiment. This experimental test was a preliminary analysis about the efficiency of a manual wheelchair, hence nobody else was involved. The limitation of having just one able-bodied male user is clear, however, each test was repeated ten times in order to obtain more fair results. Prior to the trials, the participant provided written informed consent. The ethical approval was granted, and the work was performed in accordance with the Declaration of Helsinki.

## 3.2 Equipment

Two load cells connected the handles with the traction cable to measure the user input forces $F_R$ (right hand) and $F_L$ (left hand). Two Hall sensor, one for each rear wheel, detects the frequency, $f_R$ (right wheel) and $f_L$ (left wheel), of sixteen equidistant magnets positioned on each wheel. The angular speed of right and left wheels, $\omega_R$ and $\omega_L$, were obtained by the Eq. 1. The wheelchair speed $V$ is obtained by the Eq. 2.

$$\omega_R = f_R\frac{\pi}{8};\ \omega_L = f_L\frac{\pi}{8} \tag{1}$$

$$V = \frac{\omega_R + \omega_L}{2} r_{Rw} \tag{2}$$

## 3.3 Data Acquisition

The input hand forces, $F_R$ and $F_L$, and the frequencies of rotation, $f_R$ and $f_L$, were recorded with an acquisition device, National Instrument USB-6341. The signals were synchronized with the Analogical Input Sample clock with a 1000 Hz sampling frequency.

## 3.4 Experimental Tests

The experimental tests consisted of propelling from standing start Handwheelchair.Q on a flat hallway covered with smooth flooring. The user was asked to drive the wheelchair at approximately 1 m/s. For each configuration, the test was performed 10 times. The propulsion motion is composed of two phases: an active phase and a recovery phase, as shown in Fig. 5. The active phase occurs when the force $F$, sum of $F_R$ and $F_L$, is grather than $F_{trig}$ while the recovery phase occurs when the force $F$ is less than $F_{trig}$. The value $F_{trig}$ represents the condition in which the ratchet mechanism inserted in the wheels enable the propulsion or active phase. The force $F1$ is the average force during the complete cycle and $F2$ is the average force during the active phase. Starting from a standstill, the wheelchair first has an acceleration phase, not reported, and then reaches a condition of periodic steady state phase. In the periodic steady state phase the wheelchair speed, $V$, oscillates around the average value, $V_{mean}$. In the active phase, the maximum and the minimum of $V$ has been identified to define $\Delta V$, namely the difference between $V_{max}$ and $V_{min}$. In addition, the input power $PI$ was calculated as follow:

$$PI = Fr_P \frac{V}{r_{Rw}} \quad (3)$$

In particular, *PI1* is the average power during the complete cycle while *PI2* during the active phase. Lastly, the input energy *EI* for each propulsion cycle was calculated as follow:

$$EI = PI2T_A = PI1(T_A + T_R) \quad (4)$$

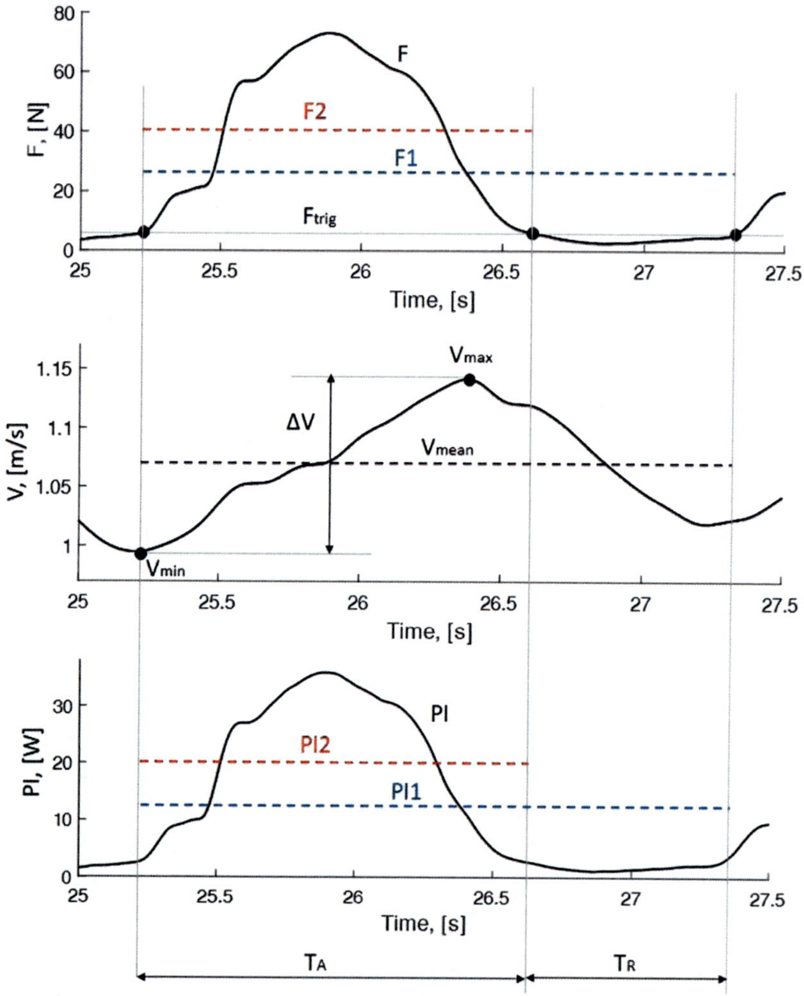


**Fig. 5.** Force, wheelchair speed and input power data in a single cycle.

## 4 Results and Discussions

For each cycle, the variables *F1*, *F2*, *V*, $T_A$, $T_R$, $\Delta V$, *PI1* and *PI2* were calculated and then the average values (and the standard deviation) were reported in Table 2 for each configuration of Handhweelchair.Q.

**Table 2.** Data (SD).

| | Freewheel | Standard | Easywheel |
|---|---|---|---|
| $V$ [m/s] | 0.98 (0.03) | 1.02 (0.02) | 1.09 (0.04) |
| $\Delta V$ [m/s] | 0.030 (0.005) | 0.033 (0.004) | 0.027 (0.005) |
| *F1* [N] | 27.13 (2.22) | 29.51 (2.76) | 25.39 (3.51) |
| *F2* [N] | 42.43 (0.81) | 43.72 (1.17) | 41.24 (1.43) |
| *TA* | 1.47 (0.05) | 1.54 (0.05) | 1.37 (0.06) |
| *TR* | 0.83 (0.02) | 0.74 (0.02) | 0.85 (0.05) |
| *PI1* [W] | 11.78 (0.58) | 13.39 (0.65) | 12.34 (0.91) |
| *PI2* [W] | 18.43 (2.07) | 19.84 (2.30) | 20.05 (3.56) |
| *EI* [J] | 27.09 | 30.55 | 27.46 |

In Fig. 6, the three configurations are compared in terms of force. As expected in the standard configuration the forces *F1* and *F2* are greater than the Freewheel and Easywheel configuration because the wheelchair efficiency is lower, due to rolling resistance.

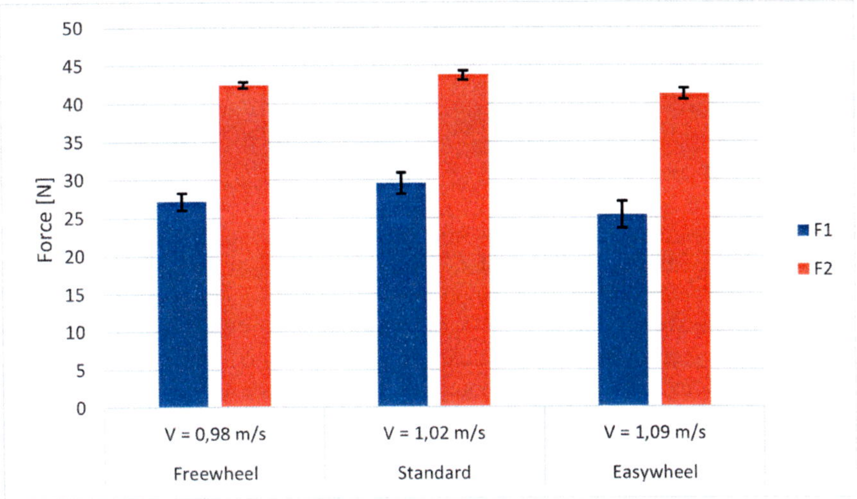


**Fig. 6.** Force comparison.

The wheelchair speed of the Easywheel configuration is the highest but the forces are the lowest.

In Fig. 7, the three configurations are compared in terms of input power. The *PI1*, namely the input power in the active phase, increases with increasing of the wheelchair speed. The input power *PI2* for Freewheel configuration is the lowest and is higher for the standard configuration.

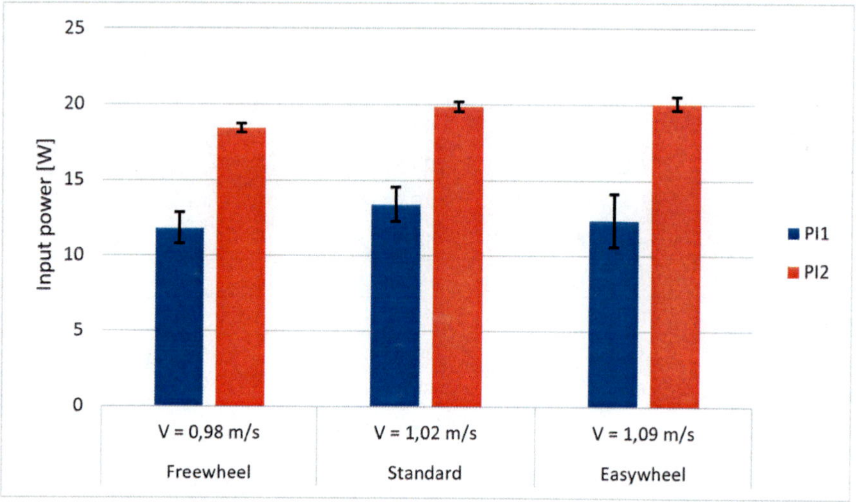


**Fig. 7.** Power comparison.

An additional comparison can be done examining the input energy, *EI*. The *EI* for Easywheel and Freewheel is approximately the same about 27 J instead of 30 J for the standard configuration. At the end, it is important to underline that the $\Delta V$ of Standard configuration is higher.

## 5 Conclusions

This preliminary test has compared the efficiency of the prototype Handwheelchair.Q in three different configurations. In absolute terms, the Easywheel configuration seems to be the best configuration in terms of efficiency in the specific conditions of the test even if it is the heaviest configuration. The inertia effect and the rolling friction of the front wheel have an effect on the global efficiency of the wheelchair. In addition, is important to point out that the technological characteristics of the front wheels have an important role in terms of efficiency of the wheelchair. Another important effect concerns the distribution of the normal force on the ground as shown in literature [10, 11].

This work represents the preliminary test of a wide experimental campaign that it will be executed in the future, where many aspects will be considered. First of all, the front wheels represent just one specific configuration, in addition the position of the return pulley, the radius of the pulley that defines the transmission ratio, and the position of the centre of mass of the user will be examined in detail. Secondly, specific tests will be carried out in order to optimise the efficiency of the transmission of motion and the biomechanical motion depending on the anthropometry characteristic of the user. A specific methodology will be defined in order to optimise the different parameters for each user. The above-mentioned aspects will be examined in different conditions as different speeds, different terrains (asphalt, dirt road, cycle path, smooth surface, …) and depending on the use: sport, light physical or recreational activities, speed up moving in around the city etc. In conclusion, the tests will involve different subject, disabled people especially.

## References


1. O'Brien, T.D., et al.: Well-being, health and fitness of children who use wheelchairs: feasibility study protocol to develop child-centred "keep-fit" exercise interventions. J. Adv. Nurs. **71**(2), 430–440 (2015)
2. Archer, T.: Physical exercise and its impact on psychology. Clin. Exp. Psychol. **2**, e104 (2016). https://doi.org/10.4172/2471-2701.1000e104
3. World Health Organization: The World Health Report 2002: Reducing Risks Promoting Healthy Life, p. 61. WHO, Geneva (2002)
4. Georgian, B., Lorand, B.: The influence of leisure sports activities on social health in adults. Springerplus **5**(1), 1–7 (2016). https://doi.org/10.1186/s40064-016-3296-9
5. https://www.globalgoals.org/3-good-health-and-well-being. Consulted on 24 May 2021
6. Cavallone, P., Bonisoli, E., Quaglia, G.: Handwheelchair.Q: new prototype of manual wheelchair for everyday life. In: Niola, V., Gasparetto, A. (eds.) IFToMM ITALY 2020. MMS, vol. 91, pp. 111–119. Springer, Cham (2021). https://doi.org/10.1007/978-3-030-55807-9_13
7. Cavallone, P., Bonisoli, E., Quaglia, G.: Prototyping of manual wheelchair with alternative propulsion system. Disabil. Rehabil. Assist. Technol. **15**(8), 945–951 (2020 Nov). https://doi.org/10.1080/17483107.2019.1629185. Epub 2019 Jun 28 PMID: 31250677
8. Quaglia, G., Bonisoli, E., Cavallone, P.: The design of a new manual wheelchair for sport. Machines **7**, 31 (2019). https://doi.org/10.3390/machines7020031
9. Quaglia, G., Bonisoli, E., Cavallone, P.: Handwheelchair.Q: innovative manual wheelchair for sport. In: Carbone, G., Gasparetto, A. (eds.) IFToMM ITALY 2018. MMS, vol. 68, pp. 370–378. Springer, Cham (2019). https://doi.org/10.1007/978-3-030-03320-0_40
10. Chan, F.H.N., Eshraghi, M., Alhazmi, M.A., Sawatzky, B.J.: The effect of caster types on global rolling resistance in manual wheelchairs on indoor and outdoor surfaces. Assist. Technol. **30**(4), 176–182 (2018). https://doi.org/10.1080/10400435.2017.1307880. Epub 7 Jun 2017 PMID: 28590160
11. Zepeda, R., Chan, F., Sawatzky, B.: The effect of caster wheel diameter and mass distribution on drag forces in manual wheelchairs. J. Rehabil. Res. Dev. **53**(6), 893–900 (2016). https://doi.org/10.1682/JRRD.2015.05.0074. PMID: 28475204

# Detecting Impulsive Movements to Increase Operators' Safety in Manufacturing

Valeria Rosso[1(✉)], Laura Gastaldi[2], and Stefano Pastorelli[1]

[1] Department of Mechanical and Aerospace Engineering, Politecnico Di Torino, Turin, Italy
valeria_rosso@polito.it
[2] Department of Mathematical Sciences, Politecnico Di Torino, Turin, Italy

**Abstract.** Interaction and collaboration of human and robot is becoming a quite common scenario, in manufacturing production. When an operator works with a robot, usually, controlled and predicable movements are used. However, the operator could execute unexpected impulsive movements, which may induce risks for his safety. The purpose of this methodological study is to identify such impulsive movements, by using an inertial sensor fixed on the wrist and by four features calculated on the gesture kinematics. The features were calculated on the measured linear acceleration and on the jerk using two epoch durations (100 ms and 250 ms). Then, features threshold that allow impulsive movements onset identification were evaluated. Results showed that the four features overall recognized impulsive movements during reaching task executed with different directions and cadences. Results showed also that calculating the features with shorter epoch duration gave better results compared to longer epoch duration. A combination of the four features calculated on short epoch duration would increase impulsive gesture detection, reducing occurrence of false positive. In conclusion, this methodology seems suitable to detect impulsive gestures and to intervene promptly with safety measures.

**Keywords:** SDG9 · Safety risks · Safety workplace · Reaching task · Unexpected gestures · Human-robot collaboration

## 1 Introduction

With the growing manufacturing production and high requirement for efficiency, human-robot collaboration is becoming a common scenario in the industry. Typical collaborative tasks are pick and place, reaching tasks, and assembly [1]. In such tasks, humans and robots share the workspace, therefore, operator's safety is mandatory. Collaborative robots requirements are ruled by standard regulations [2] and technical specifications [3]. However, risks can also come from the tasks itself [4]; therefore tasks recognition techniques have been developed [5].

When working with a robot, operator usually performs repetitive and controlled movements related to the task. Overall, controlled movements executed by the operator are characterized by predictable kinematics and dynamics. Consequently, human's


G. Quaglia et al. (Eds.): I4SDG 2021, MMS 108, pp. 174–181, 2022.
https://doi.org/10.1007/978-3-030-87383-7_19

movements can be detected using upper limbs gesture [6] and activity recognition algorithms [7] based, for example, on wrist-worn inertial sensor.

However, the operator could execute unexpected gestures, which fall outside the task nominal movements. Unexpected gestures could occur due to different factors, such as external disturbance, mental status of the operator in relation to environmental factors or distraction [8]. These movements are characterized by high variability and uncertainty in the execution pattern in terms of the number of human segments involved and range of segments kinematics and dynamics.

When performing unexpected movements operators may locate themselves on the robot trajectory, jeopardizing their own safety. Recent works present robotic systems able to anticipate human state of mind and unexpected behavior like lost attention or exhaustion [9]. However, to the best of authors' knowledge, the recognition of unexpected movements characterized as sudden and impulsive reactions, for example due to unexpected events, has not been investigated yet.

The purpose of the current study was to present a methodology to characterize operator's activity in terms of controlled movements, which are related to the task, and to detect impulsive movements, which are related to sudden perturbations of task execution. Then, this methodology would allow the identification of impulsive gestures onset, that can be used as a warning to prevent risks of injury for the operator.

## 2 Material and Methods

### 2.1 Material

To reproduce an industrial workstation, a test bench with five stations (5 × 5 cm) fixed on a rigid surface was used for the tests (Fig. 1). The stations 1 to 4 delimited the working area, whereas the station 5 is placed outside of the working space and the operator might reach it during the normal working activity. The five stations were located 40 cm apart from each other and the stations 3 to 5 were located 10 cm away from the edge of the rigid surface. The station 1 and 4 were located approximately at 100% and 30% of the participant upper limb length, respectively [10].

One healthy participant took part in the study. During the tests, the participant was seated in front of the working space, aligned with the middle line, with hip and knee joint angles equal to 90°. The participant's shoulders were relaxed and the trunk, held in a straight position, was 10 cm far from the test bench edge [11]. The test bench height was regulated to be 1 cm lower with respect to the participant's elbow to allow comfortable upper limb movements during the tests [12].

To collect the upper limb kinematics, one inertial measurement unit (Xsens MTx Motion Tracker, Xsens Technologies, The Netherlands) was firmly fixed to the left wrist of the participant (Fig. 1a) with an elastic band. The inertial unit was oriented with the X axis parallel with the longitudinal arm axis and the Y axis parallel to the medio-lateral wrist axis. The inertial unit was composed of triaxial accelerometer (full scale ± 50 m/s$^2$) and triaxial gyroscope (full scale ± 1200°/s). Data was recorded with a sample frequency of 100 Hz with the MT Manager software (Xsens Technologies, The Netherlands) via Bluetooth and was stored on a personal computer.

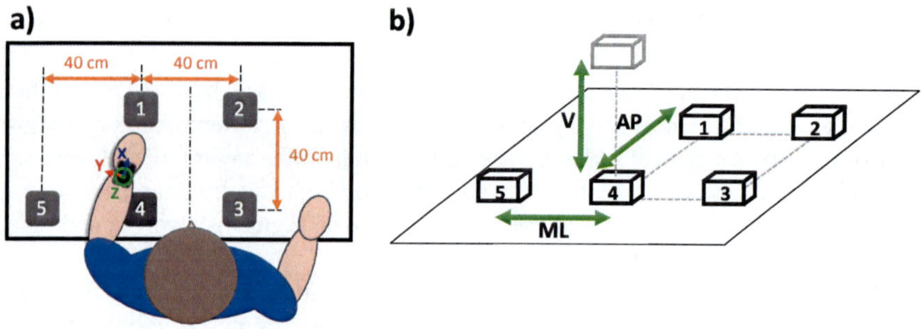


**Fig. 1.** Test bench. (a) station used for the tests and participant wearing the inertial sensor on the left wrist. (b) Reaching tasks direction: AP antero-posterior, ML medio-lateral, V vertical.

## 2.2 Protocol

The protocol executed by the participant consisted in two parts: (a) reaching task with controlled movements only; (b) reaching task with controlled movements followed by a set of impulsive movements.

In the first part of the protocol, reaching task started with both participant's hands on the surface of the test bench. The reaching task consisted in elevating the left hand, moving it between two stations with defined direction and timing, then returning in the initial position. The right arm remained in the same position during the entire test.

Three reaching task trials were executed by the participant (Fig. 1b): antero-posterior (AP) movements between the stations 4 and 1, medio-lateral (ML) movements between the stations 4 and 5, vertical (V) movements over the station 4 to reach shoulder height, which was approximately 40 cm over the test bench.

For each reaching task trial, a total of twenty movements between the two stations were performed. In-between trials the participant rested for 60 s. The trials were executed at 25 bpm and 50 bpm to simulate slow and fast execution rhythms. The cadence was paced with a metronome, so that each beat defined the movement onset.

In the second part of the protocol, the participant repeated each reaching task trial at both cadences, and, after the 20 controlled movements, 6 impulsive movements were performed in the same direction. In the controlled movements, the distance between two stations was covered according to the metronome pacing. Whereas, impulsive movements started at the metronome beat, but the distance between the two stations was covered in a shorter time. The participant was instructed to wait for the next metronome beat to perform the following impulsive movement.

## 2.3 Analysis

For this study, data from the accelerometer were considered. The analysis was computed in two steps. Firstly, the reaching task trials of the first part of the protocol were analyzed to define the threshold values that characterize operator's activity in terms of task controlled movements. Then, threshold values were used to identify the onset of impulsive movements executed in the second part of the protocol.

To characterized operator's upper limb activities, four time-domain features were used. The first two features were based on the Laban Movement Analysis [13] that had been already used to recognize human movement with a single wrist-worn accelerometer [14]. The other two features were based on the concept of Actigraphy that is usually employed to monitor long-term physical activity from acceleration signal.

From the Laban Movement Analysis, the "Time effort factor" and "Flow effort factor" were considered. "Time effort factor" is related to the movement acceleration, distinguishing between sudden and sustained movements [15]. "Flow effort factor" refers to the continuity in motion, distinguishing between movements with small or large jerk amplitude. Therefore, "Time effort factor" ($MA_{e0}$) and "Flow effort factor" ($MJ_{e0}$) were calculated from acceleration data, as the integral of the acceleration magnitude and of the jerk magnitude respectively [15].

Actigraphy has been widely used to monitor the amount of upper limb use of healthy subjects [16] and patients [17] fixing a sensor on the wrist. Traditional actigraphy counts the amount of time the acceleration remains over a defined threshold; however, innovative numerical indexes ($MA_{e1}$ and $MA_{e2}$) were proposed and validated [18]. In the current analysis, $MA_{e1}$, calculated as the standard deviation of the acceleration magnitude and $MA_{e2}$, calculated as the magnitude of the standard deviation of the acceleration components [18], were introduced.

The four time-domain features (MAe0, MJe0, MAe1 and MAe2) were evaluated over time epochs of defined length, without overlap. Since the study aims to develop a methodological approach, two different epoch lengths were evaluated: 100 ms and 250 ms. Thus, the features were calculated for the three reaching task trials and the two cadences executed with controlled movements. The maximum value for each combination was calculated and overestimated of 5% to consider a safety margin in their use as threshold. For each feature and epoch length, the threshold value representative of the operator's activity with controlled movements was defined as the maximum value among the three reaching tasks and the two cadences.

Then, the reaching task trials executed with the impulsive movements were analysed. For each epoch, the four features were calculated, and their value was compared with the threshold value to identify the onset of the impulsive movement. This analysis was repeated considering the two epoch lengths separately.

## 3 Results and Discussion

For each reaching task trial and cadence executed with controlled movements, the four features were calculated with the two epoch durations. The maximum value for each condition was identified and overestimated of 5%. As example, in Fig. 2 for each feature is reported the identification of the maximum (red dot) for the reaching task executed in antero-posterior direction at 50 bpm, analyzed using epoch length of 250 ms. The maximum value overestimated of 5% is indicated with a red dashed line.

In Table 1 are reported the maximum values overestimated of 5% for all the reaching task trials and cadences, calculated with the two epoch lengths. In addition, for each feature is reported the threshold value.

Then, the threshold values were used to identify the onset of impulsive movements. As example, in Fig. 3 for each feature are reported the results for the identification

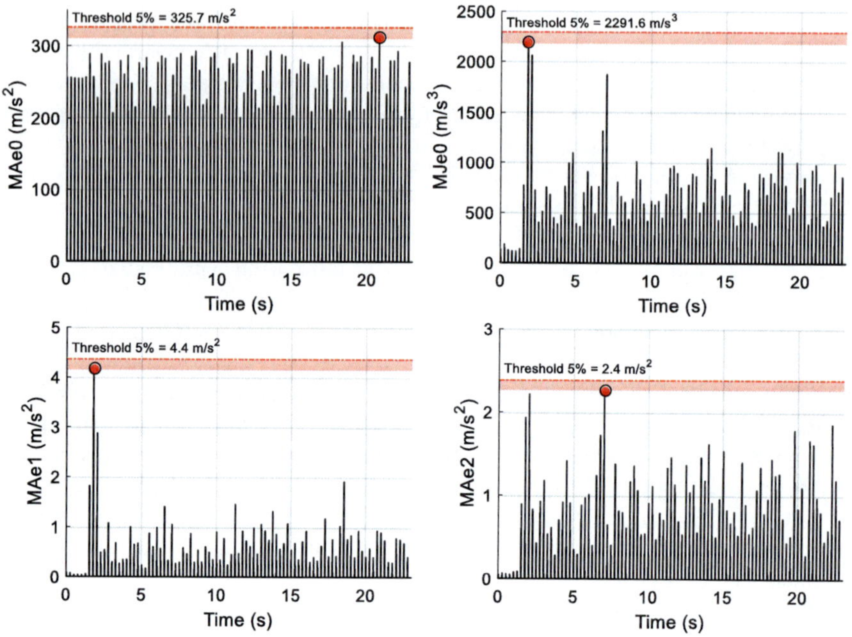


**Fig. 2.** Controlled movement. Reaching task in antero-posterior direction. Cadence 50 bpm, epoch 250 ms. Maximum values identification for the four features

**Table 1.** Features threshold values for the three reaching task trials at the two cadences calculated with epoch duration of 100 ms and 250 ms.

| Feature | Reaching task direction | Epoch duration 100 ms | | | Epoch duration 250 ms | | |
|---|---|---|---|---|---|---|---|
| | | Cadence | | Threshold | Cadence | | Threshold |
| | | 25 bpm | 50 bpm | | 25 bpm | 50 bpm | |
| $MA_{e0}$ ($m/s^2$) | Antero-posterior | 132.9 | 161.8 | 161.8 | 307.9 | 325.7 | 334.7 |
| | Medio-lateral | 124.3 | 136.9 | | 287.8 | 301.1 | |
| | Vertical | 123.9 | 151.2 | | 288.7 | 334.7 | |
| $MJ_{e0}$ ($m/s^3$) | Antero-posterior | 463.3 | 1190.8 | 1190.8 | 1006.6 | 2291.6 | 2291.6 |
| | Medio-lateral | 561.9 | 770.5 | | 936.8 | 1424.1 | |
| | Vertical | 554.4 | 1146.6 | | 1022.1 | 1850.8 | |
| MAe1 ($m/s^2$) | Antero-posterior | 0.7 | 2.2 | 2.2 | 0.9 | 4.4 | 4.4 |
| | Medio-lateral | 0.5 | 1.1 | | 0.6 | 1.5 | |
| | Vertical | 0.6 | 1.1 | | 0.6 | 1.7 | |
| MAe2 ($m/s^2$) | Antero-posterior | 1.6 | 2.5 | 2.5 | 1.5 | 2.4 | 2.4 |
| | Medio-lateral | 1.5 | 2.4 | | 1.5 | 2.4 | |
| | Vertical | 1.1 | 2.5 | | 1.1 | 2.1 | |

of the first impulsive movement onset (green dot) for the reaching tasks executed in antero-posterior direction at 50 bpm, analyzed using epoch duration of 250 ms. In black is reported the signal identified as controlled movement, in red the signal identified as impulsive movement. The threshold value is reported with the red dashed line.

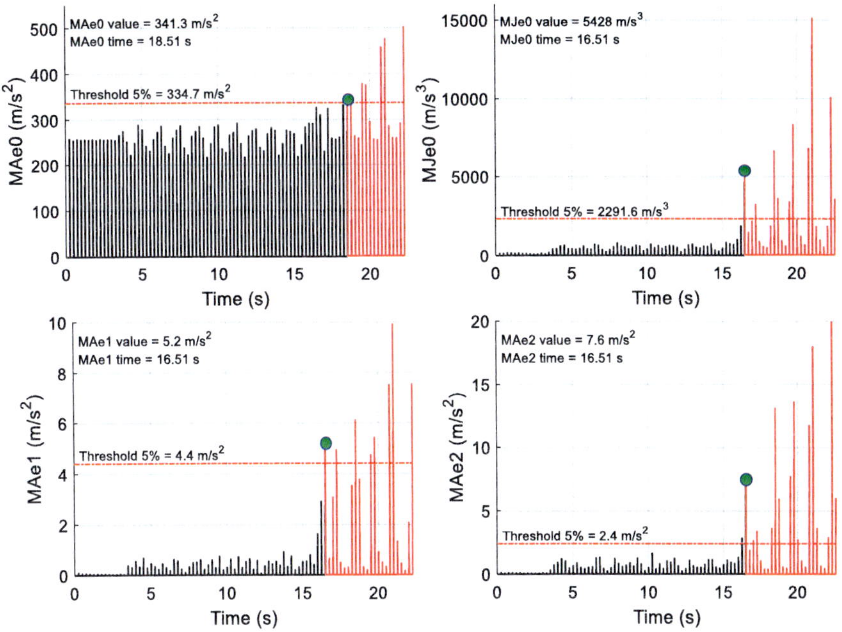


**Fig. 3.** Impulsive movement onset detection using the identified threshold for the four features. Reaching task in antero-posterior direction, Cadence 50 bpm, Epoch 250 ms

In Table 1 are reported the results for the impulsive movement onset time identification for the three reaching task trials executed at both cadences. Results for the analysis performed using 100 ms and 250 ms epoch durations are compared.

Overall, results show that the four features identified the impulsive movements for all the trials. The only exception is for the feature MAe0 for the reaching task executed in vertical direction at 50 bpm that, sharply anticipated the movement identification.

Comparing the two epoch lengths, results shows that shorter epoch length (100 ms) identified the impulsive movement onset slightly earlier compared to the longer epoch length (250 ms) for the three reaching task trials and both cadences.

When comparing the four features, both at low and high cadences, MAe0 seems to be the one that identifies the impulsive movement later with respect to the other features in the reaching task executed in antero-posterior direction. This result was found when data were analyzed with both epoch durations.

Concerning MAe1, impulsive movement were identified with a small latency in reaching task executed in medio-lateral direction at low cadence, when analyzed with epoch length of 100 ms, compared to the movement in the other direction. However, it anticipated the identification when reaching task was executed in vertical direction.

To allow quick detection limiting false positive identification, it is advisable to consider the four features analyzed with short epoch length (Table 2).

**Table 2.** Difference in impulsive movement onset between 100 ms epoch and 250 ms epoch for the three reaching task trials at the two cadences.

| Feature | Reaching task direction | Impulsive movements onset (s) | | | |
|---|---|---|---|---|---|
| | | 25 bpm | | 50 bpm | |
| | | 100 ms epoch | 250 ms epoch | 100 ms epoch | 250 ms epoch |
| $MA_{e0}$ ($m/s^2$) | Antero-posterior | 26.01 | 26.01 | 17.21 | 18.51 |
| | Medio-lateral | 27.91 | 28.01 | 15.21 | 15.26 |
| | Vertical | 27.91 | 28.01 | 15.51 | 8.51 |
| $MJ_{e0}$ ($m/s^3$) | Antero-posterior | 25.91 | 26.01 | 16.31 | 16.51 |
| | Medio-lateral | 27.91 | 28.01 | 15.21 | 15.26 |
| | Vertical | 27.91 | 28.01 | 15.51 | 15.76 |
| MAe1 ($m/s^2$) | Antero-posterior | 25.91 | 26.01 | 16.41 | 16.51 |
| | Medio-lateral | 28.01 | 28.01 | 15.21 | 15.26 |
| | Vertical | 27.81 | 28.01 | 15.51 | 15.76 |
| MAe2 ($m/s^2$) | Antero-posterior | 25.91 | 26.01 | 16.31 | 16.51 |
| | Medio-lateral | 27.81 | 28.01 | 15.21 | 15.26 |
| | Vertical | 27.91 | 28.01 | 15.61 | 15.76 |

## 4 Conclusions

In industrial field when an operator works with a robot usually performs controlled movements that can be recognized and predicted by the robot itself. However, the operator could execute impulsive movements that may collocate himself on the robot trajectory, inducing risks for safety. This methodological study aimed to assess the ability of four features to characterize operator's activity related to the task and, thus, identify the onset of impulsive movements related to perturbation of the task. Monitoring operator's activity with a single inertial sensor fixed on the wrist and analyzing motion kinematics with four features allows to identify the onset of impulsive movements. Considering all the four features and calculate them on shorter epoch lengths seems to increase the suitability of the detection. To move towards real-time operator safety monitor, it becomes mandatory to identify a system with appropriate data transmission frequency and able to analyze kinematic data with a compatible period.

## References

1. Digo, E., Antonelli, M., Cornagliotto, V., et al.: Collection and analysis of human upper limbs motion features for collaborative robotic applications. Robotics **9**, 33 (2020)
2. Robots and Robotic Devices - Safety Requirements - Part 2: Industrial Robot Systems and Integration, ISO 10218–2 (2011)
3. Robots and Robotic Devices - Industrial Safety Requirements - Collaborative Industrial Robots, ISO/TS 15066 (2016)
4. Vacca, L., Rosso, V., Gastaldi, L.: Kinematic comparison of standard judo ushiro ukemi and backward fall. Int. J. Mech. Control **22**, 153–160 (2021)
5. Marvel, J.A., Falco, J., Marstio, I.: Characterizing task-based human–robot collaboration safety in manufacturing. IEEE Trans. Syst. Man Cybern. Syst. **45**, 260–275 (2015)
6. Gupta, H.P., Chudgar, H.S., Mukherjee, S., et al.: A continuous hand gestures recognition technique for human-machine interaction using accelerometer and gyroscope sensors. IEEE Sens. J. **16**, 6425–6432 (2016)
7. Koskimaki, H., Huikari, V., Siirtola, P., et al.: Activity recognition using a wrist-worn inertial measurement unit: a case study for industrial assembly lines. In: 2009 17th Mediterranean Conference on Control and Automation, pp. 401–405 (2009)
8. Devin, S., Alami, R.: An implemented theory of mind to improve human-robot shared plans execution. In: 11th ACM/IEEE International Conference on Human-Robot Interaction (HRI), pp. 319–326 (2016)
9. Görür, O.C., Rosman, B., Sivrikaya, F., Albayrak, S.: Social cobots: anticipatory decision-making for collaborative robots incorporating unexpected human behaviors. In: Proceedings of the 2018 ACM/IEEE International Conference on Human-Robot Interaction, pp. 398–406 (2018)
10. Yang, C., Bouffard, J., Srinivasan, D., et al.: Changes in movement variability and task performance during a fatiguing repetitive pointing task. J. Biomech. **76**, 212–219 (2018)
11. Bosch, T., Mathiassen, S.E., Hallman, D., et al.: Temporal strategy and performance during a fatiguing short-cycle repetitive task. Ergonomics **55**, 863–873 (2012)
12. Qin, J., Lin, J.H., Faber, G.S., et al.: Upper extremity kinematic and kinetic adaptations during a fatiguing repetitive task. J. Electromyogr. Kinesiol. **24**, 404–411 (2014)
13. Laban, R., Ullmann, L.: The mastery of movement (2011)
14. Kikhia, B., Gomez, M., Jiménez, L.L., et al.: Analyzing body movements within the laban effort framework using a single accelerometer. Sensors **14**, 5725–5741 (2014)
15. Samadani, A.A., Burton, S.J., Gorbet. R., Kulić, D.: Laban effort and shape analysis of affective hand and arm movements. In: Proceedings - 2013 Humaine Association Conference on Affective Computing and Intelligent Interaction, ACII 2013, pp. 343–348 (2013)
16. Bailey, R.R., Klaesner, J.W., Lang, C.E.: An accelerometry-based methodology for assessment of real-world bilateral upper extremity activity. PLoS ONE **9**, e103135(2014)
17. Hayward, K.S., Eng, J.J., Boyd, L.A., et al.: Exploring the role of accelerometers in the measurement of real world upper-limb use after stroke. Brain Impair. **17**, 16–33 (2016)
18. Iacovelli, C., Caliandro, P., Rabuffetti, M., et al.: Actigraphic measurement of the upper limbs movements in acute stroke patients. J. Neuroeng. Rehabil. **16**, 1–10 (2019)

# Intercity Bus Occupant Safety

V. M. Cantor Mexquititla, M. A. Rivera-Hernández, and C. R. Torres-San Miguel(✉)

Instituto Politécnico Nacional, Escuela Superior de Ingeniería Mecánica Y Eléctrica, Unidad Zacatenco, 07738 Mexico City, México
ctorress@ipn.mx

**Abstract.** The most common means of transportation in Mexico City is the intercity bus. The frequency of transport makes it necessary to carry out measures that protect the health of the passenger. According to data from the World Health Organization (WHO), one in four of the traumatic deaths recorded in the world are related to transport. UNECE and NHTSA promote standards so that vehicles have the necessary passive safety to reduce occupant injuries in a collision event. This research shows simulations by the finite element that represent two types of collisions: frontal and rear. A bus seat and a 50-percentile Hybrid III dummy are evaluated with LS-DYNA software. Results showed the head injury criteria (HIC) and chest severity index (CSI) in a passenger when the bus crashed at a 32 km/h speed.

**Keywords:** SDG9 · HIC · CSI · Dummy · Front collision · Rear collision

## 1 Introduction

According to data from the World Health Organization (WHO) approximately one in four of the traumatic deaths that occur in vehicular accidents that are recorded in the world are related to public transport, a number that represents about 2.1% of all causes of global death Speeding is a problem adjacent to motorized vehicles and affects users of transport routes [1]. This is a widespread social problem, which afflicts the entire road network (highway, local roads, and urban roads), often at any time; about 50% of drivers exceed speed limits [2]. Currently, agencies such as the United Nations Economic Commission for Europe (UNECE) and the National Highway Traffic Safety Administration (NHTSA) are responsible for promoting bus design standards intercity between other vehicles and providing continuous improvement vehicle safety systems. Passive safety is currently a constantly growing subject due to the constant innovation that exists to build vehicles. Many old vehicles lack essential safety regulations, increasing the probability of suffering in a few years vehicles called scrap metal forcing the passenger to replace the vehicle and affect the environment due to the manufacture of these. Research focuses on studying devices such as those found in [3–5] where research focused on ISR and CRS retention systems. On the other hand, in [6, 7], shock tests of different types were performed, knowing the type of injury that for the cause of the impact may suffer, cranioencephalic and thorax injuries were analyzed in conditions of frontal shock using


G. Quaglia et al. (Eds.): I4SDG 2021, MMS 108, pp. 182–192, 2022.
https://doi.org/10.1007/978-3-030-87383-7_20

children's Dummy. Research on crash tests carried out on occupants with rare health problems such as Osteogenesis Imperfecta or some degenerative disease was analyzed in [8, 9]. These studies collaborate to realize safety vehicles by increasing protection during the manufacture of vehicles. The main objective of vehicular safety is to protect the physical integrity of the occupants during and after the accident through a risk reduction. Therefore, the development of safe and reliable buses requires optimized design methods, materials, and high-quality manufacturing procedures. For example, when an occupant is not wearing a seatbelt, when he collides, it will suffer a movement from the seat in which it is in front, producing injuries to the face, neck, legs, mainly in the knees. [10] Currently, knowing the behavior inside the body or structure lets to identify the level of safety offered by the seats to improve passive safety, reducing injuries to the occupant in the event of an accident [11, 12].

This work aims to find the deformations that occur in the seats in scenarios of a frontal collision and rear collision in an interurban bus according to UNECE standard 80 [13] and figure out the kinetic energy absorbed by the passenger during the collision by determining acceptability criteria based on the head injury criteria (HIC) and chest severity index (CSI).

## 2 Design and Modeling

The NHTSA establishes the requirements for testing seats. Standard 108 establishes the forces that the occupant seat must support, which must be 20 times more than the mass of the seat multiplied by 9.81 m/s$^2$, applied in a longitudinal direction. Acceptance criteria: The measurement of the Dummy's head for collision tests are made from the calculation of the resulting triaxial acceleration derived from the center of gravity ($\gamma$_r) measured with a CFC of 600 [14] from the following equation and expressed in g = 9.81 m/s$^2$ is calculated with Eq. (1):

$$\gamma_r^2 = \gamma_l^2 + \gamma_v^2 + \gamma_t^2 \tag{1}$$

Where $\gamma_l$ is the value of instantaneous longitudinal acceleration, ?$_v$ is the value of instantaneous vertical acceleration, and $\gamma_t$ is the value of the resulting instantaneous transverse acceleration that reflect the change in acceleration experienced by the head in the first moments after the crash, once that the impact wave reaches the base of the armchair. The analysis is carried out by selecting the maximum integration limits, taking the maximum value at 36 ms after the accident. The head injury criterion (HIC) will be calculated through Eq. (2), the acceleration data is obtained from the resulting acceleration, the values of $t_1$ and $t_2$ are expressed in milliseconds (ms) [15].

$$\left\{ (t_2 - t_1) \left[ 1/(t_2 - t_1) \int_{t1}^{t2} a\, dt \right]^{2.5} \right\} \tag{2}$$

The biomechanical criterion that must be met if the Dummy is installed in a forward-facing seat is the thorax acceptability criterion (ThAC), determined by the absolute value of the resulting acceleration. The value of this criterion is less than 30 g, except in periods less than 3 ms considering the deviation of the rib cage, this being the peak value of the deviation of any rib determined by the thoracic displacement transducers. In addition, the ThAC is measured concerning the viscous criterion (VC), which indicates that the response related to the viscera on any rib is equal to the maximum value calculated from the product of the relative compression of the normalized thoracic cage at 140 mm in length and the compression rate obtained by the compression bypass as shown in the following expression where D measured in meters to a deviation of the ribs. [16]

$$VC = max[D/0, 14\,x\,dD/dt] \tag{3}$$

The seat has a particular place and a handrail on the back, made with polypropylene. The structure is made of steel. The support is adaptable to anybody and is coated with electrostatic paint. The passenger seat commercial model for an Intercity bus was selected with the dimensions of Fig. 1. The back of the seat is made of PVC material with a steel structure [17].

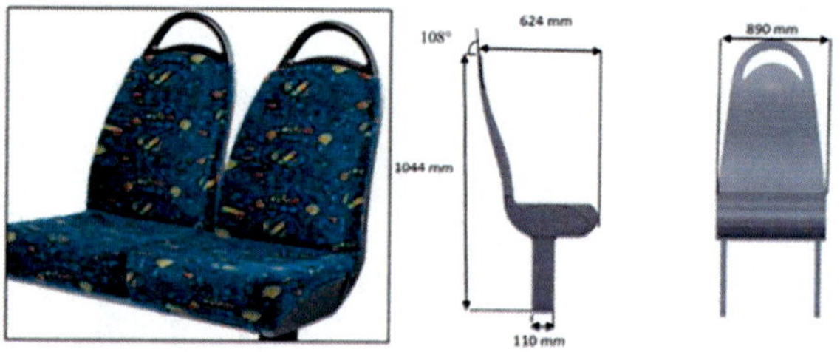


**Fig. 1.** The seat is modeled in CAD using the SolidWorks software

The discretization of the seats was carried out with HyperMesh® software, in which 102289 nodes and 434199 elements were developed in 6 volumes. The mesh that was used is dimension number 8 Fig. 2.

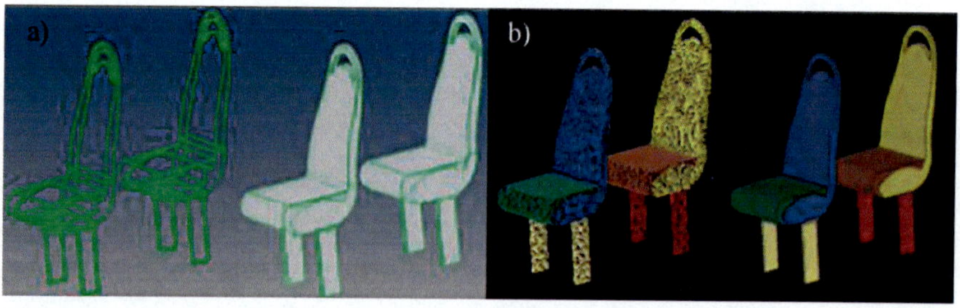


**Fig. 2.** a) Armchair discretization in the HyperMesh® software, b) the model loaded in the LS-Dyna® software.

The material used for all the seat components was Plastic-Kinematic, with the mechanical properties of Table 1.

**Table 1.** Mechanical properties of materials.

| Seat parts | Material | Density g/cm$^3$ | Modulus of elasticity kg/cm$^2$ | Poisson coefficient | Yield stretch kg/cm$^2$ |
|---|---|---|---|---|---|
| Backrest | PVC | 1.56 | 280 | 0.38 | 560 |
| Cushions | Polyurethane foam | 0.035 | 120 | 0.30 | 130 |
| Brackets | Steel SJ2340 | 7.85 | 2100000 | 0.30 | 2457.5 |

According to the UNECE 080 standard, the test is carried out at 32 km/h. The seat supports were restricted in the direction DOFX, DOFY, DOFZ. The Dummy chosen is a 50th percentile Hybrid III [18]. The contact command is defined between different parts in a simulation, and there are different contact algorithms as flexible body contact, flexible body contact with rigid body contact, surface-to-surface contact, eroded contact, attached surface, and CAD surface [19, 20]. In this simulation, the dummy-armchair contact is made, the Dummy is placed in the armchair through the * DmyPos command in the HyperMesh program to subsequently make contact between both models within the LS-DYNA software [21]. It is also essential to consider the restriction being analyzed by attaching the armchair supports to the platform to simulate the deformations suffered by the seats due to the sudden bus stop upon impact.

Conversely, a fairly standard feature in the codes used for finite element programs is that they do not contain units and yield numerical results. This leads the programmer to establish a system of compatible units, preferably based on the SI. Particular attention should be paid to this detail since the post-processor will show erroneous results if there are errors in the units.

## 3 Numerical Simulation

### 3.1 Boundary Conditions

LS-Dyna is used to analyze the deformations of the dynamic problem, and the solution methodology is based on explicit integration. The software has an impact and contact algorithm that allows complex problems to be solved and a mesh in the transition zone. The type of dummy contact with the armchair used is AUTOMATIC_SUFACE_TO_SURFACE, and the seats are connected to the platform, simulated as an urban bus floor. A simulation time of 150 ms is established because, according to the UNECE standard, the recommended time is between 100 and 200 ms. Accelerometers measure the rate of HIC and CSI [22]. The location of the accelerometer in the head is at node 133919, while that of the thorax is node 135705 Fig. 3.

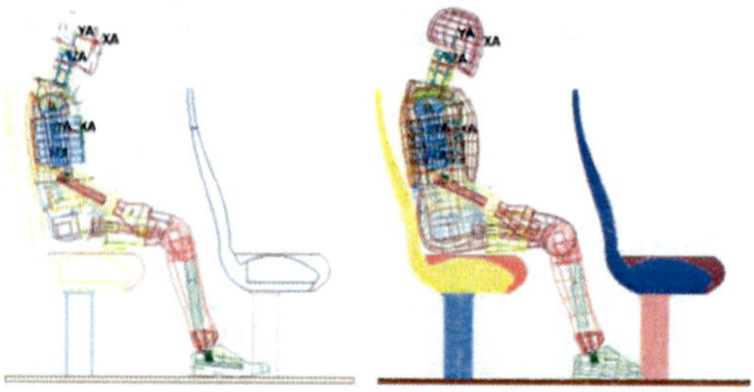

**Fig. 3.** The head and thorax accelerometer location.

## 3.2 Rear and Front Impact Simulation

The simulation was carried out with the Dummy resting position in the armchair as a passenger in an urban bus. The seat speed is about 32 km/h and is the same as the vehicle would carry in an interval of 150 ms. Figure 4 shows the dummy behavior in a range of 0 to 150 ms in steps of 30 ms during a frontal crash.

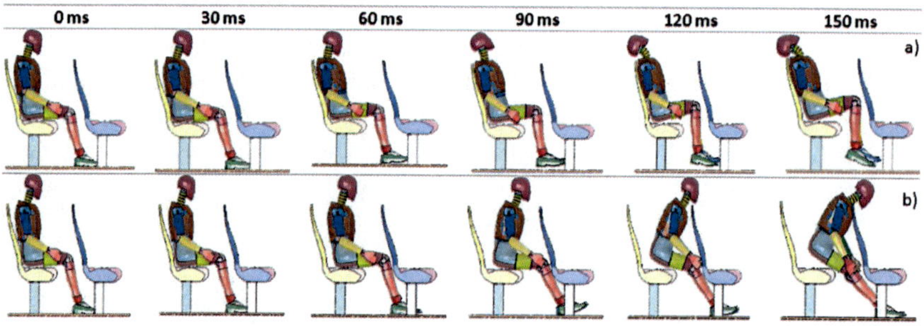


**Fig. 4.** a) Dummy simulation during a rear impact, b) Dummy simulation during a frontal impact.

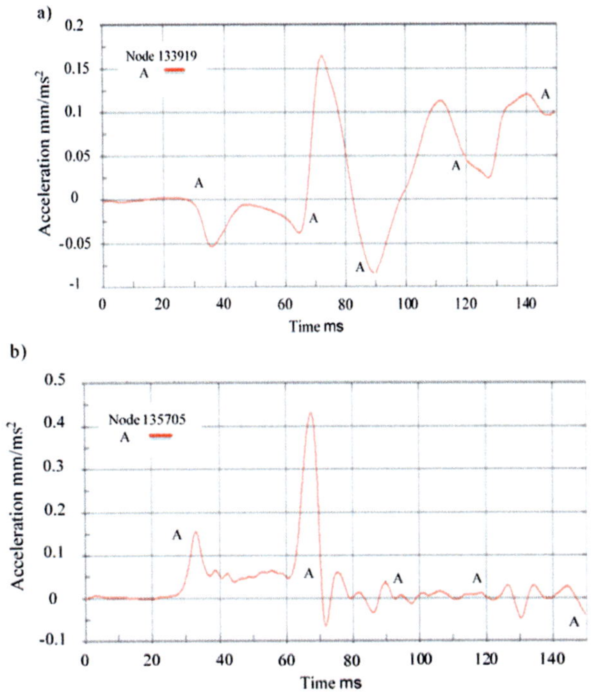


**Fig. 5.** a) Head accelerations result during a rear collision; b) Thorax accelerations result during a rear collision.

Figure 5 shows the values recorded from the accelerometers located in the head and thorax acquired during the rear collision.

Figure 6 shows the values recorded from the accelerometers located in the head and thorax acquired during a frontal collision.

### 3.3 Rear Impact Seat Stress

The main impact with the armchair is the Dummy due to its proximity. The stress of the armchair is shown, mainly in the rear armchair, in which the Dummy is placed. This one for a rear-end impact. The interval starts from 60 ms with the armchair, following a sequence of 10 ms, as shown in Fig. 7a.

### 3.4 Frontal Impact Seat Stress

Figure 7b and c show the armchair stress due to the impact of the Dummy with the seat, in time steps of 60 ms, 65 ms, 75 ms, 85 ms, 95 ms, 140 ms, 145 ms, and 150 ms, because in these periods the main stress points are generated.

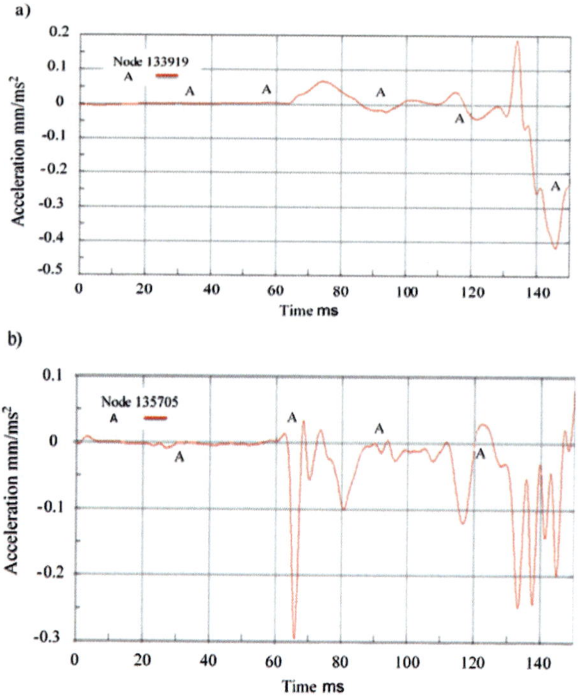


**Fig. 6.** a) Head accelerations result during a frontal collision; b) Thorax accelerations result during a frontal collision.

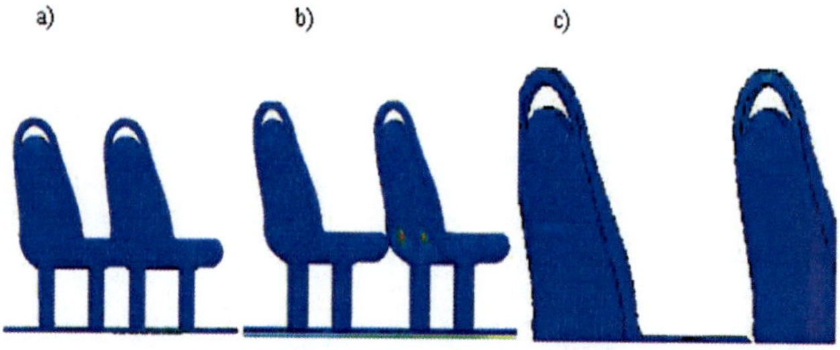


**Fig. 7.** a) Armchair stress at 70 ms interval during a rear impact. b) Armchair stress at 65 ms interval during a frontal impact. c) Contact areas between the Dummy and the rear armchair.

# 4 Results

## 4.1 Rear Impact, HIC and CSI Values

Figure 8 shows the effects of HIC in the time intervals for the calculation of 14-HIC36 injury is $t_1 = 64.76$ ms and $t_2 = 92.94$ ms giving 14-HIC3 = 183.1. While the maximum acceleration obtained was 23 G. The CSI thorax injury is equal to $t_1 = 66.68$ ms and $t_2 = 69.68$ ms obtaining in CSI = 120.6, and the maximum acceleration is approximately between 70 G [23].

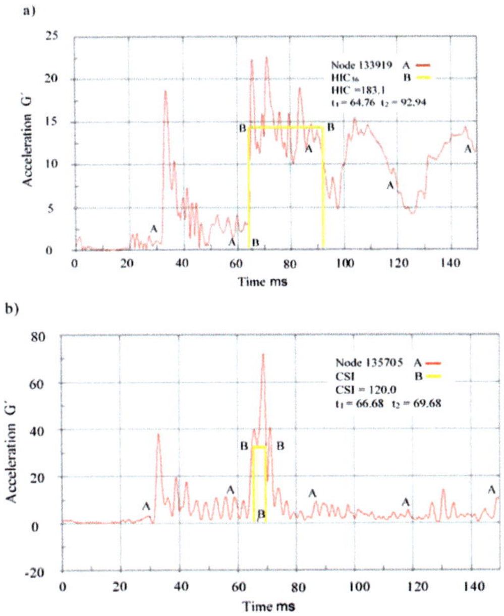


**Fig. 8.** a) HIC values during a rear impact; b) CSI values during a rear impact.

## 4.2 Frontal Impact, HIC and CSI Values

Figure 9 shows the intervals of the calculation of 14-HIC36 injury for a frontal type impact is $t_1 = 136.4$ and $t_2 = 149.0$, reaching the maximum acceleration, which is almost 100 G. Giving a value of 14-HIC36 = 291.1. For the thoracic injury, the time intervals are $t_1 = 136.8$ ms and $t_2 = 142.4$. Obtaining a CSI = 302 because it reaches the maximum acceleration of 80 G [23].

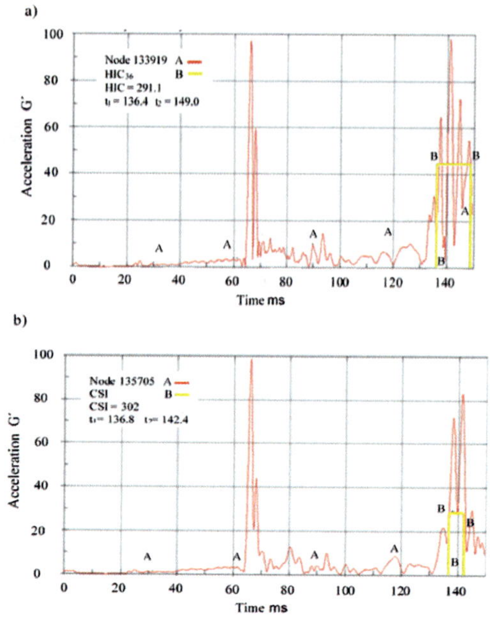


**Fig. 9.** a) HIC values during a frontal impact; b) CSI values during a frontal impact.

## 5 Conclusions

According to the data obtained from the simulations, the HIC value shows a mild concussion occurs at the level of brain damage without fracture in the skull. In the rear type, the level of brain damage was obtained as a mild shock without fracture in the skull. The thoracic deformations that occur in the two different types of impact were also analyzed, and it is concluded that the value of the chest injury for a rear impact is CSI = 120.0. Therefore, for the first case, there is a slight chest injury. Although for the frontal type, the thoracic injury index of CSI = 302.0 was obtained, but in the frontal type impact due to the index of injury that was gotten, it reflects a significant injury with severe damages causes severe injuries as fractures, costal fissure, sternal fissure, and in the part of the lung, a bruise and tracheal rupture.

**Acknowledgments.** The authors acknowledge the financial support for developing this work by the National Council of Science and Technology (CONACYT) of Mexico, the Instituto Politécnico Nacional. Authors thanks of partial support from project 20210282 and EDI grant, all provided by SIP/IPN.

## References


1. International Federation of Red Cross and Red Crescent Societies, Control de la velocidad, Global Roa. Switzerland (2008)
2. Rivers, R.W.: Evidence in Traffic Crash Investigation and Reconstruction. Charles C Thomas, Publisher, Canada (2007)
3. Martínez-Sáez, L., et al.: Design and validation of a child seat pretensor device. Rev. Fac. Ing. Univ. Antioquia **70**, 54–65 (2014)
4. Cruz-Jaramillo, I.L., et al.: Numerical child restraint system analysis in 6 years old infant during a dolly rollover test. Int. J. Crashworthiness **26**(4), 404–412 (2020). https://doi.org/10.1080/13588265.2020.1718464
5. Miranda, M., Miguel, C.R.T.S., Alejandro Flores-Campos, J., Ceccarelli, M.: Numerical simulation of a 2D harmonic oscillator as coupling system for Child Restraint Systems (CRS). In: Niola, V., Gasparetto, A. (eds.) IFToMM ITALY 2020. MMS, vol. 91, pp. 492–502. Springer, Cham (2021). https://doi.org/10.1007/978-3-030-55807-9_56
6. Cruz-Jaramillo, I.L., et al.: Numerical low-back booster analysis in a 6-year-old infant during a dolly rollover test. J. Adv. Transp. **2020**, 1–9 (2020). https://doi.org/10.1155/2020/5803623
7. Cruz-Jaramillo, I.L., Torres-San-Miguel, C.R., Cortes-Vásquez, O., Martínez-Sáez, L.: Numerical low-back booster analysis on a 6-year-old infant during a frontal crash test. Appl. Bionics Biomech. **2018**, 1–6 (2018). https://doi.org/10.1155/2018/2359262
8. Martinez, L., et al.: Crash impact dummies adapted to people affected by osteogenesis imperfecta. In: IRCOBI Conference Proceedings, p. 768–769 (2016)
9. Cuautle-Estrada, A., et al.: Lower Limb design for a child test dummy, based on osteogenesis imperfecta characteristics. J. Phys. Conf. Ser. **1723**(1), 012019 (2021). https://doi.org/10.1088/1742-6596/1723/1/012019
10. Abbas, A.K., Hefny, A.F., Abu-Zidan, F.M.: Seatbelts and road traffic collision injuries. World J. Emerg. Surg. **6**(1), 1–6 (2011). https://doi.org/10.1186/1749-7922-6-18
11. Yuce, C., et al.: A case study: designing for sustainability and reliability in an automotive seat structure. Sustainability **6**(7), 4608–4631 (2014)
12. Guler, M.A., Atahan, A.O., Bayram, B.: Effectiveness of seat belt usage on the rollover crashworthiness of an intercity coach. In: Proceedings of the International Technical Conference on the Enhanced Safety of Vehicles (ESV), Stuttgart, Germany (2009)
13. International Standard Organization, INTERNATIONAL STANDARD ISO 6487. Road vehicles: Measurement techniques in impact test Instrumentation, vol. Sixth Edi (2015)
14. Linder, A., Avery, M., Kraft, M., Kullgren, A.: Change of Velocity and pulse characteristics in rear impacts: real world and vehicle tests data. In: Proceedings of 18th Enchanced Safety of Vehicles Conference, pp. 1–9
15. Yang, H., Wang, Z., Xie, K.: Impact of connected vehicles on mitigating secondary crash risk. Int. J. Transp. Sci. Technol. **6**(3), 196–207 (2017). https://doi.org/10.1016/j.ijtst.2017.07.007
16. Hiemstra-van Mastrigt, S.: Comfortable passenger seats: recommendations for design and research (2015)
17. Rust, W., Schweizerhof, K.: Finite element limit load analysis of thin-walled structures by ANSYS (implicit), LS-DYNA (explicit) and in combination. Thin-Walled Struct. **41**(2–3), 227–244 (2003). https://doi.org/10.1016/S0263-8231(02)00089-7
18. Anghel-Vasilescu, P., Dorignac, J., Geniet, F., Leon, J., Taki, M.: Nonlinear supratransmission in multicomponent systems. Phys. Rev. Lett. **105**(7), 074101 (2010). https://doi.org/10.1103/PhysRevLett.105.074101
19. Hallquist, J.O.: LS-DYNA Keyword User's Manual, vol. I (2007)
20. Vincze-Pan, S., Csiszar, A.: Real and Simulated Crashworthiness Tests on Buses. In: International Technical Conference on the Enhanced Safety Vehicles, p. 10 (2005)

21. Li, Z., et al.: Finite element analysis of pedestrian lower limb fractures by direct force: the result of being run over or impact? Forensic Sci. Int. **229**(1–3), 43–51 (2013). https://doi.org/10.1016/j.forsciint.2013.03.027
22. Timmel, M., Kolling, S., Osterrieder, P., Du Bois, P.A.: A finite element model for impact simulation with laminated glass. Int. J. Impact Eng. **34**(8), 1465–1478 (2007)
23. UNECE, Prescripciones Uniformes Relativas a la Homologación de los Vehículos Respecto a la Protección de los Ocupantes en Caso de una Colisión Frontal, E/ECE/TRANS/505/Rev.1/Add.93/Rev.2, 2013.

# A New Mask-Based Technique for Stereotactic Brain Neurosurgery

Mirko Patuzzo(✉), Nicola Sancisi, and Vincenzo Parenti-Castelli

Department of Industrial Engineering, University of Bologna, Bologna, Italy
mirko.patuzzo@studio.unibo.it

**Abstract.** The state of the art of stereotactic surgery is reported with particular attention to robot-based surgical systems in order to highlight what are the technical limits still present. From this basic knowledge, alternative operating methods based on thermoplastic masks, already in use in radiosurgery, are proposed, which can compensate for these technical limitations and improve the experience for both the patient and the surgeon. In particular, two different solutions using the mask are presented. Finally, a comparison between the two solutions is discussed and possible future developments are outlined.

**Keywords:** SDG3 · Stereotactic surgery · Mask-based techniques · Brain surgery

## 1 Introduction

Stereotactic surgery refers to a specific operation methodology used in neurosurgery, which uses special high-precision devices to identify and reach targets in areas of the brain by the least invasive possible way. To achieve this objective, various systems have been proposed, which vary in degree of complexity, precision and costs.

At the end of the nineteenth century, the development of neuroscience was limited by the technological level of the time, in particular the absence of diagnostic techniques. It can therefore be said that neuroscience is deeply linked to intracranial identification techniques. The ability to precisely locate injuries and the development of non-invasive operating techniques were then the key to give prestige to this discipline [1, 2].

The first identification technique was developed in 1918 by Walter Danny [3] and was based on air injection into the brain ventricles (ventriculography) to evaluate the shape and displacement of the latter and identify the lesion indirectly. Since then, numerous diagnostic techniques were proposed but only in 1972 a direct image of a patient's brain tissues was obtained. This was achieved through the studies of Hounsfield on X-rays that led to Computer Tomography (CT) [4]. Similarly, in 1970 Cope and Damadian developed the basics of the actual Magnetic Resonance Imaging (MRI) [5]. These diagnostic techniques improved over the years, becoming the standards in the analysis and identification of brain structures [1, 2].

An additional key to the development of stereotactic surgery was brought by the invention of the stereotactic frame, which was used on humans in 1947. Spiegel and


G. Quaglia et al. (Eds.): I4SDG 2021, MMS 108, pp. 193–200, 2022.
https://doi.org/10.1007/978-3-030-87383-7_21

Wycis, indeed, developed a preliminary stereotactic surgery by exploiting data from ventriculography to locate particular intracranial anatomical structures [1, 2]. The frame-based procedure takes its name from the tool that is the basis of the operation [6, 7]: the stereotactic helmet or frame. This device (Fig. 1) consists in a rigid ring fixed to the patient's skull by screws under local anaesthesia. A system of guides is mounted on the ring and allows translation of a pointer, housing the surgical tool, with respect to the skull. This frame is radiopaque, i.e., it can be identified in CT. Thus, it provides a fixed and known reference system with respect to the intracranial target to reach.

Frame-based stereotactic surgery is still a standard of quality, precision and effectiveness. The stereotactic frame itself is indeed a relatively simple tool, which however guarantees a high accuracy: in vitro and in vivo studies reported errors in reaching the target of $2.23 \pm 0.14$ mm and $2.7 \pm 1.1$ mm respectively [8, 9]. However, a few negative aspects have led to devise new operating methodologies. One is the invasiveness of the process, due to fixing the frame to the patient's skull via screws. A further problem is given by the time between the preoperative and operative phases: it is necessary the patient wears the frame for the possible shortest time to avoid problematic repercussions at the anchorage points such as local infections and bleeding.

Robotics allowed reduction of these problems due to the robot ability to analyse the 3-D space and provide accurate motion. These new techniques allowed removal of the stereotactic frame (frameless techniques) with benefits for the patient but also for the surgeon. Despite these advantages, already theorized in 1985, only a few robots were marketed [6], due to their high technical complexity and costs.

The scope of this paper is to propose a solution that combines the advantages of stereotactic neurosurgery techniques, mitigating their problems. The stereotactic frame is substituted with a less invasive but equally accurate restraint system. This is based on a thermoplastic mask, which allows the patient's head mobility without requiring recalibration, also resulting in a compact system with moderate costs. The mask is the base for a mini-parallel robot with small mass and sufficient positional accuracy. The advantages of two possible solutions are discussed, together with future developments.

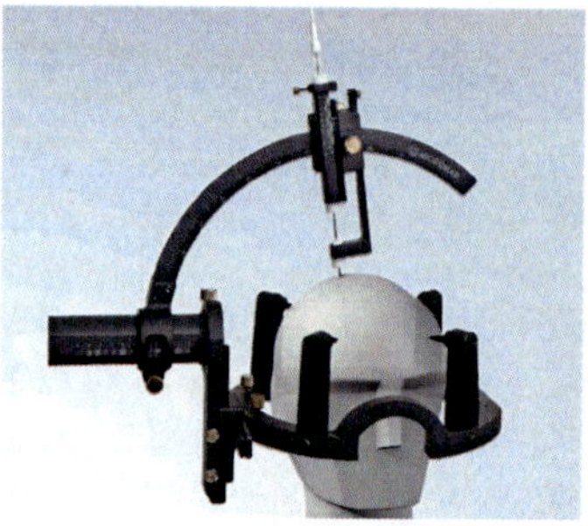

**Fig. 1.** Example of stereotactic frame with attached guides and the tool.

## 2 Stereotactic Mask

Stereotactic masks are widely used in radiosurgery [10]; however, they have never been applied in stereotactic surgery. Each mask is composed of a sheet of thermoplastic material, which is immersed in a thermostatic bath, becoming malleable. It can then be located on the patient's face and carefully formed by the operator to best match his somatic traits (Fig. 2). Once formed, the mask is left to cool until it regains its rigidity. The process takes about 15–30 min. Stereotactic masks may take different forms depending on certain characteristics [10]. A first feature is the polymeric material. It determines the main properties of the mask and in some cases, it can be integrated with Kevlar fibres. Another feature is the lightening hole scheme which can vary between 20% and 40% of the mask volume. Finally, the shape of the patient's body and consequently the areas of the patient's body covered by the mask and the anchor points may vary. All these parameters may lead to a variation in the mask stiffness and in the patient comfort.

The shape of the mask is also an interesting aspect to consider. Indeed, the thermoplastic masks on the market have different shapes depending on the areas of the body that will bind and the number of fixing points. The mask can only bind the head or reach the shoulder area (Fig. 3) with obvious advantages in the rigidity of the structure at the expense of comfort. In addition, they are distinguished by the number of anchor points, which can be two or three in the case of a face cover mask and four or five for a cover of the face and shoulders. A relatively recent evolution consists of two shells in thermoplastic material, the first that is arranged on the patient's nape and the second that is formed on the face. In this way it is possible to create a complete shell that covers the entire head ensuring better rigidity and positional accuracy [11].

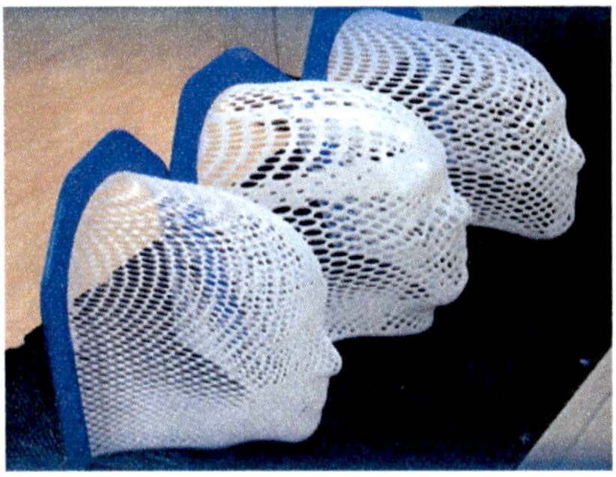

**Fig. 2.** Stereotactic mask example.

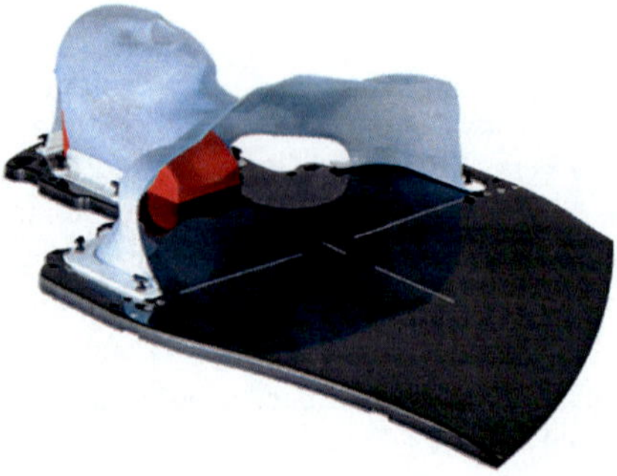

**Fig. 3.** Stereotactic mask that reaches the shoulder area.

## 3 Preoperative and Operational Procedure

The proposed system requires the use of a thermoplastic mask. However, as an element of discontinuity from the system used in radiosurgery, the mask will not be fixed to a support on the operating table but only to the patient's head thus ensuring head mobility, albeit limited, in the operating field.

In the preoperative phase, before surgery, an MRI of the patient's head is carried out to identify the target to be reached and plan the tool trajectory to reach the target itself, avoiding critical areas.

On the day of the surgery, first the thermoplastic mask is formed. For this type of application, a mask consisting of two shells is chosen since it guarantees a better positional accuracy both in the formation phase and in the repositioning phase, should the latter necessary as it will be shown in Sect. 5.

The half-shell that will cover the nape and the neck is the first to be formed (Fig. 4a). A sheet of thermoplastic material made malleable by a thermostatic bath is dried and then placed on supports. Such supports ensure a certain distance of the sheet from the cot on which the patient lies down in order to provide the necessary space to allow the half-shell deformation under the pressure of the head. During this process, the operator must take care to adhere the sheet to the patient's nape as better as possible in order to improve the positional accuracy of the system. Finally, a cooling time varying from fifteen to thirty minutes ensures a complete sheet solidification. It is important to note that, in the case of the lower half-shell, it is not necessary for the sheet to present a lightening pattern since there is no danger of clogging the patient's respiratory tract or generating a sense of claustrophobia in the predisposed patients. This improves the stiffness and the accuracy of the system.

Between the fastening headband of the half-shell in the nape and the neck and that of the face there is a support head (K in Fig. 4b), made of composite material, which has a dual function: it improves the rigidity of the system and presents an anchor zone for the registration points that will be discussed later.

The further step is the formation of the half-shell that will cover the patient's face (Fig. 4c). A second sheet of thermoplastic material, also in this case with an integrated restraint ring, is placed in the thermostatic bath. Once the required malleability is reached,

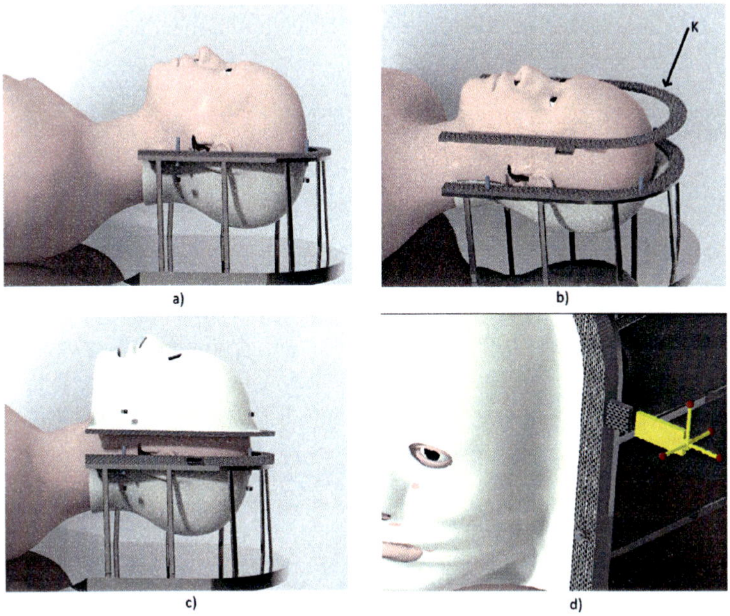


**Fig. 4.** Stereotactic mask production process: a) formation of the half-shell and fixing to the supports; b) fixing support headband K; c) superior half-shell formation and mask closure; d) reference system insertion.

it is extracted, dried, placed on the patient's face and fixed to the support headband. It is again the operator's task to make the material adhere to the patient's features, accurately tracing areas such as the nose, eyes, chin, and mouth. In this case a sheet with a lightening pattern is required, to allow patient breathing and to limit the sense of claustrophobia. Finally, the upper half-shell is solidified. To univocally define a coordinate system both in the image and in the device space, four markers (spheres with radio opaque materials suitably positioned) is fixed to the support headband (Fig. 4d). The four spheres represent the four points for the definition of the $4 \times 4$ transformation matrix between the two spaces. It is worth noting that three markers would be sufficient to define the coordinate transformation, but the fourth marker is used for redundancy and to improve accuracy. Markers shown in Fig. 4d are representative: a higher accuracy is reached by a larger distance between markers, depending on the specific space allowed.

Once the thermoplastic mask has been formed, a CT is carried out to obtain a detailed image of the patient's skull box including the external frame and coordinate system. Then, registration of the two three-dimensional models from CT and MRI is performed, to obtain an accurate representation of both the skull box and the brain. Through this procedure, the position of the target with respect to the reference system defined by the four spheres is determined.

The patient is then taken to the operating room and placed on the surgical bed. Here, a manipulator is attached to the stereotactic mask to guide and point the surgical tool.

Figure 5 shows a possible solution with a Gough-Stewart parallel mini-manipulator [6, 12] mounted on a circular guide.

## 4 Discussion

The use of thermoplastic masks in the field of stereotactic surgery would bring several advantages for the surgeon and the patient.

First, the stereotactic coordinates are fully defined as a result of the registration between CT and MRI. In this way the system is not affected by possible displacements of the patient's head during the operation. The system is also extremely compact, as it is frame-based. This ensures better mobility of the surgeon in the surgery room and the possibility of moving the patient's head as needed. A further advantage is the cost of the product. In fact, the thermoplastic mask is a widely distributed commercial product with a low price.

As for the patient, the mask guarantees a high comfort, especially when compared to frame-based systems, which require the use of screws to constrain the frame on the skull. In addition, as an additional comfort factor, it allows a slight mobility of the head with respect to the surgical bed.

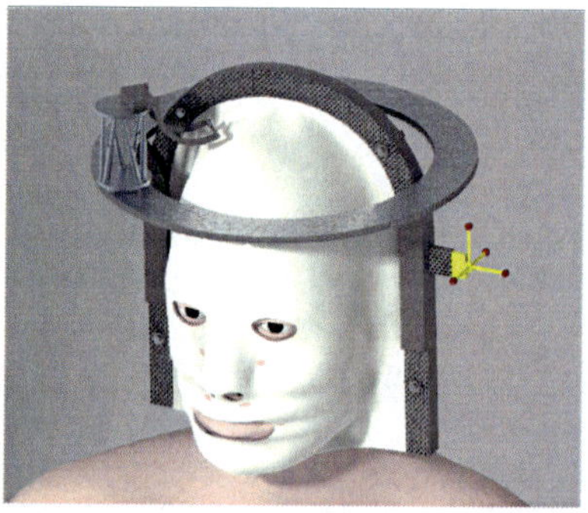

**Fig. 5.** System with stereotactic mask on which the circular guide is fixed. As an example, a mini Gough-Stewart robot is represented with a tool guide on the movable plate.

Among possible drawbacks of the proposed system, we can hypothesize the occurrence of movements of the mask in the intraoperative phase that would produce a variation of the stereotactic coordinates, introducing a positional error. Such displacements, in particular, could be due to the presence of soft tissue between the mask and the skull. However, the thermoplastic mask, if well formed around the nasal septum, chin area and other bony protrusions, should adequately avoid possible relative shifts. Therefore, assuming that the mask is sufficiently rigid to be assumed as a rigid body, these zones would introduce the six degrees of constraint necessary to ensure correct positioning in the absence of intraoperative displacements. For instance, standard thermoplastic masks for stereotactic radiotherapy showed displacements (mm) in the three orthogonal directions $0.1 \pm 0.3$, $0.2 \pm 0.4$ and $0.3 \pm 0.6$ respectively, and rotations pitch, roll and yaw 0.5° ± 0.6°, 0.2° ± 0.3° and 0.4° ± 0.3° respectively [11]. The exact amount of displacement of the current solution should be however experimentally quantified. A further possible drawback is related to neurosurgeons who may have some difficulties in stereotaxic operations to adapt to the new protocol, which is rather different from the current standard they are familiar with.

## 5 Second Preoperative and Operational Procedure

A second procedure is proposed based on the hypothesis that the relative position errors of the stereotactic mask and the skull is negligible between subsequent repositions.

The procedure requires that the stereotactic mask be formed in the first preoperative phase and followed by the MRI before disassembling the mask from the patient. In this way, during the preoperative phase it is possible to derive the stereotactic coordinates of the target with respect to the external reference system fixed to the mask. The mask is then dismounted, and the surgeon can plan the surgery within the time between the preoperative and operative phases. Then, the mask is repositioned on the patient before the operation and the CT scan is performed. If repositioning were accurate enough, it would be sufficient to register the external reference system fixed to the mask, on the MRI and CT images, to have an accurate identification of the stereotactic coordinates. The other steps are the same as reported for the first solution in the previous sections.

This intervention methodology would avoid the process of bone model registration, reducing time and costs.

## 6 Conclusions

Starting from a brief historical excursus on the birth and evolution of stereotactic surgery, an innovative operating methodology, and a variant of it, have been proposed, which exploit the use of a stereotactic mask. The proposed techniques are based on pre-existing systems but modified to integrate special characteristics. A discussion was then carried out on the positive and negative characteristics of the new procedures.

Further studies and practical tests will be carried out in order to validate the proposed techniques and the assumed hypotheses.

## References


1. Shurkhay, V.A., Goryaynov, S.A., Aleksandrova, E.V., Spallone, A., Potapov, A.A.: Navigation systems in neurosurgery. Voprosy neirokhirurgii imeni N.N. Burdenko **80**(6), 107 (2016). https://doi.org/10.17116/neiro2016806107-114
2. Grunert, P., Darabi, K., Espinosa, J., Filippi, R.: Computer-aided navigation in neurosurgery. Neurosurg. Rev. **26**(2), 73–99 (2003). https://doi.org/10.1007/s10143-003-0262-0
3. Dandy, W.E.: ventriculography following the injection of air into the cerebral ventricles. Ann. Surg. **68**(1), 5–11 (1918). https://doi.org/10.1097/00000658-191807000-00002. PMID:17863946; PMCID:PMC1426769
4. Beckmann, E.C.: CT scanning the early days. Br. J. Radiol. **79**(937), 5–8 (2006). https://doi.org/10.1259/bjr/29444122. PMID: 16421398

5. Cope, F.W., Damadian, R.: Cell potassium by 39K spin echo nuclear magnetic resonance. Nature **228**(5266), 76–77 (1970). https://doi.org/10.1038/228076a0. PMID: 5456216
6. Faria, C., Erlhagen, W., Rito, M., De Momi, E., Ferrigno, G., Bicho, E.: Review of robotic technology for stereotactic neurosurgery. IEEE Rev. Biomed. Eng. **8**, 125–137 (2015). https://doi.org/10.1109/RBME.2015.2428305. Epub 2015 Apr 30 PMID: 25955851
7. Rhodes, M.L., Glenn, W.V., Azzawi, Y.M., et al.: Stereotactic neurosurgery using 3-D image data from computed tomography. J. Med. Syst. **6**, 105–119 (1982). https://doi.org/10.1007/BF00994124
8. Quiñones-Hinojosa, A., Ware, M., Sanai, N., McDermott, M.: Assessment of image guided accuracy in a skull model: comparison of frameless stereotaxy techniques vs. frame-based localization. J. Neuro-Oncol. **76**(1), 65–70 (2005). https://doi.org/10.1007/s11060-005-2915-z
9. Bradac, O., Steklacova, A., Nebrenska, K., Vrana, J., de Lacy, P., Benes, V.: Accuracy of varioguide frameless stereotactic system against frame-based stereotaxy: prospective, randomized single-center study. World Neurosurg. **104**, 831–840 (2017). https://doi.org/10.1016/j.wneu.2017.04.104. Epub 2017 Apr 25 PMID: 28454992
10. Tunkr, A.: «Assessment of Various Thermoplastic Masks Used for Radiation Therapy Immobilization in Terms of Shrinkage and Pressure», B.Sc. McMaster Univerity thesis (2015)
11. Meer, S., et al.: EP-1506: Evaluation of cranial setup accuracy: a double shell positioning system versus an in-house 3-point mask solution. Radiother. Oncol. **115**, S819–S820 (2015). https://doi.org/10.1016/S0167-8140(15)41498-7
12. Joskowicz, L., et al.: Image-guided system with miniature robot for precise positioning and targeting in keyhole neurosurgery. Comput. Aided Surg. **11**(4), 181–193 (2006). https://doi.org/10.3109/10929080600909351

# Estimation of Force Effectiveness and Symmetry During Kranking Training

Elisa Digo[1](✉), Laura Gastaldi[2], Mattia Antonelli[1], Valerio Cornagliotto[1], and Stefano Pastorelli[1]

[1] Department of Mechanical and Aerospace Engineering, Politecnico Di Torino, Turin, Italy
elisa.digo@polito.it

[2] Department of Mathematical Sciences "G.L. Lagrange", Politecnico Di Torino, Turin, Italy

**Abstract.** The third Sustainable Development Goal of the 2030 Agenda promotes healthy lives and well-being for all people of all ages. A good way to ensure a healthy lifestyle is to perform daily physical activity. Among different exercises of cardiovascular training, kranking is a program that involves arm-cranking gesture performed on a stationary handbike. In order to correctly perform this activity, biomechanical parameters have to be monitored. The present pilot study aimed at developing a setup for the quantitative evaluation of the force effectiveness and symmetry during different conditions of upper limbs kranking. One healthy young subject performed different tasks of steady-state cycling on varying cadence, braking torque, and motion pattern. Strain gauges positioned on the handles of a commercial arm-cranking machine allowed the estimation of total and effective forces applied by the user. Moreover, an optical motion capture system was adopted to evaluate the kinematics of the upper limbs during the movement. Comparing the total and the effective forces, the effectiveness of the gesture was evaluated for all testing conditions. Overall, results suggest that the developed setup is adequate to efficaciously identify possible alterations of performance parameters during upper limbs kranking.

**Keywords:** SDG3 · Kranking · Upper limbs · Physical training · Active ageing · Inclusive fitness

## 1 Introduction

In line with the third Sustainable Development Goal (SDG3) of the 2030 Agenda, the assurance of healthy lives and the promotion of well-being play key roles for all people of all ages [1]. In this context, physical activity provides many benefits to the health condition. Among different training programs, kranking represents a suitable alternative to the most common physical exercises. Indeed, this stationary form of cycling powered by upper limbs has lots of advantages in different fields such as athletic training, physical therapy, active ageing, and inclusive fitness. More in detail, kranking helps the strengthening of upper body muscles, the increase of workout endurance, and the enhancement of cardiovascular conditions. Hence, many sports activities that involve


G. Quaglia et al. (Eds.): I4SDG 2021, MMS 108, pp. 201–208, 2022.
https://doi.org/10.1007/978-3-030-87383-7_22

upper limbs may be attributed to handcycling exercise [2, 3]. In a clinical context, repetitive movements typical of kranking favour rehabilitation and physical therapy in terms of motor learning, functional recovery, and movement coordination [4–6]. The consistent functional consequences of ageing make the elderly health condition a social issue to cope with. Accordingly, kranking might be adopted to improve mobility, balance, and postural stability of healthy elderly people [7]. Finally, handbikes have also become increasingly popular in sports, recreation, and outdoor transportation of people with lower limb impairments [8].

The biomechanical study of the arm-cranking movement from different perspectives can offer concrete help in enhancing performance during physical training. First, the kranking gesture can be analyzed evaluating the benefits on performance of the in phase (the two handles are moved to keep them paired) and antiphase (the movement of one handle has a phase of 180° with respect to the other handle) cycling patterns of motion [9, 10]. Then, starting from biomechanical musculoskeletal models of the human upper body [11], the kranking gesture can be studied considering the involvement of different muscles and identifying push and pull phases in a cycling period [12–14]. Moreover, the analysis may focus on the impact of extrinsic biofeedbacks displaying the performance metrics of work, speed, and cadence during the exercise [15, 16]. Finally, the investigation of the arm-cranking gesture can also concentrate on forces, torques, powers, and cadences at stake [17–19].

The aim of the present pilot study was to validate the tuned setup for the quantitative evaluation of the force effectiveness and symmetry during different conditions of upper limbs kranking. In detail, one healthy and young subject with no experience in hand cycling performed both in phase and antiphase steady-state cycling changing cadence, braking torque, and motion pattern. Strain gauges were positioned on the handles of a commercial arm-cranking machine to estimate the total and effective forces applied by the user. An optical motion capture system consisting of cameras and markers was adopted to evaluate the kinematics of the upper limbs during the movement. Relating total and effective forces, the effectiveness of the gesture was assessed.

## 2 Materials and Methods

### 2.1 Instruments

The testing activity involved a Krankcycle® machine (Matrix, JHT ITALIA s.p.a.), which is a stationary handbike with independent crank handles, designed for the training of a subject. The power exerted by the user is dissipated by a brake acting on a flywheel. The resistant torque of the brake can be manually adjusted through a knob on the front part of the device. The advantage offered by two independent crank handles allows the reproduction of both in phase and antiphase cycling. In the former pattern the cranks are aligned, while in the latter pattern the cranks are turned in a diametrically opposite position (Fig. 1A).

Two strain gauge sensors were placed on the pins connecting the grips with the crank handles to evaluate the radial forces exerted by the user on each handle (Fig. 1B). Each sensor was made up of four extensometers (HBM, Germany) with electric resistance type 1-LY11–0.6/120. A control unit Spider8 (HBM, Germany) with a built-in conditioning unit was adopted to convert the analogic inputs into digital signals which were processed exploiting the software Catman Easy.

A stereophotogrammetric system with two high-speed cameras was installed in the testing laboratory. Each camera was oriented with the lens orthogonal to the sagittal plane to record both sides of the Krankcycle. The cameras frame rate was set to 100 fps, thus ensuring a proper reconstruction of the cycling gesture, taking place at a comfortable self-selected rate up to 2 Hz [8]. Cameras were calibrated before and after trials using a dedicated calibration tool. Since the cranking movement can be approximately considered planar, a simplified 2D biomechanical model of the upper limbs, based on six markers per side, was adopted. Markers were arranged on the wrists, elbows, shoulders, and acromions of the participant. Two additional markers were fixed on the extremities of each grip to evaluate its instantaneous position and orientation (Fig. 1C).

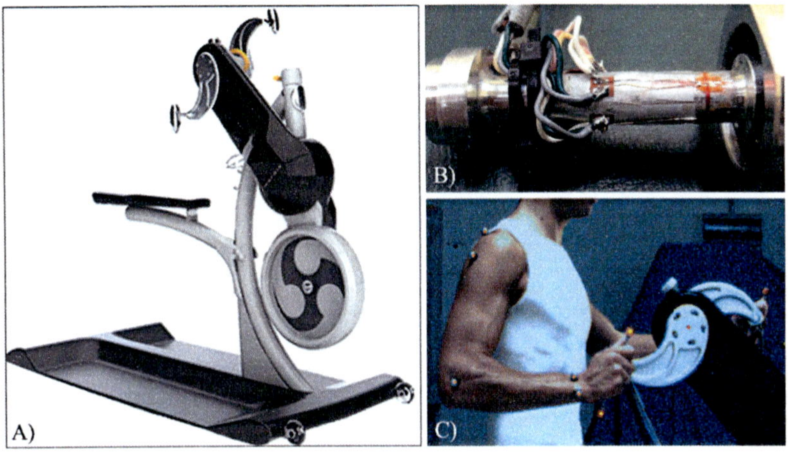


**Fig. 1.** A) The Krankcycle® machine. B) Strain gauges installed on the handle shaft. C) Right view on an antiphase movement with markers fixed on the upper limb and handle extremities.

## 2.2 Protocol

The test consisted of the acquisition of kinematic and dynamic parameters during kranking performed by a young healthy male subject. The study was approved by the Local Institutional Review Board and all procedures were conformed to the Helsinki Declaration. Before the experiment, video cameras were calibrated. A board with a grid of points was placed in the field of vision of each camera according to the plane of motion of the crank handle and the human arm. Since the distance between the points in the grid was known, the software created a correspondence between each pixel in the recorded movie and a specific bi-dimensional position. Once the two video cameras were set, the synchronization between the video recording and the signals coming from the strain gauge sensors was performed through an external triggering device.

Subsequently, the cadence, torque, and motion pattern were varied to define many kranking tests enabling qualitative considerations on the effects of these parameters on the kranking activity. In detail, low and high cadences were respectively self-selected by the subject with the goal of investigating the performances at different speeds of cycling. Then, two qualitative levels (low and high) of resistant torque on the flywheel were imposed on the Krankcycle brake. Moreover, the hand-cycling style was also explored activating the two crank handles in phase or antiphase. As Table 1 shows, the most significant tests were selected and analyzed. In particular, the test condition characterized by low self-selected cadence, high braking torque, and in phase cycling pattern was assumed as the reference for the comparison of the other conditions. Once the specific experimental setting was defined, each test was repeated three times for 10 s. Data acquisition was performed in steady-state condition (after 60 s from the start) through dedicated hardware and software devices.

**Table 1.** Kranking characteristics of the selected tests.

| | Test 1 (ref) | Test 2 | Test 3 | Test 4 |
|---|---|---|---|---|
| Cadence (Hz) | Low | High | Low | Low |
| Torque (Nm) | High | High | Low | High |
| Pattern | In phase | In phase | In phase | Antiphase |

### 2.3 Data Analysis

Once the tests were performed, all the collected data was properly handled to extract relevant kinematic and dynamic information (Fig. 2). Exploiting the optoelectronic system, the trajectories of body markers and grip extremities ($H_1$ and $H_2$) were known. The positions of shoulders (S), elbows (E), and wrists (W) were defined in the cameras global frame. Furthermore, the centres of rotation of the crank handles (C) and the centres of the grippers (H) were identified in the cameras global frame. Then, H were related to C. Since the point of application of the force action was not directly visible in the movie frames, it was individuated geometrically. The adopted methodology consisted of measuring the position of the pin along the segment connecting the grip extremities. Since the grip was a rigid body, the relative position between its extremities and the connection point with the pin was invariable. Figure 2 shows the reference frames (rf) taken into account in the analysis. A fixed rf 0 ($x_{0i}y_{0i}$) was placed corresponding with the rotation axis of the crank handle on the $i^{th}$ side, where "i" stands for the left or right side. The rf C ($x_{Ci}y_{Ci}$) was fixed to the crank and its orientation with respect to the rf 0 depends on the rotation angle $\theta_{Ci}$ of the crank itself. Then, the rf H ($x_{Hi}y_{Hi}$) was defined as linked to the handle grip. Thus, the $i^{th}$ force expressed in the rf H was referred to the rf 0, as described in Eq. 1:

$$F_0 = A_C^0 \cdot A_H^C \cdot F_H \tag{1}$$

where $F$ is the force action referred to the fixed rf ($F_0$) and the handle rf ($F_H$); $A_C^0$ is the transformation matrix defining C with respect to 0; $A_H^C$ is the transformation matrix defining H with respect to C. In the following, to lighten the notation, all the vectors are expressed in the fixed reference frame, hence omitting the subscript. The torque transmitted to the device on each side was estimated as described in Eq. 2:

$$T_i = b_i \times F_i \quad (2)$$

where $b_i$ is the vector along with the crank. Accordingly, the total torque (Nm) applied by the subject was obtained as the sum of the two contributes on the right and left sides. Moreover, cadence (Hz) and phase error (deg) were calculated for each test. Subsequently, right and left performance parameters were computed as the ratio between the effective force ($F_{Ef}$), i.e. the force component that contributes to the forward motion of the crank, and the total force ($F$). Finally, the ratio between left and right effectiveness was estimated. For each test, the results of the three attempts were averaged to compare different activity conditions in terms of kinematics, forces, and effectiveness.

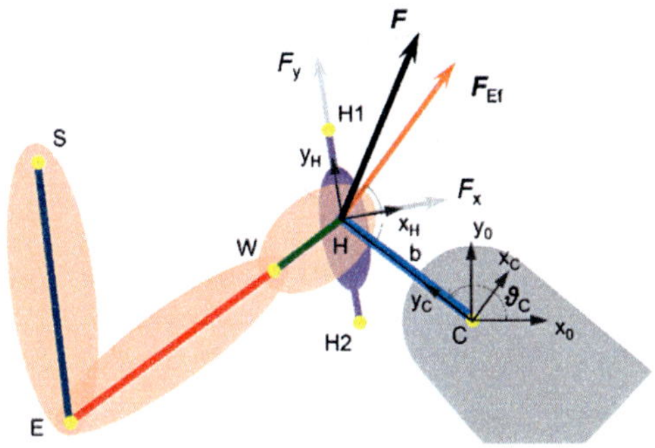


**Fig. 2.** Schematic description of reference frames adopted during the analysis.

# 3 Results and Discussions

The following results refer to the performance parameters evaluated varying the testing conditions (cadence, torque, and motion pattern) with the aim of analyzing possible differences in terms of symmetry, correctness of motor activity, and proprioception.

Figure 3 illustrates mean values of right and left effectiveness defined for all testing conditions as the ratio between the effective force $F_{Ef}$ and the total force $F$. For each test, values of all repetitions were averaged and expressed over a complete rotation. According to Fig. 3, the effectiveness is not homogeneous during a single period. In the tests with the in phase pattern (Fig. 3.1, 2, 3), the effectiveness improves when the movement occurs in both the areas proximal (PR) and distal (DS) with respect to the human trunk. On the contrary, a slight worsening of effectiveness can be observed in correspondence of the superior (UP) and inferior (DW) regions. Since the participant was right-handed, Fig. 3.4 depicts a lower left effectiveness in the anti-phase pattern.

Table 2 shows mean and standard deviation values of significant parameters estimated for all repetitions of the four reported conditions. The right (R) effectiveness was adequate (≥0.92) independently of the testing condition. However, the reduction of the braking torque (test 3) caused a slight decrease in effectiveness (−3%). Considering the left (L) weak arm effectiveness, a significant difference between in phase and antiphase gestures can be highlighted. Indeed, for the in phase conditions, both the rise of the cadence (test 2) and the reduction of the torque (test 3) produced an increase of effectiveness (+4% and + 5%, respectively). On the contrary, as stated before, the antiphase condition (test 4) caused a remarkable worsening of effectiveness (−9%).

Observing the L/R effectiveness ratios, it is evident that both the rise of the cadence (test 2) and the reduction of the torque (test 3) improved the symmetry of effectiveness (+5% and + 9%, respectively). On the contrary, the antiphase condition (test 4) led to a lower effectiveness ratio with respect to the reference test (−7%). Moreover, the phase error of in phase gestures is negligible (≤2.11°), whereas in antiphase conditions it is notable (12.32°). This confirms that, during the antiphase test, it is more difficult to maintain a constant phase delay of 180° between the two handles.

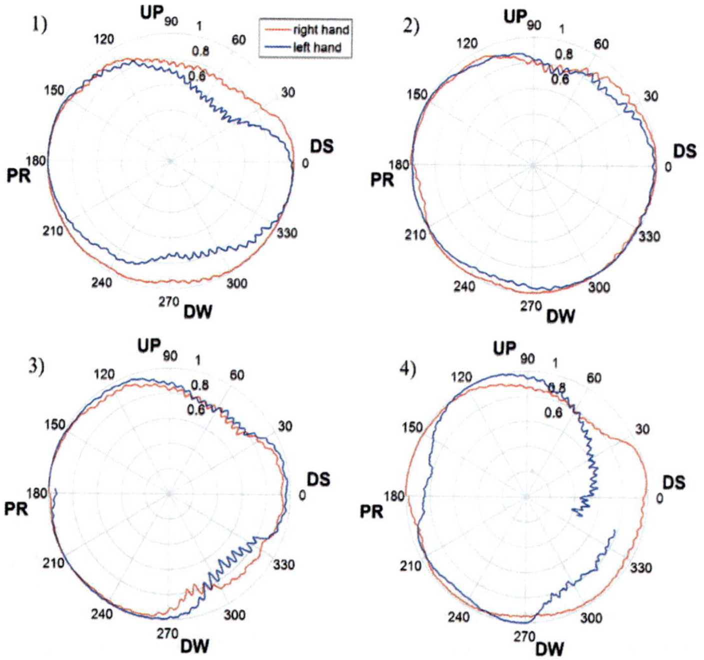


**Fig. 3.** Mean values of right and left effectiveness estimated as functions of the kranking angle. 1) Reference; 2) Cadence increase; 3) Torque reduction; 4) Antiphase handcycling pattern.

**Table 2.** Mean and standard deviation (std) values of significant parameters.

| | Test 1 (reference) | Test 2 | Test 3 | Test 4 |
|---|---|---|---|---|
| | Mean (std) | Mean (std) | Mean (std) | Mean (std) |
| Cadence (Hz) | 0.63 (0.03) | 1.31 (0.06) | 0.72 (0.04) | 0.63 (0.04) |
| Torque (Nm) | 12.82 (3.59) | 15.78 (3.98) | 6.22 (1.71) | 13.04 (1.72) |
| Phase error (deg) | −0.28 (1.52) | 2.11 (1.80) | 2.09 (1.55) | 12.32 (1.86) |
| R-effectiveness | 0.95 (0.07) | 0.95 (0.06) | 0.92 (0.07) | 0.94 (0.07) |
| L-effectiveness | 0.87 (0.12) | 0.91 (0.08) | 0.92 (0.08) | 0.78 (0.19) |
| L/R eff. ratio | 0.91 (0.09) | 0.96 (0.06) | 1.00 (0.05) | 0.84 (0.23) |

## 4 Conclusions

This pilot study aimed at validating the presented setup to quantitatively evaluate the force effectiveness and symmetry during different kranking conditions. Overall, results suggest that the current setup is suitable for highlighting deficits in terms of force effectiveness and gesture symmetry during the arm-cranking gesture. Hence, the proposed tool might be adopted to evaluate the performance through the monitoring of the effectiveness parameters. This aspect represents an improvement in different contexts such as athletic training, physical therapy, active ageing, and inclusive fitness.

The present work focused mainly on the evaluation of the force effectiveness and symmetry during the kranking activity. To provide better performance and effectiveness of training, the proposed setup has been equipped with real-time biofeedbacks of the cranking rate and left and right effectiveness and symmetry. Even additional sensors such as wearable inertial sensors and EMG sensors might be added for the extraction of other kinematic and muscular features.

**Acknowledgement.** Authors thank JHT ITALIA s.p.a. – MATRIX for providing the Krankcycle® machine.

## References

1. United Nations: Goals @ Sdgs.Un.Org. https://sdgs.un.org/goals
2. Quittmann, O.J., et al.: Kinematics and kinetics of handcycling propulsion at increasing workloads in able-bodied subjects. Sports Eng. **21**(3), 1–12 (2018). https://doi.org/10.1007/s12283-018-0269-y
3. Pearson, S., Hume, P., Cronin, J., Slyfield, D.: America's Cup Sailing: Effect of Standing Arm-Cranking ("Grinding") Direction on Muscle Activity, Kinematics, and Torque Application. Sports. **4**, 37 (2016). https://doi.org/10.3390/sports4030037
4. Oyake, K., et al.: Unilateral arm crank exercise test for assessing cardiorespiratory fitness in individuals with Hemiparetic stroke. BioMed. Res. Int. **2017**, 1–10 (2017). https://doi.org/10.1155/2017/6862041

5. Brizuela, G., Sinz, S., Aranda, R., Martínez-Navarro, I.: The effect of arm-crank exercise training on power output, spirometric and cardiac function and level of autonomy in persons with tetraplegia. Eur. J. Sport Sci. **20**, 926–934 (2020). https://doi.org/10.1080/17461391.2019.1674927
6. Diserens, K., et al.: The effect of repetitive arm cycling on post stroke spasticity and motor control. J. Neurol. Sci. **253**(1–2), 18–24 (2007). https://doi.org/10.1016/j.jns.2006.10.021
7. Hill, M., Oxford, S., Duncan, M., Price, M.: Arm-crank training improves postural stability and physical functioning in older people. Exp. Gerontol. **113**, 218–227 (2018). https://doi.org/10.1016/j.exger.2018.10.007
8. Verellen, J., Theisen, D., Vanlandewijck, Y.: Influence of crank rate in hand cycling. Med. Sci. Sports Exerc. **36**, 1826–1831 (2004). https://doi.org/10.1249/01.MSS.0000142367.04918.5A
9. Chen, H.-Y., Chen, S.-C., Chen, J.-J.J., Fu, L.L., Wang, Y.L.: Kinesiological and kinematical analysis for stroke subjects with asymmetrical cycling movement patterns. J. Electromyogr. Kinesiol. **15**(6), 587–595 (2005). https://doi.org/10.1016/j.jelekin.2005.06.001
10. Meyns, P., Van de Walle, P., Hoogkamer, W., Kiekens, C., Desloovere, K., Duysens, J.: Coordinating arms and legs on a hybrid rehabilitation tricycle: the metabolic benefit of asymmetrical compared to symmetrical arm movements. Eur. J. Appl. Physiol. **114**(4), 743–750 (2013). https://doi.org/10.1007/s00421-013-2814-5
11. Gastaldi, L., Panero, E., Rosso, V., Pastorelli, S., Vieira, T., Botter, A.: Upper Limbs Musculoskeletal OpenSim model: customization and assessment. In: Niola, V., Gasparetto, A. (eds.) IFToMM ITALY 2020. MMS, vol. 91, pp. 162–170. Springer, Cham (2021). https://doi.org/10.1007/978-3-030-55807-9_19
12. Faupin, A., Gorce, P., Watelain, E., Meyer, C., Thevenon, A.: A biomechanical analysis of handcycling: a case study. J. Appl. Biomech. **26**, 240–245 (2010). https://doi.org/10.1123/jab.26.2.240
13. Litzenberger, S., Mally, F., Sabo, A.: Biomechanics of elite recumbent handcycling: a case study. Sports Eng. **19**(3), 201–211 (2016). https://doi.org/10.1007/s12283-016-0206-x
14. Smith, P.M., Chapman, M.L., Hazlehurst, K.E., Goss-Sampson, M.A.: The influence of crank configuration on muscle activity and torque production during arm crank ergometry. J. Electromyogr. Kinesiol. **18**, 598–605 (2008). https://doi.org/10.1016/j.jelekin.2006.12.006
15. Lin, P.Y., Chen, J.J.J., Lin, S.I.: The cortical control of cycling exercise in stroke patients: an fNIRS study. Hum. Brain Mapp. **34**, 2381–2390 (2013). https://doi.org/10.1002/hbm.22072
16. Digo, E., Pastorelli, S., Vieira, T., Botter, A., Gastaldi, L.: Upper limbs cranking for post-stroke rehabilitation: a pilot study on healthy subjects. IEEE International Symposium on Medical Measurements and Applications, MeMeA 2020 - Conference Proceedings (2020). Doi: https://doi.org/10.1109/MeMeA49120.2020.9137282
17. Verellen, J., Meyer, C., Reynders, S., Van Biesen, D., Vanlandewijck, Y.: Consistency of within-cycle torque distribution pattern in hand cycling. J. Rehabil. Res. Dev. **45**, 1295–1302 (2008). https://doi.org/10.1682/JRRD.2007.12.0205
18. Goosey-Tolfrey, V.L., Alfano, H., Fowler, N.: The influence of crank length and cadence on mechanical efficiency in hand cycling. Eur. J. Appl. Physiol. **102**, 189–194 (2008). https://doi.org/10.1007/s00421-007-0576-7
19. Krämer, C., Hilker, L., Böhm, H.: Influence of crank length and crank width on maximal hand cycling power and cadence. Eur. J. Appl. Physiol. **106**, 749–757 (2009). https://doi.org/10.1007/s00421-009-1062-1

# Applications of Intermittent Pneumatic Compression for Diagnostic and Therapeutic Purposes

Carlo Ferraresi[1], Walter Franco[1], Daniela Maffiodo[1], Carlo De Benedictis[1](✉), Maria Paterna[1], Daniel Pacheco Quiñones[1], Leonardo Ermini[2], and Silvestro Roatta[2]

[1] Department of Mechanical and Aerospace Engineering, Politecnico di Torino, Turin, Italy
carlo.debenedictis@polito.it

[2] Laboratory of Integrative Physiology, Department of Neuroscience, University of Torino, Turin, Italy

**Abstract.** Intermittent Pneumatic Compression (IPC) technique is prescribed for several treatments, as the management of venous leg ulcers or the prevention of deep vein thrombosis. Commercial devices do not enable the full customization of the compressive patterns due to design specifications and low dynamics. However, IPC can be implemented in a wide scenario of clinical protocols, and not only as a therapeutic tool. In this paper, the results of the research on IPC devices conducted at the Politecnico di Torino (Turin, Italy) are presented. In particular, applications regarding the treatment of the end-diastolic volume (EDV) reduction, the investigation of vascular phenomena as hyperemia, and the assessment of venous pulse wave velocity (vPWV) are discussed. The outcomes of the research demonstrate that IPC technology can lead to the creation of widely used diagnostic, therapeutic and rehabilitative devices.

**Keywords:** SDG3 · Therapeutic and rehabilitative devices · Intermittent pneumatic compression · Hemodynamics · Hyperemia · vPWV · Pneumotronics · Human-machine interaction

## 1 Introduction

In an extensive sense, Intermittent Pneumatic Compression (IPC) represents the application of mechanical pressure stimuli on a person's body, which can be either periodic or impulsive. An IPC device consists of one or more inflatable sleeves and an inflating system that allows a controlled and well-defined pressure trend to be applied to the lower or upper limbs of patients. In a first possible application, the applied compression may mimic the action of the muscle pump and promote venous return. For this reason, it can be used for the treatment of various pathologies, such as lymphedema, venous leg ulcers [3–5], prevention of deep vein thrombosis [6, 7], and for the treatment of subjects with limited mobility due to recent surgery or pathological condition.

IPC devices come with either single or multi-chamber sleeves [5]. The latter can provide sequential compression that can push the blood to the central part of the


G. Quaglia et al. (Eds.): I4SDG 2021, MMS 108, pp. 209–218, 2022.
https://doi.org/10.1007/978-3-030-87383-7_23

patient's body; the single-chamber devices, instead, produce controlled compression that is focused to a defined point of the body. The choice of one or the other type of device depends only on the specific application. In recent years, in order to improve the venous blood return and to prevent a reduction in left ventricle end-diastolic volume (EDV) in subjects with a walking disability, a multi-chamber IPC device has been developed at Politecnico di Torino [8–10].

In another application, Ferraresi et al. [11] describe a single chamber IPC device with high dynamic performance and capable to apply a customizable pressure pattern to the limbs of the subject. The device is therefore particularly suitable for research purposes oriented to better understand the mechanism behind the hyperemic response that develops in response to compression stimuli and its attenuation following consecutive compressions [12, 13]. The same device was also employed in a different application, oriented to the non-invasive assessment of a subject's volemic status [14, 15]. It is based on the generation and measurement of the propagation velocity of a pressure wave in the veins, namely the venous Pulse Wave Velocity (vPWV), which is related to vessel stiffness and therefore to vascular filling. The PWV has been used for years to monitor the stiffness of the arteries [16–18], but it is difficult to use it to monitor the venous compartment due to the lack of natural pulsations that therefore must be generated by the IPC device.

This work presents the outcomes of the research activity on IPC devices carried out by the group of functional biomechanics at the Department of Mechanical and Aerospace Engineering (DIMEAS) of Politecnico di Torino, Italy. The activity detailed in this work was carried out in collaboration with relevant research centers, in particular the Laboratory of Integrative Physiology of the University of Torino, Italy and the Laboratory of Sports Physiology of the University of Cagliari, Italy. The diversity of the applications shown in this paper and in the literature signals the relevancy of the research theme. In fact, the improvement of treatments for cardiovascular diseases and the definition of new methodologies for the investigation of hemodynamics are well in agreement with the rationale behind SDG3 promoted by the United Nations, which aims at ensuring healthy lives and well-being.

## 2 IPC Device for the Improvement of Venous Blood Return

Patients with a walking disability are subjected to a reduction in the EDV of the left ventricle, since the blood pressure gradient between the right atrium and the postcaval vein is normally supplied by effects related to locomotion, as the rhythmic contraction of the triceps surae which compresses the veins in the calf. For this reason, a device aiming to generate intermittent stimulation to the lower limbs can be effectively used to improve the cardiac functionality in those patients.

Ferraresi et al. [8], on the basis of a mathematical modeling, created a multi-bladder IPC device designed to favor the return of venous blood in people with motor deficit in the lower limbs. The mechatronic system was provided with six inflatable bladders integrated within two shells, each mounted on a separate segment (calf and foot, see Fig. 1). The bladders were designed to direct the pneumatic energy towards the limb without dispersion related to the deformation of the materials. Thanks to compliant and air-tight material, as well as to proper shaping of each bladder, it was possible to achieve good performance of the IPC device.

The control system was based on a group of six 3-way electro-pneumatic valves, six pressure sensors and a programmable logic controller. The valves and pressure sensors were mounted on the calf shell, in order to improve the dynamic performance of the pneumatic system.

Subsequently, Manuello Bertetto et al. [9] applied this device in experimental trials, for objective evaluation of its effectiveness in the enhancement of venous blood return. To assess EDV non-invasively, an equipment able to measure the thoracic electrical bioimpedance was used.

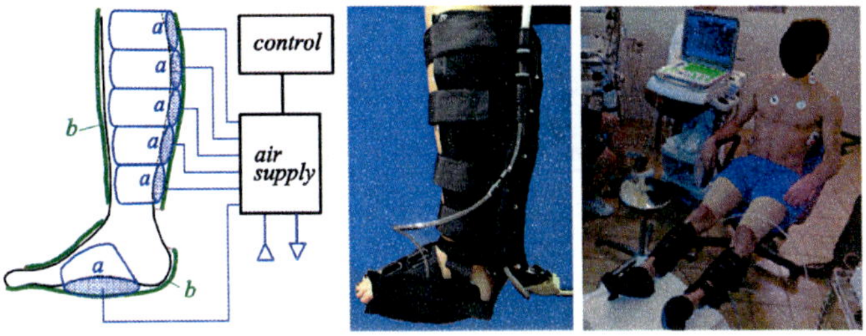


**Fig. 1.** Pneumotronic IPC device for the improvement of venous blood return; application in experimental trials for EDV assessment.

Compression-relaxation protocols were designed considering the alternating activation of two sleeves to simulate the muscle pattern during locomotion. The mechatronic device was tested on 19 healthy voluntary participants, which were initially monitored for 3 min to collect TEB baseline values. During the experimentation, which involved activation-deactivation sequences following a peristaltic compression (caudal-rostral trend), the EDV increased of about 10% with respect to preliminary observation done during initial monitoring of the subjects.

Although the study did not consider patients unable to walk or bed-ridden, it confirmed the relevance of the implementation of such IPC devices for the treatment of conditions related to reduction of venous blood return.

## 3 IPC Device for the Study of Vascular Phenomena

Besides therapeutic treatments, IPC devices can be effectively used to examine in-depth vascular phenomena as hyperemia. Ferraresi et al. [11] presented the design and modeling of an IPC device aimed at providing customizable pressure profiles to human limbs, whereas Messere et al. [12, 13] showed applications of such device on a set of healthy subjects, aimed at developing protocols for the study of the mechanisms behind hyperemia (i.e., the local increase of blood flow to tissues).

The device, shown in Fig. 2, was based on a pneumatic circuit used to inflate a blood pressure cuff wrapped around the limb to stimulate. In order to achieve appropriate responses, it was designed to control the pressure with a step reference of magnitude between 0 and 250 mmHg, with critical attention given to static and dynamic performance.

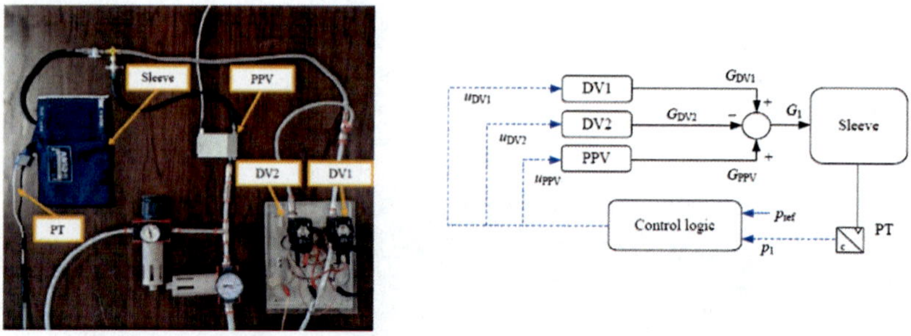


**Fig. 2.** Picture (left) and scheme (right) of the pneumotronic system [11]. DV1 and DV2 are the digital valves respectively used to charge and discharge the cuff, PPV is the pressure proportional valve, PT is the pressure transducer.

In order to fulfil the required specifications of accuracy and short time response, a pneumatic circuit based on the combination of 2/2 digital solenoid valves and a pressure proportional valve, shown in Fig. 2, was designed. The digital valves (DV1 and DV2) were used to quickly charge and discharge the cuff, while the compact pressure proportional valve (PPV) was used to accurately control the level of pressure inside the cuff. A control system, combined with a pressure transducer PT positioned inside the cuff, was used in the final architecture to drive the valves, depending on the operating conditions of the system.

In order to investigate the sensitivity of the device to the variation of some physical parameters, an analytical model was implemented. In particular, the interaction between the cuff and the biological tissues was studied to achieve an accurate matching between the model and experimental data.

The final design of the pneumotronic IPC device showed high dynamic performance and was able to provide sharp, customizable stimuli to the limb of a subject (see Fig. 3) in a wide range of pressure, necessary for the investigation of several vascular phenomena with different dynamics.

### 3.1 Investigation of Vascular Reactivity to Mechanical Stimuli

Resistance vessels in skeletal muscle exhibit a rapid dilatation in response to mechanical stimuli, such as short-lasting external compressions, whose underlying mechanisms are still unclear. Messere et al. [12] performed an analysis on 10 healthy volunteers with the IPC device described in the previous section. The subjects sat on a chair with a backrest, the leg fully extended with the cuff wrapped distally to the knee. The protocol started after 10 min of rest and provided customized pressure patterns. With respect

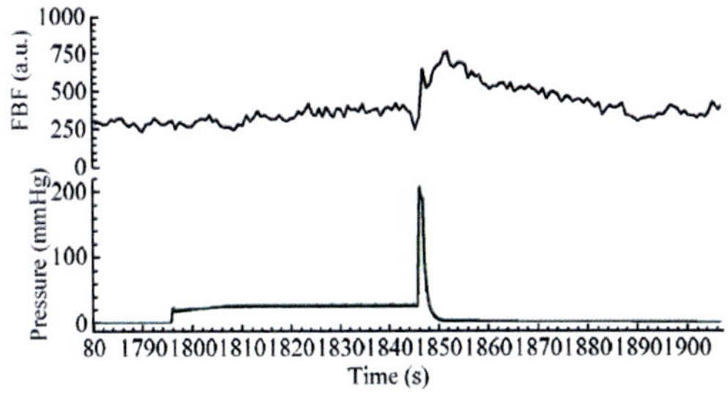


**Fig. 3.** Recordings of a post-compression hyperemia from a representative subject: femoral artery blood flow (FBF) and cuff pressure [11].

to traditional devices, that are often meant to apply fast and short-lasting stimuli, the custom-design IPC device was necessary to generate sequences of static pressure levels that can be considered to modulate vascular filling in venous compartments as well as pressure gradients across the wall of the vessels.

During the compressions, the blood flow of the femoral artery (FBF) was measured by Eco-Doppler sonography. The flow response highlighted a significant increase of FBF after the occurrence of the high-level compression. Preliminary investigations showed suitable performance of the device for the investigation of vascular phenomena behind the compression-induced rapid hyperemia that still do not have a univocal interpretation [19–24].

Another relevant issue recently investigated is the role of tissue oxygenation in shaping the magnitude of this mechanically-induced rapid hyperemia. Messere et al. [13] hypothesized that increased tissue oxygenation could explain the progressive attenuation of the hyperemia in response to repetitive stimulation. To measure local hemodynamic changes, a continuous wave near-infrared spectroscopy (NIRS) device was used, providing a tissue oxygenation index. The NIRS measurement was obtained for wrist extensor muscles and for the lateral head of the gastrocnemius muscle of the right leg, since two different stimulation points were considered (Fig. 4).

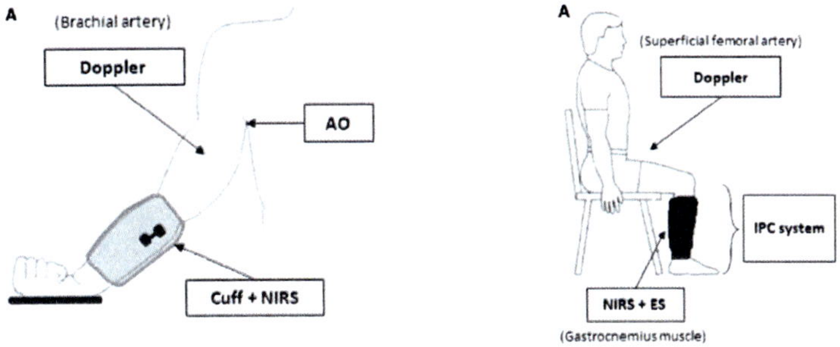


**Fig. 4.** Set-up of the 1st (left) and 2nd (right) protocol for the investigation of hyperemia [13].

In the first protocol, the IPC device was used to deliver two series of two compressive stimuli to the forearm with a time interval of 25 s between each compression, with peak pressure 250 mmHg and duration 1 s (Fig. 4, left). In the first series, the hyperemia was not limited, whereas it was prevented after the first compression in the second series. It was observed that preventing the hyperemia and the ensuing increases in oxygenation also prevented the attenuation of the hyperemic response to a subsequent stimulus.

The second protocol was designed to understand if the tissue oxygenation produced by a compression or a muscle contraction (which similarly compresses intramuscular vessels) affects the hyperemia generated by a subsequent compression or contraction of the same muscle (Fig. 4, right). During this protocol, the lateral gastrocnemius of the right leg was stimulated by means of an electrical stimulator at 20 Hz frequency. Compression of the leg was obtained with an IPC system including four bladders, with peak pressure 150 mmHg and duration of about 2–3 s. Several combinations of muscle contraction and pneumatic compression were considered.

The results of the study demonstrated that the hemodynamic responses to muscle contraction and to limb compression are very similar, suggesting a common underlying mechanism related to mechanical deformation of blood vessels. Moreover, the results showed, for the first time in humans, that the tissue hyper-oxygenation has a significant role in limiting further responses in skeletal muscle.

## 4 IPC in the Study of Venous Compartment Hemodynamics

As discussed in Sect. 3, IPC technique can be employed to investigate vascular phenomena and not only as a therapeutic tool in patients with some deficits as EDV reduction (Sect. 2). Moreover, IPC devices can also be used to generate pressure pulses in veins to allow for the measurement of Pulse Wave Velocity (vPWV).

### 4.1 IPC as a Tool in a System for the Measurement of Venous PWV

Ermini et al. [14] carried out a study on 8 healthy volunteers with a revised version of the IPC device shown in Sect. 3. The experimental set-up is shown in Fig. 5. A compressive stimulus was delivered directly to the foot by inflating a pneumatic cuff. The peak pressure considered was equal to 200 mmHg, with a duration of 1 s and inflation time of 400 ms. The foot compression was used to create a pressure pulse that propagated along the veins until it was detected by a Doppler ultrasound system at superficial femoral vein (SFV) level. The venous blood velocity was recorded with a linear probe positioned on the leg, with an incident angle of about 60°.

A pressure transducer was positioned on the cuff outlet and used to monitor the pressure inside the inflating volume (Fig. 6a, bottom), whereas leg venous pressure (LVP) was estimated as the hydrostatic load relative to the vertical distance between the leg and the venous point of collapse, which was detected by a second dedicated ultrasound system. The Doppler signal was sampled at 10 kHz (Fig. 6a, top) and used to extract the maximum velocity profile (Fig. 6c). The footprint of the latter, detected by a custom-made algorithm, was used to calculate the transit time of the pulse wave. The vPWV was then calculated as the ratio between the traveled distance (measured between the ankle and the probe positions) and transit time.

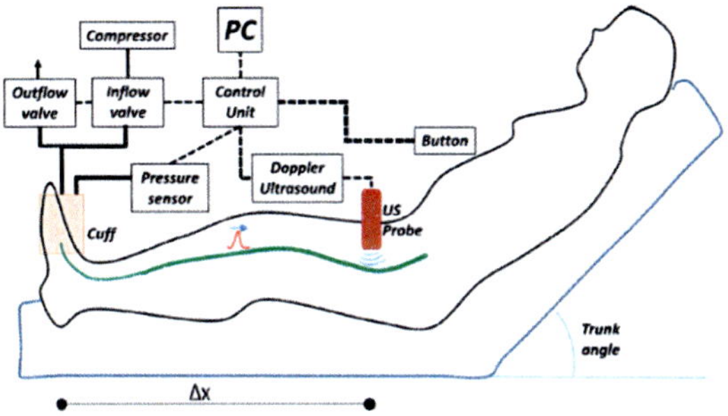


**Fig. 5.** Set-up for the measurement of vPWV in response of different trunk orientation [14].

The trigger of the stimulation was provided directly by the participant subject with a button, at the end of the expiratory phase in order to reduce the interference between the vPWV assessment and the respiratory activity and to improve the reproducibility of the measurements.

This study provided a significant proof of concept for a new methodology focused on the calculation of vPWV. Although PWV can be reliably assessed in arteries, where it is a well-established marker of cardiovascular risk, the measurement of vPWV has not been investigated with the same attention. In particular, the lack of relevant pulsation in the venous compartment, as well as the low blood pressure levels involved, represent significant limitations for a reliable measurement. For these reasons, the use of an IPC device can be really effective in generating the pressure pulse required for such measurement. The results of the study, in terms of vPWV values, were compatible with those obtained with different techniques (1–3 m/s for supine position, 4–15 m/s in the arm for venous pressures of 20–80 mmHg). Moreover, the linear relationship observed between vPWV and LVP was compatible with the literature [25–28] and confirmed that vPWV assessment could represent an indirect estimation of the venous pressure, which is directly connected to the volemic status of a patient.

### 4.2 Estimation of vPWV in Response to a Simulated Fluid Change

Ermini et al. [15] conducted an experiment on 15 healthy volunteers with a modified version of the IPC device shown in the previous section. In particular, the system was used to measure the vPWV along the arm, through a synchronous delivery of the compression (at the hand) with respiratory and cardiac cycles. The hemodynamic perturbation was realized by means of passive leg raising (PLR) maneuvers. An overview of the experimental set-up is given in Fig. 7. A fast compression (peak pressure: 400 mmHg, lasting about 1 s, inflation time: 400 ms) was delivered to the hand of each subject through the cuff.

A Doppler ultrasound system was used to detect the pressure pulse propagating proximally at the level of the basilic vein (BV). To further improve the quality of the measurements, the detection was synchronized with both respiratory and cardiac activities. In particular, the measurement was always performed at the end of the expiratory phase and at the same time instant within the cardiac cycle, corresponding to the lowest

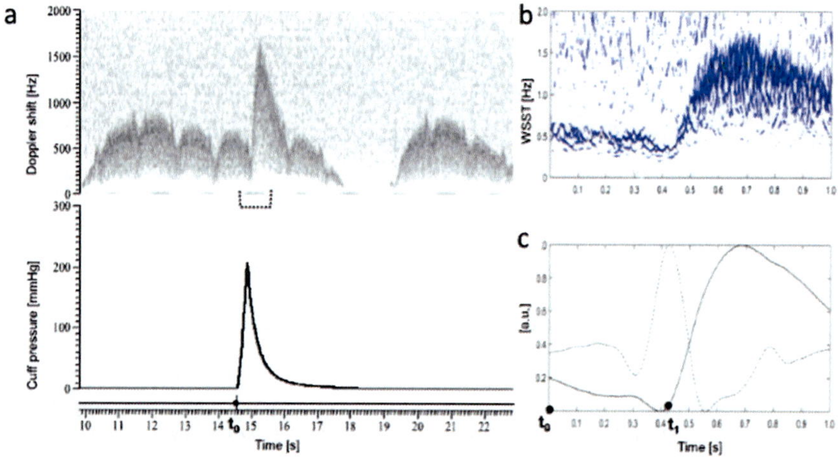


**Fig. 6.** Representative acquired and processed signals. (a) Venous pulse wave detected by Doppler ultrasound (top) and compressive stimulus profile (bottom). (b) Wavelet synchro-squeezed transform (WSST) of the Doppler signal. (c) Smoothed normalized maximum energy profile of WSST (solid line) and its normalized second derivative (dashed line), whose maximum identifies the footprint of the profile which was used to calculate the transit time of the pulse wave [14].

blood velocity. Cardiac synchronization was assessed by monitoring of the R-wave with a commercial ECG system.

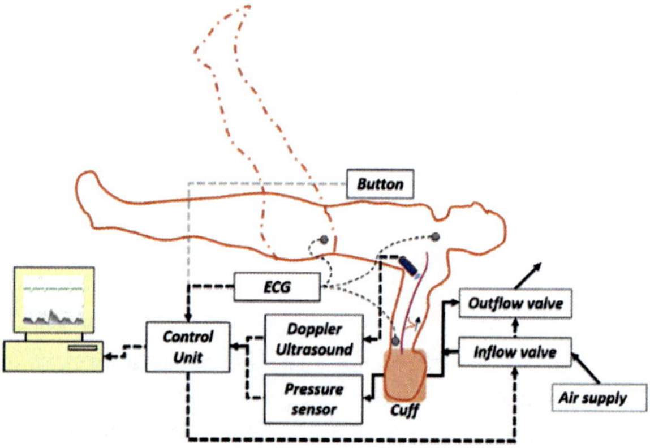


**Fig. 7.** Set-up to estimate vPWV in response of passive leg rising maneuvers [15].

This study showed for the first time that vPWV is sensitive to simulated changes in blood volume, as produced by PLR. With respect to traditional techniques, this methodology is objective. Furthermore, the measurement is sensitive even to limited hemodynamic challenges, and it is not invasive. Given the relationship between vPWV and the volemic status of a patient, this technique could be used for monitoring purposes or to provide a quantitative assessment of vascular filling during fluid administration in patients.

## 5 Conclusions

The applications of IPC technique can be of a therapeutic and rehabilitative type, such as the improvement of the return of venous blood to the heart in people with disorders to the cardiovascular system, or of a diagnostic type, such as the measure of the propagation velocity of a pressure wave in the venous compartment, which allows an indirect evaluation of the volemic status of a subject. Further applications can deepen the study of still poorly defined phenomena in the cardiovascular system.

The variety of applications from the physiological point of view requires that the specific IPC device must be designed and manufactured with non-common techniques, to meet the required precision and dynamic performance.

The applications of IPC are completely non-invasive, and this constitutes a fundamental element for the realization of devices of limited cost, easy to use and therefore very widespread. They can therefore represent an important means of spreading well-being and good health widely, in the spirit of the SDG3 of the United Nations Development Program.

## References


1. Johansson, K., Lie, E., Ekdahl, C., Lindfeldt, J.: A randomized study comparing manual lymph drainage with sequential pneumatic compression for treatment of postoperative arm lymphedema. Lymphology **31**, 56–64 (1998)
2. Zaleska, M., Olszewski, W.L., Durlik, M.: The effectiveness of intermittent pneumatic compression in long-term therapy of lymphedema of lower limbs. Lymphat Res. Biol. **12**(2), 103–109 (2014)
3. Comerota, A.J.: Intermittent pneumatic compression: physiologic and clinical basis to improve management of venous leg ulcers. J. Vasc. Surg. **53**, 1121–1129 (2011)
4. Nelson, E.A., Mani, R., Thomas, K., Vowden, K.: Intermittent pneumatic compression for treating venous leg ulcers. Cochrane Database Syst. Rev. **16**(2), CD001899 (2011)
5. Sparks-DeFriese, B.J.: Vascular Ulcers. In: Physical Rehabilitation, pp. 777–802, W.B. Saunders (2007)
6. Flam, E., Berry, S., Coyle, A., Dardik, H., Raab, L.: Blood-flow augmentation of intermittent pneumatic compression systems used for the prevention of deep vein thrombosis prior to surgery. Am. J. Surg. **171**, 312–315 (1996)
7. Zhang, D., Li, F., Li, X., Du, G.: Effect of Intermittent pneumatic compression on preventing deep vein thrombosis among stroke patients: a systematic review and meta-analysis. Worldviews Evid. Based Nurs. **15**(3), 189–196 (2018)
8. Ferraresi, C., Maffiodo, D., Hajimirzaalian, H.: A model-based method for the design of intermittent pneumatic compression systems acting on humans. Proc. Inst. Mech. Eng. PART H **228**(2), 118–126 (2014)
9. Manuello Bertetto, A., Meili, S., Ferraresi, C., Maffiodo, D., Crisafulli, A., Concu, A.: A mechatronic pneumatic device to improve diastolic function by intermittent action on lower Limbs. Int. J. Autom. Technol. **11**(3), 501–508 (2017)

10. Maffiodo, D., De Nisco, G., Gallo, D., Audenino, A., Morbiducci, U., Ferraresi, C.: A reduced-order model-based study on the effect of intermittent pneumatic compression of Limbs on the cardiovascular system. Proc. Inst. Mech. Eng. Part H **230**(4), 279–287 (2016)
11. Ferraresi, C., et al.: Design and simulation of a novel pneumotronic system aimed to the investigation of vascular phenomena induced by Limb compression. J. Bionic. Eng. **16**, 550–562 (2019)
12. Messere, A., et al.: Delivery of customizable compressive patterns to human limbs to investigate vascular reactivity. Biomed. Phys. Eng. Express **4**(6), 067003 (2018)
13. Messere, A., et al.: Hyper-oxygenation attenuates the rapid vasodilatory response to muscle contraction and compression. Front. Physiol. **9**, 1078 (2018)
14. Ermini, L., Ferraresi, C., De Benedictis, C., Roatta, S.: Objective assessment of venous pulse wave velocity in healthy humans. Ultrasound Med. Biol. **46**(3), 849–854 (2019)
15. Ermini, L., Chiarello, N., De Benedictis, C., Ferraresi, C., Roatta, S.: Venous Pulse Wave Velocity variation in response to a simulated fluid challenge in healthy subjects. Biomed. Signal Process. Control **63**, 102177 (2021). https://doi.org/10.1016/j.bspc.2020.102177
16. Boutouyrie, P., Briet, M., Vermeersch, S., Pannier, B.: Assessment of pulse wave velocity. Artery Res. **3**, 3–8 (2009)
17. Safar, M.E.: Arterial stiffness as a risk factor for clinical hypertension. Nat. Rev. Cardiol. **15**, 97–105 (2018)
18. Lin Wang, Y.Y.: Did you know developing quantitative pulse diagnosis with realistic haemodynamic theory can pave a way for future personalized health care. Acta Physiol. **227**(3), e13260 (2019). https://doi.org/10.1111/apha.13260
19. Mohrman, D.E., Sparks, H.V.: Myogenic hyperemia following brief tetanus of canine skeletal muscle. Am. J. Physiol. **227**, 531–535 (1974)
20. Tschakovsky, M.E., Sheriff, D.D.: Immediate exercise hyperemia: contributions of the muscle pump versus rapid vasodilation. J. Appl. Physiol. **97**, 739–747 (2004)
21. Clifford, P.S., Tschakovsky, M.E.: Rapid vascular responses to muscle contraction. Exerc. Sport Sci. Rev. **36**, 25–29 (2008)
22. Turturici, M., Mohammed, M., Roatta, S.: Evidence that the contraction-induced rapid hyperemia in rabbit masseter muscle is based on a mechanosensitive mechanism, not shared by cutaneous vascular beds. J. Appl. Physiol. **113**, 524–531 (2012)
23. Turturici, M., Roatta, S.: Inactivation of mechano-sensitive dilatation upon repetitive mechanical stimulation of the musculo-vascular network in the rabbit. J. Physiol. Pharmacol. **64**, 299–308 (2013)
24. Jasperse, J.L., Shoemaker, J.K., Gray, E.J., Clifford, P.S.: Positional differences in reactive hyperemia provide insight into initial phase of exercise hyperemia. J. Appl. Physiol. **1985**(119), 569–575 (2015)
25. Mackay, I., Van Loon, P., Campos, J., de Jesus, N.: A tecnique for the indirect measurement of the velocity of induced venous pulsation. Am. Heart J. **73**, 17–23 (1967)
26. Anliker, M., Wells, M.K., Ogden, E.: The transmission characteristics of large and small pressure waves in the abdominal vena cava. IEEE Trans. Biomed. Eng. **16**, 262–273 (1969)
27. Minten, J., Van De Werf, F., Auber, A., Kasteloot, H., De Geest, H.: Apparent pulse wave velocity in canine superior vena cava. Cardiovasc Res. **17**, 627–632 (1983)
28. Nippa, J., Alexander, R., Folse, R.: Pulse wave velocity in human veins. J. Appl. Physiol. **30**, 558–563 (1971)

# Intersomatic Screw Experimental Study to Treat Lumbar Disc Herniations

A. Oropeza-Osornio[1], C. R. Torres-San Miguel[2](✉), and G. M. Urriolagoitia-Calderón[2]

[1] Instituto Politécnico Nacional, Escuela Superior de Ingeniería Mecánica y Eléctrica, Unidad Ticomán, 07340 Ciudad de México, Mexico

[2] Instituto Politécnico Nacional, Escuela Superior de Ingeniería Mecánica y Eléctrica, SEPI, Unidad Zacatenco, 07738 Ciudad de México, Mexico

ctorress@ipn.mx

**Abstract.** In this work, the biomechanical behavior of an intersomatic screw located in an intervertebral disc after a discectomy is studied. For the analysis, a clamping device was designed to reproduce the rotational movements of the vertebrae. The mechanism was adapted to a Shimadzu® universal mechanical testing machine, using an axial compression load in a range of 0 to 3000 N on two types of specimens, with different properties but with similar geometries, an L2 vertebral unit –L3 from resin vertebrae and an L4-L5 vertebral unit from a porcine specimen. The deformation of the vertebrae was obtained using the Photostress technique. The results show the biomechanical evaluation of an interbody screw, which can be used as an alternative to alleviate pathologies that involve the wear of the lumbar intervertebral disc and increase the quality of life and well-being of people, which is the Third Goal for Development Sustainable (SDG3) of the 2030 Agenda of the United Nations.

**Keywords:** SDG3 · Biomechanics · Herniated discs · Intersomatic screws · Testbed

## 1 Introduction

The spinal column is a set made up of bones and connective tissues based on the spinal cord and nervous tissue, perfectly structured and articulated, which fulfill specific functions of support, protection, and movement [1–5]. In specific cases, the spine is affected by vertebrae diseases, such as the nerve roots or the spinal cord. Depending on the type of disease and its severity, the treatment is performed, ranging from taking drugs and rest to surgical interventions and implants. Surgical intervention is one of the main alternatives to treat spinal conditions to release trapped nerves or tumors that compress the spinal cord [6]. The most frequent cause of nerve root entrapment is the so-called "Vertebral Disc Herniation", but deformations of the vertebral bodies or tumors of the spine are also conditions that imply restructuring the harmony of the spine, so it is necessary the use of implants [7–9]. Implants have been based on three variables: materials, geometry and


G. Quaglia et al. (Eds.): I4SDG 2021, MMS 108, pp. 219–228, 2022.
https://doi.org/10.1007/978-3-030-87383-7_24

restriction movement [10]. Implants are made up of plates, connecting rods and bearings, and metallic and polymeric materials, which provide greater flexibility. However, these use has not been entirely satisfactory since it has been necessary to complement their application with intersomatic cages, which have helped reduce postoperative pain. However, it allows better stability and reduced movement. This additional stability can promote successful fusion, resulting in the patient being able to quickly return to all daily activities [11–14].

The research aims in an experimental study of an intersomatic screw, whose geometry is new, with the aim of knowing its biomechanical behavior and determining if its use in people is feasible, and contribute to the Third Goal for Development Sustainable (SDG3) of the United Nations 2030 Agenda, referring to the health and well-being of people, since back pain occurs in men and women, between 30 and 50 years old, of which 90% only require conservative treatment such as exercise, physical therapy, steroids and inflammatory. The remaining percentage requires surgical interventions.

## 2 Method

Experimental tests were carried out to validate the behavior of a cannulated intersomatic screw, using the L4–L5 vertebral unit of a porcine specimen due to its remarkable similarity with the geometry of human vertebrae and a resin L2–L3 vertebral unit. The cannulated intersomatic screw insert (1) is made with Titanium due to its high compatibility in humans. Its tip is blunt (3). It also has a conical external thread (2) formed by deep and flat teeth, which provide greater support and anchoring in the vertebral space to avoid any antero-posterior (AP) or medio-lateral sliding (ML). The implant has a screw head (4) at its opposite end, which adjusting and providing extra torque. An internal longitudinal hole in the body has a thread (6) at the ends. This provides an adequate positioning, placement and accommodation of the intersomatic screw. It is inserted like any screw; a suitable anchor does not accept rotational or sliding movements once released from the positioning instrument. It is placed through a cannula. When the interbody screw is placed, the lost space is restored (Fig. 1).

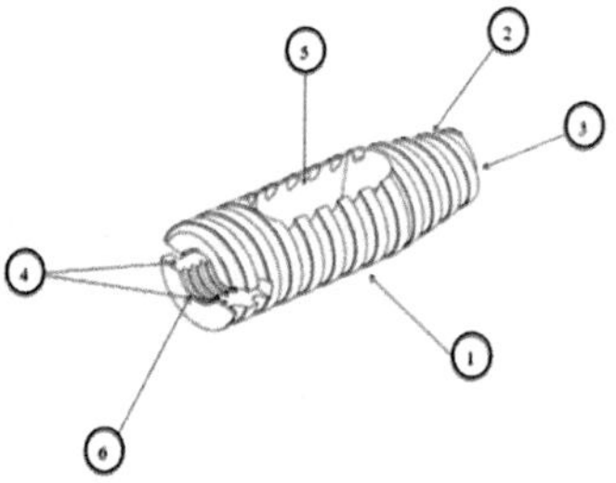

**Fig. 1.** Intersomatic implant.

### 2.1 Simplified Load Simulator (SLS)

The design of the SLS, considered many experimental studies that evaluated injuries, [15–17] to mention a few examples: test bench design, which is capable of reproducing impact scenarios by applying a compression force to the thorax produced by the seat-belt, characterization of the impact on the head in terms of acceleration and force, among others [18–20]. The vertebral movements and loads are studied by a Simplified Load Simulator (SLS) mechanism, consisting of two sections [21, 22]. The lower section generates a flexion-extension, adduction, and abduction movement in the vertebrae, moving the base to which the specimen is attached. The upper section provides a rotational movement in the L4 vertebra using gears to transmit movement. The rotation system generated by the upper section (Fig. 3) of the mechanism results from the coupling of a Dynamixel® Model MX-64 servo motor with a capacity of 6 Nm integrated to a spur gear system pinion-crown in order to increase torque generated by the motor and provide the necessary force to obtain the required movement (Fig. 2).

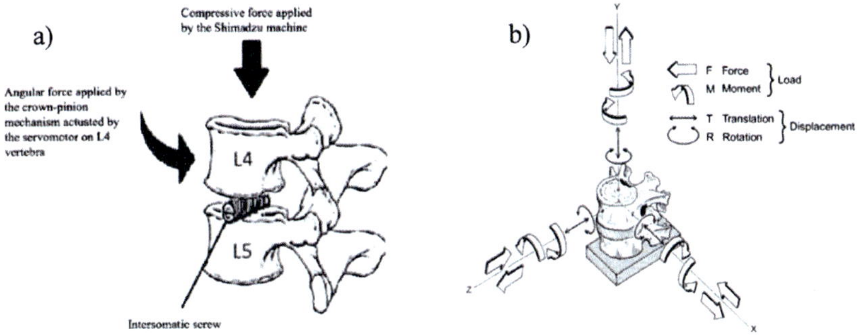


**Fig. 2.** a) Free body diagram of vertebrae and screw, b) DOF of vertebrae unit.

The movements mentioned above were created through a mobile base, manipulated by an electric linear actuator I.E.I. brand with a capacity of 110 lbs, which changes the angle of inclination of the base to produce the desired movements.

## 3 Experimental Analyses

The tests were carried out with different specimens and similar geometries. The tests applied photoelastic resin to the porcine vertebrae specimens. The human geometry vertebra was obtained through computational techniques from computed tomography. The tests have been done with the Shimadzu® universal mechanical testing machine, where the Simplified Load Simulator (SLS) was adapted (Fig. 3). The assembly of those mechanisms helps to simulate the compression movement generated by gravity and the rotational and abduction movements generated by the spine (Fig. 4).

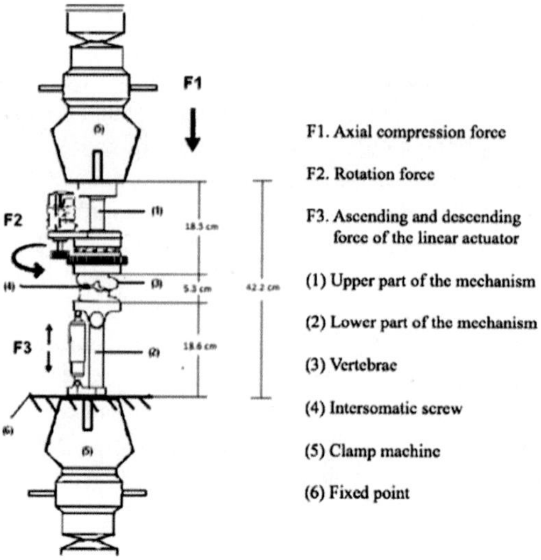


**Fig. 3.** Simplified load simulator.

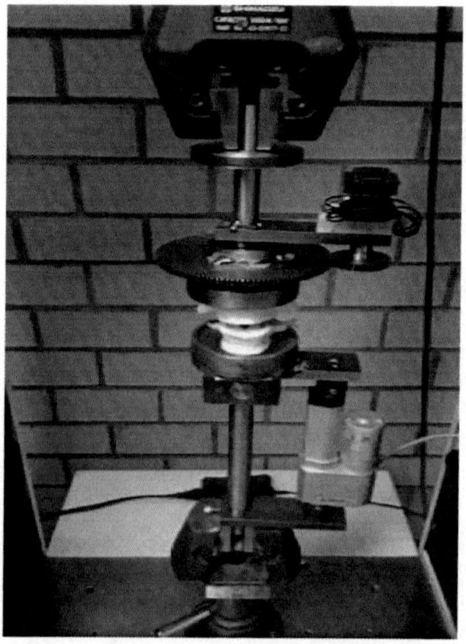

**Fig. 4.** Assembly of the compression test.

The human L2–L3 lumbar vertebrae photoelastic specimens were obtained through computational techniques. The test conditions were as follows:

- Adult patient with a 56-year-old height of 1.73 m and a weight of 89 kg.
- The patient had a medical CT scan due to a fracture of Ribs 7 and 8.
- Room temperature: 24 °C
- Relative humidity of the room: 50%
- Duration of compression test: 101 s.

The compression test was performed by including the intersomatic screw between the vertebrae. As shown in Fig. 5, the load on the photoelastic resin vertebrae was increased at a constant speed in a range from 0 to 1487.1 N. When this load value was reached, the L2 vertebra fractured.

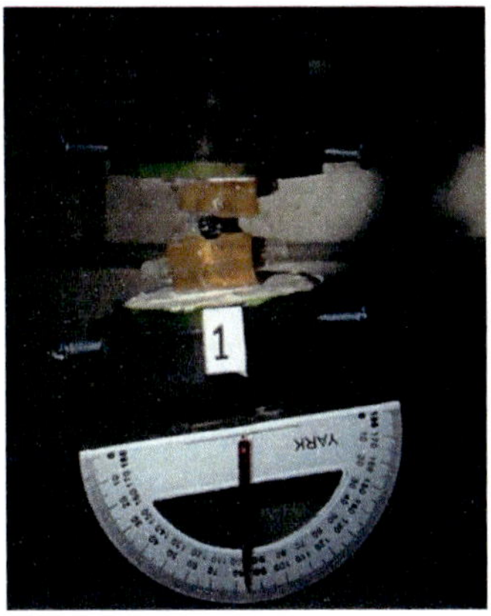


**Fig. 5.** L2 and L3 resin vertebrae under compression.

The photostress technique was used to carry out compression tests on L4 and L5 porcine vertebrae with the intersomatic screw. The test conditions were as follows:

- Duroc-Jersey swine selected with an average age of 6 months and a weight of 120 kg. The pig was slaughtered with the recognized standards for this effect. Next, the vertebral column was separated from the rest of the porcine body and the lumbar section was isolated, preparing the L4 and L5 vertebrae to be compression tested.
- Room temperature: 24 °C
- Relative humidity of the room: 50%
- Duration of compression test: 194 s.

The compression test was performed by gradually increasing the load, as shown in Fig. 6, ranging from 0 to 3017.8 N. The vertebrae began to fracture when this load value was reached.

**Fig. 6.** Porcine vertebrae L4 and L5 under Compression, observed by reflective polariscope

# 4 Results

In Fig. 7, it is possible to observe that the deformation in the vertebra remained practically linear within the range of 0.2 mm (45N) to 1.4 mm (1159N) and that at a load of 1487.1 N, the photoelastic specimen of vertebrae L2 was fractured, registering a maximum displacement of 1,684 mm.

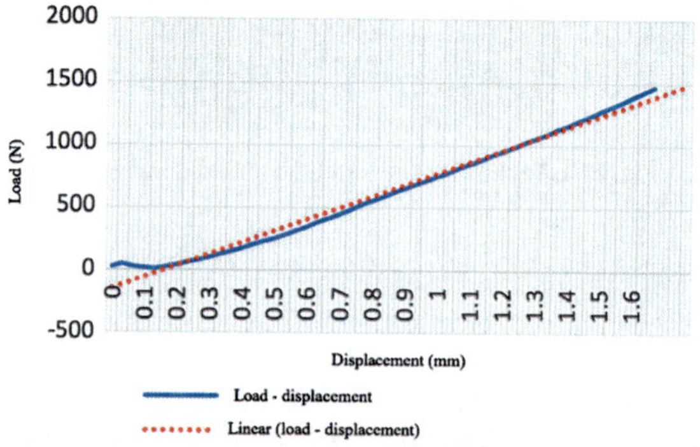


**Fig. 7.** Load-displacement graph and linear trend in displacement.

In Fig. 8, the photoelastic specimens are shown after compression testing, L2 with vertebral body fracture, L3 only with traces of the square thread of the interbody screw power.

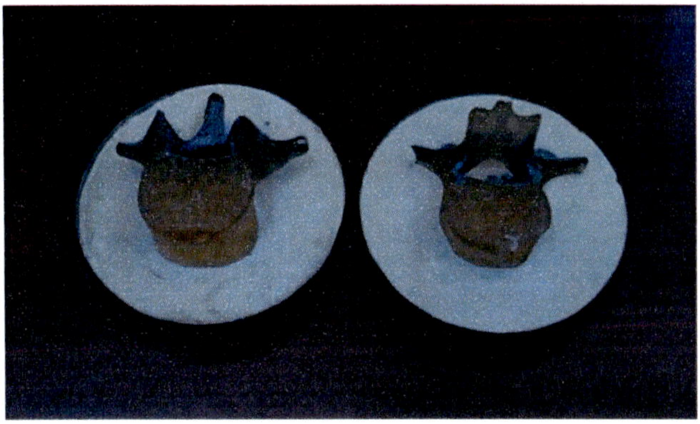

**Fig. 8.** Specimens after the compression test, L2 with vertebral body fracture (right) and L3 with traces of the intersomatic screw (left).

In Fig. 9, the deformation in the porcine vertebra remained practically linear throughout the test is observed, and from a load of 3017.8 N, the vertebrae began to fracture. A maximum displacement of 3,233 mm was recorded.

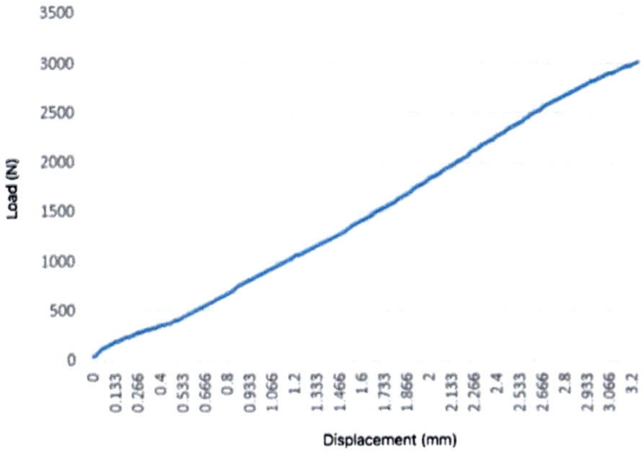


**Fig. 9.** Compression test, load-displacement.

In Fig. 10, the porcine specimens of L4 and L5 are shown after the compression test. The traces of the intersomatic screw are observed, except in the central part of the body of the vertebrae, which indicates that the screw worked as a simple beam.

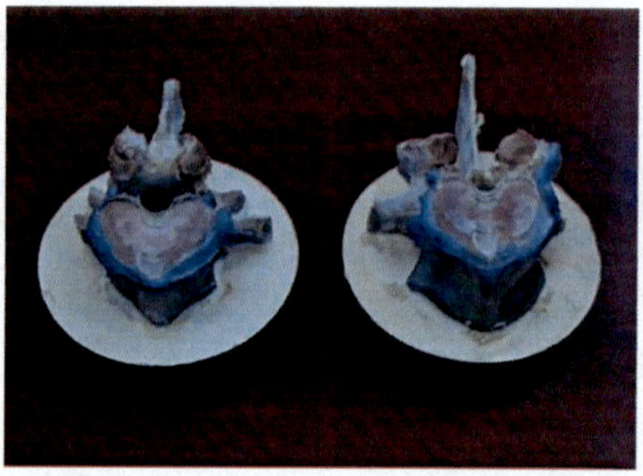

**Fig. 10.** Intersomatic screw traces in L4 and L5 specimens.

## 5 Conclusion

The Simplified Load Simulator (SLS) developed for the experimental tests is suitable since acting in conjunction with a universal mechanical testing machine. The spinal column supports the tensile and compression, bending, flexo-torsion loads, and rotational movements. Furthermore, the configuration of the intersomatic screw allows to install and ensures a perfect anchorage since it does not accept rotation or sliding movements after being implanted. In addition to the stability it provides, the proposed prosthesis was able to withstand a force of 3000 N without undergoing modifications in its shape, so we can say that if this device is implanted, the patient will be able to load approximately 200 kg. The spine stabilization depends on the prosthesis used, mainly on its shape, material and the personal characteristics in which it is implanted. The prosthesis proposed in this work is necessary to carry out other types of tests through finite element analysis, in vivo tests, and even propose new screw shapes and configurations.

**Acknowledgments.** The authors acknowledge the financial support for the development of this work by the National Council of Science and Technology (CONACYT) of Mexico, the Instituto Politécnico Nacional. Authors thanks of partial support from project 20210282 and EDI grant, all provided by SIP/IPN.

## References

1. Cloward, R.B.: The treatment of ruptured lumbar intervertebral discs by vertebral body fusion I; indications, operative technique, after care. J. Neurosurg. **10**, 154–168 (1953)
2. Bailey, R.W., Bagley, C.: Stabilization of the cervical spine by anterior fusion. J. Bone Joint Surg. **42**(4), 565–594 (1960)
3. Ramani, P.S.: Textbook of Cervical Spondylosis (Ed. by, Jaypee), p. 327 (2004)
4. Hitchin, P.W., Traynelis, V.C., Rengachary, S.: Techniques in Spinal Fusion and Stabilization (Ed. by, Thieme), p. 174 (1995)
5. Xu, H.-Z., Wang, X.-Y., Chi, Y.-L., Zhu, Q.-A., Lin, Y.: Biomechanical evaluation of a dynamic pedicle screw fixation device. Center, Department of Orthopaedic Surgery, Second Affiliated Hospital of Wenzhou Medical College, Wenzhou, China, 7 December 2005

6. Disch, A.C., Luzzati, A., Melcher, I., Schaser, K.D., Feraboli, F., Schmoelz, W.: Three-dimensional stiffness in a thoracolumbar model: a biomechanical in vitro study. Centre for Musculoskeletal Surgery, Musculoskeletal Tumour Surgery Section, Charite'-University Medicine Berlin, Augustenburger Platz Berlin, Germany, 3 July 2007
7. Gay, R.E., Ilharreborde, B., Zhao, K.D., Berglund, L.J., Bronfort, G., An, K.-N.: Stress in lumbar intervertebral discs during distraction: a cadaveric study. Department of Physical Medicine and Rehabilitation, Mayo Clinic, USA, 31 July 2007
8. Meyers, K., et al.: Use of instrumented pedicle screws to evaluate load sharing in posterior dynamic stabilization systems. Department of Biomechanics, Hospital for Special Surgery, New York, USA, 15 August 2007
9. Wheeler, D.J., et al.: Interlaboratory variability in vitro spinal segment flexibility testing. A University of Minnesota, Minneapolis, USA, 29 June 2011
10. Torres-San Miguel, C.R., Hernández-Gómez, J., Urriolagoitia-Sosa, G., Romero-Ángeles, B., Martínez-Sáez, L.: Design and manufacture of a customised temporomandibular prosthesis. Rev. Int. Metodos Numer. Calc. Diseno Ing. **35**, 1–22 (2019)
11. Tang, J.A., Scheer, J.K., Ames, C.P., Buckley, J.M.: Pure moment testing for spinal biomechanics applications: fixed versus 3D floating ring cable-driven test designs. The Taylor Collaboration, San Francisco, USA, 21 December 2011
12. Galbusera, F., Bellini, C.M., Anasetti, F., Ciavarro, C., Lovi, A., Brayda-Bruno, M.: Rigid and flexible spinal stabilization devices: a biomechanical comparison. J. Med. Eng. Phys. **33**, 490–496 (2011)
13. Rapoff, A.J., Ghanayem, A.J., Zdeblick, T.A.: Biomechanical comparison of posterior lumbar interbody fusion cages. Spine **22**, 2375–2379 (1997)
14. Calalpa-Torres, I., Torres-SanMiguel, C.R., Urriolagoitia-Sosa, G., Cuautle-Estrada, A., Romero-Angeles, B., Urriolagoitia-Calderón, G.M.: Antigravity Device for Invertebral Rehabilitation. Engineering Design Applications III, pp. 13–21, Springer, Heidelberg (2020)
15. Ramirez, O., Ceccarelli, M., Russo, M., Torres-San-Miguel, C.R., Urriolagoitia-Calderon, G.: Experimental dynamic tests of RIB implants. In: Carbone, G., Gasparetto, A. (eds.) IFToMM ITALY 2018. MMS, vol. 68, pp. 353–361. Springer, Cham (2019). https://doi.org/10.1007/978-3-030-03320-0_38
16. Ramirez, O., Torres-San-Miguel, C., Ceccarelli, M., Urriolagoitia-Calderón, G.: Experimental characterization of an osteosynthesis implant. Adv. Mech. Mach. Sci. **73**(16), 53–62 (2019)
17. Torres, C.R., García, G., Aguilar, L.A., Martínez, L.: Design of a rib impactor equipment. VIII Int. Congr. Eng. Phys. **48**, 1–6 (2016)
18. Rueda Arreguin, J.L., Ceccarelli, M., Torres San Miguel, C.R.: Design of an articulated neck for testbed mannequin. In: Niola, V., Gasparetto, A. (eds.) IFToMM ITALY 2020. MMS, vol. 91, pp. 94–101. Springer, Cham (2021). https://doi.org/10.1007/978-3-030-55807-9_11
19. Rueda Arreguín, J.L., Ceccarelli, M., Torres-San-Miguel, C.R., Morales Cruz, C.: Lab experiences on impact biomechanics of human head. In: Rauter, G., Cattin, P.C., Zam, A., Riener, R., Carbone, G., Pisla, D. (eds.) MESROB 2020. MMS, vol. 93, pp. 229–237. Springer, Cham (2021). https://doi.org/10.1007/978-3-030-58104-6_26
20. Rueda Arreguín, J.L., Ceccarelli, M., Torres San Miguél, C.R.: Design and simulation of a parallel-mechanism testbed for head impact. In: Zeghloul, S., Laribi, M., Sandoval Arevalo, J. (eds.) RAAD 2020. MMS, vol. 84, pp. 400–407. Springer, Cham (2020). https://doi.org/10.1007/978-3-030-48989-2_43

21. Caravaggi, P., Chaudary, S., Uko, L., Chen, L., Khamsi, B., Vives, M.: A novel design for application of pure moments in-vitro: application to the kinematic analysis of the cervical spine. Vives Joint Motion Lab, Orthopedics Dept., University of Medicine and Dentistry of New Jersey, Newark (NJ), USA (2013)
22. Holsgrove, T.P., Gheduzzi, S., Gill, H.S., Miles, A.W.: The development of a dynamic, six-axis spine simulator. Centre for Orthopaedic Biomechanics, Department of Mechanical Engineering, University of Bath, Bath, UK, 26 November 2013

# Robotics and Mechatronics

# A Proof of Concept of Self Replicating Robots for Space Settlements

Paolo Gallina[1,2(✉)], Nicola Scuor[1,2], Massimiliano Nolich[1,2], Lorenzo Scalera[1,2], Alessandro Gasparetto[1,2], and Stefano Seriani[1,2]

[1] Department of Engineering and Architecture, University of Trieste, Trieste, Italy
pgallina@units.it

[2] Polytechnic Department of Engineering and Architecture, University of Udine, Udine, Italy
alessandro.gasparetto@uniud.it

**Abstract.** Efficient agriculture in a hazardous environment is an essential factor for space settlement. At the same time, transporting resources for settlement is very expensive and, to a certain extent, impracticable. It is therefore necessary to find techniques and tools to exploit resources on site. The aim of the project presented in this paper consisted in realizing a partially Self Replicating mobile Robot (SRR) that can monitor and helping to cultivate plants. The fibers of the plants have been used to build the frame of another SRR. This way, except for the electronic components, it was possible to replicate the robot through cultivated plants. In particular, it was possible to replace the weight of the new robot of 30%.

**Keywords:** Responsible consumption and production · Robotics · Space settlement · Plants in space · Self replication

## 1 Introduction

Ever since the dawn of the Space Age, humans have made plans for exploring and ultimately colonizing the surface of most planets and moons in the Solar System, with the main candidates by far being the Moon and Mars.

In most concepts for moon bases that have been put forward since the beginnings, the key role is that of life-support systems. If settlement and colonization is supposed to tend to complete autonomy from Earth, means to exploit resources *in situ* to provide breathable atmosphere, water, food and shelter are the main issue [8, 10]. Both for the Moon and Mars the soil and atmosphere (in the latter) in general provide all basic chemicals to support crops [10, 14, 16]. In 2010, Maggi

Supported by organizations: Dubai Future Foundation through Guaana.com open research platform; Italian MIUR under PRIN grant "SEDUCE" no. 2017TWRCNB and Departments of Excellence grant L. 232/16; University of Trieste - University funding for scientific research projects - FRA 2018.


G. Quaglia et al. (Eds.): I4SDG 2021, MMS 108, pp. 231–240, 2022.
https://doi.org/10.1007/978-3-030-87383-7_25

et al. in [7], produced a very thorough analysis of the influence of the martian environment on a possible agricultural endeavor.

Agriculture provides excellent sustainment capabilities to a putative planetary base. However, agriculture is also labor intensive, requiring multiple tasks to perform the basic activities. Providing a base with a single robot to perform all tasks is unrealistic; however, the costs associated with sending a fleet of robots from Earth are high. This is the main reason why the concept of In-Situ Resources Utilization (ISRU) is generally proposed [9].

The idea of employing growable materials to increase the potential of robotic systems or other functional structures has been investigated in literature [13]. The term "growable robotics" refers to a type of robotic systems that self-grow either a part or all with minimal input from external production infrastructure [11]. Ideally, following the fabrication of single components, the issue is that of assembling the parts in order either to perform maintenance, or to fabricate an entire robot [12].

Partial replication and assembly can be exploited also to the aim of "growing" the robot itself, or to enhance its capabilities [17].

Given the state-of-the-art of self-replication, assembly, "growing" etc. settlement and colonization activities can greatly benefit from this kind of technology domain. In particular, in 2002 Chirikjian et al. [2] and Ellery [5] propose the use of self-replicating robots to the development of the lunar surface; in particular Chirikjian analyzes the feasibility of SRRs in the context of lunar resource development. In [3], growing robots are introduced, starting from the original concept of a self-replicating automaton to the achievements obtained by plant inspiration, which provided an alternative solution to the challenges of creating robots with self-building capabilities. The use of 3D printing machines is considered enabling towards this endeavor, as Ellery points out in 2015 [4].

Building from the work presented up to this point, we propose a methodology and framework that enables self-replication through the use of agriculture as an ISRU technique. A mobile robot is sent to the target proto-planetary surface and is instructed to tend to crops (in our examples bamboo [1,15]) which yield organic compounds like polymers and fibers, which in turn will constitute the structural components of the next generation of robots. This architecture is expected to completely automate the process.

In our study, a mobile robot is instructed to tend to crops (in our examples bamboo [1]) which yield organic compounds like polymers and fibers, which in turn will constitute the structural components of the next generation of robots. This architecture is expected to completely automate the process. We propose a conceptual framework which effectively utilizes resources available *in situ* through the chemical and physical processes associated with agriculture; the soil and $CO_2$ are transformed into fibrous plant matter, harvested and finally converted into mechanical and structural components.

## 2 Self-replicating Robot Architecture

The main objective of the project is twofold. On the one hand, it consists in demonstrating that some structural components of a robot can be made out of vegetable fibres. On the other hand, it consists in demonstrating that the robotic system is able to automate certain phases linked to the production process of the required crops. Together, the achievement of the two objectives represents a first step towards the self-generation of robotic systems. The main schema the project and the modules involved are sketched in Fig. 1.

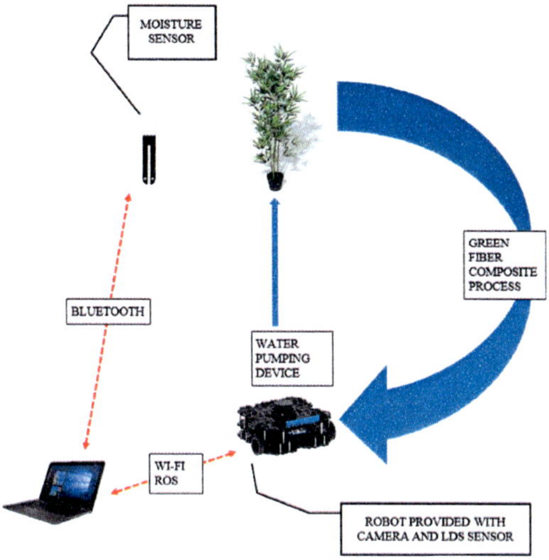


**Fig. 1.** Functional schematics of the project

A mobile robot has been employed to cultivate a plant of bamboo. There are several reasons why we decided to use bamboo. Bamboo is one of the fastest growing plants on the planet. A plant of bamboo can reach its full height in just one growing season and, over the next few years, the stem reaches maturity and hardness. Therefore the bamboo is highly sustainable since it can be harvested in few years. Moreover it possess interesting mechanical properties as far as compression and tensile strength [15]. For this reasons, it represents an excellent renewable and construction material [1,6].

The robot and the plants are located inside an "arena" made of wood panels. The robot employed is a differential wheeled mobile platform, a TurtleBot 3[1]. The robot is provided with a camera (Raspberry pi camera) and a laser distance sensor (LDS). In addition, a water tank has been built. It is positioned above the LDS and allows, through the opening of a solenoid valve, to irrigate the plant

[1] http://www.robotis.us/turtlebot-3/.

when required. Inside the arena, a single bamboo plant has been rooted in a pot. To complete the system a sensor was inserted in the plant soil. It measures moisture levels and temperature. The research project is split in three phases. In the first one the robot cultivates the plant: by means of a vision system, considering also moisture data, it plans the irrigation of the plant and assesses its health condition. During the second phase, the plant is harvested and the fibres needed to make the composite material are obtained. The third phase consists in the realization of the structural components of the robot that are replaced to the original ones.

The robot first generates a 2D map of the environment by means of SLAM (Simultaneous Localization And Mapping) process, analyzing the data recorded by the LDS.

The robot receives instructions from a master PC through ROS (Robot Operating System). The master PC feeds the robot with actions, analyses the images provided by the camera and records data related to the plant soil conditions: moisture, temperature and light.

In order to cultivate the plant, periodically, the master PC performs the following tasks: a) Moves the robot to the target position (in front of the plant); b) Takes a picture of the plant; c) Performs the image analysis in order to derive plant life conditions (see next section for details); d) Measures moisture, temperature and light intensity (Bluetooth communication); e) If required, according point to c) and d), activates the irrigation system; f) Moves the robot to its home position. When the plant has fully completed its growth, its fibers can be used to produce the composite material.

## 3 Proof of Concept

In order to monitor the plants which have been assigned to its care, the main requirement of the robot is to be able to deliver water to the soil in the pot. Additionally, the robot must monitor the status of the plant, to insure its well-being and to signal possible drastic deviations. The plant watering process can be broken down in the following sequence of events: measurement of the water content in the soil and the status of the plant; loading of water from a fixed dispenser; navigation to the location of the pot; positioning of the water faucet on top of the pot; delivery of water to the soil.

### 3.1 Plant Health Monitoring: Leaves Color

The visual status of the plant is monitored via image analysis of the leaves. The robot operates in a structured environment in which the localization system is facilitated by the lack of additional obstacles and the vision system works with uniformly colored background surfaces on which the plants stand out clearly. At regular intervals, an image is taken of the plant and is analyzed. The element which provides a quantitative measure of the status of the plant is the color of the leaves. The image analysis process can be broken down as follows: localization

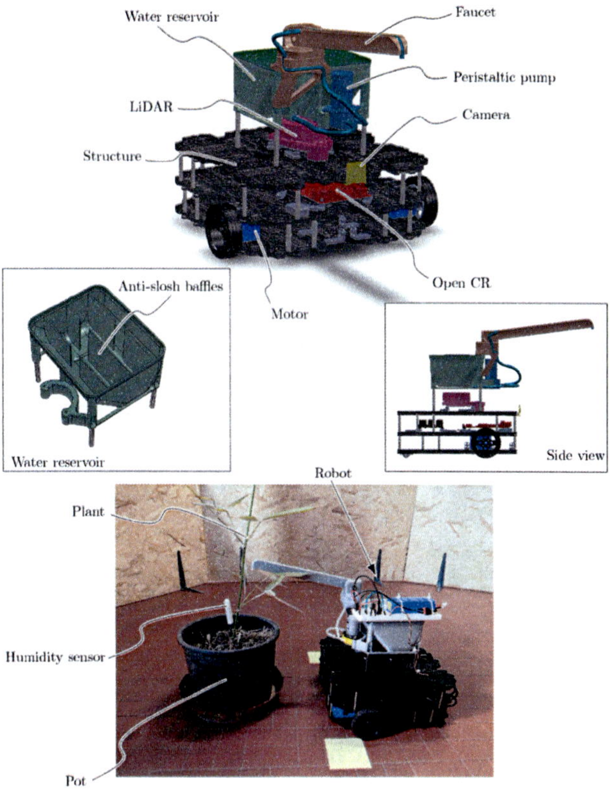


**Fig. 2.** Robotic system

of the pot; selection of the foliage Region Of Interest (ROI); recognition of the leaves and segmentation; auto-calibration of color data based on the segmented background; measurement of average leaves color in the segmented leaves data (Fig. 2).

In Fig. 3 the main stages of the analysis are visible; in a) the localization of the pot is visible. The localization is performed using a technique called cross-correlation, where a sample of the image (b)) is searched for in the base image. The position at which the pot is found by the algorithm is shown as a red crosshair in a) and is exploited to determine the ROI where the foliage is expected to be: this region is indicated by a red boundary in the same image. It should be noted that the size of the ROI is determined by the scale of the pot, which in turn is extrapolated from the cross-correlation results.

A second cross-correlation step is performed on the foliage ROI to determine the precise location of the leaves. The sample used in this case is much more complex: the baseline is shown in c), where one can see a set of single leaves; The result of this is shown in form of a map in d), which, using Otsu's thresholding

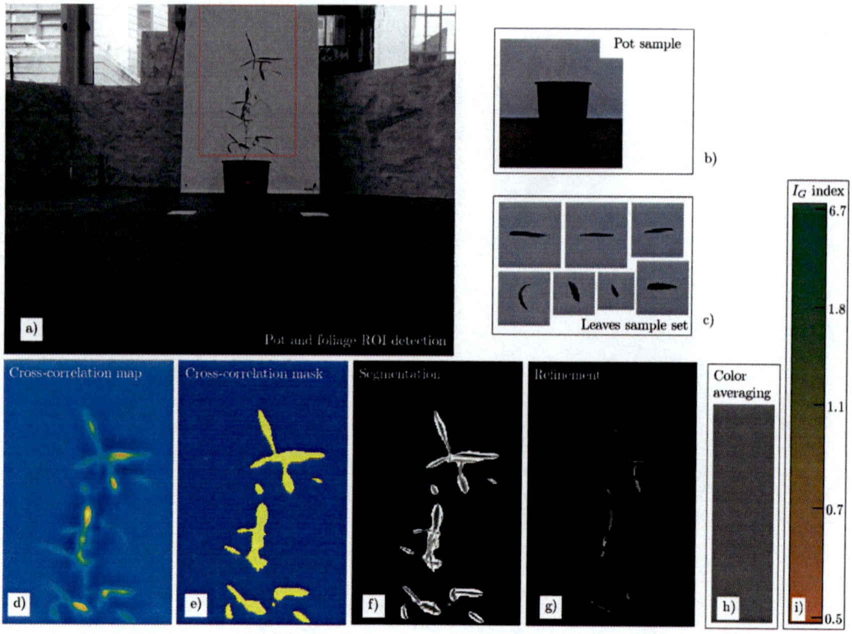


**Fig. 3.** Plant monitoring through image analysis. In a) the view of the camera is shown with the ROI highlighted in red (this view is shown in gray-scale here); in b) the sample for cross-correlation is shown; in c) the sample set is visible for the leaves detection; in d) through g) the main phases of the leaves detection are illustrated; in h) an example of resulting color is shown; in i) the scale of the index $I_G$ is reported.

technique, serves as the basis for the creation of a mask, shown in e). At this point segmentation is carried out and the approximate location of the leaves is determined; in order to completely remove the background, Otsu's thresholding is used again, thus refining the segmentation, as illustrated in g). Finally, the RGB data from this last masked image is averaged and the mean color of the leaves, extracted (h).

An index $I_G$ is shown for each color, indicating how far the detected color is in the green spectrum compared to the other color components:

$$I_G = \frac{G}{R} \tag{1}$$

where G and R are the measured intensities of the green and red color channels in the detected color, respectively.

### 3.2 Bamboo Fiber Composite Process

The goal of this sub-task was to define a manufacturing process for a polymer matrix bamboo fiber reinforced composite, suited to build the structural frame of the robots used for replacing some structural elements of the robot. This process requires human intervention because of its complexity and the many steps required.

Due to lack of time during the experiment, only one plant was grown. However, the process requires a large number of plants in order to obtain the necessary fibers. In our case, 3 trunks of plants were selected, each 1.5 m high.

In a first step, well seasoned bamboo stems were effectively crumbled in a general-purpose, two-shaft shredder. The resulting product was composed of a mixture of two prevailing fractions: internode fragments, characterized by a relatively high aspect ratio, and node fragments, characterized by an equiaxial shape. Since only the internodal fragments were considered as the suited fraction for further processing, a separation step is required. This was done using a dynamic flotation system, composed of a plastic tank, a water flow system and a collecting mesh filter (see Fig. 4).

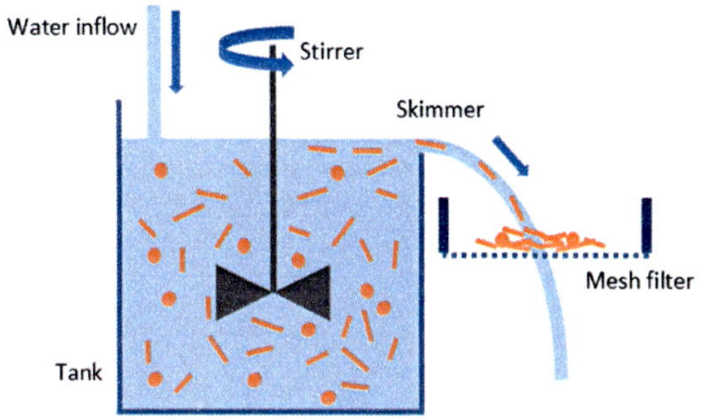


**Fig. 4.** Dynamic flotation system.

The useful fragments fraction was further processed, aiming to reduce the average fiber diameter, while preserving the fiber lenghth (i.e. improving the fibers aspect ratio).

The resulting fiber mixture was then oven-dried, to reduce humidity to an acceptable level for incorporation into the resin matrix (60 °C, 24 h). Then, a square area of approximate dimensions $400 \times 400\,\text{mm}^2$ was delimited with a vacuum sealing tape. The resin adduction tube and vacuum pulling tube were disposed along two opposite edge of the square. This area was then filled with a fiber layer, which was covered with a peel-ply tissue patch to avoid potential puncturing problems caused by the fibers against the vacuum bag (see Fig. 5).

The stack was then sealed with a polyethylene vacuum bag, and vacuum was pulled to check the air tightness of the whole system.

After 48 h of curing at room temperature, the composite board was extracted from the manufacturing bed, and post-cured at 60 °C for 12 h, to improve the mechanical properties of the resin matrix. The composite board was subsequently dry-sanded.

**Fig. 5.** Vacuum sealing process and the resin adduction device.

The mechanical properties of the composite have been determined with in-plane tensile tests, considering the material as quasi-isotropic in the board plane. The results of the tests show a value of the in-plane tensile strength of 25.6 $\pm$ 1.5 Mpa, with an initial elastic modulus of 4.9 $\pm$ 0.3 GPa. Eventually, the plate was cut to the required dimensions. It was used to replace the first structural layer of the initial robot, as it can be seen in Fig. 6.

**Fig. 6.** Robot with the modified structure: one structural layer was replaced with the bamboo fiber one.

## 4 Conclusions

This project identified a roadmap to implement self-replicating robotic systems through the cultivation of plants. The principle that has guided this research is rooted in the extraordinary properties of plant fibers: low payload for space transportation, high mechanical properties, production flexibility and scalability. To validate the basic implementation strategy, a robotic prototype capable of growing bamboo plants has been realized.

## References


1. Awalluddin, D., et al.: Mechanical properties of different bamboo species. In: MATEC Web of Conferences, vol. 138, pp. 1–12 (2017)
2. Chirikjian, G.S., Zhou, Y., Suthakorn, J.: Self-replicating robots for lunar development. IEEE/ASME Trans. Mechatron. **7**(4), 462–472 (2002)
3. Del Dottore, E., Sadeghi, A., Mondini, A., Mattoli, V., Mazzolai, B.: Toward growing robots: a historical evolution from cellular to plant-inspired robotics. Front. Robot. AI **5**, 16 (2018)
4. Ellery, A.: Notes on extraterrestrial applications of 3d-printing with regard to self-replicating machines. In: IEEE International Conference on Automation Science and Engineering, October 2015, pp. 930–935 (2015)
5. Ellery, A.: How to build a self-replicating machine on the moon. In: Earth and Space 2016: Engineering for Extreme Environments - Proceedings of the 15th Biennial International Conference on Engineering, Science, Construction, and Operations in Challenging Environments, pp. 713–724 (2016)
6. Lorenzo, R., Godina, M., Mimendi, L., Li, H.: Determination of the physical and mechanical properties of Moso, Guadua and Oldhamii bamboo assisted by robotic fabrication. J. Wood Sci. **66**(1), 1–11 (2020). https://doi.org/10.1186/s10086-020-01869-0
7. Maggi, F., Pallud, C.: Martian base agriculture: the effect of low gravity on water flow, nutrient cycles, and microbial biomass dynamics. Adv. Space Res. **46**(10), 1257–1265 (2010)
8. Mankins, J.C., Mankins, W.M., Walter, H.: Biological challenges of true space settlement. Acta Astronaut. **146**, 378–386 (2018)
9. Naser, M.: Space-native construction materials for earth-independent and sustainable infrastructure. Acta Astronaut. **155**, 264–273 (2019)
10. Nelson, M., Dempster, W., Allen, J.: Integration of lessons from recent research for "earth to mars" life support systems. Adv. Space Res. **41**(5), 675–683 (2008)
11. Pataranutaporn, P., De La Vega, J., Chowdhury, A., Ng, A., Mihaleva, G.: Toward growable robot: Eexploring and integrating flexible - biological matter with electronics (2018)
12. Suthakorn, J., Kwon, Y., Chirikjian, G.: A semi-autonomous replicating robotic system. In: Proceedings of IEEE International Symposium on Computational Intelligence in Robotics and Automation, CIRA, vol. 2, pp. 776–781 (2003)
13. Wakimoto, T., Takamori, R., Eguchi, S., Tanaka, H.: Growable robot with 'additive-additive-manufacturing', April 2018 (2018)
14. Wamelink, G., Frissel, J., Krijnen, W., Verwoert, M., Goedhart, P.: Can plants grow on mars and the moon: a growth experiment on mars and moon soil simulants. PLoS ONE **9**(8) (2014)

15. Yamaguchi, C., Serra, M., Righetto, A.: Bamboo: a great ally of the civil construction. Smart Innov. Syst. Technol. **140**, 547–562 (2019)
16. Yazawa, Y., Yamaguchi, A., Takeda, H.: Lunar minerals and their resource utilization with particular reference to solar power satellites and potential roles for humic substances for lunar agriculture. Moon Prosp. Energy Mater. Res. **9783642279690**, 105–138 (2012)
17. Yim, M., et al.: Modular self-reconfigurable robot systems [grand challenges of robotics]. IEEE Robot. Autom. Mag. **14**(1), 43–52 (2007)

# Mechanism Designs for Solar Tracking

Matteo Russo[1] (✉) and Marco Ceccarelli[2]

[1] Faculty of Engineering, University of Nottingham, Nottingham, UK
matteo.russo@nottingham.ac.uk
[2] LARM2: Laboratory of Robot Mechatronics, University of Rome "Tor Vergata", Rome, Italy
marco.ceccarelli@uniroma2.it

**Abstract.** In this paper, mechanism design for solar trackers is discussed in terms of serial and parallel architectures that are analyzed to characterize the feasible performance of mechanism solutions that can fully track the sun during its diurnal east-to-west and seasonal north-south motions. Their main operational characteristics are discussed as design criteria and used to compare existing solutions. New mechanism designs are outlined as examples of the proposed guidelines.

**Keywords:** SDG7 · Solar energy · Mechanism design · Parallel manipulators · Solar trackers

## 1 Introduction

In 2015, the United Nations adopted the 2030 Agenda for Sustainable development, a plan of action with 17 sustainable development goals and 169 targets to improve life on the planet [1]. Several of these goals and targets address the need for "*access to affordable, reliable, sustainable and modern energy.*" With Goal 7, the United Nations commit to: "*ensure universal access to affordable, reliable and modern energy services. By 2030, increase substantially the share of renewable energy in the global energy mix,* (…) [and] *enhance international cooperation to facilitate access to clean energy research and technology, including renewable energy, energy efficiency and advanced and cleaner fossil-fuel technology, and promote investment in energy infrastructure and clean energy technology.*"

Among the different sources of renewable energy, solar energy is one with the fastest growing technology, mostly obtained through small- or medium-sized installations of photovoltaic panels on the roofs of houses and business sites. The widespread access to the energy source throughout the globe and the scalability of photovoltaic technologies make solar energy particularly suited to address the global need for clean energy infrastructures.

Obtaining a high efficiency is one of the main concerns with photovoltaic technologies, since their performance is significantly influenced not only by the intrinsic efficiency of the panel but also by the incident solar radiation, which depends on the diurnal and seasonal motions of the sun, with a radiation perpendicular to the panel


G. Quaglia et al. (Eds.): I4SDG 2021, MMS 108, pp. 241–249, 2022.
https://doi.org/10.1007/978-3-030-87383-7_26

representing the optimal scenario. Therefore, a wide range of mechanisms has been proposed, and is available on the market, to adapt the orientation of photovoltaic panels to the sun.

The performance of solar panels can be drastically increased even by introducing a single axis of rotation which follows the sunrise-to-sunset motion. Single-axis trackers can be either actuated by a motor [2, 3] or passively controlled through heliotropic materials [4, 5]. Nevertheless, an optimal performance is obtained with two-axis trackers, which can follow not only the diurnal east-to-west motion of the sun, but also its seasonal north-south motion. A simple mechanical solution to obtain this motion is given by an open loop 2R serial chain with motors on both joints, and serial linkages in general are widely used for solar tracking, e.g. [6], thanks to their immediate, simple control and low mechanical complexity.

Alternatively, parallel mechanisms are characterized by an increased complexity in mechanical architecture, kinematic modelling, and control. However, they have two main advantages over serial chains when used as solar tracking mechanisms: first, parallel architectures are usually characterized by high stiffness, which is required by solar trackers as they have to resist external forces and disruption due to wind and weather; moreover, the higher positional accuracy of parallel mechanisms results in a better precision in orientation than serial mechanisms, enabling a more accurate tracking of the sun motion and thus achieving a higher efficiency.

Several lower-mobility parallel mechanisms have been proposed for solar trackers: in [7], the 3-Degree-of-Freedom (DoF) CAPAMAN is optimized to follow the motion of the sun; the 2PRRR-2PRRS in [8] is another solution with low energy consumption. Further examples of three-limbed parallel designs for double-axis tracking can be found in [9], with a U-2PUS mechanism optimized for a −90° to +90° azimuth motion and a 0° to 90° elevation motion; in [10], with an asymmetric PRS-PUS-RU chain that tracks the −90° to −31° elevation, 90° to −63° azimuth typical of the winter solstice; in [11], where the proposed 3RSR-US mechanism achieves a −90° to +90 ° azimuth motion and a 0° to 90° zenithal motion; and in [12, 13], with U-3PSS and U-PRU-PUS architectures.

While all these works introduce different designs that can achieve double-axis solar tracking, they mostly represent adaptations of existing lower-mobility parallel mechanisms that were originally designed for other applications. Thus, the mechanical architecture is often overengineered and introduces unrequired complexity in the specific task. For this reason, this research paper addresses the problem of mechanism design specifically targeted to solar tracking applications. First, the requirements for solar tracking are discussed by analysing the motion of the sun throughout days and seasons. Then, feasible mechanism solutions are identified and characterized with their kinematic model. Then, they are discussed for the required range of motion and compared in performance to identify the advantages and disadvantages of different designs.

## 2 Requirements for Solar Tracking

The position of the sun is usually recorded on a polar chart that shows the altitude and azimuth of its path, such as the one in Fig. 1. With reference to Fig. 2, the altitude of the sun can be defined with an elevation (or altitude) angle $\alpha_s$, which represents its angular height as measured from the horizontal plane. The minimum elevation is 0° at sunrise and sunset, and a maximum of 90°, representing the sun at its zenith, can be only observed between the Tropic of Capricorn and the Tropic of Cancer. The altitude is sometimes measured with the zenith angle $\psi_S$, which is the complementary of the elevation angle. The azimuth angle $\gamma_S$ completes the polar representation of the sun path by expressing the angular coordinate on the horizontal plane of the projection of the sun's position on the south-north direction.

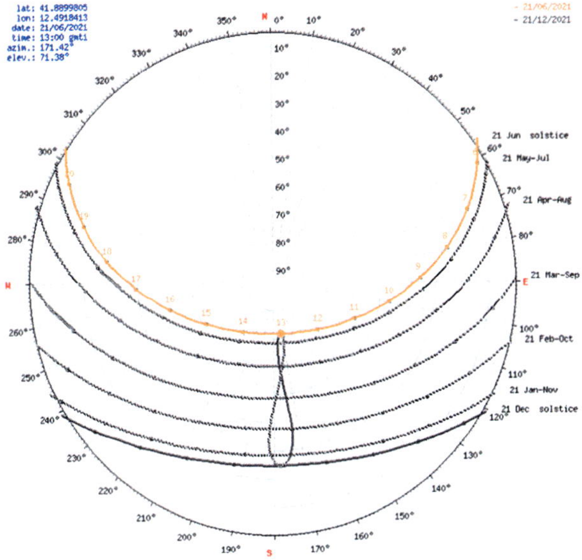


Fig. 1. Representation of the sun path on a polar chart: an example for Rome, Italy, 2021 [14].

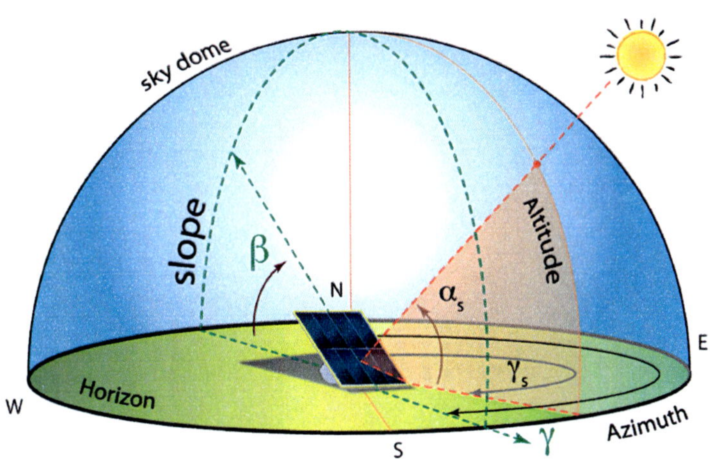


Fig. 2. Polar representation of the position of the sun with azimuth and altitude angles.

Since the solar path varies according to the location on earth, as shown in the example path charts in Fig. 3, the motion requirements for a solar tracker can significantly vary depending on where the panel is installed, with more demanding requirements closer to the Equator. Whereas most previous study consider a −90° to +90° azimuth motion and a 0° to 90° elevation motion, a range of −120° to +120° azimuth motion and a 0° to 80° elevation motion better characterizes the motion of the sun in a European location. This choice is purely exemplary; the methodology here proposed can be applied to any other location by changing the range requirement to the local one.

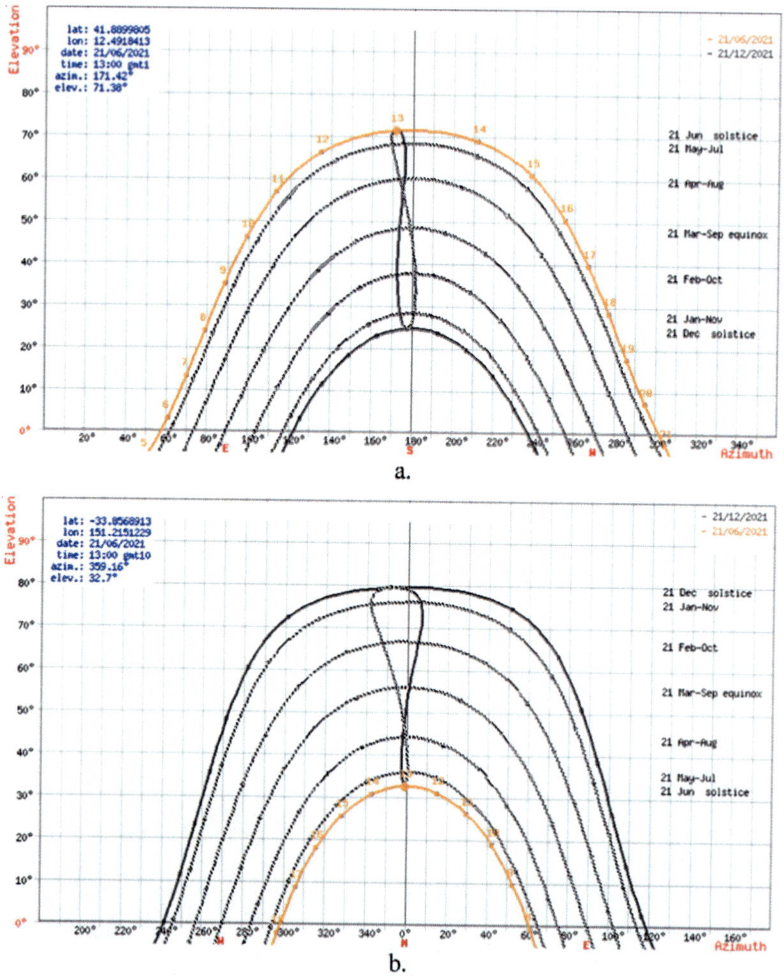


**Fig. 3.** Examples of solar paths in polar altitude-azimuth coordinates [14]: a. In Rome, Italy, 2021; b. In Sydney, Australia, 2021.

In Fig. 4, the main design requirements and expected operation capabilities are outlined referring to mechanism design for efficient use-oriented solutions. Sun tracking requires a suitable fairly slow motion of a surface facing the sun so that the sun rays are normally incident to it. The mechanism structure is expected to have a corresponding motion control and the capability to handle the high payload of the photovoltaic panels as well as external loads, such as wind. Its general features include a stiff low-cost solution both in design and operation, and a robust design able to sustain hazardous weather conditions, spanning from very cold to very hot temperatures, rain, and wind. Complementary to the above design requirements, the mechanism is expected to have a fairly simple action with a limited regular maintenance and a compact light mechanism to facilitate a modular installation of with multiple units in outdoor environments.

The main operation goal can be identified in the efficient work of the unit, including both tracking mechanism and photovoltaic cell, in terms of continuous autonomous functioning. For example, when considering a photovoltaic panel of $2 \times 1$ m size and a weight of 800 N, a tracking mechanism should be contained within the projection of the panel onto the ground and with a height of 2 m (for a $2 \times 1 \times 2$ m volume) to enable adjacent installations. Furthermore, its power consumption should be limited to a fraction of the captured solar energy significantly lower than the increase in captured energy when compared to a static panel. Thus, both requirements and goals can be formulated even as design and operation criteria to be considered for optimal solutions.

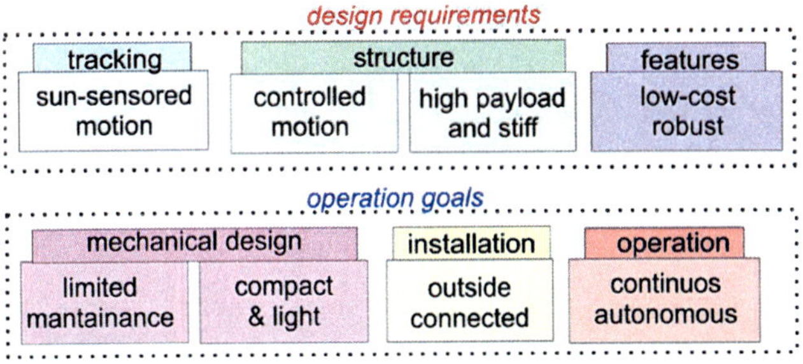


**Fig. 4.** A scheme with the main design characteristics for solar tracking mechanisms.

## 3 Existing Mechanism Solutions

The simplest solar tracking mechanisms are characterized by a single axis of rotation that follows the altitude of the sun; these designs consist of a single revolute joint actuated by a motor, as shown in the scheme in Fig. 5a. Even though a single degree of freedom significantly boosts the performance of photovoltaic panel, the seasonal motion of the sun requires a second actuator that enables an additional rotation. Whereas a serial 2R mechanism as per Fig. 5b satisfies these motion requirements, most of the other operational goals and desired features outlined in the previous section, such as high stiffness, low consumption, accurate motion, and resistance to weather hazards, make parallel mechanisms extremely interesting for solar tracking despite the added mechanical complexity.

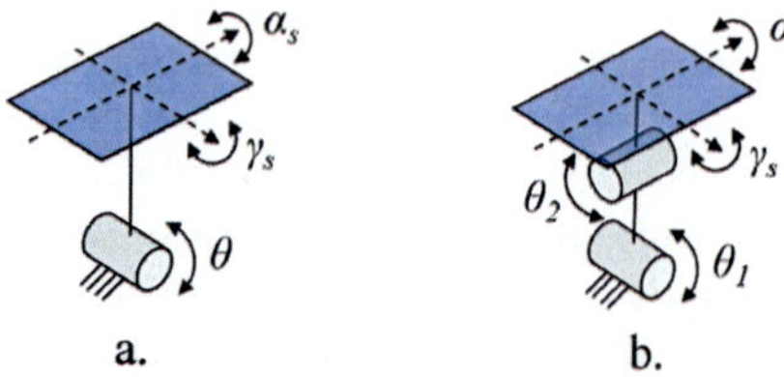


**Fig. 5.** Examples of existing mechanism solutions based on serial chains: a. A single axis tracker with an open-loop R chain [2]; b. A two-axis tracker with an open-loop RR chain [6].

Most of the existing parallel mechanism solutions are based on 2-degree-of-freedom structures with a tripod architecture, where two limbs are actuated and the third one is passively constraining the motion of the platform. In the kinematic schemes in Fig. 6, which reports two examples from literature [7–13], the main features of these tripods can be appreciated. First, they are driven by linear motors fixed on the ground, which usually present an improved motion performance when compared to the rotary motors observed in serial structures, thanks to the reduced backlash and to worm gears or equivalent transmissions preventing backdriving motion from the wind or other weather conditions. Furthermore, as the wrench due to the weight of the system is partially supported by the passive chain, the power consumption can be minimized with a proper dimensional synthesis.

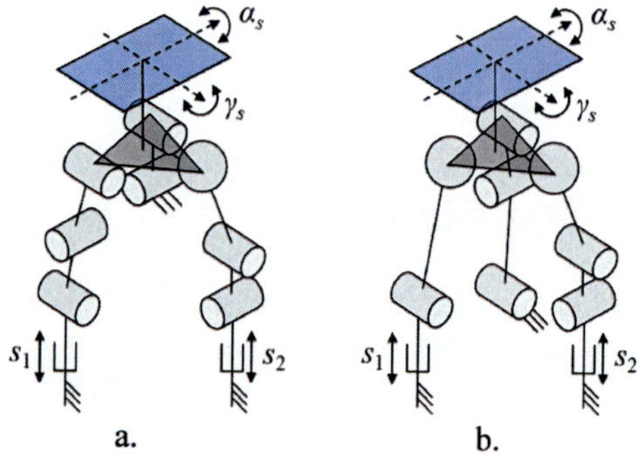


**Fig. 6.** Examples of existing mechanism solutions based on parallel chains: a. U-PRU-PUS architecture [13], also similar to the U-2PUS in [9]; b. PRS-PUS-RU architecture [10].

In conclusion, existing mechanism solutions are based on either serial chains with one or two rotational degrees of freedom, which are characterized by a simple design and motion planning, or on asymmetric parallel mechanisms with tripod structures, with better performance but added complexity.

## 4 Novel Mechanisms for Solar Tracking

In this section, the three mechanisms in Fig. 7 are introduced as alternatives to the existing solutions that have been outlined previously. These mechanisms aim at improving the performance of solar trackers by following the design guidelines that were presented in Sect. 2.

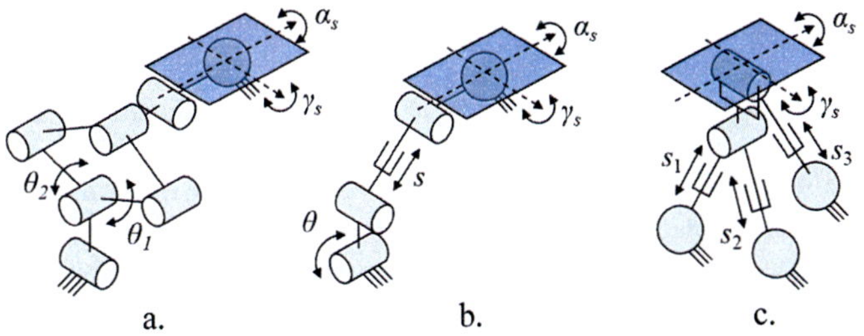


**Fig. 7.** Examples of novel mechanism solutions: a. 2RRRR-SR closed-chain mechanism; b. RRPRS closed-chain mechanism; c. 3SPR parallel mechanism.

The proposed solutions range from serial closed-chain structures, such as the one in Fig. 7a and 7b, to the parallel tripod in Fig. 7c. By using closed-chain structures, these solutions achieve a higher stiffness than the open-chain structures in Fig. 5. The mechanism in Figs. 7a and 7b are both characterized by a photovoltaic panel connected to the ground with an idle spherical joint (or equivalent) placed in the center of mass of the panel or on its vertical. In this way, the weight of the panel is supported by this idle joint and the actuation force/torque needed for solar tracking is minimized. Thus, the orientation is controlled by a 2-degree-of-freedom chain, which is a parallelogram in the mechanism in Fig. 7a and a rotational and linear motors in series in the mechanism in Fig. 7b, acting on one side of the panel. The tripod in Fig. 7c is based on a fully parallel mechanism reported in [15], which presents a wide singularity-free reachable workspace when compared to similar-sized structures while maintaining a comparable stiffness, as shown in [15]. The main features of these designs are summarized in Table 1.

When compared to the serial chains in Fig. 5, the proposed mechanisms are all characterized by a better stiffness because of their closed-chain architectures. Thus, they present a better reliability, as they are more suited to sustain weather hazards. Furthermore, they can represent an alternative to the parallel mechanisms in Fig. 6 thanks to their fairly simple kinematics, which lead to an efficient control.

**Table 1.** Summary of the features of the proposed mechanism solutions.

| Kinematic chain: | 2RRRR-SR | RRPRS | 3SPR |
|---|---|---|---|
| Diagram | Figure 7a | Figure 7b | Figure 7c |
| Structure | 1 closed-loop chain | 1 closed-loop chain | 3 parallel chains |
| Panel weight | Supported by idle spherical joint | Supported by idle spherical joint | Linear motors (non -backdrivable) |
| Degrees of Freedom | 2 (rotational) | 2 (1 rotational, 1 linear) | 3 (linear) |
| $-120°$ to $+120°$ azimuth | Yes | Yes | Yes |
| 0° to 80° elevation | Yes | Yes | Yes |

## 5 Conclusions

In this paper, the main requirements and features of solar tracking mechanisms are discussed. The motion range of the mechanism is presented in function of the solar path and its variability in different locations. Other operational features are outlined, and existing mechanism solutions are presented for both serial and parallel architectures. Finally, the proposed guidelines are used to define three potential solutions for solar tracking devices as based on closed-chain mechanisms.

## 1. References

1. United Nation General Assembly, 2015. Transforming our world: the 2030 Agenda for Sustainable Development. A/RES/70/1
2. Sallaberry, F., Pujol-Nadal, R., Larcher, M., Rittmann-Frank, M.H.: Direct tracking error characterization on a single-axis solar tracker. Ener. Conv. Manag. **105**, 1281–1290 (2015)
3. Clifford, M.J., Eastwood, D.: Design of a novel passive solar tracker. Solar Ener. **77**(3), 269–280 (2004)
4. Cezan, S.D., Baytekin, H.T., Baytekin, B.: Self-regulating plant robots: bioinspired heliotropism and nyctinasty. Soft robot. **7**(4), 444–450 (2020)
5. Li, C., Liu, Y., Huang, X., Jiang, H.: Direct sun-driven artificial heliotropism for solar energy harvesting based on a photo-thermomechanical liquid-crystal elastomer nanocomposite. Adv. Funct. Mat. **22**(24), 5166–5174 (2012)
6. Yao, Y., Hu, Y., Gao, S., Yang, G., Du, J.: A multipurpose dual-axis solar tracker with two tracking strategies. Renew. Ener. **72**, 88–98 (2014)
7. Jiménez, E., Ceccarelli, M., Carbone, G.: A dynamic analysis of the robot CaPaMan (Cassino Parallel Manipulator) as solar tracker. In: *Proceedings of IFToMM-FeIbIM International Symposium on Mechatronics and Multibody Systems MUSME, Valencia*, pp. 579–594 (2011)
8. Altuzarra, O., Macho, E., Aginaga, J., Petuya, V.: Design of a solar tracking parallel mechanism with low energy consumption. *Proc. Instit. Mech. Eng. Part C: J. Mech. Eng. Sci. 229*(3), 566–579 (2015)
9. Cammarata, A.: Optimized design of a large-workspace 2-DOF parallel robot for solar tracking systems. Mech. Mach. Theor. **83**, 175–186 (2015)
10. Wu, J., Chen, X., Wang, L.: Design and dynamics of a novel solar tracker with parallel mechanism. IEEE/ASME Trans. Mech. **21**(1), 88–97 (2015)

11. Mauro, S., Battezzato, A., Biondi, G., Scarzella, C.: Design and test of a parallel kinematic solar tracker. Adv. Mech. Eng. **7**(12), 1687814015618627 (2015)
12. Wu, J., Zhang, B., Wang, L.: Optimum design and performance comparison of a redundantly actuated solar tracker and its nonredundant counterpart. Solar Ener. **127**, 36–47 (2016)
13. Du, X., Li, Y., Wang, P., Ma, Z., Li, D., Wu, C.: Design and optimization of solar tracker with U-PRU-PUS parallel mechanism. *Mech. Mach. Theor. 155*, 104107 (2021)
14. Sun position. https://www.sunearthtools.com. Accessed 15 Apr 2021
15. Russo, M., Ceccarelli, M., Takeda, Y.: Force transmission and constraint analysis of a 3-SPR parallel manipulator. *Proc. Inst. Mech. Eng. Part C: J. Mech. Eng. Sci. 232*(23), 4399–4409 (2018)

Check for updates

# An Automated Procedure for Crack Detection

Erika Pellegrino(✉) and Tania Stathaki

Imperial College London, London, England
e.pellegrino16@imperial.ac.uk

**Abstract.** Nowadays the civil infrastructure is exposed to several challenges such as daily vehicular traffic and extreme weather conditions, i.e. ghastly winds, strong rain etc. It is well known that these may determine structural deterioration and damages, which can even cause catastrophic collapses related to significant socio-economic losses. For this reason it is evident that automatic inspection and maintenance must play a decisive role in the future. With the objective of quality assessment, cracks on concrete buildings have to be identified and monitored continuously. Due to the availability of cheap devices, techniques based on image processing have been gaining in popularity, but they require a rigorous analysis of large amounts of data. Moreover, the detection of fractures in images is still a challenging task, being these structures sensitive to noise and to changes in environmental conditions. This paper proposes an automatic procedure to detect cracks in images along with a parallel implementation on heterogeneous High Performance Architectures aiming both at automatizing the whole process and at reducing its execution time.

**Keywords:** SDG11 · Infrastructure · Innovation · Industry

## 1 Introduction

The concrete infrastructure requires a constant monitoring due to the deterioration caused by several factors. In current practice the detection of defects is absolved manually by inspectors, but this modality is time consuming and error prone [1,3]. This task is critical also for assessing the concrete quality [2]. Due to availability of cheaper devices, image-based techniques have been gaining in popularity, but they require large volumes of data to be processed as quickly as possible. The detection of fractures is still challenging in image processing. The main reasons are that they have a complex topology, a thickness similar to the image resolution and are easily corrupted by noise. One of the scopes of this paper is to propose a variational model to detect cracks in images. It is well known that classical edge detection techniques based on gradient information are not successful with these structures. The intuitive idea is that a fracture (when it is modeled as an indicator function supported by a smooth curve) can

Supported by Imperial College London.


G. Quaglia et al. (Eds.): I4SDG 2021, MMS 108, pp. 250–254, 2022.
https://doi.org/10.1007/978-3-030-87383-7_27

be approximated by a sequence whose terms have a Hessian that blows up in its perpendicular direction, while their gradient is null. Motivated by these observations, showing that a suitable model should involve higher order derivatives, we propose a crack recovery method that falls within second order variational models. It is based on the Blake-Zisserman [4] functional or, to be more precise, on its approximation by elliptic functionals defined on Sobolev spaces [5]. In this paper we propose an implementation on High Performance Computing (HPC) architectures that are designed to execute massive parallel tasks. The paper is organized as follow: in Sect. 2 we describe the mathematical model and the parallel implementation, in Sect. 3 we discuss performance results and in Sect. 4 we draw some conclusions.

## 2 Mathematical Model and Parallel Implementation

Variational methods have addressed successfully problems such as image segmentation and edge detection. They propose as solution a minimizer of a global energy. A first example is described by Mumford and Shah (MS) in their famous paper [8] where they proposed a first order functional, whose minimization determines an approximation of the image by means of a piecewise smooth function and detects edges as singularities in the image intensity. However, this model is not suitable for cracks because they do not represent singularities in the intensity function, but in its gradient instead. For this reason, we propose a second order variational model based on the Blake-Zissermann (BZ) functional [4]. This was introduced with the aim of overcoming some limitations of the MS approach, such as the over segmentation and the lack in detecting gradient discontinuities. Being the original formulation not suitable for numerical treatment, we had to work on a different approach that is based on the approximation proposed by Ambrosio and Tortorelli (AT) for the MS functional [6,7]. In their model, they replaced the unknown discontinuity set by an auxiliary function which smoothly approximates its indicator function. In our case two auxiliary functions are introduced as indicators of both intensity discontinuity and gradient discontinuity sets. As numerical minimization algorithm we chose an "inexact" block-coordinate descent scheme (BCD) in order to address the heterogeneous environment. Although the model is global, several numerical experiments have highlighted that the solutions weakly depend on boundary conditions and are energetically close to the initial data. This motivate the adoption of a tiling scheme to address very large images: the minimizer is assembled by merging together local solutions restricted to portion of images.

## 3 Numerical Results

We tried our methods on images of cracks taken in tunnels in Greece and backscattered electron images of concrete samples. In the former case the challenge was to reconstruct the whole structure avoiding the effect of the noise and the

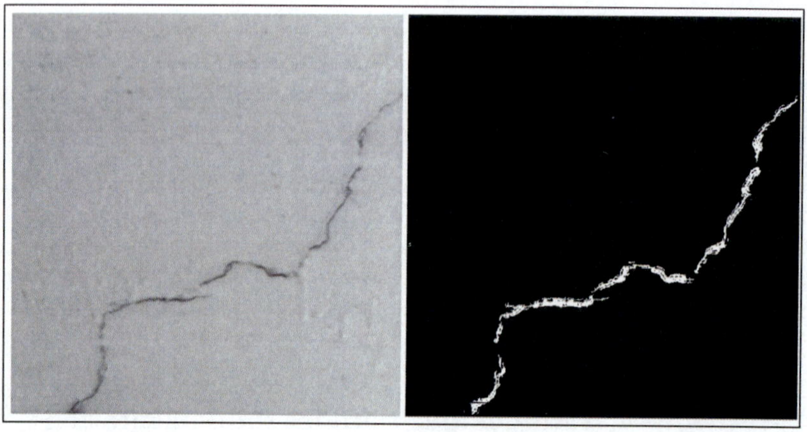

**Fig. 1.** Crack on a concrete wall

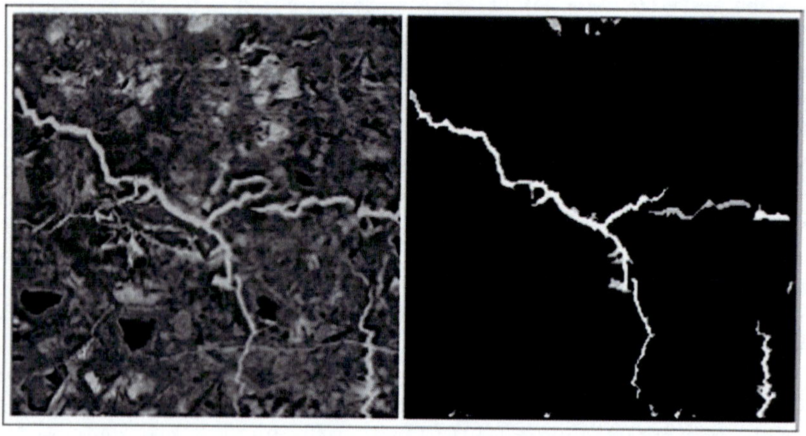

**Fig. 2.** Microcracks on a BSE microscopy of concrete materials

environmental conditions (i.e. lights). In the latter the aim was to "insulate" the structure from the complex texture in the background.

In both cases the structures have been detected correctly (Fig. 1, 2). In order to reduce the execution time and to provide a automatic procedure we tested a sequential implementation with a parallel one based on the OpenMP framework [11] that implements two strategies for collaboratively executing a program on an environment composed by devices of different types (aka heterogeneous architectures). The experiments were performed both on a commodity PC and on a High Performance Computing cluster [10]. A sequential version was executed on a workstation equipped with a processor Intel (R) Xeon CPU E6-79 at 3.40 GHz with 32 GB of RAM and total number of cores 12, running an Ubuntu 18.04

operating system. The parallel version based was executed on a heterogeneous cluster equipped with ×86-64 processors, running a CentOS 7.6 operating system. Overall, the results show a significant reduction in the execution time with respect to the sequential algorithm (Table 1).

**Table 1.** Run time comparisons for a single image

| Algorithm | Time (s) |
|---|---|
| Sequential | 13.184868 |
| Parallel 24 cores | 1.056701 |
| Parallel 48 cores | 0.5915374 |

## 4 Conclusions

In this paper we proposed a variational method to detect cracks in images and a parallel implementation targeting modern High Performance Computing architectures with the aim of automatizing the detection process and of reducing its execution time. We got promising results for both the quality of the reconstruction and the reduction of its duration. As future plan, we would like to test our procedure on real world scenarios in order to understand if it could be used as useful tool for assessing the building infrastructure and the concrete quality. We plan also to reduce further the execution time by designing a domain specific hardware accelerator tailored specifically to detect cracks in real time. Moreover, it would be interesting to understand how the algorithm work with isotropic structures such as bubbles.

**Acknowledgement.** We would like to thank the National Technical University of Athens for having provided images of cracks in the tunnels of Egnatia Motorway in Metsovo and the Concrete Durability Laboratory at Imperial College London for the back-scattered electron images.

## References

1. Protopapadakis, E., Voulodimos, A., Doulamis, A., Doulamis, N., Stathaki, T.: Automatic crack detection for tunnel inspection using deep learning and heuristic image post-processing. Appl. Intell. **49**(7), 2793–2806 (2019). https://doi.org/10.1007/s10489-018-01396-y
2. Lee, H.X.D., Wong, H.S., Buenfeld, N.R.: Self-sealing of cracks in concrete using superabsorbent polymers. Cem. Concr. Res. **79**, 194–208 (2016)
3. Dams, B., et al.: Aerial additive building manufacturing: three-dimensional printing of polymer structures using drones. In: Proceeding of the Institution of Civil Engineers: Construction Materials, pp. 1–12 (2017)
4. Blake, A., Zisserman, A.: Visual Reconstruction. MIT Press, Cambridge (1987)

5. Ambrosio, L., Faina, L., March, R.: Variational approximation of a second order free discontinuity problem in computer vision. SIAM J. Math. Anal. **32**(6), 1171–1197 (2001)
6. Tortorelli, V.M., Ambrosio, L.: Approximation of functionals depending on jumps by elliptic functionals via Gamma-convergence. Comm. Pure Appl. Math. **43**, 999–1036 (1990)
7. Ambrosio, L., Tortorelli, V.M.: On the approximation of free discontinuity problems, pp. 105–123 (1992)
8. Mumford, D., Shah, J.: Optimal approximations by piecewise smooth functions and associated variational problems. Commun. Pure Appl. Math. **42**(5), 577–685 (1989)
9. Bellettini, G., Coscia, A.: Discrete approximation of a free discontinuity problem. Numer. Funct. Anal. Optim. **15**(3-4), 201–224 (1994)
10. http://www.imperial.ac.uk/admin-services/ict/self-service/research-support/rcs/
11. https://www.openmp.org/
12. Author, F.: Article title. J. **2**(5), 99–110 (2016)
13. Author, F., Author, S.: Title of a proceedings paper. In: Editor, F., Editor, S. (eds.) CONFERENCE 2016, LNCS, vol. 9999, pp. 1–13. Springer, Heidelberg (2016). https://doi.org/10.10007/1234567890
14. Author, F., Author, S., Author, T.: Book title, 2nd edn. Publisher, Location (1999)
15. Author, A.-B.: Contribution title. In: 9th International Proceedings on Proceedings, pp. 1–2. Publisher, Location (2010)
16. LNCS Homepage. http://www.springer.com/lncs. Accessed 4 Oct 2017

# Modelling of Unmanned Aerial Vehicles with Vertical Take Off and Landing Capabilities

Armando Casazza[1], Roberto Fiorenzani[1], Alberto Mela[2], Luca Pugi[2](✉), and Alberto Reatti[2]

[1] Sky Eye Systems SRL, Cascina, Italy
a.casazza@skyeyesystems.it
[2] University of Florence, Florence, Italy
luca.pugi@unifi.it

**Abstract.** UAVs offer an affordable and sustainable way to improve monitoring of large areas contributing, as example, to gather large amount of data or to improve the surveillance of sites in which criminal activities such as clandestine immigration should harm safety of peoples. Patrolling of large areas involves the use of Unmanned Aerial Vehicles with large autonomies, a feature that is much easier to be obtained with fixed wing planes. Also, VTOL (Vertical Take Off and Landing) features are highly desirable to assure an easy and fast deployment. For these reasons, a hybrid propulsion layout is adopted: wings assure lift for prolonged missions in an efficient way, while multiple rotors with fixed axis are used to assure hovering and VTOL capabilities. So, the superposition of these two propulsion systems assures a relatively simple and robust construction and control of the UAV. However, some further cautions must be taken during the transition from cruise speed to hovering condition and vice versa. For these reasons authors propose in this work a simplified model aiming to simulate the behavior of the UAV to support a better comprehension of involved dynamics and a better design optimization of the system.

**Keywords:** Decent Work and Economic Growth (SDG 8) · Peace · Justice and Strong Institution (SDG 16) · UAVs · Modelling of dynamical systems

## 1 Introduction: Modeled Hybrid UAV

UAVs (Unmanned Aerial Vehicles) offer to perform in an affordable way a lot of different task involving the availability of a prolonged aerial support to monitor large areas or to cover large distances avoiding traditional economical and technological limitations of conventional, pre-existing technologies. UAVs technology is spreading in a wide variety of shapes corresponding to different involved applications [1].

This work is mainly focused on the development of a model able to simulate the behavior of UAVs that adopt a hybrid propulsion layout [2] to gather in a single platform the autonomy of an unmanned fixed wing plane with maneuverability and hovering performances of rotating wing systems such as quad copters. Reference layout for the


G. Quaglia et al. (Eds.): I4SDG 2021, MMS 108, pp. 255–263, 2022.
https://doi.org/10.1007/978-3-030-87383-7_28

proposed model is represented by the simplified layout of Fig. 1: a fixed wing airplane [3] is propelled by a rear propeller able to provide a longitudinal thrust ($T_0$ in Fig. 1) ailerons and rudders (on boom mounted inverted v-tail) are used to control the trajectory when cruising. Four vertical thrusters fixed on booms are used to control the UAV as a quadcopter [4] during vertical take-off and landing also assuring hovering capabilities. For what concern energy management, it is supposed the adoption of a hybrid series layout described in Fig. 2: all the propellers are moved by electrical motors fed by lithium batteries that are continuously recharged by an alternator that convert the mechanical energy provided by a two-stroke internal combustion engine, working as a range extender [5].

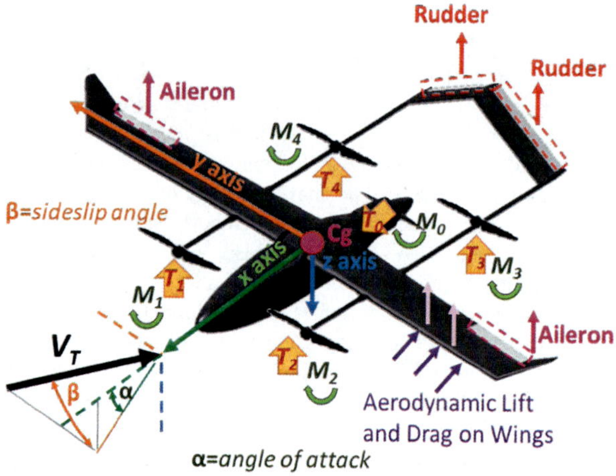


**Fig. 1.** Layout of modelled UAV, simplified scheme

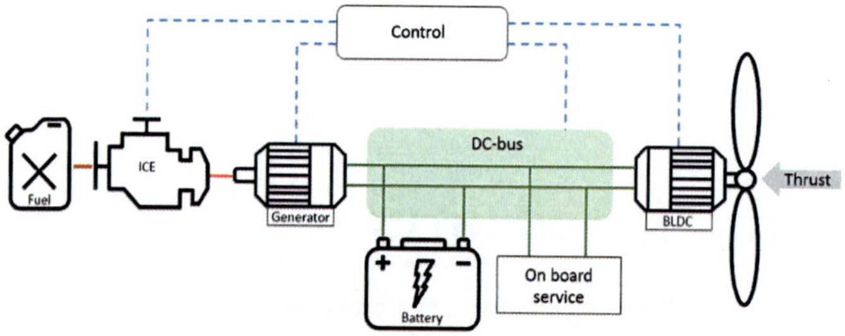


**Fig. 2.** Hybrid series Layout

## 2 AUV Dynamic Model

AUV is modelled using a Matlab-Simulink™ model as a rigid body, adopting the same notation and conventions [3] that are often used also for underwater vehicles especially when complex multi-thruster configurations are adopted [6].

UAV dynamics is described by Eq. (1) where following symbols are adopted:

- $m$ and $I_{ij}$ describe the inertial properties of the vehicle, respectively the mass and the corresponding elements of the matrix of inertia, described by the i-th and j-th pedex.
- $u,v,w$, and $p,q,r$ are respectively vehicle linear and angular vehicle speed components of the vector $\nu$ respect to a body reference system visible in Fig. 1.
- $X, Y, Z, L, M, N$, is the corresponding vector $\tau$ of external applied forces and torques respect to the same body constrained reference system.

Vector of external forces and torques is then calculated as the sum of different contributions provided by different subsystems as visible in the scheme of Fig. 3:

- Weight ($\tau_g$) and aerodynamic forces($\tau_{aer}$) from wings and navigation surfaces (ailerons and rudders).
- Thrust and reaction torques from installed propellers ($\tau_{prop}$).

$$\tau = I_M \dot{\nu} + C(\nu)\nu = \begin{cases} X = m(\dot{u} - rv + qw) \\ Y = m(\dot{v} + ru - pw) \\ Z = m(\dot{w} - qu + pv) \\ L = I_{xx}\dot{p} - I_{xy}\dot{q} - I_{xz}\dot{r} + (I_{zz} - I_{yy})qr + (I_{xy}r - I_{xz}q)p + (r^2 - q^2)I_{yz} \\ M = I_{yy}\dot{q} - I_{xy}\dot{p} - I_{yz}\dot{r} + (I_{xx} - I_{zz})pr + (I_{yz}p - I_{xy}r)q + (p^2 - r^2)I_{xz} \\ N = I_{zz}\dot{r} - I_{xz}\dot{p} - I_{yz}\dot{q} + (I_{yy} - I_{xx})pq + (I_{xz}q - I_{yz}p)r + (q^2 - p^2)I_{xy} \end{cases} \quad (1)$$

## 2.1 Weight and Aerodynamic Forces

Aerodynamic forces $\tau_{aer}$ (2) to which a generic wing/aerodynamic body is subjected are proportional to wing surface $S_w$ and to the dynamic pressure contribution $p_B$ (3) which is substantially the specific kinetic energy of the fluid expressed as the product of density $\rho$ and the squared value of the relative speed $v_T$ of air respect to the vehicle.

Coefficients of proportionality adopted in (2) are called $C_i$ where the index i-th identifies a specific coefficient: as example $C_D$ represents the proportionality coefficient to calculate longitudinal Drag force $X_{aer}$. For the calculation of aerodynamic torques/moments $L_{aer}$ $M_{aer}$ $N_{aer}$, aerodynamic coefficients are scaled respectively respect to the wingspan $b$ and the aerodynamic cord $l_{mac}$.

$$\tau_{aer} = \begin{bmatrix} X_{aer} \\ Y_{aer} \\ Z_{aer} \\ L_{aer} \\ M_{aer} \\ N_{aer} \end{bmatrix} = \begin{bmatrix} -C_D \\ C_Y \\ -C_L \\ bC_l \\ l_{mac}C_M \\ bC_N \end{bmatrix} p_B S_w \quad (2)$$

$$p_B = \frac{1}{2}\rho v_T^2 \quad (3)$$

Aerodynamics coefficients $C_i$ are tabulated functions of attack($\alpha$) and sideslip($\beta$) angles that describe the direction of the relative speed of the air that is investing the plane as described in Fig. 1. For navigation surfaces such as ailerons and rudders-elevators these coefficients are further corrected and tabulated to take count of their variable orientation respect to incoming airflow. Angles $\delta_{rl}$ $\delta_{rr}$ $\delta_{al}$ $\delta_{ar}$ are used to describe respectively the orientation of rudders (left and right) and ailerons (left and right). All these tabulated contributions are summed to calculate global cumulative aerodynamics coefficients and consequently aerodynamic forces acting on the vehicle. Tabulated functions describing the behavior of the vehicle are provided by the industrial partner that has supported this study and they have obtained from CFD calculations validated by experimental data, since most of components of the system simulated in this work are internally designed, engineered and tested by the company.

For what concern contribution of gravitational forces, the calculation is performed according (4) since adopted reference frame is constrained to vehicle center of gravity, null contributions in terms of restoring moments/torques should be calculated.

$$\tau_g = \left[0\ 0\ mg \cos\theta \cos\phi\ 0\ 0\ 0\right]^t \tag{4}$$

## 2.2 Calculation of Thrust and Reaction Torques Generated ByPropellers

For what concern propellers the thrust produced by the i-th propeller $T_i$ is calculated from tabulated values of the thrust coefficient $C_{Ti}$ [7] defined according (5). Reaction Torques $M_i$ are defined respect to tabulated dimensionless torque coefficient $C_{Mi}$ (6):

$$C_{Ti} = \frac{T_i}{\rho n_i^2 d_i^4} \Rightarrow T_i = C_{Ti} \rho n_i^2 d_i^4 \tag{5}$$

$$C_{Mi} = \frac{T_i}{\rho n_i^2 d_i^5} \Rightarrow M_i = C_{Mi} \rho n_i^2 d_i^5 \tag{6}$$

Thrust and torque coefficients are often referred respect to the so called advance coefficient $J_i$ (7) of the i-th propeller:

$$J_i = \frac{v_i}{n_i d_i} \tag{7}$$

In Eqs. (5), (6) and (7), following symbols are adopted:

- $n_i$ is the rotation speed of the *i-th* propeller in Hz;
- $d_i$ is the diameter of i-th diameter;

Data are provided by APC [8] the supplier of adopted propellers. Expected behavior of chosen propellers for vertical propellers (numbered from 1 to 4 in the scheme of Fig. 1) and the rear one is shown respectively in Figs. 3/a/b.

Electrical actuators and generators used for the proposed application are supposed to be brushless PM motors for which is assumed the simplified model described in (8), being the torque $Q_m$ produced by the motor proportional to motor current *I:*

$$V = k_\phi \omega + RI + L\dot{I} + jLI\omega \tag{8}$$

In (8) motors are modelled as equivalent DC motors in which whose armature circuit is modelled considering the voltage-velocity $k_\phi$ and losses due to internal resistance $R$ and inductance $L$. Motors are supposed to be speed controlled by a driver with a constant efficiency $\eta_D$. For multi-quadrant operations power exchanged by the drive with a constant voltage source $W_{DC}$ respect to motor one $W_M$ is calculated according (9):

$$\mathit{if}\ W_m \geq 0 \Rightarrow W_m = W_{DC}\eta_D; \quad \mathit{if}\ W_m < 0 \Rightarrow W_m = \frac{W_{DC}}{\eta_D} \tag{9}$$

Vector of efforts $\tau_{prop}$ produced by the five propellers including reaction torques is described by (10), where the coordinate $z_0$ along the body constrained vertical direction is considered to better evaluate the contribution in terms of exerted pitch torque.

$$\tau_{prop} = \begin{bmatrix} T_0 \\ 0 \\ -T_1 - T_2 - T_3 - T_4 \\ l\frac{\sqrt{2}}{2}(+T_1 + T_2 - T_3 - T_4) - M_0 \\ l\frac{\sqrt{2}}{2}(-T_1 + T_2 + T_3 - T_4) + T_0 z_0 \\ M_1 - M_2 + M_3 - M_4 \end{bmatrix} \tag{10}$$

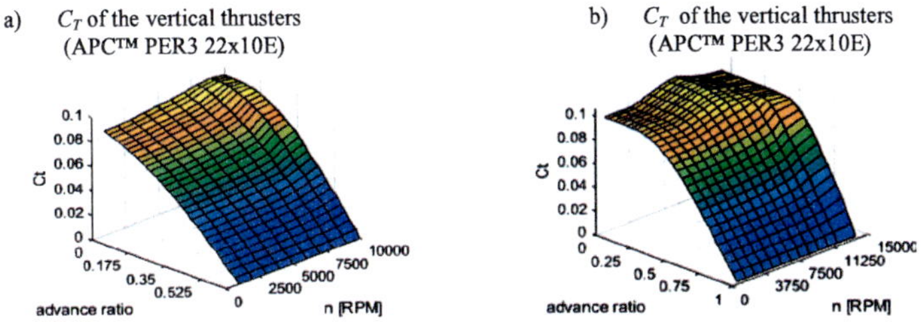


**Fig. 3.** Tabulated propeller performances, (a) for vertical propellers, (b) for the rear one

## 3 Power Management System

For what concern vehicle batteries, it is considered a simplified model in which the battery is modelled as an ideal voltage source $V_{OC}$ (SOC) followed by a constant resistance in series $R_{bat}$. According to this simplified model, power that is really consumed inside the battery $W_{soc}$ can be directly calculated knowing the power exchanged by the battery with external loads $W_{bat}$ (11):

$$W_{SOC} = \frac{V_{OC}^2}{2R_{bat}}\left(1 - \sqrt{1 - 4\frac{W_{bat}R_{bat}}{V_{OC}^2}}\right) \tag{11}$$

In (10) $V_{OC}$ is a tabulated function of the SOC (State of Charge) which is defined according as the fraction of current stored energy $E(t)$ respect to maximum one $E_{max}$ (12)

$$SOC(t) = \frac{E(t)}{E_{\max}} = \frac{E_0 + \int_0^t W_{soc} dt}{E_{\max}} \tag{12}$$

Battery is continuously recharged by a two-stroke engine: coupling between motor and alternator is simulated and optimized to exploit the maximum efficiency (expressed in term of Brake Specific Fuel Consumption) of the motor respect to power required by the alternator as visible in Fig. 4.

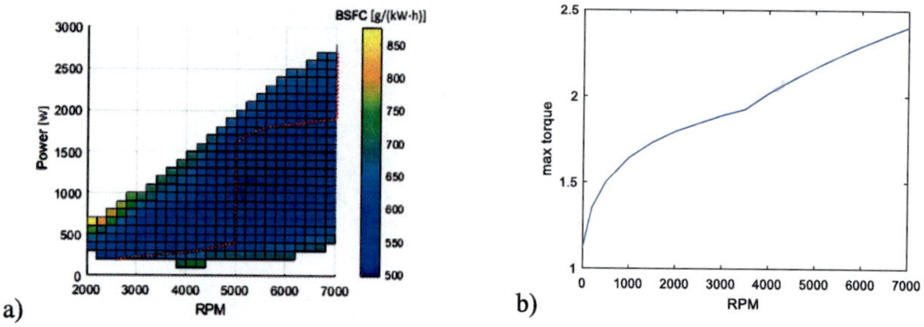


**Fig. 4.** Optimized working curve of the engine to obtain max efficiency vs. required power (a) and corresponding torque behavior of the alternator (b)

## 4 Simulation of Vertical Take Off Maneuver

Proposed model has been used to simulate a complete mission of the UAV. In this preliminary work attention is focused on the simulation of the take off maneuver described in Fig. 5: in current implementation proposed control system is based on the superposition of two controllers, the VTOL one and the CRUISE one. As Take Off starts the first controller, the VTOL one is activated to perform a vertical trajectory to gain minimum elevation $h_{min}$. As VTOL control is activated only the four vertical propellers numbered from 1 to 4 are activated. To better understand the behavior of proposed control system is adopted a Vehicle Carried Reference Frame: a reference frame constrained to body but with fixed axis aligned to North, East and center of the earth (NED). In (13) it is calculated the corresponding behavior of the vector $T_n$ of forces along the three-axis respect to vehicle orientation expressed in terms of Euler angles $\varphi.\psi.\theta$.

$$T_n = \begin{bmatrix} T_x \\ T_y \\ T_z \end{bmatrix} = \begin{bmatrix} (\sin\psi\sin\phi + \cos\psi\cos\phi\sin\theta)\left(\sum_{i=1}^{4} T_i\right) \\ (-\cos\psi\sin\phi + \sin\theta\sin\psi\cos\phi)\left(\sum_{i=1}^{4} T_i\right) \\ (\cos\theta\cos\phi)\left(\sum_{i=1}^{4} T_i\right) \end{bmatrix} \approx \begin{bmatrix} (\phi\sin\psi + \theta\cos\psi)\left(\sum_{i=1}^{4} T_i\right) \\ (-\phi\cos\psi + \theta\sin\psi)\left(\sum_{i=1}^{4} T_i\right) \\ \left(\sum_{i=1}^{4} T_i\right) \end{bmatrix} \tag{13}$$

Considering small variations of pitch and roll angles, planar forces in $x_n$ and $y_n$ directions are approximately linear dependent from $\varphi$ and $\theta$, while the vertical force is simply the sum of the thrusts assured by vertical propellers. Torques need to control vehicle orientation along the three axis can be calculated according (10) introducing negligible approximations since adopted yaw and roll angles are very small.

Once UAV reach an assigned elevation, a bump-less transition procedure starts: VTOL control is maintained only to assure a constant elevation $h_{ref}$ respect to ground and then a second controller the CRUISE one is activated also imposing longitudinal vehicle speed reference $u_{ref}$. $u_{ref}$ is a value of vehicle speed for which, according (2), aerodynamic lift generated by wings is big enough to completely balance the weight of the plane. So, transition phase ends when the plane reach a longitudinal speed that allows the plane to switch off energy consuming propellers of the VTOL system. To perform this operation CRUISE control is designed with a closed loop strategy that is often adopted in literature [2]: since lift is proportional to the squared vehicle of incoming air, altitude is smoothly controlled by adjusting the thrust $T_0$ delivered by the rear propeller. An increase of $T_0$ is associated to a relatively small increase of speed since variation of altitudes are substantially proportional to the integral of the square value of the speed, so most of the energy delivered by this actuator, once drag is balanced is converted in an increase of the potential energy of the system and consequently of the altitude. A second loop is used to control the speed of the plane by adjusting through the action of navigation surfaces (as example elevators) the pitch angle of the plane: also, in this case effectiveness of proposed control depends from the conversion of plane kinetic energy in to potential one smoothed by drag variations. Finally, a third loop controls through rudders and ailerons UAV steering. As visible in Fig. 5, proposed take-off strategy produces a smooth and stable behavior since plane maintains an almost constant altitude

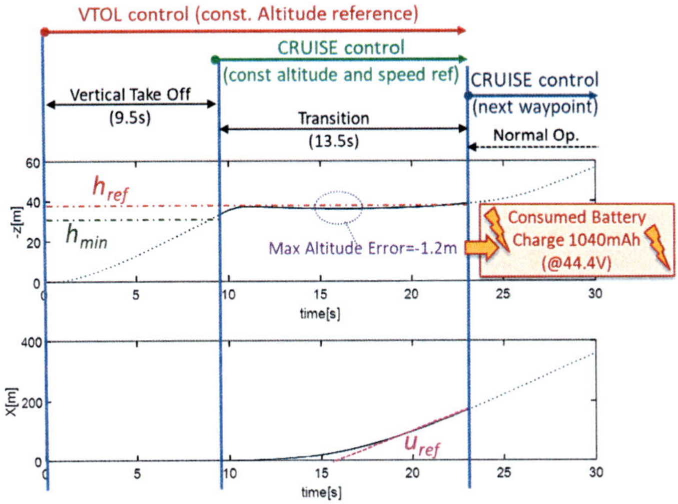


**Fig. 5.** Optimized working curve of the engine to obtain max efficiency vs. required power (a) and corresponding torque behavior of the alternator (b)

during all the performed transition, but it is also clear that energy consumption should be improved: since acceleration of the UAV is caused only by the rear propeller so duration for which vertical propellers of VTOL must work is relatively long.

## 5 Conclusions and Future Developments

In this work authors have been able to design a complete model of an AUV in which different aspects related not only to control but also to detailed design of the propulsion system are investigated. This pave the way to future investigations in which this model should be used to test, as example, different control strategies to save energy during take off. Authors are also interested to investigate the usage of a torque allocation procedures aiming to systematically optimize both energy efficiency and robustness respect to external disturbances (wind) and unmodelled dynamics also transferring some know how from automotive [9] marine [10] systems. Also future extension of this study should regard a further hybridization of motion capabilities of the vehicle on the example of some of the most recent contributions introduce in the HeritageBot project [11–13].

## References


1. Gupta, S.G., et al.: Review of unmanned aircraft system (UAS). Int. J. Adv. Res. Comput. Eng. Technol. (IJARCET) **2**(4), 1646–1658 (2013)
2. Balmer, G.R.: Modelling and Control of a Fixed-wing UAV for Landings on Mobile Landing Platforms (2015)
3. Gryte, K., et al.: Aerodynamic modeling of the skywalker x8 fixed-wing unmanned aerial vehicle. In: 2018 International Conference on Unmanned Aircraft Systems (ICUAS), pp. 826–835. IEEE (2018)
4. Luukkonen, T.: Modelling and control of quadcopter. Independent research project in applied mathematics. Espoo **22**, 22 (2011)
5. Powell, B.K., Pilutti, T.E.: A range extender hybrid electric vehicle dynamic model. In: Proceedings of 1994 33rd IEEE Conference on Decision and Control, pp. 2736–2741. IEEE (1994)
6. Pugi, L., Pagliai, M., Allotta, B.: A robust propulsion layout for underwater vehicles with enhanced manoeuvrability and reliability features. Proc. Inst. Mech. Eng. Part M J. Eng. Maritime Environ. **232**(3), 358–376 (2018). https://doi.org/10.1177/1475090217696569
7. Allotta, B., Pugi, L., Pagliai, M., Paolucci, L. Innovative redundant and reconfigurable propulsion systems: Fast prototyping through simplified simulation and design procedures. In: Marine Engineering: Emerging Developments and Global Challenges, pp. 69–95 (2017)
8. Free technical data available at official site of APC. https://www.apcprop.com/technical-information/performance-data/
9. Montani, M., et al.: ESC on in-wheel motors driven electric vehicle: handling and stability performances assessment. In: Proceedings - 2020 IEEE International Conference on Environment and Electrical Engineering and 2020 IEEE Industrial and Commercial Power Systems Europe, EEEIC/I and CPS Europe 2020, Article number. 9160768 (2020). https://doi.org/10.1109/EEEIC/ICPSEurope49358.2020.9160768
10. Pugi, L., Allotta, B., Pagliai, M.: Redundant and reconfigurable propulsion systems to improve motion capability of underwater vehicles. Ocean Eng. **148**, 376–385 (2018). https://doi.org/10.1016/j.oceaneng.2017.11.039

11. Ceccarelli, M., et al.: 2. HeritageBot service robot assisting in cultural heritage. In: Proceedings - 2017 1st IEEE International Conference on Robotic Computing, IRC 2017, pp. 440–445, Article number. 7926580 (2017)
12. Ceccarelli, M., Cafolla, D., Russo, M., Carbone, G.: HeritageBot platform for service in Cultural Heritage frames. Int. J. Adv. Robot. Syst. **15**(4), 1729 (2018)
13. Cafolla, D., Russo, M., Ceccarelli, M.: Experimental validation of HeritageBot III, a robotic platform for cultural heritage. J. Intell. Rob. Syst. **100**(1), 223–237 (2020). https://doi.org/10.1007/s10846-020-01180-6

# Underactuated Soft Gripper for Helping Humans in Harmful Works

Gabriele Maria Achilli[1], Silvia Logozzo[2], Maria Cristina Valigi[2(✉)], Gionata Salvietti[3], Domenico Prattichizzo[3], and Monica Malvezzi[3]

[1] Department of Engineering, University of Perugia, 05100 Terni, Italy
gabrielemaria.achilli@studenti.unipg.it
[2] Department of Engineering, University of Perugia, 06125 Perugia, Italy
{silvia.logozzo,mariacristina.valigi}@unipg.it
[3] Department of Information Engineering and Mathematics, University of Siena, 53100 Siena, Italy
monica.malvezzi@unisi.it

**Abstract.** Robotics and automation are essential to improve life quality by assisting and substituting humans in dangerous tasks. Starting from this idea, this paper introduces an innovative gripper that can be employed in the new industrial policy, such as Industry 4.0, in which machines and humans work together. In particular, the paper describes the study of a new soft gripper as a modification of a underactuated gripper previously designed. The new proposed gripper presents an improved design, and it employs more sustainable materials resulting more compliant and adaptable. Therefore, it can be introduced within different industrial and service contexts with the aim of improving the process efficiency and even sustainability, according to SDG9. In particular, the proposed gripper can be used in the recycling and waste disposal chain with a consequent improvement of workers' health, helping the achievement of SDG3, SDG8 and SDG12. The main feature of the gripper is the modularity, which makes it highly versatile and reconfigurable. reducing waste and obsolescence. In addition, the paper investigates the application of a new environment-friendly material for realising the phalanges comparing it with different materials in terms of structural mechanics by means of FEM analysis. Results demonstrate that both the new design and the employment of the new "green" material make the gripper more efficient and helpful in achieving different *Sustainable Development Goals*. This is also the basis of a circular perspective for the protection of the environment and the optimization of costs and resources and human lives.

**Keywords:** SDG3 · SDG8 · SDG 9 · SDG 12 · Underactuated robots · Robotic manipulation

## 1 Introduction

Grasping and the manipulating objects with robotic devices in a safe and robust way has represented an important issue, particularly in unstructured environments, due to the


G. Quaglia et al. (Eds.): I4SDG 2021, MMS 108, pp. 264–272, 2022.
https://doi.org/10.1007/978-3-030-87383-7_29

complexity of tasks [1, 2]. Robotic grippers typically are assemblies of rigid joints and links [3]. In this paper, after a brief description of a robotic gripper previously proposed in [4], some improvements are studied and designed to obtain a new device with a different structure, using a "green" material. The new gripper has a modular design, presenting flexibility and a good ability to manipulate free-form objects. The device consists of deformable elements that adapt themselves to the shapes of the objects. The elements are connected by compliant flexure joints, which resulted to increase the performance of grasping, enhancing the range of motion and compliance and dereasing internal friction, as it happens also with the contact rolling joints as demonstrated in [5]. It is a underactuated system with a great versatility and cost-effectiveness because the use of fewer actuators allows to create lighter mechanisms. A underactuated soft gripper similar to the one proposed in tis work is presented in [6] but with different configuration of fingers, phalanges and transmission and also different construction materials. All the components of the gripper can be 3D printed. In other works, the use of shape memory alloys to realize fingers and joints was investigated, as for example in [7]. Underactuated grasping systems are also studied in literature for the realization of prosthetic hands as the "Federica" hand [8]. An interesting application is also related to the improvement of grasping due to a movable palm, as in [9].

This paper proposes and investigates the manufacturing of some components of the gripper employing a more sustainable material (PLA) than the usually used ones. Finite Element Method (FEM) simulations are carried out to evaluate the behavior of the components made in the new material in working conditions. Finally, the proposed model is compared with the initial version to evaluate the improvements.

The proposed robotic gripper can help to meet some of the 17 UN's Sustainable Development Goals (SDG) in industry for achieving a healthier planet.

The use of robots in the industry is increasing together with the implementation of automated production systems, estimating an increase in self-processing processes up to 70%, so that most of the current jobs will be fully automated. Obviously, the overall trend in the use of robots leads to a reduction in manual and routine working and in an increase of conceptual activities for human workers. One of the consequences is that, in many industrial fields, the quality of life of workers will be improved together with their health. In this context, the robotic gripper developed in this work is conceived for the manipulation of objects with variable shapes and characteristics, trying to replace human hands in harmful or repetitive activities. In a prospective of circular economy, the proposed device can find applications in specific contexts such as the manipulation of waste in the recycling and waste disposal chain, where the processing of waste and the separate collection process are sometimes dangerous for the health of workers for [10]. For these reasons this technology contributes at satisfying SDG3 and SGD8.

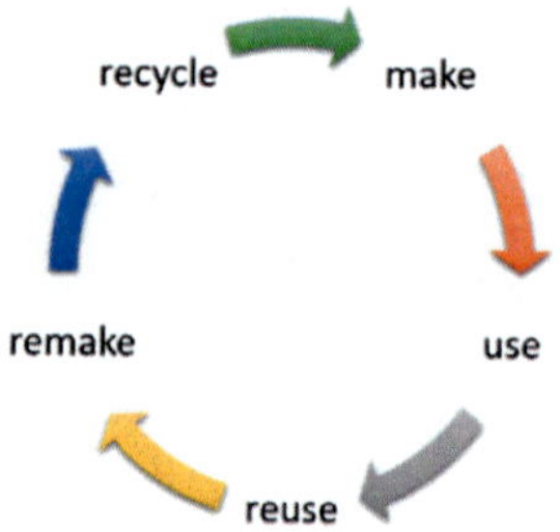


**Fig. 1.** Circular economy prospective

Moreover, thanks to the modularity, in case of damage of certain components, such as phalanges, it is possible to replace only damaged ones and not the entire finger with a project of circular economy whose aim is make, use, reuse and remake (Fig. 1). Furthermore, the use of a more sustainable material such as the PLA (Polylactide), contributes at satisfying SDG 9 and SDG 12.

## 2 Gripper's First Version

The starting gripper is a underactuated mechanism able to adapt itself to the shape of the object to grasp. It is a kind of soft robotic hand composed by two modular fingers that are actuated by a single tendon by means of a differential system [11, 12]. The flexible joints connect rigid modules to build a deformable structure that can fit the shape of the grasped object. Normally soft hands have not enough dexterity to perform a precise grip and the working environment, such as a planar surface, represents a constraint able to reduce the uncertainties that can be exploited by the robotic hand. From this observation a flat surface was inserted to reduce constrains and improve the hand manipulation capability.

**Fig. 2.** Starting gripper

This idea was applied for the device showed in Fig. 2, that is the first version of the robotic hand, newly designed for applications in the waste industry. The presence of the rigid scoop connected to the base of the gripper increases the certainty of the grasp and allows the gripper to have the capability of grasping object not necessarily placed on a regular surface. Anyway, some limits are found as reported in Table 1.

**Table 1.** Gripper's limits and reason of its redesign

| | Gripper limits |
|---|---|
| 1 | Difficulties of adapting to non-planar rigid surfaces since the scoop is rigid |
| 2 | Risk of damage for the grasped objects |
| 3 | Insufficient adaptability to the object shapes: when the object does not present a planar surface, the grasping is performed almost exclusively by the two robotic fingers |
| 4 | Low mobility of the scoop. For grasping small objects, a good dexterity is needed to prevent them from falling |
| 5 | Difficulties in selection of particular objects within a heterogeneous mix of different shapes |
| 6 | Weak versatility and reconfiguration |

## 3 Design of the Modified Solution

The development of the gripper regards the improvement of the device performances by means of an alternative solution able to make the soft hand more modular and facilitating the grasping abilities. The insertion of two new flexible modular phalanges for each finger and of a new slot with a through hole, allows the introduction of additional modules able to convert the pair of fingers in an actuated scoop as it is shown in Fig. 3.

**Fig. 3.** The new mechanism that converts the pair of fingers in a scoop. The black module is made in TPU (Thermoplastic polyurethane), the two finger modules in ASA (Acrylonitrile styrene acrylate).

The black module is made in TPU and it has cylinders for connection with the same shape of modules so that the modularity is assured. The single parts of the new device are shown in Fig. 4. To ensure stability and ease of assembly, different types of phalanx modules have been provided but the diameter of the used holes and cylinders remained unchanged: this allows further modularity in the device and makes use of a design already tested and working. The new scoop is shown in Fig. 5 and its extremity has a flat face able to contact the object being gripped. On the other extremity the scoop has a T-insert that allows the insertion of the module inside the end flange slot.

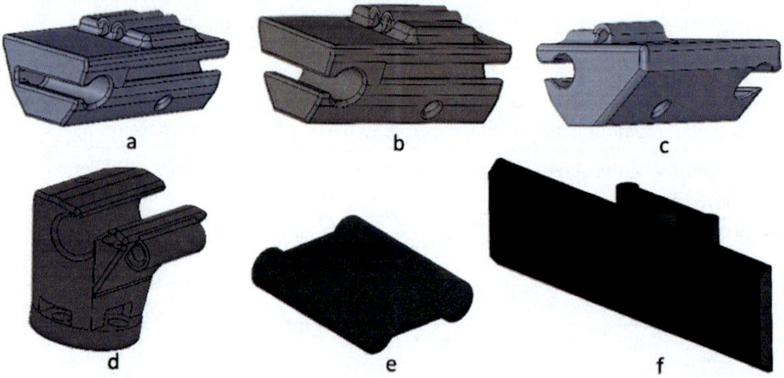


**Fig. 4.** Single parts of the new device: a. Starting module (ASA), b. Middle module (ASA), c. End Module (ASA), d. Base connection with the finger (ASA), e. Flexible joint between a pair of modules (TPU), f. Flexible module that converts the pair of fingers in a scoop (TPU).

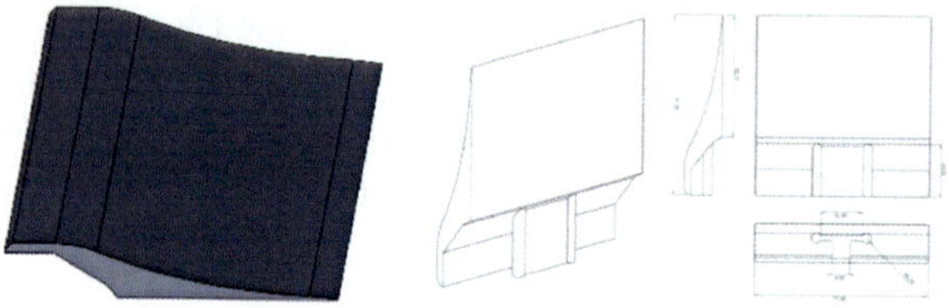

**Fig. 5.** The scoop with its end flat.

### 3.1 New Material

In its first version the gripper was made by ASA for the rigid parts and TPU for the soft parts [13], as shown in Fig. 6a. As known, TPU is an elastic polyurethane plastics and its deformability depends on the Shore Hardness scale value [14]. In this study a 98A TPU was used. The phalanx modules and the scoop were made of ASA material that is thermoplastic with rigid mechanical properties [15]. Both materials are oil-based, with a high environmental impact during production [16]. The study proposes to evaluate the characteristics of the gripper by replacing the rigid parts made in ASA with the PLA. PLA is a thermoplastic material derived from a wide variety of plant fibers, with a low environmental impact [17]. In Fig. 6b the prototype of the device made in PLA is shown.

This material is less elastic than ASA, anyway it has the mechanical properties described in Table 2 [18], compatible with the applications of the presented work. In fact, the demand for resistance to bending deformation on the entire scoop is addressed above all by the flexible joints in TPU (Fig. 4e), which remain unchanged in the two versions. The subsequent study through the FEM analysis shows damage risk areas of the rigid components printed in PLA compared to ASA.

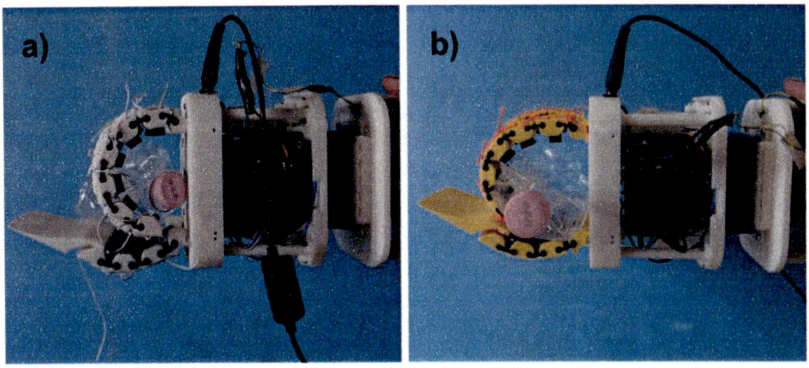


**Fig. 6.** Gripper's prototypes: a. device in ASA; b. device in PLA

**Table 2.** Mechanical properties of the polymeric filament used

| | ASA | PLA | TPU |
|---|---|---|---|
| Elastic modulus ($N/mm^2$) | 2097 | 3149 | 15.6 |
| Poisson's ratio | 0.3712 | 0.36 | 0.495 |
| Shear modulus ($N/mm^2$) | 761.7 | 1287 | 26 |
| Yield strength ($N/mm^2$) | 55 | 77 | 16 |
| Mass density ($kg/m^3$) | 1030 | 1024 | 1135 |

### 3.2 FEM Analysis

The upper part of the prototype was examined using the FEM (finite element method), to evaluate the displacements and the stress applied to the scoop. The two base connectors were blocked with fixed constraints, and a support constraint on the lower edge of the scoop was applied to simulate the presence of an applied force by the actuator through the tendons tension that close the scoop. Based on the described schematic representation, a realistic behavior of the device has been simulated by COMSOL Multiphysics. Two series of trials were done: the first with the parts in ASA, the second in PLA. A force of 19.16 N was applied, equivalent to a weight force of 2 kg; this value was chosen to compare the results obtained with the FEM analysis with a realistic trial, using objects with a maximum mass of 2 kg. The force was applied to a pair of modules. The results obtained with ASA material are shown in Fig. 7 and the ones obtained by PLA are shown in Fig. 8.

## 4 Result and Discussions

The tests carried out showed acceptable displacements in the areas affected by deformation and in accordance with the initial design conditions. Moreover, no critical concentration stress areas were highlighted (Fig. 8). There is only a stress area in the finger base connectors, near the hole for the flexible joints: this value is conservative, less than a possible failure of the component for the applied load values (Table 3). In the subsequent realization through AM (additive manufacturing), the thickness of the part should be evaluated. The choice of the PLA instead of ASA for the intended use of the studied device is more than acceptable. In addition, the use of PLA for the rigid components allows the use of a green filament with less toxicity than ASA material during manufacturing process. The highlighted deformability characteristics confirmed the possibility of building a new prototype entirely in PLA and seeing its behavior even in real scenarios.

**Table 3.** FEM results

| Plot | ASA | PLA |
|---|---|---|
| Static displacement (mm) | 0.642 | 0.622 |
| Static nodal stress ($N/mm^2$) | 3.791 | 4.377 |

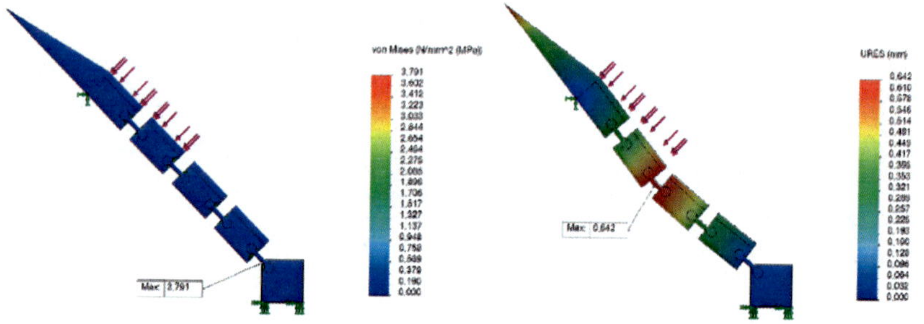

**Fig. 7.** Results obtained with ASA material

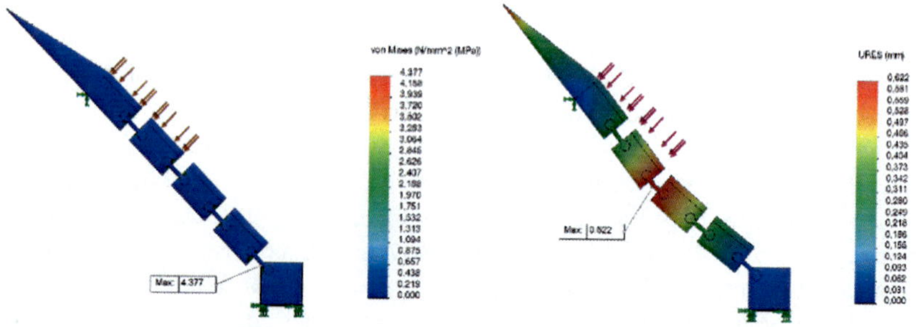

**Fig. 8.** Results obtained with PLA material

## 5 Conclusion

This paper presents the study and optimization of a soft gripper. In particular, the paper is focused on the application of a new environment-friendly material to make the gripper more efficient and helpful in achieving different Sustainable Development Goals as SDG3, SGD8, SDG9, SDG12. In fact, the proposed gripper, has the aim of reproducing the human hand functions, to help men in harmful tasks, promoting human health, decent work, and allowing human workers to be employed in more conceptual and intellectual activities, spending less time for manual repetitive tasks. Besides improving human health and life and work conditions, this project considered also environmental aspects by complying some aspects of the circular design for sustainability. In fact, low environmental impact materials were studied to realize the soft gripper, with particular attention to keep or improve all the manipulation abilities.

## References


1. Carbone, G.: Grasping in Robotics, Mechanisms and Machine Science, vol. 10. Springer, Heidelberg (2012). https://doi.org/10.1007/978-1-4471-4664-3
2. Mason, M., Salisbury, J.J.: Robot Hands and the Mechanics of Manipulation. The MIT Press, Cambridge (1985)
3. Melchiorri, C., Kaneko, M.: Robot Hands. In: Siciliano, B., Khatib, O. (eds.) Springer Handbook of Robotics, pp. 463–480. Springer, Cham (2016). https://doi.org/10.1007/978-3-319-32552-1_19
4. Salvietti, G., Malvezzi, M., Iqbal, Z., Eslami, T.: Soft hands with embodied constraints: the soft scoopgripper. In: Proceedings of the 2019 International Conference on Robotics and Automation (ICRA), Montreal, QC, Canada, 20–24 May 2019
5. Boisclair, J.M., Laliberte, T., Gosselin, C.: On the optimal design of underactuated fingers using rolling contact joints. IEEE Robot. Autom. Lett. **6**(3), 4656–4663 (2021)
6. Ma, R., Odhner, L., Dollar, A.: A modular, open-source 3D printed underactuated hand. In: Proceedings - IEEE International Conference on Robotics and Automation, Karlsruhe, Germany (2013)
7. Maffiodo, D., Raparelli, T.: Flexible fingers based on shape memory alloy actuated modules. Machines **7**(2), 40 (2019)
8. Cosenza, C., Niola, V., Savino, S.: A simplified model of a multi-jointed mechanical finger calibrated with experimental data by vision system. Proc. Inst. Mech. Eng. Part K: J. Multibody Dyn. **235**(1), 164–175 (2021)
9. Espinosa-García, F.J., Arias-Montiel, M., Ceccarelli, M., Carbone, G., Lugo-González, E.: Design and experimental characterization of a novel subactuated mechanism for robotic finger and movable palm. Int. J. Mech. Control **20**(2), 141–146 (2019)
10. Drudi, D.: Job hazards in the waste industry. Solid Waste Technol. **14**(1), 32–35 (2000)
11. Malvezzi, M., Iqbal, Z., Valigi, M.C., Pozzi, M., Prattichizzo, D., Salvietti, G.: Design of multiple wearable robotic extra fingers for human hand augmentation. Robotics **8**(4), 102 (2019)
12. Malvezzi, M., Valigi, M.C., Salvietti, G., Iqbal, Z.: Design criteria for wearable robotic extra–fingers with underactuated modular structure. Mech. Mach. Sci. **68**, 509–517 (2019)
13. Achilli, G., Valigi, M.C., Salvietti, G., Malvezzi, M.: Design of soft grippers with modular actuated embedded constraints. Robotics **9**(4), 105 (2020)
14. https://aerospace.basf.com/elastollan.html (Online)
15. Yap, L., et al.: Finite element analysis of 3D-printed acrylonitrile styrene acrylate (ASA) with ultrasonic material characterization. Int. J. Comput. Mater. Sci. Eng. **8**, 1950002 (2018)
16. Gu, J., Wensing, M., Uhde, E., Salthammer, T.: Characterization of particulate and gaseous pollutants emitted during operation of a desktop 3D printer. Eur. J. Eng. Sci. Technol. **1**(2), 59–64 (2019)

17. Pang, X., Huang, X., Tang, Z., Chen, X.: Polylactic acid (PLA): research, development and industrialization. Biotechnol. J. **5**, 1125–1136 (2010)
18. Żur, P., Żur, A., Baier, A.: Finite elements analysis of PLA 3D-printed elements and shape optimization. Eur. J. Eng. Sci. Technol. **2**(1), 59–64 (2019)

# A Control Architecture for the Hybrid Hexapod Robot R3HC

Fernando Gómez-Bravo[1(✉)], Alejandro Garrocho Cruz[1], and Giuseppe Carbone[2]

[1] Electronic, Computing and Automation Engineering Department, University of Huelva, Huelva, Spain
{fernando.gomez,alejandro.garrocho}@diesia.uhu.es

[2] Department of Mechanical Engineering, Energy and Management, University of Calabria, Rende, Italy
giuseppe.carbone@unical.it

**Abstract.** This paper describes the control architecture developed for the hybrid hexapod robot R3HC, created at the University of Huelva in collaboration with the University of Calabria. This architecture supports the communication processes that allow the robot controller to acquire data from a Lidar sensor and, at the same time, to coordinate legs and wheels motions so that the robot is capable of navigating and overcoming obstacles laying on the floor.

**Keywords:** SDG9 · Robotics and mechatronics · Hexapod robot · Hybrid locomotion · Control architecture

## 1 Introduction

At the present time, hybrid robots are one of the branches of the robotics world where the biggest and most important challenges are presented. Hybrid robots combine the benefits of walking and wheeled robots. These types of systems are especially suitable for exploring uneven ground [4], allowing them to move autonomously by rolling on wheels or walking on legs when necessary. In this way, they are faster than walkers robots on flat surfaces, thanks to its wheels, and able to overcome obstacles by using the legs, what cannot be done by traditional wheeled robots [6]. Thus, these robots are appropriate to innovate in industrial infrastructures, facing challenges related with the SDG9 United Nation Goals.

Among hybrid robots it is worth highlighting wheeled hexapods, which are mechanics vehicles that can walk or roll on six legs or wheels [4]. In this area, it should be noted the Cassino Hexapod Hybrid Robots, I, II and III, designed in the "LAboratorio di Robotica e Mecatronica" (LARM) at the Cassino University by the team of Prof. Carbone, and now further developed at the University of Calabria [7,8], see Fig. 1. Cassino hexapods are six-legged robots, comprising two joints and a wheel on each leg. Three DC motors control the motion of each leg, therefore a robots is driven by eighteen motors operated simultaneously [3].


G. Quaglia et al. (Eds.): I4SDG 2021, MMS 108, pp. 273–281, 2022.
https://doi.org/10.1007/978-3-030-87383-7_30

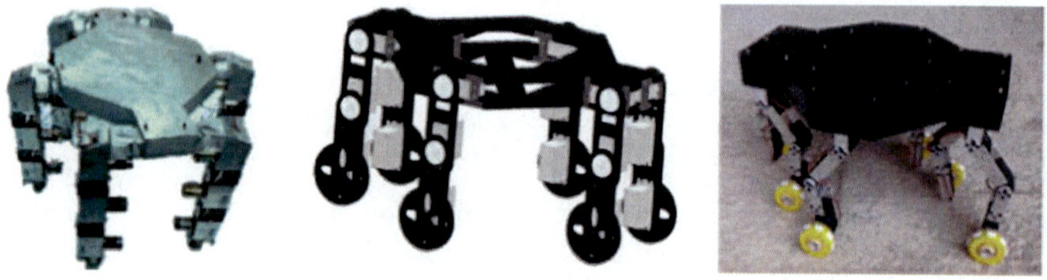

**Fig. 1.** The Cassino hexapod robots.

As an evolution of the Cassino hexapods, the robot R3HC was created in the University of Huelva in collaboration with the team of Prof. Carbone. R3HC incorporates new features in terms of locomotion capabilities and also in the control architecture, specially designed to navigate experimentally in unstructured scenarios using a Lidar sensor. This work aims to describe this architecture and the communication process particularly developed for these purposes.

The paper is structured as follow. In Sect. 2 the main features of R3HC are introduced. Section 3 describes the control architecture. The paper closes with the experimental result and conclusions.

## 2 R3HC Characteristics

R3HC robot differentiates itself from the Cassino Hexapod versions in a twisting joint at the top of each leg, as shown in Fig. 2. This characteristic allow robot's legs to rotate around their longitudinal axis, which enable a range of new and sophisticated movements, increasing their manoeuvrability abilities and control complexity [3]. In consequence, R3HC is composed by six legs with four motors each: wheel, knee, hip and twist motor, in ascending order. Thus, R3HC robot uses twenty-four motors. Another difference is the use of digital servomotors (Dynamixel motors, [1]), which requires the controller system to be capable of managing a particular communication protocol for reading and sending data.

**Fig. 2.** R3HC leg model and schematic.

By comparison, the last version of Cassino hexapod robots is equipped with omni-wheels [5] that allow it to move in any direction without leg rotation. By contrast, R3HC manoeuvrability is supported by the new twisting joint of each leg that allow the robot to describe smooth trajectories, overcoming different obstacles or rotating about itself.

Finally, different from traditional mobile robots, where range sensors are configured to visualize obstacles in the horizontal plane, R3HC incorporated a Lidar sensor in the upper part of the robot, placed on an tilted support, facing the ground, see Fig. 3. This will allow the robot to obtain information about small obstacles lying on the floor, as well as bigger obstacles that must be avoided by changing robot motion.

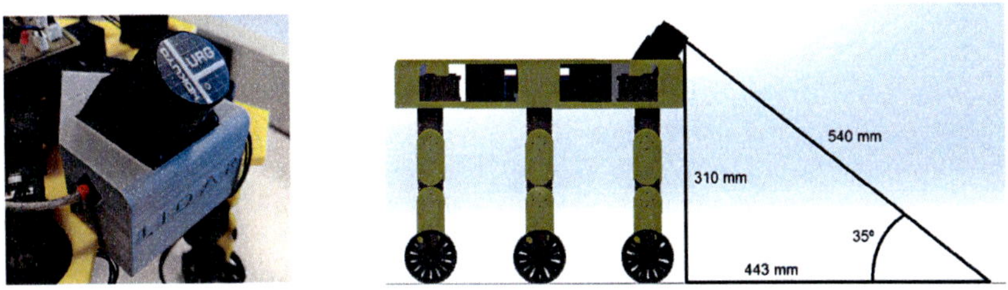


**Fig. 3.** Lidar vision on the tilted support.

## 3 Control Architecture

R3HC has been particularly designed to navigate over flat surfaces where obstacles lying on the floor can occasionally appear. The control architecture is intended to coordinate locomotion on wheels and legs, allowing the robot to overcome obstacles, smaller than the size of the robot step, while the robot is rolling. For this purpose, this architecture must support the communication processes with the motors and the Lidar, and, at the same time, take decision about the robot navigation. When the robot is moving on a flat surface and a small obstacle appears opposite to it, the robot must overcome it by simply lifting one, two or more legs. To execute this manoeuvre, the robot has to decide which legs need to be raised, what movement pattern should be done, and identify the right time to start the overcoming manoeuvre. In this sense, with the aid of the Lidar, it is possible to create obstacle analysis algorithms that will enable R3HC to have capacity to decide the type of manoeuvre depending on obstacles dimensions and relative position. On this subject, we have carried out several tests that will be exposed in the experimentation section.

### 3.1 Hardware Architecture

This architecture has been implemented by means of a PCB, an Arduino MEGA, an Ethernet Shield to connect Arduino to a local network, and a PC laptop, see Fig. 4.

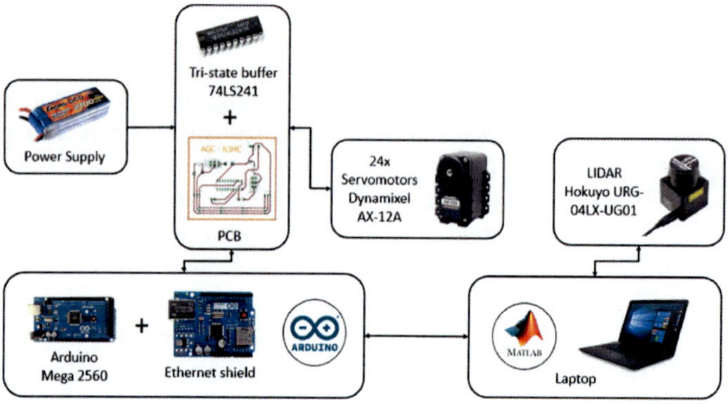


**Fig. 4.** Hardware structure.

The Arduino MEGA acts as the R3HC low-level controller that handles communication with the motors. The PCB is used for servomotors connections, through a serial bus implemented by a tri-state buffer (74LS241), and for power supply. In addition, the laptop, which can be placed on the robot, allows to acquire information from the Lidar sensor, record data from experiments and execute the major computation processes (as a decision-making algorithm). The Lidar sensor and Arduino will connect to the laptop via USB, implementing asynchronous serial communications.

Regarding the Lidar sensor characteristic, we are using the Lidar Hokuyo URG-04LX-UG01. It is a laser sensor for area scanning. The scanned area is a 240° semicircle with a maximum radius of 4000 mm (this will be limited to 1000 mm during experimentation). The use of the Lidar sensor will allow the robot to obtain a point cloud that will be further analyzed by the control algorithm for decision making purposes.

### 3.2 Software Architecture and Communication Handling

This architecture is aimed at optimising read-write messages speed, and taking decision about robot manoeuvring. The communication process that supports the robot operation is handled by an algorithm, implemented in Matlab, that runs within the laptop, and manages the decision making process, Fig. 5. The laptop will request the Lidar data through a 'callback' function. This type of functions are configured so that a task is carried out when a certain event occurs. In this case, this event happens each time that the data of the Lidar are available at the input buffer of the serial port. The task associated to the callback is the algorithm for Lidar data processing. By means of this algorithm the size of the obstacles facing the robot is estimated. Through the interpretation of these data, a decision-making about the robot motion will be made, (see Fig. 5). Particularly, the algorithm will decide about the execution of an adequate manoeuvre according to the size and position of the obstacle.

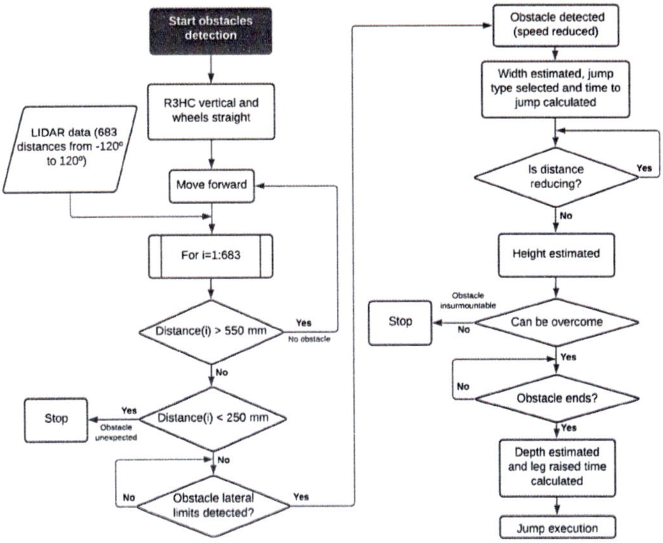


**Fig. 5.** Flow chart of the decision making algorithm.

The communication flow of the whole system is depicted in the chart of Fig. 6. Each type of manoeuvre has a particular label. When the algorithm decides that a particular manoeuvre should be executed, it will deliver to Arduino a request, containing the label associated to the selected manoeuvre. Arduino will receive the message containing the adequate movement through an interruption, which takes place every time a message is received on its serial port. At this moment, Arduino will start controlling the motion of the legs associated to the selected manoeuvre. These movements are made up of specific positions of the servomotors, which are sent from Arduino to the corresponding motors through the bus implemented in the PCB, using the Dynamixel library.

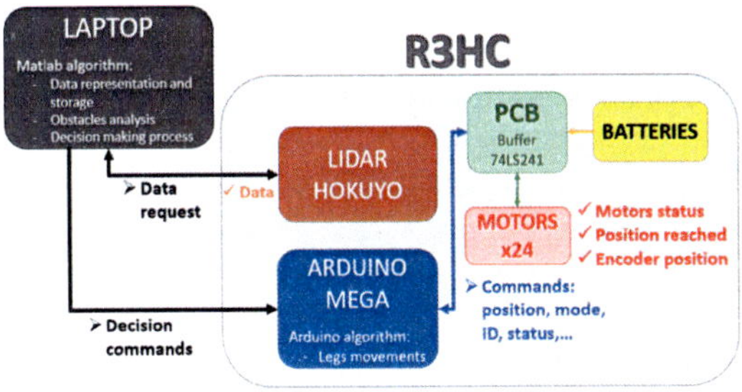


**Fig. 6.** Communication chart.

## 4 Experimental Results

The main objective of these experiments was to verify the correct operability of the proposed architecture, as well as configure the communication processes described before. Once the correct operation of the system was verified, the effectiveness of the designed overcoming strategy was tested.

First of all, as mentioned above, it is necessary to place a laptop on the robot. As a consequence, several tests were carry out to verify that the robot was capable of supporting the computer weight while moving, see Fig. 7a.

After this, the capability for sending and receiving data from motors had to be checked. Thus, initially, two basic tests were performed (see Figs. 7b and 7c). The data of the encoders position were requested in two different situations and further evaluated and represented in the laptop during robot motion. Figure 7b presents the evolution of the encoder of a wheel motor when the robot moves forward, and Fig. 7c illustrates the evolution of the encoders position during a single leg jump (hip and knee motors).

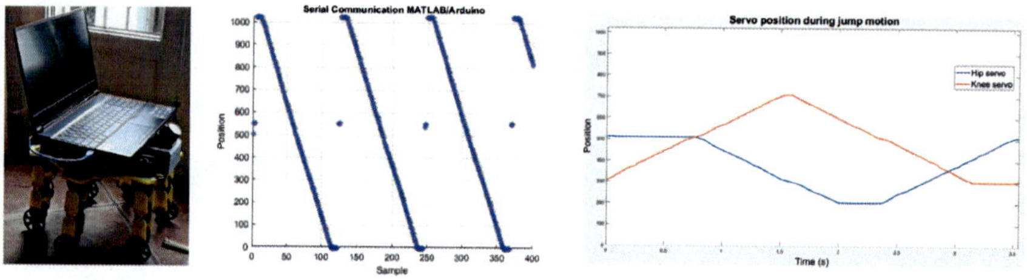


(a) (b) (c)

**Fig. 7.** Test 1. (a) laptop test; (b) values of a wheel encoder while rolling; (c) joints position values while jumping.

Likewise, the process of acquiring information from the Lidar needed to be tested. By this experiment, the correctness of the perception of the environment was verified, and an analysis of the data received while the hexapod was moving, facing one or several obstacles, was performed. Figure 8 illustrates the robot perception in different moments of this situation. These pictures represent the Lidar cloud data points in the sensor local reference frame. On the right side it is visible what would be opposite to the robot. If there are no obstacles, the measurements kept approximately constant on the X axis (vertical line, Fig. 8b), whereas the profile of the object appears or disappear as the obstacle is detected by the Lidar Fig. 8c–8f). In this way, the characteristics of the surrounding obstacles can be determined by analysing these measurements.

The last experiment illustrates the performance of two obstacle stepping manoeuvres. The robot is moving forward, and an obstacle to avoid appears opposite to the robot. To do so, it is first necessary to create particular movements that allow R3HC to overcome obstacles of different sizes and positions.

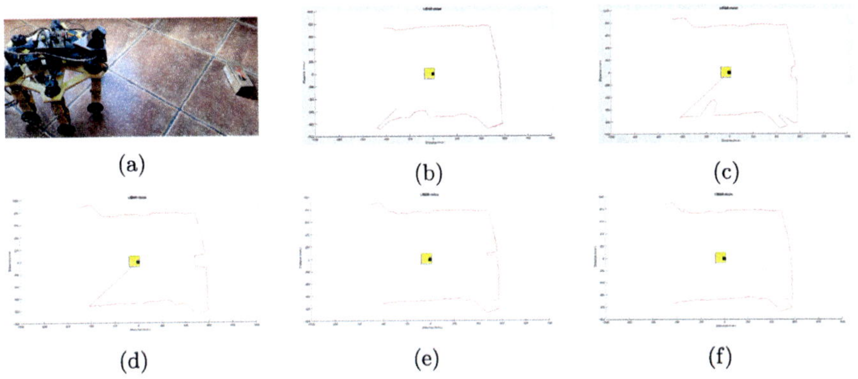

**Fig. 8.** Test 2. Lidar measurements: (a) R3HC rolling; (b) before obstacle detection; (c)–(e) obstacle detection; (f) after obstacle detection.

Two basic manoeuvres have been designed, one for jumping a lateral obstacle, which affects the movement of a single leg, and the other for jumping a central obstacle, which affects the movement of central legs, see Fig. 9. The movements for rising the legs are based on those presented in [2,3] and [4]. By using these movements and knowing the characteristics of the obstacle, the robot can choose the adequate jumping manoeuvre (according to the diagram shown above).

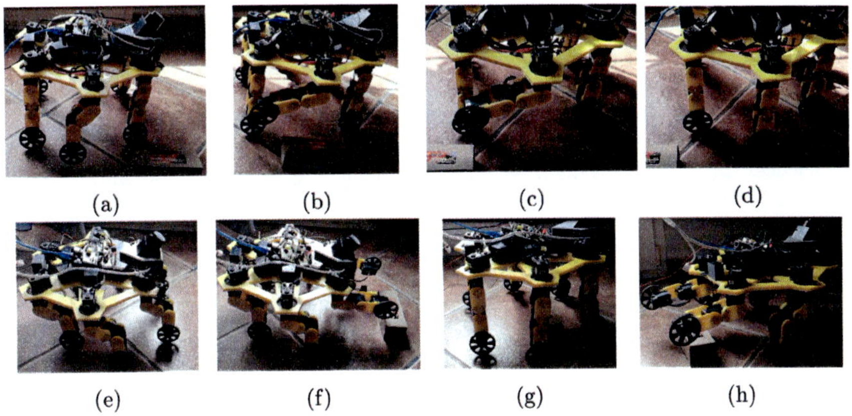

**Fig. 9.** Test 3 and 4. Lateral jump: (a)–(d); Central jump: (e)–(h).

As can be seen in these previous figures, once the robot is at the correct distance, it starts lifting the leg, without stopping the forward motion. The robot executed the step movement by turning knee and hip, making the leg rise. Once the obstacle has been overcome, the leg is lowered by returning the same joints to he initial configuration. Notice that, in order to maintain steadiness, wheels velocities and twisting joint of the other legs must receive the correspondent references, which represents a challenge for the coordination of the movements.

In order to jump an obstacle involving both front and rear legs, as shows figures from Fig. 9e to Fig. 9h, it is necessary to crouch to lower the Center of Mass (CoM), since raising front or rear legs at the same time, may cause the CoM to be outside of the of the support polygon, and the robot be at risk of falling.

## 5 Conclusion

This article proposes a control architecture developed for controlling the hybrid hexapod robot R3HC. The system is capable of managing several communication protocols that makes possible to coordinate the movements of 24 digital motors, and also to acquire data from a Lidar sensor. The presented approach allows the robot to select and implement different strategies for overcoming obstacles, based on the the data acquired by the Lidar.

Experimental results demonstrate the computational efficiency of the proposed approach. The communication strategy in this paper has been particularly designed for R3HC, while the control algorithm can be applied to other robots with similar characteristics.

## References

1. Dynamixel: Robotis e-manual. https://emanual.robotis.com/docs/en/dxl/ax/ax-12a/
2. Gómez-Bravo, F., Aznar, M.J., Carbone, G.: Planning an optimal step motion for a hybrid wheeled-legged hexapod. In: Petuya, V., Pinto, C., Lovasz, E.C. (eds.) New Advances in Mechanisms, Transmissions and Applications. Mechan. Machine Science, vol. 17, pp. 207–214. Springer, Dordrecht (2014). https://doi.org/10.1007/978-94-007-7485-8_26
3. Gomez-Bravo, F., Villadoniga, P., Carbone, G.: Design and operation of a novel hexapod robot for surveillance tasks. In: Ferraresi, C., Quaglia, G. (eds.) RAAD 2017. MMS, vol. 49, pp. 707–715. Springer, Cham (2018). https://doi.org/10.1007/978-3-319-61276-8_75
4. Gómez-Bravo, F., Aznar, M.J., Carbone, G.: Optimal motion for obstacle stepping in a hybrid wheeled-legged hexapod. In: 2014 IEEE International Conference on Autonomous Robot Systems and Competitions (ICARSC), pp. 53–58. IEEE (2014)
5. Orozco-Madaleno, E.C., Gomez-Bravo, F., Castillo, E., Carbone, G.: Evaluation of locomotion performances for a mecanum-wheeled hybrid hexapod robot. IEEE/ASME Trans. Mechatron. **26**(3), 1657–1667 (2020)
6. Orozco-Magdaleno, E.C., Cafolla, D., Castillo-Castañeda, E., Carbone, G.: Experimental validation of a gait planning for obstacle avoidance using mecanum wheels. In: Uhl, T. (ed.) IFToMM WC 2019. MMS, vol. 73, pp. 2391–2400. Springer, Cham (2019). https://doi.org/10.1007/978-3-030-20131-9_236

7. Orozco-Magdaleno, E.C., Cafolla, D., Castillo-Castaneda, E., Carbone, G.: Static balancing of wheeled-legged hexapod robots. Robotics **9**(2), 23 (2020)
8. Orozco-Magdaleno, E.C., Cafolla, D., Castillo-Castañeda, E., Carbone, G.: A hybrid legged-wheeled obstacle avoidance strategy for service operations. SN Appl. Sci. **2**(3), 1–13 (2020). https://doi.org/10.1007/s42452-020-2141-5

# Does Inertia Matching Imply Energy Efficiency?

Paolo Boscariol(✉), Roberto Caracciolo, and Dario Richiedei

DTG - Università degli Studi di Padova, Vicenza, Italy
paolo.boscariol@unipd.it

**Abstract.** The aim of this work is to analyze the problem of designing energy-efficient servo driven systems, by providing a choice of the proper motor size and the correct gear ratio for an arbitrary load. The analysis takes into account a typical test case and a large number of possible design, which are analyzed by parameterizing the design space on the basis of the motor size and reduction ratio, and on the inertia ratio as well. The analysis of the feasibility of each design, and of the resulting energy consumption for each design sample, show that the most common sizing guideline, based on inertia matching, is incapable of capturing the most energy efficient design.

**Keywords:** SDG12 · Energy efficiency · Servo system design · Brushless motor · Gearbox · Inertia matching

## 1 Introduction

Servo-controlled systems are the core elements of most mechatronic systems, automatic machines and robots, and as such are widely adopted in industry. Their relevance in countless applications is also testified by their impact on the overall energy consumption by the industry: several studies [4,6] indicate that electric motors are responsible for up to 70% of such energy consumption. The relevance of servo-driven system, and the push by government policies is now fostering the development and the adoption of 'greener' solutions also in mechatronic systems, which can help meeting the goal of more sustainable consumption and production set by the Sustainable Development Goal 12 [5]. As a result, many manufacturers now offer solutions characterized by high efficiency [7,12], which in the case of servo-drive systems means improving the efficiency of the motor, of its electronic driver, and of the gearbox, being these its key components.

Choosing 'premium' components is not the only - or possibly the best - path to a greener industry: several studies have shown that the available options for the improvement of energy efficiency are several, as different approaches can lead to non-negligible improvements [3]. A consolidated design principle supporting energy efficiency is the adoption of lightweight components or of springs to store and release, at proper time, some mechanical energy [2,8,11],


G. Quaglia et al. (Eds.): I4SDG 2021, MMS 108, pp. 282–289, 2022.
https://doi.org/10.1007/978-3-030-87383-7_31

using an approach usually referred to as 'natural motion' [13]. Alternatively, the energy consumption can be minimized by a careful motion design, since a careful tuning of the acceleration profile can be beneficial in the reduction of the inertial forces to be counteracted by the actuators, which are usually the most relevant contribution to the overall energy consumption [1,11].

Energy-efficiency can be, however, tackled starting from the initial stages of the design of a machine: in particular by carefully choosing the electric drive, the motor and the gearbox (if any), a naturally more efficient solution can be implemented. Despite the relevance of the problem, it is recognized that choosing the wrong size motor for the application is one of the most common mistake in system design [12].

The most used approach to the design of a mechatronic systems, which essentially boils down to the proper choice of a motor and a reduction gear from some catalogs, follows a procedure in which the reducer is chosen first. This choice comes first since the design is usually driven by an estimation of the load parameters - i.e. load speed and torque profiles are usually known in good detail. The choice of the motor comes after, and consists in a process which discards the unsuitable ones: among the non-discarded solutions, the cheaper options is usually referred to as the 'best one'. Such approach has several shortcomings: first of all, the procedure might require some iterations, as the initial choice of the gearbox might be non-compatible, or good enough, when considered together with the population of available motors. Moreover, in this procedure the energy consumption is generally not taken into consideration, and in most cases it does not lead to an energy efficient solution. Manuals often suggest a design procedure based on simple analytic relationships, the most common one being based on the achievement of inertia matching [9]. This work will briefly analyze the problem of designing energy efficient servo-driven systems, and the generally detrimental effect of inertia matching on energy consumption. The data gathered from datasheet will be used to test hundreds of different design, which are checked for feasibility and energy requirement.

## 2 System Modeling

A typical servo-driven system can be represented by the diagram shown in Fig. 1(a), as a cascade of a motor, a gearbox, and a load. Assuming that the load motion profile is known, it can be represented as a time-varying resistant torque profile $T_2(t)$; hence, assuming that both Coulomb and viscous friction act on the motor shaft, the motor should provide the following torque:

$$T_m(t) = (J_m + J_r)\,\ddot{\vartheta}_l(t)i + k_v\dot{\vartheta}_l(t)i + T_f + \frac{T_2(t)}{i} \tag{1}$$

being $J_m$ and $J_r$ the motor and reduced moment of inertias, $k_v$ the viscous friction constant, $T_f$ the Coulomb friction torque, and $i$ the gear ratio. Assuming an equivalent DC-motor description of the brushless motor that drives the system, the voltage drop across the motor leads and the current absorption can be estimated as:

$$V(t) = RI(t) + k_b \vartheta_m(t); \qquad I(t) = \frac{T_m(t)}{k_t}; \tag{2}$$

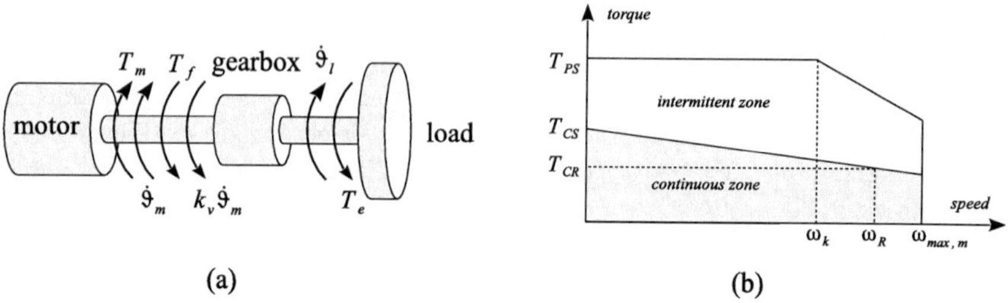


**Fig. 1.** System modeling (a) and motor characteristic curve (b)

in which $k_t$ is the motor torque constant and $k_b$ is the back-emf constant. This simple, but generally accurate model allows to predict the motor electric power absorption as a voltage to current product, which, being then integrated over time, provides the overall energy consumption over a time frame $[0, T]$ as:

$$E_m = \int_0^T VIdt = \int_0^T RI^2dt + \int_0^T k_b\dot{\vartheta}_m Idt = \underbrace{\frac{R}{k_t^2}\int_0^T T_m^2 dt}_{\text{Joule losses}} + \underbrace{\frac{k_b}{k_t}\int_0^T \dot{\vartheta}_m T_m dt}_{\text{mechanical losses}} \tag{3}$$

Equation (3), together with (1) provide an energy estimation model that can be used to forecast the energy consumption of a given design, which is directly usable if the proper motor and gearbox characteristics can be drawn from catalog data. Each design must however be checked for feasibility, i.e. it must be checked that the maximum and average rating listed in datasheets are not exceeded. In order to ensure feasibility, the gear ratio must be checked against the resulting maximum gearbox speed, $\omega_{1B}$, against the gearbox rated speed, $\omega_{1r}$, and against maximum motor speed, $\omega_{m,max}$. The last data is also represented in the motor speed-torque characteristics as shown in Fig. 1(b). The gearbox torque capability suggested by the manufacturer should not be exceeded as well, hence the impact of the peak load toque, $T_2^{max}$, and of the root-mean-cubic value of the load torque, $T_2$, should be checked as well. For a detailed treatment of such design boundaries please refer to [10]. The motor size is to be checked as well, as to ensure that no overheating conditions are encountered during normal operation: to do so, the motor torque-speed pairs should be checked to ensure that the motor RMS operative point lies in the continuous operative zone, which is shown in Fig. 1(b). Without going into details, for which, again, the reader should refer to [10], a correct operation of the motor can be ensured by a sufficiently large motor size, i.e. by ensuring that its continuous torque at stall, $T_{CS}$, is large enough, ad for an adequately rated gearbox. Within this limitations, and according to the energy requirements set by Eqs. (1,3), the relationship between the two main parameters of the design, $i$ and $T_{CS}$, and the energy consumption can be investigated. An analysis of this kind is provided in the next section, by choosing a benchmark load and by testing a large number of design using catalogs data. A list of 45 motors and of 17 reducers, which leads to 765 possible design, is tested and feasibility, together with energy efficiency, is verified for each possible choice.

### 2.1 Inertia Matching Conditions

The concept of inertia matching is well established in engineering practice, and has been, for a long time, suggested in several application books and manufacturers' datasheet. The Inertia Ratio (IR) is defined as the ratio between the reflected load moment of inertia (plus the reducer inertia) and the motor inertia: as shown in the work [9], in such condition the motor effort, for a given load acceleration, is minimized. IR has gained a noticeable popularity as it has been shown to minimize the risk of instability, which can occur if the control tuning cannot compensate the effects of backlash and elasticity, which are mostly due to the non-ideal characteristics of the gearbox. The development of modern control systems, and the general increase of the bandwidth of commercially available servo control systems, has however limited the impact of such problem, and rarely closed-loop instability is caused by an inertia mismatch. In order to pursuit inertia matching, the transmission ratio is set as $i_{IM} = \sqrt{J_m/J_l}$. This choice simplifies the design procedure, allowing to choose first the motor and then the gear ratio. The problem is however, and in general, more complex, as both the motor choice and the gear ratio $i$ do affect the performance of the servo-drive system to be designed. Hence, both the choice of the gearbox and of the motor should be made within a coherent and possibly concurrent design approach. If energy optimality is sought, the traditional design approaches should be revisited, in particular, as it will be shown with a simple test-case, aiming for minimum (or close to minimum) motor sizing or to unitary Inertia Ratio is not, in most cases, the best energy-wise option. Usually a minimum size motor is chosen to reduce the impact of its inertia on the overall motor torque, however the effects of a larger motor inertia can be compensated by an improved torque generation, i.e. by a higher motor torque constant.

## 3 Design Analysis

The test-case under consideration is built by considering a load with inertia $J_l = 2$ kgm$^2$, on which act a constant resistant torque $T_e = 0.5$ Nm and a viscous friction torque, represented by the load-side viscous friction constant $k_v = 1 \times 10^{-2}$ Nms/rad. The load is moved according to a trapezoidal motion profile, with $\lambda = .2$, over a duration $T = 3$ s. Each motion sequence, which provides a load displacement equal to $h = 15\pi$ rad, is followed by a pause interval whose duration is $T_p = 1$ s. The resulting load-side speed and torque pairs are shown in Fig. 2.

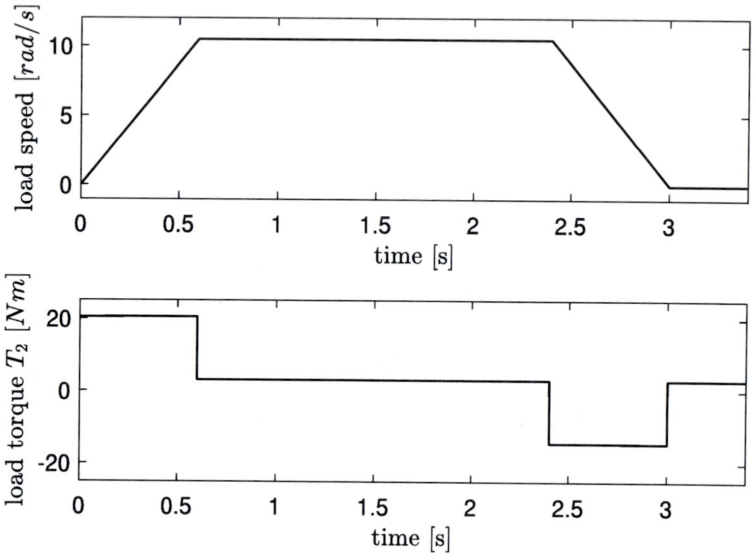


**Fig. 2.** Load speed $\omega_l$ and load torque $T_2$

The aforementioned 765 design solutions are then verified for feasibility, and the respective energy requirements are represented in Fig. 3, which plots the overall energy consumption, $E_m$, as a function of $i$ and of the motor size $T_{CS}$. The best energy-wise design is highlighted by the black circle.

The motors on right-side of the figure are the ones with the smallest continuous stall torque, hence the 'smallest' ones, i.e. the energy-optimal solution is found for an oversized motor, and for a relatively small reduction ratio. The main features of the energy-optimal design are shown in Table 1, which also highlights that the energy improvement over the minimum size design is rather relevant, being equal to 25%. The latter design requires a motor with $T_{CS} = 0.84$ Nm and a 40:1 reduction: the resulting Inertia Ratios are, respectively, equal to 12.11 and 41.76. Hence both design deviate radically form the inertia matching condition.

**Table 1.** Comparison of motor-gearbox designs

| Design objective | Gear ratio $i$ | $T_{cs}$ $[Nm]$ | IR | Energy consumption $[J]$ | % variation |
|---|---|---|---|---|---|
| Minimum energy | 7 | 11.93 | 12.11 | 66.1 | - |
| Unitary IR | 25 | 11.93 | 0.98 | 111.4 | +68.5% |
| Minimum motor size | 40 | 0.84 | 41.86 | 88.8 | +34.3% |

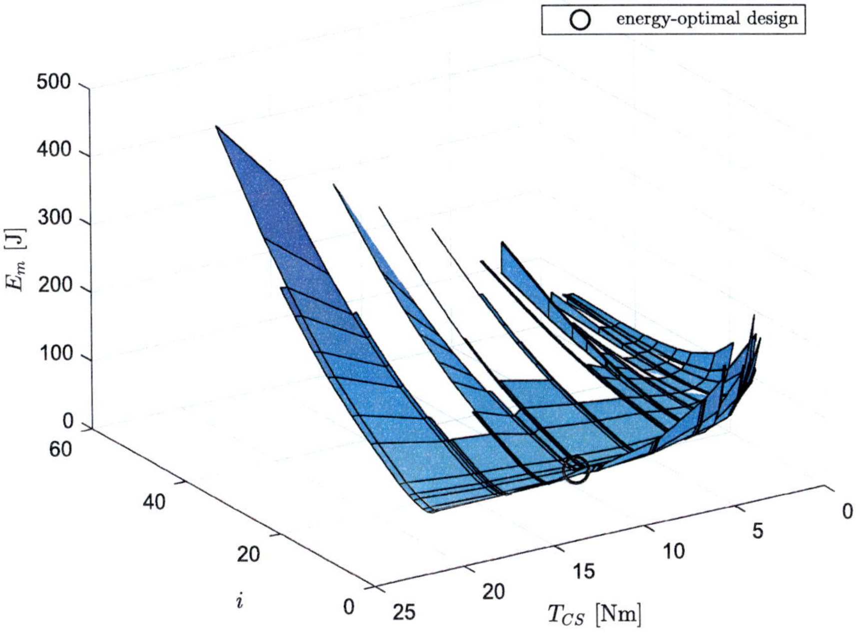


**Fig. 3.** Energy consumption vs. $T_{CS}$ and $i$: the energy-optimal design is shown by the black circle

Plotting the total energy consumption for the 501 feasible designs among the 765 possible ones versus the inertia ratios results in the logarithmic plot in Fig. 4: the analysis suggests that the inertia ratio should be chosen between 5 and 25. The inertia matching condition does not appear to be energetically sound, as a unitary Inertia Ratio design requires, in the best case, an energy equal to 111.4 J: this figure is almost twice the minimum energy figure.

This analysis, whose results are also corroborated by several other test-cases not reported here, suggest that an energy-efficient design is generally found for oversized motors and for inertia ratios larger than 5, as designs with unitary IR and close to minimum size motors are rather inefficient. The rather generous oversizing suggested by the energetic analysis can be explained by the results provided in Fig. 5, which shows how the two most-relevant motor parameters are scaled with the motor size. The ratio $R/k_t^2$, whose reciprocal is the square value of the motor constant, falls rapidly for any increase of $T_{CS}$: hence the energy associated by the Joule losses, as highlighted in Eq. (3), can be reduced by choosing larger motors. Certainly a larger motor implies a larger inertial load, which affects both the magnitude of both Joule and mechanical losses, however this effect has a limited impact, since it is usually overcompensated by the change in the ratio $R/k_t^2$. The analysis of the nameplate data of the 45 motors under consideration shows that the proportionality between $J_m$ and $T_{CS}$ is almost linear, as it can be seen in Fig. 5. Hence, the effects of the decrease of

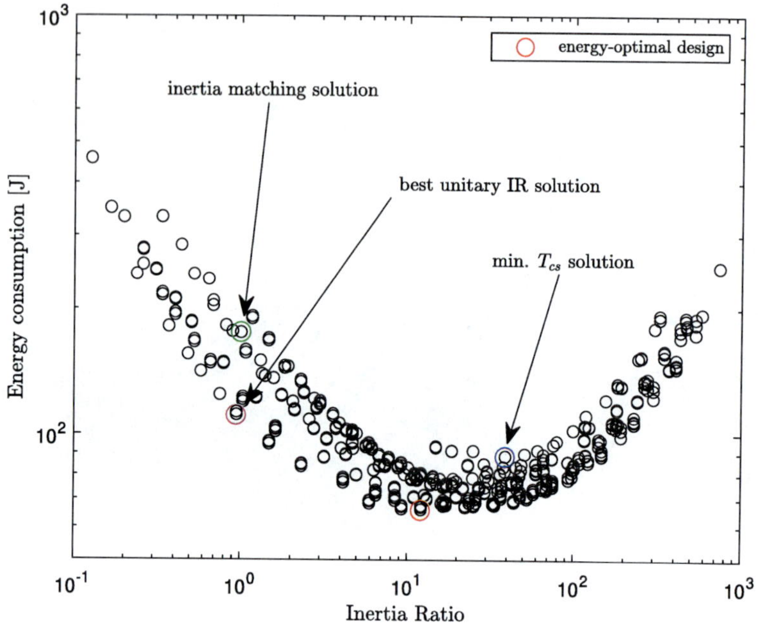


**Fig. 4.** Energy consumption vs. inertia ratio for test case II: the energy-optimal design is shown by the red circle

$R/k_t^2$ with the increase of $T_{CS}$ prevails over the effects of a larger $T_m$, and as a results a larger (but not too large) motor can be energetically more efficient.

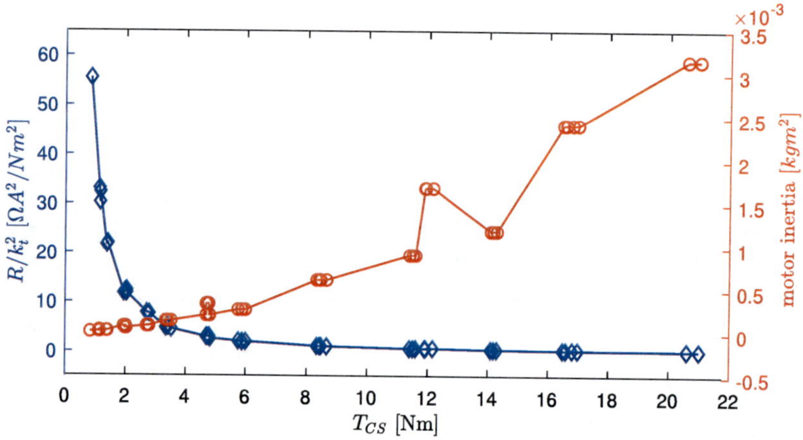


**Fig. 5.** $R/k_t^2$ ratio vs. $T_{CS}$ for the 45 motor samples

## 4 Conclusion

In this work an analysis on the impact of a classic design rule, i.e. the inertia matching condition, is re-considered from the energy consumption point of view. Starting form an estimation model, which is used to forecast the energy consumption of a servo-driven system, a large number of possible designs are tested. The results highlight the complex relationship between design parameters and energy consumption, which calls for an integrated motor-gearbox design procedure. Moreover, it is found that targeting for inertia matching and minimum motor size are both, energy-wise, inefficient choices. In general, is it found that efficiency can be boosted by using large inertia ratios and oversized motors.

## References


1. Boscariol, P., Caracciolo, R., Richiedei, D., Trevisani, A.: Energy optimization of functionally redundant robots through motion design. Appl. Sci. **10**(9), 3022 (2020)
2. Carabin, G., Palomba, I., Wehrle, E., Vidoni, R.: Energy expenditure minimization for a delta-2 robot through a mixed approach. In: Kecskeméthy, A., Geu Flores, F. (eds.) Multibody Dynamics 2019. ECCoMAS 2019, vol. 53. pp. 383–390. Springer, Cham (2019). https://doi.org/10.1007/978-3-030-23132-3_46
3. Carabin, G., Wehrle, E., Vidoni, R.: A review on energy-saving optimization methods for robotic and automatic systems. Robotics **6**(4), 39 (2017)
4. De Almeida, A., Fong, J., Brunner, C., Werle, R., Van Werkhoven, M.: New technology trends and policy needs in energy efficient motor systems - a major opportunity for energy and carbon savings. Renew. Sustain. Energy Rev. **115**, 109384 (2019)
5. Guevara, S., Julián, I.P.: Sustainable consumption and production: a crucial goal for sustainable development - reflections on the Spanish SDG implementation report. J. Sustain. Res. **1**(2) (2019)
6. Lu, S.M.: A review of high-efficiency motors: specification, policy, and technology. Renew. Sustain. Energy Rev. **59**, 1–12 (2016)
7. Lukaszczyk, M.: Improving efficiency in electric motors. World Pumps **4**, 34–41 (2014)
8. Palomba, I., Wehrle, E., Carabin, G., Vidoni, R.: Minimization of the energy consumption in industrial robots through regenerative drives and optimally designed compliant elements. Appl. Sci. **10**(21), 7475 (2020)
9. Pasch, K., Seering, W.: On the drive systems for high-performance machines. J. Mech. Transm. Autom. Des. **106**(1), 102–108 (1984)
10. Richiedei, D.: Integrated selection of gearbox, gear ratio, and motor through scaling rules. Mech. Based Des. Struct. Mach. **46**(6), 712–729 (2018)
11. Richiedei, D., Trevisani, A.: Analytical computation of the energy-efficient optimal planning in rest-to-rest motion of constant inertia systems. Mechatronics **39**, 147–159 (2016)
12. Saidur, R.: A review on electrical motors energy use and energy savings. Renew. Sustain. Energy Rev. **14**(3), 877–898 (2010)
13. Scalera, L., Carabin, G., Vidoni, R., Wongratanaphisan, T.: Energy efficiency in a 4-DoF parallel robot featuring compliant elements. Int. J. Mech. Control **20**(02), 49–57 (2019)

# Preliminary Comparison Between Handheld and Mobile Robotic Mapping Systems

Eleonora Maset[1(✉)], Lorenzo Scalera[2], Alberto Beinat[2], Federico Cazorzi[1], Fabio Crosilla[2], Andrea Fusiello[2], and Alessandro Gasparetto[2]

[1] Department of Agricultural, Food, Environmental and Animal Sciences (DI4A), University of Udine, Udine, Italy
{eleonora.maset,federico.cazorzi}@uniud.it

[2] Polytechnic Department of Engineering and Architecture (DPIA), University of Udine, Udine, Italy
{lorenzo.scalera,alberto.beinat,fabio.crosilla,andrea.fusiello, alessandro.gasparetto}@uniud.it

**Abstract.** *Make cities and human settlements inclusive, safe, resilient and sustainable* is one of the Sustainable Development Goals drawn by the United Nations. To support renovation projects, a first essential step is to gain an updated and precise knowledge of indoor and outdoor public spaces and urban areas, which requires accurate and fast mapping systems. In this context, in addition to well-established manual topographic techniques, mobile robotics could play a crucial role in the perspective of acquiring automatic surveys of such environments. In this paper, we present a preliminary comparison between handheld and mobile robotic mapping platforms, using a 3D laser scanner that realizes real-time mapping on the field. Experimental results show that the robotic system could be an efficient alternative to the handheld-mode survey, providing precise point clouds with uniform density.

**Keywords:** Sustainable cities and communities (SDG11) · Mobile robotics · Laser scanning · Mapping systems · Simultaneous Localization and Mapping (SLAM)

## 1 Introduction

The United Nation 2030 Development Agenda has drawn the importance of making cities and human settlements inclusive, safe, resilient and sustainable [17]. In particular, the Sustainable Development Goal 11 (SDG11) indicates the target to enhance inclusive and sustainable urbanization, as well as to protect and safeguard the world's cultural and natural heritage. At the same time, the SDG11 outlines the aim to substantially increase the number of cities and human settlements adopting and implementing integrated policies and plans towards resource efficiency, adaptation to climate change, and resilience to disasters.


G. Quaglia et al. (Eds.): I4SDG 2021, MMS 108, pp. 290–298, 2022.
https://doi.org/10.1007/978-3-030-87383-7_32

In this context, acquiring and recording updated high-resolution 3D information of internal and external environments is crucial in the study and analysis of both buildings and human settlements [9]. The survey of civil structures, usually performed by means of classical surveying technologies such as Photogrammetry and Terrestrial Laser Scanning (TLS), has been revolutionised in recent years by portable Mobile Mapping Systems (MMSs). These devices can be easily carried by a person who acquires accurate 3D data of the surrounding environments by simply walking through the area of interest, as shown in [13]. Nowadays, several studies have been proposed that examine advantages and disadvantages of portable MMSs in diversified test fields, as demonstrated by a flourishing literature on the topic [16]. For example, in [6] a wearable mobile laser system is evaluated for the indoor 3D mapping of a complex historical site, whereas in [10] the performance of a handheld laser scanner is investigated in different outdoor scenarios, such as the survey of a building facade and a mountain torrent. Mapping systems are used also to inspect constructions or manufactured parts and identify possible discrepancies between the as-built workpieces and their nominal specifications [12]. The data acquired by portable mapping devices are essential to create a Building Information Model (BIM) according to the as-built condition of the structure [11], and recently proved to be fundamental to support functional and occupancy analysis of buildings in challenging situations, such as the COVID-19 pandemic [5].

The acquisition of 3D information on buildings and human settlements could be improved and automated by mounting the mapping devices (e.g., laser scanners or RGB-D cameras) on a robotic platform, which can be steered from remote through the area of interest or can autonomously perform the required survey. Nowadays, mobile robots have been increasingly applied for mapping and surveying in several different fields, such as proximal sensing and precision farming in the agricultural field [14,15], the exploration and inspection in hazardous and challenging environments [4,18], as well as inspections in disaster situations. Mobile robots are also employed for the mapping of archaeological and cultural heritage sites, as shown in [3]. A review on autonomous mobile scanning systems for the digitization of buildings can be found in [1].

In this paper, we present a preliminary comparison and performance evaluation between handheld and robotic mapping systems. In particular, we used a 3D laser scanner, coupled with an Inertial Measurement Unit (IMU), that realizes high-resolution mapping of the surrounding environment and can provide real-time results. Experimental tests are carried out performing the survey of the ground floor of the Rizzi building of University of Udine (Italy) and considering two different scenarios: in the first case, referred to as *handheld*, the device is attached to a telescopic pole and carried manually, while in the second case (*robotic*) it is mounted on a mobile robot. The analysis of the results shows that the robotic mapping platform compares favourably with respect to the handheld modality, and the feasibility of the robotic system for future automatic surveying with the goal of supporting renovation projects for sustainable cities.

The remainder of the paper is organized as follows: in Sect. 2 the materials and methods used in this work are described, whereas in Sect. 3 the experimental results are presented. Finally, the paper is concluded in Sect. 4.

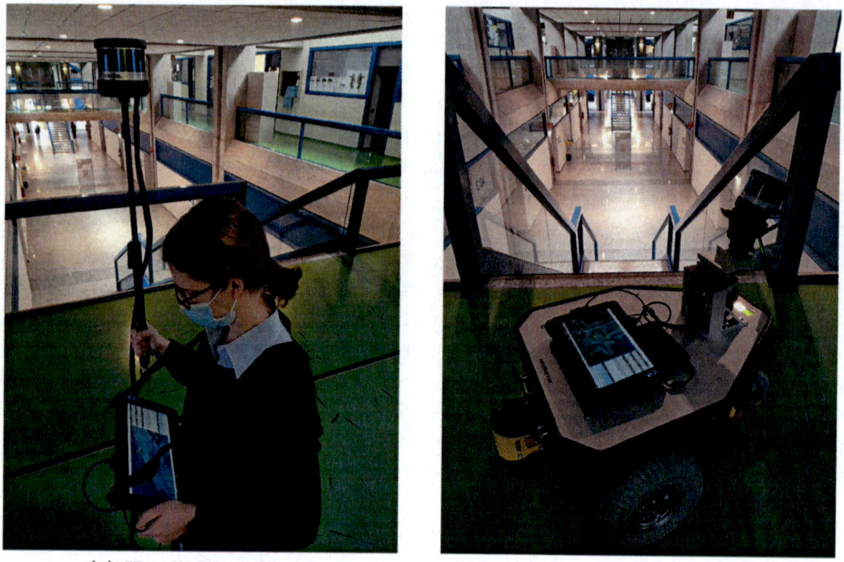

(a) Handheld platform (b) Robotic platform

**Fig. 1.** Laser scanner attached to a telescopic pole and carried manually (a), and mounted on the mobile robot (b).

## 2 Materials and Methods

The experimental tests were performed using the HERON Lite system by Gexcel srl [7]. This instrument is composed of a Velodyne Puck LITE laser scanner coupled with a XSens MTI inertial sensor. The laser head is characterized by 16 channels emitting infrared laser beams, allowing to cover a 360° horizontal field of view and a 30° vertical field of view. In its standard configuration, the HERON Lite device is thought to be attached to a telescopic carbon fiber pole, which is manually held during the mapping procedure (Fig. 1(a)). In our tests the HERON Lite system was mounted on a mobile robot, with an angle of approximately 30° between the vertical direction and the laser rotation axis (Fig. 1(b)). In particular, the MP-500 mobile system by Neobotix was adopted, which is equipped with two drive wheels, an undriven castor wheel, and a 2D laser scanner used for safety purposes. The robot has dimensions equal to $814 \times 592 \times 361$ mm and a maximum speed of 1.5 m/s. In this first work, the mobile robot has been steered using a joystick connected by a wi-fi link.

The HERON Lite system provides a map of the scanned environment in real-time, using an on-line Simultaneous Localization and Mapping (SLAM) algorithm. Thanks to this approach, new scans are incrementally added to the current point cloud of the surveyed area as soon as they are available, simultaneously estimating both the system trajectory and the map. More in detail, the procedure is carried out exploiting IMU data to compute a rough estimate of the sensor position and attitude, which is refined through a registration process that aligns each cloud (a single 360° scan) to the previous ones using the well-known Iterative Closest Point (ICP) algorithm [2]. However, this real-time method tends to drift on long paths, downgrading the accuracy of the produced map. So, the final point cloud of the environment is defined via an off-line post-processing, carried out through the HERON Desktop software, that implements a full SLAM approach [8]. In this case, the map is computed simultaneously exploiting all the measurements acquired during the entire survey: a global registration among all the scans is performed, minimizing misalignment errors among the clouds and closing the loops (if any) in the trajectory. These refinements usually lead to a 3D point cloud characterized by a precision of 3 cm and a global accuracy of 5–20 cm, depending on the geometric characteristics of the surveyed area and the presence of closed-loops on the trajectory, which can significantly affect the performance of the SLAM algorithms [7].

To provide a preliminary comparison between handheld and mobile robotic mapping platforms, we surveyed the ground floor of the west-wing corridor of the Rizzi building of University of Udine (Italy). In both cases the trajectory followed a closed loop around the square-based plant. The main characteristics of the data acquisitions are reported in Table 1. Please note that both surveys were processed assuming the same parameter values for the SLAM algorithms previously described. Figure 2 shows the point cloud obtained with the robotic mapping system, after the manual removal of outliers (points falling outside the area of interest).

**Table 1.** Characteristics of the surveys performed with the handheld and the robotic mapping platforms.

| Survey platform | Acquisition time | Trajectory length | Points number |
|---|---|---|---|
| Handheld | 5 min 30 s | 342 m | 35,755,046 |
| Robotic | 13 min 40 s | 339 m | 47,694,577 |

## 3 Results and Discussion

To evaluate the results provided by the mobile robotic mapping platform, a comparison with the map produced through the handheld system was performed, investigating at first the possible presence of relative deformations between the two models. In particular, we focused the attention on the south part of the surveyed corridor (Fig. 2(b)) and we preliminary registered the two point clouds via the ICP algorithm. In fact, the coordinates of each model are expressed in an

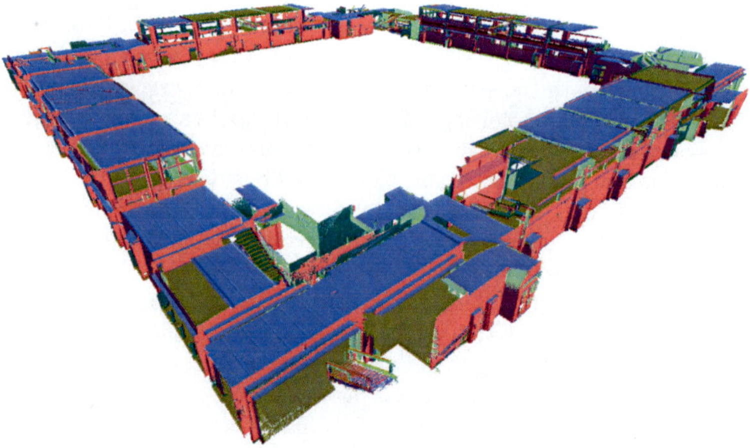

(a)

(b)

**Fig. 2.** Point cloud of the whole corridor of the university building (a), and a detailed view of the south part (b). Points are colored according to the local plane inclination.

arbitrary reference frame and a relative alignment is therefore needed to subsequently evaluate the differences between the two point clouds. The point-to-point absolute distances between the models were estimated using CloudCompare software[1], obtaining a mean distance of 2.2 cm and a standard deviation of 1.5 cm. Figure 3 reports the results for the floor and a wall of the analysed corridor. It can be noticed that higher discrepancies are located on the floor, whereas the wall shows tight correspondence (distances $< 2$ cm) between the two models. In any case, the computed differences are compatible with the accuracy of the instrument, according to the manufacturer's specifications.

Observing instead the point cloud density that characterizes the two models, it is interesting to highlight that the handheld survey modality provides a bimodal distribution (Fig. 4(a)) with higher density values on the walls

[1] https://www.danielgm.net/cc/.

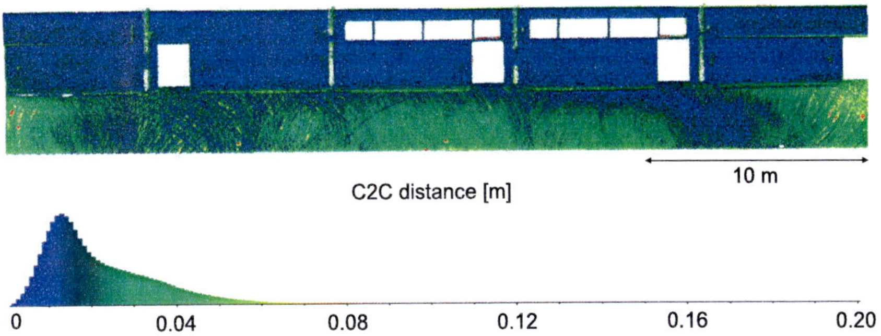


**Fig. 3.** Cloud-to-cloud absolute distances computed between the point clouds obtained by the handheld and the robotic mapping platforms.

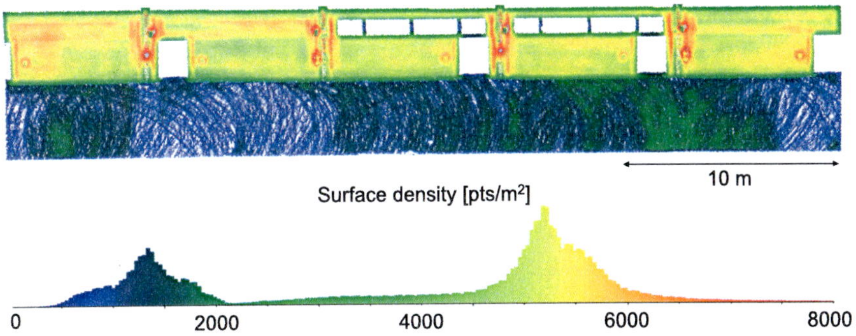


(a) Handheld mapping platform

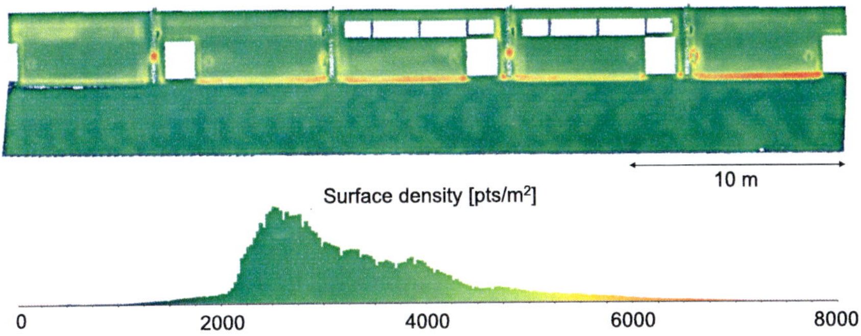


(b) Robotic mapping platform

**Fig. 4.** Surface density characterizing the point cloud obtained from the handheld-mode survey (a) and the robotic mapping system (b).

**Table 2.** Results of the plane fitting procedure. The RMS of the distances between the points and the corresponding estimated plane are reported for each considered patch.

| Plane | Handheld mapping | Robotic mapping |
|---|---|---|
| Wall #1 | 2.5 cm | 1.6 cm |
| Wall #2 | 3.0 cm | 1.6 cm |
| Wall #3 | 2.6 cm | 2.9 cm |
| Wall #4 | 2.2 cm | 2.0 cm |
| Floor #1 | 1.4 cm | 1.4 cm |
| Floor #2 | 1.2 cm | 1.3 cm |

(on average, approximately 5000 points/m$^2$), that decreases to values below 2000 points/m$^2$ on the floor. On the contrary, the survey performed through the mobile robotic platform guarantees a more uniform point distribution, with an average density of 3300 points/m$^2$ that characterizes all the surveyed surfaces.

In addition to a more consistent point density, the point cloud acquired with the robotic system shows also a slightly lower level of noise. This feature was quantitatively verified extracting six patches of points from the walls and floor and fitting a plane to each subset. Table 2 reports the Root Mean Square of the distances of the points from the corresponding fitted plane, showing on average a RMS of 2.2 cm for the model obtained from the handheld-mode survey and 1.8 cm for the robotic mapping case. A possible explanation of this behavior may lie in the fact that the robot performs the survey at lower speed and, above all, avoids the oscillations and sudden movements to which the laser scanner can be subjected when it is mounted on a pole and carried by a walking person.

## 4 Conclusion

Acquiring updated and detailed 3D models of buildings is an essential preliminary step in any renovation project that can lead to more sustainable and resilient cities, which is the aim of the Sustainable Development Goal 11. In this context, robotic mapping systems could represent a fundamental aid to reduce and automate the time-consuming manual work usually required for the surveying operations. In this paper, we presented preliminary results relating to indoor surveys carried out by mounting a 3D laser scanner on a remote-controlled robot. Quantitative evaluations demonstrate a tight correspondence between the point cloud obtained from the robotic mapping and the one retrieved via the handheld platform, with the former showing a more uniform point distribution and lower noise level: robotic mapping could be therefore a viable alternative to the well-established surveying mode.

**Acknowledgments.** We gratefully thank Prof. Paolo Gallina, Dr. Stefano Seriani and Matteo Caruso for their help in setting up the mobile robot.

## References

1. Adán, A., Quintana, B., Prieto, S.A.: Autonomous mobile scanning systems for the digitization of buildings: a review. Remote Sens. **11**(3), 306 (2019)
2. Besl, P., McKay, N.: A method for registration of 3-D shapes. IEEE Trans. Pattern Anal. Mach. Intell. **14**(2), 239–256 (1992)
3. Borrmann, D., Hess, R., Eck, D., Houshiar, H., Nüchter, A., Schilling, K.: Evaluation of methods for robotic mapping of cultural heritage sites. IFAC-Papers OnLine **48**(10), 105–110 (2015)
4. Chen, J., Cho, Y.K.: Detection of damaged infrastructure on disaster sites using mobile robots. In: 16th International Conference on Ubiquitous Robots, pp. 648–653. IEEE (2019)
5. Comai, S., et al.: Indoor mobile mapping system and crowd simulation to support school reopening because of COVID-19: a case study. ISPRS Arch. **44**, 29–36 (2020)
6. Di Filippo, A., Sánchez-Aparicio, L.J., Barba, S., Martín-Jiménez, J.A., Mora, R., González Aguilera, D.: Use of a wearable mobile laser system in seamless indoor 3D mapping of a complex historical site. Remote Sens. **10**(12), 1897 (2018)
7. Gexcel SRL: HERON lite (2021). https://gexcel.it/en/solutions/heron-mobile-mapping/heron-lite. Accessed 14 May 2021
8. Grisetti, G., Kümmerle, R., Stachniss, C., Burgard, W.: A tutorial on graph-based slam. IEEE Intell. Transp. Syst. Mag. **2**(4), 31–43 (2010)
9. Gupta, T., Li, H.: Indoor mapping for smart cities–an affordable approach: using kinect sensor and zed stereo camera. In: International Conference on Indoor Position and Indoor Navigation, pp. 1–8. IEEE (2017)
10. Maset, E., Cucchiaro, S., Cazorzi, F., Crosilla, F., Fusiello, A., Beinat, A.: Investigating the performance of a handheld mobile mapping system in different outdoor scenarios. Int. Arch. Photogramm. Remote Sens. Spatial Inf. Sci. XLIII-B1-2021, 103–109 (2021)
11. Maset, E., Magri, L., Fusiello, A.: Improving automatic reconstruction of interior walls from point cloud data. ISPRS Arch. XLII-2/W13, 849–855 (2019)
12. Maset, E., Scalera, L., Zonta, D., Alba, I., Crosilla, F., Fusiello, A.: Procrustes analysis for the virtual trial assembly of large-size elements. Robot. Comput. Integrated Manuf. **62**, 101885 (2020)
13. Nocerino, E., Menna, F., Remondino, F., Toschi, I., Rodríguez-Gonzálvez, P.: Investigation of indoor and outdoor performance of two portable mobile mapping systems. In: Videometrics, Range Imaging, and Applications XIV, vol. 10332 (2017)
14. Ristorto, G., Gallo, R., Gasparetto, A., Scalera, L., Vidoni, R., Mazzetto, F.: A mobile laboratory for orchard health status monitoring in precision farming. Chem. Eng. Trans. **58**, 661–666 (2017)
15. Shalal, N., Low, T., McCarthy, C., Hancock, N.: Orchard mapping and mobile robot localisation using on-board camera and laser scanner data fusion-part a: tree detection. Comput. Electron. Agric. **119**, 254–266 (2015)
16. Tucci, G., Visintini, D., Bonora, V., Parisi, E.I.: Examination of indoor mobile mapping systems in a diversified internal/external test field. Appl. Sci. **8**(3), 401 (2018)

17. United Nations: The Sustainable Development Goals (2015). https://sdgs.un.org/goals. Accessed 14 May 2021
18. Zimroz, R., Hutter, M., Mistry, M., Stefaniak, P., Walas, K., Wodecki, J.: Why should inspection robots be used in deep underground mines? In: Widzyk-Capehart, E., Hekmat, A., Singhal, R. (eds.) 27th International Symposium on Mine Planning and Equipment Selection - MPES 2018, pp. 497–507. Springer, Cham (2019). https://doi.org/10.1007/978-3-319-99220-4_42

# Design and Implementation the Robotics System for Glass Containers Depalletizing

Irina I. Kalapyshina[1,2](✉) and Konstantin A. Nuzhdin[1]

[1] ITMO University, St.-Petersburg, Russia
[2] SPbF IZMIRAN, St.-Petersburg, Russia

**Abstract.** The paper describes the original robotic system for depalletizing glass containers DP3, which works as a part of an automated food packaging line. This system was developed during the implementation of the project on the introduction of robotics technologies for the food industry in order to improve working conditions, increase the productivity of the enterprise through the modernization of production. The developed robotic system and the method of separation of glass containers implemented in it allows you to automatically select a number of layers of containers standing on transport pallets. The introduction of the described robotics system increased productivity and allowed to eliminate the human factor. The article describes the existing methods of separating the layers of packaging, their comparative analysis is carried out, and the original method used for depalletization of glass containers is proposed. The process of debugging a robotics systemis described and attention is paid to the impact of the work performed on economic processes.

**Keywords:** Industry · Innovation and infrastructure robotics system · Decent work and economic growth · Glass container · Pallet disassembly · Conveyor · Automatic line

## 1 Introduction

In the modern food industry, glass containers are widely used. In the technological process of packaging finished products, there is a problem of forming a number of tare from a layer, for feeding into automatic product packaging lines. The problem is caused by the way glass containers are transported by suppliers. Glass tare is delivered on pallets of standard sizes, it is organized in layers (10–13 layers), in each of which 13 × 15 pieces are placed, between the layers there are sheets or pallets of cardboard for the stability of the structure. Figure 1 shows the developed robotics system and pallets (Fig. 1, position 3) consisting of layers of packaging separated by cardboard. In the Fig. 1 shows the physical configuration of the robotic system. Using a vacuum gripper, the robot manipulator moves the container layer from one of the pallets to the corresponding storage conveyor, eliminates the intermediate packaging material. Next, the container is moved to the discharge conveyor using a developed separation device.

The problem is that in order to feed the packaging to the automatic packaging lines, it must be separated from the layers and fed "in a rank" to a narrow single-row conveyor.


G. Quaglia et al. (Eds.): I4SDG 2021, MMS 108, pp. 299–306, 2022.
https://doi.org/10.1007/978-3-030-87383-7_33

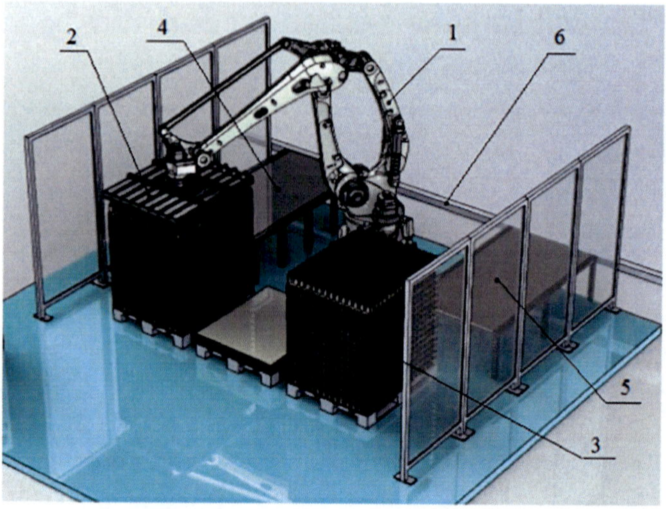


**Fig. 1.** General view of DP-3 robotics system with glass container depalletization, where: 1 - robot manipulator, 2 - gripper devices for the tare layer, 3 - pallets with tare layers, 4, 5 - storage conveyors, 6 - discharge conveyor.

Currently, there are various ways to separate tare, which are listed below. However, none of the existing solutions can be used in this automated cell. The main reasons are the cramped dimensions allocated for the placement of the entire robotics system, the need to minimize the cost of the system and eliminate the use of workers for depalletization and unloading of glass containers on the discharge conveyor.

### 1.1 General Overview

At the moment, there are technical solutions to the problem of forming a number of containers from a layer for feeding into automatic lines, but the proposed system differs significantly in technical characteristics. Let's consider some of the most popular analogues. An overview of the analogs is presented in Table 1.

The main criteria for evaluating the described systems:

- Performance (the speed of automatic lines is usually more than 60 tare units per minute)
- Overall size (the space that the system occupies in the production room, preferably small overall size)
- Reliability (uninterrupted operation of the system for a long time)
- Cost (the result of comparing the tare separation systems according to the main criteria is presented in Table 2)

The point 2 and point 3 systems have high performance, but are not suitable due to their large overall dimensions. The point 4 system has small dimensions, but low productivity, clearly visible in the presence of gaps between the discharged tare.

**Table 1.** General overview of tare separation systems

| Description | Structural scheme |
|---|---|
| **Semi-automatic tare separation device** [1]. This device implements a mechanism that works on the principle of a centrifuge. In the structural scheme of the semi-automatic tare separation device, the disk of the centrifuge with the container installed on it moves at a constant angular velocity, due to the forces of inertia, the container is pressed against the board. At the edge of the disc, a separator in the form of a wedge element is installed to organize the exit of the tare to the discharge conveyor tangentially to the disc. The tare is placed by the workers in the central part of the disk. Structural scheme: where: 1-centrifuge, 2-wedge separator, 3-discharge conveyor, 4-tare | 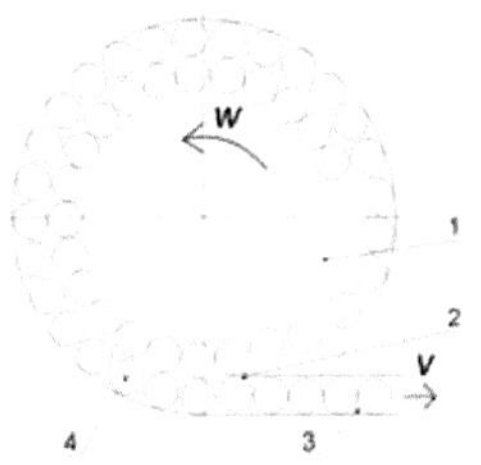  |
| **«DillinAutomation» company separation System.** Structural scheme demonstrates how it works. In this system, the moving flow of the tare is narrowed by means of the oblique side. To prevent the possibility of jamming of tare, several narrow conveyors are used, installed side by side to each other and moving at different speeds [2]. The conveyor 1 moves at the minimum speed, the discharge conveyor 3 moves at the maximum speed. In the "DillinAutomation" system, a structural protrusion is additionally placed on the discharge conveyor near the row formation area. Structural scheme «DillinAutomation»: where: 1-the oblique side, 2 – the first," slow " conveyor, 3-the diverting conveyor, 4-the structural protrusion. | 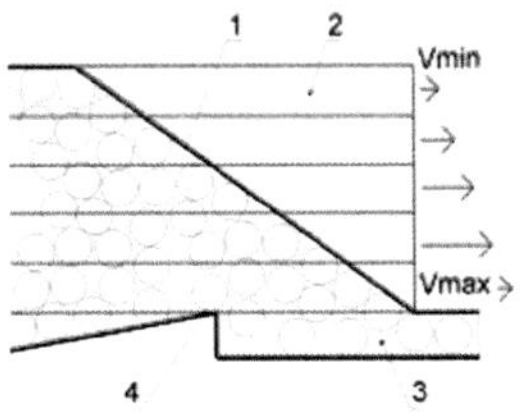  |
| **«Australis Engineering» company separation System.** The separation device developed by Australis Engineering also uses several narrow conveyors, as does the DillinAutomation container separation system [3]. However, in this system, the container is not pressed against the sides. The difference in the speed of the conveyors is not used for pushing, but for stretching the containers in one row. The separation system of the company "Australis Engineering": where: 1-the oblique side, 2-the first," slow " conveyor, 3-the diverting conveyor | 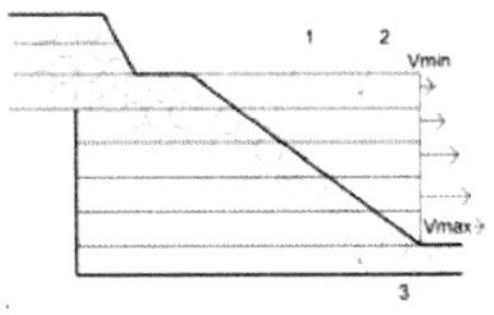  |
| **«Ska Fabricating» company separation System** [4]. It feeds the container to a single-row conveyor located in the center. On both sides of the single-row conveyor are two wider ribbons that move at opposite speeds to each other. The container is pressed against the oblique side, and sent to the discharge conveyor. The third conveyor counteracts the jamming of the tare in the exit area. Ska Fabricating packaging separation system: 1-discharge conveyor, 2-supply conveyor, 3-counter conveyor | 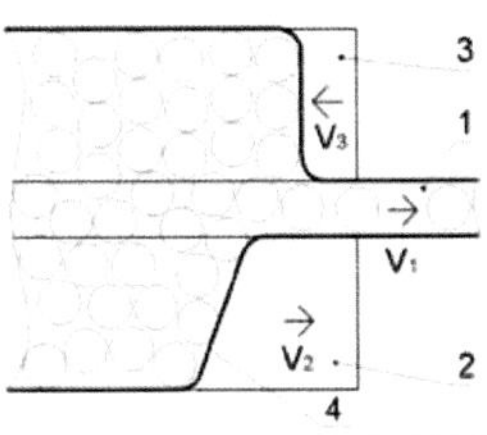  |

**Table 2.** Comparison table of tare separation systems

| | Performance | Overall size | Reliability | Cost | Minimizing the human factor |
|---|---|---|---|---|---|
| Semi-automatic tare separation device | ●●●○○ | ●○○○○ | ●●●●○ | ●○○○○ | ○○○○○ |
| DillinAutomation | ●●●●● | ●●●●○ | ●●●●○ | ●●●●○ | ●●●●● |
| Australis engineering | ●●●●○ | ●●●●● | ●●●●○ | ●●●●○ | ●●●●● |
| Ska fabricating | ●●●○○ | ●●●●○ | ●●●●○ | ●●●●● | ●●●●● |

Thus, to solve the problem of depalletization of glass containers in a robotics system, a semi-automatic separation device is best suited point 1 Table 2. This system successfully solves the problem of continuous delivery of containers, but due to its shape is unsuitable for automatic feeding from pallets [5–7]. The possibility of feeding a rectangular layer to the centrifuge disc is complicated by the fact that the disc is constantly moving and the position of the tare on it is difficult to predict. Feeding from the side of the board is impossible due to the presence of moving cans there, and the supply of tare to the central part significantly increases the overall dimensions and weight of this device. Also, the need to manually load the tare into the device does not allow you to exclude manual labor for using this system in automatic mode.

As a result, to solve the problem, it was decided to use the principle of dividing the tare on the basis of an analog with a change in the shape of the table from round to square and the exclusion of the human factor.

## 2 Synthesis of a Glass Tare Depalletization System for a Robotics System

To completely exclude people on the work site of the robotics system, it was necessary to solve a number of problems. One of the main tasks is to develop a depalletization system that will meet all the requirements specified in Table 2. Figure 2 shows the separation unit: Fig. 2b shows the schematic diagram of the developed glass tare depalletization system, where the storage conveyor 1 receives the depalletized container in layers of 12 × 12 pieces as it moves along the discharge conveyor 2. The separation device 3 is a rigid stop with an angle of 90° [8, 9]. To increase the force at the outlet from the storage to the discharge conveyor, a side conveyor 4 is installed.

General view of the robotics system with depalletization shown in Fig. 3. One of the important problems is the removal of the intermediate packaging separating the layers of tare [10]. This technical difficulty can be solved using the additional four pneumatic grippers, in Fig. 4 position 1. These structural units have additional degrees of freedom, and their own engines, which allows them to be effectively included in the scenarios of the entire robotics system with the depalletization.

The movement of a container unit at the stage of the discharge conveyor can be divided into two movements, one of which is translational and the other is rotational.

a)

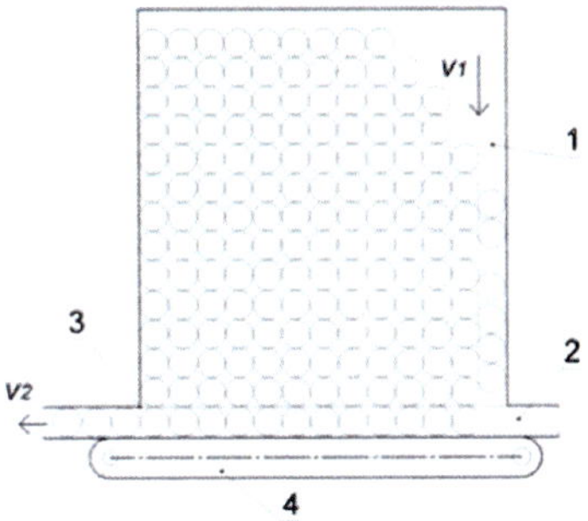


b)

**Fig. 2.** Separation installation: a) - physical configuration; b) - structural scheme, where: 1-feed conveyor, 2-discharge conveyor, 3-separation device, 4-side conveyor

**Fig. 3.** Debugging the work of the robotics system with the depalletization, general view;

**Fig. 4.** Debugging the work of the robotics system with the DP-3 depalletization system, removal of intermediate packaging separating the layers of tare.

Thus, for a complete description of the movement of the tare unit in the global coordinate system OXY, it is sufficient to define three independent coordinates x0(t), y0(t), φ0(t) [11, 12]. In this case, there are only two independent parameters, since the movement of a container unit along the discharge conveyor towards the separating device (along the OX axis) gives a change in only the X coordinate, while Y remains unchanged. This system has two degrees of freedom and Eq. 1 in the form of Lagrange can be used to solve it:

$$\begin{cases} x_0 = f_1(t) \\ \phi_0 = f_2(t) \end{cases}$$

$$\begin{cases} \dfrac{d}{dt}\left(\dfrac{dQ}{d\dot{x}_0}\right) - \dfrac{dT}{dx_0} = d(\sum A_1(P_i))dx_0 \\ \dfrac{d}{dt}\left(\dfrac{dQ}{d\dot{\phi}_0}\right) - \dfrac{dT}{d\phi_0} = d(\sum A_1(P_j))d\phi_0 \end{cases} \tag{1}$$

Where: $\varphi_0$ - rotation of the container, $x_0$ - movement along the conveyor, Q – generalized force acting on the container, T – kinetic energy, $A_{1,2}$ – work of force.

Equation 1 can be used to describe the movement of a unit container along a discharge conveyor. Rotational motion, characterized by the angle of rotation φ0, is a consequence of the forces acting on the container. Figure 4, shows a diagram for a container unit at the time of its movement along the outgoing conveyor. Friction force Ffr2 - between the bottom of the container and the discharge conveyor is much less than the friction force Ffr4 between the side surface of the container and the side conveyor Ffr2 « Ffr4, which is an important design feature of the described DP-3 system. The container is rolled along the side conveyor, experiencing pressure - H from the side of the container layers located on the storage conveyor, H ≠ const, since the number of container layers and elements in them is constantly changing. Vs - conveyor speed, Fv - force moving container (Fig. 5).

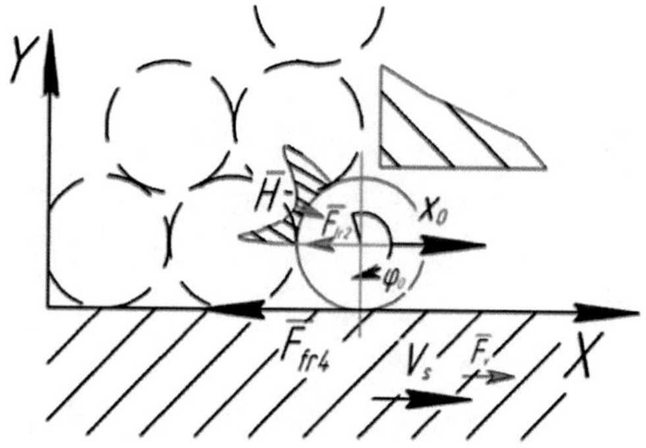


**Fig. 5.** Separation device, power circuit acting on the unit of the tare output from the conveyor.

Suppose that the container moves without slipping, then its movement can be dscribed by Eq. (2), based on system (1):

$$\begin{cases} m\ddot{q}_1 = (H(t))\cos(\alpha) - F_{fr4} - F_{fr2} + F_v \\ J\ddot{q}_2 = R(F_{fr4} + F_{fr2}) - (H(t)) \end{cases} \tag{2}$$

Where: $q_1 = x_0$ – container movement, $q_2 = \varphi_0$ – container rotation, $\alpha$ - angle between the resultant force H and the axis X.

Using Eq. 2, you can investigate the movement of the container, calculating, on the basis of statistical data, the variable force H.

## 3 Conclusion

The results of the comparison of the main characteristics of the analog and the DP3 system are presented in Table 3.

**Table 3.** Comparative table of the analog and the DP-3 system

| | Performance | Overall size | Reliability | Cost | Minimizing the human factor |
|---|---|---|---|---|---|
| Semi-automatic tare separation device | ●●●○○ | ●○○○○ | ●●●●○ | ●○○○○ | ○○○○○ |
| Depalletization system DP-3 | ●●●●● | ●●○○○ | ●●●●○ | ●●○○○ | ●●●●● |

The developed robotics system with depalletization has a low cost and allows depalletization of containers with high performance and reliability. It is important that it allows you to use the system as part of a robotics system, because it is an automatic system that minimizes the human factor. As a result of the introduction of the technological line, the depalletization time was significantly reduced, approximately 6 times, and the defects associated with the work of low-skilled personnel were completely eliminated. A continuous flow of containers along the conveyor to the packaging shop is provided.

## References


1. Gentile, S.S., Gladieux, M.J., Cechner, S.L.: Adjustable topguide assembly for air driven conveyor [txt]: patent № US 6,843,367 B1, USA, 18 Jan 2005
2. Cronin, B., Lemmer, N., Lowrey, D.: The applicant listed for this patent is Ska Fabricating Inc. Invention is credited to [txt]: patent №17/014840 A1 USA, 11 Mar 2021
3. EZY-Change Mk2 Guide Rails [txt]: patent №: 2015100359, Australis Engineering Pty Limited (2018)
4. «Ska Fabricating» Company Homepage. https://skafabricating.com/. Accessed 15 Oct 2017

5. Boso, D.P., et al.: An integrated DEM–FEM approach to study breakage in packing of glass cartridges on a conveyor belt. Granular Matter **22**(4), 1–16 (2020). https://doi.org/10.1007/s10035-020-01030-0
6. Mans, J.: Contract packing carbonated beverages. Packag. Dig. **48**(2), 30–32 (2011)
7. Bogoyavlensky, A.V., Khramov, I.M.: Transport and loading devices of automated production; Ministry of Science and Higher Education of the Russian Federation, Ural Federal University named after the first President of Russia B. N. Yeltsin. - Yekaterinburg: Ural University Press, 25 may 2020. https://elar.urfu.ru/handle/10995/94363. (rus)
8. Volkov, R.A., Gnutov, A.N., Dyachkov, V.K., et al.: Under the general editorship of Yu. A. Pertena. L.; Conveyors: Handbook/Mashinostroenie, Leningr, otd-nie, p. 367. s. il (1984). (rus)
9. Nuzhdin, K.A., Musalimov, V.M., Kalapyshina, I.I.: Modelling of nonlinear dynamic of mechanic systems with the force tribological interaction. Tribol. Ind. **37**(3), 360–365 (2015)
10. Kozhushko, G.G., Lukashuk O.A.: Calculation and design of belt conveyors. Ministry of Education and Science of the Russian Federation, Ural Federal University. - Yekaterinburg: Ural University Press, 232 (2016). (rus)
11. Kalapyshina, I.I., Perechesova, A.D., Nuzhdin, K.A., Musalimov, V.M., Zamoruev, G.B.: Modeling of mechatronic systems in MATLAB (Simulink/Simmechanics). In: 26th European Modeling and Simulation Symposium, EMSS, pp. 106–111 (2014)
12. Bock, F., Bruhn, H.: Case study on scheduling cyclic conveyor belts. Omega **102**, 102339 (2021)
13. Yang, C., Bu, L., Chen, B.: Measur. J. Int. Measur. Confed. **178**, 109342 (2021)
14. Tojiyev, R., Isomidinov, A., Alizafarov, B.: Strength and fatigue of multilayer conveyor belts under cyclic loads. Turk. J. Comput. Math. Educ. **12**(7), 2050–2068 (2021)

# Experimental Characterization of an Electrical Propulsion Unit for Service UAVs

Cinzia Amici[1], Federico Ceresoli[1], Matteo Saponi[2], Marco Pasetti[3], Simone Zanoni[4], Alberto Borboni[1], Monica Tiboni[1](✉), and Rodolfo Faglia[1]

[1] Department of Mechanical and Industrial Engineering, University of Brescia, via Branze, 38, 25123 Brescia, Italy
monica.tiboni@unibs.it

[2] Imbal Carton s.r.l., Via Gardesana, 54, 25080 Prevalle, BS, Italy
m.saponi@imbalcarton.com

[3] Department of Information Engineering, University of Brescia, via Branze, 38, 25123 Brescia, Italy

[4] Department of Civil, Environmental, Architectural Engineering and Mathematics, University of Brescia, via Branze, 43, 25123 Brescia, Italy

**Abstract.** The increasing attention to the critical environmental situation leads towards the greater diffusion of vehicles based on affordable and clean energy, like electric energy. The electrification of traction thus favours the development of sustainable cities. This also applies to the Unmanned Aerial Vehicles (UAVs), for which the electric solution in most cases is also the optimal solution. This paper presents the experimental characterization of the propulsion unit of a medium-sized drone (within the mass limit of 25 kg at the take-off) for service applications. The characterization process primary aims to integrate the power consumption data provided by the manufacturer, and evaluate the actuators performance in different conditions: i) considering two alternative configurations, i.e. motor with and without propellers, and ii) evaluating the contribute of the gravitational acceleration $g$ on the actuation systems performance, i.e. with the supporting frame of the test bench at different angles with respect to the ground plane.

**Keywords:** SDG 7 (affordable and clean energy) · SDG11 (sustainable cities and communities) · Electrical propulsion · Unmanned Aerial Vehicles (UAVs) · Experimental characterization

## 1 Introduction

The impact of vehicles emissions on the environment is a global problem. The vehicles electrification process aims to combat it, in step with at least two of the Sustainable Development Goals (SDG) stated by the United Nations, i.e. SDG 7, Affordable and Clean Energy, and SDG 11, Sustainable Cities and Communities. In the field of Unmanned Aerial Vehicles (UAVs), the use of electric propulsion is widespread [1]. The type of adopted propulsion greatly varies depending on the size of the UAV [2]: for small or


G. Quaglia et al. (Eds.): I4SDG 2021, MMS 108, pp. 307–314, 2022.
https://doi.org/10.1007/978-3-030-87383-7_34

medium-sized drones, electric motors are almost universally used, therefore these vehicles assure a null environmental impact in terms of emissions. For large drones, the adopted solution is significantly influenced by the performance required to the aircraft especially in terms of autonomy and load capacity. In fact, the purely electric solution may not allow these performances to be always achieved, and in those cases fuel or hybrid solutions are generally adopted.

In the segment of medium-sized UAVs (with a mass at the take-off of the order of 20–25 kg and with an autonomy of several kilometres), projects for devices with social purposes are increasing. Unlike smaller UAVs, mainly dedicated to applications requiring limited payloads (e.g. surveillance, shooting, monitoring) or last-mile delivery tasks, these drones can be used for more challenging missions, such as delivery requiring wider roundtrips or specific civil service applications demanding for heavier payloads, such as cameras, sensors and instrumentation for agriculture change detection [3, 4]. These solutions have an important indirect impact on the environment, because they ease the reduction of the road mobility and the consequent impact on the air, thus contributing to make cities and communities more sustainable.

Specific application field, purpose and expected operative conditions strongly affect the design requirements of devices [5–7], and the need to use affordable and clean energy for vehicle power generation has important consequences in the design of drones. The characterization of the electric motors used for these UAVs represents a crucial phase of the design/construction process [8, 9], also considering that these systems must guarantee a degree of safety compatible with the social context within which they interact. Furthermore, the components characterization is an important step to develop condition monitoring of the systems [10–12], that could improve the reliability in operation. Besides, according to literature the design of many UAV is currently based on legacy-driven selections or manufacturer data, which in some cases, especially for less common components, are limited or inaccurate [13].

Within this context, the paper presents the experimental characterization of an electric propulsion unit, based on a DC brushless motor and on a propeller, which will be part of the propulsion system of a medium-sized drone. The vehicle is designed to not overcome the mass limit of 25 kg defined by the Italian and European regulation framework as one of the class thresholds for civil drones, will be used for service purposes, such as deliveries, and therefore must guarantee the highest level of reliability. Since data concerning the energy consumption demanded by the propulsion units are required for the following design stages, like the sizing of the power supply system, the experimental campaign on the selected actuator became necessary to gather the needed data. Different tests have been performed, with the dual purpose of assessing the motors characteristics for throttle percentages lower than 40%, to integrate the datasheet provided by the manufacturer with the missing values, and of validating the provided data. Initially, a set of tests was carried out to calibrate the measurement system, investigating the correspondence between values assigned to the radio-controller and input voltage for the actuator. Then, the actuators have been studied considering two alternative configurations: i) motor without external loads (M) and ii) actuator with propellers (MP). Different working conditions have been also analysed for both the actuation configurations, evaluating the contribute of the gravitational acceleration $g$ on the actuation systems performance.

# 2 Materials and Methods

## 2.1 Experimental Setup

As Fig. 1 depicts, a double-configuration test bench was arranged for the performance analysis of the actuation system.

The experimental setup was composed of: i) the actuation unit, i.e. T-MOTOR MN801-S KV150 motor, and T-MOTOR G26x8.5 propellers in case, ii) a T-MOTOR FLAME 80A HV V2.0 ESC module, iii) two SPEKTRUM SPMX50006530 LiPo batteries, connected in series, iv) a SANWA EXES PLUS radio-controller, equipped with an FHSS 2.4 GHz receiver module, v) the supporting frame, adjustable to the inclination $\delta$ of 0° and 57° with respect to the ground plane, and vi) the measurement system, composed of an ammeter and a voltmeter.

Since the radio-controller was used in an unconventional way, i.e. as part of the measurement chain, a preliminary calibration of the system was performed. Tests compared the voltage measured at the radio-controller output at different values of imposed setpoint, in two conditions: increasing the setpoint values from 0% to 100% of the radio-controller full scale, and decreasing it from 100% to 0%.

Figure 2 describes the schematic layout of the experimental tests. Table 1 and Table 2 present a synthesis of the technical data provided by the manufacturer, describing the performance of the evaluated actuators, and actuator with propellers units, respectively.

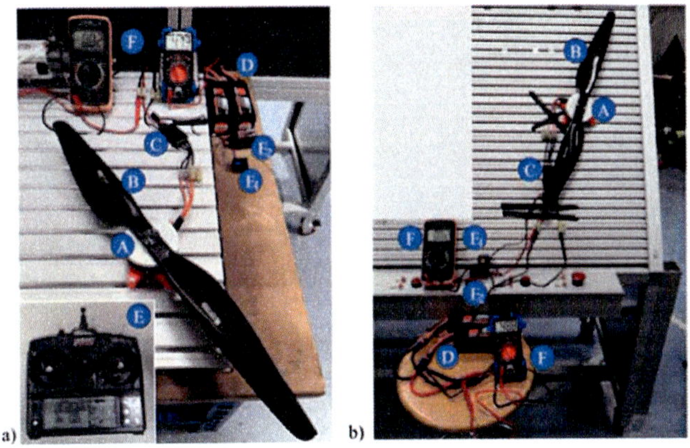


**Fig. 1.** The experimental setup, with the supporting frame at a) 0° and b) 57°. In the pictures, A indicates the motor, B the propeller, C the ESC unit, D the batteries and E the radio-controller, with $E_1$ and $E_2$ its dedicated receiver and battery respectively; F indicates the ammeter and voltmeter realizing the measurement system.

## 2.2 Data Analysis

For data acquired from the measurement system calibration, root mean square (RMS), standard deviation (SD) and maximum value of the residual between measuring cycles were computed.

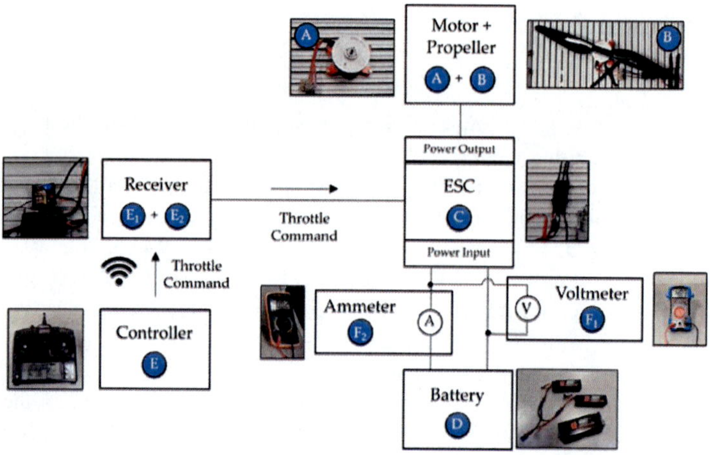


**Fig. 2.** Schematic of the experimental setup. The user block can describe the motor or both the motor and propeller elements according to the protocol of the specific test.

**Table 1.** Synthesis of technical data for the T-MOTOR MN801-S KV150 motor.

| Parameter | Value |
|---|---|
| Size | $\phi$ 89·39.4 mm |
| Mass | 480 g |
| Idle current (15 V) | 1.1 A |
| Peak current (180 s) | 45 A |
| Max power (180 s) | 2800 W |

**Table 2.** Synthesis of technical data provided by the manufacturer for the propulsion unit composed by a T-MOTOR MN801-S KV150 motor and T-MOTOR G26x8.5 propeller. Data are presented at 60% of the maximum Throttle, and the extreme values of the range provided by the manufacturer (i.e. 40 and 10% of the maximum Throttle).

| Parameter | Throttle 60% | Range (Throttle 40%–100%) |
|---|---|---|
| Thrust (g) | 6383 | 3509–12076 |
| Torque (Nm) | 1.69 | 0.83–3.75 |
| Current (A) | 17.07 | 7.03–5247 |
| RPM | 3798 | 2612–5247 |
| Power (W) | 817 | 337–2738 |
| Efficiency (g/W) | 7.81 | 10.41–4.41 |

For the actuators, the power consumption in the M and MP configurations were analyzed, and the trendlines for i) the MP data with $\delta$ at 0°, ii) the values stated by the manufacturer, and iii) the set of both these data subsets, were evaluated respectively.

In the M configuration, the power consumptions with $\delta$ at 0° and 57° were investigated, computing RMS, SD and maximum value of the residual between the values of the two conditions.

## 3 Results

The results of the calibration process performed on the radio-controller are reported in Fig. 3. The residual evaluated as difference between decreasing and increasing setpoints presented a maximum absolute value of 1 mV, rms of 0.67 mV and SD of 0.50 mV.

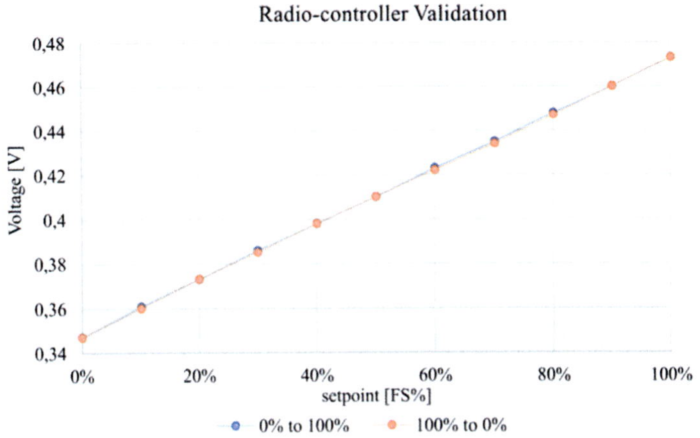


**Fig. 3.** Measured voltage with respect to the imposed setpoint; values acquired for increasing values of the setpoint are reported in blue, and for decreasing in orange

Figure 4 collects the results of the tests performed with $\delta$ at 0°, for both the M and MP configurations, i.e. motor without and with propellers respectively. Data acquired in MP configuration allow integrating the characteristic profile of power consumption provided by the manufacturer, and Fig. 4 depicts as trendline the best fitting third order spline approximating the data for both the subsets of experimentally measured data and manufacturer's values, as well as for the combined set of the available data.

The results of the tests for the analysis of the gravity contribution to the actuator performance are presented in Fig. 5. The residual evaluated as difference between the two series of data presents a maximum absolute value of 5.77 W, a RMS of 3.29 W and a SD of 1.18 W.

## 4 Discussion

Literature presents few examples of propulsion system design for medium-sized UAVs, and still few products are suitable as components for this drone class. Moreover, also when components are available on the market, the potential issue of incomplete datasheets or missing information remains [13]. Therefore, to enable the sizing of the power supply system of a drone, a proper characterization of the components in the propulsion unit is needed. In the current case, the data provided by the manufacturer for the selected pair of actuator and propellers were partially incomplete. The performed experimental campaign allowed us to gather the missing data and evaluate the system performance in some specific working conditions.

The initial tests on the measurement system confirm that multiple repetitions detect negligible result variations (SD of 0.50 mV) also under those operational conditions which would emphasize systematic errors, like different pitch angles.

The analysis of the actuator performance with respect to the presence of external loads offers data suitable for a further step of the design process, i.e. the components modelling required by the design optimization study. As Fig. 4 depicts, the actuator performance remarkably changes whether or not the propellers are connected. Both measured data for the MP condition and power consumption data provided by the manufacturer depict a trend that can be described by a third order spline. Evaluating the data subsets independently, the acquired data suggest a faster increasing trend than manufacturer values. This phenomenon could simply indicate different conditions in the experimental setup that could affect the measurement process, like different sensitivity of the systems to non-idealities (e.g. the ground effect), or reflect the lower reliability of the interpolating curve of measured data for values higher than 40% throttle, since no data are available for the fitting in that range. The latter interpretation is also supported by the analysis of the red trendline, defined considering the set of both measured values and manufacturer data, which better overlaps with the expected trend suggested by the manufacturer indications.

Besides, the assessment of the propulsion unit sensitivity to the $\delta$ angle provides a preliminary indication of the system behavior for different pitch values of the UAV. As described in Fig. 5 the actuator performance results little affected by variations in the value of the $\delta$ angle, also in the worst working condition, i.e. with the actuator in configuration M. In fact, the maximum detected variation in the power consumption between these two cases results lower than the 4.7% of the maximum measured power (131.25 W for $\delta$ null, at 100% throttle).

Further investigations were also performed, like the analysis of the stability in the power consumption of each actuation system. In this case, current and voltage variations were monitored at different working velocity, along at least 30 s of use, but no remarkable variations were detected.

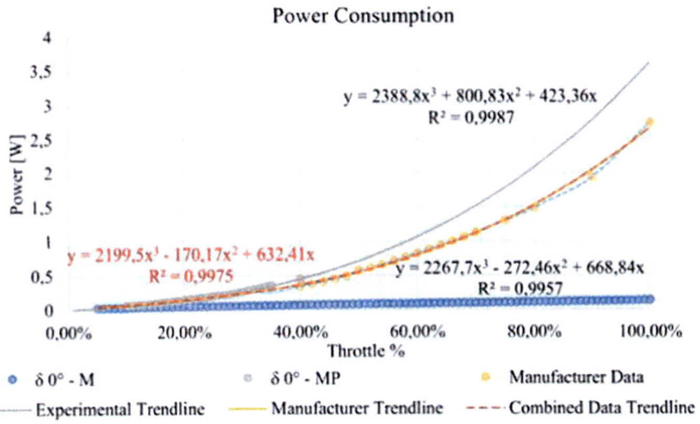


**Fig. 4.** Power consumption for $\delta$ at 0°: motor without external loads (M) and with propellers (MP) configurations are reported in blue and gray respectively; in yellow the data provided by the manufacturer. In continuous line the trendlines of the measured data and manufacturer values subsets: in dashed red line the trendline for the set of the combined data.

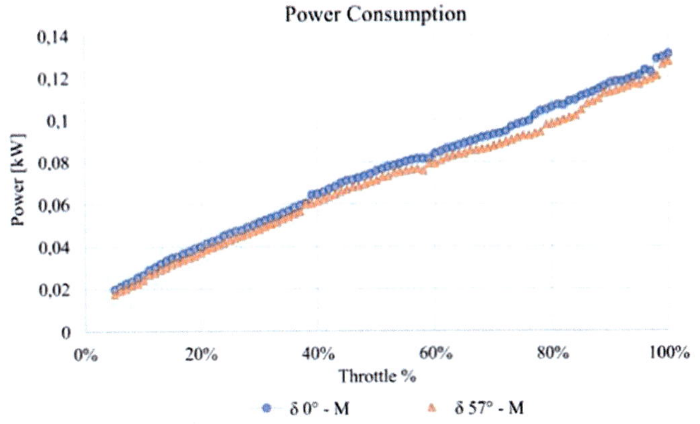


**Fig. 5.** Power consumption of the motor without external loads (M), with blue circles for $\delta$ at 0°, and orange triangles for $\delta$ at 57°.

## 5 Conclusions

In this paper the experimental characterization of the propulsion unit for a medium-sized service UAV (total mass at take-off lower than 25 kg) was presented. A test bench was arranged, and an experimental campaign was designed and carried out, to complete the power consumption data provided by the actuator manufacturer and obtain the data required for the further design steps. Experimental tests revealed that the selected actuators present a robust response to pose variations, but are sensitive to the presence of propellers. Further analyses would investigate the characteristic profile thrust-velocity of the motor, the response of the actuator to external variable loads and ground effect, or conditions for the arising of vibratory phenomena.

**Fundings.** This research was partially funded by Regione Lombardia, Call Hub Ricerca e Innovazione, within the project "MoSoRe@Unibs - Infrastrutture e servizi per la Mobilità Sostenibile e Resiliente", 2020–2022.

## References

1. Amici, C., Ceresoli, F., Pasetti, M., Saponi, M., Tiboni, M., Zanoni, S.: Review of propulsion system design strategies for unmanned aerial vehicles. Appl. Sci. **11**(11), 5209 (2021). https://doi.org/10.3390/app11115209
2. Hassanalian, M., Abdelkefi, A.: Classifications, applications, and design challenges of drones: a review. Prog. Aerosp. Sci. **91**, 99–131 (2017). https://doi.org/10.1016/j.paerosci.2017.04.003
3. Stöcker, C., Bennett, R., Nex, F., Gerke, M., Zevenbergen, J.: Review of the current state of UAV regulations. Remote Sens. **9**(5), 33–35 (2017). https://doi.org/10.3390/rs9050459
4. Stroppiana, D., et al.: Estimating crop density from multi-spectral UAV imagery in maize crop. ISPRS - Int. Arch. Photogramm Remote Sens. Spat. Inf. Sci., XLII-2/W13(2/W13), 619–624 (2019). https://doi.org/10.5194/isprs-archives-XLII-2-W13-619-2019
5. Quaglia, G., Oderio, R., Bruzzone, L., Razzoli, R.: A modular approach for a family of ground mobile robots. Int. J. Adv. Robotic Syst. **10** (2013). https://doi.org/10.5772/56086
6. Amici, C., Pellegrini, N., Tiboni, M.: The robot selection problem for mini-parallel kinematic machines: a task-driven approach to the selection attributes identification. Micromachines **11**(8) (2020). https://doi.org/10.3390/MI11080711
7. Formicola, R., Ragni, F., Mor, M., Bissolotti, L., Amici, C.: Design approach of medical devices for regulation compatibility: a robotic rehabilitation case study. In: Proceedings of the 7th International Conference on Information and Communication Technologies for Ageing Well and e-Health, pp. 146–153 (2021). https://doi.org/10.5220/0010402801460153
8. Zulkipli, A.H., Raj, T., Hashim, F.H., Huddin, A.B.: Characterization of DC brushless motor for an efficient multicopter design. In: 2016 International Conference on Advances in Electrical, Electronic and Systems Engineering, ICAEES 2016, pp. 586–591 (2017). https://doi.org/10.1109/ICAEES.2016.7888114
9. Muehlebach, M., D'Andrea, R.: The flying platform – a testbed for ducted fan actuation and control design. Mechatronics **42**, 52–68 (2017). https://doi.org/10.1016/j.mechatronics.2017.01.001
10. Tiboni, M., Remino, C.: Condition monitoring of a mechanical indexing system with artificial neural networks (2017). https://doi.org/10.1007/978-3-030-11220-2_19
11. Tiboni, M., Incerti, G., Remino, C., Lancini, M.: Comparison of signal processing techniques for condition monitoring based on artificial neural networks. In: Fernandez Del Rincon, A., Viadero Rueda, F., Chaari, F., Zimroz, R., Haddar, M. (eds.) CMMNO 2018. ACM, vol. 15, pp. 179–188 (2019). Springer, Cham. https://doi.org/10.1007/978-3-030-11220-2_19
12. Tiboni, M., Bussola, R., Aggogeri, F., Amici, C.: Experimental and model-based study of the vibrations in the load cell response of automatic weight fillers. Electronics **9**(6), 1–14 (2020). https://doi.org/10.3390/electronics9060995
13. Bershadsky, D., Haviland, S., Johnson, E.N.N.: Electric multirotor propulsion system sizing for performance prediction and design optimization (2015)

# Study of Cooling Systems for Electronic Components Used in Robot Applications

Narcis Gratian Crăciun, Erwin Christian Lovasz, and Carmen Sticlaru(✉)

Politehnica University of Timisoara, Timişoara, Romania
narcis.craciun@student.upt.ro, {erwin.lovasz, carmen.sticlaru}@upt.ro

**Abstract.** A cooling system for a printed circuit board used in robots or mechatronic systems is a device that has the role of dissipating the heat generated by electronic components in the external environment. The two most efficient types of cooling systems are heatsink-based and heat pipe-based. To determine which of the two types is more efficient, 3D models are made for both systems. For the heatsink-based system, 5 models are made to determine how the fin geometry influences the efficiency of the system. Thermal simulations are performed for all models and the results are compared. The results of the best heatsink system are compared with the results of the heat-pipe based system to determine which of the two has the highest efficiency.

**Keywords:** SDG4 · Cooling system · Heatsink · Heat pipe

## 1 Introduction

The expansion of robotics in recent years is overwhelming, robots are used in many fields, from the automotive industry, to medicine, home help, entertainment, etc. [1]. Along with this expansion, the current trend towards the development of medium to small robots can be observed. This means that the electronic components are also smaller due to lack of space. Agglomeration of components in combination with their increasing required conversion powers leads to a major problem in both operation and lifespan, namely the increase in temperature [2]. Another factor that aggravates the problem is the increasing use of plastic materials that are known to have poor heat transfer properties [2].

The book Fundamentals of Electronic Systems Design [3], shows how lowering the operating temperature of a component by up to 50% has the effect of lowering the failure rate from 5 to 10 times. Decreasing the operating temperature of a semiconductor by 10 degrees Celsius at the junction of the temperature leads to halving its failure rate [3].

For these reasons, the use of a system to dissipate the heat generated by electronic components is imperative in Robotics and Mechatronics industry. Cooling systems take the heat generated and release it into the environment by heat transfer. Heat transfer can be done in 3 ways, namely, convection, conduction and radiation [2]. The most common cooling systems are those based on heat sink and heat pipes and can use all of these 3 processes to dissipate heat [4].


G. Quaglia et al. (Eds.): I4SDG 2021, MMS 108, pp. 315–324, 2022.
https://doi.org/10.1007/978-3-030-87383-7_35

The radiator of the heat sink system is made of a material with a high thermal conductivity, generally aluminum or copper and is placed in contact with the components that generate heat. Heat is absorbed by the base of the radiator by thermal conduction and is dissipated into the environment by the process of convection or thermal radiation [2].

The heat pipe system consists of a pipe made of a material with good thermal properties, generally copper, inside which there is a working environment (water, glycol), the system also has a wick and a radiator [2]. One end of the pipe (evaporator) absorbs the heat generated by the electronic components leading to the vaporization of the working environment, then the heat is transferred to the other end (condenser) in contact with the radiator. Here the working environment condenses, the heat is dissipated and transferred to the environment by the radiator by convection and radiation [2].

These systems are found especially in devices that have printed circuit boards (PCBs). This means that they are found in industrial robots, stepping robots, robots in the medical industry, domestic robots, robots used in agriculture and farms [5], practically in all robots that have circuits with a tendency to overheat. The Big Dog stepper robot, for example, uses 2 such heatsink-based cooling systems, one for the oil pump PCB and one for the engine [6]. These heatsink-based systems can also be used to transfer heat from the final effector of a robotic arm or to cool the engines of nuclear rollers [7, 8].

Because these types of systems have proven their efficiency and applicability in many applications, interest in their development and improvement has been directly proportional. There have been numerous studies over the years on how to increase the efficiency of these types of systems [9, 10]. Parameters such as material, system size, wall thickness, contact surface with electronic components, fins geometry influence both the efficiency and cost of a heat sink system [17, 18]. For the heat pipe, beside the geometry, size of the contact zone with electronics and material, the wick structure and the working fluid type are very important when designing a heat pipe cooling system [19, 20].

This study follows two research directions of cooling systems for electronic components. In the first phase, the influence of the geometry of the fins from the heat sink systems is studied by making five models of radiators with the same base but different geometries of the fins. The five models are thermally simulated to determine how much their performance is influenced by the geometry of the fins.

The second phase of the study seeks to compare a heat sink cooling system with a heat pipe cooling system to determine which of the two systems has the highest heat dissipation performance generated by the electronic components of a PCB. The two 3D models of the systems are subjected to a thermal analysis to determine which system is more efficient.

## 2 Design of Models for Both Types of Systems

### 2.1 Design for the Five Heat Sink Models

First the study involves comparing 5 models of radiators made of aluminum to see how the geometry of the fins affect the efficiency. All 5 prototypes presented in figures from 1 to 5 have the same radiator base, diameter of 10 cm and height of 4 cm, but the geometry of the fins is different. The design of the fins respects the current knowledge about increasing the efficiency of the heatsink, namely: the fins are positioned vertically on, the channels between the fins have a greater distance and the area of the fins is as large as possible (Figs. 1, 2, 3, 4 and 5).

**Fig. 1.** First heat sink prototype

**Fig. 2.** Prototype number 2

**Fig. 3.** Prototype number 3

## 2.2 Design of the Heat Pipe and the Heat Sink System

For the second path of the study, a comparation between two passive cooling systems is made, one based on heat sink and one on heat pipes.

**Fig. 4.** Prototype number 4

**Fig. 5.** Prototype number 5

The heatsink model, Fig. 6 is made of an aluminum block by machining. It has a complex geometry of the wings and the space between them is large enough to favor the convection process.

The heat pipe system consists of a heat pipe made of copper and a radiator made of aluminum. The pipe is flattened and is placed in a horizontal position. The working liquid used is water. Because the heat pipe system also needs a radiator that takes thermal energy from the heat pipe and dissipates it outside by thermal convection, we made an aluminum radiator, but unlike the studies conducted in this direction, the radiator in the heat pipe system has no fins. The top of the radiator is flat, to reduce the manufacturing cost. In Fig. 7 is presented the cooper heat pipe and in Fig. 8 the aluminum heat sink.

The heat pipe is flattened precisely to increase the contact surface with the components that emit heat. The radiator has a geometry that facilitates the takeover of thermal energy from the pipe through the large contact surface between the two. A single heat pipe was used for this study, but depending on the case, several pipes can be installed. The pipe has no areas of curvature, because those areas reduce the efficiency rate of the device.

## 3 Performing Thermal Simulations

In order to perform the simulations, all the prototypes, with heat sink and heat pipe, were assembled with the PCB. Between the PCB and the surface of the heatsink, a thermal paste is applied which has the role of facilitating the heat transfer between the components of the PCB and the heatsink.

**Fig. 6.** Heat sink model

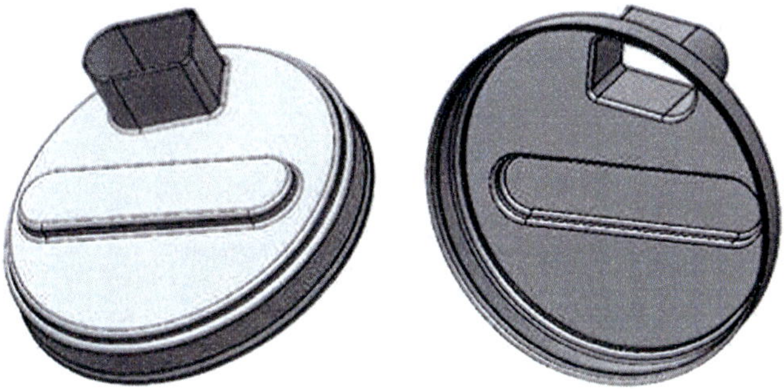

**Fig. 7.** Heat sink of the heat pipe system

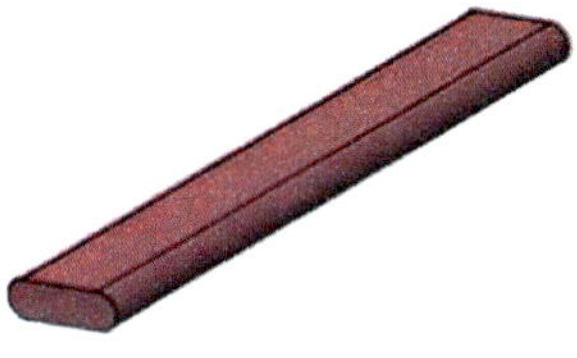

**Fig. 8.** Heat pipe

The PCB is made of FR408 with the following parameters: material density 1.85 g/cm$^3$, specific heat capacity 950 J/(kg · K) and thermal conductivity of 0.29 W/(kg · K).

Thermal paste is made of aluminum nitrate with parameters for simulation: density of 3.26 g/cm$^3$ thermal conductivity of 0.3 W/(m · K) and specific thermal capacity of 740 J/(kg · K).

### 3.1 Thermal Simulations of Heat Sink Prototypes

The heat sink prototypes are made of aluminum with the following parameters for simulation: material density 2.7 g/cm$^3$, specific heat capacity 910 J/(kg · K) and thermal conductivity of 205 W/(m · K).

The prototypes are thermally simulated (figures from 9 to 13) in 3D cad software Creo 4 in order to determine how much heat can be dissipated, how heat is distributed in the area of interest and what is the maximum temperature reached.

For this purpose, the simulation conditions are presented: 4W load for the PCB surface on which the components to be cooled are located (the 4 W rating was determined by simulations and previous experiments), the value of 7 W/(m$^2$ · K) is a conservative estimate of natural convection on the outer surface of the radiator in the environment. The results are expressed in Celsius degrees (Figs. 9, 10, 11, 12 and 13).

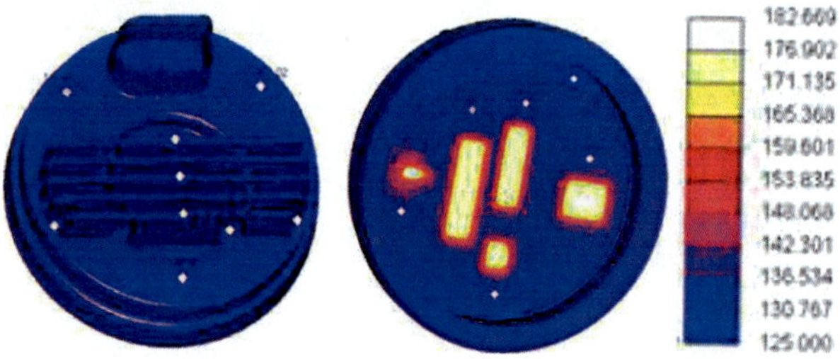


**Fig. 9.** Thermal simulation for prototype number 1

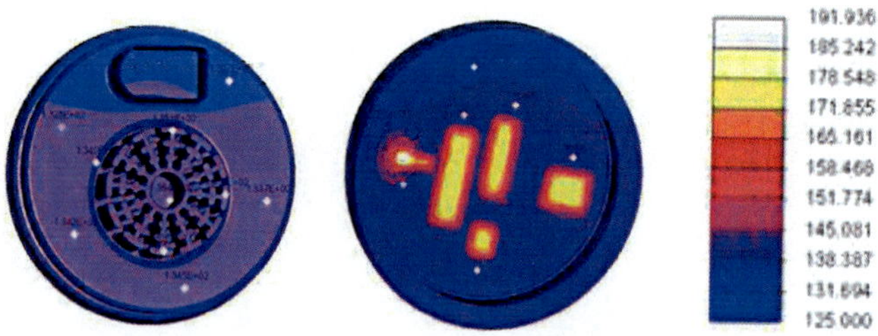


**Fig. 10.** Thermal simulations for prototype number 2

The results of the thermal simulations for the 5 prototypes are presented in Table 1. As it can observed, the best results are obtained by the prototype numer 5, and on

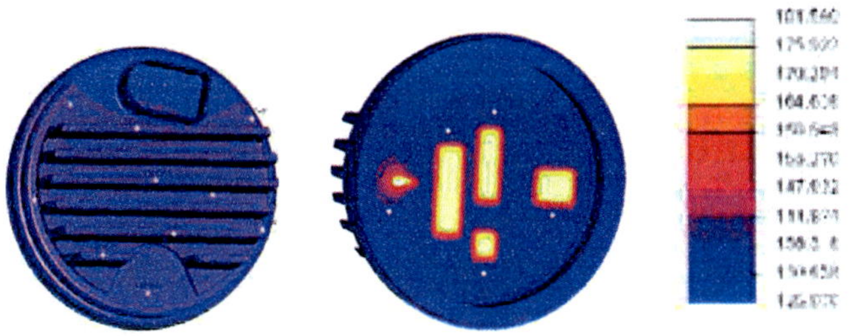

**Fig. 11.** Thermal simulations for prototype number 3

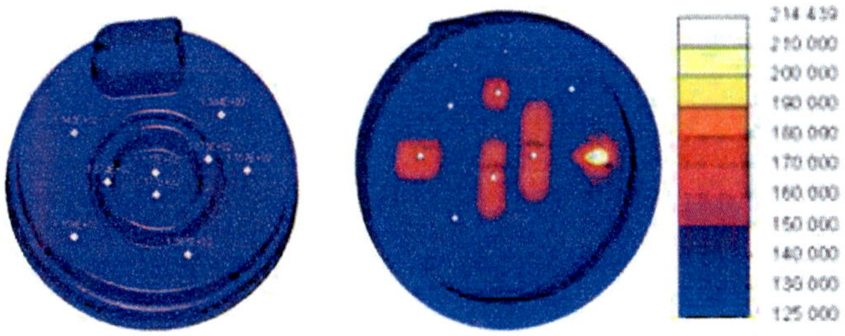


**Fig. 12.** Thermal simulations of prototype number 4

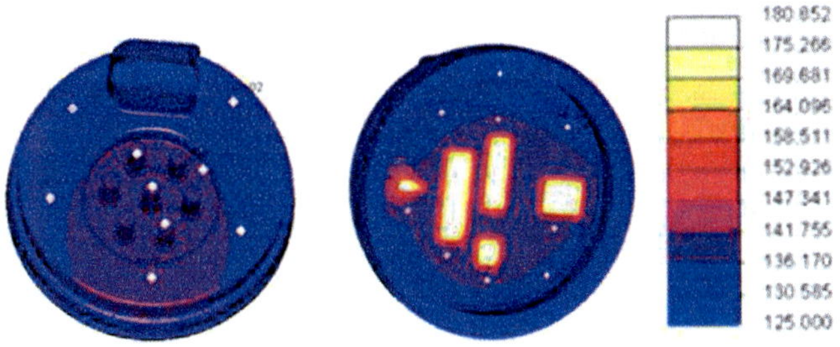


**Fig. 13.** Thermal simulations of prototype number 5

the second place in prototype number 3. These results are another confirmation of the studies on fin geometry.

**Table 1.** Thermal simulations for the 5 prototypes.

| Prototype | Maximum Temperature [°C] | Minimum Temperature [°C] |
|---|---|---|
| 1 | 182,669 | 136,534 |
| 2 | 191,936 | 138,387 |
| 3 | 181,58 | 136,316 |
| 4 | 214,439 | 130 |
| 5 | 180,852 | 136,17 |

## 3.2 Thermal Simulations for Heat Pipe and Heat Sink Prototype

The heatsink prototype is made of aluminum and has the following parameters for simulation: material density 2.7 g/cm$^3$, specific heat capacity 910 J/(kg · K) and thermal conductivity of 205 W/(m · K).

The prototype is thermally simulated in 3D software tool named Creo in order to determine how much heat can be dissipated, how heat is distributed in the area of interest and what is the maximum temperature reached as can be seen in Fig. 14.

For this purpose, the simulation conditions are presented: 4W load for the PCB surface on which the components to be cooled are located (the 4 W rating was determined by simulations and previous experiments), the value of 7 W/(m$^2$·K) is a conservative estimate of natural convection on the outer surface of the radiator in the environment.

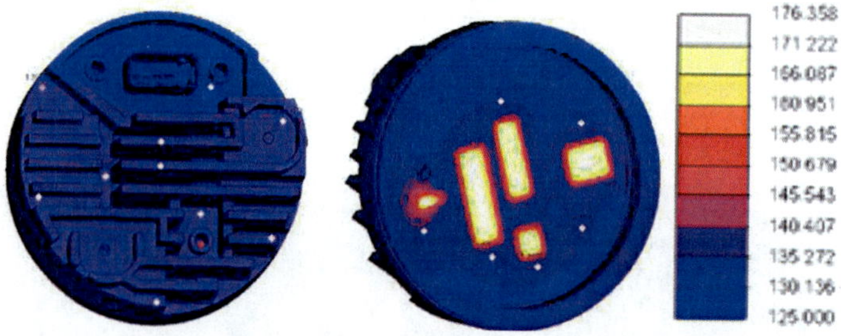


**Fig. 14.** Thermal simulations for heat sink system

The heat pipe prototype is simulated in Ansys program because it has a higher complexity than the heatsink one, and the water inside the pipe must also be taken into account. The volume of water was considered strictly conductive. The heat transfer between solid and water was done with the help of a transfer coefficient corresponding to a heat transfer with stationary liquid (Fig. 15). The loads of the model were: the thermal conductivity of aluminum of 237 [W/m · K], because the radiator is made of aluminum, the density of aluminum of 2.7 g/cm$^3$ and the density of copper of 8.96 g/cm$^3$ and the thermal conductivity of the pipe of heat of 4,382 [W/m · K].

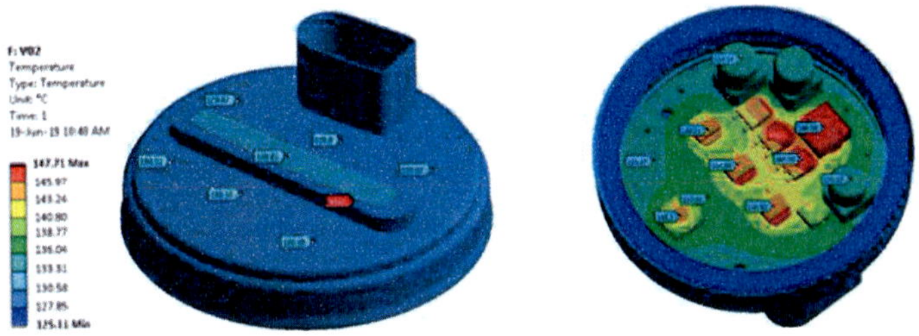

**Fig. 15.** Thermal simulations for the heat pipe system

The results from thermal simulations made in CREO and Ansys for the two models are presented in Table 2. The simulations are performed in two programs because for heat pipe, the working fluid must be taken into account, and the Ansys program is more suitable for such cases. The results show that the heat-pipe based system is more efficient to transfer heat from the PCB components than the aluminum heat sink system. The temperature requirements differ depending on the project, in this case, the maximum temperature allowed was 150 °C.

**Table 2.** Thermal simulations made in CREO and Ansys.

| Prototype | Maximum Temperature [°C] | Minimum Temperature [°C] |
|---|---|---|
| Heat sink | 176,358 | 135,252 |
| Heat pipe | 147,91 | 129,8 |

## 4 Conclusions

This paper presents a study on the cooling systems for the heat generated by the electronic components of printed circuit boards used in robot applications. The study presents 2 research directions on heat sink systems and heat pipe systems. In the first phase, the influence of the geometry of the fins from the radiators of the heat sink systems is studied by making 5 prototypes that are thermally simulated. The obtained results demonstrate that the geometry of the fins has an influence on the system performance and as a consequence, the performance of a heat sink system can be increased if the design of the fins is suitable for the application in question. The second direction of study presents the comparison of a model of heat sink system with a heat pipe cooling system. Thermal simulations show that the heat pipe system has a higher efficiency than the heat sink system. These results confirm the efficiency of heat pipe systems and open a new path in the case of cooling systems for electronic components in robots. Heat pipe systems take up less space, allow access to components in hard to reach places, are lighter than radiator systems and are therefore ideal candidates for robots where low weight and high speed are required.

## References

1. Ben-Ari, M., Mondada, F.: Robots and their applications. Elem. Robot., 1–20 (2018)
2. Lienig, J., Bruemmer, H.: Fundamentals of Electronic Systems Design, Chapter 5, Thermal Management and Cooling (2017)
3. Lienig, J., Bruemmer, H.: Fundamentals of Electronic Systems Design, Chapter 4, Reliability Analysis (2017)
4. Edmunds, L.: Heatsink Characteristics. International Rectifier
5. https://dspace.emu.ee/xmlui/bitstream/handle/10492/6453/AR2021_98_Galins_V_Tehnika_doi_054.pdf?sequence=1&isAllowed=y. Accessed 22 Mar 2021
6. Raibert, M., Blankespoor, K., Nelson, G., Playter, R.: BigDog team: BigDog, the rough-terrain quaduped robot. Boston Dynamics, Waltham
7. Pollak, M., Drobransky, J.: Structural design and material cutting using a laser end effector on a robot arm. TEM J. (2020)
8. Han, Y., Luan, W., Juang, Y., Zhang, X.: Thermal control of electronics for nuclear robots via phase change materials. Energy Procedia **75**, 3301–3306 (2015)
9. Staliulioni, Z., Zhang, Z., Pittini, R., Andersen, M.A.E.: Investigation of heat sink efficiency for electronic component cooling applications. Elektronika ir Elektrotechnika **20**, 49–54 (2014)
10. Kumar, A., Londhe, K., Kaushik, V.R., Khamkar, N.: Heat sink design for optimal performance of compact electronic appliances-a review (2017)
11. Oguntala, G., Sobamowo, G., Ahmed, Y., Abd-Alhameed, R.: Thermal prediction of convective-radiative porous fin heatsink of functionally graded material using adomian decomposition method. Computation **7**, 19 (2019)
12. Khan, W.A., Culham, J.R., Yovanovich, M.M.: The role of fin geometry in heatsink performance
13. Zohuri, B.: Design guide and heat pipe selection. In: Zohuri, B. (ed.) Heat Pipe Applications in Fission Driven Nuclear Power Plants, pp. 265–308. Springer, Cham (2019). https://doi.org/10.1007/978-3-030-05882-1_9
14. https://core.ac.uk/download/pdf/286834033.pdf. Accessed 16 May 2021

# Genetic Algorithm Based PID Parameter Optimization for Longitudinal Dynamics of a Fixed Wing Mini UAV

Şaban Ulus(✉) and Ikbal Eski

Engineering Faculty, Mechatronics Engineering Department, Erciyes University, Kayseri, Turkey
{sabanulus,ikbal}@erciyes.edu.tr

**Abstract.** Unmanned aerial vehicles (UAVs) have gained very much importance in both civilian and industrial areas. Especially, in military applications, there are growing demands on the UAVs' usage for reconnaissance, surveillance, defence, and attack purposes. Therefore, autonomous control of the UAVs has great importance to achieve stable and successful missions. Most of the control techniques in the aircraft control systems use the classical PID (proportional + integral + derivative) control method but this method may have some insufficient response characteristics in terms of changing weather conditions and disturbances. At this point, because of the PID usage in most UAV systems, PID controller parameters must have been obtained for optimum control performance of any state dynamics of the UAV. In this study, longitudinal dynamics are obtained for pitch angle and altitude dynamics of a mini fixed-wing Cessna-182 UAV at 60 km/h flight speed. PID parameters are investigated for the best performance of evolutional genetic algorithms (GA) in terms of different performance indexes such as integrated square error (ISE), integrated absolute error (IAE), time multiply of square error and absolute errors (ITSE, ITAE) and linear quadratic regulator (LQR) cost functions. The simulation results indicate the best performance obtained for the LQR index in the point of overshoot and settling time. In addition, proportioned hybrid ITSE and ITAE cost also developed their performance either for overshoot or settling time. This paper contributes to sustainable development goal (SDG9) by investigating aerial vehicle innovation for a stable and reliable flight purpose in the industrial and civilian applications.

**Keywords:** SDG9 · UAV · Genetic algorithm · PID control · Longitudinal dynamic

## 1 Introduction

UAV systems and application areas are very popular and gained more attraction in both civilian and military applications. UAVs have very many investigations and developments in agricultural areas such as spraying pesticides, photogrammetry etc. Also, especially, they are used for reconnaissance, surveillance, and armed UAV for military


G. Quaglia et al. (Eds.): I4SDG 2021, MMS 108, pp. 325–334, 2022.
https://doi.org/10.1007/978-3-030-87383-7_36

applications. By the increase of the technological advancements and usage of the UAVs, their control systems are investigated in terms of autonomy and stability with reliability purposes. UAV autopilot systems are a very important effect on the controllability, navigation, and stability of the control of the aircraft [1]. The autopilot applications for aircrafts can have a possibility of stable and effective flight missions without human controls [2].

In the previous studies, the controller design implementations on UAVs for both fixed-wing and multi-rotor systems have been investigated by using modern approaches. Most of the control algorithms deal with the nonlinearities to achieve system stability and control. Even if the modern control approaches achieve stability and optimization, they have complexity and more cost for some situations [3]. Besides, especially, for linear systems, the PID controller algorithm is mostly chosen for its simplicity and performance. Öner et al. [4] have been investigated their UAV altitude dynamics by using the PID controller methods. Wagner et al. [5] used numerical feedback control and PI control by using 6 degrees of freedom Monte-Carlo simulation model for the digital landing control. It is noted that the numerical feedback control has valuable results for robust control. Bayraktar et al. [6] implemented a UAV test by analyzing hybrid algorithms on the hardware design and real flight experiments.

In the research, many different modern control methods about UAV dynamics such as fuzzy control [7, 8] adaptive control [9, 10] artificial neural networks (ANN) [11], genetic algorithms (GA) [12] have been investigated. Kurnaz et al. [13] proposed an autonomous flight control structure based on the artificial neuro-fuzzy inference system (ANFIS) for a UAV. They developed 3 fuzzy logic control (FLC) modules for altitude, lateral and longitudinal movements in three-dimensional space. Ulus and Eski [8] proposed ANFIS, FLC, hybrid ANFIS-PID, PD-Fuzzy-PI control approaches and compared the results with PID Ziegler-Nichols (PID-ZN) for a mini "Ultra-stick 25e" fixed-wing UAV model. The authors noted that hybrid approaches have more acceptable results. Sun et al. [14] studied an adaptive FLC by considering a full state nonlinear system. The implemented model was acceptable with the Lyapunov function.

Kim et al. [15] analyzed the LQR control. The disturbance and changing conditions for the model are implemented by the modern LQR control approach. Pang et al. [16] studied the tracking performance of an Osprey aircraft which is related to model uncertainties. Kalman Filter (KF) was applied to estimate system states. According to the results, the variance was decreased compared with the PID controller. Selma [17] proposed an ANFIS and particle swarm optimization (PSO) technique for trajectory planning of a quadrotor air vehicle and results were compared to the classical PID controller.

This paper presents a Cessna-182 mini-UAV's longitudinal dynamics control for classical PID-ZN and evolutional GA methods for optimizing PID controller's objective functions such as integrated square error, absolute error, time-multiplied square error, time-multiplied absolute error, (ISE, IAE, ITSE, ITAE), LQR cost function and proportioned ITSE + ITAE cost functions. The PID for pitch angle and altitude dynamics of the model are obtained by the ZN method roughly. Finally, the optimized GA-PID results were given for the eight objective functions to analyze performance criteria. The simulations show that the LQR performance has more advantage in terms of the overshoot and settling time especially. The controller effort and feedback error are optimized by noise and disturbance covariances in LQR. The other acceptable results can be improved using ITSE + ITAE with a different ratio. Overshoot or settling time can be decreased by using different a, b ratios as in Table 2.

## 2 UAV Longitudinal Dynamics and Control

In this study, to design a UAV control system, Cessna-182 mini-UAV is used and shown in Fig. 1. The UAV has ailerons, an elevator, rudder, flaps, and electric motor. Aircraft longitudinal dynamics include pitch angle ($\theta$), pitch rate ($\dot{\theta}$), forward (u), downward (w) airspeed and altitude (h) dynamics. The equations are obtained by linearizing at 60 km/h cruise speed. Table 1 shows some specifications of the UAV.

**Table 1.** The mini-Cessna-182 model specific parameters

| Parameter | Specification | Value |
|---|---|---|
| S | Wing reference area | 0.23 $m^2$ |
| b | Wingspan | 1.3 m |
| $\overline{c}$ | Chord of the wing | 0.177 m |
| m | Weight | 1.21 kg |
| Ixx | Roll moment of inertia | 0.0424 $kg.m^2$ |
| $I_{yy}$ | Pitch moment of inertia | 0.046 $kg.m^2$ |
| Izz | Yaw moment of inertia | 0.079 $kg.m^2$ |
| $I_{xz}$ | Product moment of inertia | 0.0012 $kg.m^2$ |

**Fig. 1.** Cessna-182 mini- UAV model

The longitudinal dynamics of the Cessna-182 can be presented in Eqs. (1), (2).

$$\dot{x}(t) = A.x(t) + B.u(t) + E.\mathrm{u(t)} \tag{1}$$

$$y(t) = C.x(t) + D.u(t) \tag{2}$$

A, B, C, D, E can be defined as state matrix, control input, output, feed-forward, and uncertainty matrixes, respectively. The state, input, and output vectors are x(t), u(t), y(t), respectively. The state-space form for the Cessna-182 has given in Eq. (3) [1].

$$\overbrace{\begin{Bmatrix}\Delta\dot{u}\\ \Delta\dot{w}\\ \Delta\dot{q}\\ \Delta\dot{\theta}\\ \Delta\dot{h}\end{Bmatrix}}^{\dot{x}_{long}}=\overbrace{\begin{bmatrix}X_u & X_w & 0 & -g & 0\\ Z_u & Z_w & u_0 & 0 & 0\\ M_u+M_{\dot{w}}Z_u & M_w+M_{\dot{w}}Z_w & M_q+M_{\dot{w}}u_0 & 0 & 0\\ 0 & 0 & 1 & 0 & 0\\ -\sin(\theta_0) & -\cos(\theta_0) & 0 & -u_0.\cos(\theta_0) & 0\end{bmatrix}}^{A_{long}}\overbrace{\begin{Bmatrix}\Delta u\\ \Delta w\\ \Delta q\\ \Delta\theta\\ \Delta h\end{Bmatrix}}^{x_{long}}+\overbrace{\begin{bmatrix}X_{\delta_e} & X_{\delta_T}\\ Z_{\delta_e} & Z_{\delta_T}\\ M_{\delta_e}+M_{\dot{w}}Z_{\delta_e} & M_{\delta_T}+M_{\dot{w}}Z_{\delta_T}\\ 0 & 0\\ 0 & 0\end{bmatrix}}^{B_{long}}\overbrace{\begin{Bmatrix}\Delta\delta_e\\ \Delta\delta_T\end{Bmatrix}}^{u_{long}} \tag{3}$$

In Eq. (3), state variables are given as u, w, q, θ, h. These are known as velocity components in the x and z-axis, pitch rate, pitch angle and altitude of the aircraft, respectively. $X_u$, $X_w$, $Z_u$, $Z_w$, $M_u$, $M_w$, $M_q$, $M_{\dot{w}}$ are the state derivatives. $u_0$,$\theta_0$ are airspeed, pitch angle at the steady-state. $X_{\delta_e}$, $X_{\delta_T}$, $Z_{\delta_e}$, $Z_{\delta_T}$, $M_{\delta_e}$, $M_{\delta_T}$, $M_{\dot{w}}$ are stability derivatives in $B_{long}$ matrix. $\Delta\delta_e$,$\Delta\delta_T$ are the elevator and throttle inputs. Equation (4) gives numerical values obtained from the UAV to analyse stability and control.

$$\overbrace{\begin{Bmatrix}\Delta\dot{u}\\ \Delta\dot{w}\\ \Delta\dot{q}\\ \Delta\dot{\theta}\\ \Delta\dot{h}\end{Bmatrix}}^{\dot{x}_{long}}=\overbrace{\begin{bmatrix}-0.106 & 0.243 & 0 & -9.81 & 0\\ -0.801 & -5.79 & 16.67 & 0 & 0\\ 2.017 & -3.52 & -3.52 & 0 & 0\\ 0 & 0 & 1 & 0 & 0\\ -0.052 & -0.997 & 0 & -16.65 & 0\end{bmatrix}}^{A_{long}}\overbrace{\begin{Bmatrix}\Delta u\\ \Delta w\\ \Delta q\\ \Delta\theta\\ \Delta h\end{Bmatrix}}^{x_{long}}+\overbrace{\begin{bmatrix}7.153 & 0\\ 0 & -9.38\\ 0.072 & -169.3\\ 0 & 0\\ 0 & 0\end{bmatrix}}^{B_{long}}\overbrace{\begin{Bmatrix}\Delta\delta_e\\ \Delta\delta_T\end{Bmatrix}}^{u_{long}} \tag{4}$$

Altitude (h) and pitch angle (θ) transfer functions against elevator deflection can be obtained from Eqs. (4) for Cessna-182 aircraft.

$$\frac{\delta\theta}{\delta e}=\frac{-169.3s^2-964.8s-137.6}{s^4+21.43s^3+151.1s^2+30.36s+142.2} \tag{5}$$

$$\frac{\delta h}{\delta e}=\frac{9.37s^3+5783s^2+1.633x10^4s+3318}{s^5+21.43s^4+151.1s^3+30.36s^2+142.2s} \tag{6}$$

Transfer functions for altitude and pitch angle dynamics obtained from the state space using MATLAB. Transfer functions used for analyzing PID Ziegler-Nichols (PID-ZN) results and comparing performance criteria using GA optimization and objective functions.

## 3 The PID Controller Structure

PID control is the well-known and most widely used structure in the control system design because of its simplicity and low-cost effect. However, under some external disturbance conditions, it may not be a proper design for nonlinear systems if the parameters are not chosen optimally. The PID structure can be defined below as shown in Eq. (7).

$$u(t)=K_p e(t)+K_I\int_0^t e(t)dt+K_D\frac{de(t)}{dt} \tag{7}$$

In Eq. (7), e(t) is defined as the error of the reference and feedback signal. The PID controller gains in this study found by using ZN setting at MATLAB.

### 3.1 Performance Evaluation of the Different Cost Functions

In this paper, different performance criteria for each dynamic have investigated and simulation results for optimization of PID parameters presented. Eight different cost functions experimented for the PID fitness function (J) given in Eq. (8).

$$J = parm(1) + parm(2)\frac{1}{s} + parm(3)\frac{s}{0.001s + 1} \tag{8}$$

In Eq. (8), parm(1), parm(2), parm(3) variables indicate $K_p$, $K_i$, $K_d$ gains which searched for the minimum cost value. Here, a high pass filter approach used for the derivative block instead of just using the s transfer function block. In Eqs. (9), (10), (11), (12), (13) give the integrated squared error, absolute error, time multiplied square error, time multiply absolute error and LQR cost functions, respectively.

$$ISE = \sum_{t=0}^{k} (x_{\mathrm{m}} - \hat{x})^2 \tag{9}$$

$$IAE = \sum_{t=0}^{k} \left|(x_{\mathrm{m}} - \hat{x})\right| \tag{10}$$

$$ITSE = \sum_{t=0}^{k} t'.(x_{\mathrm{m}} - \hat{x})^2 \tag{11}$$

$$ITAE = \sum_{t=0}^{k} t'.\left|(x_{\mathrm{m}} - \hat{x})\right| \tag{12}$$

$$LQR = \sum_{t=0}^{k} (Q.\left|(x_{\mathrm{m}}(t) - \hat{x}(t))\right|^2 + R.u(t)^2) \tag{13}$$

$x_m$ is the measured signal and $\hat{x}$ is desired input value. Q, R can be defined as disturbance and noise covariances and selected as $Q = 1, R = 0.001$ in this study. "u(t)" is the control input signal. The balance between the control input and disturbance error can be set by Q and R values before GA optimization.

## 4 PID Optimization with the Genetic Algorithm Method

GA is used for the best PID parameter selection for altitude and pitch angle dynamics. The core of the GA is related to the selection of the population, crossover, and mutation parameters. In each generation, the procedure follows the selection, crossover, mutation approach. In the selection, individuals having strong adaptation capability to the problem are selected for the next generation. According to the fitness value of the individuals, having a low-cost value of the individuals may be selected while high-cost values are not used. In MATLAB, some GA options are selected as follows. (population = 100, Generations = 50, Selection = Uniform, Crossover = scattered, CrossoverFraction = 0.8, Mutation = Gaussian, CreationFcn = Uniform, Const.Tol. = 1e−3, EliteCount = 0.05 * population, FunctionTol. = 1e−6).

Fitness functions evaluated for Eqs. (9), (10), (11), (12), (13) and proportioned ITSE + ITAE values. Figure 2 shows the flow chart of the GA method. GA simulations have done until fitness value expectations and max. generations are met.

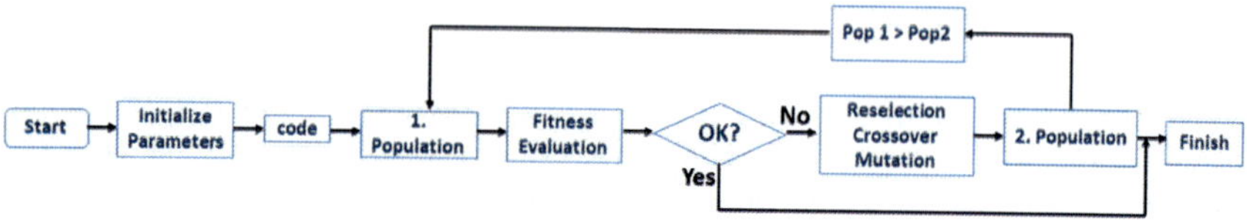


**Fig. 2.** Flow diagram of the GA

## 5 Simulation Results

In this section, GA optimization results for the different fitness functions are shown for comparison. The structure of the PID optimization with GA is depicted in Fig. 3. Table 2 gives the optimum PID parameter values and PID-ZN values for each dynamic and each cost function.

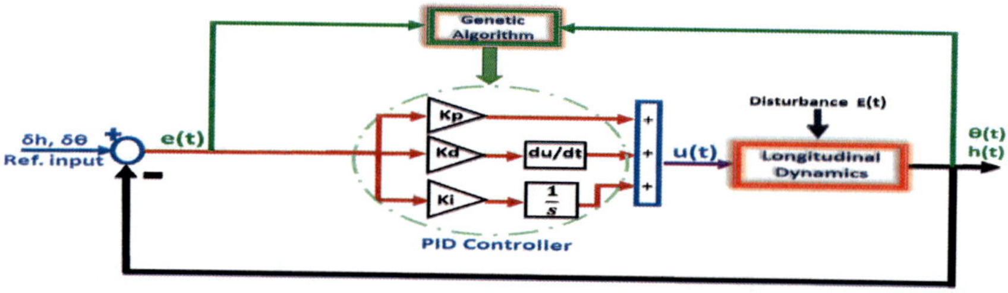


**Fig. 3.** Structure of the PID parameter optimization with GA

Figure 4 and Fig. 5 represent the pitch angle and altitude step response plots for different cost function approaches.

The transient response results for pitch angle are listed in the Table 3. As seen in the results, LQR cost and ITAE have very less overshoot than classical PID and others. Also, ITSE + ITAE proportional values improved either overshoot or settling time. Even if, ISE, IAE have less settling time, they are not acceptable with high overshoot.

**Table 2.** Numerical values of the controller parameters and cost functions

| | | PID (ZN) ISE | ISE | IAE | ITSE | ITAE | LQR | J = a*ITSE + b*ITAE | | |
|---|---|---|---|---|---|---|---|---|---|---|
| | | | | | | | | a/b | | |
| | | | | | | | | 1 | 3 | 0.33 |
| θ (pitch) | Kp | −5 | −1.6 | −5.01 | −6.16 | −3.44 | −5.13 | −3.27 | −5.99 | −5.13 |
| | Ki | −25 | −16.15 | −15.6 | −17.3 | −20.7 | −14.4 | −19.0 | −20.55 | −20.8 |
| | Kd | 0.016 | −13.27 | −5.64 | −1.01 | −0.17 | −0.25 | −0.3 | −0.281 | −0.07 |
| | Cost | 137.1 | 5.3 | 60.73 | 0.89 | 157.3 | 23.76 | 86.9 | 39.69 | 117.9 |
| h (altitude) | Kp | 0.648 | 0.016 | 0.408 | 0.247 | 0.607 | 0.021 | 0.59 | 0.56 | 0.599 |
| | Ki | 0.623 | 0.127 | 0.486 | 0.209 | 4.242 | 0.129 | 3.82 | 2.792 | 4.059 |
| | Kd | 0.039 | 0.126 | 0.103 | 0.102 | 0.096 | 0.123 | 0.096 | 0.097 | 0.095 |
| | Cost | 329.1 | 50.7 | 160.6 | 4.67 | 51.43 | 51.98 | 30.5 | 19.64 | 41.04 |

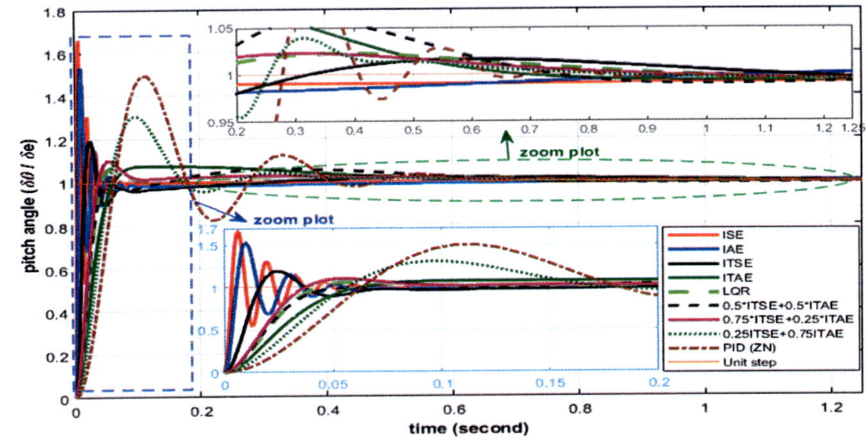


**Fig. 4.** Step responses of the pitch angle dynamics of the Cessna-182 UAV

Altitude dynamic step response of each controller setting are given in Fig. 5. The classical PID has a very big overshoot and slow settling time. In contrast, GA based optimization results give more stability and less overshoot. Table 4 indicates the altitude transient response of each criterion. The simulation results show that LQR, ISE and ITSE have very acceptable results comparing to others.

**Table 3.** Transient response parameters for pitch angle dynamic

| Controller Perf. Criteria | Pitch angle [θ] (°) – [δθ/δe] | | | | | |
|---|---|---|---|---|---|---|
| | Rise T. (s) | Peak T. (s) | Peak Val. | Overshoot (%) | Settling T. (s) | Steady-Error (%) |
| PID (ZN) | 0.0042 | 0.117 | 1.489 | 48.9 | 0.57 | 0.0016 |
| ISE | 0.0025 | 0.007 | 1.659 | 65.9 | 0.068 | 0.0022 |
| IAE | 0.004 | 0.01 | 1.5308 | 53.08 | 0.086 | 0.0023 |
| ITSE | 0.0125 | 0.025 | 1.1935 | 19.35 | 0.198 | 0.0022 |
| ITAE | 0.0413 | 0.1 | 1.072 | 7.2 | 0.486 | 0.0018 |
| LQR | 0.0295 | 0.06 | 1.065 | 6.5 | 0.468 | 0.0022 |
| 0.5*(ITSE+ITAE) | 0.004 | 0.336 | 1.057 | 5.7 | 0.586 | 0.0083 |
| 0.75*ITSE+0.25*ITAE | 0.025 | 0.057 | 1.0953 | 9.53 | 0.408 | 0.0019 |
| 0.25*ITSE+0.75*ITAE | 0.038 | 0.097 | 1.304 | 30.4 | 0.387 | 0.0019 |

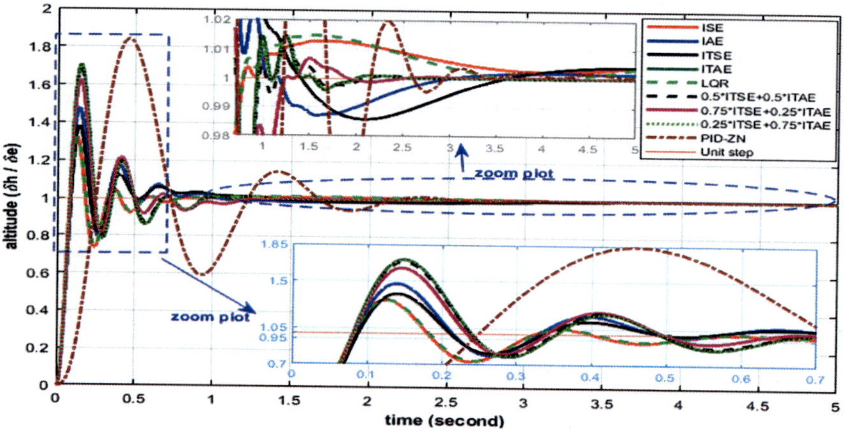


**Fig. 5.** Step responses of the altitude dynamics of the Cessna-182 UAV

**Table 4.** Transient response parameters for altitude dynamic

| Controller Perf. Criteria | Altitude [m] – [δh/δe] | | | | | |
|---|---|---|---|---|---|---|
| | Rise T. (s) | Peak T. (s) | Peak Val. | Overshoot (%) | Settling T. (s) | Steady-Error (%) |
| PID (ZN) | 0.015 | 0.4547 | 1.843 | 84.3 | 2.05 | 0.002 |
| ISE | 0.051 | 0.126 | 1.322 | 32.2 | 0.54 | 0.0013 |
| IAE | 0.05 | 0.14 | 1.4756 | 47.56 | 0.94 | 0.006 |
| ITSE | 0.055 | 0.142 | 1.378 | 37.8 | 1.01 | 0.0017 |
| ITAE | 0.049 | 0.152 | 1.704 | 70.4 | 0.903 | 0.0006 |
| LQR | 0.051 | 0.126 | 1.3166 | 31.66 | 0.54 | 0.0024 |
| 0.5*(ITSE+ITAE) | 0.05 | 0.15 | 1.6814 | 68.14 | 0.915 | 0.0007 |
| 0.75*ITSE+0.25*ITAE | 0.054 | 0.147 | 1.6288 | 62.88 | 1.121 | 0.0011 |
| 0.25*ITSE+0.75*ITAE | 0.051 | 0.15 | 1.6983 | 69.83 | 0.913 | 0.0008 |

## 6 Conclusion

In this study, PID (ZN) controller parameters improved using the GA optimization algorithm for Cessna-182 fixed-wing UAV which will be in real-time experiments. Eight different fitness functions evaluated in terms of step responses. Transient response characteristics like overshoot, peak time, settling time and steady error of each dynamic obtained. GA method improved the performance of the classic PID-ZN settings. Moreover, LQR performance has more developments on overshoot and settling time in general perspective. Noise and disturbance covariance parameters selected for less controller effort in the cost function. The proportioned ITSE + ITAE also improved either overshoot or settling time according to their rate. Results show that LQR gives the best PID optimal parameters for this study.

**Acknowledgement.** This study is a part of the research project FBA-2018-8391. The authors appreciate for financial support being supplied by the Scientific Research Project Coordination Unit of Erciyes University.

## References

1. Nelson, R.C.: Flight Stability and Automatic Control. 2nd. edn. Mc. Graw Hill International Editions Singapore (1998)
2. Ribeiro, L.R., Oliveira, N.: UAV autopilot controllers test platform using Matlab/Simulink and X-Plane. In: IEEE Frontiers in Education Conference, pp. S2H-1–S2H-6. Arlington, USA (2010)
3. Kada, B., Ghazzawi, Y.: Robust PID controller design for an UAV flight control system. In: Proceedings of the World Congress on Engineering and Computer Science, San Francisco, USA, pp. 1–6 (2011)
4. Öner, K.T., Çetinsoy, E., et al.: Mathematical modeling and vertical flight control of a tilt-wing UAV. Turkish J. Electr. Eng. Comput. Sci. **20**(1), 149–157 (2012)
5. Wagner, T., Valasek, J.: Digital autoland control laws using direct digital design and quantitative feedback theory (2006)
6. Bayraktar, S., Fainekos, G.E., et al.: Experimental cooperative control of fixed-wing unmanned aerial vehicles. In: 43rd IEEE Conference on Decision and Control, pp. 4292–4298 (2004)
7. Wu, J., Zhou, Z., et al.: Fast linear quaternion attitude estimator using vector observations. IEEE Trans. Autom. Sci. Eng. **15**, 307–319 (2018)
8. Ulus, Ş., Eski, İ.: Neural network and fuzzy logic-based hybrid attitude controller designs of a fixed-wing UAV. Neural Comput. Appl. **33**(14), 8821–8843 (2021). https://doi.org/10.1007/s00521-020-05629-5
9. Aliyu, B., Petinrin, A., et al.: PID control design of sideslip angle for a fixed-wing mini-UAV. Adv. Res. **6**(3), 1–14 (2016)
10. Mystkowski, A.: Robust control of the micro UAV dynamics with an autopilot. J. Theor. Appl. Mech. **51,** 751–761 (2013). 01/13
11. Poksawat, P., Wang, L., et al.: Gain scheduled attitude control of fixed-wing UAV With automatic controller tuning. IEEE Trans. Control Syst. Technol., 1–12 (2017)
12. Sarhan, A., Qin, S.: Adaptive PID control of UAV altitude dynamics based on parameter optimization with fuzzy inference. Int. J. Model. Optim. **6**(4), 246–251 (2016)

13. Kurnaz, S., Cetin, O., et al.: Adaptive neuro-fuzzy inference system based autonomous flight control of unmanned air vehicles. Expert Syst. Appl. **37**(2), 1229–1234 (2010)
14. Sun, K., Mou, S., et al.: Adaptive fuzzy control for non-triangular structural stochastic switched nonlinear systems with full state constraints. IEEE Tran. Fuzzy Syst. (99), 1–1, (2018). 11/26
15. Kim, A.R., Keshmiri, S., Blevins, A., Shukla, D., Huang, W.: Control of multi-agent collaborative fixed-wing UASs in unstructured environment. J. Intell. Robotic Syst. **97**(1), 205–225 (2019). https://doi.org/10.1007/s10846-019-01057-3
16. Pang, Y., MacKunis, W.: Asymptotic tracking for an osprey fixed-wing aircraft via nonlinear generalized minimum variance control. J. Aerosp. Eng. **33**(3), 04020010 (2020)
17. Selma, B., Chouraqui, S., Abouaïssa, H.: Optimal trajectory tracking control of unmanned aerial vehicle using ANFIS-IPSO system. Int. J. Inf. Technol. **12**(2), 383–395 (2020). https://doi.org/10.1007/s41870-020-00436-6

# Performance Analysis of Localization System for Wireless Robotic Capsule Endoscopy Based on 5 DOF

Memduh Suveren(✉) and Muzaffer Kanaan

Department of the Mechatronic Engineering, Erciyes University, Kayseri 38039, Turkey
{msuveren,mkanaan}@erciyes.edu.tr

**Abstract.** The wireless robotic capsule endoscopy technique is a relatively painless and invasive medical imaging technique. Most diseases in the gastrointestinal (GI) tract can be diagnosed with robotic capsule endoscopy, even in areas that cannot be reached with conventional colonoscopy. Knowing the position of the robot in robotic capsule endoscopy both speeds up the treatment process and gives the opportunity to control the robot from the outside during the procedure (active capsule endoscopy). In this study, magnetic positioning technique was used to obtain the positions of the robotic capsule in the small intestine model. With a ring-shaped permanent magnet placed around the capsule, the Magnetic Flux Density (MFD) equations were calculated analytically using two different techniques: Biot-Savart and Charge model. The positioning performances of both magnetic models were compared, and the Artificial Bee Colony (ABC) optimization algorithm and the Levenberg-Marquardt (LM) method were used together while calculating the nonlinear equations. As a result, we found that the Charge model was 61% faster than the Biot-Savart model under the same simulation conditions, and the position and angle errors of the Charge model were at least 87% less on average than the Biot-Savart model. Under noisy simulation conditions, the performance of the Charge model was observed to be either better or very close to that of Biot-Savart.

**Keywords:** SDG9 · Robotic capsule endoscopy · In body localization · Annular magnet · Artificial bee colony algorithm · Levenberg-Marquardt method

## 1 Introduction

Wireless robotic capsule endoscopy has been popularly used in the diagnosis of digestive system disorders in recent years [1]. With this technique, which is used as an alternative to classical endoscopy and colonoscopy techniques, many digestive system diseases such as Crohn's disease, colon cancer, intestinal bleeding and polyps can be diagnosed [2]. Capsule endoscopy has great advantages, considering the physical (risk of perforation and bleeding in the intestinal wall, mostly anesthesia requirement, etc.) and psychological effects (embarrassment and discomfort of the process) created by classical methods on the patient [3].

Wireless robotic capsule endoscopy basically starts with the swallowing of the small capsule and works by recording the images in the gastrointestinal tract until it is thrown


G. Quaglia et al. (Eds.): I4SDG 2021, MMS 108, pp. 335–344, 2022.
https://doi.org/10.1007/978-3-030-87383-7_37

out of the body. A capsule basically consists of 4 main parts as can be seen in Fig. 1a. These are image receiving camera, data processing processor, transmitter antenna that communicate with the external environment, and battery providing energy to the capsule. Also, Fig. 1b shows a real commercial wireless capsule Fig. 1b.

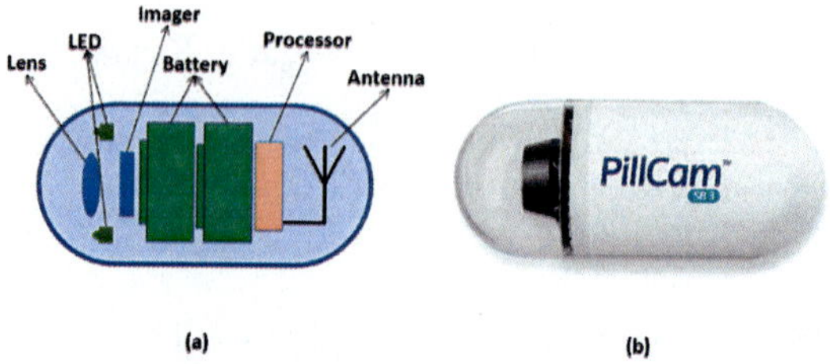


**Fig. 1.** (a) Wireless capsule structure, (b) Real endoscopy capsule [4]

Robotic capsule endoscopy, besides being a painless and non-invasive technique, is also a suitable method for positioning. There are many studies on obtaining the location of the disease diagnosed in the intestine by capsule endoscopy method. Because knowing the location of the disease both facilitates the work of the specialist who will perform the operation and accelerates the healing process of the patient. When studies on capsule localization are examined in the literature, there are a wide range of studies in scopes. These can be summarized as localization with Electromagnetic waves, magnetic positioning and positioning with other techniques (MRI, X-ray, Gamma ray ultrasound). Since X-ray and Gamma ray techniques have harmful effects on human health [5, 6], MR and Ultrasound methods require extra space in the capsule and their power consumption is high, so these techniques cannot be used widely. In radio wave positioning, positioning errors are relatively higher in RF positioning, as high frequency signals are subjected to significant attenuation as they pass through different body tissues [7, 8].

In the magnetic positioning process performed inside the body, since the relative permeability of the tissues is very close to that of the air, it is not affected by the body tissues, that is, it does not suffer a fading caused by the tissues [9]. For this reason, studies have been carried out on magnetic capsule localization. When the previous studies on magnetic capsule localization are examined in the literature, magnetic positioning basically takes place under three main headings [10]. These are the scenario where a current stimulated coil is used in the capsule [11], the scenario where the magnetic sensor is placed in the capsule [12] and the scenario with the permanent magnet in or on the capsule. The disadvantage of the first two scenarios is that placing a magnetic sensor or coil in a capsule with limited space increases both energy consumption and capsule size. For this reason, in this study, the localization process was made with permanent magnets. To avoid space constraints in the capsule, annular magnets were placed on the capsule to wrap around it.

In order to position a magnet, the magnetic flux model of the magnet must first be obtained. For this, the four different models are mostly used in the literature: Dipole model, Finite element technique, Current model (Biot-Savart) and Charge model. In this paper, we provide magnetic localization of a capsule inside the small intestine model with two different analytical flux density models (Biot-Savart, Charge Model) of an annular magnet. The overall goal is to comparatively show the magnetic field models that are optimal in terms of localization performance. The contributions of this paper can be summarized as follows. We explore the issue of positioning the robotic wireless capsule using a realistic 3-D model of the human body and intestine. We also compare two magnetic models used in the literature in terms of 5-D positioning performance for both noisy and noiseless situations.

In this paper, we gave information about the simulation environment in Sect. 2, we obtained the localization algorithm of the magnetic capsule in Sect. 3 and discussed the results in Sect. 4.

## 2 Simulation Environment

In order to localize the robotic capsule following the GI tract, the simulation environment was created first. The Real body related data of the human body in the simulations were obtained with Remcom™ and VariPose® software within the scope of the Visible Human Project [13]. With these body data, sensor locations to be placed around the abdomen and possible locations where the capsule can be found in the small intestine model were determined. Figure 2 shows the magnetic sensors and magnetic capsule positions obtained from the body model, as well as the body-related coordinate system.

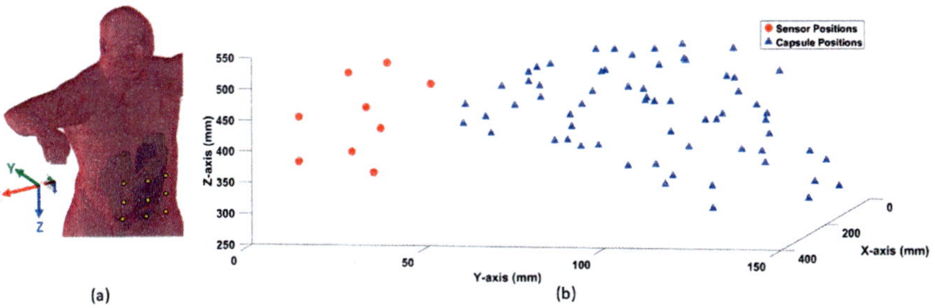


**Fig. 2.** (a) Body coordinate system (b) Magnetic capsule and sensor positions in the body

In Fig. 2, the dots represent the sensor locations, and the triangles the locations of the capsule with a magnet around it. Also, 9 hall effect sensors are placed on the body surface and uniformly distributed in the XZ plane. In addition, 64 different position and angle sets were determined in the robotic capsule to be in the intestine. In the simulations NdFeB35 magnet type was selected and the parameters of this magnet data were used [14]. Considering the actual capsule dimensions, the permanent annular magnet to be used in the analysis was determined as 15 mm in outer diameter, 10 mm in inner diameter and 12 mm in length. In addition, it is assumed that the magnet has uniform magnetization in the x-axis direction in the diameter direction, and its value is determined as 430.000 (A/m).

# 3 Localization Algorithms

The localization algorithm used to calculate the 5-D pose (3-D positions and 2-D angles) of the robotic capsule consists of 3 parts as shown in Fig. 3. These are analytical model in which flux density expressions are calculated, sensor measurements in which real magnetic values are measured, and optimization algorithms in which nonlinear equations are solved.

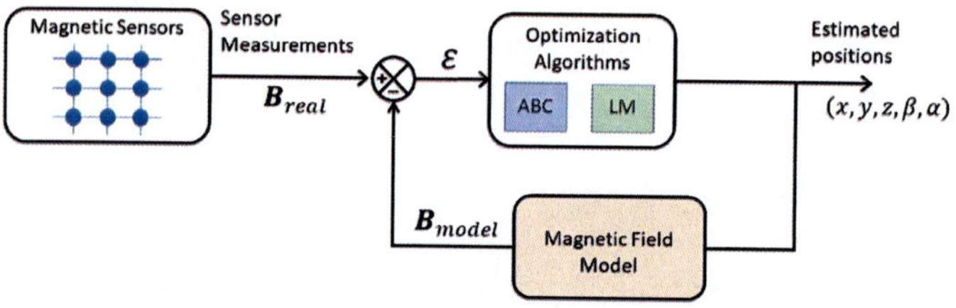


**Fig. 3.** General structure of the magnetic localization

## 3.1 Biot-Savart Model of Annular Magnet

In the Biot-Savart model, the permanent magnet is reduced to a distribution of equivalent current. The current following the A-B-C-D-A path is used to obtain the MFD formed at a generic P point by the cylindrical magnet, whose magnetization direction is z-axis, given in Fig. 4a.

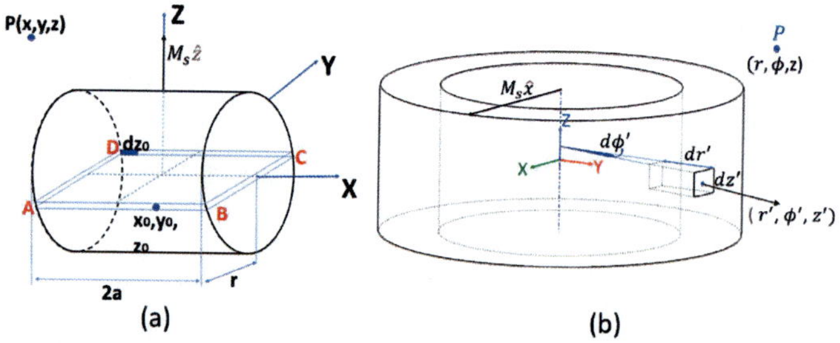


**Fig. 4.** (a) Biot-Savart law, (b) Charge model on the permanent magnet

where, $M_s$ is magnetization of the magnet, P points are observation points in space, $(x_0, y_0, z_0, r^1, \phi^1, z^1)$ are the source coordinates of the magnet and $z_0$, differential thickness at the path. The surface current of cylindrical magnet is calculated with $\mathbf{I} = J_s dz_0 \mathbf{dl}$. Here $J_s$ is the current density (A/m) and **dl** is the direction of the current. The previous study [15] is used when calculating the MFD created by any $(x_0, y_0, z_0)$ point on the ABCD path at the P point. The MFD differential equations obtained in the ABCD path are converted to the flux density equation with Eqs. (1) and (2).

$$\mathbf{dB}_{ABCDA} = \mathbf{dB}_{AB} + \mathbf{dB}_{BC} + \mathbf{dB}_{CD} + \mathbf{dB}_{DA} \tag{1}$$

$$\mathbf{B} = (B_x, B_y, B_z) = \left( \int_{-r}^{r} \mathbf{dB}_{ABCDA}\hat{\mathbf{x}}, \int_{-r}^{r} \mathbf{dB}_{ABCDA}\hat{\mathbf{y}}, \int_{-r}^{r} \mathbf{dB}_{ABCDA}\hat{\mathbf{z}} \right) \quad (2)$$

where, **B** is the MFD vector and $B$ is the MFD value. In order to limit the computation time, the MFD expressions were numerically integrated around the $z_0$ and Eq. (3) calculated by Simpson's method.

$$B_M = \frac{r}{3n}\left[ f_M(-r) + 4\sum_{k=0}^{n-1} f_M(-r + \frac{2kr + r}{n}) + 2\sum_{k=0}^{n-1} f_M(-r + \frac{2kr}{n}) + f_M(r) \right], M = x\|y\|z \quad (3)$$

In these equations, $n$ is the number of iterations, $M$ is the coordinate indicator of x, y, or z, and $r$ is the radius of the magnet. Here the used $f_x, f_y, f_z$ functions are given in detail in study [15]. It should be noted that these functions take up too much space to share here.

### 3.2 Charge Model of Annular Magnet

In this model, a magnet is reduced to a distribution of equivalent "Magnetic Charge". The flux density expressions of the cylindrical magnet given in Fig. 4b in cylindrical coordinates can be solved with the general expression in Eq. (4).

$$\mathbf{B}(\mathbf{x}) = \frac{\mu_0}{4\pi}\oint_S \frac{M_s\cos(\phi)(\mathbf{x}-\mathbf{x}')}{\left|\mathbf{x}-\mathbf{x}'\right|^3} ds' = -\frac{\mu_0}{4\pi}\int_{z_1}^{z_2}\int_0^{2\pi} \nabla\left(\frac{M_s\cos(\phi)}{\left|\mathbf{x}-\mathbf{x}'\right|}\right) Rd\phi' dz' \quad (4)$$

where, $M_s$ is magnetization of the magnet, $\mu_0$ is magnetic permeability of the space, $R$ is radius of the magnet, $\nabla$ is gradient operator on the cylindrical coordinates and $|\mathbf{x} - \mathbf{x}'|$ vector is the Euclidean distance between the source point on the magnet and the P point. When the solution techniques given in the [16] study is used, the numerical solution of the MFD expressions is obtained by Simpson's method as in Eq. (5).

$$B(r,\phi,z) = \frac{\mu_0 M_s R}{2n}\sum_{j=0}^{n}\sum_{k=1}^{2}(-1)^{k+1} S_\phi(j)\cos(\phi'(j)) I\Big[\big(r - R\cos(\phi - \phi'(j))\big)\hat{\boldsymbol{r}} + R\sin(\phi - \phi'(j))\hat{\boldsymbol{\phi}}$$

$$-\frac{1}{\sqrt{r^2 + r'^2 - 2rr'\cos(\phi-\phi') + (z-z')^2}}\hat{z}\Big] \qquad S_\phi(j) = \begin{cases} \frac{1}{3}, & j = 0, j = n \\ \frac{4}{3}, & j = 1,3,5... \\ \frac{2}{3}, & j = 2,4,6... \end{cases}$$

$$\phi'(j) = \frac{j}{n}2\pi,\ I = \frac{(z - z_k)(r^2 + r'^2 - 2rr'\cos(\phi-\phi') + (z-z')^2)^{-\frac{1}{2}}}{r^2 + r'^2 - 2rr'\cos(\phi-\phi')} \quad (5)$$

where, $n$ is the number of iterations, $R$ is radius of the magnet, $(r, \phi, z)$ are the cylindrical coordinates of the P point and $(r', \phi', z')$ are the source coordinates of the magnet as shown in Fig. 4b.

### 3.3 Coordinate System of the Permanent Magnet

The MFD values calculated in the previous sections were calculating the flux density values at a specified point in space according to the magnet's own local coordinate. However, if the magnet is rotated or translated as in Fig. 5, these MFD values will change.

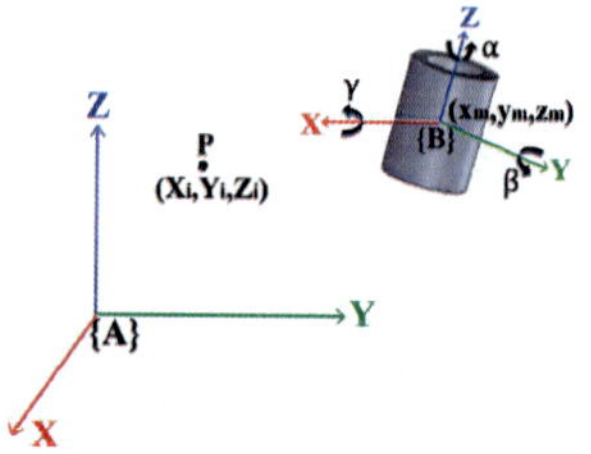


**Fig. 5.** Coordinates and rotation angles of the annular magnet

where, $(\gamma, \beta, \alpha)$ are the rotation angles about x, y and z axes respectively, $(x_m, y_m, z_m)$ are the magnet positions with respect to global {A} coordinate system, $(X_i, Y_i, Z_i)$ are the i-th sensor positions, {A} is the global coordinate system and {B}is the local coordinate system of the magnet. The rotation matrix of system {B} with respect to {A} is calculated by Eq. (6). Since the magnet magnetization vector is in the x-axis direction, rotation is performed in the x, y and z axes, respectively.

$$ {}_{B}^{A}\mathbf{R} = \mathrm{R}_{XYZ}(\gamma, \beta, \alpha) = \begin{bmatrix} c\alpha c\beta & c\alpha.s\beta.s\gamma - s\alpha.c\gamma & c\alpha.s\beta.c\gamma + s\alpha.s\gamma \\ s\alpha.c\beta & s\alpha.s\beta.s\gamma + c\alpha.c\gamma & s\alpha.s\beta.c\gamma - c\alpha.s\gamma \\ -s\beta & c\beta s\gamma & c\beta c\gamma \end{bmatrix} \tag{6} $$

The **P** sensor point which is defined according to the global coordinate system, is reduced to the magnet's own coordinate system (${}^{B}\mathbf{P}$) with the ${}_{B}^{A}\mathbf{R}$ transformation matrix and the MFD is calculated according to the obtained positions as in Eq. (7). As shown in Fig. 4, the magnetization vector of the Biot-Savart model is in the z direction, while the Charge model is in the x direction. In order to eliminate this mismatch, the Biot-Savart model is also rotated 90° in the y direction in addition to ${}_{B}^{A}\mathbf{R}$.

$$ {}^{B}\mathbf{P} = {}_{B}^{A}\mathbf{R}^{-1}.\left[(X_i, Y_i, Z_i) - (x_m, y_m, z_m)\right] \quad \Rightarrow \quad \mathbf{B}_i = {}_{B}^{A}\mathbf{R} f_{annular}\left({}^{B}\mathbf{P}\right) \tag{7} $$

Here, $(x_m, y_m, z_m)$ are the magnet positions, $(X_i, Y_i, Z_i)$ are the i-th sensor positions, $f_{annular}$ is a function that converts cylindrical magnet equations of the Biot-Savart and Charge model at the Eqs. (3) and (5) into annular magnet expressions. The $f_{annular}$ function can be found by the difference of the MFD values calculated for the outer radius and the inner radius using the superposition technique: $f_{annular} = B(x, y, z, r_{outer}) - B(x, y, z, r_{inner})$.

### 3.4 Optimization Algorithms

The MFD equations calculated in the previous section are nonlinear with 6 unknowns $(x, y, z, \gamma, \beta, \alpha)$. For this reason, optimization algorithms have been used to solve these nonlinear equations. With the experience we gained in previous study [17], ABC (Artificial Bee Colony) and LM (Levenberg-Marquardt) algorithms were run as hybrid in simulations. The ABC algorithm is basically a population-based optimization algorithm that adopts the swarm behavior of bees in nature. The job of the ABC algorithm in simulations is to solve the MFD equations, predict 6 unknown parameters of the magnet and present them to the LM algorithm. The LM algorithm provides a more precise position by using the positions taken from ABC as the initial value. The aim of both optimization algorithms is to minimize the error value ($\varepsilon$) seen in Fig. 3. For this, Eq. (8) is used as cost function.

$$\varepsilon = \sum_{i=1}^{N} \left[ (B_{ix} - \hat{B}_{ix})^2 + (B_{iy} - \hat{B}_{iy})^2 + (B_{iz} - \hat{B}_{iz})^2 \right] \tag{8}$$

where $(\hat{B}_{ix}, \hat{B}_{iy}, \hat{B}_{iz})$ represents the estimated MFD values, $(B_{ix}, B_{iy}, B_{iz})$ represents the known actual MFD values measured by the sensor. As a result, when the positioning process is performed, the positioning performance in the simulations is calculated with Eq. (9).

$$\begin{aligned} E_P &= \sqrt{(x - \hat{x})^2 + (y - \hat{y})^2 + (z - \hat{z})^2} \\ E_\beta &= \left|\beta - \hat{\beta}\right|, E_\alpha = \left|\alpha - \hat{\alpha}\right| \end{aligned} \tag{9}$$

$(E_P, E_\beta, E_\alpha)$ are localization error, absolute $\beta$ angle error, and absolute $\alpha$ angle error, respectively. As mentioned before, the $\gamma$ angle cannot be calculated due to the singularity problem, and therefore the degree of freedom of the system is 5D.

## 4 Results

As mentioned in the previous sections, simulations were made within the framework of the diagram given in Fig. 3. The performance of the two different MFD models was obtained in the same code with the Matlab software under the same conditions. In the simulations, firstly, the sensitivity of the MFD values calculated by the Biot-Savart and Charge model was examined. As described in the previous sections, the integral equations of both models have been discretized and the number of solutions $n$ has become an important factor. Figure 6a shows the convergence graph of both methods depending on $n$. When the graph is examined, it is seen that the Biot-Savart Model converges more slowly than the Charge model.

When $n = 10^6$ is selected, both models calculate the MFD values very close to each other ($10^{-15}T$ error). Therefore, the actual MFD values are calculated using this value of $n$. However, $n = 20$ value was used while calculating the MFD value during the localization stage in the simulations. In short, $n = 10^6$ was set when calculating the actual MFD values, and $n = 20$ was used when calculating the estimated MFD values.

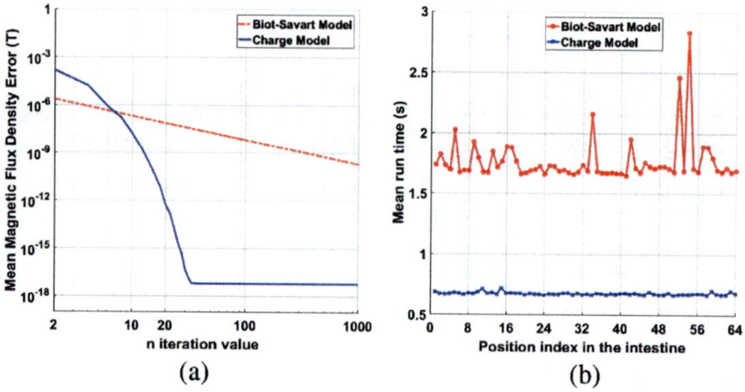


**Fig. 6.** (a) Convergence graph depending on $n$, (b) Run time comparison for $n = 20$

In simulations, the population number of the ABC algorithm was set to 30 and the number of iterations to 45. Under these conditions, the localization and angle errors obtained in 64 different positions when analyzed for both models are given in Fig. 7 and Fig. 8, respectively. When calculating these errors Eq. (9) is used. In order to eliminate the randomness while obtaining these values, the analysis was run at each location 100 times and averaged. In addition, the average calculation time of each point is in Fig. 6b.

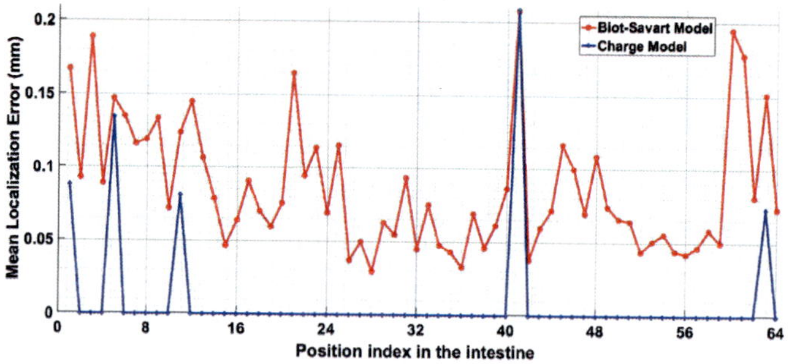


**Fig. 7.** The mean localization error comparison of the two models

The data calculated for both scenarios were also analyzed in terms of maximum, average and standard deviation error and Table 1 was drawn.

When the given figures and Table 1 are examined, it is seen that the Charge model is better than Biot-Savart in all aspects. 89.5% and 87.4% difference are observed in terms of localization error and angle errors, respectively. Also, under equal simulation conditions, the Charge model is 61.6% faster than Biot-Savart. For this reason, the Charge model has many advantages over the Biot-Savart model, especially in real-time robotic capsule endoscopy applications.

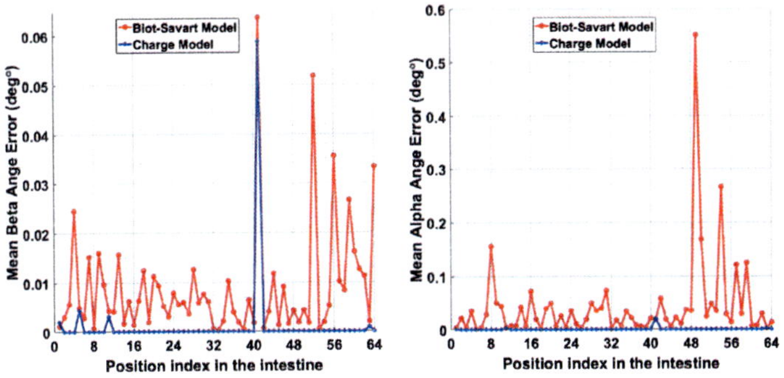


**Fig. 8.** The mean angle estimation errors of the magnet

**Table 1.** Performance comparison of Biot-Savart and charge models

| | | Biot-Savart | Charge model | % Difference |
|---|---|---|---|---|
| Loc. error (mm) | Max | 0,2082 | 0,2078 | 0,2 |
| | Mean | 0,0870 | 0,0091 | 89,5 |
| | SD | 0,0438 | 0,0347 | 20,8 |
| Beta angle error ($\beta^o$) | Max | 0,0635 | 0,0588 | 7,4 |
| | Mean | 0,0087 | 0,0011 | 87,4 |
| | SD | 0,0116 | 0,0074 | 36,2 |
| Alpha angle error ($\alpha^o$) | Max | 0,5496 | 0,0200 | 96,4 |
| | Mean | 0,0408 | 0,0004 | 99,0 |
| | SD | 0,0790 | 0,0025 | 96,8 |
| Run time per point (ms) | Max | 2830,6 | 723,3 | 74,4 |
| | Mean | 1768,2 | 679,1 | 61,6 |

The performance of both models is also studied in noisy situations. For this, the noise was added to the actual MFD values depending on the noise level. In order to provide real system conditions, independent noises are added to the sensor calculations at each point with Eq. (10).

$$B = B + B \times randn[-1, 1] \times {}^{nl}/_{100} \tag{10}$$

The positioning error investigated for different noise levels is given in Fig. 9. As the error level increases, it is seen that both models converge and there is no difference between them after a point. As a result, it was concluded that the Charge model gave more accurate results than the Biot-Savart model in every way and gave similar results with the Biot-Savart in noisy conditions. But in terms of speed, the Charge model is always faster than Biot-Savart.

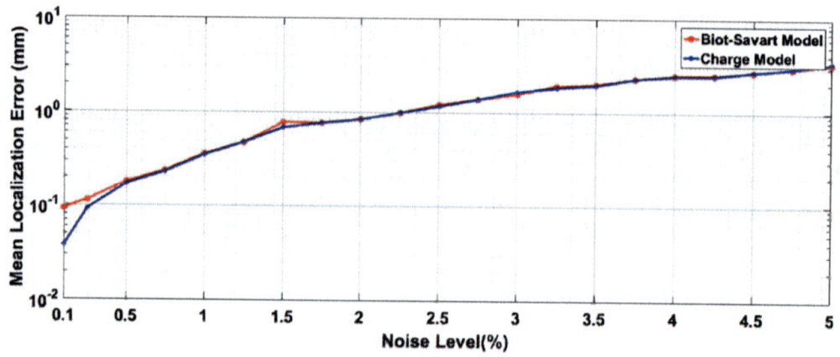


**Fig. 9.** Mean localization error at the noisy simulation condition.

## References


1. Iddan, G., Meron, G., et al.: Wireless capsule endoscopy. Nature **405**(6785), 417 (2000)
2. Silva, J., Histace, A., et al.: Toward embedded detection of polyps in WCE images for early diagnosis of colorectal cancer. Int. J. Comput. Assist. Radiol. Surg. **9**(2), 283–293 (2014)
3. Riccioni, M.E., Urgesi, R., et al.: Colon capsule endoscopy: advantages, limitations and expectations. Which novelties? World J. Gastrointest. Endosc. **4**(4), 99 (2012)
4. https://www.medtronic.com/covidien/en-us/products/capsule-endoscopy/pillcam-sb-3-system.html. Accessed 22 May 2021
5. Kuth, R., Reinschke, J., et al.: Method for determining the position and orientation of an endoscopy capsule guided through an examination object by using a navigating magnetic field generated by means of a navigation device. Google Patents (2007)
6. Wilding, I., Hirst, P., et al.: Development of a new engineering-based capsule for human drug absorption studies. Pharm. Sci. Technol. Today **3**(11), 385–392 (2000)
7. Kanaan, M., Akay, R., et al.: In-body ranging for ultra-wide band wireless capsule endoscopy using neural networks based on particle swarm optimization. SUJEST **6**(2), 207–217 (2018)
8. Suveren, M.: Ultra Wide Band (UWB) Wireless Systems in Implant Body Area Networks and Modelling of the Distance Measurement Error, MSc Master thesis, Erciyes University, Kayseri (2015)
9. Aziz, S.M., Grcic, M., Vaithianathan, T.: A real-time tracking system for an endoscopic capsule using multiple magnetic sensors. In: Mukhopadhyay, S.C., Gupta, G.S. (eds.) Smart Sensors and Sensing Technology. LNEE, vol. 20, pp. 201–218 (2008). Springer, Heidelberg. https://doi.org/10.1007/978-3-540-79590-2_14
10. Than, T.D., Alici, G., et al.: A review of localization systems for robotic endoscopic capsules. IEEE Trans. Biomed. Eng. **59**(9), 2387–2399 (2012)
11. Nagaoka, T., Uchiyama, A.: Development of a small wireless position sensor for medical capsule devices. In: The 26th IEEE EMBS, pp. 2137–2140. IEEE (2004)
12. Guo, X., Yan, G., et al.: A novel method of three-dimensional localization based on a neural network algorithm. J. Med. Eng. Technol. **33**(3), 192–198 (2009)
13. Ackerman, M.J.: The visible human project. Proc. IEEE **86**(3), 504–511 (1998)
14. Alliance. https://allianceorg.com/pdfs/Magnet_Tutorial_v85_1.pdf. Accessed 21 May 2021
15. Song, S., Li, B., et al.: 6-D magnetic localization and orientation method for an annular magnet based on a closed-form analytical model. IEEE Trans. Magn. **50**(9), 1–11 (2014)
16. Furlani, E.P.: Permanent Magnet and Electromechanical Devices: Materials, Analysis, and Applications. Academic Press (2001)
17. Suveren, M., Kanaan, M.: 5D magnetic localization for wireless capsule endoscopy using the Levenberg-Marquardt method and artificial bee colony algorithm. In: IEEE 30th PIMRC (2019)

# Digital Twin of a Servo Driver of a Servo Motor as a First Step Towards a Digital Twin of a Robot Mechanism


Sefa Furkan Küçükoğlu[1(✉)], Giuseppe Carbone[2], and Mehmet İsmet Can Dede[1]

[1] Izmir Institute of Technology, Urla, Izmir, Turkey
{sefakucukoglu,candede}@iyte.edu.tr
[2] University of Calabria, Cosenza, Italy
giuseppe.carbone@unical.it



**Abstract.** Digital Twin (DT) offers us to acquire actual system's critical information and hence, it may be possible to develop and produce more suitable systems in terms of low energy consumption and effectiveness. In this way, responsible consumption and production systems can be designed and the system's parameters can be tuned via DT. In this study, the model of a servomotor system that is used for industrial purposes is experimentally obtained. This study consists of two steps. In the first step, within the embedded control of the system, position and velocity control loops are deactivated. Then through the servo driver, currents with sinusoidal waveforms at various frequencies are applied to the servomotor. The resultant angular velocity of the motor is monitored and recorded. The amplitude of the current is kept constant during this study. The frequency of the current, however, is increased logarithmically. By using these data, a first-order transfer function (TF) is identified for the motor model. In the second step, all control loops are activated. Consequently, the total servomotor system could be represented in a digital environment. Furthermore, the static friction issue is overcome by using a Coulomb friction model with stiction effect. Finally, several experiments are conducted and then results are compared with the digital model of the servomotor system. The results clearly show that digital model can fairly represent the physical system.



**Keywords:** Responsible consumption and production · Transfer function · Servo system · Static friction · Coulomb friction model · Stiction effect


## 1 Introduction

Over the last few decades, there has been remarkable attention to automation systems. More recently, Industry 4.0 has changed our daily life and so it makes us develop various point of view about technology. With the aid of this, a good deal of product has been produced in a short time, with low-cost and low manpower. Consequently, there have been numerous researches on systems' control, flexibility, efficiency etc.



G. Quaglia et al. (Eds.): I4SDG 2021, MMS 108, pp. 345–352, 2022.
https://doi.org/10.1007/978-3-030-87383-7_38

Since servo motor systems yield satisfactory results in position and velocity control applications, they are one of the highly preferred actuation systems in industrial applications. For example, in [1] a macro – micro mechanism for laser cutting process has been designed with servo motor system so that the end – effector of the mechanism can follow a desired trajectory with wondrous precision under high acceleration. An ultra-precision positioning technique has been developed with the combination of a global stage and micro stage in [2] in order to achieve sub-micrometer accuracy.

When it comes to control and model of the servo motors, modeling, and identification of the parameters of a servo motor have been investigated in [3–5]. A Genetic Algorithm (GA) method is used in order to improve the transient response of the system and define the optimal parameters of the PID controller for the desired control system design specifications in [6]. Other than the traditional control approaches, there are studies with fuzzy neural network controller [7], a cascade controller with feedforward robust adaptive fuzzy compensator [8], and a robust internal model control (IMC) based on sliding mode control (SMC) [9] in literature, in order to compensate the modeling uncertainties. When we look at the current studies related to control and model of the servo motors, it is seen that several methods are implemented. Generally, there is a tendency to identify the parameters of the servo motors by conducting simulations and/or experiments.

As a first step to create a DT of a robotic mechanism presented in [1], the servo motor system of this mechanism is considered in this work. Two servo motors are used in this planar parallel mechanism. The aim of this machine is to shorten the task completion by reaching 5g acceleration while it preserves the positioning accuracy at less than ±0,100 mm/m, and repeatability at less than ±0,050 mm. To achieve this aim, trajectory planning algorithms are improved with the inclusion of stiffness model of the mechanism. Nevertheless, another critical factor in precise control of the mechanism under high accelerations is the servomotor control, which is main focus of this paper.

To study the control of a servo motor which is coupled with a gearbox, the initial step is to obtain its model. Firstly, frequency domain studies are carried out by generating the Bode plot of the servo motor system that is subjected to current control loop via experiments. Consequently, a transfer function is fitted to the obtained Bode plot. Thus, the motor model is obtained via experimental identification techniques. Coulomb friction model with stiction effect is used as an additional part to the obtained linear model of the motor [10]. The cascade control parameters were previously tunned with the aid of servo driver's auto-tuning capability. Then, the exact replica of this controller with its control parameters are included in the simulation model as the digital twin of the servo driver. This part is accomplished without any experimental identification procedure. Consequently, the controller and its parameters in both servo driver and the simulation model are exactly same. Thereby, the system including the servo motor and the embedded controller in the servo driver are represented in the digital environment. The sampling time of the system is kept constant as 2 ms throughout the study. The final outcome of this study is considered to be satisfactory, albeit there are minuscule differences between digital and experimental system. This error is investigated by quantifying it with Mean Absolute Error (MAE) and R-Squared ($R^2$). As a result of this work, the first step of designing a DT of the laser - cutting mechanism is realized. Kinematic, dynamic and stiffness models of the laser - cutting mechanism have already been developed. This mechanism consists of two servo motors, and servo system model obtained will be used along with the kinematic, dynamic and stiffness models so that the whole system can be represented in the digital environment.

## 2 System Description and Experimental Procedure

System operation is summarized in Fig. 1. A Kollmorgen servo motor is used in this study [11]. Servo driver is connected to a computer via ethernet connection, and the parameters of the controller embedded in the servo driver is changed via Kollmorgen AKD Graphical User Interface (AKD GUI). With the help of this GUI, controller mode and enable operation can be selected. On the other hand, Quanser Q8 Data Acquisition Card (DAQ) is used in order to supply the required analog input and evaluate the related output. Then these signals are monitored and recorded in another computer via MATLAB/Simulink Software.

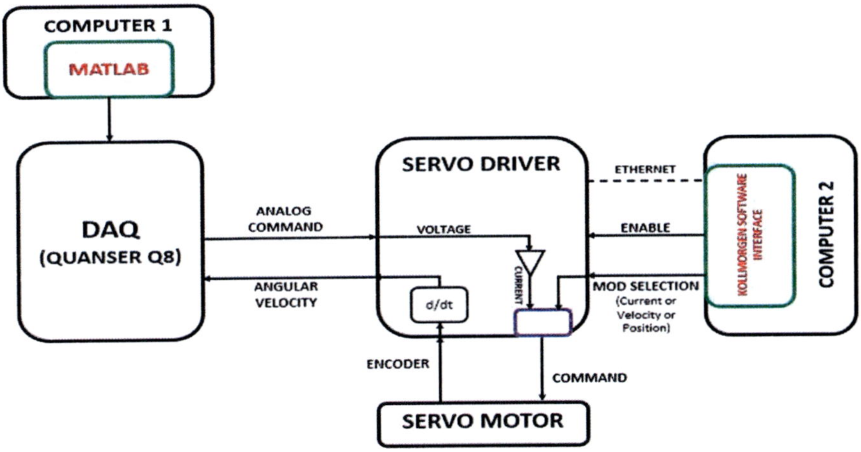


**Fig. 1.** System operation

### 2.1 Obtaining the Transfer Function of the System with Current Controller

The whole system can be represented as a block diagram in Fig. 2. This figure is taken from the AKD GUI. The system consists of three cascade control loops, position, velocity, and current control loops, from left to right, respectively. As a first step, while the position and velocity loops are deactivated, the current loop is activated.

During the experiments, the amplitude of the sinusoidal current input is kept constant (0.2 A peak to peak). However, when this sinusoidal signal with this amplitude is suddenly applied to the motor, various problems, such as overheating problems can occur, and this might result in nonlinear behavior of the system. Therefore, the input signal's amplitude (command) is risen to the designated value with a ramp function.

The frequency of the sinusoidal current input is applied within the interval 0.2 rad/s–60 rad/s. During the tests, for each designated frequency of the sinusoidal input, the

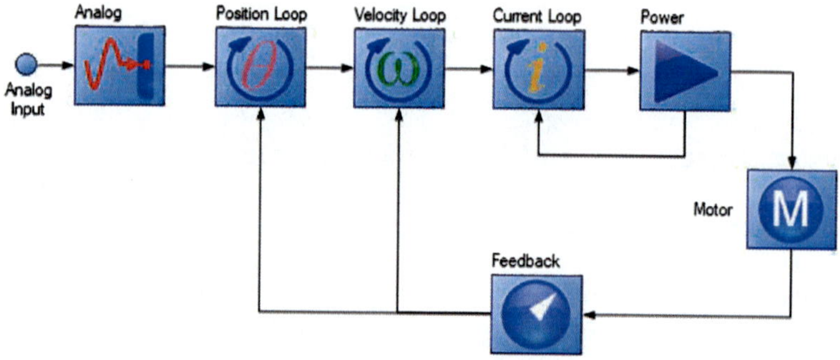


**Fig. 2.** System block diagram

corresponding angular velocity is recorded as an output of the servo motor system. The recording of the angular velocity is performed when the system's output reaches a steady peak value.

The obtained data are plotted in Fig. 3a. The frequency axis of this plot is in logarithmic scale. Red asterisks denote the measured data for the designated frequency. In this graph, the gain (magnitude ratio) is calculated by dividing the peak of the measured output velocity by the peak of the measured input current. The initial gain (75.8959 dB) and the gain value where the line intersects the cut-off frequency are important to find the TF of the investigated system. These values are encircled and emphasized in Fig. 3a.

The expectation is to receive a first-order delay transfer function due to the relation between the input current and the output velocity. Since the input, current is directly related to the input torque by the help of the torque constant of the motor, this relationship is an obvious one by neglecting the faster dynamics of the system such as the servo driver's circuitry. The necessary equations to determine the TF of the system are presented in Eqs. (1), (2), (3), and (4).

$$\mathrm{T(s)} = \frac{\mathrm{a}}{\mathrm{s+b}} = \frac{\left(\frac{\mathrm{a}}{\mathrm{b}}\right)}{\left(\frac{\mathrm{s}}{\mathrm{b}}+1\right)} \tag{1}$$

These two parameters, a, b, are to be determined from the measured data's response. In Eq. (1), the "b" value represents the cut–off frequency of the first-order TF, i.e., b = 1.6 rad/s. The remaining part of the TF is calculated as follows by considering the initial gain value.

$$75.8959 = 20\log_{10}\left(\frac{a}{b}\right) = 20\log_{10}\left(\frac{a}{1.6}\right) \tag{2}$$

$$3.7948 = \log_{10}\left(\frac{\mathrm{a}}{1.6}\right) \Rightarrow \mathrm{a} = \left(10^{3.7948}\right)(1.6) = 9975 \tag{3}$$

Therefore, TF of the current controlled servo motor system is identified to be,

$$\mathrm{T(s)} = \frac{9975}{\mathrm{s}+1.6} \tag{4}$$

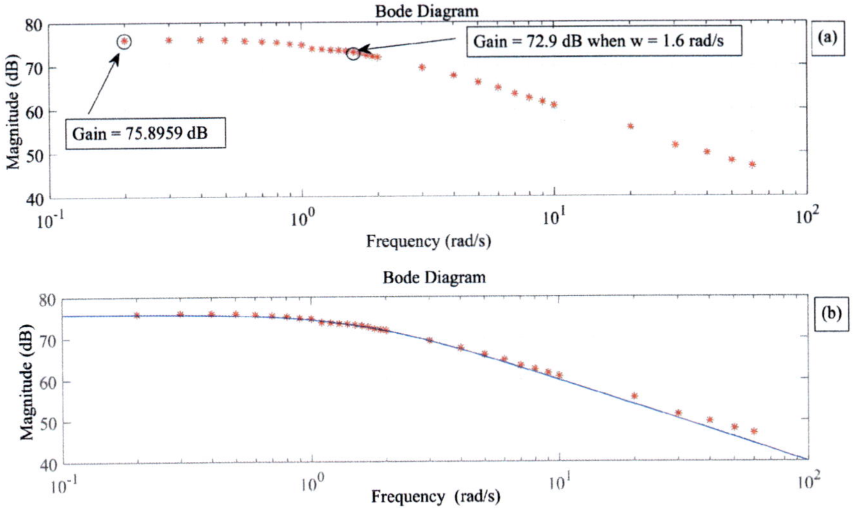


**Fig. 3.** Experimental results for identifying the current-controlled servo motor system (a) Measured Data (red) (b) Bode Diagram of the fitted TF (blue) and measured data (red).

The frequency response of this identified TF and measured data are illustrated together in Fig. 3b. It is seen that the response of the TF is to a great extent in harmony with measured data.

In addition to multiple experiments, random signals are generated and supplied to the inputs of both the identified TF and the actual servo motor system, simultaneously. In doing so, the behavior of the identified TF and the response of the servo motor system are

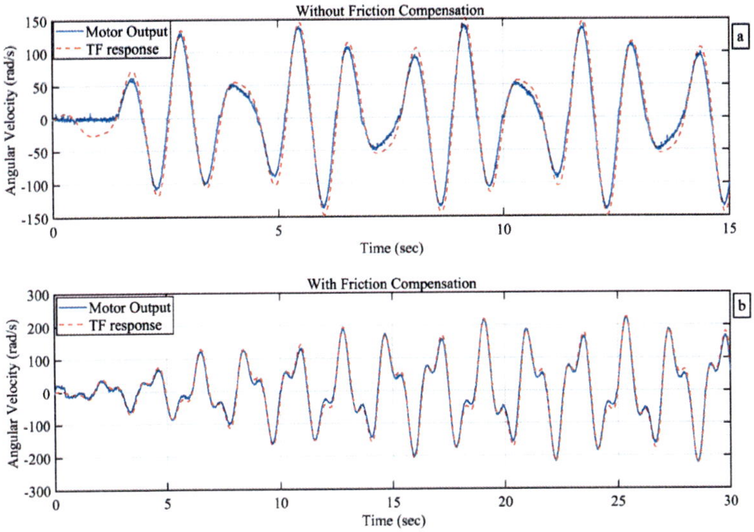


**Fig. 4.** TF (red) and servo motor's measured (blue) outputs when (a) there is no friction compensation. (b) there is friction compensation.

observed and evaluated at various frequencies and amplitudes. An example is illustrated in Fig. 4a and Fig. 4b.

In the first few seconds, because of the static friction, the motor could not initiate its motion. Since the identified TF is a linear model of the system, it does not contain the nonlinearity feature that the static friction brings to the actual system. Consequently, a coulomb friction model with stiction effect is implemented on the system's model. The result of the system with the static friction model is given in Fig. 4b.

## 3 Position Control of the Servo Motor System

As a next step for generating the DT of the servo motor system, the embedded controller of the servo driver is studied. Initially, the position and velocity control loop parameters are tuned and adjusted automatically by the servo driver's software. The same control loops with the same parameters that is embedded in the servo driver are used in the simulation model of the servo motor system. Consequently, the DT of the total servo driver with all the control loops is generated. Proportional (P) and Proportional and Integral (PI) controllers are implemented for position and velocity loops, respectively. To validate the accuracy of the DT, same random inputs are supplied to both the DT and the actual servo motor system. The results of an experimental study against the results obtained from the DT of the system are given in Fig. 5.

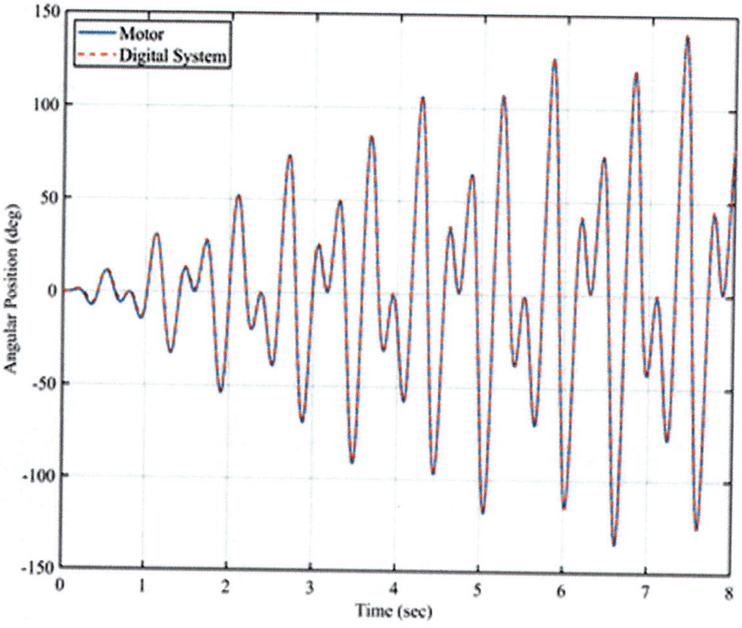


**Fig. 5.** Response of both the actual system and its DT for a random input

**Table 1.** Performance of the obtained system.

| Input signal | MAE | R2 |
|---|---|---|
| 4sin (t) | 0.3002° | 1 |
| 7sin (5t) | 0.3624° | 1 |
| 7sin (t) | 0.395° | 1 |
| 7sin (3t) | 0.3283° | 1 |
| 3sin (t) | 0.2725° | 1 |
| 3sin (10t) | 0.2441° | 0.9998 |
| [3sin (6t + 2.2864) + 3sin (10t)] | 0.2707° | 0.9998 |

## 4 Conclusion and Discussion

As it is observed from Fig. 5, the DT of the actual servo motor system produces similar results with the actual system. Multiple experiments with different amplitudes and frequencies are performed which could not be presented in this paper due to space limitations. The performance of the DT is evaluated quantitatively with two performance metrics, Mean Absolute Error (MAE) and R-Squared ($R^2$). The evaluations based on these metrics are presented in Table 1. The input signal presented in this Table is in radians and t variable is in seconds.

As a result of this study, the DT of the servo motor system of a mechanism is generated. After implementing the kinematics, dynamics and stiffness model of the mechanism, the controller parameters of the system can be tuned without the need of using the actual system.

**Acknowledgments.** This study is founded by The Scientific and Technological Research Council of Turkey (grant number 116M272). Also, we would like to thank Dr. Emre Uzunoğlu, İbrahimcan Görgülü and Assoc. Prof. Dr. Gökhan Kiper.

## References

1. Dede, M.İ.C., Kiper, G., Uzunoğlu, E.: A macro-micro mechanism design for laser cutting process. In: The 17th International Conference on Machine Design and Production, UMTIK, Bursa (2016)
2. Pahk, H.J., Lee, D.S., Park, J.H.: Ultra precision positioning system for servo motor-piezo actuator using the servo loop and digital filter implementation. Int. J. Mach. Tools Manuf. **41**, 51–63 (2001)
3. Garrido, R., Miranda, R.: DC servomechanism parameter identification: a closed loop input error approach. ISA Trans. **51**, 42–49 (2012)
4. Ishak, N., Abdullah, N.I., Rahiman, M.H.F., Samad, A.M., Adnan, R.: Model identification and controller design for servomotor. In: 2010 6th International Colloquium on Signal Processing & Its Applications, pp. 1–4 (2010). https://doi.org/10.1109/CSPA.2010.5545294

5. Wu, Y., Wang, Z., Li, Y., Chen, W., Du, R., Chen, Q.: Characteristic modeling and control of servo systems with backlash and friction. Math. Prob. Eng. (2014). Article ID: 328450
6. Bindu, R., Namboothiripad, M.K.: Tuning of PID controller for DC servo motor using genetic algorithm. Int. J. Emerg. Technol. Adv. Eng. **2**(3) (2012)
7. Lin, F.J., Wai, R.J., Chen, H.P.: A PM synchronous servo motor drive with an on-line trained fuzzy neural network controller. IEEE Trans. Energy Convers. **13**(4), 319–325 (1998). https://doi.org/10.1109/60.736317
8. Liu, Y., Wang, Z.Z., Wang, Y.F., Wang, D.H., Xu, J.F.: Cascade tracking control of servo motor with robust adaptive fuzzy compensation. Inf. Sci. **569**, 450–468 (2021)
9. Li, P., Zhu, G.: Robust internal model control of servo motor based on sliding mode control approach. ISA Trans. **93**, 199–208 (2019)
10. Niranjan, P., Karinka, S., Sairam, K.V.S.S.S.S., Upadhya, A., Shetty, S.: Friction modeling in servo machines: a review. Int. J. Dyn. Control **6**, 893–906 (2018)
11. AKM Series Servo Motor Specifications. https://www.kollmorgen.com/en-us/products/motors/servo/akm-series/akm-series-ac-synchronous-motors/ac-synchronous-servo-motors/

# Experimental Evaluation of the Behaviour of Nitinol Bars for a Compliant Parallel Trunk Rehabilitation Device

F. J. Campa(✉), D. Diaz-Caneja, and O. Altuzarra

University of the Basque Country UPV/EHU, Escuela de Ingeniería de Bilbao, Plaza Ingeniero Torres Quevedo 1, 48013 Bilbao, Spain
fran.campa@ehu.eus

**Abstract.** The objective of the present work is to improve a device for rehabilitating the trunk balance and motion of patients who have suffered a stroke. The device is based on a parallel continuum manipulator coupled to a standing frame, where a wearable jacket is actuated by four flexible limbs connected to rotary motors. It has been designed to perform basic trunk exercises achieving translation coupled to rotation around the anteroposterior, medio-lateral and longitudinal axis. After developing and testing a first version, it was decided that the design of the flexible limbs had be improved to allow a better behaviour under compressive forces and high flexural deformations, avoiding the permanent deformations that the Nylon 66 bars suffered. Here, the tests carried out with Nitinol bars are shown. Experimental testing has been done and the results have been compared with a model of the bars deflection. The results show that the Nitinol bars are a good solution for the device, although a good characterization of its nonlinear behaviour must be done.

**Keywords:** SDG3 · Rehabilitation · Stroke · Parallel continuum manipulator · Nitinol

## 1 Introduction

The purpose of the present work project is to analyze the viability of using Nitinol slender bars for the flexible limbs of a mechatronic device for the rehabilitation of the trunk motion in patients who have suffered a stroke, losing the proprioception in half the body. The prototype is based on a jacket that the patient wears and is actuated by flexible limbs to reproduce the exercises proposed by physiotherapists, also measuring data that can be useful for the medical staff to monitor the rehabilitation process. The device is a 'parallel continuum mechanism' [1], using very flexible elements instead of kinematic joints to create the movement. The flexibility of the slender elements on these mechanisms brings several advantages: they are easier to wear, they can absorb external disturbances, the flexibility of the device is psychologically less intimidating for the patient, naturally absorbing possible perturbations or spasms, and, they naturally act as a lowpass filter


G. Quaglia et al. (Eds.): I4SDG 2021, MMS 108, pp. 353–361, 2022.
https://doi.org/10.1007/978-3-030-87383-7_39

of the motors motion, performing a more progressive and physiological motion without complex control. On the other hand, these mechanisms are relatively recent, with complex kinematics and dynamics, so not only there is room for improvement in their design and control but also in their manufacturing techniques and selection of optimum materials.

To justify why such development can have a great importance in our society, it must be said that, at present, aging in population has increased age-related disorders as central nervous system diseases, affecting the quality of life of adults [2, 3]. For example, in some countries, cerebral vascular accident is the first cause of mortality in women, and the second in men, only surpassed by heart attack [4]. Although nowadays the mortality rate stands at 11% [5], it has increased the number of patients who survive but then, a rehabilitation process is required. Unfortunately, as a stroke can cause very disabilities depending on the area damaged, the rehabilitation process must be custom designed for each patient necessities. In early stages, the rehabilitation process focuses on the motor recovery of the torso and the upper and lower limbs, doing passive mobilizations, re-education of balance, or active assisted movements and muscular strengthening of the affected area, by means of anterior-posterior, rotational, lateral or manipulation movements [6]. To avoid falling or other accidents, assistive devices as standing frames are generally used for safe positioning of the patient.

On the other hand, the increase in patients with cerebral vascular accident has meant an increase in the number of patients per physiotherapist. As a result, this has led to a decrease in the patient's hospital re-habilitation time [7]. In addition, there is still a lack of means to measure objectively the patient evolution and determine the duration of their rehabilitation. In the recent years, the use of robotics devices has been promoted in the search for more effective methods for post stroke rehabilitation, as they allow mitigating the large amount of labour required, in addition to the possibility of measuring the performance of the patient, providing very useful information about the recovery state. Many of the current robots designed for physical therapy are focused on rehabilitation of upper and lower limbs. The MIT-MANUS, the ARM-GUIDE, or the MIME system, are some of the more successful devices [7]. Regarding the lower limb rehabilitation, the Lokomat [8] or the Gait Trainer [9] are two examples of these research areas. However, the recovery of mobility in the torso is another important neuromotor task that involves a significant physical effort for the physiotherapist and has an impact on the stability, mobility and quality of life of the patient. Hence, it can be said that trunk rehabilitation is currently a field of interest in the area of robotic rehabilitation devices.

## 2 Prototype Design

### 2.1 Prototype Concept

The use of the medical device prototype developed takes place in one of the first steps in the rehabilitation, that is, to recover the ability of balancing the trunk, as a previous step to train the balance in standing position and the gait. This stage takes place in a standing frame, which fixes the legs and hip of the patient, so it is possible to perform specific of trunk motion and balance training with no risk.

The rehabilitation device is a parallel continuum manipulator, as seen in Fig. 1 [10]. It consists on a jacket, which is the mobile platform of the parallel mechanism, and four flexible limbs linked in one end to the jacket by means of spherical joints (S-joints, $B_i$) and in the other end to the rotary motors that act as a revolute joint (R-joints, $A_i$). The patient keeps a standing pose up to the hip due to the standing frame that fixes the patients knees and hip. The motors are fixed to the standing frame, and rotate the limbs, whose deflection generate translational and rotational motion in the jacket. The contact surface between patient and jacket is located on his back. The patient wears the jacket using adjustable straps as in a backpack, keeping a proper contact between patient and jacket, to transmit the motion properly.

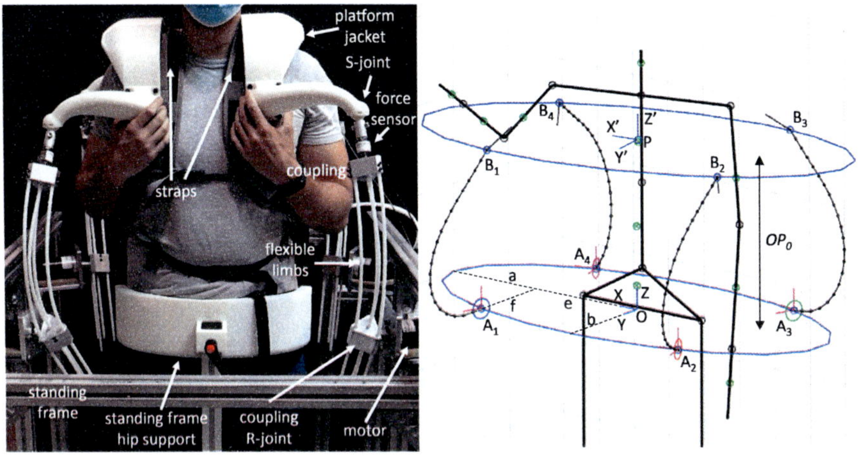


**Fig. 1.** Elements of the trunk motion rehabilitation device and kinematic diagram.

The transmission of the motors rotary motion to the jacket or platform is achieved thanks to the deflection of the limbs. The flexible limbs were designed to reach a balance between flexibility and rigidity. On one hand, the objective is to have a workspace big enough to meet the kinematic requirements of a person with minimal capabilities. On the other hand, it must be also capable of applying enough force to the patient. In the preliminary design, they are formed by several elements in series: an aluminum coupling to the motor axis, a set of four slender bars, another aluminum coupling where the bars are fixed, an uniaxial force sensor and the final aluminum coupling to the S-joint, see Fig. 1. The four HBM U9C uniaxial force sensors were introduced to be able to check the normal tensile or compressive forces in the limbs and use that data for the hybrid force and position control of the device, which adapts the device mechatronics stiffness to the patient capabilities.

The central bar is fixed in both ends, while the three outer bars can slide through the R-joint coupling, so they should remain parallel to the central one during the deflection. In the preliminary design, each bar was made of Nylon 6.6 with 6 mm of diameter and the length of the bars with no deflection was of 300 mm. The outer bars were placed in a circumference of 50 mm of diameter with the center in the central bar.

## 2.2 Problems Detected in the Flexible Limbs

During the experimental testing, the prototype demonstrated the capability of pushing and pulling the person who wears it achieving rotations up to 10° around the longitudinal and anteroposterior axis, and up to 15° around the medio-lateral axis. However, there were to problems with the Nylon 6.6 bars. The first is the lack of parallelism between the flexible bars. This is due to the compressive forces that, in some positions, some of the outer bars suffer. As a result, they slide trough the R-joint aluminum coupling losing the parallelism, even colliding between them, and thus not working as expected. This could be solved fixing all the bars in both ends, but in that case no parallelism could be expected, so collisions would be expected, apart from being unesthetic.

The second problem with the bars was the appearance of plastic deformations after several cycles of testing, which would mean that the bars should be substituted frequently. This is not a problem as the Nylon 6.6 can be recycled, but it is not the ideal solution. To solve this second problem, as an alternative to Nylon 6.6, in the current work the Nitinol alloy is going to be tested. It is a superelastic alloy that it is very interesting for the application here presented as it can strain up to 8% without permanent deformation and has a long fatigue life. To solve the first problem, as Nitinol has a higher strength, only one bar of Nitinol as a backbone is going to be tested. That would simplify the design avoiding having to use parallel bars to achieve limbs with the required stiffness. Plus, Nylon 6.6 is not biodegradable, so Nitinol could be a more environmentally friendly alternative.

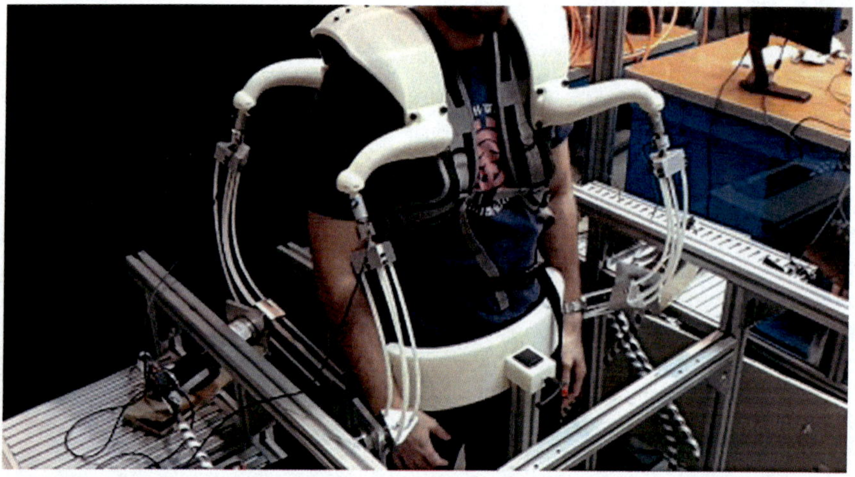

**Fig. 2.** Lack of paralelism in the flexible limbs during a longitudinal (vertical axis) rotation test.

# 3 Experimental Methodology for Testing the Nitinol Bars

## 3.1 Test Bench

A test bench has been developed to test the new Nitinol bars of 3 mm of diameter and 500 mm length, which is the same as in the prototype. The Nitinol or Nickel-Titanium alloy used is the Nitinol Grade 1 alloy from the manufacturer Fort Wayne. Its mechanical behaviour is characterized by a hysteretic stress-strain curve, where two plateaus are present. In the batch received, the upper plateau stress appears at 516 MPa and the lower plateau stress at 169 MPa. The linear behaviour up to the upper plateau is characterized for a Young modulus of 63 GPa.

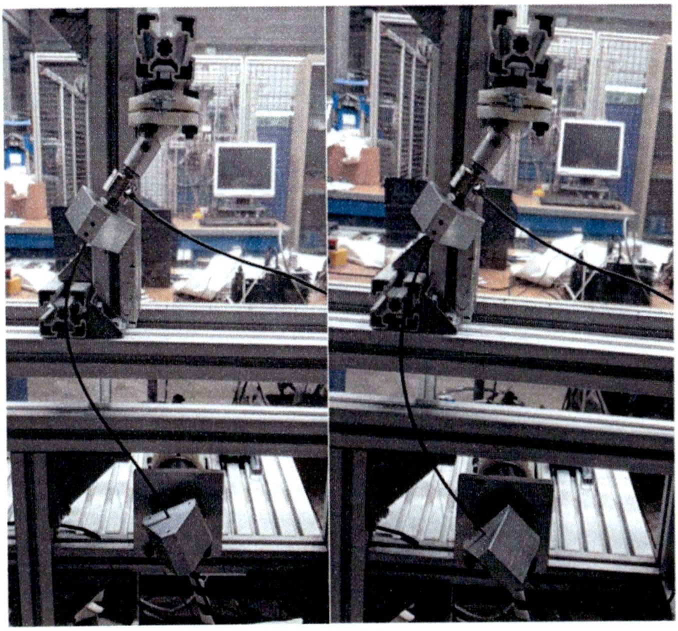

**Fig. 3.** Two captions of the Nitinol bar deflection during the tests: Left) Second deflection mode. Right) First deflection mode.

The test bench has been designed to reproduce the kinematic chain of the limbs of the rehabilitation device but fixing the S-joints. In the test bench the actuator is a Maxon RE 50 DC motor with an encoder HEDL 5540 and GP52B reductor of 81:1, coupled to the Nitinol bar, which is coupled in the other end to a spherical joint, see Fig. 3. The motor axis and the spherical joint are aligned vertically and separated by a distance of 390 mm. The total length of the bar from the R-joint to the S-joint is 439.3 mm, that is distributed as follows: 22.8 mm of the R-joint, 300 mm of the bar and 116.5 mm for the ensemble joint- HBM U9C uniaxial force sensor-spherical joint. Instead of using 4 parallel bars as in the case of Nylon 6.6, only one bar will be tested due to the higher stiffness of Nitinol.

The purpose of the experiment is to rotate the motor inducing a deflection on the bar, while measuring the motor torque and position and the axial force in the sensors. The signals have been sampled with a frequency of 500 Hz and filtered by means of a low pass first-order filter with a cut-off frequency of 20 Hz. Those measurements have been compared with a quasistatic model of the deformation of the bar which takes into account the large non-linear deformations [10]. Also, the capability of the bar to recover the initial form with no permanent deformations after several cycles will be visually evaluated. The range of the motion goes from 5° to 30°, which is the angle of the bar at the R-joint with respect to the vertical. To avoid the influence of the dynamics on the values measured, a smooth motion has been programmed with a spline profile of order 7 with a maximum angular velocity of 0.24 rpm, see Fig. 4.

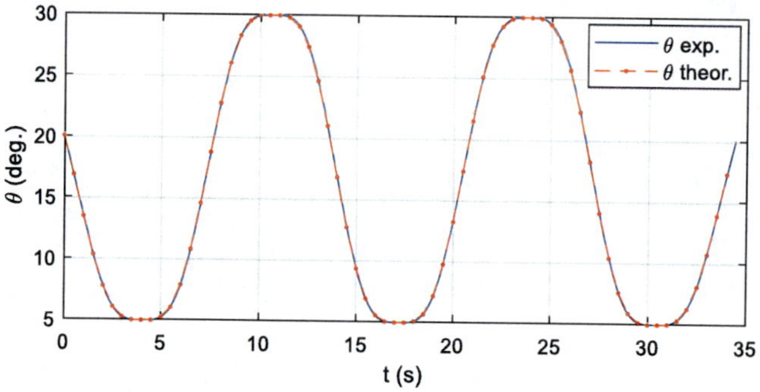


**Fig. 4.** Motor motion profile in the test: programmed (theor.) vs. measured (exp.).

The position control algorithm for the motor is a cascaded PID controller where the input is the initial and final angular positions and the velocity. An EPOS2 50/5 controller has been used to control current and velocity loop, whereas the position loop has been implemented in Labview RT using a CompactRio system. The cycle time for the current loop is 0.1 ms, for the velocity loop is 1 ms and for the position loop is 2 ms. In addition, the motor has been characterized to separate inertial and frictional effects on the motor torque from the torque required to deform the bars.

## 4 Results

In Fig. 5 it is shown how the motor torque and the force measured in the sensor change during the motion. As a reference, the expected values using the current model which assumes a linear behavior of the material is shown. As it can be seen, there is a sudden step in both values in the maximum and minimum positions of the cycle, when the inversion of the motion happens. Such steps can occur due to Coulomb friction in the motor, backlash in the spherical joint, or due to the deflection of the structure of the test bench. However, all these effects were measured and considered negligible or corrected, as in the case of the Coulomb friction.

Hence, it is clear that the nonlinear behavior is due to the Nitinol hysteretic characteristics. The linear model was then adapted to calculate the stress along the bar, revealing that in all the positions of the bar from 5° to 30° there are several sections with a normal stress higher than the 516 MPa of the upper plateau, so the hysteretic behaviour jumps in.

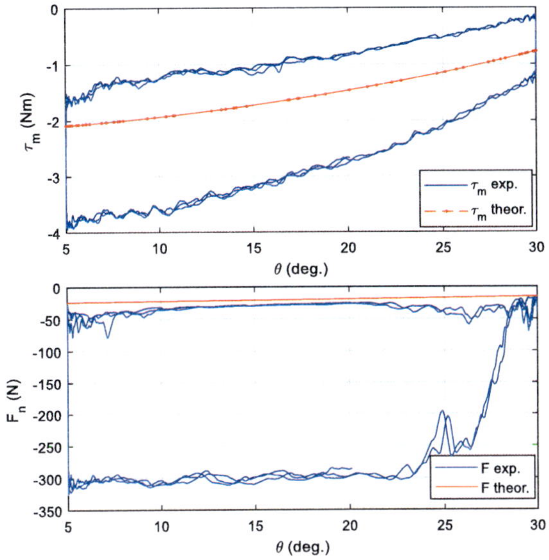


**Fig. 5.** Top) Motor torque vs. angular position. Bottom) Sensor force vs. angular position. Experimental values (exp.) vs. linear model of the bar (theor.).

This behavior means that the models and the control must be adapted to introduce the dependency of the motion direction, which is not an easy task. However, the behavior of the Nitinol bars in practice was better than the one of Nylon bars. As an example, of its capability to recover its initial shape with less permanent deformation, in Fig. 6 it is seen the shape of both kind of bars after 50 cycles, where the deformation of the Nitinol bar is approximately a third of that of the Nylon bar.

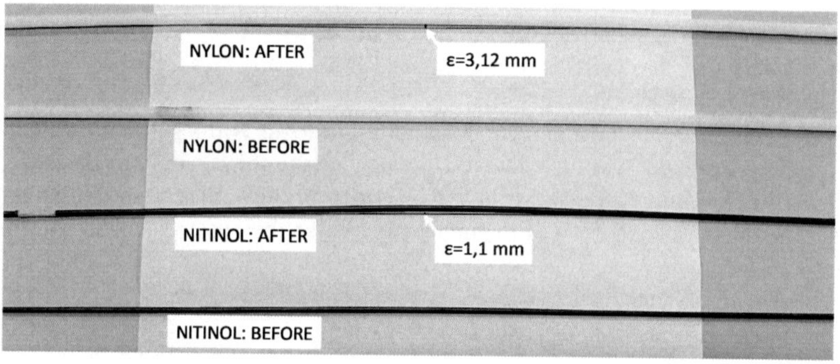


**Fig. 6.** Shape of the Nitinol and Nylon bars after 50 cycles.

## 5 Conclusions

In the present work, Nitinol bars of 3 mm of diameter have been tested as an alternative to the current limbs of Nylon 6.6 used in a prototype for a parallel continuum medical device for trunk rehabilitation. The new material has proven to be a good candidate as the higher stiffness allows using only one bar per limb, it is more durable regarding the recovery of the initial shape, and avoids the use of a not biodegradable material. On the other, hand, the hysteretic behavior must be taken into account in order to be able to accurately predict their behavior and to implement it in the control of the device.

**Acknowledgements.** Authors want to acknowledge the support given by the Spanish Government (Ministerio de Economía y Competitividad y Fondo Europeo de Desarrollo Regional, FEDER, DPI2015–64450-R), and the Basque Government for the project IT949–16. Thanks are addressed to I. Lascurain and L. Santisteban, from the Faculty of Medicine & Infirmary of the UPV/EHU, A. Bengoetxea from Université Libre de Bruxelles, and Hospital Gorliz for their valuable advice.

## References

1. Bryson, C.E., Rucker, D.C.: Toward parallel continuum manipulators. In: 2014 IEEE International Conference on Robotics and Automation (ICRA), Hong Kong, pp. 778–785 (2014)
2. Atashzar, S.F., Shahbazi, M., Patel, R.V.: Haptics-enabled interactive neurorehabilitation mechatronics: classification, functionality Challenges and Ongoing Research. Mechatronics **57**, 1–19 (2019)
3. Reinkensmeyer, D.J., Emken, J.L., Cramer, S.: Robotics, motor learning, and neurologic recovery. Annu. Rev. Biomed. Eng. **6**(1), 497–525 (2004)
4. Ringleb, P.A.: Guidelines for management of ischaemic stroke and transient ischemic attack 2008. Cerebrovasc. Dis. **25**(5), 457–507 (2008)
5. Domínguez, D., Grau, M.: Entrenamiento aeróbico y de fuerza en la rehabilitación del ictus. Fisioterapia **33**(5), 210–216 (2011)
6. Lum, P., Reinkensmeyer, D., Mahoney, R., Rymer, W.Z., Burgar, C.: Robotic devices for movement therapy after stroke current status and challenges to clinical acceptance. Top Stroke Rehabil. **8**(4), 40–53 (2002)

7. Xie, S.: Advanced Robotics for Medical Rehabilitation. Springer, Cham (2016). https://doi.org/10.1007/978-3-319-19896-5
8. Nam, K.Y., Kim, H.J., Kwon, B.S., Park, J.W., Lee, H.J., Yoo, A.: Robot-assisted gait training (Lokomat) improves walking function and activity in people with spinal cord injury: a systematic review. J. Neuroeng. Rehabil. **14**(1), 1–13 (2017)
9. Schmidt, H., Werner, C., Bernhardt, R., Hesse, S., Krüger, J.: Gait rehabilitation machines based on programmable footplates. J. Neuroeng. Rehabil. **4**, 1–7 (2007)
10. Diaz-Caneja, D., Campa, F.J., Altuzarra, O.: Design and modeling of a parallel continuum manipulator for trunk motion rehabilitation. J. Med. Devices **15**(1), 011109 (2021)

# Model Based Estimation of Passive Joint Clearance in Parallel Robot and Its Application to Failure Detection

Masumi Ohno and Yukio Takeda(✉)

Tokyo Institute of Technology, Tokyo 152-8552, Japan
takeda.y.aa@m.titech.ac.jp

**Abstract.** This paper focuses on the detection of failure of parallel robot caused by the existence of excessive joint clearance due to the wear of joint elements during a long use under high load at high speed. In order to enable the estimation of passive joint clearance values in parallel robot without specially designed trajectories and any additional equipment, a linearized model to formulate the relationship between the clearance and actuation torque difference was presented introducing the stochastic variables. Solution process of this formulation by using the torque data obtained by experiments and dynamic analyses for a set of trajectories was proposed. Experiments were conducted for an industrial Delta robot in which spherical joints having clearance ranging from 0 to 0.15 mm were used. Performance of the estimation of joint clearance was stochastically evaluated, and the results showed the feasibility of the failure detection based on the proposed method.

**Keywords:** SDG9 · Parallel robot · Failure detection · Estimation of joint clearance · Dynamic model · Stochastic approach

## 1 Introduction

For a sustainable industrialization in the global level, realization of efficient and systematical automated production system based on the zero-downtime concept is demanded. To this end, each robot and machine in the production system should be always kept mechanically healthy so that it can properly work. For such a purpose, an automated preventive and diagnosis function is required to be implemented in the controller which enables continuous health monitoring. Apart from failures in huge systems, for responding to the mechanical deterioration and failure of mechanical parts such as bearings, gears, bolts and links according to the long-time usage, periodical inspection or continuous health monitoring are effective. Once failure is found, replacement and repair of the elements can be efficiently planned and easily done without emergent downtime. Continuous health monitoring function can be implemented in the control software if it does not require high computational cost nor large memory. Use of the cloud computing using the IoT technology will be also applicable for sharing the data with the suppliers.


G. Quaglia et al. (Eds.): I4SDG 2021, MMS 108, pp. 362–371, 2022.
https://doi.org/10.1007/978-3-030-87383-7_40

Utilization of big data and/or artificial intelligence technology would be considered effective for the health monitoring of robot [1–3]. For a better failure detection performance, data obtained under some specific conditions will be considered useful. Even though such special data might be included in the big data, its effective use would be almost impossible since the effectiveness of such data is not clearly understood without preliminary knowledge as far as only the "big data" is used. Once the big data approach had been already deployed, it would be unlikely to find such special data, and there would be a big barrier to obtain such effective data by the users or suppliers. Another approach for the fault detection based on the physical model of robot including the phenomena caused by the faults needs less data as far as the model is sufficiently precise and will be helpful to find better conditions for the detection [4–7]. Application of fault detection method based on physical model will be rather supportive in robot, the operation conditions of which always change in trajectories, load and temperature according to the users than automation machines repeatedly conducting the same tasks. In order to make the quality of the products and manufacturing cost be independent of the manufacturing place all over the world, function for fault detection should be implemented in all robots as the standard equipment and fault detection should be automatically conducted by the robot itself. For such a purpose, implementation of the model based fault detection function is considered practical.

Delta robot is one of the translational parallel robots with 3 degrees of freedom (DOF). It and its extension version having one additional rotational DOF around the vertical axis are widely spread in the food packaging industries for performing high speed pick-and-place tasks. Such a Delta robot has 12 spherical joints, each of which is usually composed of a housing made by plastic material and a steel ball. Due to the relative motion between these bearing elements under high load during the high speed/high acceleration motions, the plastic material suffers wear, which results in the existence of clearance. Such clearance is developed according to the repeated use. Once clearance size exceeds allowable limit, vibrating behavior and big noise during operation will occur, and at last the bearing will be broken. Once such breakage happens, the production line including the robot with failure should be stopped, which means the existence of downtime. Therefore, in order to avoid such a situation, estimation of the joint clearance should be always done. To this end, the authors have been trying to develop a method for detecting the excessive clearance of spherical joints of Delta robot. We focused on the change of the actuation torque caused by the clearance of joints, and proposed a method to design the special trajectories used in the estimation. Its effectiveness has been confirmed through experiments [7]. From the standing point of view of the users of robots, daily use of such special trajectories for failure detection is considered inconvenient and inefficient. Therefore, another approach in which trajectories prepared for manufacturing are utilized is demanded.

The present paper proposes a new method for estimating the clearance of spherical joints of Delta robot, which does not use specially designed trajectories, based on the physical model of Delta robot with joint clearance. Experimental results are presented, and the effectiveness of the proposed method and the feasibility of the fault detection are discussed. We considered the cases where one or two joints in a robot have excessive clearance because existence of more than two joints having excessive clearance is practically considered rather rare.

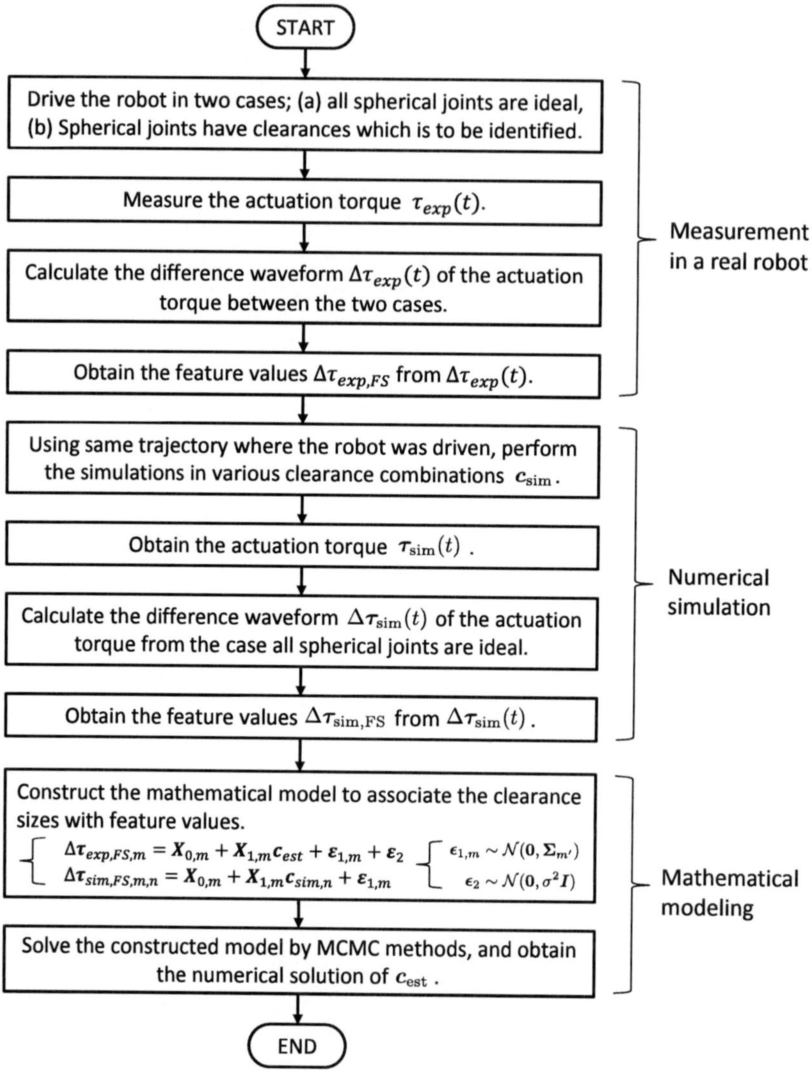


**Fig. 1.** Flowchart of the proposed method

## 2 Method

Figure 1 shows the process for estimating the joint clearance of robot proposed in this paper. The details are explained below.

### 2.1 Formulation

A vector $\boldsymbol{c}$, hereafter called as "joint clearance" is composed of clearance values of joints in the robot. The effect of joint clearance $\boldsymbol{c}$ on the behavior of the robot is evaluated by the difference of actuation torque $\Delta\boldsymbol{\tau}_{FS}$ in this paper. We assume that the following relationship between $\boldsymbol{c}$ and $\Delta\boldsymbol{\tau}_{FS}$ with respect to a trajectory $TJ_m$($m$ is the trajectory number) holds:

$$\Delta\boldsymbol{\tau}_{FS,m} = \mathbf{X}_{0,m} + \mathbf{X}_{1,m}\boldsymbol{c} + \boldsymbol{\varepsilon}_{1,m} \tag{1}$$

Here, $\boldsymbol{X}_{0,m}$ and $\boldsymbol{X}_{1,m}$ are coefficient matrices dependent on the trajectory. The torque difference with respect to time $\Delta\boldsymbol{\tau}(t)$ can be obtained as the difference between the actual torque under the existence of joint clearance $\boldsymbol{c}$,$\boldsymbol{\tau}_m(t)$, and the actuation torque when $\boldsymbol{c} = 0$, $\Delta\boldsymbol{\tau}_{ideal,m}(t)$, as

$$\boldsymbol{\tau}_m(t) = \boldsymbol{\tau}_m(t) - \boldsymbol{\tau}_{ideal,m}(t). \tag{2}$$

$\Delta\boldsymbol{\tau}_{FS,m}$ in Eq. (1) is defined as a feature value extracted from $\Delta\boldsymbol{\tau}_m(t)$. Though the effect of joint clearance on the actuation torque is expected nonlinear, we assume a linear relationship as shown in Eq. (1) for simplicity. To consider the error by linearization, a stochastic variable $\boldsymbol{\varepsilon}_{1,m}$ is introduced in Eq. (1).

Through extraction of feature values from the actuation torque differences obtained by the theoretical analysis based on the dynamic model of the robot under several conditions of joint clearance $\boldsymbol{c}_n$ ($n = 1, \cdots, N$; $N$ is the total number of sets of joint clearance) and trajectories $TJ_m$ ($m = 1, \cdots, M$), a variety of set of torque difference and joint clearance can be obtained. This set is represented as $(\Delta\boldsymbol{\tau}_{sim,FS,m,n}, \boldsymbol{c}_{sim,n})$ ($n = 1, \cdots, N$; $m = 1, \cdots, M$).

Through torque measurement of the real robot under unknown joint clearance for a trajectory $TJ_m$ and data processing for feature extraction, a feature value corresponding to $\Delta\boldsymbol{\tau}_{FS,m}$ in Eq. (1) can be experimentally obtained. This is denoted as $\Delta\boldsymbol{\tau}_{exp,FS,m}$.

The following mathematical model, in which results obtained by the theoretical analysis and experimental measurement are used, to represent the relationship between the joint clearance and torque difference is considered:

$$\Delta\boldsymbol{\tau}_{exp,FS,m} = \boldsymbol{X}_{0,m} + \boldsymbol{X}_{1,m}\boldsymbol{c}_{est} + \boldsymbol{\varepsilon}_{1,m} + \boldsymbol{\varepsilon}_2 \tag{3a}$$

$$\Delta\boldsymbol{\tau}_{sim,FS,m,n} = \boldsymbol{X}_{0,m} + \boldsymbol{X}_{1,m}\boldsymbol{c}_{sim,n} + \boldsymbol{\varepsilon}_{1,m} \tag{3b}$$

A stochastic variable $\boldsymbol{\varepsilon}_2$ is introduced as an error term to compensate for the error in the dynamic model and experimental measurement. We assume that $\boldsymbol{\varepsilon}_{1,m}$ and $\boldsymbol{\varepsilon}_2$ are normally distributed, and represented as follows:

$$\boldsymbol{\varepsilon}_{1,m} \sim N(0,\ \boldsymbol{\Sigma}_{\mathrm{m}}) \tag{4a}$$

$$\boldsymbol{\varepsilon}_2 \sim N(0, \sigma^2 \boldsymbol{I}) \tag{4b}$$

$N(\boldsymbol{\mu}, \boldsymbol{\Sigma}_m)$ represents a multi-variable normal distribution of average $\boldsymbol{\mu}$ and variance-covariance matrix $\boldsymbol{\Sigma}_m$. $\boldsymbol{I}$ represents a unit matrix.

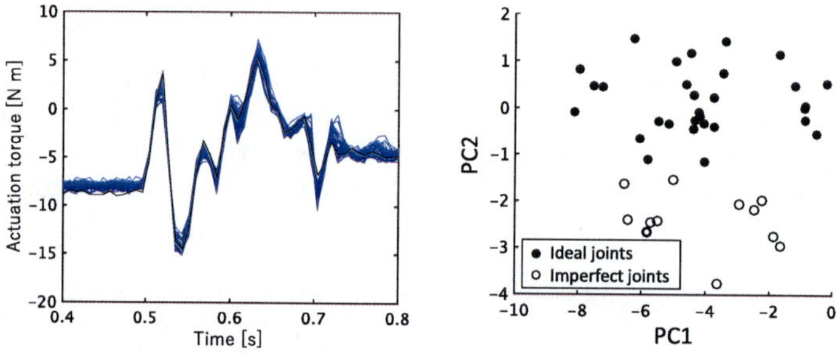


(a)Measured torque (b) PCA result

**Fig. 2.** Effect of joint clearance on driving torque and its feature values

As can be found above, the proposed method needs torque measurements under $\boldsymbol{c} = 0$. This may lead to inconvenience in the use of data for new trajectories after the robot has been used for a long period ($\boldsymbol{c} \neq 0$). In order to enable it to adapt to such a situation, the improvement of the formulation presented above may be needed, however, we believe that it will be enabled based on the current formulation by adding some adjustment. On the other hand, since $\boldsymbol{\varepsilon}_2$ quantitatively represents the difference between the dynamic model and real robot, it will be useful as an indicator of accuracy of the model. Therefore, its value can be utilized as a measure of individual difference of each robot from the standard model and change over time of each robot.

## 2.2 Feature Extraction by Principal Component Analysis

A method to obtain the feature value $\Delta\boldsymbol{\tau}_{FS,m}$ from the time-series value $\Delta\boldsymbol{\tau}_m(t)$ is considered. Among the time-series data obtained by torque measurement/theoretical analysis based on the model, there should be some feature values corresponding to the joint clearance. Even if we have huge data, it is not easy directly to extract such feature from the data itself or huge amount of calculation might be needed. Therefore, we applied the principal component analysis to extract the feature values which are hidden in the time-series values. Principal component scores are taken as the feature values.

Principal component vectors respectively obtained using the torque values as the time-series data by theoretical analysis and experimental measurement for a trajectory are slightly different from each other due to the modeling error. Since the proposed method to estimate the joint clearance is based on the comparison between the theoretical model and the data obtained by experiment as represented by Eq. (3), the feature values for theoretical analysis and experiment should correspond to a same principal component vector for a trajectory. In this paper, the scalar projection values of the experimental values on to the principal component vector obtained by theoretical analysis are used.

As an example, measured torque values for several conditions of joint clearance for a trajectory and their principal component scores obtained according to the process mentioned above are shown in Fig. 2. In this case, it can be observed that the effect of joint clearance can be extracted by the second principal component scores.

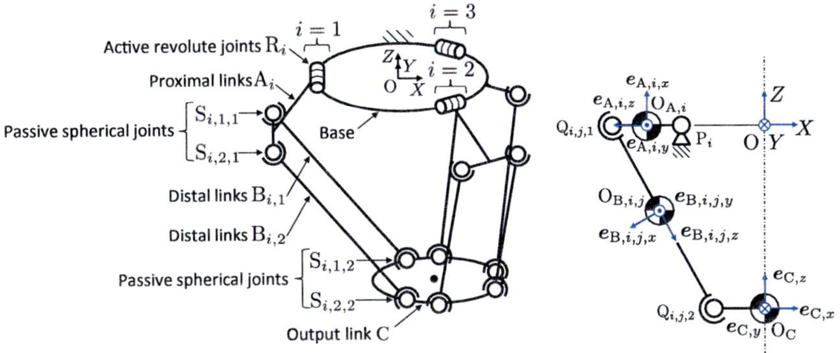


(a) Kinematic structure (b) Dynamic model

**Fig. 3.** Delta robot

### 2.3 Estimation of Joint Clearance

Calculation process to estimate the joint clearance by solving Eq. (3) is as follows:

(1) Torque values as time-series data for trajectories $TJ_m(m = 1, \cdots, M)$, $\Delta\boldsymbol{\tau}_{exp,m}(t)$, are obtained by experiments.
(2) Torque values for the trajectories $TJ_m(m = 1, \cdots, M)$ under a variety of joint clearances $\boldsymbol{c}_{sim,n}(n = 1, \cdots, N)$, $\Delta\boldsymbol{\tau}_{sim,m,n}(t)$, are obtained by theoretical analysis.
(3) Sets of feature values of torque and joint clearance, $(\Delta\boldsymbol{\tau}_{exp,FS,m}, \boldsymbol{c}_{est})$ and $(\Delta\boldsymbol{\tau}_{sim,FS,m,n}, \boldsymbol{c}_{sim,n})$ are obtained by the principal component analysis and scalar projection, where $\boldsymbol{c}_{est}$ is unknown and $\boldsymbol{c}_{sim,1} = 0$.
(4) A set of $M \times N$ simultaneous equations constructed according to the process above are solved for $J + 3M + 1$ unknowns ($J$ is the number of joints with clearance). The numbers of unknowns are $J$ for clearance values of joints, $M$ for $\boldsymbol{X}_{0,m}$, $M$ for $\boldsymbol{X}_{1,m}$, $M$ for $\boldsymbol{\varepsilon}_{1,m}$ and 1 for $\boldsymbol{\varepsilon}_2$, respectively. Since this equation is a stochastic equation, all unknown variables are solved as stochastic variables having probability distribution by applying the Bayesian inference.

## 3 Application to Delta Robot

### 3.1 Delta Robot and Preparation for Experiments

Figure 3(a) shows the mechanism configuration of the Delta robot. It is a translational parallel robot with 3 DOF, in which three identical chains composed of a spatial parallelogram having 4 spherical joints and a crank are symmetrically arranged between the

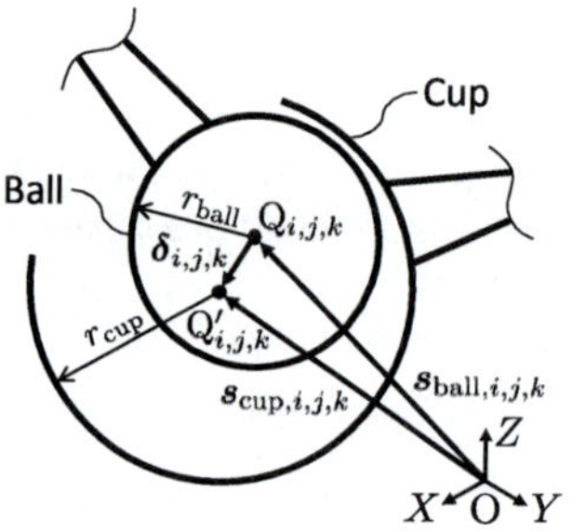


**Fig. 4.** Spherical joint with clearance

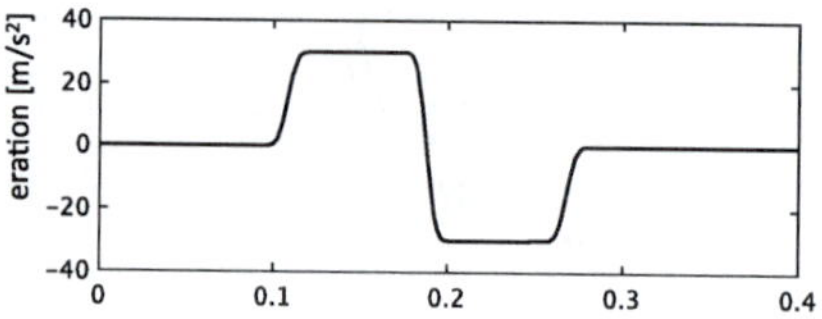


**Fig. 5.** Acceleration profile

base and the output link. This paper considers its extended version with 4 DOF, in which a drive chain to generate the rotational motion around the vertical axis is attached to one of the chains ($i = 1$). Figure 3(b) shows the dynamic model of the Delta robot. Values of kinematic and dynamic parameters used in the theoretical analysis are determined so as to correspond to the industrial Delta robot used in the experiments. As mentioned above, the values of the dynamic parameters of chain 1 are different from those of the other chains because of the effect of the drive chain for rotational motion.

Spherical joint with clearance $S_{i,j,k}(i = 1, 2, 3; j = 1, 2; k = 1, 2)$ is modelled as shown in Fig. 4. Clearance is defined as $c_{i,j,k} = r_{cup,i,j,k} - r_{ball,i,j,k}$ where $r_{cup,i,j,k}$ and $r_{ball,i,j,k}$ are radii of the cup and ball of a joint, respectively.

Trajectories of the output link used in the experiments were prepared such that they are all straight-line ones of 200 mm displacement in the cartesian space and of the same acceleration profile shown in Fig. 5. By randomly combining the start point and direction of motion within the workspace of the experimental robot, 100 trajectories were generated. We prepared three kinds of spherical joints of $c = 0, 0.14$ mm and 0.15 mm. By changing the location of spherical joint(s) having excessive clearance ($c =$ 0.14 or 0.15 mm) among the six joints located on the output link $S_{i,j,2}(i = 1, 2, 3; j = 1, 2)$ ($J = 6$) while the maximum number of joints having excessive clearance in one robot is 2, we obtained 43 cases of joint clearances in total. This means $N = 43$. Among them, 12 cases were prepared for one joint having excessive clearance ($J_{AC} = 1$), and 30 cases for 2 joints having excessive clearance ($J_{AC} = 2$), while only one case for $c = 0$ ($J_{AC} = 0$). For all 100 trajectories, we conducted theoretical analysis based on the dynamic analysis method for Delta robot with joint clearance [8] combined with all 43 cases of joint clearances. The number of trajectories used for identification of joint clearance was set as 10 ($M = 10$) for all experiments.

### 3.2 Results and Discussions

For a case of joint clearance $c_{1,1,2} = 0.14\,\mathrm{mm}$, $c_{2,1,2} = 0.15\,\mathrm{mm}$ where $J_{AC} = 2$, the estimated joint clearance values are shown as a result obtained for a set of $M (= 10)$ trajectories in Fig. 6. In this estimation, all joints were considered as joints having non-zero clearance. Even though the estimated values of clearance are smaller than the actual ones, it is confirmed that estimated values of clearances of the two joints in which excessive clearances were actually given are apparently larger than the others. This shows the effectiveness of the proposed method.

**Table 1.** Calculated AUC values of each condition

| $J_{AC}$ | $S_{1,1,2}$ | $S_{2,1,2}$ | $S_{3,1,2}$ | $S_{1,2,2}$ | $S_{2,1,2}$ | $S_{3,2,2}$ | Total |
|---|---|---|---|---|---|---|---|
| 1 | 0.986 | 0.911 | 0.899 | 0.969 | 0.959 | 1.000 | 0.954 |
| 2 | 0.937 | 0.827 | 0.942 | 0.942 | 0.907 | 0.982 | 0.923 |

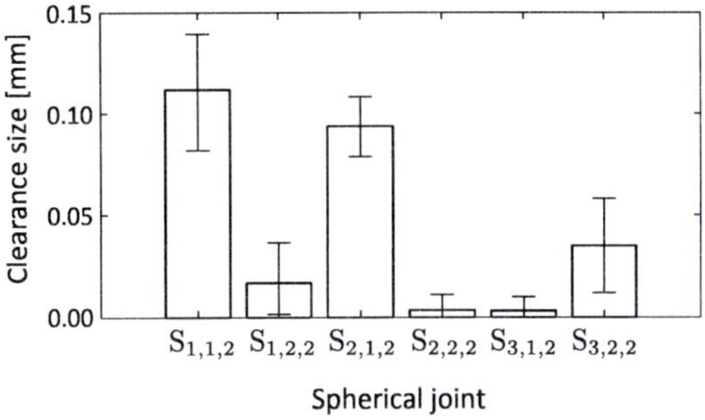


**Fig. 6.** Estimated clearance values

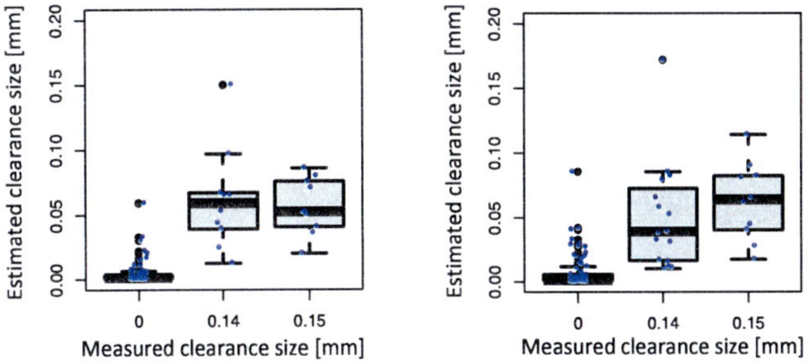


(a) $J_{AC} = 1$ (b) $J_{AC} = 2$

**Fig. 7.** Estimated joint clearance for $S_{1,1,2}$

For all cases of joint clearances, joint clearances were repeatedly estimated 100 times by changing the combination of trajectories. An example result is shown in Fig. 7 as

Box Plots in which the distribution of the estimated clearances among the 100 trials with different combination of trajectories is shown. In Fig. 7, the results for the cases where the number of joints having excessive clearance is 1 and 2 ($J_{AC} = 1, 2$) are separately shown. All of the results obtained are investigated by drawing ROC (Receiver Operating Characteristic) curves and evaluated by means of AUC (Area Under the Curve) values for all 6 joints. The results are summarized in Table 1. It is known from these results that it is statistically possible to estimate the joint clearance and to distinguish between the ideal joint and the one having excessive clearance around $c \geq 0.15$ mm by the proposed method without using specially designed trajectories under the condition of $J_{AC} \leq 2$ among the total 12 joints though the estimated values are rather smaller than the actual ones. It is also known that the probability to successfully estimate the clearance decreases as the number of joints having excessive clearance increases.

## 4 Conclusions

The relationship between the joint clearance value and the torque difference in a Delta robot was formulated by a linear equation introducing stochastic variables for the estimation of the joint clearance. Solution procedure using the data by experiment and theoretical analysis was presented. Through experiments, the effectiveness of the proposed method has been confirmed. The proposed method uses the torque measurement data, which is easily measured by the standard controller, when the robot operates under trajectories for manufacturing. Therefore, there are no needs for additional equipment and specially designed trajectories. In addition, thanks to the use of the dynamic model of the robot, huge number of preliminary experiments for learning or tuning of the estimation program are not necessary. The authors believe that such a failure detection system can be implemented as a standard equipment in the robot system, which will contribute to the global development and foundation of manufacturing technology.

## 1. References


1. Soualhi, M., Nguyen, K.T.P., Medjaher, K.: Pattern recognition method of fault diagnostics based on a new health indicator for smart manufacturing. Mech. Syst. Signal Process. **142** (2020)
2. Bittencourt, A.C., Saarinen, K., Sander-Tavallaey, S., Gunnarsson, S.: A data-driven approach to diagnostics of repetitive processes in the distribution domain – application to gear box diagnostics in industrial robots and rotating machines, Mechatronics **24** (2014)
3. Eski, I., Erkaya, S., Savas, S., Yildirim, S.: Fault detection on robot manipulators using artificial neural networks. Robot. Comput. Integrated Manuf. **27** (2011)
4. Sabry, A.H., Nordin, F.H., Sarby, A.H., Kadir, M.Z.A.A.: Fault detection and diagnosis of industrial robot based on power consumption modeling. IEEE Trans. Ind. Electron. **67** (2020)
5. Li, K.-L., Tsai, Y.-K., Chan K.-Y.: Identifying joint clearance via robot manipulation. Prof. IMechE Part C, J. Mech. Eng. Sci. **232** (2018)
6. Ahmad, S., Zhang, H., Liu, G.: Distributed fault detection for modular and reconfigurable robots with joint torque sensing: a prediction error based approach. Mechatronics **23** (2013)

7. Ohno, M., Takeda, Y.: Design of target trajectories for the detection of joint clearances in parallel robot based on the actuation torque measurement. Mech. Machine Theory **155** (2021)
8. Ohno, M.: Dynamic analysis of parallel robot with spherical joints considering joint clearances and clearance identification based on actuation torque measurement. Dissertation, Tokyo Institute of Technology (2021)

# Mobile Robotics for Sustainable Development: Two Case Studies

Rocco Galati, Giacomo Mantriota, and Giulio Reina(✉)

Politecnico di Bari, Department of Mechanics, Mathematics and Management, Bari 70126, Italy
{rocco.galati,giacomo.mantriota,giulio.reina}@poliba.it

**Abstract.** The use of robotic solutions for sustainability is critical to reduce pollution and waste of materials during the manufacturing processes, and it can significantly reduce the amount of energy employed compared with traditional methods. This paper presents two case studies in the development of mobile robotic solutions used to increase the manufacturing sustainability by avoiding waste of materials and physical stress for the operators. Both mobile robots can improve the manufacturing performance by reducing costs and time with special attention to the environment and to the safety.

**Keywords:** SDG9 · Mobile robotics · Sustainable innovation · Service robotics

## 1 Introduction

Over the past few years, more and more companies have been trying to find new and innovative methods to integrate robotic solutions into their manufacturing line to improve their processes and increase their products quality. However, despite these efforts, the optimization of manufacturing processes is currently only partially addressed using industrial mobile robots. Increasing the autonomy and precision of robots can be beneficial in many ways: for example, by reducing the labor costs or limiting the waste of raw materials [1]. The increasing adoption of artificial intelligence has increased the capabilities of mobile robots to solve more complex problems with important fall outs in real-world applications including monitoring of water, air and ground, and precision farming [2–6]. Considering all these benefits, it is pretty evident that robotics can help to improve sustainability solutions and can provide a very useful tool to fight climate change, support recycling activities and make manufacturing more sustainable by reducing waste since they can be extremely accurate, i.e., they can measure the quantity of herbicide to spray on each plant by saving time and money, and the total amount of energy needed to complete a single task. Moreover, robots can accomplish assignments that can be dangerous for human health. This paper presents two case studies about the development of robotic platforms used to increase the manufacturing sustainability by avoiding waste of


G. Quaglia et al. (Eds.): I4SDG 2021, MMS 108, pp. 372–382, 2022.
https://doi.org/10.1007/978-3-030-87383-7_41

materials and physical stress for the human operators. The first section describes the designing solutions adopted to develop a tracked mobile robot used to compress flax fibers inside containers by increasing their bulk density while the second one reports the study carried on to build a holonomic mobile robotic platform to automatically mark lines on the floor of wide exhibition halls in order to outline the area where each stand should be placed by saving time, human efforts and environmentally friendly paints. Final results are provided at the end of each section.

## 2 Mobile Robotics for the Compaction of Linen Fibers

Flax is also known under the name of Linen, cellulosic fibers obtained from the stem of the flax plant, which are much stronger and more lustrous than other natural fibers like cotton; the fiber is derived by the stalk of a plant, Linum Usitatissimum, which typically is from 80 to 120 cm high with few branches and small flowers with a colour that can change from white to intense blue. The individual fiber cell is long, typically 27 mm but ranging from 9 to 70 mm in length, and has an average diameter of 23 $\mu$m. Fiber cell bundles are located in the outer parts of the stem, just under the epidermis and cortical parenchyma and are derived from phloem cells. It is possible to see how this type of flax fibers is composed by a large quantity of Cellulose and Lignin, with a bulk density of 133 kg/m$^3$ and a pH equal to 7.7. Linen is incredibly easy to grow requiring a very little amount of water and most of the farmers let nature do the work. Typically, a flax plant requires just a quarter of the water that its cotton counterpart would. Another interesting aspect related to the flax plants is that they are able to trap massive amounts of $CO_2$ in the atmosphere. The European Confederation of Flax and Hemp (CELC) state that one hectare of flax retains 3.7 tons of $CO_2$ every year and it is equivalent to the emissions from an average car left it running for nearly six days without stopping. In France, that is one of countries with the widest flax plantations in Europe, the total amount of $CO_2$ absorbed each year is equal to the one generated by a light bulb being lit for over 2 million years. France, Belgium, and the Netherlands account for 80% of the world's flax linen production covering 97,000 hectares which equates to 133,000 tins of long fibers and, usually, all the farmers respect their ecosystems using zero irrigation, GMO-free and zero-waste methods.

However, the increased demand for linen fibers is leading the suppliers to intensify their manufacturing processes and to organize frequent and recurring ground shipments all over the world. Although, recently, lots of improvements have been introduced on the production line, only very few have been made concerning the storage of the linen fibers into the truck containers for the transportation. The flax fibers are manually and gradually distributed inside the container by two or more workers placed at the top of the container by using long pitchforks until the top of the container is reached; usually, a vibrator attached to the container distributes the material in a more uniform way by making it vibrate to try to maximize the used space inside the container. No other mechanised process has been designed or proposed to be used to perform this kind of operation, so all the major flax fibers producers still rely on traditional techniques. However, this procedure does not correctly compress and compact the various layers of flax fibers, and so this leads to a non-optimised occupied space for the same shipping volume by increasing the number of shipments and the overall time needed to deliver a specified amount of material. It is for this reason that most companies involved in flax production are investing their resources to find new ways to fill the available volume in the container with as much material as possible in order to increase the transported flax at the same shipping price. Moreover, the process of filling the container by using a mechanical hopper releases small fiber particles into the air by potentially generating an explosive environment that can be very dangerous for all the human operators close to the container.

### 2.1 A Service Compactor Vehicle for Flax Fibers

A service vehicle has been developed to address the problems related to the safety of workers involved in the filling process and to the optimization of the available volume in each container. Full details of the robot can be found in [7], and they are not repeated here. To accomplish these two tasks, the vehicle has been equipped with large rubber tracks and a navigation system capable to make it autonomously move inside the container by rolling over the flax material to compress it, without destroying its fibers, by actually increasing the material transported in each shipment. Basically, the main purpose of the vehicle is to increase the overall bulk density seen as the measure of the bulk of the material and it is used to express how much weight of a material could be packed per unit area in a container typically used for storage or transportation purposes as it is without compress or compact it. Bulk density is typically expressed as kg per $m^3$ ($kg/m^3$) and it is about $133\,kg/m^3$ for Linen fibers.

At the top of each container there is a mechanical hopper that pushes the material down to the center of the container floor which is made up by a metal plate. The material is clustered only at the center forming up a mound, which needs to be manually scattered by human operators with the aid of pitchforks to gradually fill up all the container volume. It is very clear that two main aspects must be taken into account while designing the robot locomotion unit: the vehicle should be able to gradually spread the flax fibers from the center of the container along its entire length and, in addition, it needs to float over the fibers layers from down to top as quickly as the material quantity increases inside the container itself. Therefore, a correct balance between the vehicle ground pressure and its traction force over a very uneven surface must be found in order to make it have enough power to smooth, level and compress the material by moving it from the mound, and an adequate ground footprint to be able to float over the flax fibers layers.

For all these reasons, a locomotion unit based on large tracks with skid-steering system has been chosen over a solution based on wheel to overcome the rolling motion resistance and to increase the floating capacity of the vehicle [8]; since the standard containers used for flax transportation have a minimum width of about 2470 mm, a maximum overall width of 1900 mm was set for the frame of the vehicle with the idea of making it move straight forward and backward through the container to continuously roll on the material.

To avoid destroying the flax fibers, the belt of the two tracks has been made by 100% natural rubber with four reinforcing layers with no treads, and with a total length of 4700 mm, thickness of 11 mm and width of 860 mm. The corresponding ground contact length results in about 1970 mm, which offers a total contact footprint surface of 3.4 $m^2$. In Fig. 3, on the left, it is possible to see a CAD representation of the whole vehicle as it has been designed; as showed, both left (5) and right (6) tracks have an inverse trapezoidal shape to easily move the flax fibers towards the containers corners and to create enough space inside the frame to place the electronics and the motors. There is no roof or protection shield on the top of the vehicle since the belts act as conveyor for the material that fall from the hopper placed above it: this solution helps the vehicle to spread the material not only by using its tractive force, but also by using the tracks as a real belt conveyor. Front (7) and back (8) bumpers are used as limit switches to detect when the vehicle is approaching the container walls in both directions while the side wheels (9) made in soft rubber are used to avoid the vehicle to scratch the container's walls in case of trajectory deviations and support it to come back on track since they are mounted on springs. The right side of Fig. 3 shows a semi-assembled and powder coated version of the robot's frame, made of Fe360 iron, with all the already galvanized drive and castor wheels and the gearboxes. Since the vehicle is fully electric and it operates in a flammable dusty environoment, all the electronics components are designed to agree with the ATEX standard used for systems intended for use in potentially explosive atmospheres (Fig. 1).

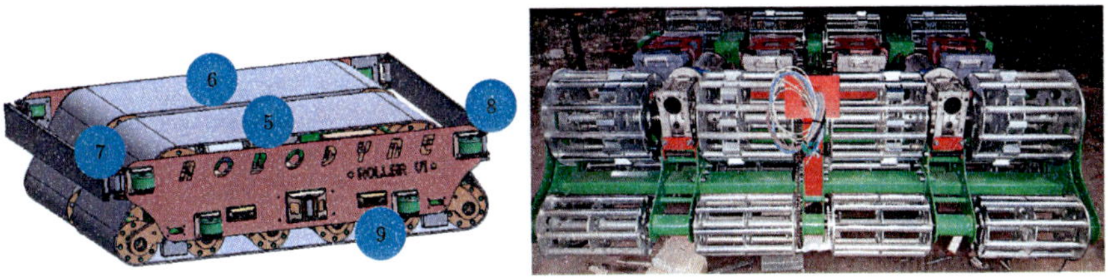


**Fig. 1.** The whole vehicle assembly with tracks (5 and 6), bumpers (7 and 8) and side wheels (9), on the left, a robot semi-assembled frame, on the right

Before performing the tests with the real container, several runs were conducted in a mock-up test bed made up by several chipboards with a thickness of 10 mm to simulate a small container filled with flax fibers, as showed in Fig. 2.

Afterwards, the vehicle has been loaded inside a real container to check its performance in a real work case; at the beginning, the unit has been positioned at the center of an empty container, as showed on the left picture in Fig. 3, and then the hopper has been operated to gradually release the flax fibers into the container from top to down.

**Fig. 2.** The service unit during the first tests made in a chipboards box

The vehicle has been remotely controlled to move straight forward and backward by a human operator in order to roll over the flax shives deposited on the container metal floor. After having filled all the container with flax fibers, the vehicle has been removed and the total measured weight was about $W_f = 19000\,\text{kg}$, which is greater than the weight $W_i = 12000\,\text{kg}$ measured with standard methods. This results in an improved bulk density equal to $b_s = 211\,\text{kg/m}^3$ with an increase of about 58% with respect to the standard value ($b_s = 133\,\text{kg/m}^3$). It is recommended not to exceed the new bulk density value in order to avoid to stress too much the fibers since the maximum density for the specific material used for the experiments is $b_{s,max} = 245\,\text{kg/m}^3$ with a security buffer about 15% greater than the obtained $b_s$.

**Fig. 3.** On the left, the service vehicle at the beginning of the compression process, while, on the right, the compressed layer

## 3 Mobile Robotics for Floor Marking

Globally, there are approximately 32,000 exhibitions each year, featuring 4.5 million exhibiting companies and attracting over 303 million visitors. Exhibitors and visitors combined spend around €116 billion every year on exhibitions, making exhibitions a significant global industry. As stated by the Global Association of the Exhibition Industry (UFI), in 2018, the total number of venues with a minimum of 5,000 $m^2$ of gross indoor exhibition space around the world was 1,217 and the gross exhibition space available is 34.7 million $m^2$ [9]. Additional venue space has become available in almost all regions of the world and substantial venue investments and upgrades have taken place to increase venue capacities. Reducing the impact businesses have on the environment and ensuring that they contribute positively to the regions they operate in is high priority for most industries, including exhibitions. Exhibitions can contribute to a sustainable world as, in addition to the positive economic and educational impact they have at regional levels, they can reduce multiple travel. By hosting market players at an exhibition, professionals can schedule meetings and develop business relationships with multiple contacts all under one roof.

Almost all the fair companies use a totally empty hall for the exhibitions since each fair is based on its own and unique layout. Moreover, in this way, the exhibitors can choose a different shape for their stands starting from a free area that typically has a rectangular plan, 100 m wide and 200 m long. After having sold all the available spaces, it is mandatory, for the company that hosts the fair, to mark on the floor the position where each stand is located with the highest possible accuracy; the markers are used to establish the boundaries where each stand must be built. Since there can be several exhibitions in succession in order to optimize space, time, and money, the marking operation needs to be as fast as possible and it is usually carried on during night and without breaks by resulting in a high physical effort for the human operators. Usually, this kind of process takes about 8 to 12 h to be completed depending on the total square meters to be marked and it is performed by four to eight workers that manually place long adhesive tapes on the floor by measuring each side of every single stand while bending over their back on floor for all the time. A team of workers is able to fix up to several kilometers of tape since a large exhibition can have several hundred stands with hundreds of coordinates to be marked on the floor. The adhesive tape used in this process is not recyclable and so a lot of waste is produced after each exhibition given that the tape cannot be used again for the next exhibition. The only alternative to the use of the adhesive tape is the application of non-permanent paints that can be dangerous for the human health and for the environment. Moreover, since the marking operation is done manually, under stress and without the correct tools, it is possible to find measurement errors that lead to wrong area dimensions and to misalignment between multiple stands.

Considering that it is mandatory to minimize the position error for each marker, a holonomic locomotion system based on four mecanum wheels has been preferred over other solutions since the vehicle needs to move instantaneously in

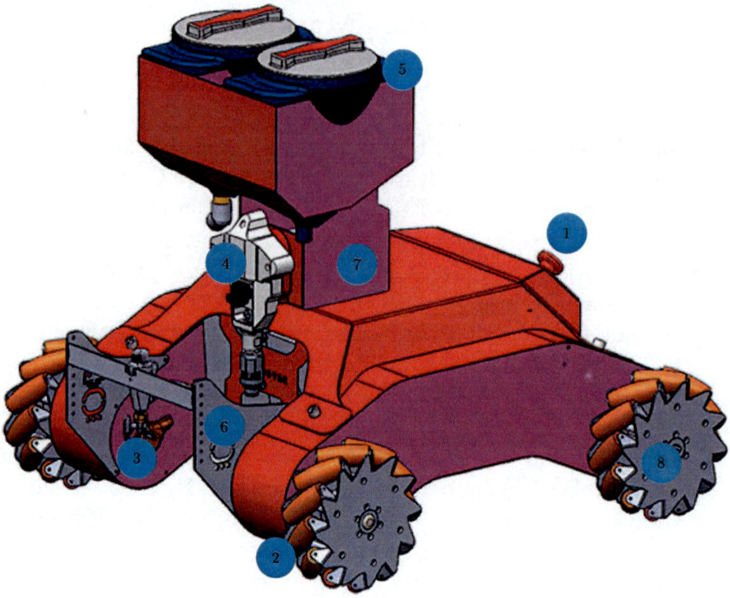


**Fig. 4.** 3D rendering of the robot

any direction and also on the spot without introducing significant slipping effects. As it is possible to see in Fig. 4, the structure of mecanum wheels (2) relies on a central wheel with twelve rollers placed at an angle of 45° around the periphery of the wheel itself. The rollers are shaped such that the silhouette of the mecanum wheel is circular and they are made up by soft nylon to better adapt their profile to the ground surface. The angled rollers are able to translate a portion of the force in the rotational direction of the wheel to a force normal to the wheel direction. Depending on each individual wheel direction and speed, the resulting combination of all these forces allow the vehicle to move easily in any direction. Each mecanum wheel is connected to a motor and a gearbox (8) equipped with a high resolution encoder and they are independently controlled by a ROS Node that runs on the onboard computer. Instead of using an automatic tape dispenser device as a tool to place the maker on the floor, the vehicle use an airless spraying system powered by an hydraulic pump (4), commanded by a motor driver, that suck out the paint stored in two upper tanks (5) and feed it to a bottom nozzle (3) usually placed at 25 cm away from the floor even if its height can be adjusted by using the positioning holes available on the nozzle support (6). Finally, an internal 12 VDC air compressor (7) is used to enable or disable the nozzle. All the system can be powered on and off by using the main control panel (1) which also provides a series of power connectors for additional devices.

### 3.1 Mapping

Usually, stands are square or rectangular-shaped and can have different dimensions depending on how much square meters have been reserved by the exhibitor and it is not unusual to have nested plants when two or more companies opt to share their areas. Figure 5 highlights a typical example of an exhibition hall where it is possible to see all the stands marked with red lines whose perimeter needs to be reported on the floor. The black squares are the pillars that can have different dimensions that range from $1 \times 1$ m up to $2 \times 2$ m and they are the only fixed structure that it is possible to find in each hall.

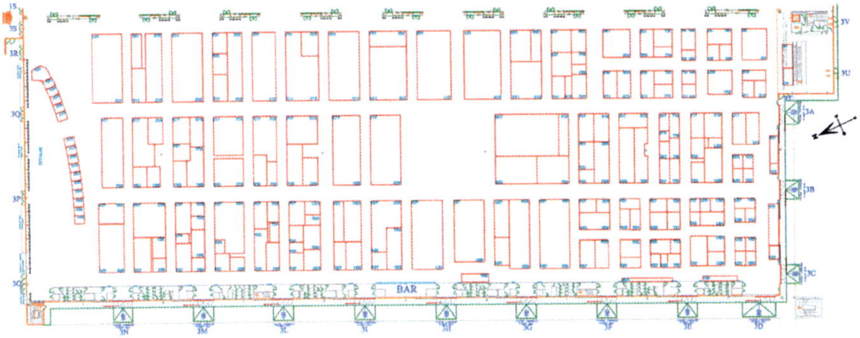


**Fig. 5.** A representation of a common exhibition hall where the inner area for each stand is marked with red lines while the black squared boxes are the pillars

One of the most critical aspects to take into account during the designing of such autonomous marking system is that the area to cover is too wide and most of the commercial sensors are not able to accurately sense the environment at distances greater than 40–50 m from the vehicle especially considering that the air can be dusty and the walls are covered by non-reflective paint. Since the exhibition halls are $100 \times 200$ m wide at maximum and each one has at least four pillars distanced about 50 m one from the other, the localization problem can be solved by mounting two LMS531 laser sensors, with a maximum range of 80 m and a maximum horizontal aperture angle of 180°, on the front and on the rear side of the vehicle with the idea to use them to always calculate the distance between the vehicle and the fixed pillars and the perimetral walls. In additional, four wheel encoders with 1024 pulses per revolution have been used to retrieve the overall velocity of the system and to provide another way to compute the total distance traveled. By comparing the position of the vehicle from these fixed structures and by reporting it on the hall plan, based on a DWG format, it is possible to accurately detect the vehicle pose in the space.

As reported in Fig. 6, a configuration with two laser sensors has been used to avoid empty spots and to maximize the number of anchor points with the

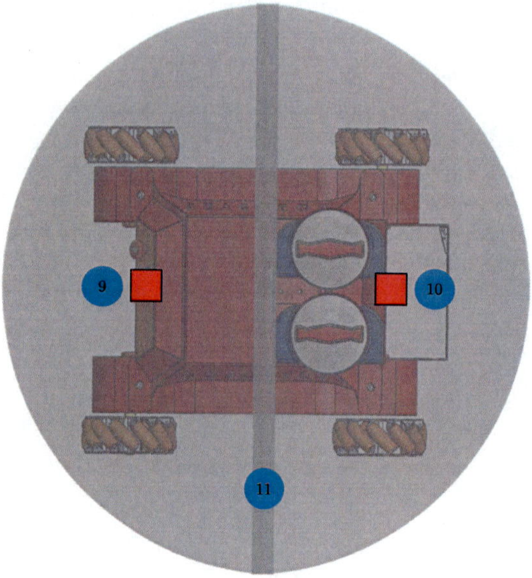


**Fig. 6.** Both front (9) and back (10) laser sensors have been positioned in order to overlap their 180° aperture angles by removing any empty spot.

pillars and the side walls; in particular, the 180° field of view of both lasers (9) and (10) overlaps in (11) by covering all the vehicle's body.

The navigation software, Fig. 7, is based on ROS (Robot Operating System) and it is made up by several nodes to accomplish different tasks. The onboard computer, equipped with Intel i7 CPU, 16 GB RAM DDR3 and a solid state disk, runs Ubuntu 18.04 and ROS Melodic and elaborates the data coming from the four wheel encoders and from both lasers in order to generate continuous odometry data; these information are encapsulated in a ROS topic and sent to the localization node that compares the distances reported by the laser sensors with the DWG map file that contains the coordinates of each fixed structure (i.e. the pillars and the walls). If a match if found between these two data set, then the x,y coordinates about the current vehicle position are published. An additional node checks if the current position relies on a point that needs to be marked on the floor and proceed to activate the air compress to enable the sprayer.

Final application is showed in Fig. 8 where it is possible to see the vehicle performing a rounded square perimeter (13) while facing pillars (14) and (15) with its front laser and running alongside the walls (16). The paint used for the marking process is a liquid temporary peelable protective coating which is environmentally friendly since it is based on pure water and it can be manually and easily removed from the floor and it can be collected to be shipped to paint-recycling facilities. First tests demonstrated how the vehicle is able to mark a

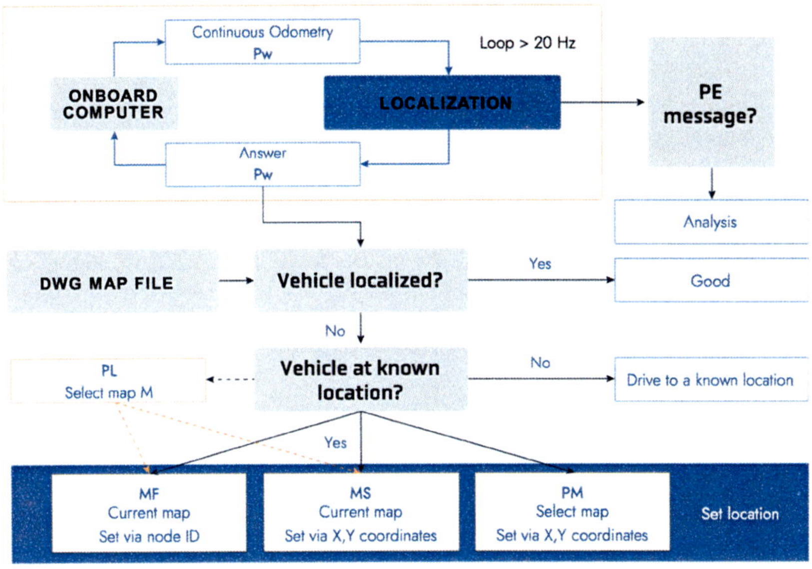


**Fig. 7.** Basic workflow for the navigation software

full hall in 180 min saving about the 75% of the time required by the human operators for the same task.

**Fig. 8.** The vehicle while marking a stand perimeter with round corners

## 4 Conclusions

Two robotics platforms have been presented as an example of sustainable solutions to improve traditional manufacturing processes by limiting the waste of material and the psychical stress for the human operators. The service compactor vehicle has been proved to increase the total amount of flax fibers available for each container by reducing the air pollution caused by the need to have multiple long-distance ground transportation. The results obtained by the line marker vehicle showed how this solution is able to drastically decrease the time required to mark an entire exhibition hall by minimizing also the waste of materials since it is based on a environmentally friendly paint instead of adhesive tape as it happens with the traditional marking methods.

**Acknowledgment.** The financial support of the projects: Agricultural inTeroperabiLity and Analysis System (ATLAS), H2020 (Grant No. 857125), and multimodal sensing for individual plANT phenOtyping in agriculture robotics (ANTONIO), ICT-AGRI-FOOD COFUND (Grant No. 41946) is gratefully acknowledged.

## References

1. Bugmann, G., Siegel, M., Burcin, R.: A role for robotics in sustainable development. In: Proceedings of the IEEE African'11, Livingstone, Zambia, pp. 1–4 (2011)
2. Bausys, R., Cavallaro, F., Semenas, R.: Application of sustainability principles for harsh environment exploration by autonomous robot. Sustainability **11**, 2518 (2019)
3. Galati, R., Reina, G.: Terrain awareness using a tracked skid-steering vehicle with passive independent suspensions. Front. Robot. AI **6**, 46 (2019)
4. Distante, C., Indiveri, G., Reina, G.: An application of mobile robotics for olfactory monitoring of hazardous industrial sites. Indust. Robot. **1**, 51–59 (2019)
5. Carbonari, L., Botta, A., Cavallone, p., Tagliavini, L., Quaglia, G.: Data-driven analysis of locomotion for a class of articulated mobile robots, J. Mech. Robot. **13**(5), 050905 (2021)
6. Reina, G., Foglia, M.: Modelling and handling dynamics of a wind-driven vehicle. Vehicle Syst. Dynam. **57**(5), 697–720 (2019)
7. Galati, R., Mantriota, G., Reina, G.: Design and development of a tracked robot to increase bulk density of flax fibers. J. Mech. Robot. **13**(5), 050903 (2021)
8. Wong, J., Huang, W.: Wheels vs. tracks"– A fundamental evaluation from the traction perspective. J. Terramechan. **43**, 27–42 (2006). https://doi.org/10.1016/j.jterra.2004.08.003
9. The Global Association of the Exhibition Industry (UFI): The global exhibition industry is of major economic importance. https://www.ufi.org/industry-resources/global-exhibitions-day/exhibitions-are-a-large-global-industry/. Accessed 12 July 2021

# Balancing of a Haptic Device with a Set of Springs

Mattia Persampieri, Gianluca Gatti, and Giuseppe Carbone(✉)

Department of Mechanical, Energy, and Management Engineering, University of Calabria, Rende, Italy
giuseppe.carbone@unical.it

**Abstract.** This paper addresses the balancing of a haptic device by using a set of springs. The proposed haptic device is based on a purely translational parallel manipulator. Given the specific application as haptic device, it is necessary to eliminate or minimize the effect of gravity for achieving a suitable positioning and force feedback. Given the peculiarities of this haptic device, this paper reports the design, construction and preliminary testing of a simple and low-cost balancing solution based on an external set of springs. Simulation results and preliminary tests are reported for showing the engineering feasibility and effectiveness of the proposed solution.

**Keywords:** Robotics · Automation · Static balancing · Manufacturing. SDG4

## 1 Introduction

Over the decades, the use of haptic interfaces has become more frequent and important especially in the medical field regarding surgical operations and in the engineering field for reverse engineering. Several works in the literature show a wide range of possible solutions, where it is of fundamental importance the accuracy of positioning and the ability to return a feeling of the forces of interaction. In this paper, we have dealt with the self-balancing of a haptic device to improve performance in terms of positioning accuracy and interaction feeling. The results are analyzed in detail to identify the solutions that allow the best balancing, minimizing the effect of the gravitational field on the haptic interface. Finally, a prototype of the proposed balancing solution was created, and a preliminary experimental validation is reported to demonstrate the effectiveness of the proposed solution.

Haptic interfaces allow human-robot communication through touch, from which feedback is obtained simultaneously according to the user's movements, generally through sensors. More adequately, it is defined as the association of gesture with touch and kinesthesia for providing communication between human and machine [4]. Haptic interfaces differ mainly in their structural type and degrees of freedom. As far as structural type there are serial robot with an open kinematic chain and parallel robot with closed kinematic chain. The degrees of freedom (dofs) are variable and ranging from three to six or more. The minimum number of dofs is three because the systems provide at least three translations.


G. Quaglia et al. (Eds.): I4SDG 2021, MMS 108, pp. 383–392, 2022.
https://doi.org/10.1007/978-3-030-87383-7_42

## 2 R-CUBE Parallel Architecture as Haptic Device

R-Cube (Fig. 1) is a manipulator with a translational parallel structure with three dofs, where the movements along the x, y, z axes have been decoupled using only rotational joints. The name of the manipulator is such because R stands for "Revolute Joints" while Cube is linked to the cubic workspace. The simplicity of this structure is also due to the use of only one actuator for each kinematic chain: a 180° rotation is sufficient to create the desired translation [14]. The remarkable fact of this device is the presence of many positive aspects that make an interesting comparison with other analyzed devices. The main positive aspects of R-Cube are easy manoeuvrability, simple geometry, 3D motion, low costs, and high structural stiffness.

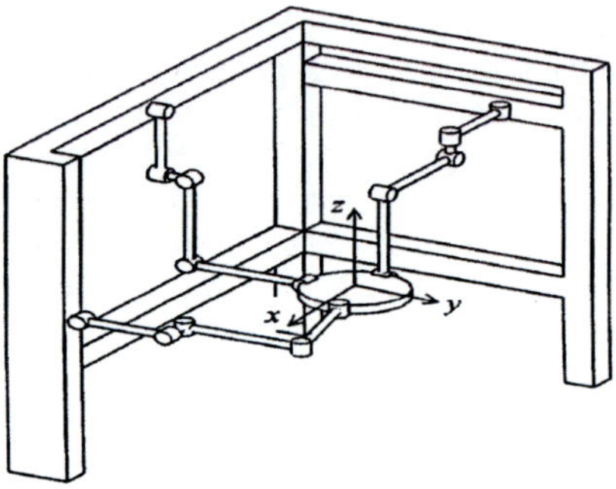


**Fig. 1.** R-Cube Model [14].

## 3 Balancing Models and Simulations

Static balancing is used to reduce the effect of gravity forces, caused by the masses of the mechanism structure and payload, on the motorized joints of the robotic device. Often the device mass can be ten times greater than the mass of the payload [7]. A robotic manipulator is statically balanced if no torque or force is required to the actuators to keep the system in balance [8, 9].

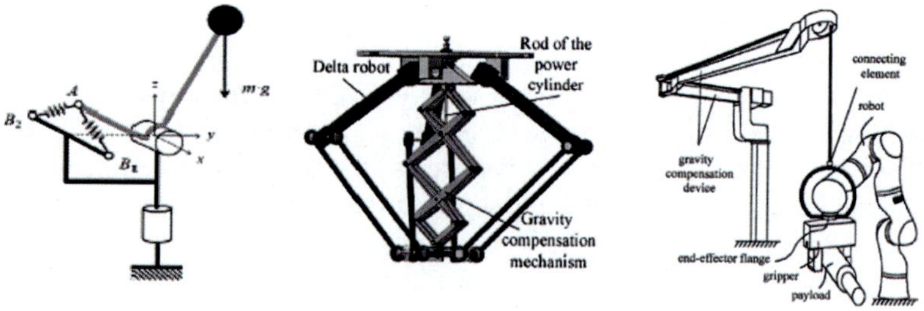


(A) (B) (C)

**Fig. 2.** Examples of balancing systems [11, 12].

The most important types of balancing system (Fig. 2) are:

- Balancing with counterweights.
- Balancing with spring (Fig. 2a).
- Balancing with auxiliary actuator (Fig. 2b).
- Balancing with cables and pulleys (Fig. 2c).

The main features of these types of balancing systems are summarized in Table 1. To balance the R-Cube manipulator a normal traction spring was used [12].

The preliminary study of balancing was made on a two-dimensional model built on Matlab (Student Version) and then compared with a multibody model made on Working Model (Student Version). The parameters for both simulations are visible in Table 2 and relate to stiffness, springs attachment, joystick positions, initial spring length and weight force. In Table 2 with the symbols k1 and k2, we refer to the stiffness of a spring, with L0 we refer to the initial spring length, with the symbols O1 and O2 we refer the attachments of the springs, with the symbols Px and Py we refer to the joystick positions along X and Y directions, respectively, and with WF we refer to the weight force. In Fig. 3 we can see the model built on Matlab and in Fig. 4 there is the model made on Working Model. Finally, in Table 3 there is the comparison between Matlab and Working Model with relative results showing the degree of balancing in the system. The imbalance forces marked with the letter E are given by the difference between the force along the various directions and the weight force.

**Table 1.** Comparison table of Balancing typology.

| **Static Balancing** | 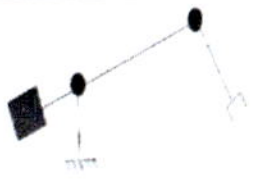 |  | 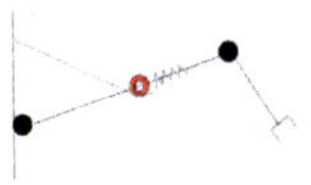 | 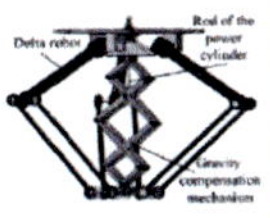 |
|---|---|---|---|---|
| Balancing typology | counterweights | springs | cables and pulleys | Auxiliary actuators |
| Advantages | Easy application | Irrelevant weights gain | Excellent versatility | Balancing accuracy and stability |
| Disad-vantages | Weights gain and size in-crease | Difficult planning and difficult find-ing suitable springs | Keep low stress in the cables, possible cable interference | Weight, price, and computa-tional costs |
| Application | Industrial serial robot | Any robot | Serial and parallel robots | Spatial Robots |

To verify the results simulations implemented in Matlab (Fig. 5) are compared to those using the multibody software Adams (Student Version) (Fig. 6). The parameters for both simulations are listed in Table 4. The main start parameters are stiffness, springs attachment, joystick position, weight force and initial spring length.

**Table 2.** Simulation parameters used for Working Model and Matlab

| k1 [N/mm] | k2 [N/mm] | L0 [mm] | O1 [mm] | O2 [mm] | Px [mm] | Py [mm] | WF [N] |
|---|---|---|---|---|---|---|---|
| 0.15 | 0.15 | 310 | (360; 470) | (0; 470) | (180; 191; 201; 211; 221) | (190; 180; 170; 160; 150) | 5 |

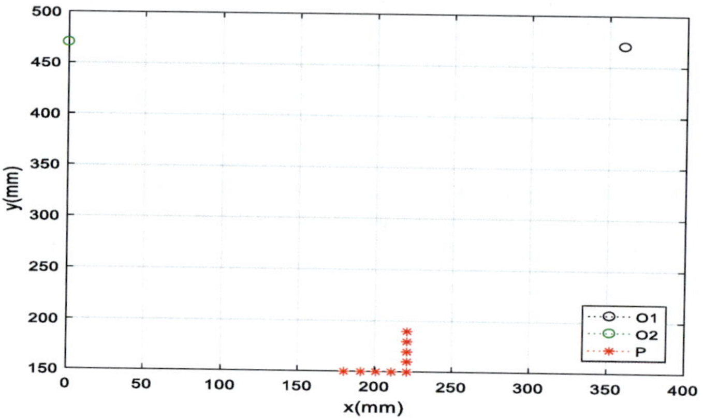


**Fig. 3.** Spring attachment and joystick position, with P we refer to Joystick Position and with O1 and O2 we refer springs attachments.

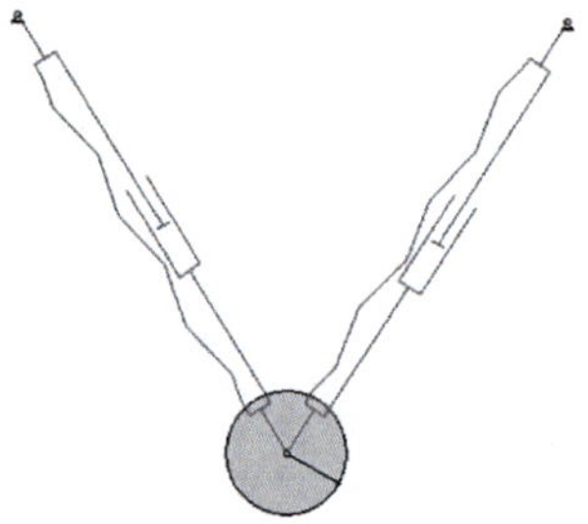

**Fig. 4.** Two-dimensional model on Working Model.

**Table 3.** Simulation results.

| Spring attachment coordinates O1= (360 mm; 470 mm) O2= (0 mm; 470mm) K=0,15 N/mm L0=310 mm | | | | | | | | | | | |
|---|---|---|---|---|---|---|---|---|---|---|---|
| Joystick position [mm] | | Matlab [N] | | | | Working Model [N] | | | | Differences between Matlab and WM [N] | |
| Px | Py | Fy | Fx | Ey | Ex | Fy | Fx | Ey | Ex | ΔFy | ΔFx |
| 180 | 190 | 5.77 | 3.71 | 0.77 | 3.71 | 5.77 | 3.71 | 0.77 | 3.71 | 0.00 | 0.00 |
| 190 | 180 | 7.99 | 5.01 | 2.99 | 5.01 | 7.99 | 5.00 | 2.99 | 5.00 | 0.00 | 0.01 |
| 200 | 170 | 10.28 | 6.34 | 5.28 | 6.34 | 10.28 | 6.33 | 5.28 | 6.33 | 0.00 | 0.02 |
| 210 | 160 | 12.65 | 7.69 | 7.65 | 7.69 | 12.64 | 7.67 | 7.64 | 7.67 | 0.01 | 0.02 |
| 220 | 150 | 15.09 | 9.05 | 10.09 | 9.05 | 15.08 | 9.02 | 10.08 | 9.02 | 0.01 | 0.03 |
| The minimum unbalance values are highlighted in green, and the maximum value are highlighted in red | | | | | | | | | | | |

**Table 4.** Simulation parameters used for Adams and Matlab.

| k1 [N/mm] | k2 [N/mm] | O1 [mm] | O2 [mm] | O3 [mm] | Px [mm] | Py [mm] | Pz [mm] | WF [N] |
|---|---|---|---|---|---|---|---|---|
| 0.15 | 0.15 | (399; 399; 298) | (0; 399; 298) | (399; 0; 298) | (159; 180; 187) | (164; 185; 189) | (130; 180; 193) | 5 |

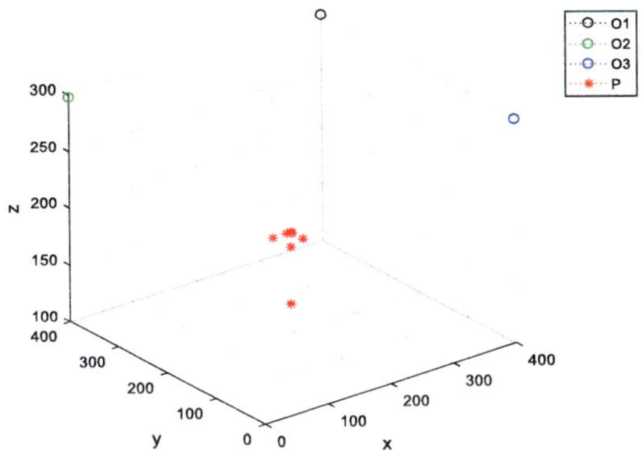


**Fig. 5.** With letter P we refer to the joystick positons and with O1, O2, O3 we refer to the spring's attachments.

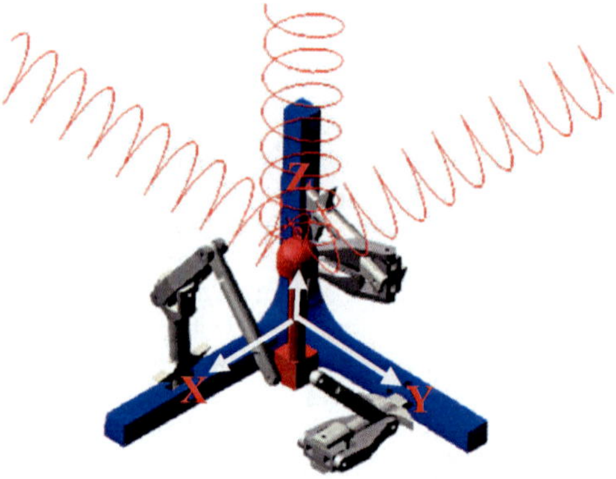


**Fig. 6.** Three-dimensional model of the proposed system within MSC.Adams.

In Table 5 we can see the forces along Z-, Y-, and Z-axes, and the comparison between Matlab and Adams.

In Table 6 there are the unbalancing forces comparison between Matlab, and Adams (Student Version) and we can see there is little difference between both software.

**Table 5.** Force comparison table between Matlab and MSC.Adams.

| Spring attachment coordinates O1 = (399 mm; 399 mm; 298 mm) O2 = (0 mm; 399 mm; 298 mm) O3 = (399 mm; 0 mm; 298 mm) k1 = k2 = 0.15 N/mm | | | | | | | | | | | | |
|---|---|---|---|---|---|---|---|---|---|---|---|---|
| Joystick position [mm] | | | | Matlab [N] | | | Adams [N] | | | Differences between Matlab and Adams [N] | | |
| N° | Px | Py | Pz | Fx | Fy | Fz | Fx | Fy | Fz | ΔFx | ΔFy | ΔFz |
| 0 | 159 | 164 | 130 | 0.00 | 0.00 | 000 | 0.00 | 0.00 | 0.00 | 0.00 | 0.00 | 0.00 |
| 1 | 180 | 185 | 180 | 2.22 | 2.25 | 1.36 | 2.10 | 2.16 | 1.38 | 0.12 | 0.09 | −0.02 |
| 2 | 187 | 189 | 193 | 2.65 | 2.67 | 1.43 | 2.34 | 2.44 | 1.42 | 0.31 | 0.23 | 0.01 |
| | | | | F1 | F2 | F3 | F1 | F2 | F3 | ΔF1 | ΔF2 | ΔF3 |
| Joystick position 1 | | | | −1.35 | 2.40 | 2.40 | −1.35 | 2.43 | 2.39 | 0.00 | −0.03 | 0.01 |
| Joystick position 2 | | | | −1.65 | 2.85 | 2.83 | −1.75 | 2.88 | 2.76 | 0.10 | −0.03 | 0.07 |
| | | | | ΔL1 [mm] | | ΔL2 [mm] | | ΔL3 [mm] | | | | |
| Joystick position 1 | | | | −9.00 | | 16.00 | | 16.00 | | Δz of + 50 mm | | |
| Joystick position 2 | | | | −11.00 | | 19.00 | | 19.00 | | Δz of + 13 mm | | |

**Table 6.** Unbalancing force comparison table between Matlab and Adams.

| Spring attachment coordinates O1 = (399 mm; 399 mm; 298 mm) O2 = (0 mm; 399 mm; 298 mm) O3 = (399 mm; 0 mm; 298 mm) k1 = k2 = 0.15 N/mm | | | | | | | | | | | |
|---|---|---|---|---|---|---|---|---|---|---|---|
| Joystick Position [mm] | | | Matlab [N] | | | Adams [N] | | | Differences between Matlab and Adams [N] | | |
| Px | Py | Pz | Ex | Ey | Ez | Ex | Ey | Ez | ΔEx | ΔEy | ΔEz |
| 159 | 164 | 130 | 0.00 | 0.00 | 0.00 | 0.00 | 0.00 | 0,00 | 0.00 | 0.00 | 0.00 |
| 180 | 185 | 180 | 2.22 | 2.25 | −3.64 | 2.10 | 2.16 | −3.62 | 0.12 | 0.09 | −0.02 |
| 187 | 189 | 193 | 2.65 | 2.67 | −3.57 | 2.34 | 2.44 | −3.58 | 0.31 | 0.23 | 0.01 |
| Average | | | 2.44 | 2.46 | −3.61 | 2.22 | 2.30 | −3.60 | | | |
| Standard Dev. | | | 0.30 | 0.30 | 0.05 | 0.17 | 0.20 | 0.03 | | | |

## 4 Preliminary Testing on a Built Prototype

A CAD model of R-Cube is illustrated in Fig. 7a) and the physical 3D printed prototype is illustrated in Fig. 7b).

Figure 8 shows the R-Cube manipulator balanced with normal springs and the use of IMU sensor for detecting displacements (Fig. 9) and accelerations (Fig. 10). To have more spring attachment, was used a metallic net (Fig. 7b) with mesh 1 cm × 1 cm. In Fig. 9 and Fig. 10 the movement is a translation along the three directions X- Y- and Z- axes. The motion is obtained manually by an operator with measured accelerations ranging value from 0 to 0.2 g after performing translations ranging from 0 to 100 mm. Very small accelerations have been measured also after releasing the end-effector. After its release, the end-effector stands still with full balancing of the gravity forces.

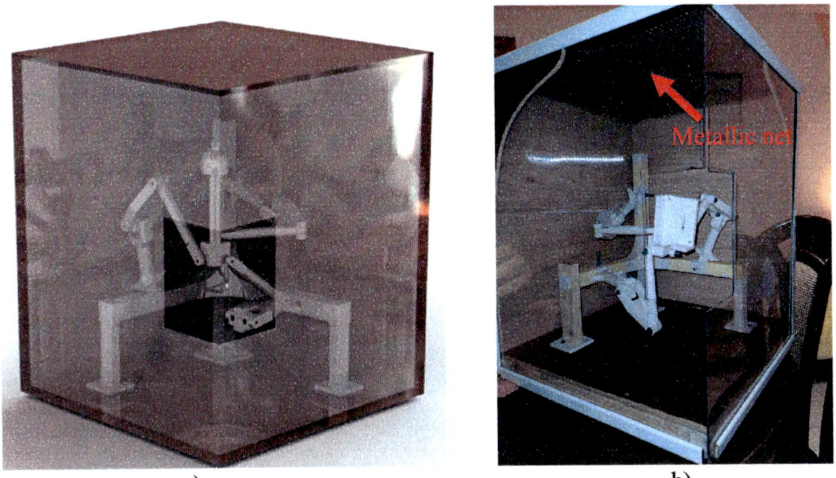


a) b)

**Fig. 7.** The proposed design solution: a) a CAD model; b) a built prototype.

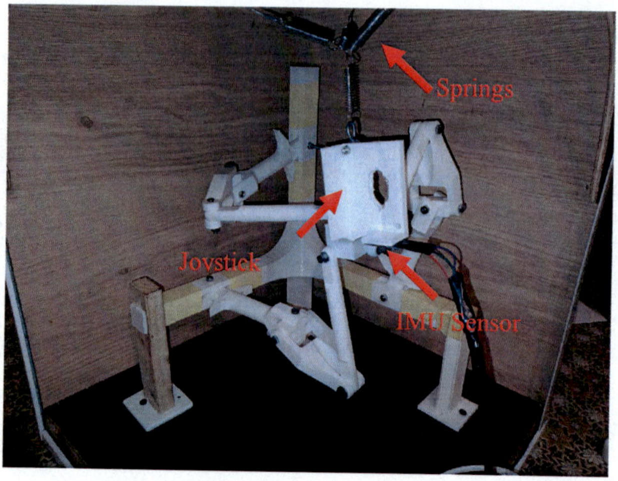


**Fig. 8.** Built prototype with balancing system and IMU sensor.

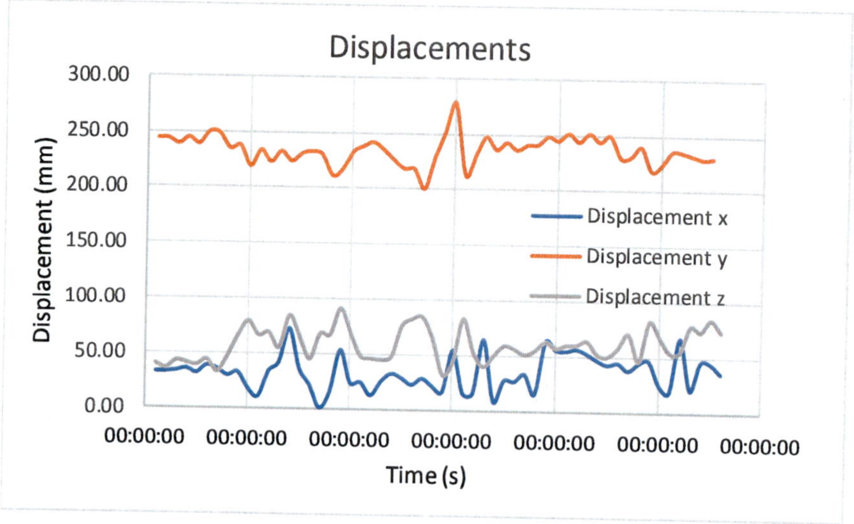


**Fig. 9.** Joystick displacement variation measured with the IMU sensor for the motion indicated in Table 2.

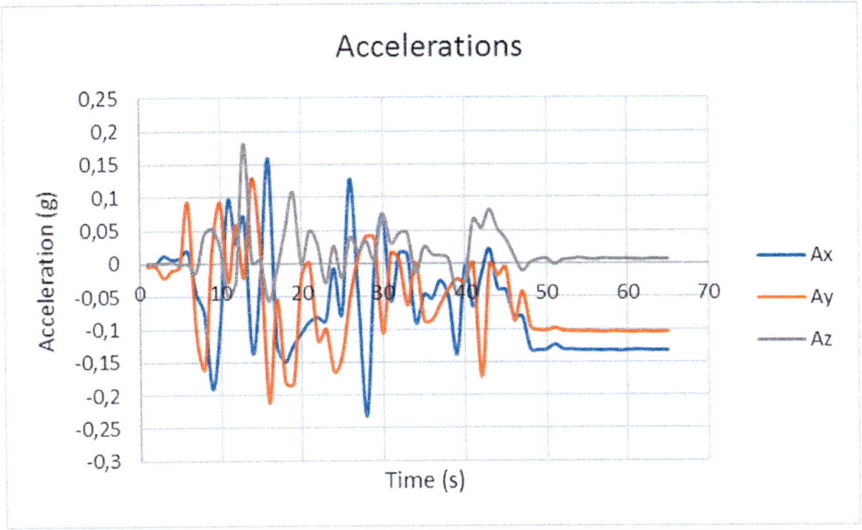


**Fig. 10.** Joystick acceleration variation measured with the IMU sensor.

## 5 Conclusions

This paper has investigated a balancing approach for a parallel manipulator used as haptic device, based on a simple set of linear springs. Numerical simulations have been carried out to properly size and locate springs. Numerical simulations have been used to validate a possible design solution with a suitable balancing behavior in various configurations and along various paths within the feasible workspace. A prototype was built and tested to prove the simplicity and effectiveness of the proposed approach applied to the specific case of study. Future developments will address the optimization of the device.

## References

1. Hamza-Lup, F.G., Popovici, D.M., Bogdan, C.M.: Haptic feedback systems in medical education. IEEE J. Adv. Distribut. Learn. Technol. **1**(2), 7–16 (2013)
2. Shakhovska, N. (ed.): Advances in Intelligent Systems and Computing. AISC, vol. 512. Springer, Cham (2017). https://doi.org/10.1007/978-3-319-45991-2
3. Vulliez, M., Zeghloul, S., Khatib, O.: Design strategy and issues of the Delthaptic, a new 6-DOF parallel haptic device. Mech. Mach. Theory, Elsevier **128**, 395–411 (2018)
4. Hayward, V., Astley, O.R., Cruz-Hernandez, M., Grant, D., Robles-De-La-Torre, G.: Haptic interfaces and devices. Sens. Rev. **24**(1), 16–29 (2004)
5. Görgülü, İ., Carbone, G., Dede, M.İ.C.: Time efficient stiffness model computation for a parallel haptic mechanism via the virtual joint method. Mech. Mach. Theory **143** (2020)
6. Perreault, S., Cardou, P., Gosselin, C.: Approximate static balancing of a planar parallel cable-driven mechanism based on four-bar linkages and springs. Mech. Mach. Theory **79**, 64–79 (2014). https://doi.org/10.1016/j.mechmachtheory.2014.04.008. ISSN 0094-114X
7. Stefan, S.: Static balancing of robot mechanisms and manipulation devices. J. Mech. Eng. Strojnícky časopis **68**(2), 77–90 (2018)

8. Agrawal, S.K., Fattah, A.: Gravity-balancing of spatial robotic manipulators. Mech. Mach. Theory **39**(12), 1331–1344 (2004)
9. Yang, Z.-W., Lan, C.-C.: An adjustable gravity-balancing mechanism using planar extension and compression springs. Mech. Mach. Theory **92**, 314–329 (2015)
10. Cafolla, D., Carbone, G., Ceccarelli, M.: Balancing of a 3-DOFs Parallel Manipulator. Dynamic Balancing of Mechanisms and Synthesizing of Parallel Robots Zhang D., Wei B. (eds), Springer, Cham, pp. 173–191 (2016). https://doi.org/10.1007/978-3-319-17683-3
11. Gatti, G., Carbone, G.: Gravity compensation of robotic manipulators using non-linear spring configurations (2020)
12. Arakelian, V.: Gravity compensation in robotics. Adv. Robot. **30**(2), 79–96 (2016)
13. Ciupitu, L.: Adaptive balancing of robots and mechatronic systems. Robotics **7**(68) (2018). https://doi.org/10.3390/robotics7040068
14. Li, W., Gao, F., Zhang, J.: R-CUBE, a decoupled parallel manipulator only with revolute joints. Mech. Mach. Theory **40**(4), 467–473 (2005)
15. Nikoobin, A., Moradi, M.: Optimal balancing of robot manipulators in point-to-point motion. Robotica **29**(2), 233–244 (2011). https://doi.org/10.1017/S0263574710000093
16. Nikoobin, A., Moradi, M., Esmaili, A.: Optimal spring balancing of robot manipulators in point-to-point motion. Robotica **31**(4), 611–621 (2013). https://doi.org/10.1017/S026357471200063X

# Green Tribology

# Wear Assessment and Reduction for Sustainability: Some Applications

Silvia Logozzo(✉) and Maria Cristina Valigi

Department of Engineering, University of Perugia, Perugia, Italy
{silvia.logozzo,mariacristina.valigi}@unipg.it

**Abstract.** Wear assessment and reduction are topics of great significance in green tribology as part of the *Sustainable Development Goals*. In fact, minimization of wear plays a fundamental role in tribology and sustainability, because it promotes enhancement of useful life of components, limitation of waste material and dangerous debris and energy saving. When these goals are achieved in biomechanical applications, they result in improvements of health and life quality of human beings (SDG3). In the industrial field the main result of wear assessment and optimized design of components and processes is the reduction of pollution and minimization of energy dissipation and costs (SDG9, SDG12). In this paper the analysis of different wear assessment workflows and techniques is presented, and advantages in terms of sustainability are discussed. Some case studies are presented to demonstrate how wear assessment helped achieving sustainable products and applications. An example of redesign of mixing blades, based on digital wear evaluation techniques and application of a power consumption prediction model is presented. Experimental results show that the new design results in a longer durability of the blades and in a lower power consumption. Another case study regarding the evaluation of damage of teeth due to friction is presented. Results were used to make a diagnosis and treat the patient to prevent further teeth damages.

**Keywords:** SDG12 · SDG3 · SDG9 · Green tribology · Wear reduction · Energy saving

## 1 Introduction

Mechanical Engineering is a discipline focused on design, analysis, manufacturing, and maintenance of mechanical systems, based on the knowledge of multiple sciences. Mechanical engineers have an important role in achieving *Sustainable Development Goals* (SDG) [1] as the main objectives of their work are the optimization of machines, products, and processes in terms of reduction of power consumption, pollution, and waste materials. Furthermore, when mechanical engineers design new mechanical or biomechanical components, they aim at improving resistance and lifecycle, bringing social-economic advancement and a contributing to a healthier life. Thus, sustainability for humans and management of resource responsibly should naturally be the leading idea of all the mechanical applications. Among the SDG in mechanical engineering


G. Quaglia et al. (Eds.): I4SDG 2021, MMS 108, pp. 395–402, 2022.
https://doi.org/10.1007/978-3-030-87383-7_43

green tribology is a key discipline, with 12 principles oriented towards sustainability [2]. Green tribology's aim is to lead all the aspects resulting from the interaction of surfaces in relative motion towards environment-social-friendly repercussions. From this perspective, wear reduction and friction control are among the 12 goals of green tribology, as they contribute to the reduction of energy consumption and heat dissipation (SDG12), limitation of pollution and extension of lifecycle of mechanical (SDG9) or biomechanical (SDG3) components, with evident enhancement of sustainbility. Wear reduction is obviously a consequence of the optimization of the wear resistance of components, which in turn depends on accurate design based on the application of wear prediction models and on the execution of specific wear assessment campaigns [3, 4]. Wear prediction models are developed and applied to different fields of mechanics, as for example to evaluate losses of material and possible failures of biomedical prostheses, as hip bearings [5], total shoulder prostheses [6] or total knee replacements. In biotribology, sometimes the wear characterization is performed directly by collecting data from clinical trials and explants (in-vivo characterization) [7], or from laboratory simulations (in-vitro characterization) [8] and analyzing them by statistical approaches. Both in industrial and biomedical applications, experimental wear assessment is fundamental both for validating mathematical models [9], and for characterizing the wear resistance of components to enhance their useful life [10, 11], to redesign them with the purpose of minimizing material and energy consumption [12, 13] or to plan more effective maintenance interventions for human safety purposes [14]. Wear assessment methods based on mathematical models are indirect methods, as they evaluate wear based on mathematical theories, while wear evaluation techniques based on experimental approaches are direct methods, as they allow to directly measure the real volume of material loss. Among the experimental techniques, the use of digital measuring instruments like 3D optical scanners as well as dimensional control software is becoming more and more important as they allow to get a more accurate and complete scenario of the wear distribution and progress, reducing waste of time, costs and enhancing the quality of the results [3, 15]. This paper aims at presenting and discussing different wear assessment workflows oriented to achieve SDG. The described case studies demonstrate how wear assessment can help increasing sustainability.

## 2 Materials and Methods

### 2.1 Wear Assessment Workflows for Achieving SDG

Wear assessment can be included in many applications as part of a complete workflow oriented to sustainability. The main SDG achievable by wear study are SDG12, SDG9 and SDG3, by reduction of pollution and material/energy waste, enhancement of diagnostics and safety, increase of long-term resistance of biomechanical implants for improvements of life and health. All these goals can be reached by applying different workflows, three of which are presented and analyzed in this paper.

The first wear assessment workflow for sustainability (Fig. 1) is aimed at the experimental or simulated characterization of wear progress to analyze how a specific component can be redesigned and optimized to show a longer lifecycle, to reduce the material loss and save energy during its exercise. The goal of the first wear characterization is to evaluate the distribution and quantification of wear over the component's surface.

This step can be done both by simulation with wear prediction models, already validated by experiments, or by experimental campaigns on the worn component after a certain period of use. The experimental characterization in this case is preferably as it can give result also to phenomena occurred during exercise which could not be predictable by simulations.

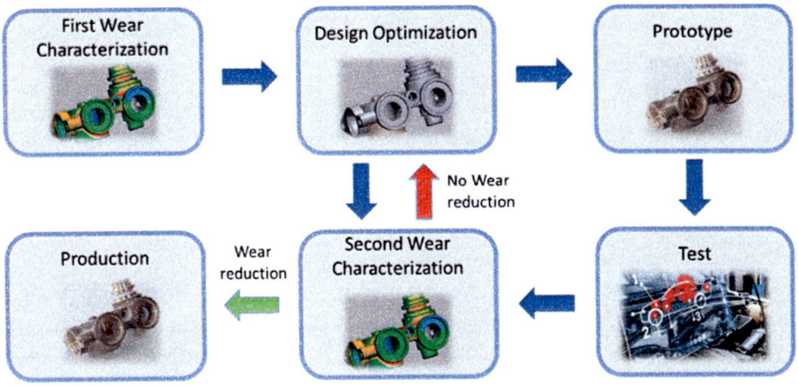


**Fig. 1.** First wear assessment workflow for sustainability

The second step of the workflow is much simpler if the wear characterization is complete as much as possible. In fact, design optimization operations depend on the deep knowledge of how and where wear occurs and for what reasons. It is often useful to set different optimized designs to check their effectiveness in the next step.

The second wear assessment after design optimization is needed to check if the component has been effectively improved in terms of wear resistance. This step can be performed directly using the CAD projects if wear simulation software is used or it must be performed on new optimized test components. If physical prototypes are used and installed in the real machine or exercise environment the evaluation of wear resistance is much better, even because in the real working environment other aspects of the component effectiveness can be considered and evaluated, as for example if they are still fast or sufficiently efficient in accomplishing their work.

This workflow can be applied both to mechanical components and orthopedic prostheses with obvious advantages. In the case of biomechanical prostheses as hip or knee replacements, the test phase can be performed using specific laboratory wear simulators which replicate the dynamic and kinematic conditions of the human walk or of other joint movements, like shoulder inserts.

The second wear assessment workflow for sustainability is represented in Fig. 2. In this case the wear characterization allows to analyze malfunctioning both in mechanical and biomedical environments. The knowledge of how wear is distributed over the component's surface gives a feedback on possible interventions to set in order to restore a healthy situation. For example, throughout the wear characterization of teeth the dental practitioner can diagnose bruxism phenomena or malocclusions and he can intervene by designing a proper bite to help patient to recover and improve his situation.

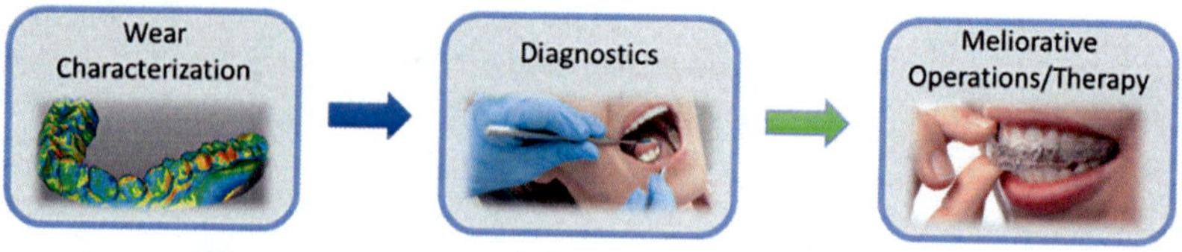


**Fig. 2.** Second wear assessment workflow for sustainability

If for instance this workflow is applied to orthopedic endo-prostheses, it enables researchers to evaluate if debris were released in the patient joint and allows the medical doctor to set proper treatments. In the mechanical, civil or industrial fields, this workflow helps enhancing production processes or optimizing machinery exercise by changing working parameters like geometrical configurations or kinematic and dynamic factors, rheological and resistance properties of processed materials, etc.

The third wear assessment workflow is mainly applied to machinery and infrastructures. In this case the wear assessment process is useful to characterize problems due to cracks, geometrical changes, and loss of material. Based on dimensional control and diagnostics, proper ordinary or extraordinary maintenance interventions can be planned to assure safety and correct functioning. Some examples of this workflow can be found in applications related to infrastructures like railways or in production plants.

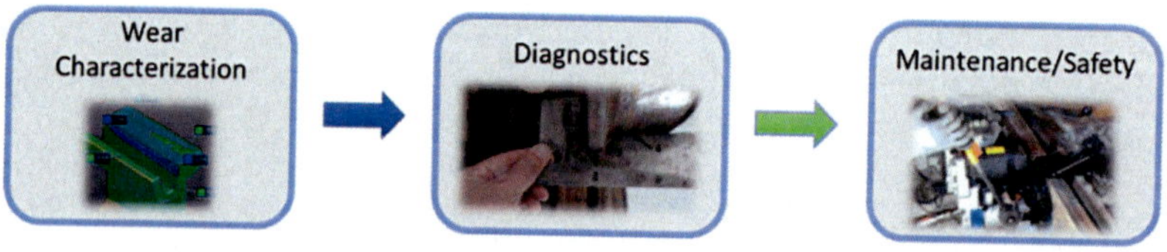


**Fig. 3.** Third wear assessment workflow for sustainability

In the case of railways, the wear characterization process allows to detect critical situations where tracks need to be restored or it can be used to investigate eventual damages of rolling components like train wheels. In production plants, wear assessment allows to evaluate the entity of eventual damages and to plan replacements.

## 2.2 Digital Wear Assessment Methods and Advantages for Sustainability

Wear assessment can be very helpful in finding all the needed information to operate an effective redesign or to detect damages in machinery, infrastructures, and medical applications, especially if it is performed by 3D digital methods [3]. Those methods include the use of 3D optical or contact digitizers and mesh to CAD or mesh to mesh compare software, as cutting-edge inspection software. 3D scanners allow to reconstruct complete 3D digital models of the worn components or damaged human body parts. 3D inspection software enables to compare the worn or damaged part with a reference one (CAD project of 3D scan of the unworn part), displaying 3D wear maps. In this way a complete scenario of the distribution and amount of material loss can be achieved, overcoming the main disadvantages of traditional methods like weighing. Moreover, the use of portable 3D digitizers gives the opportunity of measuring and analyzing components in their working environment without dismounting them from the machine.

# 3 Results

## 3.1 Applications of the First Wear Assessment Workflow and Relative SDG

One of the applications of the first wear assessment workflow regards the design optimization of components like mixing blades of concrete twin shaft mixers. The aim of this study was to redesign those blades in order to achieve two main goals oriented to sustainability: longer useful life and reduction of power consumption. The considered machine was a twin shaft mixer (S.I.Co.Ma. srl, Perugia, Italy) processing 4500 $m^3$ of concrete, with 8 blades for each shaft (Fig. 4a). The analyzed blade was the fourth of the left shaft and the measurements were done directly on the machine when it was new and after 6 and 12 months, named $t_1$ and $t_2$, respectively. The analyzed blade has a typical T shape, as represented in Fig. 4b.

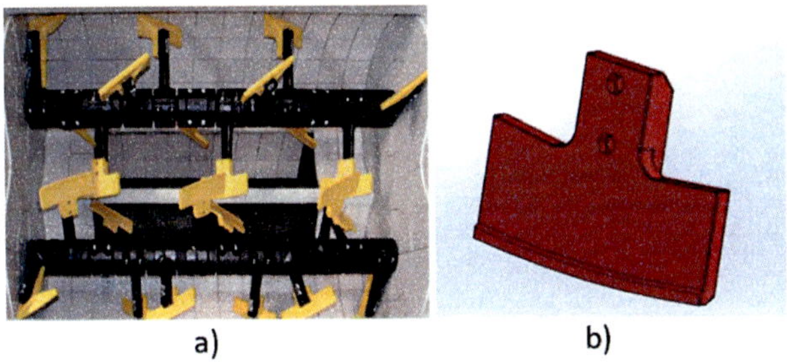


**Fig. 4.** a) Twin shaft mixer; b) T blade.

The first wear characterization was aimed at studying wear progress after different working periods, to characterize the most worn areas and analyze how the blade design could be modified to reduce wear but also to save energy. Wear mechanisms affecting mixing blades durability are mainly erosive and abrasive. Erosive wear is typical of the mixing wet phase, as it is due to slurry flow. Abrasive wear is typical of the dry phase, as it occurs due to the mechanical impact between blades and solid aggregates. Corrosive wear is normally negligible with respect to abrasive and erosive wear as passivation protects blades from chemical aggression. The wear evaluation was performed by using a metrology grade portable 3D optical laser scanner (FreeScan X7, Shining 3D Tech. Co., Ltd., Hangzhou, China) and a mesh compare software (Geomagic Wrap 2017, 3D Systems, Inc., Rock Hill, South Carolina, USA). After the first evaluation the bottom section of the blade resulted as the most worn region and wear was mostly concentrated on the right and external wing. Furthermore, thanks to a validated power consumption model, the power spent with a typical cycle with T blades was calculated in the transient and regime phase of the mixing operation [16–18]. With these results a new blade design was studied with a L shape, that was an asymmetric T with a longer wing on the outer part of the blade. This new shape processed by the power consumption model resulted to guarantee an energy saving of about 9% both in the transient and regime phase. Then sixteen L new L blades were prototyped and tested with the same twin shaft mixer, with the same mixture composition. The fourth blade of the left shaft was measured after the same time periods $t_1$ and $t_2$, to check its wear resistance. The L design and results of the final wear evaluation at the end of $t_2$ are reported in Fig. 5.

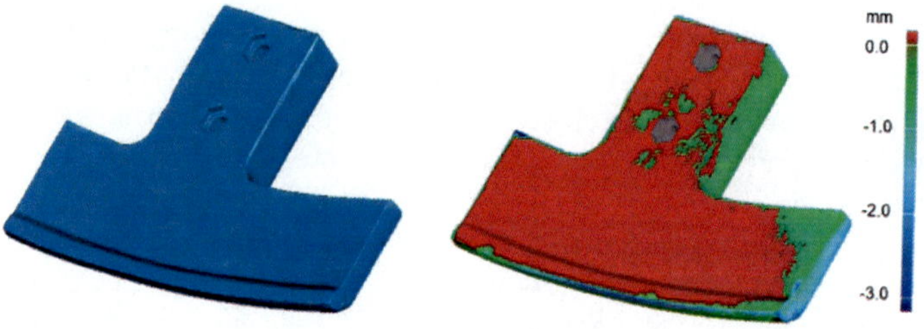


**Fig. 5.** Final wear characterization

Considering the blade bottom section, the initial volume $V_0$ and the current volume $V_i$, after each test period $t_i$, the wear rate can be calculated according to

$$\chi_i = \frac{V_0 - V_i}{V_0}\% \quad (1)$$

Results of both first and final wear assessment are reported in Table 1. One can observe that with the new design the wear resistance of the blade is improved, allowing a longer lifecycle of the component.

**Table 1.** Wear results

| Time period | Material loss $mm^3$ | Wear rate |
|---|---|---|
| $t_1$ (T blade) | 6180 | 3.1% |
| $t_2$ (T blade) | 8519 | 4.4% |
| $t_1$ (L blade) | 4538 | 2.3% |
| $t_2$ (L blade) | 7140 | 3.7% |

## 3.2 Applications of the Second Wear Assessment Workflow and Relative SDG

An application of the second wear assessment workflow for sustainability is the case of the wear characterization of teeth of a 50 years old female patient. The aim of the application was to investigate the damage of the patient's teeth to identify the cause and plan a treatment. The patient went to the dental practitioner after three years from the previous visit and the doctor found the teeth were damaged and consumed. To determine the distribution of losses and to find the cause of the damage the dies generated by the old and new dental impressions were acquired by a dental desktop 3D scanner (AutoScan, Shining 3D Tech. Co., Ltd., Hangzhou, China) and superimposed with a mesh compare software (Geomagic Wrap 2017, 3D Systems, Inc., Rock Hill, South Carolina, USA). Results (Fig. 6) revealed a bruxism condition, as the material loss was randomly distributed over the teeth surface. If the wear was concentrated in a specific region, malocclusion issues could be considered as the cause of the damage. From results the technician was able to design a proper bite for the patient. Subsequent follow-up with further wear assessments demonstrated that the damage was not increasing and that the therapy was effective.

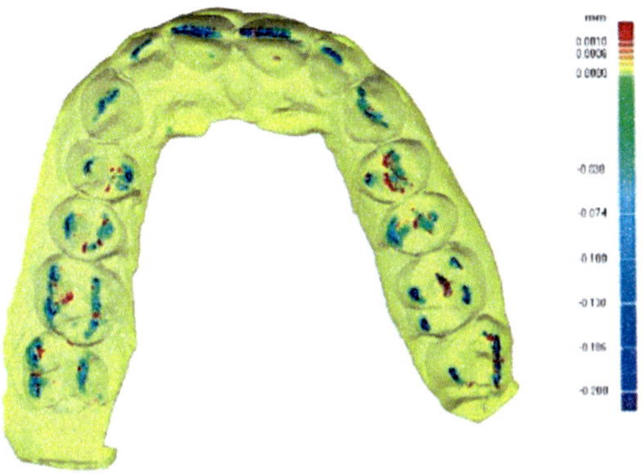


**Fig. 6.** Wear characterization of teeth

## 4 Conclusion

This paper presented three wear assessment workflows as part of the green tribology discipline and oriented to achieve three of the *Sustainable Development Goals*: SDG12, SDG9 and SDG3. In fact, the three workflows, throughout various wear evaluation processes, can promote energy and material saving, reduction of pollution, extension of the lifecycle of mechanical and biomedical components, enhancement of the infrastructures' safety, improvement of health and life of people. The three workflows were described in detail and two case studies were presented. The first one described the improvement of a mechanical component of a concrete mixing machine in terms of wear resistance, with consequent extension of useful life, reduction of material waste and pollution and power consumption during the mixing cycle. The second case study regarded a dental application, where the 3D wear assessment procedure enabled to diagnose the disease and to plan the treatment.

# 1. References

1. Okokpujie, I.P., Fayomi, O.S.I., Oyedepo, S.O.: The role of mechanical engineers in achieving sustainable development goals. In: Procedia Manufacturing 35 (Proceedings of the 2nd International Conference on Sustainable Materials Processing and Manufacturing, SMPM 2019), pp. 782–788 (2019)
2. Nosonovsky, M., Bhushan, B.: Green tribology: principles, research areas and challenges. Phil. Trans. R. Soc. A **368**, 4677–4694 (2010)
3. Valigi, M.C., Logozzo, S., Affatato, S.: New challenges in tribology: wear assessment using 3D optical scanners. Materials **10**(5), 548 (2017)
4. Mattei, L., Di Puccio, F.: Influence of the wear partition factor on wear evolution modelling of sliding surfaces. Int. J. Mech. Sci. **99**, 72–88 (2015)
5. Di Puccio, F., Mattei, L.: Biotribology of artificial hip joints. World J. Orthopedics **6**(1), 77–94 (2015)
6. Mattei, L., Di Puccio, F., Joyce, T.J., Ciulli, E.: Numerical and experimental investigations for the evaluation of the wear coefficient of reverse total shoulder prostheses. J. Mech. Behav. Biomed. Mater. **55**, 53–66 (2016)
7. Affatato, S., Caputo, D., Bordini, B.: Does the body mass index influence the long-term survival of unicompartmental knee prostheses? A retrospective multi-centre study. . Int. Orthopaedics (SICOT) **43**, 1365–1370 (2019)
8. Affatato, S., Ruggiero, A.: A critical analysis of TKR in vitro wear tests considering predicted knee joint loads. Materials **12**(10), 1597 (2019)
9. Butini, E., Marini, L., Meli, E., Rindi, A., Valigi, M.C., Logozzo, S.: Development and validation of wear models by using innovative three-dimensional laser scanners. Adv. Mech. Eng. **11**(8), (2019)
10. Valigi, M.C., Logozzo, S., Affatato, S.: In vitro 3D wear characterization of knee joint prostheses. Mech. Mach. Sci. **73**, 3855–3863 (2019)
11. Mattei, L., Di Puccio, F., Ciulli, E., Pauschitz, A.: Experimental investigation on wear map evolution of ceramic-on-UHMWPE hip prosthesis. Tribol. Int. **143**, 106068 (2020)
12. Valigi, M.C., Logozzo, S., Rinchi, M.: Wear resistance of blades in planetary concrete mixers. Design of a new improved blade shape and 2D validation. Tribol. Int. **96**, 191–201 (2016)
13. Valigi, M.C., Fabi, L., Gasperini, I.: Wear resistance of new blade for planetary concrete mixer. In: 5th World Tribology Congress, WTC 2013, vol. 2, pp. 1208–1211 (2013)
14. Valigi, M.C., Logozzo, S., Butini, E., Meli, E., Marini, L., Rindi, A.: Experimental evaluation of tramway track wear by means of 3D metrological optical scanners. In: Proceedings of the 11th International Conference on Contact Mechanics and Wear of Rail/Wheel Systems (CM2018), pp. 1007–1012 (2018)
15. Affatato, S., Valigi, M.C., Logozzo, S.: Knee wear assessment: 3D scanners used as a consolidated procedure. Materials **13**(10), 2349 (2020)
16. Valigi, M.C., Logozzo, S., Landi, L., Braccesi, C., Galletti, L.: Twin-shaft mixers' mechanical behavior numerical simulations of the mix and phases. Machines **7**(2), 39 (2019)
17. Valigi, M.C., Logozzo, S., Galletti, L.: Effect of design parameters and operating conditions in concrete twin shaft mixers. Mech. Mach. Sci. **68**, 424–431 (2019)
18. Valigi, M.C., Logozzo, S., Landi, L., Galletti, L.: Power absorption and experimental results of concrete twin shaft mixers [Leistungsabsorption und Versuchsergebnisse bei Doppelwellenmischern für Beton]. Betonwerk und Fertigteil-Technik/Concrete Plant and Precast Technol. **3**, 40–44 (2020)

# The Challenges of Oil Free Bearings in Micro-turbomachinery

Federico Colombo(✉), Luigi Lentini, Terenziano Raparelli, and Andrea Trivella

Politecnico di Torino, Turin, Italy
{federico.colombo,luigi.lentini,terenziano.raparelli, andrea.trivella}@polito.it

**Abstract.** Among all types of bearings, gas bearings are the most environmental friendly supports as they do not need to be lubricated with mineral oil. The air or the gaseous working fluid used in micro-turbomachinery are used to lubricate the bearings and realize the supporting task. Due to their characteristics, they cannot replace the other bearings in all applications; however, the research is always working in order to broaden the boundaries of gas bearings. This paper is aimed at describing the most promising application domains in which gas bearings are challenging and gives a survey on all types of gas bearings, discussing their pros and cons.

**Keywords:** SDG7 · Oil-free bearings · Micro-turbomachinery · Gas bearings

## 1 Introduction

In micro-turbomachinery higher and higher rotational speeds are asked today in order to accomplish the miniaturization demands. The trend is towards a downscaling to obtain more compact devices with higher energy to weight ratios.

A broad range of applications can be listed, such as air management systems for aircrafts, auxiliary power units, micro gas turbines for power generation or fuel cell hybrid systems, industrial turbo-compressors and blowers, turbo expanders, cryogenic pumps, cryo-coolers and automotive turbochargers.

Generally, micro-turbomachines are supported by fluid film journal bearings, which can bear high loads at speeds much higher respect to the maximum speeds tolerated by rolling element bearings. Anyway, oil-lubricated hydrodynamic bearings are not much more appealing than conventional rolling element bearings. Firstly because cavitation of the liquid lubricant prevents their applicability at very high speeds. Secondly, unless the process fluid is used as a lubricant [1], an external system is needed to provide the lubricant. Thirdly, a sealing system with inherent losses is necessary to prevent leakage of the lubricant fluid.

In summary, in the comparison between rolling element bearings and fluid film bearings, external lubrication systems and sealing are usually necessary for both types


G. Quaglia et al. (Eds.): I4SDG 2021, MMS 108, pp. 403–411, 2022.
https://doi.org/10.1007/978-3-030-87383-7_44

of bearings. The situation changes in case a gaseous fluid or ambient air is used as a lubricant.

Gas bearings offer advantages respect to oil lubricated bearings such as low friction and low heat generation. The reduced power losses are due to the low viscosity of the gaseous lubricant, which is at least two orders of magnitude lower than oil.

In addition, they do not require complex and costly systems for the lubricant circulation. Eliminating the oil-lubricated bearings, the system is simplified and can be more compact. However, gas bearings disadvantages include lower load capacity, less fluid damping and smaller clearances/tight tolerances. Moreover, the efforts spent in scaling the systems to obtain more compact designs bring to reach higher operating temperatures.

With the increasing demand of economical, more compact, light and environmental friendly energy conversion systems, gas bearing represent a challenging sector in oil-free turbomachinery.

The challenges of gas bearings in the aforementioned applications concern i) the environmental friendliness, ii) the low power losses, iii) the possibility of working at extreme temperature and last but not least iv) the possibility of reaching very high rotational speeds.

There are different types of gas journal and thrust bearings that could be suitable for oil free lubricated systems. In the following, the application domains in which gas bearings are employed are described. Then, a state of the art on the bearing types is given and pros and cons of each typology are discussed. Finally, the evolution undertaken by researchers in this field in the last years is mentioned and future trend is conjectured.

## 2 The Application Domains

For high speed applications, gas bearings represent a clean oil-free solution. They are extensively adopted in high precision electrospindles and very high speed spindles for micromachining. Another important sector involves several micro-turbomachinery applications of power up to 200 kW. A brief list of these applications is here below presented.

### 2.1 Organic Rankine Cycle Machines

The Organic Rankine cycle (ORC) is employed to convert heat at low temperature to electricity [2]. The heat can be wasted heat, geothermal heat, solar energy or heat from biomass combustion. Example of ORC are domestic cogeneration plants, industrial or farming processes.

Micro-cogeneration ORC have become popular in the last years. To eliminate the need of auxiliaries for providing the lubricant, a low-boiling working fluid could be chosen and a small amount of this used to lubricate the bearings. This is an advantage of the so-called oil-free bearings, which don't need an external oil supply system.

## 2.2 Air Cycle Machines

The Air Cycle Machines (ACM) are high speed components employed in the environmental control system of aircrafts for managing cooling, heating and pressurization of the environment. The ACM has a turbine, a compressor and a fan mounted on a single shaft rotating at about 40 krpm. Since 80s rolling element bearings have been converted to foil air bearings, which do not need externally pressurized air.

## 2.3 Heat Pumps

In EU countries, there is a high demand of low temperature heat in range 60 °C to 100 °C. A more efficient way of providing this heat respect to the combustion of fossil fuels is by using heat pumps.

Heat pumps are used in buildings for space heating and hot water production and in industry for waste heat recovery, cooling and refrigeration [3]. In industry, there are large quantities of waste heat, which can be converted in process heat through heat pumps. The heat source, however, has a significant impact on system volumes and efficiency.

The simplest heat pump system consists of a single stage compressor, a condenser, an expansion valve and an evaporator. More complex cycles have been developed for different applications in order to increase performance.

Small scale oil free turbocompressor are expected to spread out also in this setting. In this case, the working fluid operates near the saturation point and it is important to avoid condensation in bearings.

## 2.4 Helium and Hydrogen Refrigerators

Dynamic gas bearings have been employed in helium [4, 5] and more recently hydrogen refrigerators, progressively replacing the oil bearings and the externally pressurized gas bearings. As the dynamic viscosity of hydrogen is about two orders of magnitude lower than the viscosity of helium, initially the dynamic gas bearings employed helium as a lubricant in order to reach the required load capacity [6]. More recently, these bearings were improved and redesigned to work successfully with hydrogen [7].

## 2.5 Turbo Blowers

An important setting of turbo blowers is the treatment of wastewater and sewage. Large volumes of air are required to be pumped through the water. This process reduces the levels of volatile organic compounds and removes dissolved gases.

Turbo blowers are high-speed compressors providing high air flow rates with a low pressure ratio. They consist of a single impeller connected with an electric motor, a thrust bearing and a cooling fan. Oil free turbo blowers have been investigated up to 75 kW power and 30 krpm rotational speed [8].

### 2.6 Automotive Turbochargers

Turbochargers are employed to improve the efficiency and the power output of internal combustion engines [5]. Heat energy is recovered from the exhaust gases and used to supply an additional air flow into the engine. Foil bearings have been employed to support turbochargers of diesel vehicle engines [9]. These bearings demonstrated to operate up to 90 krpm.

## 3 Overview on Gas Bearings

Gas bearings can be categorized into three main groups: aerostatic, aerodynamic and hybrid. The first ones need an external system to provide pressurized air; the second ones do not need compressed air supply systems as the load capacity is generated by the dragging effect of rotation. The third category involves both the external pressurization and the dynamic effect of rotation.

Bearings that exploit the aerodynamic effect are tilting pad bearings, herringbone grooved bearings and bearings with a modified geometry respect to plain bearings. A description of all these typologies for journal bearings follows (see Fig. 1), with also a discussion about pros and cons of each solution.

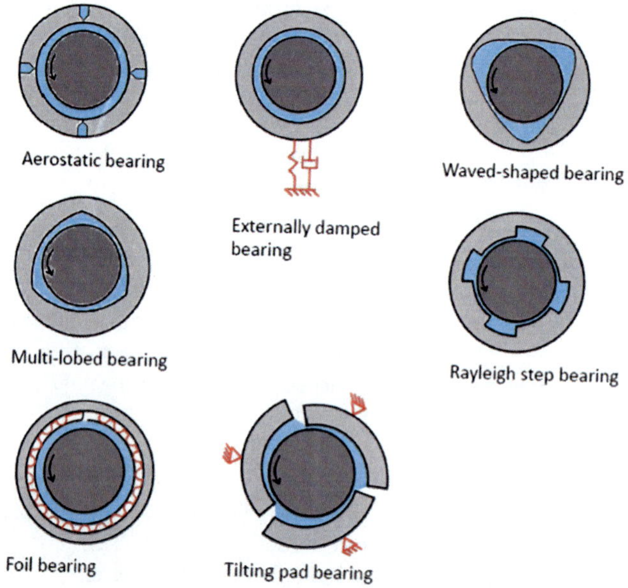


**Fig. 1.** Different typologies of gas journal bearings

### 3.1 Aerostatic Bearings

Aerostatic bearings are endowed of a series of orifices through which the externally pressurized air flows into the radial gap between the shaft and the bushing. The pressure distribution that is established generates the load capacity when the shaft is eccentric, as it creates a restoring force when the shaft is moved away from the center of the bushing. The direct coefficients of stiffness of such bearings are high and proportional to the supply pressure. For this reason, also the stability threshold is high.

A shortcoming of such bearings, which penalize this typology especially in turbomachinery, is the need of the supply system, which represents a power-consuming device and decreases the total efficiency of the machine. For this reason, aerostatic bearings are generally used in precision application, where the complexity of the system justifies an auxiliary equipment.

Porous bearings are a particular typology of aerostatic bearings. They are composed of a porous material that is crossed by the supply air flow. They are equivalent to aerostatic bearings endowed with lots of very small supply orifices. The pressure distribution encountered in these bearings is uniform and little affected by form errors of the bushing internal profile. Consequently, they are particularly suitable to precision applications.

The limitations of aerostatic bearings are connected to the maximum pressure at which the air is compressed in industrial environments. Compared to fluid bearings lubricated with mineral oil, the pressure is at least a tenth, therefore also the load capacity is limited.

### 3.2 Bearings with Grooves

Herringbone grooved journal bearings and spiral grooved thrust bearings are self-acting gas bearings. They are provided of proper shallow grooves machined on the shaft which facilitate the pumping of the lubricant fluid inside the bearing and generate the load capacity above a minimum speed. The grooves form an angle with the circumferential direction and through them, the fluid is dragged from the bearing edges into the bearing during rotation.

The grooves in journal bearings contribute to reduce the attitude angle and increase the stability threshold speed [10]. This type of bearings is not compliant and the film heights must be quite small in order to assure stability. Due to this, stringent manufacturing tolerances and good surface roughness are required. A shortcoming of grooved bearings is associated to possible thermal distortions, which could cause a seizure or a significant decrease of load capacity. Moreover, the small clearance causes high frictional losses.

Spiral groove gas thrust bearings technology has been considered from 50s and modern applications are limited to small size rotors. Their load capacity is high but this requires tight tolerances to achieve. They are quite sensitive to thermal distortions that in practice reduce their load capacity [11].

### 3.3 Externally Damped Bearings

As gas bearings suffer at very high rotational speeds of self-excited instability due to the low damping and cross-coupled stiffness coefficients, they can be mounted flexibly on a damping supporting structure. The damping introduced by this structure is helpful in raising the threshold speed of journal bearings [12, 13]. Investigations on the stiffness and damping characteristics are essential in designing the supporting parts [14, 15].

### 3.4 Bearings with Modified Geometry

The increased stability can also be assured by a modification of the plain bearing profile: the so-called noncircular bearings try to break the symmetry of pressure and reduce the cross coupling behavior in order to postpone the instability onset. It is the case of waved-shaped, multi-lobed and Rayleigh step bearings.

Waved bearings are very similar to multi-lobed bearings; the first ones present a sinusoidal clearance in the circumferential direction and are quite easy to manufacture, while the second ones, together with the Rayleigh step bearings, are more difficult to fabricate due to the more complex profile and the required strict tolerances.

The main shortcoming of such bearings is due to air compressibility, which limits their maximum pressure and load capacity.

### 3.5 Tilting Pad Bearings

These bearings are composed of multiple pads surrounding the shaft [16]. The pads can accommodate their attitude rotating about a pivot thus decreasing the attitude angle and increasing stability respect to fixed geometry bearings. It is known that they have excellent stability characteristics thanks to their low cross-coupled stiffness [17].

The pads can be pivot-less, they can be connected by a bolt and nut mechanism to the external carter or they can be obtained from the external structure by EDM technology. By use of compliant elements, they can easily compensate for the shaft centrifugal expansion.

Due to the conformable gap geometry, tilting-pad bearings, together with foil bearings, show superior stability properties and can in part compensate thermal distortions.

### 3.6 Foil Bearings

Air foil bearings makes use of metal foils surrounding the shaft and preloaded against it [18]. When the shaft rotates, an aerodynamic pressure is generated between shaft and foils separating them with a thin film of air. As the shaft expands, the foils get pushed further away keeping the radial gap quite constant. The aerodynamic pressure distribution is fully developed above a threshold speed named liftoff speed. Below this rotational speed, sliding contact occurs between bearing and shaft, so adequate solid lubrication is important not to damage the surfaces.

The radial compliance of the bearing provides tolerance to shaft misalignment, shock loading, centrifugal and thermal distortion.

Multiple foils can be arranged at discrete circumferential positions like in the multi-pad foil bearing or a single top foil can be wound around the shaft. In this last case, a bump foil is positioned outside the top foil in order to provide an elastic foundation which can compensate for radial vibrations of the shaft [1].

A similar principle is used in the thrust bearings [19, 20], where the top foil is supported by axial springs or corrugated foils which form the elastic foundation.

Initially employed in air cycle machines (ACMs) [21], due to low load capacity requirements and low operating temperatures, their performance has increased in the last three decades: more advanced solid lubricants have been developed and foil bearings with higher load capacity and dynamic characteristics have been produced. These advances have broadened the applications of foil bearings to include turbochargers, power generators [22] and small gas turbine engines [23, 24].

### 3.7 Squeeze Film Bearings

In these bearings, the load capacity is generated by the radial motion of the shaft (squeeze), imposed by an external exciter, such as a piezoelectric actuator or an electromagnetic device. The squeeze motion provides also viscous damping but generally, the stiffness is small [25]. Generally, they are independent of external gas supply.

## 4 The Evolution of the Last years and Future Trends

As well explained in [26], the research on gas bearings began in the end of the 19th century and the first industrial applications began to develop after the second World War. From 50s up to 90s gas bearings technology played a key role in several settings. Initially the research efforts were aimed at increasing stiffness and optimizing the steady-state characteristics. Then attention was put in dynamic stability problems, which are still a challenging research field. Since 70s the interest in this field began to slow down. In the last years it has been experienced a turnaround: despite gas lubrication is considered a full developed applied science, modern technology demands higher and higher requirements, especially in high precision and in high rotational speed applications.

It is in these two application domains that gas bearings are expected to consolidate their presence. In particular, the micro-turbomachinery sector offers a challenge to further develop these components as the requirements of long life, low losses and miniaturization needs matches well with the peculiarities of gas bearings.

## 5 Conclusions

This paper shows a state of the art of the gas bearings technology and an overview on their potential applications.

Being oil free components, these devices are environmentally friendly and have gradually replaced oil-lubricated bearings in numerous applications.

Depending on the application domain, a different type of gas bearings is more suitable. The most promising ones, due to the self-acting load capacity, are aerodynamic bearings: foil bearings, grooved bearings and titling pad bearings.

Research on these components is still very challenging especially in micro-turbomachinery due to the market demands in terms of miniaturization, stability at increased rotational speeds, increased load capacity and high temperature operation.

## References


1. Heshmat, H., Hryniewicz, P., Walton, J.F., Willis, J.P., Jahanmir, S., DellaCorte, C.: Low-friction wear-resistant coatings for high-temperature foil bearings. Tribol. Int. **38**, 1059–1075 (2005). https://doi.org/10.1016/j.triboint.2005.07.036
2. Lecompte, S., Huisseune, H., Van Den Broek, M., Vanslambrouck, B., De Paepe, M.: Review of organic Rankine cycle (ORC) architectures for waste heat recovery. Renew. Sustain. Energy Rev. **47**, 448–461 (2015). https://doi.org/10.1016/j.rser.2015.03.089
3. Arpagaus, C., Bless, F., Schiffmann, J., Bertsch, S.S.: Pompes à chaleur à multiples températures: Une synthèse de la littérature. Int. J. Refrig. **69**, 437–465 (2016). https://doi.org/10.1016/j.ijrefrig.2016.05.014
4. Jain, A., Jadhav, M.M., Karimulla, S., Chakravarty, A.: Parametric studies on floating pad journal bearing for high speed cryogenic turboexpanders. IOP Conf. Ser. Mater. Sci. Eng. **278** (2017). https://doi.org/10.1088/1757-899X/278/1/012026
5. Lee, Y.B., Kwon, S.B., Kim, T.H., Sim, K.: Feasibility study of an oil-free turbocharger supported on gas foil bearings via on-road tests of a two-liter class diesel vehicle. J. Eng. Gas Turbines Power. **135**, 1 (2013). https://doi.org/10.1115/1.4007883
6. Eber, N., Quack, H., Schmid, C.: Gas bearing turbines with dynamic gas bearings and their application in helium refrigerators. Cryogenics **18**(11), 585–588 (1978)
7. Ohlig, K., Bischoff, S.: Dynamic gas bearing turbine technology in hydrogen plants. AIP Conf. Proc. **1434**, 814–819 (2012). https://doi.org/10.1063/1.4706994
8. Kim, T.H., Lee, Y.B., Kim, T.Y., Jeong, K.H.: Rotordynamic performance of an oil-free turbo blower focusing on load capacity of gas foil thrust bearings. J. Eng. Gas Turbines Power. **134**, 1–7 (2012). https://doi.org/10.1115/1.4004143
9. Lee, Y.B., Park, D.J., Kim, T.H., Sim, K.: Development and performance measurement of oil-free turbocharger supported on gas foil bearings. J. Eng. Gas Turbines Power. **134**, 32506–1–32506–11 (2012). https://doi.org/10.1115/1.4004719
10. Gu, L., Guenat, E., Schiffmann, J.: A review of grooved dynamic gas bearings. Appl. Mech. Rev. **72** (2020). https://doi.org/10.1115/1.4044191
11. Rimpel, A.M., Vannini, G., Kim, J.: A rotordynamic, thermal, and thrust load performance gas bearing test rig and test results for tilting pad . J. Bearings Spiral Groove Thrust Bearings J. Eng. Gas Turbines Power. **139**, 1–11 (2017). https://doi.org/10.1115/1.4037315
12. Waumans, T., Peirs, J., Al-Bender, F., Reynaerts, D.: Aerodynamic journal bearing with a flexible, damped support operating at 7.2 million DN. J. Micromech. Microeng. **21** (2011). https://doi.org/10.1088/0960-1317/21/10/104014
13. Belforte, G., Colombo, F., Raparelli, T., Viktorov, V.: High-speed rotor with air bearings mounted on flexible supports: test bench and experimental results. J. Tribol. **130**, (2008). https://doi.org/10.1115/1.2908905
14. Battig, P., Schiffmann, J.: Data-driven model for the dynamic characteristics of o-rings for gas bearing supported rotors. J. Appl. Mech. Trans. ASME. **86**, (2019). https://doi.org/10.1115/1.4043473
15. Al-Bender, F., Colombo, F., Reynaerts, D., Villavicencio, R., Waumans, T.: Dynamic characterization of rubber o-rings: squeeze and size effects. Adv. Tribol. (2017). https://doi.org/10.1155/2017/2509879
16. San Andrés, L.: Hybrid Tilting Pad Gas Bearings: Analysis and Experimental Validation. TRC-B&C-1-05, Turbomachinery Laboratory, Texas A&M University College Station, TX (2005)
17. Lihua, Y., Huiguang, L., Lie, Y.: Dynamic stiffness and damping coefficients of aerodynamic tilting-pad journal bearings. Tribol. Int. **40**, 1399–1410 (2007). https://doi.org/10.1016/j.triboint.2007.03.007

18. Agrawal, G.L.: Foil air/gas bearing technology — an overview (1997). https://doi.org/10.1115/97-GT-347
19. Park, D.J., Kim, C.H., Jang, G.H., Lee, Y.B.: Theoretical considerations of static and dynamic characteristics of air foil thrust bearing with tilt and slip flow. Tribol. Int. **41**, 282–295 (2008). https://doi.org/10.1016/j.triboint.2007.08.001
20. Yong-Bok, L., Tae, Y.K., Chang, H.K., Tae, H.K.: Thrust bump air foil bearings with variable axial load: theoretical predictions and experiments. Tribol. Trans. **54**, 902–910 (2011). https://doi.org/10.1080/10402004.2011.606957
21. Walton, J.F., Heshmat, H.: Application of foil bearings to turbomachinery including vertical operation. In: Turbo Expo: Power for Land, Sea, and Air, vol. 78583, p. V001T04A004. American Society of Mechanical Engineers (1999)
22. Bruckner, R.J.: Windage power loss in gas foil bearings and the rotor-stator clearance of high speed generators operating in high pressure environments. Proc. ASME Turbo Expo. **5**, 263–270 (2009). https://doi.org/10.1115/GT2009-60118
23. Walton, J.F., Heshmat, H., Tomaszewski, M.J.: Testing of a small turbocharger/turbojet sized simulator rotor supported on foil bearings. J. Eng. Gas Turbines Power. **130** (2008). https://doi.org/10.1115/1.2830855
24. Bruckner, R.J.: An assessment of gas foil bearing scalability and the potential benefits to civilian turbofan engines. Proc. ASME Turbo Expo. **1**, 29–35 (2010). https://doi.org/10.1115/GT2010-22118
25. Yoshimoto, S.: Floating characteristics of squeeze-film gas bearings with vibration absorber for linear motion guide. J. Tribol. **119**, 531–536 (1997). https://doi.org/10.1115/1.2833533
26. Al-Bender, F.: Air Bearings. Theory, Design and Appications. Wiley (2021)

# Contact Models and Reduction of Instabilities in Sealing Rings for Sustainability Purposes

Maria Cristina Valigi(✉), Silvia Logozzo, and Claudio Braccesi

Department of Engineering, University of Perugia, Perugia, Italy
{mariacristina.valigi,silvia.logozzo,claudio.braccesi}@unipg.it

**Abstract.** Instability phenomena in mechanical seals are responsible of malfunction, fluid leakage and noise generation, which in turn lead to disadvantages in terms of *Sustainable Development Goals* (SDG). Mechanical face seals are mainly employed in pumps used in the automotive and appliance industries. The aim of these seals is to separate two chambers containing fluids, assuring the containment, during the rotation of a shaft. Tribological instabilities can cause a stick-slip behavior, which leads to failures, wear, and fluid leakage with consequent waste of material, increase of costs and pollution, release of fluids potentially dangerous for human health. Thus, reduction of these phenomena will help to achieve SDG12 - Responsible Consumption and Production, SDG9 - Industry, Innovation and Infrastructure, and SDG3 - Good Health and Wellbeing. Proper mathematical models are very useful to predict instabilities and optimize the service conditions. In this paper, tribological models based on different contact theories are compared in terms of stick-slip predictivity, with respect to experimental data gathered with a dedicated tribometer. Results allow to discuss the accuracy of different theories in predicting the critical speed of the seals corresponding to the manifestation of instabilities and leakages.

**Keywords:** SDG12 · SDG9 · SDG3 · Green tribology · Stick-slip · Seals

## 1 Introduction

Mechanical face seals are essential components in several engineering applications, as they are widely used in the automotive and appliance fields, in pumps, compressors, mixers and hydraulic machines. Although they have been used for years, currently the issues related to their exercise need to be further studied and understood, to predict and avoid undesired dysfunctions.

Malfunctions in mechanical seals are mainly related to tribological issues, like frictional instabilities which cause an intermittent contact between the two sealing rings, named stick-slip. Stick-slip is responsible of phenomena which reduce the sustainability because it determines the occurrence of failures, wear, and fluid leakage with consequent waste of material, increase of costs, pollution and contamination, release of fluids potentially dangerous for human health. From this perspective, prediction and limitation of instabilities and stick-slip in mechanical seals is an application of green tribology


G. Quaglia et al. (Eds.): I4SDG 2021, MMS 108, pp. 412–420, 2022.
https://doi.org/10.1007/978-3-030-87383-7_45

[1] and can contribute to achieve *Sustainable Development Goals* (SDG) [2], as in particular SDG12-Responsible Consumption and Production, SDG9-Industry, Innovation and Infrastructure, and SDG3-Good Health and Wellbeing. A mechanical face seal is designed to be mounted on a rotating shaft working between two different zones or chambers full of fluids, with the aim of keeping the two chambers separated, avoiding fluid leakage, loss of pressure or contamination (Fig. 1).

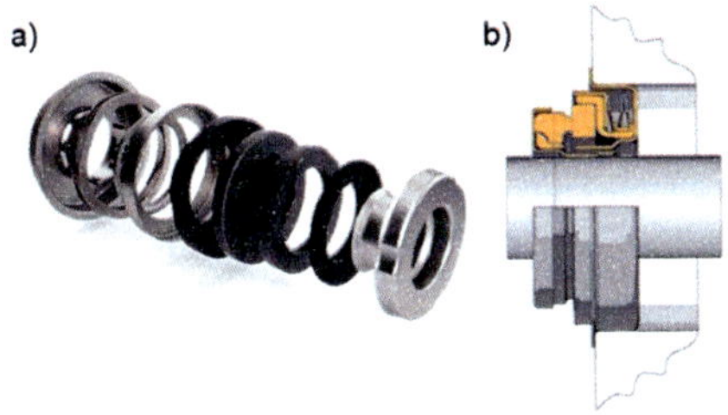


**Fig. 1.** Mechanical face seal: a) exploded view; b) section

The sealing effect is obtained with two rings, one integral with the rotating shaft and one stationary, which are pushed one against the other but avoiding the contact between their surfaces. The gap between the rings is filled with a lubricant under pressure, which assures the sealing effect.

During exercise, the mechanical seal can be continuously subjected to vibrations and instabilities due to misalignments of the rings or to friction fluctuations caused by the sporadic contact between the rings. Two of the main phenomena due to instabilities are stick-slip and ringing [3, 4], which in turn can cause fluid leakage, deformations and fatigue. The stick and slip behavior can arise when the lubrication regime is mixed or boundary and the friction torque fluctuates during the shaft rotation, due to an intermittent mechanical contact between the rotating and stationary rings [5]. This phenomenon is relevant in mechanical seals with no anti-rotation pins. In particular, stick and slip was observed when the rotating shaft decelerates [6, 7]. Stick-slip often causes friction induced vibrations and fatigue failures [8, 9].

Ringing is strictly related to fluid leakage, often due to closure pressure gradients and spring stiffnesses [3, 10, 11]. As crucial components for the operation of several machines, mechanical seals deserve scientific efforts to study the influence of service conditions and design parameters on their dynamic behavior. In this scenario, the development and validation of prediction mathematical models is a topic of great significance. In particular, the study and modelling of the mechanical contact between rings is very useful to allow an accurate prediction of stick-slip and fluid leakage [12, 13]. Contact theories must be combined with lubrication and dynamic models to get a complete scenario of the mechanical seals behavior [14].

Some new studies are focused on the wear behavior prediction of mechanical face seals [15]; other authors investigate the dynamic behavior of seals based on misalignments and different geometries to study the stick-slip vibrations and the sealing effect [16, 17]; another field of research deals with study and limitation of generation of acoustic emissions due to instabilities [18]. This paper is an extension of a previous study [4], where a tribo-dynamic model was developed to predict the friction instability arising conditions in terms of shaft critical speeds. In that paper, the lubrication, dynamics, friction, and contact theories were combined to generate a mathematical model of the seal's behavior. In this paper the previous model is used again, but combining different contact theories, to analyze the prediction accuracy of the model, with respect to experimental results.

## 2 Friction Instabilities Prediction Model

Mechanical face seals are necessarily lubricated at the interface between the two rings. Infact, lubrication is responsible of the sealing effect by creating a load-bearing fluid film, with lift function. Furthermore, lubricants are useful to limit friction, wear, vibrations, and misalignments. Sometimes the lubrication is realized by means of the exercise fluid itself. The sealing effect is guaranteed when the film thickness is small, but at the same time a thin film favors the occurrence of instabilities, especially in high pressure and velocity regimes. So, a compromise is always necessary to identify the right lubricating condition. In fact, the film thickness influences the lubrication regime which can be hydrodynamic, mixed or boundary. In hydrodynamic lubrication (HL) the fluid film is thick, and the load is supported solely by the lubricant; in mixed lubrication (ML) most of the load is supported by the fluid but a small part is supported by mechanical contact between the sealing surfaces; in boundary lubrication (BL) film thickness is very small and the load is mainly supported by mechanical contact. In this last case, overheating, friction, and stick-slip can destroy the seal or at least cause dysfunctions. Normally the switch from a lubrication regime to another is not instantaneous, but there is a transient phase where two regimes coexist at the same time. The right service compromise for a mechanical face seal is the transition between the HL and ML regimes. Therefore, in the lubrication regime of a mechanical face seal, the pressure in the gap between the two sealing rings is the sum of two terms, $p_l$ due to the lubricant and $p_c$ due to the mechanical contact. In this paper the lubrication regime of a mechanical face seal was studied by a tribo-dynamic model previously developed [3, 4] but modified by adding different contact models.

### 2.1 Lubrication Model

The lubrication model is used to determine the $p_l$ in the ML regime and it is based on the Reynolds' theory and on the Patir and Cheng model [12]. According to these theories the lubrication equation can be written as:

$$\frac{1}{r}\frac{\partial}{\partial \alpha}\left(\emptyset_\theta \frac{h_n^3}{12\mu}\frac{\partial p_l}{\partial \alpha}\right)+\frac{\partial}{\partial r}\left(\emptyset_r \frac{r h_n^3}{12\mu}\frac{\partial p_l}{\partial r}\right)=\frac{r\omega}{2}\left(\frac{\partial \bar{h}}{\partial \alpha}-\sigma\frac{\partial \emptyset_s}{\partial \alpha}\right)+r\frac{\partial \bar{h}}{\partial t} \quad (1)$$

where $p_l$ is the mean hydrodynamic pressure, $r$ is the radial direction, $\mu$ is the dynamic viscosity of the lubricant, $\alpha$ is the angular coordinate, $\omega$ is the relative angular speed, $h$ is the local film thickness for rough surfaces, $\bar{h}$ is the mean value of $h$, $h_n$ is the nominal film thickness. The subscript $n$ refers to nominal quantities. $\emptyset_\theta$ and $\emptyset_r$ are the pressure flow factors introduced by Patir and Cheng which correlate the average pressure flow for rough seals to smooth ones. $\emptyset_s$ is the factor considering the additional flow transport due to sliding. The ratio $h_n/\sigma$ is a measure of the effect of surface roughness on the flow. $\sigma$ is the standard deviation of roughness. For seals $h_n/\sigma > 2$ and $\overline{h} \sim h_n$. When $h_n/\sigma$ tends to $\infty$ $\emptyset_\theta$, $\emptyset_r$ tend to 1 and Eq. (1) results in the Reynolds equation for smooth surfaces. For $2 < h_n/\sigma < 3$ the effect of roughness on the total interfacial pressure is considered in $p_c$ and for $h_n/\sigma > 3$ the effect of roughness is neglected. So $p_l$ can be evaluated considering the Reynolds equation for smooth surfaces. Thus, $\emptyset_\theta$, $\emptyset_r = 1$, $\emptyset_s = 0$ and $\overline{h} \sim h_n$.

In the condition of the studied mechanical face seal, and in ML regime, the squeeze effect is neglected, and the problem is axisymmetric. From integration of Eq. (1).

$$p_l(r) = \frac{p_2 - p_1}{\int_{r_1}^{r_2} \frac{1}{rh_n^3} dr} \int_{r_1}^{r} \frac{1}{rh_n^3} dr + p_1 \tag{2}$$

## 2.2 Contact Models

Contact models allow to determine $P_c$, due to the mechanical contact between the rings. Three different contact models were considered.

According to the first and second contact model $p_c$ can be calculated as [4]:

$$p_c = b_m S_C \tag{3}$$

where $S_c$ is the compressive yield stress of the softest material of the pair, and $b_m$ represents the area of mechanical contact which can be calculated as a normal (first contact model) and exponential (second contact model) cumulative distribution function of roughness. The third contact model is based on the Greenwood-Williamson (GW) theory [19], where the contact pressure is:

$$p_c = \frac{4}{3} \eta E^* \sqrt{\beta} \sqrt{\sigma^3} F_{3/2} \tag{4}$$

where $E^*$ is the composite Young's modulus, $\eta$ is the asperity density, $\beta$ is the asperity radius, $\sigma$ is the standard deviation of roughness, $F_{3/2}$ is a function of the film thickness $h$ introduced by GW and representing a probability density function of roughness.

## 2.3 Dynamic Model

The mechanical face seal can be treated as a 2-dof (degree of freedom) system, where the Lagrangian coordinates are the axial translation $h_n$ and the rotation $\theta$ of the stationary ring around the axis of the seal, when stick slip occurs. Equation of dynamics of the seal are:

$$m\ddot{h}_n + C\dot{h}_n + Kh_n = \int_{r_1}^{r_2} 2\pi Prdr \tag{5}$$

$$J\ddot{\theta} + C_t\dot{\theta} + K_t\theta = M_b + M_h$$

$$= \int_{r_1}^{r_2} 2\pi f(\omega) p_c r^2 dr + \int_{r_1}^{r_2} 2\pi \mu\omega \frac{r^3}{h_{1n} + n(r - r_1)} dr \quad (6)$$

$m$ and $J$ are mass and moment of inertia of the seal and f is the friction coefficient. The mass is connected to the frame by an axial stiffness $K$ and damping $C$, and a torsional stiffness $K_t$ and damping $C_t$. P is the total interfacial pressure as the sum of $p_l$ and $p_c$. The fluid film is considered with a tapered shape. The total moment is the sum of the the friction torque acting on the seal in BL conditions ($M_b$) and the viscous torque acting in HL regime ($M_h$).

## 3 Experimental Setup and Results

Experimental tests were performed to set the right dynamic viscosity for the simulations, by measuring the service temperature. The experimental apparatus used in this work was a tribological test rig (Fig. 2) equipped with the following sensors: load cell Lebow 3167 for axial load, load cell cantilever Burster 2N 205 mm for friction torque, K-thermocouple for measuring ring's temperatures. The test rig was also provided with a wireless thermometric chain for measuring temperatures of the rotating ring [20], so that the thermocouple can rotate integrally with the rotating ring. Experimental tests were performed imposing a gradual deceleration to the shaft, starting from a velocity of 2000 rpm. Temperature and velocity of the rotating ring were measured during the deceleration. Results of temperature (Fig. 3) were used as input in the tribological model to characterize the viscosity and the friction coefficient.

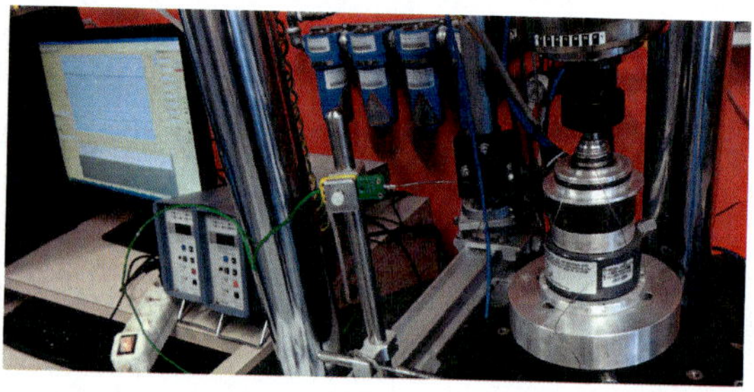

**Fig. 2.** Tribological test rig

Furthermore, during the tests, the seal showed a critical speed of 135 rad/s (1290 rpm), in the time instant corresponding to the occurrence of ringing. Simulations based on the model reproduced the friction torque characterizing the ringing and the critical speed was obtained with the same condition of experimental test.

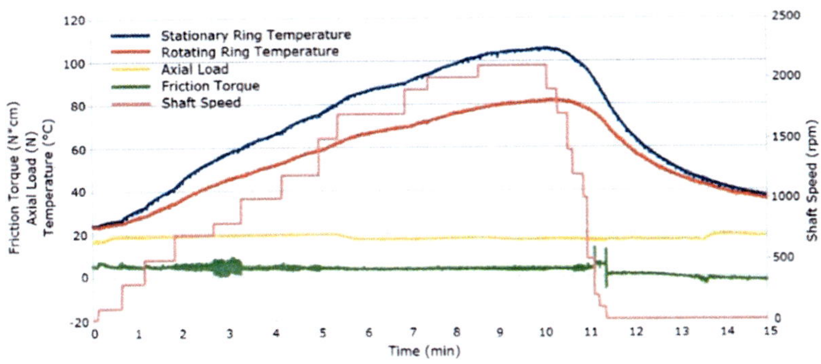


**Fig. 3.** Experimental results

# 4 Results and Discussion

All the design and service parameters used in the simulations are reported in Table 1.

**Table 1.** Design and exercise parameters

| | | | |
|---|---|---|---|
| Inner Radius $r_1$ | $9.3 * 10^{-3}$ m | Asperity Density $\eta$ | $2 * 10^{11}$ |
| Outer Radius $r_2$ | $11.1 * 10^{-3}$ m | Asperity Radius $\beta$ | $0.16 * 10^{-3}$ m |
| Curvature | $8.33 * 10^{-4}$ rad | Roughness std.dev. $\sigma$ | $0.152 * 10^{-6}$ m |
| Nominal film thickness $h_n$ | $0.1 * 10^{-3}$ m | Axial Stiffness $K$ | 42142 N/m |
| Inner Pressure $p_1$ | $10^5$ Pa | Torsional Stiffness $Kt$ | 4.3090 Nm/rad |
| Outer Pressure $p_2$ | $2 * 10^5$ Pa | Axial Damping $C$ | 2.0529 Ns/m |
| Initial shaft speed | 500 rpm | Torsional Damping $Ct$ | $1.2104 * 10^{-4}$ Nms/rad |
| Dynamic viscosity $\mu$ | $700 * 10^{-6}$ Pa * s | Mass $m$ | 0.01 kg |
| Young's Modulus $E*$ | $234 * 10^6$ Pa | Moment of Inertia $J$ | $3.4 * 10^{-7}$ kg * m$^2$ |

The simulations were done considering a deceleration curve with the same trend of the deceleration regime applied during the experimental tests. Figure 4 reports the results of the contact pressures for the three applied contact theories.

As it can be seen from Fig. 4, the contact pressure resulting from the application of the model with an exponential distribution of roughness, gives a very high contact pressure which overestimates the stick-slip phenomenon. In fact, considering this contact pressure, the lubrication regime results to be always boundary with prevalence of mechanical contact between the two rings and not enough sealing effect. For this reason, results of the critical speed corresponding to the contact model with exponential distribution of roughness were not taken into consideration. Table 2 reports the results of

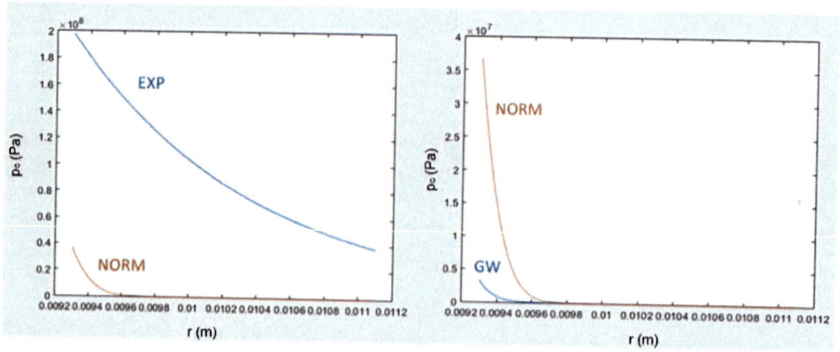


**Fig. 4.** Results of contact pressures $p_C$. NORM: normal distribution of roughness, EXP: exponential distribution of roughness, GW: Greenwood-Williamson model

the critical speeds, comparing the experimental value with the simulated results applying the different contact theories. The critical speed was found as the value of speed corresponding to the initial occurrence of vibrations and rotations of the stationary ring.

**Table 2.** Results of critical speeds

| Experimental | Normal distribution | GW model |
|---|---|---|
| 135 rad/s | 149 rad/s | 144 rad/s |

These results demonstrate that the contact model based on the GW theory provides a more accurate result, and therefore it allows a better prediction of friction instabilities in mechanical face seals. The critical speed is slightly overestimated with respect to the experimental value, but it represents a safety factor. In fact, the overestimation allows to predict instabilities before they really happen, during deceleration. This means that seals designed by using this model will be made to work in a safe range of speeds.

## 5 Conclusion

This paper presented and described different tribological models to predict friction instabilities in mechanical face seals working in mixed lubrication regimes. Different contact theories were applied and discussed with the aim of predicting the instability arising conditions represented by a critical speed of the rotating shaft corresponding to the initial vibration of the stationary ring. Experimental tests were performed to get a real value of the critical speed in the same conditions applied during the simulations and with the same deceleration regime of the shaft. Results demonstrate that the contact theory based on the Greenwood-Williamson theory provides a more accurate prediction of the stick-slip phenomenon. The improved model can be extended to other exercise parameters. These achievements are useful to enhance the sustainability of both the development of mechanical seals and the exercise of devices mounting seals. In fact, a better prediction of the occurrence of stick-slip and instabilities allows to avoid dangerous fluid leakage, failure and wear which can cause waste of material, increase of costs and pollution, and damages for human health. From this perspective, this paper provides good results in the field of green tribology for achieving *Sustainable Development Goals* and, in particular, SDG12, SDG9, and SDG3.

## References

1. Nosonovsky, M., Bhushan, B.: Green tribology: principles, research areas and challenges. Phil. Trans. R. Soc. A **368**, 4677–4694 (2010)
2. Okokpujie, I.P., Fayomi, O.S.I., Oyedepo, S.O.: The role of mechanical engineers in achieving sustainable development goals. Proc. Manufact. **35**, 782–788 (2019). https://doi.org/10.1016/j.promfg.2019.06.023
3. Braccesi, C., Valigi, M.C.: Undesired acoustic emissions of mechanical face seals: model and simulations. Tribol. Int. **71**, 125–131 (2014). https://doi.org/10.1016/j.triboint.2013.11.014
4. Valigi, M., Braccesi, C., Logozzo, S.: A Parametric Study On Friction Instabilities In Mechanical Face Seals. Tribol. Trans. **59**(5), 911–922 (2016). https://doi.org/10.1080/10402004.2015.1121311
5. Nakano, K.: Two dimensionless parameters controlling the occurrence of stick-slip motion in a 1-DOF system with Coulomb friction. Tribol. Lett. **24**(2), 91–98 (2006). https://doi.org/10.1007/s11249-006-9107-7
6. Logozzo, S., Valigi, M.: Investigation of instabilities in mechanical face seals: prediction of critical speed values. In: Uhl, Tadeusz (ed.) Advances in Mechanism and Machine Science: Proceedings of the 15th IFToMM World Congress on Mechanism and Machine Science, pp. 3865–3872. Springer International Publishing, Cham (2019). https://doi.org/10.1007/978-3-030-20131-9_383
7. Valigi, M.C., Braccesi, C., Cianetti, F., Logozzo, S.: Stick-slip simulation and detection in mechanical face seals. In: Proceedings of the ASME 2015 International Mechanical Engineering Congress & Exposition, 13–19 November, Houston, Texas, USA., Houston, Texas, USA, 15–2015 (2015)
8. Di Bartolomeo, M., Meziane, A., Massi, F., Baillet, L., Fregolent, A.: Dynamic rupture at a frictional interface between dissimilar materials with asperities. Tribol. Int. **43**, 1620–1630 (2010)
9. Brunetti, J., D'Ambrogio, W., Fregolent, A.: Friction-induced vibrations in the framework of dynamic substructuring. Nonlinear Dyn. **103**(4), 3301–3314 (2020). https://doi.org/10.1007/s11071-020-06081-9
10. Valigi, M.C., Papini, S.: Analysis of chattering phenomenon in industrial S6-high rolling mill. Diagnostyka **14**(3), 3–8 (2013)
11. Papini, S., Valigi, M.C.: Analysis of chattering phenomenon in industrial S6-high rolling mill Part II: experimental study. Diagnostyka **17**(4), 27–32 (2016)
12. Patir, N., Cheng, H.S.: An average flow model for determining effects of three-dimensional roughness on partial hydrodynamic lubrication. J. Lubr. Technol. **100**, 12–17 (1978)
13. Leine, R.I., Van Campen, D.H., De Kreker, A., Van De Steen, L.: Stick-slip vibrations induced by alternate friction model. Nonlinear Dyn. **16**, 41–54 (1998)
14. Green, I., Barnsby, R.M.: A simultaneous numerical solution for the lubrication and dynamic stability of noncontacting gas face seals. Trans. ASME **123**, 388–394 (2001)
15. He, W., Wang, S., Zhang, C., Wang, X., Liu, D.: Lubrication and wear characteristics of mechanical face seals under random vibration loading. Materials **13**(6), 1285 (2020)

16. Ma, Y., Li, Y., Xie, F., Wang, L., Zhu, K.: Numerical investigation on sealing behaviors of an extremely high-speed two-stage impellers structure in cryogenic rockets. Asia-Pacific J. Chem. Eng. **13**(5), e2229 (2018)
17. Varney, P., Green, I.: Dynamic modeling of an eccentric face seal including coupled rotor-dynamics, face contact, and inertial maneuver loads. Proc. Inst. Mech. Eng. Part J: J. Eng. Tribol. **232**(6), 732–748 (2018)
18. Hu, S., Huang, W., Shi, X., Peng, Z., Liu, X., Wang, Y.: Bi-Gaussian stratified effect of rough surfaces on acoustic emission under a dry sliding friction. Tribol. Int. **119**, 308–315 (2018)
19. Greenwood, J.W., Williamson, J.B.P.: Contact of nominally flat surfaces. Proc. Roy. Soc. **295**(1442), 300–319 (1966)
20. Valigi, M.C., Braccesi, C., Logozzo, S., Conti, L., Borasso, M.: A new telemetry system for measuring the rotating ring's temperature in a tribological test rig for mechanical face seals. Tribol. Int. **106**, 71–77 (2017)

# Green Tribology for Sustainable Development Goals

Nikolai K. Myshkin and Feodor A. Grigoriev(✉)

Metal-Polymer Research Institute of Belarus National Academy of Sciences, Gomel, Belarus

**Abstract.** Sustainable Development Goals include very important topic of industry and innovation (SDG9) where tribology, defined as the science and technology of interacting surfaces in relative motion, can play very important role. It can be explained by the losses connected with friction and wear in machines estimated by significant portion of the gross national product and energy consumption, as well as great social significance of tribology in its ecological and medical aspects. The concept of "green tribology" has been formulated meeting the SDG strategy. Green tribology can provide design solutions, better materials and lubricants for many types of tribosystems. The efficient way to decrease losses in energy and materials is to replace the metal parts of machinery and vehicles with composite components based on polymers. Combination of low mass-density, resistance to corrosion, self-lubrication, and relatively easy processing make polymers very prospective due to the use of different fillers greatly improving their tribological behavior. Recent approval of EU directives has initiated the promotion of plant oils instead of the mineral ones and it is important to find the efficient ways of such a replacement.

**Keywords:** SDG9 · Sustainable Development · Green tribology · Electrical Transport · Eco-friendly lubricants

## 1 Introduction

Friction is one of the most serious problems of the contemporary engineering and it brings enormous losses of energy. The great share of the fuel consumed by cars, locomotives, and other vehicles is spent on friction in machines and their contact with the road or rail. In general, the losses connected with friction and wear of machines are estimated by significant portion of the gross national product and energy in industry, transportation, agriculture, and house keeping [1].

There is a great social significance of tribology in its ecological and medical aspects. Reduction of losses in energy and wear of materials in machines together with the elimination of harmful emissions, use of ecologically friendly materials and lubricants are very important. In Peter Jost's speech at the Tribology Jubilee Reception in Buckingham Palace (March 2 2017) he has mentioned "Tribology has grown into nanotribology, green tribology and biotribology …and I hope it will benefit industry, the environment, the reduction of the use of materials – all leading to the security and the quality of life for this and future generations…". A general scheme presented in Fig. 1 illustrates this conclusion [1].


G. Quaglia et al. (Eds.): I4SDG 2021, MMS 108, pp. 421–428, 2022.
https://doi.org/10.1007/978-3-030-87383-7_46

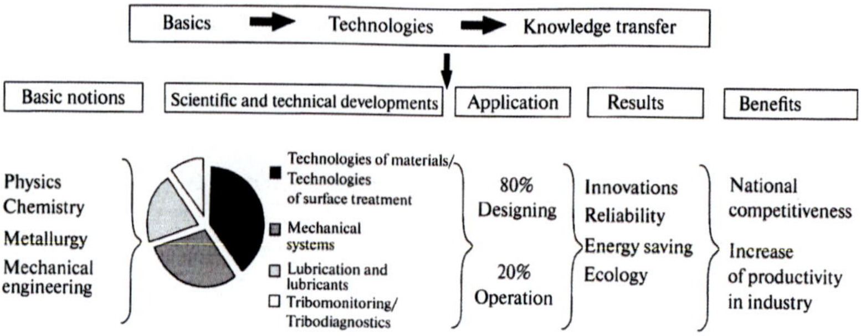


**Fig. 1.** Transfer of tribological knowledge into industry [1].

## 2 Tribology and Sustainable Development Goals

Energy and environmental sustainability have initiated interest in more efficient and cleaner transportation and industry. About one-fifth of all global industrial energy and one-third of energy used in transportation are spent for overcoming friction, not mentioning the environmental impacts of transportation affected by wear losses in moving mechanical parts and components. The recent advances in tribology could reduce energy losses due to friction and wear by 18–40%, which means up to 8.7% of the total global energy use and 1.4% of the Gross National Product (GNP) in developed countries. For example, 30% of the fuel in a passenger car goes to overcome friction with the half of consumption in the engine and transmission (Fig. 2 [2]).

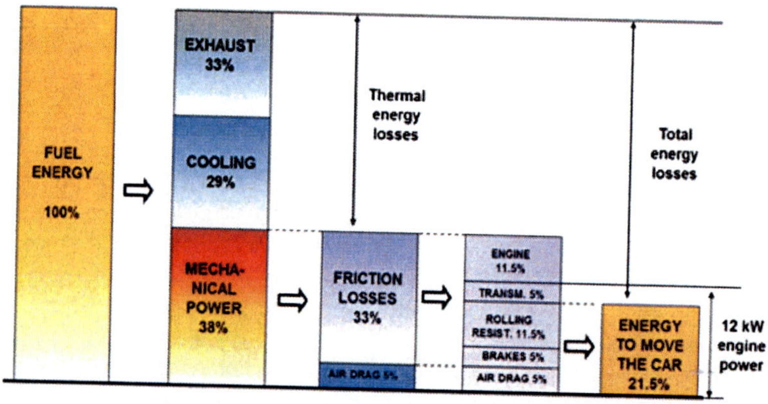


**Fig. 2.** Consumption of fuel energy in passenger car [2].

Taking in mind a quantity of cars we can expect a huge economy even at small reduction of this energy consumption in hundreds millions of cars. There is a variety of methods to solve the problems of saving the materials and energy lost due to friction and wear. Design in mechanical engineering can provide decreasing of these negative effects due to implementation of electrical drives instead of combustion engines and transmissions, thus saving power with simultaneous positive ecological effect due to decrease in harmful emissions. The energy losses of electric cars are about 3.4 times lower compared to cars with combustion engine and the $CO_2$ emissions are 4.5 times lower when the electricity comes from the green energy sources [2, 3].

Tribology can provide design solutions, better materials and lubricants for many types of tribosystems. Surface engineering which can provide efficient surface treatment and coatings for various machine components is another beneficial way in improving design efficiency. The optimistic forecast of the annual savings due to implementation of tribology advances into various segments of the global economy is presented in Fig. 3 [2].

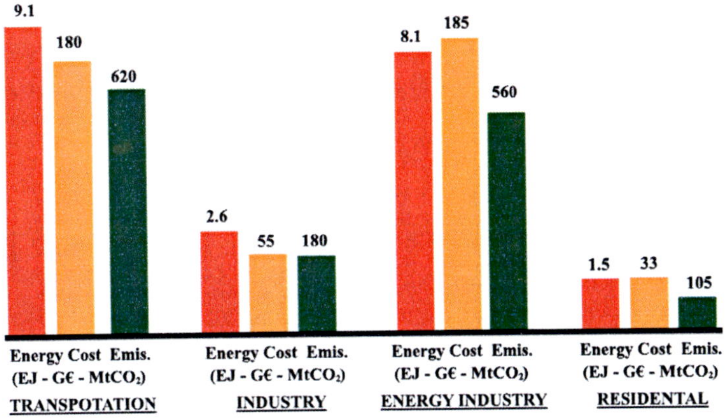


**Fig. 3.** Data on energy, cost, and emissions level potential savings for 2025: EJ $-10^{18}$ J, G€ -Giga Euro, $MtCO_2$–Megatons of $CO_2$ [2]

## 3 Prospective Materials For Green Tribology

Another efficient way to decrease losses is to replace the metal parts of machinery and vehicles with less heavy composite components based on polymers and polymer composites. It can bring a great effect due to lower friction and energy losses for tribosystems with lower mass of their components. This effect has become the most evident in transportation and aerospace which are eager to decrease the mass of vehicles.

Polymer tribology has made a dramatic progress in development of antifriction and brake materials, composites for bearings, tires, dampers and many other components of modern machinery [4]. Such polymers as polytetrafluoroethylene are often used as solid lubricants. The mechanism of PTFE friction is explained by the transfer of a very thin film to the counter face playing the role of lubricant. Solid lubricating polymers are applied as a friction additive to various composite materials, e.g., sintered powder metals or ceramic materials and coatings. Their matrix provides for strength and heat removal from the friction surfaces, while the polymer ensures low friction.

The composites with polymer matrix have a versatility of properties due to various fillers, additives, and reinforcing components. Polymer films and coatings were applied efficiently for prevention of sticking and severe wear of components in micromechanical systems and memory storage devices. By introduction of dispersed fillers it is possible to vary their physico-mechanical properties, as well as reduce or increase the friction coefficient. The tribological characteristics and advantages of polymers are illustrated in the Table 1 [4].

**Table 1.** Tribological characteristics of polymers.

| Material | Friction coefficient | Operation temperature, °C | Advantages | Disadvantages |
|---|---|---|---|---|
| Aliphatic polyamides | 0.2 ÷ 0.5 | −40 ÷ +85 | Wear and fatigue resistance | Water sorption, high coefficient of friction |
| Aromatic polyamides | 0.1 ÷ 0.3 | −100 ÷ +200 | Wear, fatigue, and heat resistance | High cost; water sorption |
| Fluoroplastics | 0.01 ÷ 0.05 | −269 ÷ +260 | Low friction, corrosion resistance | Creep and low strength |
| Polycarbonate | 0.2 ÷ 0.5 | −60 ÷ +125 | Rigidity and corrosion resistance | Low fatigue strength |
| Polyacetals | 0.1 ÷ 0.3 | −50 ÷ +120 | Wear and fatigue resistance | Abrasive effect on counterbody |
| Polyolefins | 0.1 ÷ 0.3 | −100 ÷ +100 | High resistance to aggressive media | Low mechanical strength |
| Thermoplastic elastomers | 0.3 ÷ 0.6 | −60 ÷ +120 | High elasticity and resistance to ambience | High friction and low strength |
| Polyetherether ketone | 0.2 ÷ 0.4 | −30 ÷ +250 | High heat and ambient resistance | High cost |
| Polyphenylene sulfide | 0.2 ÷ 0.5 | −30 ÷ +220 | Wear and fatigue resistance | High cost |

When the dispersion of the fillers equals to nanometer range they are found in the matrix at a distance from each other comparable to their size and material presents a layer reinforced by molecular forces. A significant progress has been achieved in materials science of composites with adoption of nanotechnologies.

Among polymer nanocomposites are those containing metal particles and their compounds with size comparable to macromolecules of polymers.

Great interest is attracted nowadays to inexpensive polymer-clay nanocomposites since their raw material is extracted in the natural clay fields. Fullerenes and carbon nanotubes are employed to monitor the properties of the composites; discovery of graphene has stimulated a new wave of investigations in this field. Polymer composites with nanofillers are taking over the areas traditionally occupied by other polymer-based materials. The preparation of polymer composites by nanofiller dispersion in molten thermoplastics is one of the most advanced and fast-developing methods. Polymer-clay nano-composites are very prospective due to their low price and availability of mass-production [5]. But the laminated silicates used as nano-fillers need increasing compatibility with the polymer matrix, so their surface should be modified by surface-active agents. The scheme of forming polymer-clay composite is presented in Fig. 4.

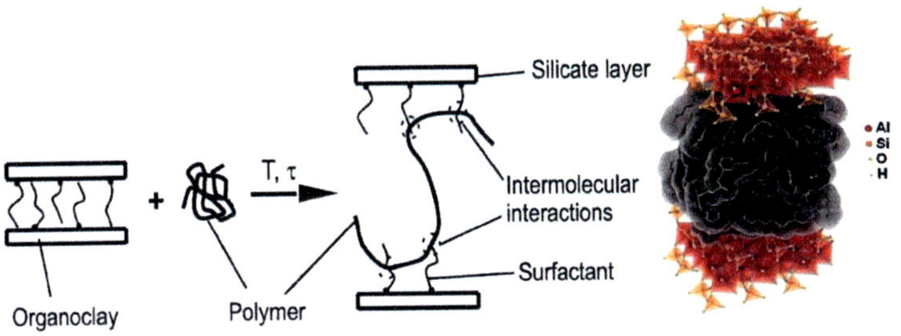


**Fig. 4.** Scheme of polymer-clay composite formation [4]

The data on properties of polymer-clay composites based on polyamide PA-6 with montmorillonite (MMT) and octadecylamine (ODA) are given in the Table 2 [4, 5].

**Table 2.** Properties of polymer-clay composites.

| Composition, wt% | Processing | $\sigma_T$, MPa | ε, (%) | $I_{abr} \times 10^3$, g/m |
|---|---|---|---|---|
| PA 6/nanoclay (MMT)–3% | in melt | 75 | 76 | 2.8 |
| PA 6 /nanoclay (MMT + ODA)–3% | in melt | 86 | 53 | 4.0 |

Abrasive wear test conditions: p = 2 MPa, v = 0.35 m/s; pin-on-ring with abrasive paper; $\sigma_T$ – tensile strength; ε – relative elongation; $I_{abr}$ – abrasive wear rate.

Polymers and polymer composites need the proper techniques and equipment for tribotests. Methodology of testing the polymers is developing with taking account of the important part of heating in their tribological performance. Development of the internationally approved standard schemes and procedures of tribotests has becoming the urgent necessity. Conformable block-on-ring is one of the efficient schemes for friction test of polymers because it proceeds at constant nominal contact area and pressure. ISO has adopted such scheme describing the design size of the polymer samples and procedure of increasing contact pressure and sliding velocity step by steps with pressure-velocity data plotting [6].

Good prospects are open in green tribology for the biodegradable polymer composites and nanocomposites with natural fillers as clays. These polymers are promising from the point of view of ecology. There is a progress in polymer recycling technologies which is important taking in mind the growing volume of polymer production and using in all the areas of industry and common life.

## 4 Lubricants for Green Tribology

One of the most common methods of friction and wear reduction is the use of a lubricant separating the rubbing bodies by a thin layer of the material with a low resistance to shear. In a number of tribosystems in machines it is possible to realize a hydrodynamic mode of lubrication exercising sliding in a liquid layer, so resistance to sliding is determined mainly by its viscosity. The development of the theory of lubrication has built the basis for design methods of the hydrodynamic and elasto-hydrodynamic lubrication [7]. In material side it is promising today to create new lubricants and additives that combine high efficiency with a possibility to use different materials in the friction pairs and provide ecological safety.

Green tribology and biotribology are now the most rapidly developing fields of tribology [1]. Nature has created perfect tribosystems and has fit the man and animals with the joints functioning in fact without any wear for many years. The development of artificial joints in orthopedics, use of implants, reproduction of natural preferences in technical devices, application of the materials imitating surface layers of the plants and animals, and etc. are all the achievements of biotribology [8].

Green lubricants and their applications are also developing quickly. Until the XX-th century vegetable oils and animal fat were most common lubricants. Development of the additives to mineral oils in the XX-th century has made them dominant in industry. In the 1990-ies ecological laws start to limit the use of mineral oils in agriculture and forestry. In 2003–2009 the approval of EU directives and road maps in "green technology" has initiated the promotion of vegetable oils. Nowadays more than 150 brands of lubricating oils and working fluids are produced commercially on the base of vegetable raw materials.

Vegetable oils provide better tribological behavior than common mineral industrial oil I-20A [9]. These oils are characterized by good anti-wear properties and protective action (wear scar diameter is much smaller for them compare to that for mineral oil, Fig. 5 a). Activation of non-saturated bonds as a result of oxidation and thermal-mechanical effect of friction can favor the tribological behavior of oils due to tribo-polymerization and formation of the protective films in the contact (Fig. 5 b, c) [9, 10].

Content of non-saturated fatty acids increases the resistance of the oils to wear with the average number of non-saturated bonds of the alkyl radicals in triglycerides contained in the oils. The score index for linseed oil is 4 times higher than that of the industrial oil I-20 at 2 times lower wear and 2 times higher critical load (Fig. 6) [9].

The test data have shown that the tribological performance of plant oils is promising for applications in the areas of industry, transportation and agriculture where the ecological safety is very important.

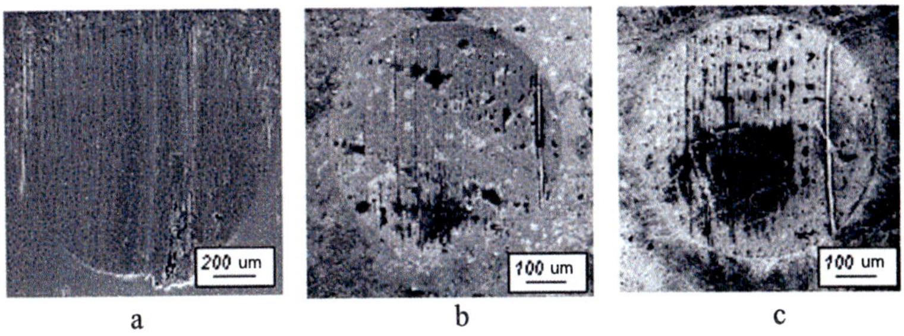


**Fig. 5.** Friction surfaces of balls after tests on a four-ball machine in a medium of various oils (a - industrial; b - rapeseed; c - linseed)

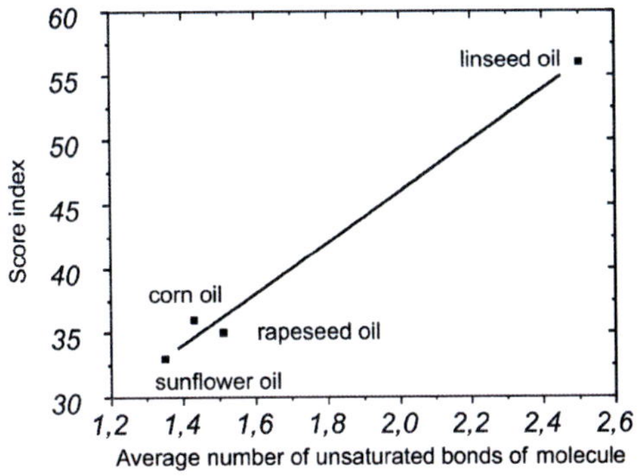


**Fig. 6.** The score index for vegetable oils of various origins. This index is equal 15 for mineral oil I-20

## 5 Conclusions

Science of friction, lubrication, and wear has grown from the ancient and medieval roots to industrial revolution and current technological transformations. Being formulated as tribology it has got a new impulse as a multidisciplinary concept in 1966 by Peter Jost's report. We have an evidence of its economic value for industrial society with the great signs of social and ecological impact. The engineering tribology has achieved remarkable results in saving energy, materials, and labor equal to some percent of the GNP in global economy.

Nowadays we need to implement the advances of green tribology to economy:

- To decrease the mass of the machines and their tribosystems due to better design and lighter materials thus reducing energy losses in friction, in achieving these purpose polymers and their composites can help.
- To use surface engineering for reducing the loss of materials due to wear, in such case coatings, implantation, texturing and patterning can be a choice.

- To use better lubricants and additives with less ecological damage; lubricants and additives of plant and animal origin can help, as well as films and coatings based on biodegradable polymers.

Tribology has done great in joining the mechanics, surface physics, and chemistry with analytical tools and computational resources. Green tribology can make an important influence on achievement of SDG in variety of areas and we should be optimistic in possibility of such achievement.

**Acknowledgements.** The authors are grateful for support by the projects of Belarus Fundamental Science Foundation T20KITG-004, T21ET-018.

## References

1. Jost, P.: Brief notes. In: Int. conference "50th anniversary of the "Jost report", p. 1–2. IMech, London (2016)
2. Holmberg, K., Erdemir, A.: Influence of tribology on global energy consumption, cost and emission. Friction **5**(3), 263–284 (2017)
3. Holmberg, K., Erdemir, A.: The impact of tribology on energy use and $CO_2$ emission globally and in combustion engine and electric cars. Tribol. Int. **135**, 389–396 (2019)
4. Myshkin, N.K., Pesetskii, S.S., Grigoriev, A.Y.: Polymer tribology: current state and applications. Tribol. Ind. **37**(2), 284–290 (2015)
5. Pesetskii, S.S., Bogdanovich, S.P., Myshkin, N.K.: Tribological behavior of polymer nanocomposites produced by dispersion of nanofillers in molten thermoplastics. In: Friedrich, K. (ed.) Tribology of Polymeric Nanocomposites, 2nd edn., pp. 119–162. Elsevier, Amsterdam (2013)
6. ISO 6601: Plastics — Friction and wear by sliding — Identification of test parameters (2002)
7. Spikes, H.: Liquid Lubrication Research; 1966 to the Present Day. In: International Conference on "50th anniversary of the "Jost report", p. 8–9. IMech, London (2016)
8. Cann, P.: Biotribology: opportunities and challenges. In: International Conference on "50th anniversary of the "Jost report", p. 10—11. IMech, London (2016)
9. Myshkin, N.K., Grigoriev, A.Y., Kavaliova, I.N.: Influence of composition of plant oils on their tribological properties. Tribol. Ind. **39**(2), 207–210 (2017)
10. Grigoriev, F.A., Kovaleva, I.N.: Corrosion activity and oxidative stability of vegetable oils used in lubricants. Friction Wear **39**(6), 535–538 (2018)

# Green Ferritic Stainless Steel: A Sustainable, Advanced and Cost-Effective Solution for the Concrete Mixer Drum

Wilian S. Labiapari[1], Ricardo J. Gonçalves Costa[1], Claudio M. de Alcântara[1], Vitor Pagani[1], Julio C. Di Cunto[1], and Jose D. B. de Mello[2(✉)]

[1] Aperam South America, Timóteo, MG, Brazil
[2] Universidade Federal de Uberlândia, Uberlândia, MG, Brazil
ltm-demello@ufu.br

**Abstract.** The use of stainless steel has been increasing in sectors where mechanical wear acts in conjunction with corrosion, such as mining and agribusiness, particularly biofuels. This replacement has led to an increase in the service life and a reduction in the weight of the structures, significantly improving the sustainability of these sectors. When using materials having a longer functional life, the production chain is more environmentally friendly because less coke and less iron ore are needed. In the case of green steels, i.e., steel produced with charcoal instead of coke and entirely recyclable, the Life Cycle Assessment (LCA) is even more favourable. Due to the unique supply chain, certified by the Forest Stewardship Council, Aperam South America is the only global producer of steels that uses 100% charcoal from proprietary eucalyptus forests instead of mineral coke. Laboratory and field tests were performed that compared abrasion-corrosion of carbon steels, advanced high-strength steels, weathering steel and nine different green stainless steels. Initially, rubber wheel and free-ball microabrasion tests were performed. In addition, a laboratory device approaching actual service conditions was built. In the field, samples were placed in sand processing equipment and inside a concrete mixer drum in a truck. Both the field and laboratory samples were collected and extensively analysed using SEM and hardness tests. Finally, a complete green stainless-steel concrete mixer drum entirely manufactured from green 410M steel was built, adapted to a truck and field-tested for five years (beginning in 2015). The useful life of the equipment increased up to three times compared to that made with standard carbon steel. It is worth noting that the concrete mixer made of green steel still preserves 70% of its thickness, predicting production of 72,000 $m^3$ of concrete and work for 12 years without the need for maintenance, which makes it the most sustainable, advanced, and cost-effective solution for the concrete mixer drum.

**Keywords:** SDG 13 and SDG 15 · Sustainability · Green stainless steel · Abrasion-corrosion; synergy; ferritic stainless steel · Concrete mixer drum


G. Quaglia et al. (Eds.): I4SDG 2021, MMS 108, pp. 429–437, 2022.
https://doi.org/10.1007/978-3-030-87383-7_47

## 1 Introduction

The iron and steel sector is, outside the power generation, the largest industrial producer of $CO_2$. It accounts for 7–9% of all direct fossil fuel emissions and 24% of the total industry $CO_2$ emissions [1].

The steelmaking process starts from iron ore for pig iron production using blast furnace processes. The standard raw material for reducing iron ore and energy source is coke, of fossil origin. The mineral coal emits 1.8 tons of $CO_2$ into the atmosphere, in addition to 7 kg of $SO_2$ per ton of steel produced. The green steel uses charcoal, of vegetable origin, instead of coke as a source of reducing agent and energy for pig iron production. The use of charcoal rescues 2.9 tons of $CO_2$ from the atmosphere for each ton of pig iron produced, generating carbon credit. Aperam South America is the only global producer of steels that uses 100% charcoal from proprietary eucalyptus forests instead of mineral coke, assuring outstanding sustainability and eco-friendliness to its Green steels portfolio. The harvesting, furnace loading and unloading, the transportation, among other carbonization stages, are entirely mechanized and use the most advanced equipment in the forestry industry. Using fume burners and gas recirculation (wood drying and coal cooling) in the coal production furnaces also helps to improve the environment and the wood transformation yield. The use of state-of-the-art technology in both eucalypt plantations and the carbonization process ensures a highly favourable Life Cycle Assessment for components made of Green stainless steels.

Stainless steel is rarely considered for wear resistance and is primarily used in applications where corrosion resistance is a dominant factor. High-strength steels have been developed for wear resistance and are widely used in various sectors, such as mining [2, 3] and agribusiness. Mechanical components in the industry are simultaneously subjected to mechanical wear and corrosion [4, 5]. Numerous studies have been conducted to emphasize the interaction between these phenomena whose importance and complexity varies according to the system: in some cases, the synergistic effect intensifies corrosion [4, 6–12], but, in others, it can cause it to severely drop [9, 10]. In the same way, the synergistic effect influences the wear phenomenon, increasing [4, 8, 13] or reducing [11, 14, 15] the total damage. In general, the range of steels studied is usually restricted to the more traditional ones: AISI 304, AISI 316, and in some cases, the duplex grade.

From this perspective, green ferritic stainless steels may provide a sustainable, cost-effective solution in less corrosive systems, particularly when compared to traditional stainless steels (owing to their low cost) or even commonly used carbon steels once the use of materials having a longer useful life present a production chain more environmentally friendly because less coke and less iron ore are needed. Labiapari et al. [16] used a free-ball microabrasion test to compare carbon steel and ferritic stainless steel for the sugar cane biofuel industry. When the test was mainly abrasive, carbon steel performed better. However, when the samples were tested after remaining 60 min in a humid environment, which was sufficient to accelerate the corrosion of the carbon steel, the ferritic stainless steel performed better. The same result was observed in the initial stages of sugar cane biofuel production. The ferritic stainless steel performed better under mechanical wear in a moderately corrosive environment (due to the humidity), considerably improving the equipment life. Based on these results, ferritic stainless steel may benefit other systems, such as concrete processing and mining. In this study, which focussed specifically on the concrete industry, laboratory and field tests were performed to compare the performance of standard carbon steels, advanced high-strength steels, and green stainless steels.

## 2 Methodology

The methodological approach used in this study is summarised in Fig. 1 which attempts to establish and organise the primary steps of an intricate combination of field and laboratory tests. A detailed description of the procedures used can be found in a recent publication by our research group [17]. The first step involved collecting carbon steel samples in actual operating conditions from concrete mixer truck drums, which were then analysed by scanning electron microscopy (SEM) to identify the relevant wear mechanisms.

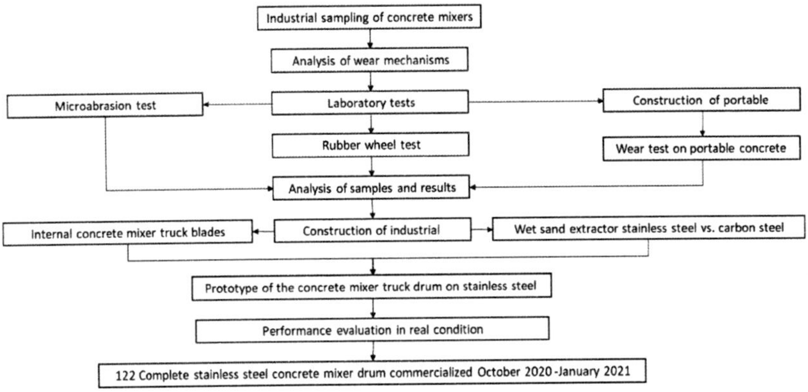


**Fig. 1.** Schematic of the methodological approach.

As a first approach, aimed to establish initial criteria allowing selecting materials in terms of abrasion resistance, conventional laboratory wear tests, including free-ball microabrasion and rubber wheel tests, were performed. Besides, a portable concrete mixer (PCM) was adapted to simulate the conditions of a concrete mixer truck drum, bringing laboratory conditions closer to actual operating situations. In this step, ferritic (410M), austenitic (AISI 304), and martensitic (AISI 420) green stainless steels were studied, as well as mild (A36) and high-strength (AHSS) carbon steels. Two industrial prototypes were built in parallel: the internal blades of a concrete mixer truck drum and a wet sand extractor. The use of the second prototype (Fig. 2) a sorter whose function was the classification of wet sand for use in concrete blends, was to extend the analysis for a more significant number of potential candidates (12 different steel types to be tested: three-carbon steels and nine green stainless steel) in actual operating conditions where abrasive and corrosive compounds certainly act.

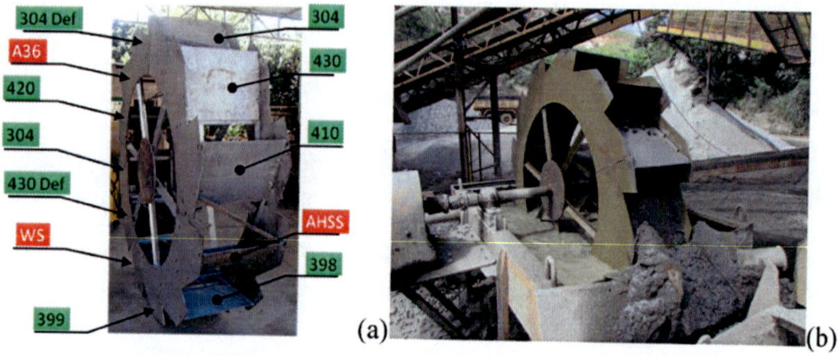


Fig. 2. (a) Sand classifier built with green stainless-steel blades made from 12 different materials, and (b) this equipment working in the field.

The last studied prototype was a full and real concrete mixer drum manufactured using 410M steel, as shown in Fig. 3. The thickness of the drum wall and internal blades was measured every three months. In addition, the total volume of the processed concrete was controlled. For comparison purposes, the amount of concrete processed in another concrete mixer drum working at the same plant but made of standard carbon steel was also monitored.

Fig. 3. (a) Real concrete mixer drum during fabrication, (b) internal blades, and (c) final prototype.

## 3 Results and Discussion

Table 1 summarises the main characteristics (chemical composition and hardness) of the tested materials. Of the most prominent features in Table 1, the high Cr content in relation to carbon steels stands out, showing the expectation of greater resistance to corrosion in this medium. On the other hand, the higher the content of Cr, and other alloy elements such as Ni and Mo, the higher the alloy cost, making the project economically unfeasible in real life. Under this perspective, our criterion will be to choose, among the steels of better performance, the one with the lowest cost or the lowest content of alloy elements.

SEM analysis of the carbon steel samples from the actual concrete mixer truck blade identified grooving and rolling abrasive phenomena as well as oxidation due to the corrosive process. In this sense, it is emphasized that, contrary to what is reported by

**Table 1.** Chemical composition and hardness of the tested steels.

| | | [wt. %] | | | | | | | | | [ppm] | | Hardness [$HV_{10}$] |
|---|---|---|---|---|---|---|---|---|---|---|---|---|---|
| | | C | Mn | Si | Cr | Ni | Mo | Al | Cu | Nb | B | N | |
| Carbon Steels | A36 | 0.138 | 1.06 | 0.01 | 0.01 | 0.01 | 0.00 | 0.033 | 0.014 | 0.001 | - | 26 | 109 ± 1 |
| | AHSS | 0.168 | 1.28 | 0.26 | 0.15 | 0.09 | 0.03 | 0.054 | 0.150 | 0.016 | 6 | 36 | 551 ± 12 |
| | WS | 0.160 | 1.36 | 0.34 | 0.64 | 0.02 | 0.01 | 0.038 | 0.320 | - | - | 62 | 170 ± 2 |
| Stainless Steels | 304 | 0.033 | 1.10 | 0.33 | 17.69 | 8.01 | 0.08 | 0.004 | 0.127 | 0.011 | - | 431 | 222 ± 3 |
| | 304 (30%) | | | | | | | | | | - | | 346 ± 2 |
| | 430Nb | 0.018 | 0.24 | 0.30 | 16.14 | 0.27 | 0.03 | 0.002 | 0.024 | 0.316 | - | 212 | 176 ± 2 |
| | 430 | 0.053 | 0.16 | 0.32 | 16.17 | 0.16 | 0.01 | 0.002 | 0.015 | 0.004 | - | 480 | 185 ± 4 |
| | 430 (30%) | | | | | | | | | | - | | 243 ± 5 |
| | 399 | 0.018 | 1.74 | 0.40 | 22.59 | 5.30 | 2.62 | 0.005 | 0.230 | 0.009 | - | 1548 | 310 ± 3 |
| | 410M | 0.015 | 0.52 | 0.50 | 11.11 | 0.35 | 0.03 | 0.002 | 0.052 | 0.027 | - | 184 | 173 ± 3 |
| | 398 | 0.012 | 1.37 | 0.34 | 22.60 | 3.51 | 0.21 | 0.003 | 0.401 | 0.006 | - | 1071 | 268 ± 2 |
| | 420A | 0.310 | 0.27 | 0.52 | 12.34 | 0.25 | 0.04 | - | 0.035 | 0.005 | - | 142 | 584 ± 22 |

concrete manufacturers and concrete mixer truck manufacturers, the moisture present, which is necessary for the manufacture of concrete, promotes the corrosive process in carbon steel. The strong presence of oxidation motivated the introduction of green ferritic stainless steel as a potential sustainable, technical and economical solution for replacing standard carbon steel.

If in the standard laboratory tests, microabrasion and rubber wheel, the greater hardness governed the performance, in the PCM test, the corrosive effect separates the materials into two distinct groups, green stainless steel, and carbon steel, with better and worse performance. For the carbon steel group, the harder was better than the softer. In the green stainless steel group, the best performance associated with higher hardness is also obeyed. The best overall performance was for green stainless steel with higher hardness, e.g., greater resistance to corrosion and wear clearly indicating the need to search for experimental conditions closer to reality to guide the subsequent industrial tests.

The internal blades of a concrete mixer truck drum produced in green ferritic stainless steel (410M) worked for about 33 months and processed approximately 10,000 m3 of concrete. Another concrete mixer, whose inner blades were made of standard carbon steel, worked for the same period on the same construction. Although the amount of concrete processed by the blades made of carbon steel was slightly higher (17%), their initial thickness was also thicker (20%), and they were wholly consumed, reaching a thickness of almost zero at the end of the test. In contrast, the thickness of the blades produced from 410M was reduced by only 55%, still presenting a residual thickness of 2.15 mm. Corrosion still accelerated wear because the inner part of the drum remained damp at all times, including during short shutdowns (e.g., overnight) or more extended shutdowns (e.g., truck maintenance). It is important to note that these effects were visible even to the naked eye for daily stops and even more severe for longer stops. It is reasonable to conclude that 410M stainless steel suffers only the mechanical effects of abrasive wear. In contrast, carbon steel suffers the combined effects of abrasion-corrosion, directly affecting its performance in real life.

Figure 4 shows the thickness reduction after 36 months of operation under severe abrasion-corrosion conditions acting in the sand classifier prototype. Regardless of the hardness, carbon steels showed a high thickness reduction rate (3.4–17 times higher), and that all the green stainless steels, regardless of composition and hardness, performed better than the carbon steels. Once more, the A36 steel (the traditional material) was the least performant, corroborating the need for a new solution. Some concrete mixer truck drum manufacturers have been testing the other two carbon steels. However, our results clearly show that green stainless steel has superior performance. Because the difference in performance between the stainless steels was not remarkable, economic considerations (summarized in Fig. 5 which was obtained from queries to metal price reference sites such as www.asianmetal.com and www.metalbulletin.com as well as confidential production cost data) and ease of production, particularly bending, cutting and weldability, were used to select the candidate to manufacture the third industrial prototype: 410M, a sustainable, ferritic, low cost, green stainless steel.

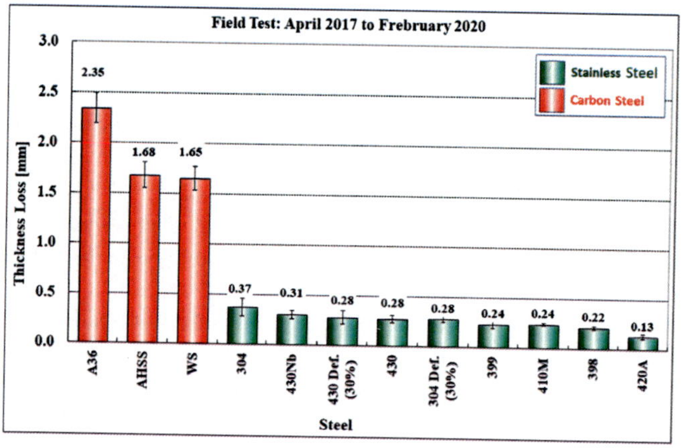


**Fig. 4.** Total average thickness loss after 36 months of the blades in a wet sand sorter under real conditions.

The third prototype was a complete full-size concrete mixer drum manufactured entirely from 410M. Both the blades and the drum thicknesses were monitored for 34 months. Figure 6 shows the wear of the drum and blades in the regions where the wear was more substantial for the prototype manufactured from 410M and for another built from A36 operating at the same construction site.

The figure also shows the accumulated amount of processed concrete. In the same period and for a similar volume of processed concrete, the wear, of both the drum and the internal blades, was considerably larger for the reference in carbon steel.

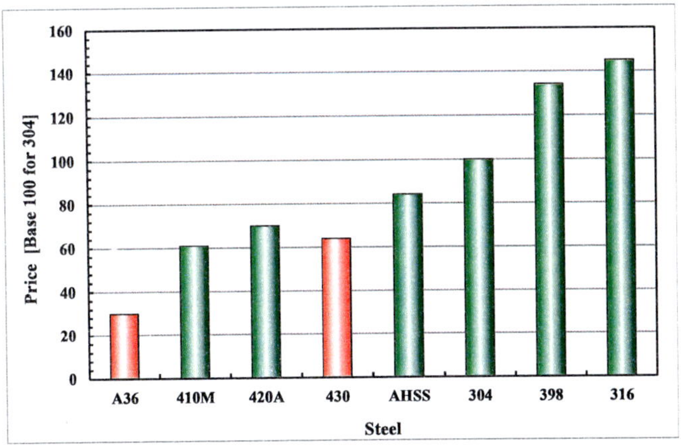


**Fig. 5.** Relative price. Steel 304 was taken as a reference since it is the most used stainless steel.

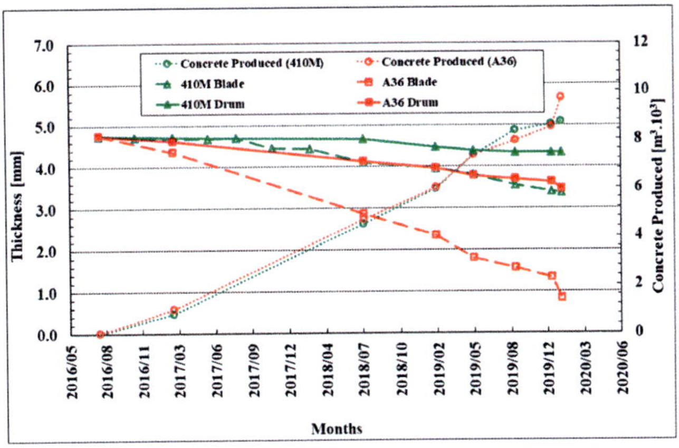


**Fig. 6.** Thickness loss measured at the most critical points of the drum and the blade of the concrete mixer truck made from A36 and 410M stainless steels.

The internal blades and the drum had a projected service life of approximately three times that of the carbon steel components. This means that while carbon steel blades/drums can produce an average of 8,000 $m^3$ of concrete, the device built from 410M can reach values as high as 25,000 $m^3$. The expectation is that the 410M drums will be able to produce 72,000 $m^3$ of concrete, working for 12 years without the need for maintenance, in contrast with the maximum four years for those produced from carbon steel. This time of use, 12 years, is often the lifetime of the truck itself. Therefore, all the equipment (truck and drum) may have the same life cycle, avoiding additional downtime for maintenance.

A 410M steel drum is approximately 55% more expensive than carbon steel, considering material and labor. However, it lasts 300% longer and avoids about 40 more downtime days throughout service life with maintenance which, according to the partner company (Pedreira Um Valemix Ltda), represents 50% savings in replacement materials and approximately US$ 85,000.00 in effective working time.

In this sense, to achieve a longer service life and shorter downtime for maintenance, some companies started to produce a concrete mixer truck using 410M as the most advanced, cost-effective solution for this application, which resulted that between December 2020 and January 2021, 122 fully stainless steel concrete mixers were sold to concrete manufacturing companies in Brazil.

## 4 Conclusions

In this study, laboratory and field tests were performed to compare carbon steels and green stainless steels aimed at determining the most advanced, cost-effective solution for achieving a longer service life and shorter downtime for concrete mixing drums and internal blades. The main conclusions are as follows:

1. SEM analysis of carbon steel samples from the original concrete mixer truck blade identified grooving and rolling abrasive phenomena as well as oxidation due to corrosion, indicating that green stainless steel was a potential technical and economical solution for replacing standard carbon steel.
2. Traditional laboratory tests, such as rubber wheel and microabrasion tests, presented degradation mechanisms that were considerably different from those of the original concrete mixer.
3. The sand classifier, allowed the simultaneous comparison of 12 materials in real conditions. Regardless of hardness, the carbon steels showed a high thickness reduction rate (3.4–17 times), and all the green stainless steels, regardless of composition and hardness, performed better than the carbon steels.
4. Because the difference in performance between the stainless steels was not notable, economic aspects and ease of production, particularly weldability, were used to select the candidate for the manufacture of the third industrial prototype: 410M, a ferritic, low cost, stainless steel.
5. Because of the longer service life and shorter downtime for maintenance, a concrete mixer truck manufactured from 410M is, at present, the most sustainable, advanced, cost-effective solution for this application.

**Acknowledgements.** The authors acknowledge the Pedreira Um Valemix Ltda for supporting this work.

## References

1. Financial Times. https://www.ft.com/content/46d4727c-761d-43ee-8084-ee46edba491a. Accessed on 25 May 2021
2. Mazur, M., Mikova, K.: Impact resistance of high strength steels. Mater. Today Proc **3**(4), 1060–1063 (2016)

3. Ramos, P.A., Brito, P.P., dos Santos, C.E., Vieira, L., da Silva, E.R.: Abrasive wear characteristics of complex carbides overlay (CCO) in comparison to low-alloy boron steels. Int. J. Refract. Met. Hard Mater. **86**, 105119 (2020)
4. Stack, M.M., Jawan, H., Mathew, M.T.: On the construction of micro-abrasion maps for a steel/polymer couple in corrosive environments. Tribol Int. **38**(9), 848–856 (2005)
5. Stack, M.M., Chi, K.: Mapping sliding wear of steels in aqueous conditions. Wear **255**(1–6), 456–465 (2003)
6. Wood, R.J.K., Sun, D., Thakare, M.R., deFrutosRozas, A., Wharton, J.A.: Interpretation of electrochemical measurements made during micro-scale abrasion-corrosion. Tribol Int. **43**(7), 1218–1227 (2010)
7. Bello, J.O., Wood, R.J.K.: Wharton, J, A: Synergistic effects of micro-abrasion-corrosion of UNS S30403, S31603 and S32760 stainless steels. Wear **263**(1–6), 149–159 (2007)
8. Sun, D., Wharton, J.A., Wood, R.J.K.: Micro-abrasion–corrosion of cast CoCrMo—Effects of micron and sub-micron sized abrasives. Wear **267**(1–4), 52–60 (2009). https://doi.org/10.1016/j.wear.2009.01.011
9. Thakare, M.R., Wharton, J.A., Wood, R.J.K., Menger, C.: Exposure effects of strong alkaline conditions on the microscale abrasion-corrosion of D-gun sprayed WC-10Co-4Cr coating. Tribol Int. **41**(7), 629–639 (2008)
10. Thakare, M.R., Wharton, J.A., Wood, R.J.K., Menger, C.: Investigation of micro-scale abrasion-corrosion of WC-based sintered hardmetal and sprayed coating using in situ electrochemical current-noise measurements. Wear **267**(11), 1967–1977 (2009)
11. Labiapari, W.S., Ardila, M.A.N., Costa, H.L., de Mello, J.D.B.: Micro abrasion-corrosion of Ferritic stainless steels. Wear **376–377**, 1298–1306 (2017). https://doi.org/10.1016/j.wear.2017.01.083
12. Marques, A., de Oliveira, M., Houmard, W., Labiapari, C.: Erosive wear study of the AISI 201LN stainless steel: a comparison with the AISI 304 and AISI 410 stainless steels. Metall. Mater. Trans. A **50**(4), 1663–1671 (2019). https://doi.org/10.1007/s11661-019-05123-1
13. Gant, A.J., Gee, M.G., May, A.T.: The evaluation of tribo-corrosion synergy for WC-Co hardmetals in low stress abrasion. Wear **256**(5), 500–516 (2004). https://doi.org/10.1016/j.wear.2003.04.001
14. SinnettJones, P.E., Wharton, J.A., Wood, R.J.K.: Micro-abrasion-corrosion of a CoCrMo alloy in simulated artificial hip joint environments. Wear **259**(7–12), 898–909 (2005)
15. Bateni, M.R., Szpunar, J.A., Wang, X., Li, D.Y.: The effect of wear and corrosion on internal crystalline texture of carbon steel and stainless steel. Wear **259**(1–6), 400–404 (2005). https://doi.org/10.1016/j.wear.2005.02.009
16. Labiapari, W.S., Alcântara, C.M., Costa, H.L., de Mello, J.D.B.: Stainless steel as an antiwear material for the bio-fuel industry. Wear **302**(1–2), 1536–1545 (2013). https://doi.org/10.1016/j.wear.2012.12.023
17. Labiapari, W.S., Gonçalves, R.J., de Alcântara, C.M., Pagani, V., di Cunto, J.C., De Mello, J.D.B.: Understanding abrasion-corrosion to improve concrete mixer drum performance: a laboratory and field approach. Wear **477**, 203830 (2021). https://doi.org/10.1016/j.wear.2021.203830

# Tribology and Sustainable Development Goals

Enrico Ciulli(✉)

University of Pisa, Pisa, Italy
ciulli@ing.unipi.it

**Abstract.** Tribology, which deals with all problems related to friction, wear and lubrication, is intrinsically connected to sustainability. In this work, after a very brief historical overview of Tribology and Sustainable Development Goals, the main tribological research fields are described. A possible organization in six tribological branches is proposed: Fundamental Tribology, Tribology of materials and lubricants, Micro and Nanotribology, Industrial Tribology, Biotribology and Ecotribology. Several aspects of all branches are related to sustainability and can be included in another specific recent discipline termed Green Tribology. Different tribological aspects that contribute to reduce energy losses, and therefore $CO_2$ emissions, and material waste by reducing friction and wear, are reported in the paper. Tribology affects many targets of the Goals. Goal 12, Responsible Consumption and Production, is probably the most affected by tribological findings, but also Goals 3, 6, 7, 8, 9, 11, 13, 14 and 15 are directly influenced. The remaining Goals are indirectly affected in different ways.

**Keywords:** SDG12 · Tribology · Sustainable Development Goals · Green Tribology · Friction · Wear

## 1 Introduction

Tribology and Sustainable Development Goals are strictly connected. In this section a very brief historical overview of both is presented showing their fundamental aspects.

### 1.1 Tribology

Tribology is an interdisciplinary science that deals with every problem related to friction, wear and lubrication. Despite its name officially appeared in 1966 [1], the application of tribological practices comes back even to the prehistoric epoch (let's think for instance about the generation of fire by friction between two pieces of wood and about the way to reduce friction using rolling elements and fluids [2]). Some historical famous milestones of Tribology are schematically shown in Fig. 1 [3]. In the figure, starting from the left, an Egyptian bas-relief shows the transports of a stone colossus on a slide with a man pouring some liquid in front of the slide. Beside, a rolling bearing for a rotating platforms of a Roman ship. A ball bearing drawn by Leonardo da Vinci, by someone considered as the first well-known tribologist, and an apparatus for friction tests are then shown. Studies on lubrication are almost always based on the use of the well-known


G. Quaglia et al. (Eds.): I4SDG 2021, MMS 108, pp. 438–447, 2022.
https://doi.org/10.1007/978-3-030-87383-7_48

Reynolds equation; the principle of the hydrodynamic lubrication is shown in the middle of the figure. A picture shows then a meeting of the committee that produced the report where Tribology was mentioned officially for the first time. More recently Tribology has taken advantage from new techniques such as particularly electronics, information and communications technology. A schematic drawing of a computer connected to an electronic system for measuring gear tooth temperature and a smart experimental system for recording experimental data of tests in bearings for turbomachinery are finally shown on the right.

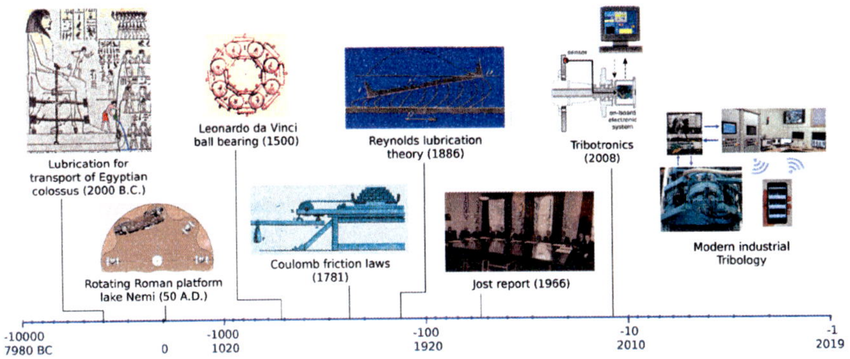


**Fig. 1.** The Tribology evolution. From [3].

## 1.2 Sustainable Development Goals

The 2030 Agenda for Sustainable Development was adopted by the General Assembly of the United Nations on 25 September 2015 [4]. It includes 17 Sustainable Development Goals (SDGs) with 169 targets for Prosperity and Peace for People and Planet based on a global Partnership (sometimes called "the five P").

The 17 SDGs [5], reported in Table 1, aim for a better quality of life for everybody with improved health, education and economic situation in strict connection with a sustainable management of the planet's natural resources. Subjects as climate, energy, science and technology are therefore also of fundamental importance for sustainable progress and life.

**Table 1.** The 17 UN Sustainable Development Goals.

| | |
|---|---|
| 1. No Poverty | 10. Reduced Inequalities |
| 2. Zero Hunger | 11. Sustainable Cities and Communities |
| 3. Good Health and Well-being | 12. Responsible Consumption and Production |
| 4. Quality education | 13. Climate Action |
| 5. Gender Equality | 14. Life below Water |
| 6. Clean Water and Sanitation | 15. Life on Land |
| 7. Affordable and Clean Energy | 16. Peace, Justice and Strong Institution |
| 8. Decent Work and Economic Growth | 17. Partnership for the Goals |
| 9. Industry, Innovation and Infrastructure | |

# 2 From Tribology to the Green Tribology Concept

## 2.1 Tribology Branches

An overview of the many subjects of the tribological studies performed today can be a useful starting point for better understanding how Tribology is connected with sustainability. In fact Tribology is often not well known among common people and even among the engineering world [6], mainly because its complexity. Tribology is a really interdisciplinary science that involves many research fields as physics, mathematics, chemistry, materials science, mechanical engineering, chemical engineering, biomedical engineering, manufacturing, and computer science. Not only industrial activities, but almost every activity of our life is related to tribological phenomena. When walking the friction between our feet and the soil is essential to avoid slipping, and when washing our teeth or when combing our hair friction and wear aspects are involved. In addition, tribological studies range from theoretical to experimental and from basic to applied ones, covering aspects from the macroscopic to the nano world. Friction, wear and lubrication aspects are involved in many different fields interconnected in different ways so that Tribology could be considered like a web, Fig. 2.

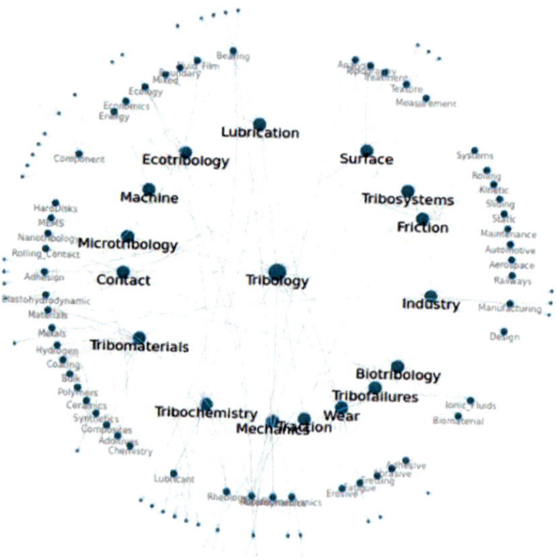


**Fig. 2.** The Tribology web.

Because of the many interconnections, it is not simple to classify the different subjects of Tribology. Usually, when the number and the importance of the studies in a specific research area have become significant, a new specific branch (or discipline) of Tribology has been created. Several different classifications in branches have been proposed during the years, all generally acceptable, but with some unavoidable overlapping among the subjects. In this paper the following list of Tribology disciplines is proposed:

Fundamental Tribology, Tribology of materials and lubricants, Micro and Nanotribology, Industrial Tribology, Biotribology, Ecotribology. The first three branches are more general, the second three more specific.

**Fundamental Tribology** involves all basic studies on friction, wear and lubrication. Theoretical and basic experimental researches are carried out that are used by all other branches. It furnishes the tools to investigate how to reduce friction (and therefore energy losses) and wear (with the related saving of materials), both in dry and lubricated contacts.

**Tribology of Materials and Lubricants** studies materials, lubricants, additives, surface treatments and coatings. Among new materials are ceramics, self-lubricating materials, and nano- and biomaterials. Tribological researches on new materials have influence on material footprint and material efficiency, resource consumption and recycling. Beside studies to reduce friction and wear in lubricated contacts with the more classical lubricants (mineral oils) and additives, there are studies to increase their durability. New ecofriendly lubricants and additives have been also developed (see below Ecotribology). Aqueous lubrication together with new aspects as superlubricity, and smart fluids, as electro-rheological (ER) and magneto-rheological (MR) fluids, able to vary their viscosity when an electric or magnetic field is applied, are also investigated. Different surface coatings for several applications are under study for further reductions of friction and wear.

**Micro and Nanotribology** performs studies at micro- and nano level (nano-reinforced materials, nanocoatings, 2D materials like graphene, fullerene-like hydrogenated carbon film, hydrogenated diamond like carbon coatings) till to the atomic scale level. The main aim is again the reduction of friction and wear. Such investigations are important for the device miniaturization and have several micro and nano mechanics applications in medicine, biotechnology, optics, electronics, aviation, etc. (micro sensors and micro machines).

**Industrial Tribology** is related to applications of tribological researches to the industrial products. It ranges from manufacturing to the machines and their components. Tribology in manufacturing involves all machining (as grinding and turning) and cold/hot forming processes. Tool coatings for extension of tool life and higher machining quality are investigated. Minimum Quantity Lubrication (MQL) has been introduced for a sustainable manufacturing for reducing the amount of cutting fluids in machining processes. Efficiency and maintainability of machines are increased by a correct tribological design of their components. Improvements in design can reduce size, friction and the amount of lubricant necessary, and produce longer life of common machine components as bearings and gears. Oil-free solutions are also studied such as magnetic bearings. In the field of transportation, in addition to the reduction of emissions related to lower energy losses, tribological studies allow also to reduce the quantity of particles emitted from brakes, tires-asphalt contacts for cars, pantographs-catenary and wheel-rail contacts for trains. Efficiency and maintainability of machines and implants can be increased also implementing corrective actions when friction-wear-lubrication problems are evidenced by monitoring. New lubrication systems, bearings and surface treatments have been developed for applications where longevity of operation is requested and maintenance is significantly costly and complicated, as for instance in wind turbines.

**Biotribology** in a broad sense deals with all tribological aspects related to living organism, animal and plants. Most studies are focused on humans. Some tribological researches involves skin and teeth friction and wear, saliva lubrication, food taste (tongue). Biomedical applications are deeply investigated, such as cardiovascular devices, tissue substitutes, synovial joints (articular cartilage and synovial fluid), artificial joints, new biomaterials. Biomimetics is another important field that can help to avoid potential toxic new materials by taking inspiration from nature. Bio-inspired surfaces and systems can be manufactured in order to replicate natural structures. Some examples are non-adhesive surfaces that simulate the lotus leaves, specially structured surfaces mimicking the gecko toes for controlling adhesion, water-resistant and waterproof materials replicating the nanostructure of cigada wings, special micro-textures (bumps and grooves) of functional surfaces used for sand erosion resistance by the desert scorpion.

**Ecotribology** is specifically focused on environmental effects of tribological studies. Environmentally friendly (ecological) lubricants and additives are important for reducing health and pollution problems. Vegetable oil-based lubricants, bio-based hydraulic and metal working fluids, gelatin based soft lubricants, ionic liquids, additives derived from natural resources as vegetable oils, coatings made by biofilms are some examples that could also provide an important friction reduction. Reconditioning of lubricants is another ecological practice to reduce the impact of non-renewable fossil liquids potentially harmful for the environment. A correct tribological design can take into account the possibility of remanufacturing of machine components as bearings and gears that do no more deliver optimum performances when worn. Remanufacturing reduces waste and emissions. Renewable energy plants are part of this branch too.

## 2.2 Green Tribology

Tribological studies can be put directly or indirectly in relation with sustainability. At the beginning of the 2000s the new term "Green Tribology" was introduced by Zhang as reported in [7] that includes all subjects of the above disciplines more directly related to sustainability, as material and energy efficiency, eco-friendly lubricants, reduced emissions. The names environmental friendly Tribology and Ecotribology were also introduced before, while the term sustainable started to be associated to Tribology successively [7, 8]. However, in a broad conception, Green Tribology can include all subjects of the tribological disciplines. Tribological theories and applications are just watched from a different point of view as before and indications on the environmental and ecological impacts of the tribological studies are explicitly mentioned. In a book edited by Nosonovsky and Bhushan [9] three areas of Green Tribology are identified: biomimetic Tribology, eco-friendly lubricants and materials, and tribological aspects of sustainable applications.

One of the main effects of Tribology on the ambient is related to reduction of energy losses and materials consumption. Obviously the first aspect is more related to friction, the second one to wear. However, a reduction of wear reduces also energy needs because it is always necessary some energy to extract materials and to manufacture machine components. Transportation, power generation, manufacturing and residential are the main sources of energy losses [10]. Friction reduction and the extension of product life by an increase of the wear resistance contribute to reduce $CO_2$ emissions. Wear protection also reduces pollution due to wear particles and fine dust. Inspiration can be also taken from the nature: biomimetic studies can help to develop controlled friction and low wear surfaces. Wear control can be obtained both with a good tribological design of the components and with condition monitoring of the machine behavior. Based on the indications from monitoring, corrective actions can be taken without delay by minimizing energy losses related to the machine malfunctioning and avoiding catastrophic damages. Tribotronics is a concept that combines Tribology and Electronics [11]. It allows to monitor and control the systems in real time. Quantities as vibrations, temperatures and sound emissions are continuously monitored by suitable sensors connected to control units and computers that can analyze results also using tribological databases and can modify some working parameters through suitable actuators. New lubricants with reduced friction (connected to lower $CO_2$ emissions) and extended life, oils and additives based on renewable materials and biodegradable and low toxicity fluids contribute to reduce pollution. Renewable clean energy plants as wind turbines and marine wave energy collecting devices are important for sustainability. Their correct working conditions are related to suitable tribological design of components for long life and reduced need of maintenance, while Tribotronics helps the plants work for long time without human intervention.

## 3 Tribology and Sustainability

Tribological studies are of fundamental importance for a sustainable development that has to combine human needs and environment's integrity. The term sustainability appears today together with Tribology in many conferences, books and papers, and the economic, environmental and social impact of Tribology is also quantified (see for instance [7, 8, 10, 12]). The enormous impact of Tribology in energy, materials, economy, $CO_2$ emission reduction has been evidenced in several studies starting from the Jost report [1]. According to [13], "*as the humanity progresses, tribology continue to make vital contributions in addressing the demands for advanced technological developments, resulting in, for example, reducing the fuel consumption and greenhouse gases emission, increasing machine durability and improving the quality of life through artificial implants, among the others*".

Tribology affects directly or indirectly several targets of almost all the 17 SDGs. The SDGs are surely related to Green Tribology that, as stated before, includes aspects of all branches described in the previous section. Two lists of Goals and targets directly and indirectly affected by Tribology are reported below. Some connections with the tribological subjects and branches are synthetically reported. They are just some of the possible connections because of the many links among the branches.

### 3.1 Goals Directly Affected by Tribology

**Goal 3-Good Health and Well-Being.** Target 3.8 mentions the *access to quality essential health-care services* that can be put in relation with Biotribology, dealing with prosthetic implants, heart valves and other medical devices for which tribological investigations are fundamental. Target 3.9 is addressed to a substantial reduction of *the number of deaths and illnesses from hazardous chemicals and air, water and soil pollution and contamination.* More efficient system with lower emissions, less material wastes through reduced wear and dimensions, reuse, environmental friendly lubricants and reduced use of lubricants are relevant for this target. Studies of Ecotribology, Industrial Tribology, Tribology of materials and lubricants are particularly involved.

**Goal 6-Clean Water and Sanitation.** Targets 6.3 and 6.6 are respectively addressed to *improve water quality by reducing pollution, eliminating dumping and minimizing release of hazardous chemicals and materials,…increasing recycling and safe reuse globally* and to *protect and restore water-related ecosystems, including mountains, forests, wetlands, rivers, aquifers and lakes*. Target 6.a mentions *water efficiency, wastewater treatment, recycling and reuse technologies*. They are again related to similar tribological fields as for Goal 3.

**Goal 7-Affordable and Clean Energy.** Target 7.3 reports the necessity to *double the global rate of improvement in energy efficiency* and is particularly related to Industrial Tribology (friction reduction, wear control, Tribotronics, efficiency and maintainability). Target 7.a mentions the *access to clean energy research and technology, including renewable energy, energy efficiency* and the promotion of *investment in energy infrastructure and clean energy technology*. This is mainly related to Ecotribology applications, as eolics and hydropower.

**Goal 8-Decent Work and Economic Growth.** Target 8.2 pushes towards *higher levels of economic productivity through diversification, technological upgrading and innovation*. Target 8.4 hopes for a progressive improvement of the *global resource efficiency in consumption and production and endeavour to decouple economic growth from environmental degradation*. Both targets can be again related to Industrial Tribology and Ecotribology (Tribotronics, friction and wear control, new coatings, ecological lubricants, reuse, minimum quantity lubrication) but also to innovative studies of Micro and Nanotribology.

**Goal 9-Industry, Innovation and Infrastructure.** Target 9.4 mentions the necessity to make infrastructure and industries *sustainable, with increased resource-use efficiency and greater adoption of clean and environmentally sound technologies and industrial processes* and is primarily related to Ecotribology and Industrial Tribology. Target 9.5 is for enhancing *scientific research, upgrade the technological capabilities of industrial sectors in all countries* and is related to Industrial Tribology and Tribology of materials and lubricants (researches for more durable and efficient components, new materials).

**Goal 11-Sustainable Cities and Communities.** Target 11.6 reports the necessity to *reduce the adverse per capita environmental impact of cities, including by paying special attention to air quality and municipal and other waste management*, while in target 11.b *resource efficiency* is mentioned again. These targets are related to transports and buildings. If the importance of Tribology in transports is clear (there is a huge number of vehicles around the world, with several tribological components, as bearings and gears), maybe less evident is the possible reduction of energy losses in buildings, where losses are nested in several common machines as electric motors, pumps and fans, present in devices such as washing machines, heating and air conditioning systems, teeming with tribological contacts.

**Goal 12-Responsible Consumption and Production.** Target 12.1 mentions again *sustainable consumption and production*, target 12.2 would like to *achieve the sustainable management and efficient use of natural resources*, and target 12.4 *the environmentally sound management of chemicals and all wastes* and the significant reduction of *their release to air, water and soil.* Target 12.5 mentions a substantial reduction of *waste generation through prevention, reduction, recycling and reuse.* Target 12.a strengths the *scientific and technological capacity to move towards more sustainable patterns of consumption and production.* All targets are again related to studies on friction and wear for reducing impact on ambient, new lubricants, biolubricants and additives, surface treatments and coatings, Tribology of machine elements.

**Goal 13-Climate Action.** Target 13.2 and 13.3 state respectively: *Integrate climate change measures into national policies, strategies and planning*, and *Improve education, awareness-raising and human and institutional capacity on climate change mitigation, adaptation, impact reduction.* Reduced $CO_2$ emissions related to lower friction losses, materials waste reduction, reduced pollution of small wear particles such as the ones produced by wear of tires, brakes and road surfaces, lubricants with longer life, lubricant reconditioning, and ecofriendly lubricants are examples of how Tribology can impact the ambient and therefore the climate. Knowledge of the tribological problems is fundamental.

**Goal 14-Life Below Water.** Target 14.1 reports about to *prevent and significantly reduce marine pollution of all kinds*, target 14.3 about to *Minimize and address the impacts of ocean acidification, including through enhanced scientific cooperation at all levels.* They are surely related to a reduction of pollution due to exhausted lubricants and to wear particles. Oil reconditioning, ecofriendly lubricants, oil-free or water lubrication and wear control techniques are tribological practices that contribute to reduce marine pollution.

**Goal 15-Life on Land.** Target 15.1 aims to *ensure the conservation, restoration and sustainable use of terrestrial and inland freshwater ecosystems*, target 15.4 *ensure the conservation of mountain ecosystems.* Oil reconditioning, remanufaturing, tribological design for longer life of the components and biomimetic applications contribute to a reduced consumption of the natural resources.

## 3.2 Goals Indirectly Affected by Tribology

**Goal 1-No Poverty.**Target 1.4 would like to *ensure that all men and women* can *have equal rights to economic resources, as well as access to... natural resources, appropriate new technology.* To spare energy and materials contributes to save money and natural resources with beneficial for all people.

**Goal 2-Zero Hunger.** Target 2.a mentions *technology development* also *in order to enhance agricultural productive capacity.* Tribological studies contribute to realize machines and components with better performances and minor downtime, also for agricultural applications, allowing to increase production.

**Goal 4-Quality Education.** Target 4.3 hopes for equal access of all women and men to *affordable and quality technical, vocational and tertiary education*, target 4.4 for a substantial increase of *the number of youth and adults who have relevant skills, including technical and vocational skills*, target 4.b mentions *training and information* *and communications technology, technical, engineering and scientific programmes.* The knowledge of the tribological concepts and their implications is important for a high quality education.

**Goal 17-Partnership for the Goals.** Target 17.6 mentions enhancement of *cooperation on and access to science, technology and innovation and enhance knowledge,* target 17.7 states *Promote the development, transfer, dissemination and diffusion of environmentally sound technologies*, target 17.16 *Enhance the global partnership for sustainable development.* Diffusion of the Green Tribology results is fundamental for sustainability. Tribological databases are available and results can be disseminated.

Tribological studies contribute to a better environmental situation and therefore to a better life for both men and women. So we could say that also **Goal 5-Gender Equality**, **Goal 10-Reduced Inequalities**, and **Goal 16-Peace, Justice and Strong Institution** are in a certain way influenced by Tribology. In particular Target 5.b mentions *Enhance the use of enabling technology, in particular information and communications technology, to promote the empowerment of women.* Computer are involved and tribological studies on hard disk technology are important.

By summarizing, Goals 3, 6, 7, 8, 9, 11, 12, 13, 14 and 15 are directly affected by tribological applications. Most probably Goal 12, aimed to Responsible Consumption and Production, is the most affected. Goals 1, 2, 4 and 17 are indirectly affected by Tribology, while its influence on the remaining Goals 5, 10 and 16 can be seen by considering the effects of Tribology in a broad sense. The different influence of Tribology on the 17 SDGs is evidenced in Fig. 3 through the different gray levels of the icons' shadows. The influence increases with the darkness of the gray.

**Fig. 3.** Icons of the 17 UN Sustainable Development Goals.

## 4 Conclusions

Several aspects of all branches of Tribology are related to sustainability and can be included in a specific quite new discipline, the Green Tribology. One of the main effects of tribological studies on the ambient is related to reduction of energy losses and materials consumption, both connected to reduced $CO_2$ emissions. Wear and friction losses and the related wastes and pollution can be drastically reduced by correct tribological design as well as by condition monitoring that also reduce machine downtime. New materials and environmental friendly lubricants can furnish an essential contribute in the field of ecological sustainability and climate changes. Tribological studies on prostheses and other medical devices contribute to the humans' well-being too.

Tribology affects several targets of the Sustainable Development Goals, in particular of Goal 12, Responsible Consumption and Production. Goals 3, 6, 7, 8, 9, 11, 13, 14 and 15 are also directly influenced by tribological findings and applications, while the remaining Goals are generally indirectly affected.

## References


1. Jost, H.P.: Lubrication (Tribology) Education and Research, A Report on the Present Position and Industry's Needs. London, Her Majesty's Stationary Office, Department of Education and Science, 80 pp. (1966)
2. Dowson, D.: History of Tribology, 2nd edn. Professional Engineering Publishing Limited, London and Bury St Edmunds, UK, 768 pp. (1998)
3. Ciulli, E.: Tribology and Industry: From the Origins to 4.0. Front. Mech. Eng. **5**(55), 1–12 (2019). https://doi.org/10.3389/fmech.2019.00055
4. United Nations General Assembly: A/RES/70/1 - Transforming our world: the 2030 Agenda for Sustainable Development, 1–35 (2015). https://sdgs.un.org/2030agenda. Accessed on 12 May 2021
5. THE 17 GOALS Homepage, https://sdgs.un.org/goals. Accessed on 21 May 2021
6. Popov, V.L.: Is tribology approaching its golden age? Grand challenges in engineering education and tribological research. Front. Mech. Eng. **4**, 1–6 (2018)
7. Zhang, S.-W.: Green tribology: fundamentals and future development. Friction **1**(2), 186–194 (2013)
8. Tzanakis, I., Hadfield, M., Thomas, B., Noya, S.M., Henshaw, I., Austen, S.: Future perspectives on sustainable tribology. Renew. Sustain. Energy Rev. **16**, 4126–4140 (2012)
9. Nosonovsky, Michael, Bhushan, Bharat (eds.): Green Tribology: Biomimetics, Energy Conservation and Sustainability, p. 632. Springer, Berlin, Heidelberg (2012). https://doi.org/10.1007/978-3-642-23681-5
10. Holmberg, K., Erdemir, A.: Influence of tribology on global energy consumption, costs and emissions. Friction **5**(3), 263–284 (2017). https://doi.org/10.1007/s40544-017-0183-5
11. Glavatskih, S., Höglund, E.: Tribotronics – Towards active tribology. Trib Int. **41**(9), 934–939 (2008)
12. Shah, R., Woydt, M., Huq, N., Rosenkranz, A.: Tribology meets sustainability. Industrial Lubrication and Tribology **73**(3), 430–435 (2021)
13. Stachowiak, G.W.: How tribology has been helping us to advance and to survive. Friction **5**(3), 233–247 (2017). https://doi.org/10.1007/s40544-017-0173-7

# Micromachines, Linkages and Mechanical Control

# Application of Apodized Fiber Bragg Gratings in Information Security Systems

Guljan Kashaganova[1], Sandugash Orazaliyeva[1], Gani Balbayev[1](✉), Zhuldyz Alimseitova[1], and Giuseppe Carbone[2]

[1] Almaty University of Power Engineering and Telecommunications, Almaty, Kazakhstan
[2] Department of Mechanical, Energy, and Management Engineering (DIMEG), University of Calabria, Rende, Italy
giuseppe.carbone@unical.it

**Abstract.** Fiber-optic sensors have several advantages, more important of which include the immunity to electromagnetic disturbances, little weight and possibility to be included into the structure being measured. The most perspective are the sensors based on the Bragg fiber gratings. Bragg fiber gratings have several advantages, for instance, they allow creating the distributed measuring massifs, which contain several sensors. As well, they are insensitive to the optic power source vibrations. Variety of using the fiber sensors based on the Bragg fiber gratings has led to producing the Bragg fiber gratings with different spectral characteristics. Homogeneous Bragg fiber gratings have the spectra with solid side lobes, which can influence at the temperature sensor processing characteristics. To level the side lobes there is applied the apodization method, which is one of the means to affect the spectral form. This paper discusses the use of fiber sensors in a security system. Namely fiber sensors based on fiber Bragg gratings. Advantages and disadvantages of fiber sensors. The principle of operation of a security system based on fiber Bragg gratings.

**Keywords:** Low power consumption · High-performance · Protection solution · Electromagnetic interference resistance · Lightning interference resistance · RF interference resistance · Reliable operation · Fiber sensors · Fiber Bragg gratings · Security system · Apodization

## 1 Introduction

Fiber optic sensors have significant advantages over traditional devices. They have high mechanical strength, resistance to elevated temperatures, vibrations and other environmental influences, the absence of electromagnetic interference, they are chemically inert, allow contactless and remote measurements.

Fiber-optic sensors can conditionally be divided into three types: point, distributed and quasi-distributed [1, 3–5, 11]. Point sensors allow you to measure and control parameters at a specific point in the object, like most non-fiber sensors. As a rule, such sensors are small in size and have high measurement accuracy. Most often, point sensors based


G. Quaglia et al. (Eds.): I4SDG 2021, MMS 108, pp. 451–460, 2022.
https://doi.org/10.1007/978-3-030-87383-7_49

on Bragg fiber gratings, long-period gratings and interferometers are used. They can be used as local thermometers, strain gauges, pressure sensors, accelerometers, etc. The indisputable advantage of distributed sensors is the ability to constantly monitor the parameters along the length (volume) of the object at any point where the fiber sensor is installed. The principle of operation of such sensor systems is based on the analysis of parameter changes along the waveguide and non-linear effects. The disadvantage of the distribution of the measured parameter along the length is the relatively low accuracy of determining the localization of disturbances (several meters long) and the relatively low accuracy of measuring the value. Distributed sensor systems can be used to monitor large areas, such as radiation and temperature sensors, allow you to analyze temperature gradients, etc.

## 2 Fiber-Optic Sensors

Currently, the following primary transducers (sensors) are used in various monitoring systems of the structure: strain gauges; acoustic transducers; molecular-electronic; fiber-optic. Strain gauges convert deformations of building structures into an electrical signal. This is due to the change in the electrical resistance of the sensor when the deformation of the measured device occurs [6]. Load cells can have different versions: wire load cells, foil load cells, film load cells.

Acoustic transducers are sensitive elements with spectral analysis of signals from stress waves in stressed structures, based on measurements of a number of acoustic parameters of building materials. Installation of these converters is possible both inside the structure during construction, and on the surfaces of structures during operation. Molecular-electronic transducers are stationary sensors for spatiotemporal analysis. They allow you to accurately record the displacement of building structures, vibrations, normal and tangential deformations, corrosion changes, etc. The effect of fiber-optic sensors is based on the dependence of the light reflection coefficient in the sensor element, rigidly connected to the controlled structural element, on the amount of deformation of this element. Fiber-optic sensors integrated into a single information and measurement system allow monitoring not only deformations, but also changes in loads, temperature, humidity, vibration of building structures, as well as penetration and destruction of protected objects, etc.

Fiber-optic sensors are a relatively new and extremely multifunctional technical application of optoelectronics, fiber and integrated optics. The reasons for this lie in the multiparametric nature of the optical signal itself – the simultaneous presence of information about changes in phase, amplitude, wavelength, and polarization in time and space [7]. Fiber-optic sensors and fiber-optic measuring systems can have distributed structures [8], operate steadily for a long time in unfavorable conditions: extreme temperatures, pressures, electromagnetic influences, in radiation, toxic or aggressive environments that have little effect on optical fibers.

The main advantage of any Fiber-optic sensors is the possibility of remote monitoring at a distance of several kilometers. In this case, the optical fiber can act as a cable for transmitting optical radiation to a sensor element located in the measurement zone, and at the same time be a sensor element. Made of quartz glass, the light guide is not afraid of the influence of weather conditions and aggressive components of the urban atmosphere. The absence of power supply and any electrical circuits in the light guide eliminates the occurrence of a fire-hazardous situation when operating fiber-optic sensors. Fiber-optic sensors are a class of unique elements that have a wide range of measured parameters, structural mobility and adaptability to various installation conditions, small overall dimensions, as well as the ability to perform measurements under the influence of powerful electromagnetic fields without losing the specified accuracy.

## 3 Fiber Sensors for Security Systems

Currently, such sensors are increasingly used in modern security systems. A security system is a set of means and methods for maintaining the safe condition of an object, preventing, detecting and eliminating threats to life, health and the environment, property and information. Today, technical means of protection are very diverse and generally very effective. However, almost all of them have one significant drawback: they detect the intrusion signal only after the intruder enters the territory of the object. Today, another type of security system has appeared and is actively developing, which uses a more modern and promising method of registering signals, based on the use of Bragg gratings used in fiber sensors. Fiber Bragg gratings (FBGs) are a region with periodic modulation of the refractive index of the fiber core. When light propagates in such a structure, it is reflected. The grating period determines the wavelength at which the constructive interference of the reflected light will occur.

Fiber Bragg gratings are a cross-section of a fiber waveguide (usually a single-mode fiber), in the structure of which a periodic structure of the refractive index with a period Λ with a certain spatial distribution is induced. As a rule, fiber Bragg gratings have a length of about a few millimeters or centimeters [9].

Figure 1 shows the basic operating principle of the FBG with an indication of the typical profile refractive index.

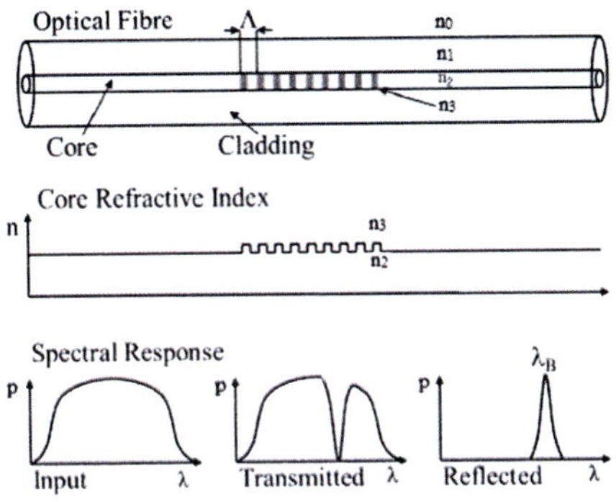


**Fig. 1.** The principle of operation of FBG

The structure of the FBG is chosen in such a way as to provide the necessary resonant interaction between the modes of the optical fiber. The interaction of the waveguide modes is usually described using the theory of coupled modes, in which it is assumed that at a certain wavelength only two modes satisfy the condition of phase synchronism and thus can effectively transfer energy to each other.

Fiber Bragg gratings bind the main mode of the light guide to the same mode propagating in the opposite direction. This means that at a certain wavelength, the radiation propagating through the optical fiber is reflected from the grating completely or partially-depending on the grating parameters. The lasers used for recording the FBG can be either continuous or pulsed, with a wavelength of radiation from the infrared to the ultraviolet range of the spectrum. Among the main methods of recording FBG, there are the step-by-step method, the phase mask method, and the interferometric method.

The Bragg fiber array has a period, in order of magnitude comparable to the wavelength of the sensor reference signal. This grating has the unique property of reflecting light in a narrow spectral range with a maximum at the wavelength λ determined by Bragg's law. One of the main advantages of fiber Bragg gratings is a unique way to convert the measured value into a change in the wavelength of radiation passing through and/or reflected from the grating, as well as ease of manufacture. FBG are well established and widely used in construction, oil production, energy, aerospace engineering, etc.

Fiber Bragg gratings have become a key component for fiber optic telecommunications, such as wavelength division multiplexing systems, fiber laser reflectors, selective reflectors and dispersion compensators, as well as for temperature, strain, pressure, ultrasound, acceleration, strong magnetic fields, chemical elements, etc.

There are three types of gratings that differ in photosensitivity and, consequently, index changes. They are known as types I, II, and IIA with a detailed description of each type in [10]. In addition, there are various physical types of lattices, such as inclined, apodized, long-period, and twitter.

The structure and profile of the refractive index of some different types are shown in Fig. 2.

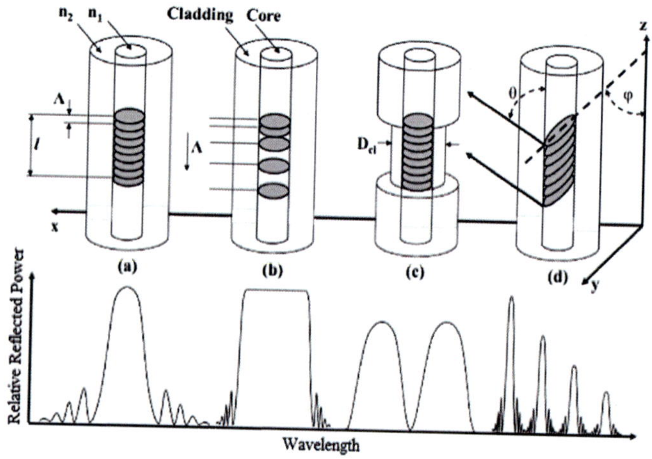


**Fig. 2.** Various types of gratings and their spectral characteristics

The structure and profile of the refractive index of some different types are shown in Fig. 2.

Fiber Bragg gratings with a variable modulation period, which are widely used in telecommunications, i.e., to compensate for the dispersion of fiber-optic communication lines and for dispersive compression of laser pulses. Typically, these applications require long gratings that are written through a phase mask. Currently, high-quality gratings with a length of more than 1 m have already been obtained. The achieved value of dispersion in gratings with a variable period of more than 1000 ps/nm makes it possible to compensate for the dispersion of a 100-km section of a standard fiber-optic line at a wavelength of 1.55 μm. Tilted fiber Bragg gratings. The tilt of the groove plane with respect to the fiber axis leads to effective coupling with cladding and emitting modes and thus forms a narrow-band absorption filter, which is used to suppress unwanted radiation in fiber laser cavities and smooth out the luminescence and gain spectra. It is a structure in which the Bragg grating plane is not perpendicular to the fiber axis but forms an angle of typically 45°.

The range and efficiency of light leakage depends on the angle formed by the plane of the grid with the axis of the fiber, and the amplitude of the refractive index. This action causes a tilted grating, it can be used for mode separation and conversion, separation of different wavelengths, and in EDFA (Erbium Doped Fiber Amplifier), improving their amplification characteristics [13].

Apodized Bragg grating. The spectrum of a uniform Bragg grating, in addition to the main peak, also has unwanted high spurious side lobes that generate crosstalk between adjacent channels in DWDM systems. The appearance of side lobes is due to the fact that the structure of the Bragg lattice, which has a finite length, quickly begins and ends. The influence of the side lobes can be limited using apodization. Apodization is the change in the amplitude of the modulation of the refractive index along the grating.

Chirped grilles. Gratings that have a period that vary according to position along the fiber are called chirped gratings. The period can be changed in various ways: for example, symmetrically (ascending or descending) to a point half the length of the lattice. It can also vary linearly with the lattice length or can be square. Another alternative is to change the average effective refractive index along the grating length, which can be achieved by changing the amplitude of the modulation refractive index profile, or by converging the fiber in the grating length region. Different wavelengths are reflected from different sections of the grating with different periods: long waves are reflected from a section with a long period, short waves travel a longer path and are reflected from an area with a short period. As a result, despite the different speeds of long and short waves, as well as the distances traveled by them, the time of their arrival at their destination is the same. Due to their characteristics, chirped gratings can be used mainly in telecommunications where they are used to compensate for dispersion in order to compress pulses that have expanded during propagation time. Broadband chirped gratings can also be used to suppress unabsorbed pumping in an erbium amplifier and to flatten its spectrum.

Long-period gratings. This term refers to a Bragg grating that has been produced by irradiation by varying the intensity of UV rays along the grating. A long-period grating contains many grating components ΔL in length and dividing them sections without modulation of the refractive index of the length of the index l, the so-called "dead zones". It should be noted that the length of the grating components and the length of the "dead zones" may vary. The spectrum of light reflected from such a structure has a number of narrow peaks.

Long-period fiber gratings have a number of important applications in various fiber-optic devices, such as fiber sensors of physical quantities, spectrally selective absorbing filters, including tunable filters, modulators of optical radiation, and others. One of the important applications of long-period gratings in telecommunications systems: smoothing the gain spectrum of fiber amplifiers used in systems with wavelength division multiplexing, as well as the luminescence spectra of broadband sources. The main advantages of this method are the simplicity of grating fabrication and their use, a wide spectral range and high gain after smoothing, low losses at the pump wavelength, and the absence of a reflected wave.

Many factors contribute to the active use of FBG in various fields: miniature – individual structures have an outer diameter of the working part of less than 100 μ, which allows them to be used in hard-to-reach places without significant structural changes. In the devices in which they operate, the temperature response is a multiplicative response, humidity, mechanical loads, etc., the absence of electromagnetic interference, the preservation of performance, in the presence of various coatings at a temperature not ie 100 to +300 °C., etc.

## 4 The Principle of the System

Consider the principle of operation of a fiber optic system based on the fiber Bragg grating.

The principle of the system is as follows. The Bragg reflective gratings in the core of a single-mode optical fiber can be created by ultraviolet radiation of an excimer laser by irradiation through an appropriate mask or by a holographic method (under the action of two interfering rays). The optical fiber segment between the two Bragg gratings is an interferometer whose reflection (and transmission) depends on the optical phase difference reflected from the first and second gratings of the optical signal.

Under the influence of deformation and acoustic vibrations, the phase difference of two rays reflected from two neighboring gratings changes. As a result, the interference pattern of two reflected rays changes (Fig. 3).

Interferometric sensors are most sensitive to changes in fiber segment length due to external factors. A single-frequency single-mode semiconductor laser operating in a pulsed mode is used as a radiation source. The pulses from each lattice system come with a different delay. To separate the signals from each section, time multiplexing is used. For signal demodulation, synchronous detection is used; for this, a phase modulator is introduced into the circuit. The optical delay line forms a series of time-shifted pulses, each of which interferes with the pulse reflected from the corresponding part of the fiber. An important role in the real application of such fiber-optic security systems is played by the fiber coating material, which provides an increase in the sensitivity of the fiber by 30 dB compared to uncoated fiber.

Structurally, the fiber-optic security system based on Bragg gratings consists of the active part (radiation input-output unit and information processing board) and passive.

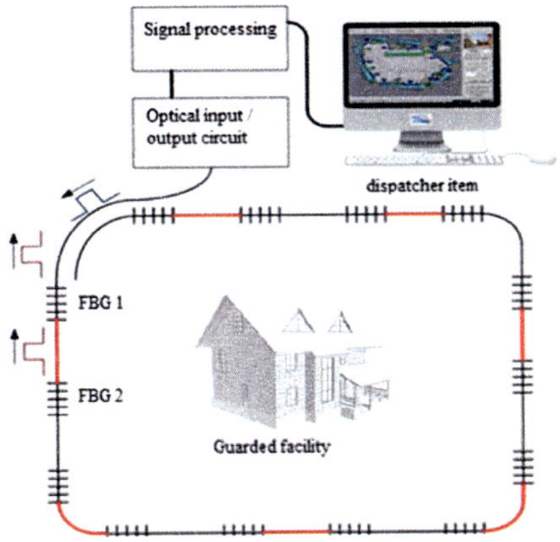


**Fig. 3.** The principle of operation of a fiber-optic security system based on Bragg gratings

The fiber-optic security system based on Bragg gratings is universal, because it is a single cable and does not contain active elements in the multi-zone section. Therefore, this system can be used both covertly (underground) and on existing fences along the perimeter of the protected area.

A feature of such security and monitoring systems is not only the ability to determine the exact location of the violation of the protected border, but also the identification and identification of the intruder by analyzing the acoustic spectrum of the intruder's noise.

## 5 Experimental Results

The growing interest in fiber-optic Bragg lattices has led to the emergence of many varieties of them. The most important property of fiber Bragg gratings is the narrow-band reflection of optical radiation, the relative spectral width of which can be $10^{-6}$ or less. The advantages of fiber gratings in comparison with alternative technologies (for example, interference mirrors and volumetric diffraction gratings) are obvious: a wide variety of obtained spectral and dispersion characteristics, many of which can be realized only on the basis of fiber gratings of RI; full fiber execution; low optical loss; relative ease of manufacture and a number of others. It can be applied various configurations of fiber Bragg gratings. More effective for security systems, we consider apodized fiber Bragg's gratings. The parameter, making the certain influence at the Bragg grating spectrum is apodization. In the simplest case we differentiate homogeneous gratings, in which the modulation depth of interference fringes fracturing factor is the same along the structure's whole length. Multitude of the fiber-optic homogeneous structures definite applications made introducing the apodised function into their production methods, which led to modulation variable depth of grating fringes refraction index. One of the most frequently used interrogation methods of temperature sensor based on the Bragg fiber gratings is the filtration by means of the second grating with the same wave length, having been created in the identical primary conditions [12]. In such system an important parameter, conditioning the given cyclical structure's practicality is the minimization of the so called side lobes [13]. One of the means to reach the effect thereof is the apodization through changing the modulation depth of refraction factor alterations in the core of the optical fiber along its axis. Cyclical structures production with any apodization functions often linked with the necessity to redesign the system and therefore the possibility to use mathematical models for simulating grating spectrum with a denoted apodization is justified (Fig. 4).

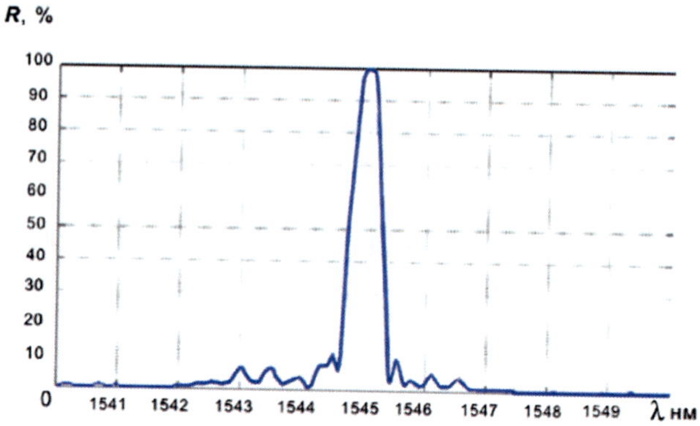


**Fig. 4.** The reflection spectrum of the apodized Bragg grating

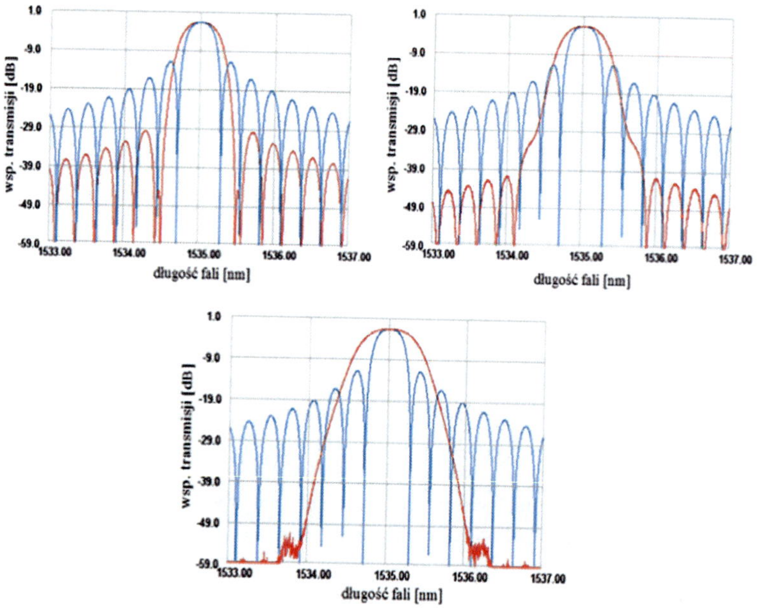


**Fig. 5.** Comparison of the spectral characteristics of fiber Bragg gratings without apodization (blue line) and apodized Gaussian function for $a = 2$, $a = 3$, $a = 5$ - (red line)

Figure 5 shows the differences in the spectral characteristics of fiber Bragg gratings without apodization and with apodization using the Gaussian function with different values of the parameter a. It can be seen from the plots that with apodized fiber Bragg gratings, the side lobes are reduced [14].

In our case, for security systems, we can use apodized fiber Bragg gratings with different periods, capable of reflecting radiation at different wavelengths. In our case, for security systems, we can use apodized fiber Bragg gratings with different periods, capable of reflecting radiation at different wavelengths. So, for the central resonant reflection wavelength $\lambda w = 1550$ nm, the width of the FBG reflection spectrum is 1–2 nm at the base for single-mode wavelengths and can reach values up to 10 nm when forming the FBG in multimode wavelengths, and the reflection coefficient at the resonant wavelength is 0.92. In this case, the attenuation of neighboring lobes in the diagram of the relative spectral reflectance of the FBG is at least 30 dB. The intrusion detection system is the most cost-effective and high-performance solution for protecting extended perimeters and other infrastructure. The system uses a fiber-optic cable, making it the most cost-competitive technology currently available for long-distance operation. Resistant to electromagnetic, lightning and radio frequency interference, it provides reliable operation in difficult conditions. The lack of power requirements in the field reduces infrastructure costs. Low power consumption - generates rated heat.

## 6 Conclusion

Fiber sensors are widely used for security systems. The undoubted advantages of fiber-optic systems include their immunity to electromagnetic radiation and electrical safety. Fiber sensors built from dielectric elements can be used not only on fences or walls, but also on explosive objects or under water. Namely, the use of fiber sensors based on Bragg gratings. The maximum length of one security zone can reach tens of kilometers. An attractive feature of the systems is the lack of active electronic equipment; this reduces the cost of installing and maintaining the security system. The security system on the Bragg gratings is devoid of most of the shortcomings of the sensors and surpasses most of the existing analogues in all its characteristics. At the same time, the most interesting is the use of this system as an underground system, which will allow secretly monitoring all objects crossing the security zone, as well as creating multi-level security systems.

## References

1. Gulyaev, Yu.V., Nikitov, S.A., Potapov, V.T.: Chamorovsky, Yu.K.: Fiber-optic technologies, devices, sensors and systems. Photon-Express Sci. **6**, 114–127 (2005)
2. Garmash, V.B., Egorov, F.A., Kolomiyets, L.N., Neugodnikov, A.P., Pospelov, V.I.: Opportunities, tasks and prospects of fiber-optic measuring systems in modern instrumentation. Photon Express Sci. **6**, 128–140 (2005)
3. Connolly, C.: Structural monitoring with fibre optics. Europhotonics **2–3**, 16–18 (2009)
4. Udd, E.: Fiber Optic Sensors. Introductory Course for Engineers and Scientists. Technosphere, Moscow (2008). 520 pgs
5. Inaudi, D.A.: Fiber optic sensors for structural control. In: Proceedings of the 14th World Conference on Earthquake Engineering, Beijing, China, No. S 25-013 (2008)

6. Wójcik, W., Kalizhanova, A.U., Kisala, P., Kashaganova, G.B., Kozbakova, A.Kh.: Measuring system using fiber-optic sensors with Bragg's fiber grating. Bull. KazNRTU **2**(132), 517–523 (2019). ISSN 1680 - 9211
7. Vuytsik, V., Kashaganova, G.B.: Investigation of the spectral characteristics of fiber Bragg gratings. In: Proceedings of the II International Scientific and Practical Conference "Information and Telecommunication Technologies: Education, Science, Practice", Almaty, Kazakhstan, 3–4 December 2015, vol. II, pp. 77–81 (2015)
8. Amirgalieva, S., Kashaganova, G., Kalizhanova, A., Kartbaev, T.: Fiber-optic sensors based on Bragg fiber gratings. Bull. KazNRTU **3**(127), 347–355 (2018). ISSN 1680 - 9211
9. Eggleton, B.J.: Novel waveguide structures for enhanced fiber grating devices. IEEE J. Selec. Top. Quantum Electron. **7**(3), 409–424 (2001)
10. Kashyup, R.: Fiber Bragg Gratings. Academic Press, San Diego (1999)
11. Tultayev, B., Balbayev, G., Orynbay, A.: Experimental study of the quality of led driver control signal broadcast through wireless networks. IOP Conf. Ser. Mater. Sci. Eng. **439**(3), 1–8 (2018)
12. Lopez-Amo, M., Lopez-Higuera, J.M.: Fiber Bragg Graing Sensors: Research Advancements, Industrial Applications and Market Expoitation, chapter Multiplexing Techniques for FBG Sensors (2014)
13. Crickmore, R., Nash, P., Wooler, J.: Fibre optic security systems for land and sea based applications. In: Proceedings SPIE, vol. 5611, pp. 79–86 (2004)
14. Buymistruk, G.Y.: Information-Measuring Equipment and Technology Based on Fiber-Optic Sensors and Systems. Publishing House IVA, GROTS Minatom, St. Petersburg (2005). 191 pgs

# Kinematic Performance Analysis of a 3-R(RPRGR)RR Planar Parallel Robot

Alexandru Oarcea[1], Antonio Marius Flavius Luputi[2], Victor Cobilean[1], Sergiu-Dan Stan[1](✉), and Erwin-Christian Lovasz[2]

[1] Technical University of Cluj-Napoca, Cluj-Napoca, Romania
{alexandru.oarcea,victor.cobilean,sergiu.stan}@mdm.utcluj.ro
[2] Politehnica University of Timisoara, Timişoara, Romania
antonio.luputi@student.upt.ro, erwin.lovasz@upt.ro

**Abstract.** The study deals with a parallel robot structure, which is actuated by three linear actuators using a geared linkage. The main advantages of this structures having geared linkages with linear actuation give the manipulator a large rotation angle with proper transmission angle while offering the possibility of avoiding the first type singularities. The main goal of this paper is to analyze the kinematic performance of a 3-R(RPRGR)RR planar parallel robot. By computing the performance indices, the whole space of the robot can be restrained, to leave the regions in which robot can move easily. MathWorks MATLAB was used for computing the performance indices and for visualization of the results.

**Keywords:** SDG9 · Performance indices · Manipulability index · Local conditioning number · MATLAB

## 1 Introduction

Parallel manipulators present a high interest for the community due their advantages over the serial configurations as higher velocities and higher accuracy of positioning of the end-effector, higher manipulated loads due to the structural stiffness. At the same time, the parallel manipulators have the disadvantages of lower workspace, which is constrained by the configurations of the robot (collisions between links) or singularity points [1, 2].

The parallel structure analyzed in this paper eliminates the first type singularities by using a geared linkage which is actuated by a linear motor. The first type singularities are described by the positions of the robot, where the actuators of the drive elements can not move the mobile platform. This type of singularities can be identified by satisfying the condition that the Jacobian matrix determinant must be equal to zero [3, 4].

## 2 Description of 3-R(RPRGR)RR

The robot presented in this paper consists of three R(RPRGR)RR kinematic chains that are attached to a triangular shaped moving platform. This configuration has 3 degrees of


G. Quaglia et al. (Eds.): I4SDG 2021, MMS 108, pp. 461–470, 2022.
https://doi.org/10.1007/978-3-030-87383-7_50

freedom (3DOF): two translations for the positioning of the platform and one rotation for the orientation of the platform [5].

The planar parallel manipulator by having linear actuation with geared linkages gives the structure a wide range of a non-zero velocity vector, avoiding the singularities in a wide range.

The structure of the mechanism gives the manipulator a wide rotational angle and a proper transmission angle. The transmission function is approximately constant for a large range.

Figure 1 shows the chain connections of the planar parallel manipulator with geared linkages. Due to the converting of the movement by the linear actuator the transmission ratio is significant, and the self-locking condition is fulfilled.

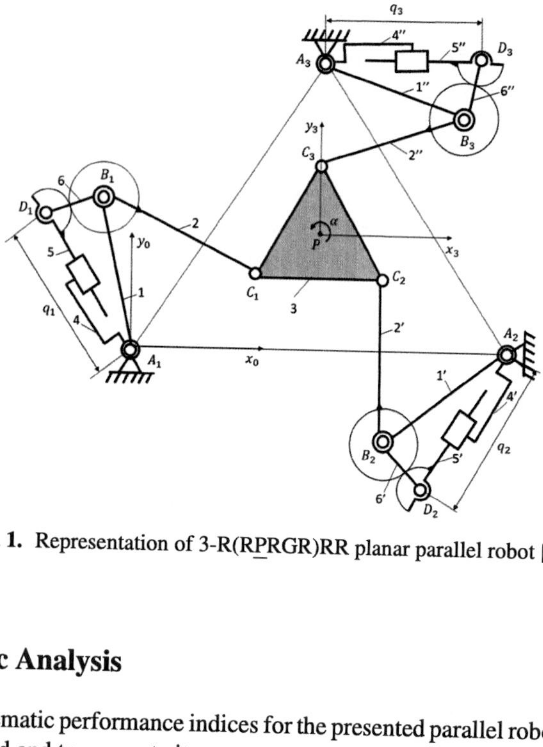


**Fig. 1.** Representation of 3-R(RPRGR)RR planar parallel robot [5].

## 3 Kinematic Analysis

To study the kinematic performance indices for the presented parallel robot, the Jacobian matrix is required and to compute it.

Corresponding to Fig. 2, the matrix form of the Chasles vector equation is [5]:

$$\begin{bmatrix} x_{A_i} \\ y_{A_i} \end{bmatrix} + \begin{bmatrix} l_1 \cos\varphi_{1i} + l_5 \cos\varphi_{2i}(q_i) \\ l_1 \sin\varphi_{1i} + l_5 \sin\varphi_{2i}(q_i) \end{bmatrix} = \begin{bmatrix} x_P \\ y_P \end{bmatrix} + \begin{bmatrix} \cos\alpha & -\sin\alpha \\ \sin\alpha & \cos\alpha \end{bmatrix} \begin{bmatrix} x_{C_i}^{(3)} \\ y_{C_i}^{(3)} \end{bmatrix} \mathrm{i} = \overline{1,3} \quad (1)$$

The slider of the linear actuator is the drive, which generates the stroke $q_i$ and the transmission function of the gear assembly is:

$$\varphi_{2i}(q_i) = (1 - i) \cdot \psi_i(q_i) + i \cdot \theta_i(q_i) + \Delta\varphi_2 \quad (2)$$

where

$$i = \pm \frac{r}{R} \tag{3}$$

$$\theta_i(q_i) = acos\left(\frac{l_6^2 - l_1^2 - q_i^2}{2 \cdot l_1 \cdot q_i}\right) \tag{4}$$

$$\psi_i(q_i) = acos\left(\frac{q_i^2 - l_1^2 - l_6^2}{2 \cdot l_1 \cdot l_6}\right) \tag{5}$$

$\Delta\varphi_2$ – the initial angle of the gear transmission.

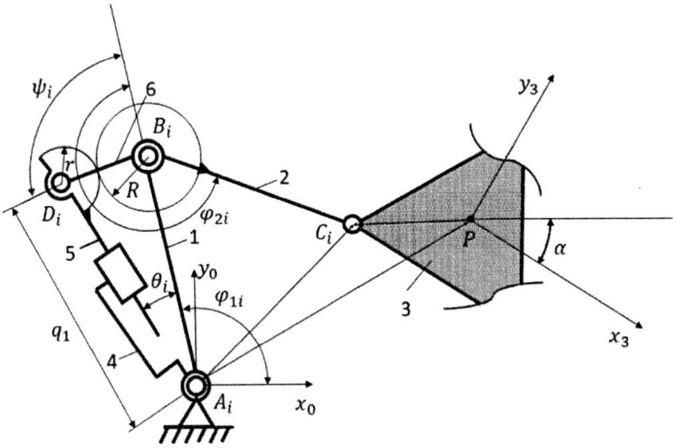


**Fig. 2.** Representation of a chain of the 3-R(RPRGR)RR planar parallel robot.

The implicit function $F_i(x_M, y_M, \alpha_M)$ is obtained from the matrix form the Chasles equations (Eq. (1)), by grouping the elements that contain the terms $\theta_{1i}$, $i = \overline{1,3}$.

The function $F_i(x_M, y_M, \alpha_M)$ can be written [5]:

$$F_i(X, Q) = 0 \tag{6}$$

where $X = \begin{bmatrix} x_M & y_M & \alpha_M \end{bmatrix}$ and $Q = \begin{bmatrix} q_1 & q_2 & q_3 \end{bmatrix}$.

In order to solve the inverse kinematic problem, the implicit function $F_i(x_M, y_M, \alpha)$ is needed:

$$F_i(x_M, y_M, \alpha_M) = (x_M + x_{C_i}^{(3)} \cos\alpha_M - y_{C_i}^{(3)} \sin\alpha_M - x_{A_i} - l_2 \cos\varphi_{2i}(q_i))^2 + (y_M + +x_{C_i}^{(3)} \sin\alpha_M - y_{C_i}^{(3)} \cos\alpha_M - y_{A_i} - l_2 \sin\varphi_{2i}(q_i))^2 - l_1^2 = 0 \tag{7}$$

## 4 Kinematic Performance Indices

Performance indices represent metrics that embody the measurement and quantification of various performance traits of the robots regarding their workspace and most performance indices are usually classified into one of the next groups [6]:

- Local indices – metrics that indicate assets of the robot which are related to a certain structure or position of the robot.
- Global indices – metrics that indicate assets of the robot separately from the structure or the position of the robot.
- Kinematic indices – metrics that present the kinematic performance of the robot.
- Dynamic indices - metrics that present the dynamic performance of the robot.
- Intrinsic indices - metrics that present performance traits of the robot while not being related to a particular application or task.
- Extrinsic indices - metrics that present performance traits of the robot while being directly associated to the task or application.

The performance indices analyzed for the 3-R(RPRGR)RR planar parallel robot are:

- Manipulability Index ($\mu$)
- Normalized Manipulability Index ($\mu_n$)
- Condition Number (k)
- Local Conditioning Index (LCI)
- Kinematic Conditioning Index (KCI)
- Global Conditioning Index (GCI)
- Global Isotropy Index (GII).

To determine the previously presented indices of the robotic system, the determination of the Jacobian matrix is required and in order to determine it, the two Jacobian matrix that are resulting from the following differential equation are combined:

$$J_x\dot{X} + J_q\dot{Q} = 0, \tag{8}$$

$$J_q = \begin{bmatrix} \frac{\partial F_1(X,Q)}{\partial q_1} & \frac{\partial F_1(X,Q)}{\partial q_2} & \frac{\partial F_1(X,Q)}{\partial q_3} \\ \frac{\partial F_2(X,Q)}{\partial q_1} & \frac{\partial F_2(X,Q)}{\partial q_2} & \frac{\partial F_2(X,Q)}{\partial q_3} \\ \frac{\partial F_3(X,Q)}{\partial q_1} & \frac{\partial F_3(X,Q)}{\partial q_2} & \frac{\partial F_3(X,Q)}{\partial q_3} \end{bmatrix} \tag{9}$$

$$J_x = \begin{bmatrix} \frac{\partial F_1(X,Q)}{\partial x_M} & \frac{\partial F_1(X,Q)}{\partial y_M} & \frac{\partial F_1(X,Q)}{\partial \alpha_M} \\ \frac{\partial F_2(X,Q)}{\partial x_M} & \frac{\partial F_2(X,Q)}{\partial y_M} & \frac{\partial F_2(X,Q)}{\partial \alpha_M} \\ \frac{\partial F_3(X,Q)}{\partial x_M} & \frac{\partial F_3(X,Q)}{\partial y_M} & \frac{\partial F_3(X,Q)}{\partial \alpha_M} \end{bmatrix} \tag{10}$$

$$J = J_q^{-1}J_x \tag{11}$$

The Manipulability Index is a performance index introduced by Yoshikawa and its computation is strictly based on the Jacobian matrix [7].

$$\mu = |det(J)| \tag{12}$$

For the determination of the Condition Number, the square roots of the values of the eigenvectors of the matrix product $JJ^T$ is required.

Furthermore, the value of the condition number of the Jacobian matrix is used as a measure of dexterity for the analyzed structure. The condition number can be also defined as the ratio of the maximum and the minimum singular values of the Jacobian matrix and it takes the following form [8]:

$$k = \sqrt{\frac{\lambda_{max}}{\lambda_{min}}} \tag{13}$$

The Local Conditioning Index (LCI) is defined as the reciprocal value of the condition number, and it is determined as following [9]:

$$LCI = \frac{1}{k} \tag{14}$$

The Minimum Singular Value represents the performance index that presents the minimal change of the speed of the end-effector based on a unit length movement of the joints of the robotic structure [7].

$$MSV = \min(\sigma_i) \tag{15}$$

The Kinematic Conditioning Index is a global index that represents the lowest Local Conditioning index over the whole reachable workspace and represents the worst performance inside the workspace for the robot [7].

$$KCI = \frac{1}{\min(k)} \cdot 100 \tag{16}$$

The Global Conditioning Index is a global index that represents the distribution of the Condition Number over the whole reachable workspace, and it is defined as following [7]:

$$GCI = \frac{1}{n} \sum\nolimits_{j=1}^{n} \frac{1}{k} \tag{17}$$

where $n$ is the number of nodes inside the workspace.

The Global Isotropy Index represents the worst performance of the robot inside its reachable workspace based on the maximum and minimum singular values of the Jacobian matrix over the whole workspace [7].

$$GII = \frac{\sigma_{min}}{\sigma_{max}} \tag{18}$$

## 5 Results

The performance indices mentioned above were analyzed for the 3-R(RPRGR)RR planar parallel robot which construction parameters are presented in Table 1.

**Table 1.** Construction parameters.

| Parameter | Symbol | Unit | Value |
|---|---|---|---|
| Outer radius of the triangular fixed platform | $r_p$ | mm | 900 |
| Outer radius of the triangular moving platform | $r_m$ | mm | 255 |
| Length of the first link | $l_1$ | mm | 300 |
| Length of the second link | $l_2$ | mm | 200 |
| Length of the carrier link | $l_6$ | mm | 50 |
| Transmission ratio | $i$ | – | 2 |

By using the constructive parameters from previous table, the performance indices presented above were computed using MATLAB, in order to iterate through the positions that can be achieved by the characteristic point (center of platform).

According to the formulation introduced by Yoshikawa for the manipulability index, the distribution in Fig. 3 was achieved (Fig. 4):

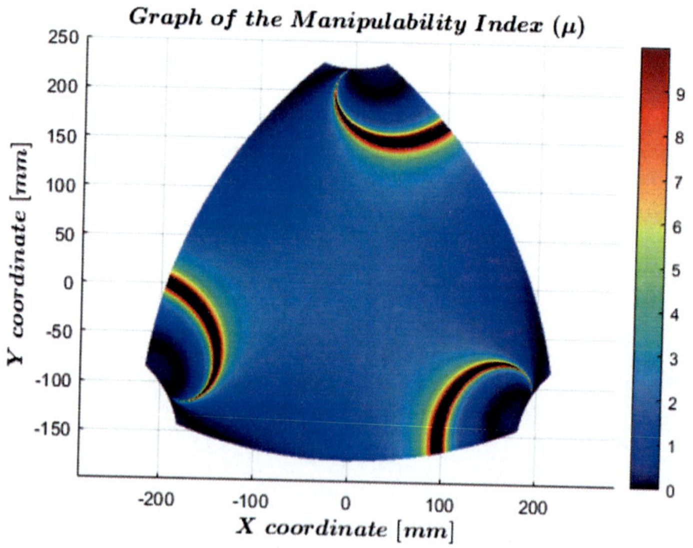


**Fig. 3.** Representation of the manipulability index for the 3-R(RPRGR)RR planar parallel robot

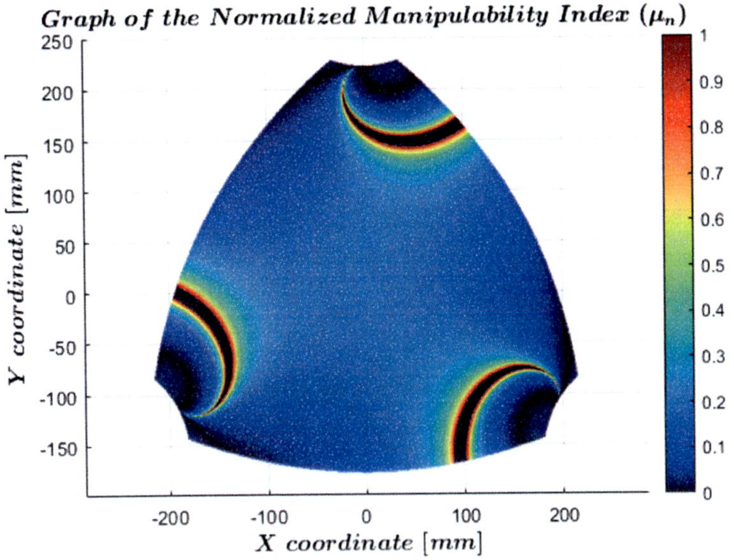


**Fig. 4.** Representation of the normalized manipulability index for the 3-R(R$\underline{P}$RGR)RR robot

From the representations of the manipulability index and the representation of the normalized mobility index, the region of the first and second order singularities of the workspace is further represented as dark regions, and it is observed that the reachable workspace is reduced near the corners of the workspace.

By the representation of the values of the condition number inside the whole workspace of the robot, it can be noticed that the reachable workspace is constrained by the regions in which the robot achieves singularity positions. The regions near the revolute joints (A1, A2, A3) (Fig. 1) cannot be achieved by the characteristic point and are represented by the empty regions in Fig. 5. In the center regions of the figure where the values of condition number are near 0, the robot is approaching an isotropic configuration.

Because the condition number is not bounded on the superior interval, more suggestive values will be given by the local conditioning index (Fig. 6), which is bounded between 0 and 1, where 0 represents singularity positions and 1 the isotropic configurations.

The representation of the Minimum Singular Value (Fig. 7) suggests that in the center region of the total workspace the platform can easily move with high velocity and high force transmission, while on the sides of the reachable workspace the value of the Minimum Singular Value approaches 0 meaning the presence of singularities. Concerning the global indices, the results are presented in the following table:

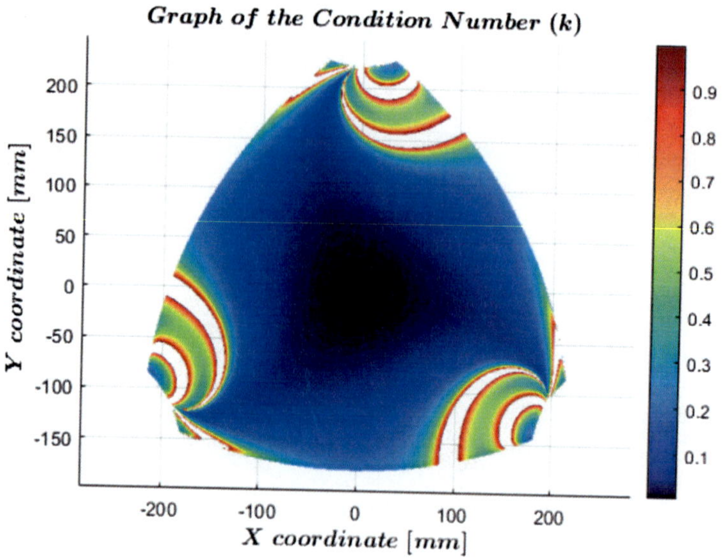


**Fig. 5.** Representation of the condition number for the 3-R(RPRGR)RR robot

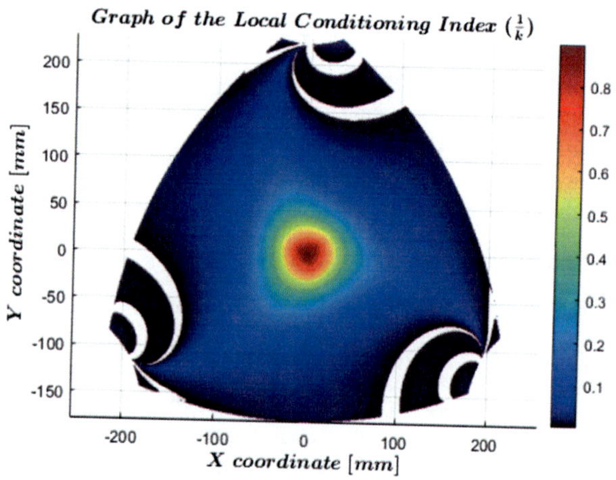


**Fig. 6.** Representation of the local conditioning index for the 3-R(RPRGR)RR robot

From the result presented in the Table 2, it can be seen that the presence of the singularity positions inside analyzed workspace conducts to the very low values of the global performance indices and it is desired to limit the analysis workspace to the workspace region that has values for the condition number of the Jacobian less then an imposed maximal value resulting in the definition of dexterous workspace in order to achieve precise values for the global indices.

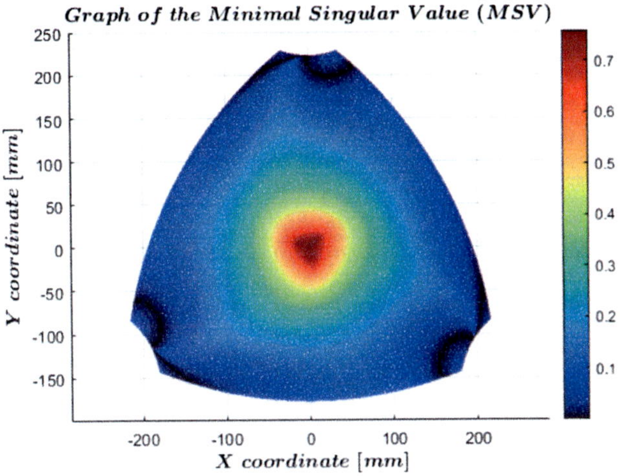


**Fig. 7.** Representation of the Minimum Singular Value for the 3-R(RPRGR)RR robot

**Table 2.** Global indices.

| Parameter | Symbol | Value |
|---|---|---|
| Kinematic conditioning index | KCI | 0.4% |
| Global conditioning index | GCI | 0.0514 |
| Global isotropy index | GII | 0 |

## 6 Conclusion

In this paper, seven performances indices of the 3-R(RPRGR)RR planar parallel robot were studied: Manipulability Index (μ), Normalized Manipulability Index ($\mu_n$), Condition Number (k), Local Conditioning Index (LCI), Kinematic Conditioning Index (KCI), Global Conditioning Index (GCI), Global Isotropy Index (GII).

For the analysis of the performance indices of the 3-R(RPRGR)RR MathWorks MATLAB was used to implement the kinematics of the robot and numerically compute the Jacobian matrix required for the determination of the performance indices.

From the results obtained, further research is proposed in order to restrict the reachable workspace to a dexterous one and reanalyze the global indices for determining more precise values for the global indices.

**Acknowledgements.** This article is supported by European Union's Erasmus+ KA203 programme, project reference: 2019-1-RO01-KA203-06315 ("Development of mechatronics skills and innovative learning methods for Industry 4.0").

## References

1. Merlet, J.P.: Determination of the orientation workspace of parallel manipulators. J. Intell. Rob. Syst. **13**, 143–160 (1995)
2. Stan, S-D.: Diplomarbeit, Analyse und Optimierung der strukturellen Abmessungen von Werkzeugmaschinen mit Parallelstruktur, IWF TU Braunschweig, Germany (2003)
3. Stan, S.-D., Manic, M., et al.: Evolutionary approach to optimal design of 3 DOF translation exoskeleton and medical parallel robots. In: 2008 Conference on Human System Interactions, pp. 720–725 (2008). https://doi.org/10.1109/HSI.2008.4581530
4. Stan, S.-D., Manic, M., Szep, C., Balan, R.: Performance analysis of 3 DOF Delta parallel robot. In: 2011 4th International Conference on Human System Interactions, HSI 2011, pp. 215–220 (2011). https://doi.org/10.1109/HSI.2011.5937369.
5. Lovasz, E.-C., et al.: Geared linkages with linear actuation used as kinematic chains of a planar parallel manipulator. In: Corves, B., Lovasz, E.-C., Hüsing, M. (eds.) Mechanisms, Transmissions and Applications. MMS, vol. 31, pp. 21–31. Springer, Cham (2015). https://doi.org/10.1007/978-3-319-17067-1_3
6. Cecarelli, M.: A synthesis algorithm for three-revolute manipulators by using an algebraic formulation of workspace boundary. ASME J. Mech. Des. **117**(2(A)), 298–302 (1995)
7. Patel, S., Sobh, T.: Manipulator performance measures - a comprehensive literature survey. J. Intell. Rob. Syst. **77**(3–4), 547–570 (2014). https://doi.org/10.1007/s10846-014-0024-y
8. Tondu, B.: A zonotope-based approach for manipulability study of redundant robot limbs. Int. J. Hum. Robot. **10**, 1350023 (2013). https://doi.org/10.1142/s0219843613500230
9. Angeles, J., LópezCajún, C.: Kinematic isotropy and the conditioning index of serial robotic manipulators. Int. J. Robot. Res. IJRR. **11**, 560–571 (1992). https://doi.org/10.1177/027836499201100605

# Control Process for an Industrial Bending Machine Designed for Metal Processing

Alexandru Oarcea[1], Victor Cobilean[1], Sergiu-Dan Stan[1](✉), and Adrian Boziean[2]

[1] Technical University of Cluj-Napoca, Cluj-Napoca, Romania
{alexandru.oarcea,sergiu.stan}@mdm.utcluj.ro
[2] Eco Sere Modulare S.R.L, Răscruci, Romania
office@ecosere.ro

**Abstract.** In this paper it was presented the working sequence process and control algorithm for an industrial bending machine, designed to comply with industrial requirements regarding metal bending, in such a way that the machine can be incorporated in an automated process, out of which results processed metal beams or profiles. The control implementation was achieved by using the Siemens LOGO! Programable relay and using Function Block Diagram (FBD) programming language. The machine was designed and build as part of Technical University of Cluj-Napoca (TUCN) agreement 19947/20.08.2020 "Research regarding development of a mechatronic system used for industrial production".

**Keywords:** SDG9 · Bending machine · Automation process · PLC · Hydraulics · Pneumatics

## 1 Introduction

The requested process represents the final step in an already existing process in which circular steel tubes of different sizes are deformed by pressing their ends to achieve a flat surface, step that is followed by rounding the corners and making a center hole by stamping (Fig. 1).

In this stage we are supposed to design and build a machine that can bend the flat surface mentioned above at an imposed angle. Having into account the required operation and the future expendability of the entire process such that it can be automated and executed by robots we decided to develop a machine that bends the flat surface by using a hydraulic cylinder that acts upon a bending knife to deform the flat surface against the surfaces of a bending prism that achieves the final form that was imposed.

To achieve multiple angles and steel tube sizes, the machine is designed so that the user can quickly swap out the bending knife from the cylinder quick mount block and the fixed bending prism according to the angle that is needed to be obtained [1, 2].

Furthermore, as a safety feature, we need to firmly secure the steel tube on the tabletop, so we decided to use two pneumatic clamps with self-blocking levers that can easily be adjusted to grip a wide variety of steel tube dimensions.


G. Quaglia et al. (Eds.): I4SDG 2021, MMS 108, pp. 471–480, 2022.
https://doi.org/10.1007/978-3-030-87383-7_51

**Fig. 1.** Actual state of the existing process

To fully control the bending process and subsequently be able to externally command and control the machine in the scope of fully automated process we decided to implement a supervising system that consist of a programable relay (Siemens LOGO!).

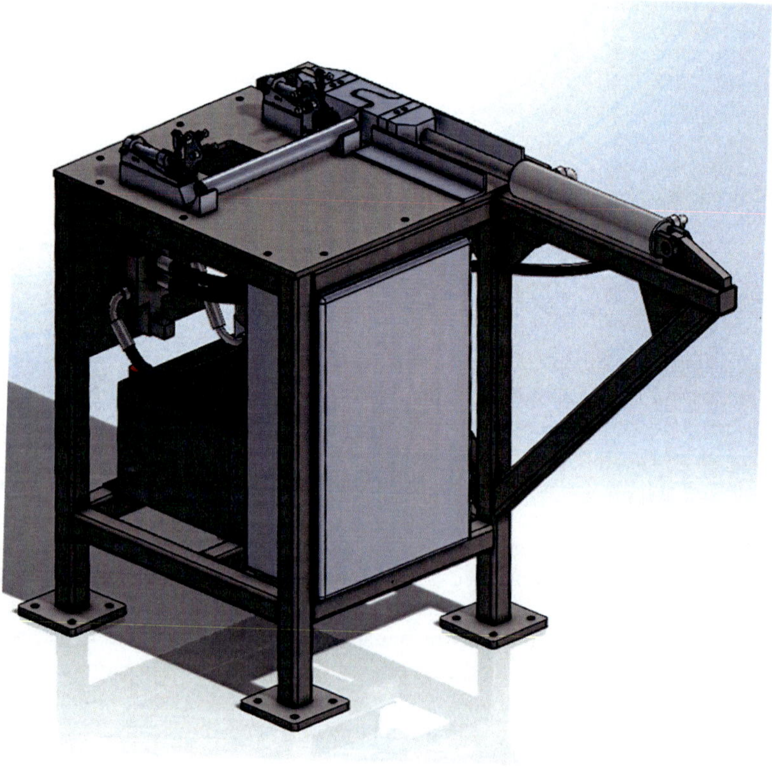

**Fig. 2.** 3D model of the proposed machine

The control system can fully observe and command the machine based on an imposed processing sequence that in the first step will be manually triggered by a foot pedal in order start the process and that can be externally triggered by the control system of the automated production line [3, 4] (Fig. 2).

## 2 Working Sequence Concept

To achieve the imposed control method, we decided to fully automate the process and develop a control method in which the machine can be used by a human operator or alternatively be used in a robotic bending application. To develop the proposed control method, we decided to use a programable logic controller (PLC) to observe and command the machine that will gather data from sensors mounted so that every moving element is detected [5].

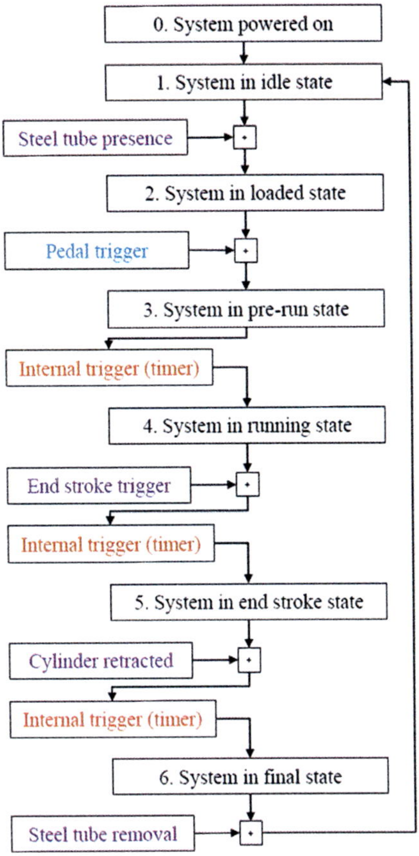


**Fig. 3.** Representation of the proposed control sequence and the triggers for each step

Furthermore, the PLC will command the steps of the machine, by enabling actuators, based on an imposed sequences in which each step is triggered by external sensors or internal events. Using this proposed method of control will assure the future integrability of the machine inside a fully automated production line (Fig. 3).

# 3 Control

To implement the control method proposed previously, we defined an internal subsequence for each step involved in the whole process because it is needed to comply with safety regulations so during the whole sequence there is a master alarm check that involves checking main power to the hydraulic system, checking the 24 V power supply and checking that the emergency button (E-STOP) that it is not triggered. All the checks from the above will generate a main alarm memory bit in case any of the previous conditions are not fulfilled [6, 7].

Furthermore, we implemented a sensor alarm memory bit in the case there is no input from the expected sensors for a certain time to proceed to the following step during the sequence (over-time sensor error).

Besides the over-time sensor error, we implemented another sensor error memory bit that is triggered by abnormal sensor input (clamping system closed sensor is on and clamping system open sensor is on at the same time, hydraulic cylinder max stroke is on, and hydraulic cylinder zero stroke is on at the same time).

The main control sequence is divided in six main steps (step zero, assigned for powering on and initializing) that control the machine, each step being designed to enable only if the triggering event is present and only if the step conditions are achieved. In this way we can reliably control each step and observe the behavior of the machine during the whole process, identifying the amount of time needed to successfully end each step to optimize the whole production time by removing waiting times implemented as a safety measure before acting the fourth, fifth and sixth step [8, 9].

The six main steps involved in the sequence are:

## 3.1 Step 0. System Powered on

During this automatic step, the camping system is commanded to open after a short foot pedal trigger, and the hydraulic cylinder is commanded to withdraw so that the initial conditions for the following step are achieved.

## 3.2 Step 1. System in Idle State

This step was needed because the main sequence needs a state in which the machine does nothing because we need to define a state that is safe for operations around the machine without having it running.

This step is triggered automatically after powering on the machine, and it is the step that the machine retunes after a whole sequence, transition being triggered by the removal of the finished product from the machine.

The step is finished by actuating on the yellow lamp that indicates that the operator needs to insert an unprocessed product inside the machine (Fig. 4a).

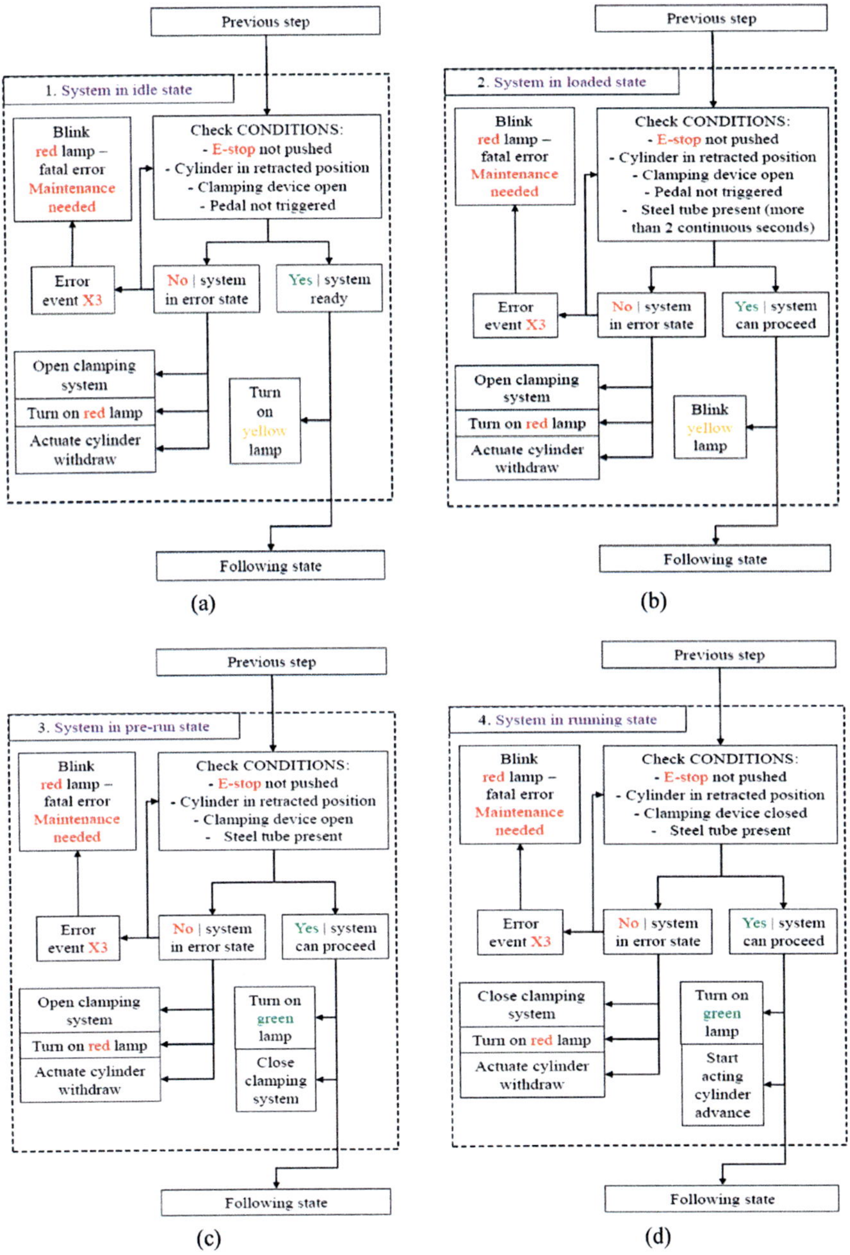


**Fig. 4.** Representation of the subsequence implemented for the step: System in idle state (a), System in loaded state (b), System in pre-run state (c), and System in running state (d)

### 3.3 Step 2. System in Loaded State

This step is trigged by inserting an unprocessed product inside the machine and upon successfully checking the initial conditions for the step (hydraulic cylinder fully retracted and the clamping device is open), and the only actuation done is blinking the yellow light, indicating that the machine is ready to start the bending process (Fig. 4b).

### 3.4 Step 3. System in Pre-run State

This step is trigged by the external foot pedal, or the external start trigger and upon successfully checking the initial conditions for the step (hydraulic cylinder fully retracted, the clamping device is open, and the unprocessed product is present for an imposed amount of time), the clamping system will be closed, the green light will turn on and the yellow light will be turned off.

The reasoning behind the implementation of this step is to separate the actions involved in the starting of the process (closing clamping system and start cylinder advance) to safely detect the clamping system state before the actuation of the hydraulic cylinder and implement a delay timer as the trigger for the following step that can be modified accordingly to the type of the processed product (Fig. 4c).

### 3.5 Step 4. System in Running State

This step is trigged by the internal timer and upon successfully checking the initial conditions for the step (hydraulic cylinder fully retracted, the clamping device is closed, and the unprocessed product is still present for an imposed amount of time), step in which the hydraulic cylinder will be commanded to advance while maintaining the green light and the clamping system turned on (Fig. 4d).

### 3.6 Step 5. System in End Stroke State

This step is triggered by the internal timer that delays the signal received from the sensor used for the cylinder achieving maximum stroke and upon successfully checking the initial conditions for the step (clamping device is closed and the unprocessed product is still present for an imposed amount of time), step in which the withdraw of the hydraulic cylinder is commended.

The reasoning behind the implementation of this step is the need for a small delay between the changing of the direction of the cylinder (advance and withdraw) to avoid mechanical and hydraulic shocks in the system.

Furthermore, we need to guarantee that the deformation has been done by letting the advance command on after the end stroke trigger for a short period of time because the sensor that detects the movement of the bending knife might induce position errors based on the type of the bending knife used during the bending process (Fig. 5a).

### 3.7 Step 6. System in Final State

This step is triggered by the internal timer that delays the signal received from the sensor used for the cylinder achieving zero stroke and upon successfully checking the initial conditions for the step (clamping device is closed and the unprocessed product is still present for an imposed amount of time), step in which the opening of the clamping device and powering on of the yellow lam is commanded.

After the removal of the processed product, the machine will return to Step 1 reinitializing the sequence. This transition is monitored to count the total cycles of the machine and the total processing time (Fig. 5b).

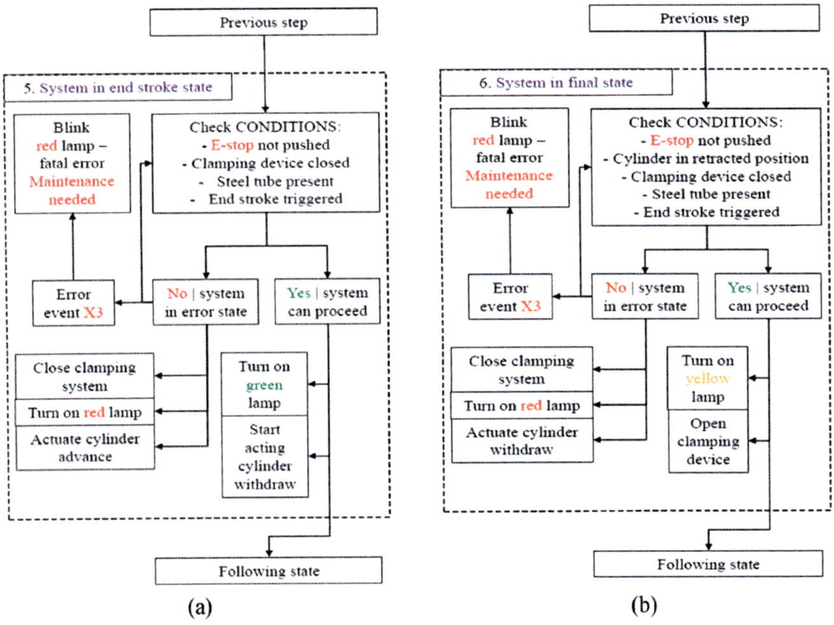


**Fig. 5.** Representation of the subsequence implemented for the step: System in end stroke state (a) and System in final state (b)

## 4 Software Implementation

Implementation of the main sequence described above was achieved by using the software Siemens LOGO! Soft Comfort, programing software designed for LOGO module series, that can be programed by using Leader (LAD) programming language or alternatively program by Function Block Diagram programming language (FBD).

During our implementation we used FBD to implement the main sequence and we divided the program in four main diagrams.

## 4.1 Input Network Diagram

Inside this network we defined all the links between the input signals and the memory bits assigned for alarms checks and sequence triggers accordingly to each step initial conditions.

## 4.2 Sequencer Diagram

The main role of this network is to exactly execute the steps in their imposed order and transition from one to another only when the accordingly trigger has been activated (Fig. 6).

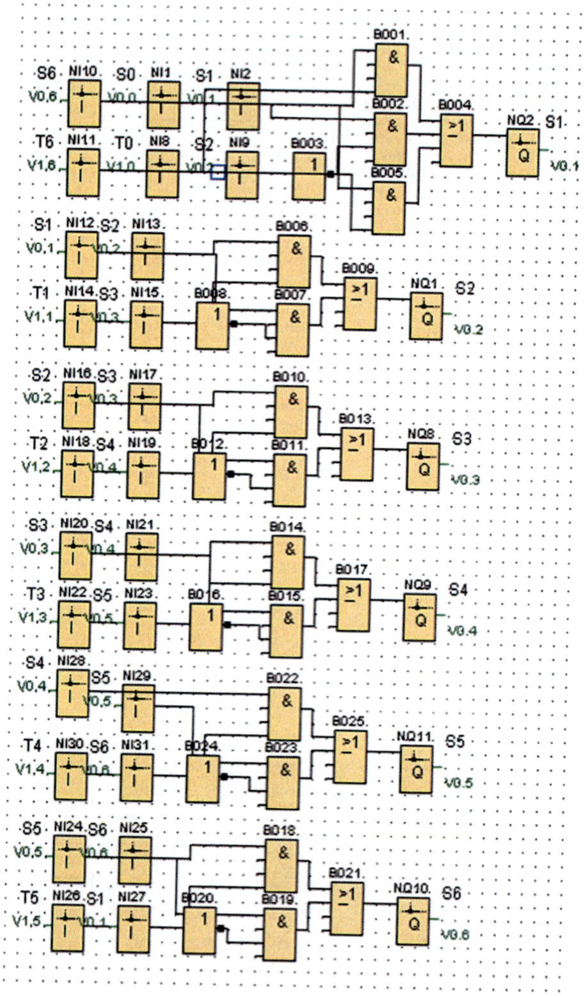

**Fig. 6.** Representation of the diagram implemented for the sequencer network

### 4.3 Alarm Network Diagram

Inside this network we defined all the memory bits and operations to define the alarm memory bits mentioned previously and the network is presented in the following figure.

### 4.4 Output Network Diagram

Inside this network we defined all the links between the output ports assigned to actuating the machine or signalizing the state of the machine and the memory bits assigned for each step so that the actuation is accordingly to the main sequence. Furthermore, each output also has a condition that assures the fact that the output would not be enabled if any error has occurred, and also if the step actuation has been enabled and an error occurs, the output will be turned off.

In the case of an emergency stop, the process will be on hold and after the cause of the stop will be resolved, the system will continue from the actual stop point by acting upon the foot pedal or sending another start external signal, without the need for a manual reset (Fig. 7).

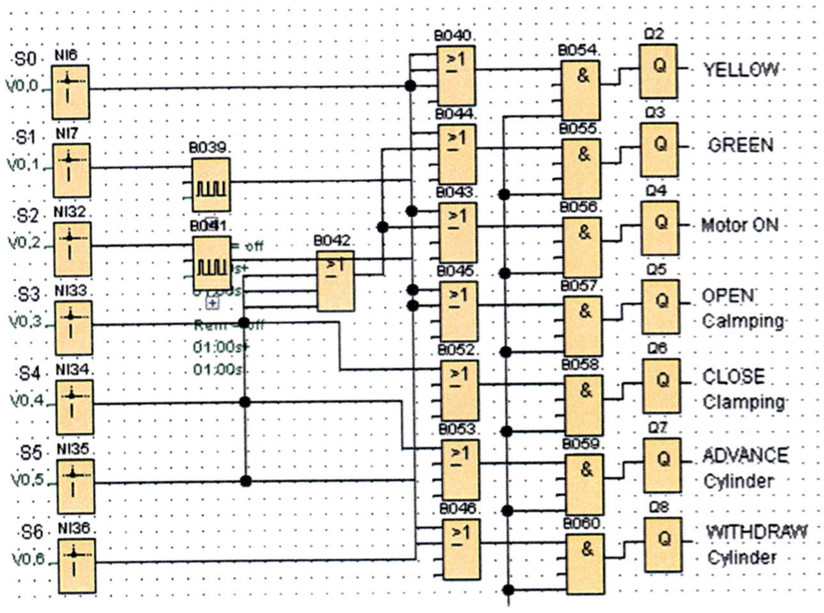


**Fig. 7.** Representation of the diagram implemented for the output network

## 5 Conclusion

In this paper it was proposed a solution involving the control of an automated bending machine composed of a hydraulic unit (motor, pump, cylinder, valves, etc.) by using the control system (Siemens LOGO!) to fulfil the imposed procedure and successfully control all the steps involved in this process.

Furthermore, the whole control sequence was virtually implemented and simulated with great success, but we propose further research regarding establishing a communication through Ethernet/IP protocol to manipulate all the record data and gain the ability to modify process parameters in real time.

**Acknowledgements.** This article has received funding and has been established in collaboration with the company ECO SERE MODULARE SRL under the TUCN agreement: 19947/20.08.2020 "Research regarding development of a mechatronic system used for industrial production."

## References

1. Rafa, V., Teutan, E.: Functioning analysis of cathode protection of pipelines for transporting of natural gas through fuzzy logic. Mechanika **20**(5), 512–515 (2014)
2. Teutan, E., Bara, M., Ardelean, I.: Generating of digitised with the aid of Tetrobot module. Mechanika **18**(6), 666–670 (2012)
3. Teutan, E., Bara, M., Ardelean, I.: Theoretical considerations regarding the virtual modelling of surfaces used in topography. Mechanika **85**(5), 61–65 (2010)
4. Bara, M., Stan, S.-D., Teutan, E., Verdes, D.: Design and virtual reality simulation of frontal - Sinusoidal ball transmission. Mechanika **78**(4), 63–68 (2009)
5. Rat, N.R., Neagoe, M., Diaconescu, D., Stan, S.D.: Dynamic simulations regarding the influence of the load-rigidity correlation on the orking accuracy of a medical Triglide parallel robot. Mechanika **17**(2), 178–181 (2011)
6. Szep, C., Stan, S.-D., Csibi, V.: Design, workspace analysis and inverse kinematics problem of delta parallel robot. Mechanika **17**(3), 296–299 (2011)
7. Verdeş, D., Stan, S.-D., Bălan, R., Coman, M.: Study of design, kinematics and virtual control of 4 degrees of freedom parallel robot. Mechanika **82**(2), 70–74 (2010)
8. Szep, C., Stan, S.-D., Csibi, V., Bălan, R.: Study of design, kinematics and accuracy modelling of 3 degrees of freedom robot. Mechanika **77**(3), 58–61 (2009)
9. Stan, S.-D., Bălan, R., Măties, V., Rad, C.: Kinematics and fuzzy control of ISOGLIDE3 medical parallel robot. Mechanika **75**(1), 62–66 (2009)

# High-Speed Dimensional Processing of Metallic Materials with an Environmentally Friendly Jet Electrolyte-Plasma Method

Alexander Popov(✉), Alla Popova, Alexandr Fumin, Mikhail Novoselov, Sergey Zakharov, and Mikhail Radkevich

Peter the Great St. Petersburg Polytechnic University, Polytehnicheskaya Street, 29, 195251 Saint-Petersburg, Russia
popov_ai@spbstu.ru

**Abstract.** This article presents the results of a study of an environmentally friendly method of surface treatment with an electric discharge formed by an electrolyte jet and a workpiece. The study of the effect of electrolytic plasma on the surface of the samples was carried out in various technological modes to obtain the maximum rate of removal from the surface of the material. As a representative material, corrosion-resistant heat-resistant steel 20X13 and stainless steel AISI 304 were used. Computer modeling of the volume and mass of the defect layer appearing on the surface of parts after various types of mechanical processing requiring removal to obtain a surface with a low roughness parameter was performed. An analysis of the surface morphology was carried out, showing the effect of electrolyte plasma on the surface of the product. A technological device has been developed that allows high-speed dimensional surface treatment with a high initial level of surface roughness inherent in blanking operations. In this case, the treated surface after the electrolytic-plasma treatment has a low surface roughness corresponding to the operations of grinding and polishing. It is shown that at small distances, high concentrations of electrolyte, the process of electrolytic-plasma treatment proceeds at a very high speed and can be used in industry to replace manual grinding and polishing operations.

**Keywords:** SDG9 · Industry · Innovation and infrastructure · Ganged current lead · Electrolyte · Surface morphology · Electrolyte-plasma discharge · Jet electrolyte-plasma treatment · Roughing

## 1 Introduction

The issues of creating promising environmentally friendly methods for processing structural materials attract considerable attention of researchers and are a vector of innovative development of the economies of states and technology companies [1, 2]. The practical application of many traditional technological processes for processing metal products has a significant impact on the environment. Among the promising and environmentally friendly technological processes, a special place is occupied by the technology of


G. Quaglia et al. (Eds.): I4SDG 2021, MMS 108, pp. 481–489, 2022.
https://doi.org/10.1007/978-3-030-87383-7_52

electrolytic - plasma processing [3–10]. This technology includes several competing methods such as electrolytic bath treatment using flowing and stationary electrolytes. Treatment in a bath using an electrolytic jet combined with an inert gas jet, a current lead that is buried and raised above the surface of the bath [11, 12]. These methods have the ability to simultaneously process a large number of products and areas. However, they require powerful technological power supplies, have a large electrolytic bath and a significant volume of vapors, the need to utilize significant volumes of electrolyte [13, 14]. Electrolytic jet processing methods have a comparable processing speed with processing in an electrolytic bath, but a significantly higher accuracy of making the dimensions of spline surfaces, the ability to carry out alloying and application of technological coatings [15–19]. In addition, the technology of jet electrolytic-plasma treatment is an environmentally friendly method even in comparison with the application in an electrolytic bath. It is more attractive due to the small amount of electrical discharge, power and volume [20]. In comparison with the processing of parts on metal-cutting machines of a turning, milling, grinding group, treatment with an electrolyte jet is a much more environmentally friendly process (Table 1). This is consistent with the reducing effect of this equipment on the environment and the service personnel. This equipment has 1.5–30 times lower consumption, 1.5 times lower noise from equipment and 100 times lower than vibration due to work without cutting. When it is operating, there is no need to dispose of oils and lubricating liquids. The security of personnel is provided by the low temperature of the applied pure environmentally electrolyte. This electrolyte does not negatively affect staff respiration. With its work there are no vibrations of equipment compared to manual grinding and prevents the development of diseases of the hands. There are no aerosols and pairs of abrasive, metal and organic dust and lubricating - cooling fluids affected by bacteria and fungi. These factors make it possible to evaluate this technology as a more environmentally friendly process compared with cutting technologies.

**Table 1.** Comparative environmental assessment.

| Machine tools | Consumed power, kWt | Noise, Db | Vibration, Hz | Lubricating and cooling liquid, MPC, mg/m$^3$ | Abrasive dust MPC, mg/m$^3$ | Water vapor, %/l | Chips/suspension, % | NaCl, MPC, mg/m$^3$ | Emulsion/electrolyte temperature, °C |
|---|---|---|---|---|---|---|---|---|---|
| Metal cutting | 2–64 | 76–110 | 2–63 | 300 | 5 | – | 100 | – | 18–56 |
| Electrolytic-plasma | 1–2 | 40–70 | 0,5–2 | – | – | 3–5 | 100 | 5 | 18–23 |

To date, the processes of jet processing are still insufficiently studied and this leads to a significant amount of research in this area. The practical application of these methods is complicated by the lack of developed technological processes that allow for high-speed dimensional electrolytic-plasma processing and reliable devices for them.

The purpose of our work is to analyze the use of various technological devices to increase the rate of specific removal of the mass of material when processing stainless steels with an electrolyte jet - plasma. The tasks of the work are the development of a digital surface model for calculating the mass of metal required for removal after the previous mechanical treatment, analysis of the operation of devices for high-speed processing, assessment of the rate of change in the mass of samples.

## 2 Methods

The experimental technique consists in establishing the relationship between the design of the ganged current lead and the parameters of the technological process. First of all, the mass of the substance removed from the anode is of interest. In accordance with Faraday's law, the mass of a substance decreasing from the anode into the electrolyte solution will be equal to:

$$m = (Q/F) * (M/z); \quad (1)$$

Where,

Q - the charge that has passed through the current lead during the electric discharge;
F - Faraday constant;
M - molar mass of the substance;
z -valence.

$$Q = I * t; \quad (2)$$

Where,

I - electric discharge current strength, A;
t - time of action of an electric discharge on the surface, s;

With fixed values of M/z, F and t, the rate of removal of the mass of the substance depends on the magnitude of the current. It is assumed that this can be achieved by increasing the current density per unit surface and creating a uniform electric field strength in the interelectrode gap. An increase in the conductivity of the electrolyte in a significant proportion will increase the current generated during the discharge. The use of sodium chloride salts makes it possible to obtain an environmentally friendly electrolyte. This electrolyte can be easily disposed of without additional requirements. In our opinion, this should open up new technological possibilities of jet electrolytic machining for high-speed dimensional machining.

## 3 Materials and Experimental Setup

As equipment for the research, we used a pilot plant for jet electrolytic-plasma treatment developed at the Higher School of Mechanical Engineering of SPbPU [21]. The principle of its operation is shown in Fig. 1. Heat-resistant corrosion steel 20X13, which is used in the manufacture of turbine blades, was chosen as the material under study. The dimensions of the marked samples are 5 × 45 × 60 mm. The samples were processed with a movable hollow cathode and a ganged current lead [22]. The travel speed of the current lead was set in the range of 150–300 mm/min. In our case, the anode was the

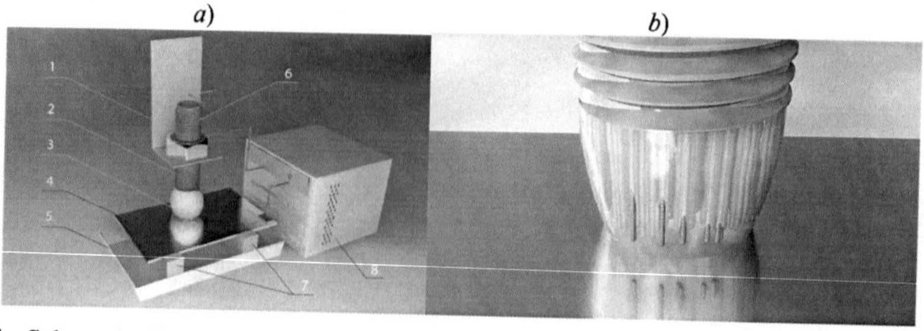


**Fig. 1.** Schematic diagrams of plasma electrolyte surface treatment: a) 1 - current lead holder, 2 - hollow current lead, 3 - electrolytic cathode, 4 - sample, 5 - electrolyte receiving bath, 6 - electrolyte supply pipeline, 7 - insulated supports, 8 - DC power supply current; b) - ganged current lead.

workpiece. The operating voltage was set at 100–125 V. The parameters of the volt-ampere characteristics were recorded with a panel ammeter and an OMIX voltmeter of the DV-1-0.5 series. Voltage and current measurement error ± 0.5%. The electrolyte used was a solution of sodium chloride (NaCl) in tap water. The limiting concentration of the solution was - 300 g/L. The electrolyte was fed through a tube with a diameter of 3 mm with a volumetric flow rate of the electrolyte of 28–90 l/h. The volumetric flow rate of the electrolyte was calculated by the formula:

$$Gv = \frac{3600 * 1}{t}; \ (1\,/\,h) \tag{3}$$

Where,

t – the time of filling a volume equal to one liter with electrolyte pumped through the hollow current lead. The electrolyte volume was measured with a volumetric glassware. The filling time of 1 L of electrolyte was recorded with a stopwatch. The surface removal depth before and after treatment was measured at least 7 times with a micrometer with a scale division of 0.01 μm. Measurement of the Rmax value was carried out after the manufacture of thin sections. The end surfaces of the samples were studied using a METAM LV-31 optical microscope (LOMO, Russia). The microscope measurement error in the range of 0–30 mm is ±0.003 mm. To simulate the microrelief of the surface after machining, the Solidworks software was used.

## 4 Modeling of the Initial Surface

As examples of various types of mechanical processing, samples were selected after sawing with a band saw, finishing milling, processing with a vulcanite wheel, and finishing grinding. The initial state of the surface after these types of processing is shown in Fig. 2 to obtain a surface with a low surface roughness parameter, it is required to remove a significant thickness of the surface layer. This layer is defective and contains a significant number of surface discontinuities. These include pores, risks, foreign inclusions. The height of this layer is equal to the value between the maximum height of the vertices and the maximum depth of the valleys at the base length. It corresponds to a value equal to Rmax. Therefore, for the formation of a layer on the surface with a roughness parameter less than Ra ≥ 0.6 μm, in some cases, it is necessary to completely remove, and in others, to partially remove the defective layer (Table 2). To estimate the mass of the defect layer for different types of technological processing, we performed modeling of the surface profile in CAD. The assessment of the maximum height of the profile irregularities Rmax and the average pitch of the profile irregularities Sm were measured on the end sections of the samples.

a) b)

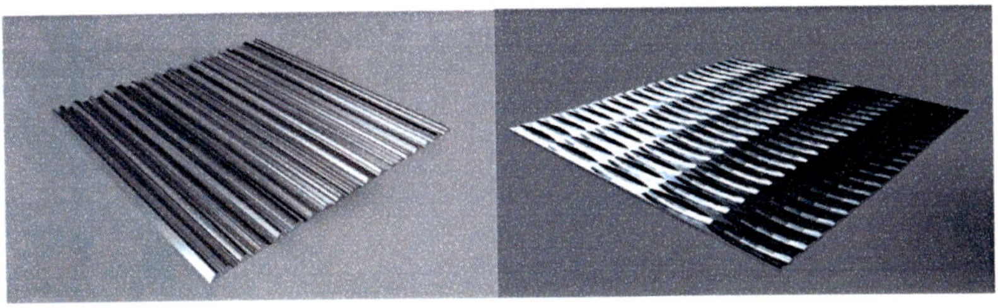

**Fig. 2.** Digital model of the surface morphology in the Solidworks program: a) - after sawing with a band saw; b) - after finishing with an end mill.

Based on these data, we have created a digital surface model in the Solidworks program.

**Table 2.** The change in the mass of the samples required to obtain a flat surface.

| Surface condition | Sawing with a band saw | Rough grinding | Finishing with a turning tool | Finishing with an end mill | Vulcanite circle treatment |
|---|---|---|---|---|---|
| Height of micro-roughness, mm | 0.61 | 0.32 | 0.17 | 0.1 | 0.11 |
| Volume, $mm^3$ | 0.31 | 0.15 | 0.09 | 0.06 | 0.05 |
| Weight, mg | 2.41 | 1.15 | 0.68 | 4.42 | 0.41 |
| Calculation area, $mm^2$ | 1 | 1 | 1 | 100 | 1 |
| Calculation of mass, mg for 1 $cm^2$ | 241 | 115 | 68 | 44 | 41 |

## 5 Experiment Results

Preliminary studies have shown that the removal of the surface layer of the material using a hollow cathode is very slow. This is due to the fact that with a hollow current supply in the form of a tube, it is not possible to equalize the electric field strength at the cathode. It will be maximal at the edges of the tube and minimal at the center of the flow and at the periphery of the electric discharge.Therefore, in order to obtain high discharge current densities and, consequently, a high rate of dissolution of the anode material, we need to additionally introduce metal conductors into the region of the electrolytic flow. This will help to increase the electrical discharge conductivity and increase the current density in the treatment area. We have conducted studies on the change in the value of the electric discharge current due to the addition of copper conductors (Fig. 3). Studies show that with an increase in the number of copper conductors Ø 1 mm, a proportional increase in the value of the electric current is observed. This shows that by increasing the number of conductors, we increase the amount of current and additionally equalize the electric field strength of the current supply. This leads to an increase in the rate of removal of the metal mass from the anode and a decrease in the surface roughness parameter to Ra 0.6 μm. In many cases, this is sufficient for sizing. If finishing is required, the surface roughness parameter can be adjusted to Ra 0.034 μm. Data on the parameters of the electric discharge are given in Table 2. The calculation of the mass of the removed metal shows that the volume corresponding to the volume of the defective layer for the coarsest technological processes is removed from the surface in a period of 18 min (Table 3).

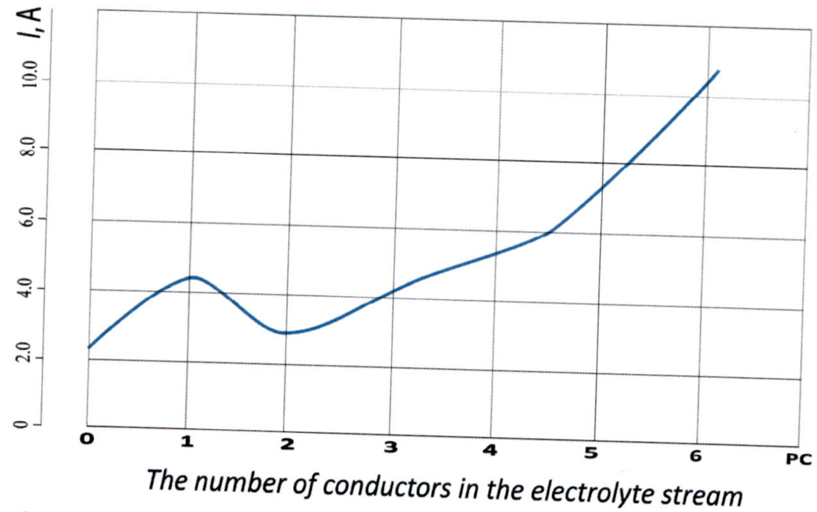


**Fig. 3.** The dependence of the value of the electric discharge current on the number of conductors introduced into the electrolytic jet.

The study of sample processing after sawing with a band saw and processing with a jet of electrolytic plasma with a ganged current supply is shown in Fig. 4.

**Table 3.** Experimental conditions and calculation of the mass of removed material.

| Parameter/Sample no. | 1 | 2 | 3 |
|---|---|---|---|
| Current strength, A | 10.3 | 8.7 | 10.6 |
| Voltage, V | 112 | 125 | 103 |
| Feed rate, mm/min | 278 | 278 | 278 |
| Interelectrode gap, mm | 5 | 5 | 3 |
| Volume flow rate of the electrolyte, l/h | 21 | 41 | 41 |
| Calculation of the removal of mass, mg per 1 $cm^2$ | 212 | 228 | 264 |

**Fig. 4.** Appearance of a specimen made of 20X13 coozion-resistant steel after sawing with a band saw and subsequent treatment with an electrolyte plasma jet.

## 6 Conclusions

1. An increase in the electrolyte feed rate when using a hollow current lead leads to a slight increase in the current in the discharge. This leads to a low rate of material removal from the surface of the product.
2. The addition of metal conductors to the electrolyte flow proportionally increases the discharge current and the rate of material removal from the product surface.
3. An increase in the number of conductors and their cross-section determines the growth of the discharge current and the rate of removal.
4. An increase in the current supply leads to a decrease in the removal speed and an increase in the quality of the treated surface.

5. The use of an electrolyte based on sodium chloride ensures the ecological purity of the process.

**Acknowledgment.** The authors would like to thank graphic designer Diana Popova for preparing illustrations.

# 1. References


1. Popova, A.I.: Osnovnye faktory innovacionnotekhnologicheskogo razvitiya (The main factors of innovative and technological development) V sbornike: Integraciya ekonomiki v sistemu mirohozyajstvennyh svyazej. Sbornik nauchnyh trudov XVII Mezhdunarodnoj nauchno-prakticheskoj konferencii, pp. 123–125 (2012). (in Russian)
2. Popova, A.I.: Voprosy integracii rossii v mirovoe hozyajstvo (Issues of Russia's integration into the world economy) V sbornike: Integraciya ekonomiki v sistemu mirohozyajstvennyh svyazej. Sbornik nauchnyh trudov XVII Mezhdunarodnoj nauchno-prakticheskoj konferencii, pp. 24–26 (2012). (in Russian)
3. Ushomirskaya, L.A., Veselovskiy, A.P.: Intensifikatsiya tekhnologicheskikh protsessov izgotovleniya detaley mashin pri ispolzovanii razlichnykh vidov energii. Metalloobrabotka **2**(56), 46 (2010). (in Russian)
4. Kulikov, I.S.: Elektrolitno-plazmennaya obrabotka materialov. In: Kulikov, I.S., Vashchenko, S.V., Kamenev, A.Ya. (eds.) NAN Belarusi, Obyedinennyy in-t energeticheskikh issledovaniy - Sosny. Belaruskaya navuka, Minsk (2010). 232s. ISBN 978-985-08-1215. (in Russian)
5. Nestler, K., Böttger-Hiller, F., Adamitzki, W., Glowa, G., Zeidler, H., Schubert, A.: Plasma electrolytic polishing - an overview of applied technologies and current challenges to extend the polishable material range. Procedia CIRP **42**, 503–507 (2016)
6. Alekseyev, Yu.G., Korolev, A.Yu., Parshuto, A.E., Niss, V.S.: Elektrolitno-plazmennaya obrabotka pri nestatsionarnykh rezhimakh v usloviyakh vysokogradiyentnogo elektricheskogo polya. Nauka i tekhnika **16**(5), 391 (2017). (in Russian)
7. Danilov, I., Hackert-Oschätzchen, M., Zinecker, M., Meichsner, G., Edelmann, J., Schubert, A.: Process understanding of plasma electrolytic polishing through multiphysics simulation and inline metrology. Micromachines **10**, 214 (2019). https://doi.org/10.3390/mi10030214
8. Zakharov, S.V., Korotkikh, M.T.: Electrolyte-Plasma Polishing Ionization Model. Lecture Notes in Mechanical Engineering, pp. 193–208 (2020)
9. Danilov, I., Paul, R., Hackert-Oschätzchen, M., Zinecker, M., Quitzke, S., Schubert, A.: Random sequential simulation of the resulting surface roughness in plasma electrolytic polishing of stainless steel. Procedia CIRP **95**, 981–986 (2020). 20th CIRP Conference on Electro Physical and Chemical Machining, ISEM 2020; Zurich, Online; Switzerland; 19 January 2021 до 21 January 2021
10. Gaysin, F.M., Son, E.Ye.: Elektricheskiye razryady v parogazovoy srede s netraditsionnymi elektrodami. Entsiklopediya nizkotemperaturnoy plazmy. Pod red. Fortova V.Ye. M.: Nauka, p. 241 (2000). (in Russian)
11. Witzke, M., Rumbach, P., Go, D.B., Sankaran, R.M.: Evidence for the electrolysis of water by atmospheric-pressure plasmas formed at the surface of aqueous solutions. J. Phys. D Appl. Phys. **45**, 5 (2012)
12. Krevsun, E.P., Kulikov, I.S.: Ustroystvo dlya elektrolitno-plazmennoy obrabotki tokoprovodyashchego izdeliya Patent na izobreteniye. Respubliki Belarus №16101 Kl. MPK: B 23H3/00, 30 August 2012. (in Russian)

13. Slovetskiy, D.I., Terentyev, S.D., Plekhanov, V.G.: Mekhanizm plazmenno-elektrolitnogo nagreva metallov. TVT. **24**(2), 353 (1986). (in Russian)
14. Grigoryev, A.I.: O perenose energii i formirovanii elektricheskogo toka v o krestnosti opushchennogo v elektrolit, silno nagretogo protekayushchim tokom elektroda. Zhurnal tekhnicheskoy fiziki, tom 74, vyp.5- S. 38–43 (2004). (in Russian)
15. Popov. A.I., Popova, A.I., Popova, D.A.: Process aspects of jet electrolytic plasma processing. St. Petersburg Polytech. Univ. J. Eng. Sci. Technol. **25** (2019)
16. Ablyaz, T.R., Muratov, K.R., Ushomirskaya, L.A., Zarubin, D.A., Sidhu, S.S.: Electrolytic plasma polishing technique for improved surface finish of Ed machined components. Eng. Solid Mech. **7**(2), 131–136 (2019)
17. Gaisin, A.F., Son, E.E.: Vapor-air discharges between electrolytic cathode and metal anode at atmospheric pressure. High Temp. **43**, 1–7 (2005). https://doi.org/10.1007/s10740-005-0040-5
18. Bagautdinova, L.N., Gaisin, F.M.: A multichannel discharge in conducting liquid at atmospheric pressure. High Temp. **48**, 126–128 (2010). https://doi.org/10.1134/S0018151X10010153
19. Popov, A.I., Novikov, V.I., Radkevich, M.M.: Characteristics of the development of electric discharge between the jet electrolyte cathode and the metal anode at atmospheric pressure. High Temp. **57**, 447–457 (2019). https://doi.org/10.1134/S0018151X19030118
20. Popov, A.I., Radkevich, M.M., Teplukhin, V.G.: Thinnest finishing treatment with a focused jet of electrolytic plasma. Lecture Notes in Mechanical Engineering (см. в книгах), pp. 139–149 (2020)
21. Popov, A.I., Radkevich, M.M., Kudryavtsev, V.N., Zakharov, S.V., Kuzmichev, I.S.: Ustanovka dlya elektrolitno-plazmennoy obrabotki turbinnykh lopatok. Patent na izobreteniye RF № 2623555. Kl. MPK: C25F7/00, 27 June 2017. (in Russian)
22. Popov, A.I., Radkevich, M.M., Medko, V.S., Novoselov, M.V.: Ustrojstvo dlya elektrolitno-plazmennoj obrabotki metallicheskih izdelij. Patent na izobretenie RF (Device for electrolytic-plasma processing of metal products. Patent for invention of the Russian Federation ) No. 2681239. Cl. IPC: C25F7 / 00, 13 June 2018. (in Russian)

# Transportation Machinery, Engines and Powertrains

# Quantum Chemical Study of Mechanochemical Reactive Mechanisms of Engine Oil Antiwear Additives

Chao Zhang(✉)

School of Mechatronic Engineering and Automation, Shanghai University, No. 99 Shangda Road, Shanghai 200444, China

**Abstract.** The life of tribopairs is often limited by the ability of their friction reduction and wear resistance. The ability to predict friction and wear, particularly for the comprehensive effects of lubricant, additive and coating characteristics, would greatly aid the development process of both tribopairs and lubricants, especially, advanced environment friendly lubricants. The purpose of this discussion is to study mechanochemical reactive mechanisms of engine oil antiwear additives via $DMol^3$ module in Materials Studio package, which is a first-principles-based quantum chemistry density functional theory (DFT) program. The S and P atoms, and PO molecules were used as species, Fe(100), Al(100) and Cu(100) surfaces were employed as the adsorption surfaces. Effects of species, adsorption position and surface on mechanochemical reactive were investigated. S atom adsorbed on Fe(100), Al(100), and Cu(100) surfaces unconditionally, while PO molecule adsorbed on Fe(100) surface unconditionally and on Al(100) and Cu(100) surfaces conditionally. The most stable S atom and PO molecule adorption surface order was Fe, Al, Cu. The most stable adsorption site of S atom order was hollow, bridge, top. The most stable adsorption site of PO molecule on Fe(100) surface order was hollow, bridge, top. Lateral repulsion among atoms and molecules on surfaces influenced adsorption configurations and energies. The coverage of the species adsorption on surfaces depended on adsorption surface and position site.

**Keywords:** SDG9 · Mechanochemical · Lubricant additives · Tribofilm · Wear · DFT · Model

## 1 Introduction

Lubricant antiwear additives are used to control the wear of rubbing parts and thus provide acceptable engine durability. There has recently been a renewed interest in antiwear additives, such as ZDDP, as a consequence of heightened environmental regulations relating to diesel exhaust to reduce that emission either through a new implementation of ZDDP, or through some substitute additive. Environmentally friendly base oils and additives are developed to meet the requirements of environmental protection [1–11]. Lubricant additive antiwear film formation is believed to be accelerated by multiple factors such as heat, load, and shear force in a sliding contact. The individual contributions of


G. Quaglia et al. (Eds.): I4SDG 2021, MMS 108, pp. 493–501, 2022.
https://doi.org/10.1007/978-3-030-87383-7_53

these factors are not fully understood because the elementary reaction pathways cannot readily be observed experimentally. The use of simulations at the quantum scale can capture reaction pathways leading to film formation [12–16]. Lubricant-chemistry reactions involve molecules from oil lubricants, additives, plating, coatings, substrate metal materials, mixtures, compounds, intermittent products, and final products. Additive antiwear film formation progresses involve isomerization, decomposition, adsorption, desorption, oxidation, restortion, and polymerization reactions.

In the current study mechanochemical reactive mechanisms of engine oil antiwear additives were examined with quantum chemical methods. The S and P atoms, and PO molecule are common compositions of the antiwear additives, and the iron, aluminium and copper materials are usually used for engine components. The S and P atoms, PO molecule, Fe(100), Al(100), and Cu(100) surfaces were employed here as examples.

## 2 Methods

In the DFT calculations, the adsorption energy was obtained for S, P and PO species on bcc Fe(100), Al(100), and Cu(100) slabs consisting of three and four layers. The three-layer slabs were modeled using a 3 × 3 supercell and vacuum region of 12 Å, in which the two bottom layers was fixed, while the upper layer combined with the adsorbtes were relaxed during optimization. The four-layer slabs were modeled using a 3 × 4 supercell and vacuum region of 12 Å, in which the two bottom layers were fixed, while the two upper layers combined with the adsorbtes were relaxed during optimization.

DFT calculations were carried out using DMol$^3$ module in Materials Studio package. The exchange-correlation energy functions were deduced in terms of spin-polarizaed Perdew-Wang (PW91) of the generalized gradient approximation (GGA). The frozen-core interaction was described by the projector-augmented wave (PAW) method. Cutoff energy of 550 eV was set as the convergence of the plane-wave expansion. The Brillouin zones were sampled with the 3 × 3 × 1 and 3 × 2 × 1 Monkhorst-Pack k-point meshes for three and four layers, respectively.

The adsorption energy, $E_{ads}$, for single atom or molecule is written as follows [17–19]

$$E_{ads} = \left(E_{surface} + E_{adsorbate} - E_{total}\right) - \Delta ZPE_{ads} - \Delta G_{abs} \quad (1)$$

where $E_{surface}$ is the total energy of fundus surface, $E_{adsorbate}$ is the total energy of free adsorbate in gas phase, $E_{total}$ represents the total energy of free adsorbate and fundus surface, $\Delta ZPE_{ads}$, corrected by the $ZPE$, is corresponding to $E_{ads}$ and determined as follows

$$\Delta ZPE_{ads} = \left(\sum_{i=1}^{vibration} \frac{h\nu_i}{2}\right)_{adsorbed} - \left(\sum_{i=1}^{vibrations} \frac{h\nu_i}{2}\right)_{gas} \quad (2)$$

where $\nu_i$ represents the vibrational frequency, $h$ is Planck's constant, $\Delta G_{abs}$ is the modified value of $E_{ads}$ at reaction temperature $T$ and reaction pressure $P$, $\Delta G$ is defined as follows

$$\Delta G = \Delta H + T\Delta S \quad (3)$$

$$\Delta S = \gamma RT - E_{ads} \tag{4}$$

where $R$ represents the ideal gas constant, $\gamma = 0(adsorptive\ species)\ or\ 1$(gaseous molecule), $\Delta U$ and $\Delta S$ are the change of internal energy and entropy of each species under reaction conditions (T, P).

A positive $E_{ads}$ shows the adsorption system is exothermic, and the more positive $E_{ads}$ is, the more likely the structure is.

The desorption energy, $E_{des}$, for single atom or molecule is written as follows

$$E_{des} = -E_{ads} \tag{5}$$

The adsorption energy of the $N(N \geq 2)$ th atom or molecule is defined by

$$\Delta E_{ads} = E_{ntotal} + E_{adsorbate} - E_{(n+1)total} \tag{6}$$

where $E_{ntotal}$ and $E_{(n+1)total}$ are the total energy of the n and n + 1 same atoms or molecules and fundus surface, respectively. A negative $\Delta E_{ads}$ for the n + 1 th same atoms or molecules shows that the fundus surface is saturated when adsorbing n same atoms or molecules.

The desorption energy, $\Delta E_{des}$, of the $N(N \geq 2)$ th atom or molecule is defined by

$$\Delta E_{des} = -\Delta E_{ads} \tag{7}$$

The lateral repulsive energies ($\Delta E_{late}$) between an added atom or molecule and the previous adsorbate is defined by:

$$\Delta E_{late} = \Delta E_{ads(n=1)} - \Delta E_{ads} \tag{8}$$

The more positive $\Delta E_{late}$ is, the more strongly the lateral repulsive energies between an added atom or molecule and the previous adsorbate.

The activation energy, $E_a$, and the reaction energy, $\Delta E_{dis}$, are defined by

$$E_a = E_{TS} - E_{IS} + \Delta ZPE_{barrier} + \Delta G_{barrier} \tag{9}$$

$$\Delta E_{dis} = E_{FS} - E_{IS} + \Delta ZPE_{reaction} + \Delta G_{reaction} \tag{10}$$

where $E_{IS}$, $E_{TS}$, and $E_{FS}$ are the potential energy of the initial (reactant), transition, and finial (product) state, respectively.

$$\Delta ZPE_{barrier} = \left( \sum_{i=1}^{vibration} \frac{h\nu_i}{2} \right)_{TS} - \left( \sum_{i=1}^{vibrations} \frac{h\nu_i}{2} \right)_{IS} \tag{11}$$

$$\Delta ZPE_{reaction} = \left( \sum_{i=1}^{vibration} \frac{h\nu_i}{2} \right)_{FS} - \left( \sum_{i=1}^{vibrations} \frac{h\nu_i}{2} \right)_{IS} \tag{12}$$

where $\nu_i$ represents the vibrational frequency, $h$ is Planck's constant, $\Delta G_{barrier}$ and $\Delta G_{reaction}$ are the modified values of $E_a$ and $\Delta E_{dis}$ at reaction temperature $T$ and reaction pressure $P$, respectively. $\Delta G$ is defined same as Eqs. (3–4).

## 3 Results and Discussion

Figure 1 shows that for S atom initially adsorbed at the top, bridge and 4F hollow positions, optimized adsorption positions kept the same sites; S adsorption energies were smallest in the top site, S was the most unstable at the top site; S adsorption energies were biggest in the 4F hollow site, S is the most stable at the 4F hollow site; S atom tended to migrate and diffuse between the bridge and 4F hollow sites on Fe(100) and Al(100) surfaces due to the small difference of S adsorption energies between the two sites on Fe(100) and Al(100) surfaces, respectively; S adsorption energies were biggest on Fe(100) surface, smallest on Cu(100) surface, corresponding to the most stable adsorption surface order: Fe(100), Al(100), Cu(100).

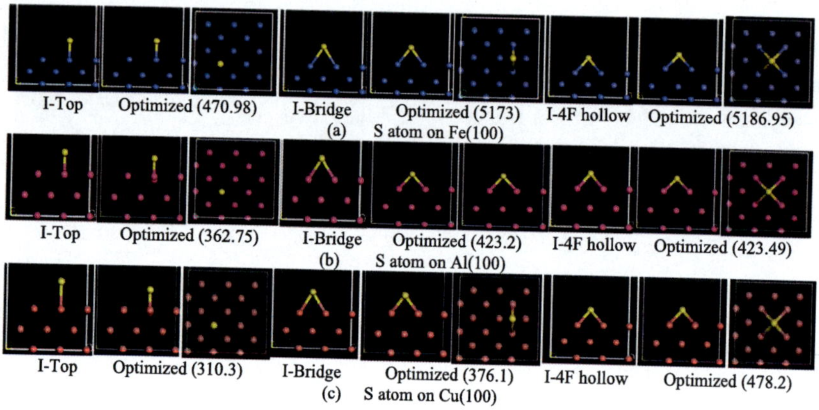


**Fig. 1.** Optimized adsorption configurations and energies of S atom on Fe(100), Al(100), and Cu(100) surfaces in gas phase (I-Top: Initial-Top; I-Bridge: Initial-Bridge; I-4F hollow: Initial-4F hollow; yellow: S; brown: Fe; orange: Cu; pink: Al; value in parenthesis is adsorption energy (unit: $kJ{\cdot}mol^{-1}$)).

Figure 2 shows that for PO molecule initially adsorbed at the top, bridge and 4F hollow positions on Fe(100) surface in Fe-O-P angle, optimized adsorption positions kept the same sites and Fe-O-P angle; for Fe(100) surface, PO adsorption energy (positive value) was smallest at the top site, biggest in the 4F hollow site, corresponding to the most stable adsorption site order: 4F hollow, bridge, top; for PO molecule initially adsorbed at the bridge site on Al(100) surface in Fe-O-P angle, optimized adsorption sites kept the same sites while the adsorption system was absorption of heat, indicating the PO molecule desorption from Al(100) surface; for PO molecule initially adsorbed at the top and 4F hollow sites on Fe(100) surface in Fe-O-P angle, respectively, optimized adsorption positions kept the same sites and Fe-O-P angle, while the PO molecule more stably adsorbed at the 4F hollow site than the top site; for PO molecule initially adsorbed at the top, bridge, 4F hollow sites on Cu(100) surfaces, the adsorption system was absorption of heat, indicating the PO molecule desorption from Cu(100) surface.

Figure 3 shows that for initially adsorbed at the top site on Fe(100) surface, PO molecule adsorption energy was smaller with Fe-O-P 180° angle than with 90° angle no

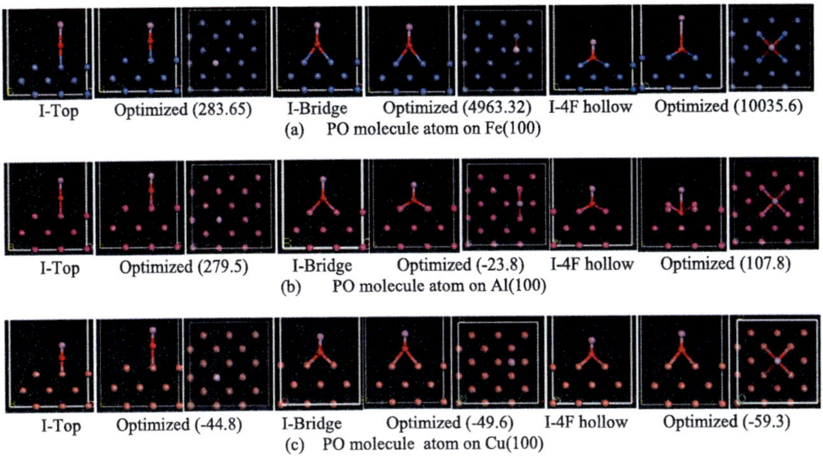


**Fig. 2.** Optimized adsorption configurations and energies of PO molecule on Fe(100), Al(100), and Cu(100) surfaces in gas phase (I-Top: Initial-Top; I-Bridge: Initial-Bridge; I-4F hollow: Initial-4F hollow; red: O; cmyk: P; brown: Fe; orange: Cu; pink: Al; value in parenthesis is adsorption energy (unit: kJ·mol$^{-1}$)).

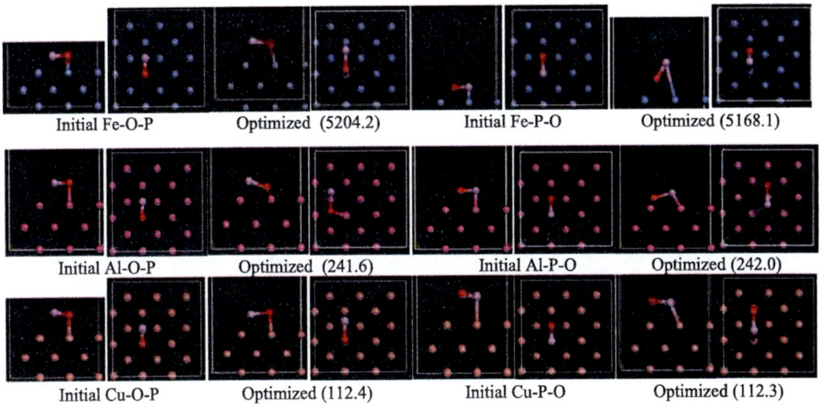


**Fig. 3.** Optimized adsorption configurations and energies of PO molecule at top sites on Fe(100), Al(100), and Cu(100) surfaces with the same initial angle in gas phase (red-O, cmyk-P, brown-Fe, orange-Cu, pink-Al, value in parenthesis is adsorption energy (unit: kJ·mol$^{-1}$)).

matter Fe-O-P or Fe-P-O; for initially adsorbed at the top sites on Cu(100), Fe(100) and Al(100) surfaces, PO molecule adsorption energies were almost same with 90° angle of Cu-O-P and Cu-P-O. Same trend was for with 90° angle of Al-O-P and Al-P-O; for with 90° angle, PO molecule adsorption energies were smallest on Cu(100) surface, biggest on Fe(100) surface, corresponding to the most stable adsorption surface order: Fe(100), Al(100), Cu(100); for initially adsorbed at the top on Cu(100) and Al(100) surfaces with angles of Cu-O-P or Al-O-P less than 180°, PO molecule could be adsorbed on Cu(100) and Al(100) surfaces.

Figure 4 shows that lateral repulsion between S atom and PO molecule made S-Fe, S-Cu, and S-Al bond tilting; after optimization, vertical mode of P-O bond at the bridge site kept same on Fe(100) surface and tilting on Cu(100) surface, respectively; adsorption energy of S atom and PO molecule was bigger on Fe(100) surface than on Cu(100) surface; after optimization and on Al(100) surface, vertical mode of Al-P bond became tilting, while vertical mode of P-O bond kept same.

Figures 5, 6 and Table 1 show the effects of the different coverage of S adsorption at the top and bridge sites on Al(100) surface and at the top site on Cu(100) surface in gas phase on the most stable configurations and energies. Al(100) surface was saturated

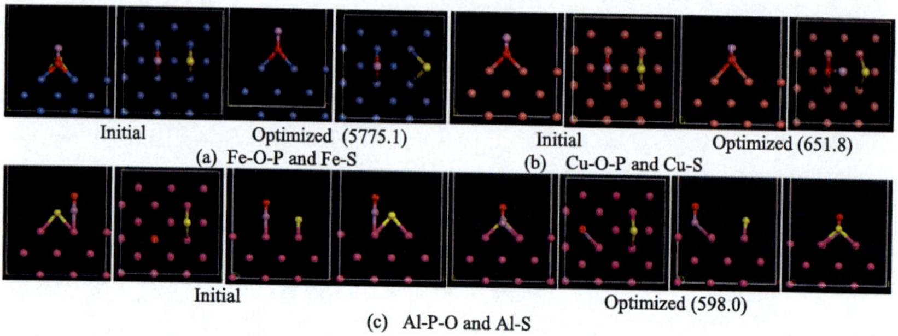


**Fig. 4.** Optimized adsorption configurations and energies of S atom and PO molecule on Fe(100), Al(100), and Cu(100) surfaces in gas phase (red-O, yellow-S, cmyk-P, brown-Fe, orange-Cu, pink-Al, value in parenthesis is adsorption energy (unit: kJ·mol$^{-1}$)).

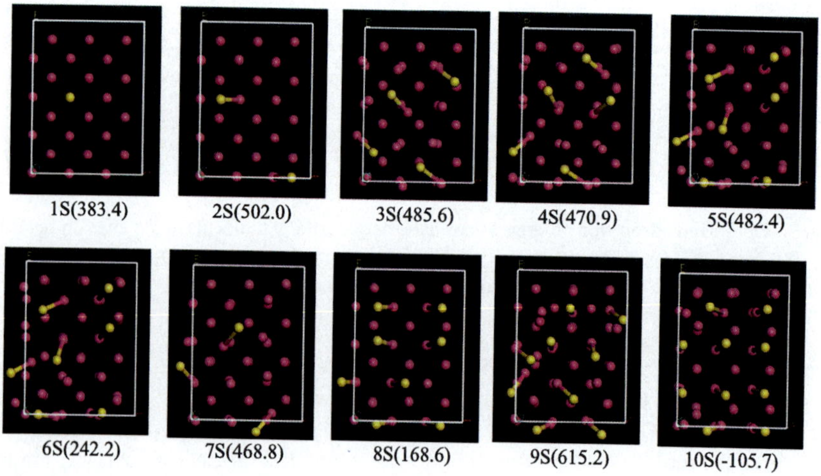


**Fig. 5.** The most stable configurations and energies of the stepwise S adsorption at the top site on Al(100) surface in gas phase (yellow: S; pink: Al; value in parenthesis is adsorption energy (unit: kJ·mol$^{-1}$)).

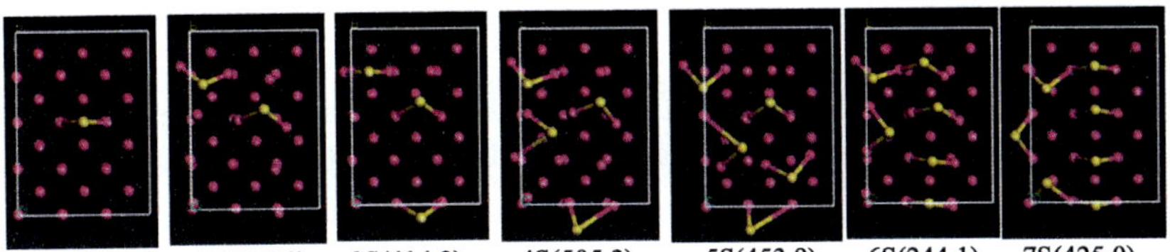

1S(461.8) 2S(487.3) 3S(414.2) 4S(505.2) 5S(453.9) 6S(244.1) 7S(425.0)

**Fig. 6.** The most stable configurations and energies of the stepwise S adsorption at the bridge site on Al(100) surface in gas phase (yellow: S; pink: Al; value in parenthesis is adsorption energy (unit: kJ·mol$^{-1}$)).

**Table 1.** Adsorption energy ($\Delta E_{ads}$), reaction energy ($\Delta E_{dis}$) and lateral repulsive energy ($\Delta E_{late}$) between an added S atom and the previous adsorption configuration of S atoms at different coverage on Al(100) and Cu(100) surfaces.

| | Number of S atom adsorbed at bridge position on Al(100) surface | | | | | | |
|---|---|---|---|---|---|---|---|
| | 1 | 2 | 3 | 4 | 5 | 6 | 7 |
| $\Delta E_{ads}$ (kJ/mol) | 461.8 | 487.3 | 414.2 | 505.2 | 453.9 | 244.1 | 425.0 |
| $\Delta E_{dis}$(kJ/mol) | -178.7 | -289.4 | -331.6 | -467.4 | -638.2 | -649.4 | -789.8 |
| $\Delta E_{late}$ (kJ/mol) | 0 | 25.5 | -73.1 | 91 | -51.3 | -209.8 | 180.9 |
| | Number of S atom adsorbed at top position on Al(100) surface | | | | | | |
| | 1 | 2 | 3 | 4 | 5 | 6 | 7 |
| $\Delta E_{ads}$ (kJ/mol) | 383.4 | 502.0 | 485.6 | 470.9 | 482.4 | 242.2 | 468.8 |
| $\Delta E_{dis}$(kJ/mol) | -122.8 | -280.1 | -432.4 | -600.3 | -782.0 | -731.4 | -937.2 |
| $\Delta E_{late}$ (kJ/mol) | 0 | 118.6 | -16.4 | -14.7 | -99 | 141.2 | -85.4 |
| | 8 | 9 | 10 | | | | |
| $\Delta E_{ads}$ (kJ/mol) | 168.6 | 615.2 | -105.7 | | | | |
| $\Delta E_{dis}$(kJ/mol) | -937.2 | -1142.1 | -727.0 | | | | |
| $\Delta E_{late}$ (kJ/mol) | 214.8 | -231.8 | 489.1 | | | | |
| | Number of S atom adsorbed at top position on Cu(100) surface | | | | | | |
| | 1 | 2 | 3 | 4 | 5 | 6 | 7 |
| $\Delta E_{ads}$ (kJ/mol) | 349.2 | 454.7 | 539.6 | 267.0 | 301.8 | 400.4 | 227.5 |
| $\Delta E_{dis}$(kJ/mol) | -92.8 | -244.4 | -516.8 | -475.9 | -464.1 | -568.9 | -536.4 |
| $\Delta E_{dis}$ (kJ/mol) | 0 | -105.5 | -190.4 | 82.2 | 47.4 | -51.2 | 121.7 |
| | 8 | 9 | 10 | 11 | 12 | | |
| $\Delta E_{ads}$ (kJ/mol) | 268.5 | 248.0 | 261.6 | 199.0 | 202.1 | | |
| $\Delta E_{dis}$(kJ/mol) | -548.6 | -533.6 | -512.6 | -397.2 | -322.6 | | |
| $\Delta E_{dis}$ (kJ/mol) | 80.7 | 101.2 | 87.6 | 150.2 | 147.1 | | |

when adsorbing nine S atoms at the top site ($\Delta E_{9ads} < 0$ *and* $\Delta E_{10ads} > 0$), while Cu(100) surface can adsorb all S atom at top site ($\Delta E_{nads} > 0, n = 1 \sim 12$), indicating Cu(100) surface could adsorb more S atoms than Al(100) surface.

## 4 Conclusions

This study used DMol$^3$ module in Materials Studio package to explore mechanochemical reactive mechanisms of engine oil antiwear additives, the following conclusions are obtained:

- Chemisorption of S atom could occur on Fe(100), Al(100), and Cu(100) surfaces.
- Chemisorption of PO molecule could occur on Fe(100) surface unconditionally, and on Al(100) and Cu(100) surfaces conditionally.
- The most stable S atom and PO molecule chemisorption surface order was Fe(100), Al(100), Cu(100).
- The most stable adsorption site of S atom order was 4F hollow, bridge, top.
- The most stable adsorption site of PO molecule on Fe(100) surface order was 4F hollow, bridge, top.
- Molecule bond angle influenced adsorption configurations and energies. Lateral repulsion among atoms and molecules on Fe(100), Al(100), and Cu(100) surfaces influenced adsorption configurations and energies.
- The coverage of the species adsorption on surfaces depended on adsorption surface and site.
- Adsorption configurations and energies of other atoms and molecules of antiwear film from ZDDP could be predicted by the present simulation approach.
- The results of this study should provide insight on the relative significance of the various theoretical mechanisms that are believed to be at work. This approach could be a useful tool to design tribopairs, lubricants and additives.

## References

1. Thomas, B.C., Wong, V.W.: Engine wear modeling with sensitivity to lubricant chemistry: a theoretical framework. Society of Automotive Engineers, number 2007-01-1566 (2007)
2. Thomas, B.C., Wong, V.W.: Engine-component wear modeling with consideration of growth and removal of surface films: computational framework and preliminary results. In: Proceedings of the ASME Internal Combustion Engine Division 2008 Spring Technical Conference, Chicago, Illinois, USA, 27–30 April 2008, Paper no. ICES2008-1693 (2008)
3. Fujita, H., Glovnea, R.P., Spikes, H.A.: Study of zinc dialkyldithiophosphate antiwear film formation and removal processes. Part I: experimental. Tribol. Trans. **48**, 558–566 (2005)
4. Fujita, H., Spikes, H.A.: Study of zinc dialkyldithiophosph-ate anti wear film formation and removal processes. Part II: kinetic model. Tribol. Trans **48**, 558–566 (2005)
5. Spikes, H.A.: Friction modifier additives. Tribol. Lett. **60** (2015). Article number: 5. https://doi.org/10.1007/s11249-015-0589-z
6. Ksenija, T.M., Lockwood, F., Spikes, H.A.: Behaviour of boundary lubricating additives on DLC coatings. Tribol. Int. **265**, 1893–1901 (2008)
7. Vengudusamy, B., Green, J.H., Lamb, G.D., Spikes, H.A.: Behavior of MoDTC in DLC/DLC and DLC/steel contacts. Tribol. Int. **54**, 68–76 (2012)
8. Quann, R.J., Jaffe, S.B.: Building useful models of complex reaction systems in petroleum refining. Chem. Eng. Sci. **51**(10), 1615–1635 (1996)

9. Stempfe, E.M.: Practical experience with highly biodegradable lubricants, especially hydraulic oils and lubricating greases. NLGI Spokesman **62**(1), 8–23 (1998)
10. Goyan, R., Melley, R.E., Wissner, P.A., et al.: Biodegradable lubricants. Lubr. Eng. **54**(7), 10–17 (1998)
11. Jackson, M.: Environmentally compatible lubricants: focusing on the long-term. NLGI Spokesman **9**, 233–239 (1993)
12. Erhan, S.Z.: USDA is designing vegetable basestocks. Lubr. Eng. **19**(2), 181–186 (1999)
13. Mosey, N.J., Woo, T.K.: A quantum chemical study of the unimolecular decomposition mechanisms of zinc dialkyldithiophosphate antiwear additives. J. Phys. Chem. A **108**, 6001–6016 (2004)
14. Khajeh, A., Hu, X., Mohammadtabar, K., Shin, Y.K., van Duin, A.C.T., Martini, A.: Statistical analysis of tri-cresyl phosphate conversion on an iron oxide surface using reactive molecular dynamics simulations. J. Phys. Chem. C **118**, 11820–11826 (2014)
15. Mohammadtabar, K., Eder, S.J., Dorr, N., Martini, A.: Heat-, load-, and shear-driven reactions of Di-tert-butyl disulfifide on Fe(100). J. Phys. Chem. C **123**, 19688–19692 (2019)
16. Shin, Y.K., Kwak, H., Vasenkov, A.V., Sengupta, D., van Duin, A.C.T.: Development of a ReaxFF reactive force field for Fe/Cr/O/S and application to oxidation of butane over a pyrite-covered $Cr_2O_3$ catalyst. ACS Catal. **5**, 7226–7236 (2015)
17. Catapan, R.C., Oliveira, A.A.M., Chen, Y., Vlachos, D.G.: DFT study of the water-gas shift reaction and coke formation on Ni(111) and Ni(211) surfaces. J. Phys. Chem. C **116**(38), 20281–20291 (2012)
18. Kapur, N., Hyun, J., Shan, B., Nicholas, J.B., Cho, K.: Ab initio study of CO hydrogenation to oxygenates on reduced Rh terraces and stepped surfaces. J. Phys. Chem. C **114**(22), 10171–10182 (2010)
19. Cao, X.M., Burch, R., Hardacre, C., Hu, P.: An understanding of chemoselective hydrogenation on crotonaldehyde over Pt(111) in the free energy landscape: the microkinetics study based on first-principles calculations. J. Catal. Today **165**(1), 71–79 (2011)

# Global Efficiency of Power-Split Hybrid Electric Powertrain

Antonella Castellano and Marco Cammalleri(✉)

Department of Engineering, University of Palermo, Palermo, Italy
{antonella.castellano,marco.cammalleri}@unipa.it

**Abstract.** The development of a proper energy management strategy in hybrid electric vehicles is crucial to pursue the best balance between emissions reduction and performance. In this respect, this paper addresses a straightforward method to investigate the behaviour of the hybrid powertrain in steady-state operation in terms of speeds and torques both of the thermal engine and the electric machines. The procedure takes into account the conversion power losses in the internal combustion engine and the electric unit and the mechanical power losses occurring in the power-split transmission. This enables the calculation of a set of maps describing the global efficiency of the hybrid powertrain and the power supplied or gathered by the batteries, which can be exploited in the implementation of a proper control strategy. The proposed approach is based on a unified parametric model that enables a comprehensive analysis of any power-split transmission.

**Keywords:** SDG13 · Hybrid electric vehicles · Power-split CVTs · Powertrain global efficiency

## 1 Introduction

The alarming aggravation of global warming has led world leaders to undertake a common commitment by adopting the United Nations Framework Convention on Climate Change (UNFCCC), whose main aim is to decrease the concentration of greenhouse gases in the atmosphere. To accomplish the goals established in the Paris Agreement (2015), relevant environmental policies have been adopted to take urgent action on climate change, by promoting the uptake of zero- and low-emission vehicles, i.e. electric and hybrid electric vehicles.

Among the Hybrid Electric Vehicles (HEVs) architectures, the power-split layout is the most promising one [1–4]. Like any other HEV, the power-split hybrid electric powertrain relies on two different power sources for motion, which are an Internal Combustion Engine (ICE) and an electric unit, which operates as a Continuously Variable Unit (CVU) and consists of two electric machines, power converters and batteries. The deployment of Planetary Gear sets (PGs) and Ordinary Gear sets (OGs), properly arranged in a Power-Split Unit (PSU), enables the decoupling of the thermal engine operation from the wheels speed. Furthermore, not only the ICE power can be used for battery charging as well as for vehicle propulsion, but also the battery pack can supply traction power by cooperating with the ICE or operating in full electric functioning mode.


G. Quaglia et al. (Eds.): I4SDG 2021, MMS 108, pp. 502–511, 2022.
https://doi.org/10.1007/978-3-030-87383-7_54

Typically, low noise, fuel-saving and reduction of $CO_2$ emissions make the pure electric mode ideal for urban driving, while the involvement of the Internal Combustion Engine (ICE) is more suitable in extra-urban driving. Nonetheless, the system operating point for given vehicle speed and demanded torque should be the one that minimises fuel consumption and optimises global efficiency, according to the current capability of the battery pack of supplying or gathering electric energy. Therefore, proper Energy Management Strategies (EMSs) based on the battery State Of Charge (SOC) have to be implemented to reduce emissions while maintaining driving comfort and high performance [5–7]. For this purpose, the power-split system has to be analysed under various operating conditions.

In this paper, a unified parametric model for Power-Split Continuously Variable Transmissions (PS-CVTs) is applied to perform a comprehensive evaluation of the overall power losses occurring in the powertrain and provide some guidelines to maximise the global efficiency. Any PS-CVT can be preliminarily analysed by the dimensionless method proposed in [8–10], which was expanded and upgraded in [11] and briefly summarised in Sect. 2. This leads to the calculation of the PSU mechanical power losses and of the subsequent real power flowing into the electric machines. Then, the ICE and electric machines efficiency maps can be introduced to investigate the powertrain behaviour for specific output speed and torque, by assuming proper constraints on the power flowing from or to the battery pack (Sect. 3). In this way, a global efficiency map can be obtained for every vehicle speed in steady-state operation. Therefore, it is possible to detect the best operating point of the whole system among the feasible ones according to the power that batteries can supply or receive. As a case study, the output-split transmission designed in [12] is analysed in Sect. 4.

## 2 Unified Parametric Model for PS-CVTs

According to the parametric model addressed in [8–11], any PSU can be modelled as a black box with two Degrees Of Freedom (DOFs) and four external ports linked to the ICE (shaft *in*), the wheels (*out*), and the two electric machines ($i$ and $o$), as it is shown in Fig. 1a. The PSU can consist of any number of PGs and OGs, arranged in one or more basic structure called Three-Port Mechanisms (TPMs). As Fig. 1b shows, each TPM consists of a PG and up to three OGs, which connect the PG branches ($R$, $S$, $C$, indicated by $X$, $Y$, $Z$ in a general way) and the PSU main shafts (*in*, *out*, $i$, $o$, …, indicated by $x$, $y$, $z$). This schematisation enables the utilization of some general equations (not reported here for brevity) based on basic functional parameters, which can be derived from the transmission constructive layout in an analysis phase [11] or properly selected during the design [8, 12].

The dimensionless approach outlined in this section is focused on analysis purpose and is suitable even for addressing multi-PG and multi-mode PS-CVTs. All the relationships describing the two-DOF system are normalised to ICE speed, torque and power, so as to reduce the mathematical treatment to a one-DOF problem. For brevity, only the main, necessary equations are reported in the following. Further explanations can be found in the previous papers [8–11]. It is worth noting that the model can be swiftly rearranged to study also the full electric operation, as it is addressed in [11].

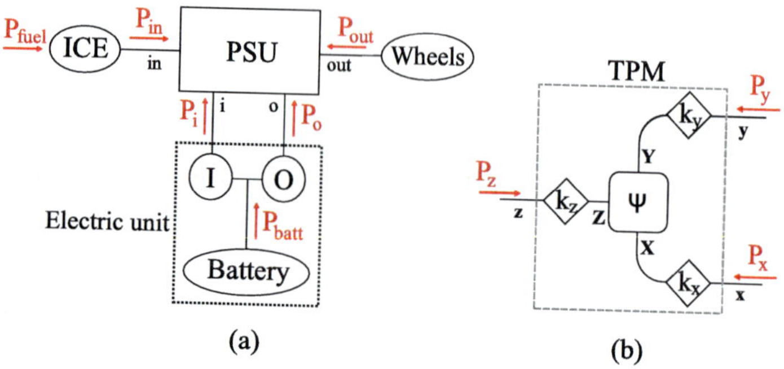


**Fig. 1.** Scheme of the power-split powertrain (a) and a generic TPM (b) with positive power flows direction.

The whole model relies on some functional parameters that rule PSU speeds, torques and power flows, as well as mechanical power losses. These functional parameters are the nodal ratios and the corresponding speed ratios. The former, indicated as $\tau_{\#k}$, are the overall transmission ratio $\tau = \omega_{out}/\omega_{in}$ achieved when a generic $k$th shaft is motionless. The latter, indicated as $\tau_{j\#k}$, are the $j$th speed ratio $\tau_j = \omega_j/\omega_{in}$ calculated for the $k$th nodal ratio. The procedure for identifying these functional parameters starting from the PSU constructive arrangement is described in [11] and relies on some simple matrix operations performed on a constraints matrix whose rows encloses all the kinematic constraints due to PGs and OGs.

One of the strengths of this model is the simplicity in the calculation of the mechanical power losses occurring in the PSU, applicable to any PS-CVT, constructive layout notwithstanding [10, 11]. Once known the constructive ratio and the efficiency of PGs and OGs and calculated the functional parameters, the mechanical power losses occurring in the PSU can be computed as a function of the overall transmission ratio $\tau$ and the opposite of the power ratio $\eta = -P_{out}/P_{in}$ (note that $\eta$ is not the global efficiency, but it considers the possibility for batteries to be charged or discharged). Figure 2 shows the map of the total mechanical power losses $\overline{p}_L$ normalised to the ICE power calculated by applying the procedure described in [10, 11] to the transmission analysed in Sect. 4.

Once computed the map of the PSU power losses, the real mechanical power flowing through the shaft $o$ can be calculated by the following equation:

$$\overline{p}_o = p_o - \frac{\tau_o}{\tau_{o\#i}}\left[\overline{p}_L + \left(\frac{\partial \overline{p}_L}{\partial \tau} + \frac{\partial \overline{p}_L}{\partial \eta}\frac{\eta}{\tau}\right)(\tau_{\#i} - \tau)\right] \tag{1}$$

where:

$$\tau_o = \frac{\omega_o}{\omega_{in}} = \tau_{o\#i}\frac{\tau - \tau_{\#o}}{\tau_{\#i} - \tau_{\#o}} \tag{2}$$

$$p_o = \frac{P_o}{P_{in}} = \frac{(\tau - \tau_{\#o})(\tau - \eta\,\tau_{\#i})}{\tau(\tau_{\#o} - \tau_{\#i})} \tag{3}$$

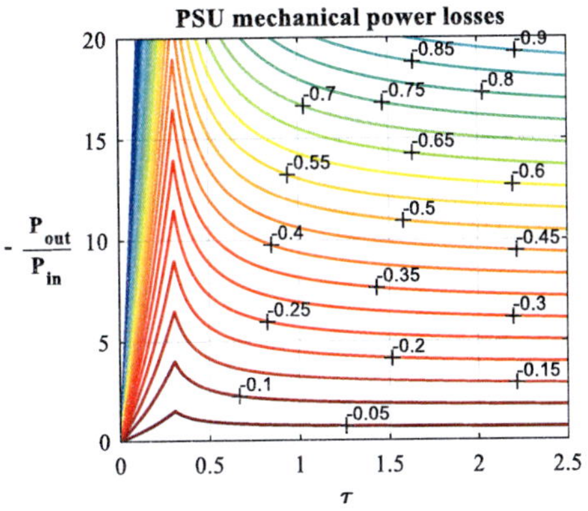


**Fig. 2.** Example of PSU mechanical power losses $\overline{p}_L$ as a fraction of the ICE power.

From Eq. (1) the real torque on the shaft $o$ normalised to $T_{in}$ can be obtained by dividing it by Eq. (2). Analogous relationships are obtained for the shaft $i$ by switching the subscripts $i$ and $o$ in Eqs. (1), (2) and (3).

## 3 Global Efficiency Evaluation

The dimensionless approach summarised in Sect. 2 enables the comprehensive analysis of the PSU response once freely assumed a speed ratio and a power or torque ratio between any two external PSU shafts. Nonetheless, the overall speed ratio $\tau$ and the overall power ratio $\eta$ were chosen as independent variables for convenience. Indeed, the output speed $\omega_{out}$ is directly related to the vehicle speed and the output torque supplied to the wheels results in acceleration, deceleration or steady-state operation. In the latter case, $P_{out}$ is dependent on the vehicle speed $V_{veh}$ as follows:

$$P_{out} = -\left(mg \sin\alpha + f_r mg \cos\alpha + 0.5 \cdot C_d A_f \rho_a V_{veh}^2\right) V_{veh} \quad (4)$$

$P_{out}$ is negative because it is delivered by the PSU. In Eq. (4), $m$ is the vehicle mass, $g$ is the gravitational acceleration, $\alpha$ is the road slope expressed in radians, $f_r$ is the rolling resistance coefficient, $c_x$ is the drag coefficient, $A_f$ is the vehicle frontal area and $\rho_a$ is the air density.

For each vehicle speed, the PSU can be analysed by imposing proper values for $\tau$ and $\eta$. In other words, the rotational speed and the real mechanical power of the electric machines can be calculated once fixed the ICE functioning point in terms of $\omega_{in}$ and $T_{in}$. Since the aim of the study is to find the system operating point that maximises the global efficiency, for fixed $V_{veh}$ it is necessary to explore all the feasible ICE functioning points, which are represented in the ICE efficiency map (Fig. 3a). Thus, for each couple of $\omega_{in}$ and $T_{in}$, the calculation of $P_{in}$ is straightforward, along with the identification of the ICE efficiency $\eta_{ICE}$. Therefore, the power that has to be provided by fuel combustion is:

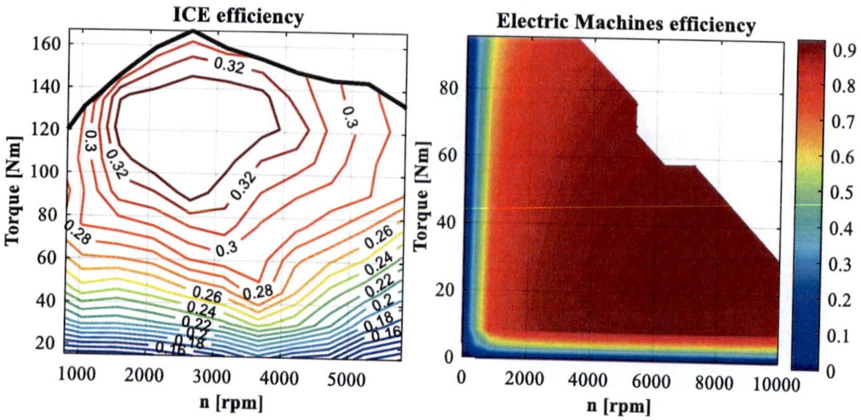


**Fig. 3.** Typical ICE efficiency map (a) and Electric Machines efficiency map (b).

$$P_{fuel}(\omega_{in}, T_{in}) = P_{in}/\eta_{ICE} \tag{5}$$

The rotational speed of the shafts $i$ and $o$ can be computed from Eq. (2) by imposing $\tau = V_{veh}/(R_{wheel} \cdot \omega_{in})$, where $R_{wheel}$ is the wheel radius. Similarly, the powers $\overline{P}_o$ and $\overline{P}_i$ flowing through the shafts $i$ and $o$ can be derived from Eq. (1) and the real torques $\overline{T}_o$ and $\overline{T}_i$ can be obtained by dividing them by $\omega_o$ and $\omega_i$. Thus, the respective electric powers $\overline{P}_{o,el}$ and $\overline{P}_{i,el}$ can be computed from the electric machines efficiency map, which indicates the efficiency of the electric machines as a function of torque and rotational speed (Fig. 3b). The net power flow that involves the battery pack is the sum of $\overline{P}_{o,el}$ and $\overline{P}_{i,el}$.

$$P_{batt}(V_{veh}, \omega_{in}, T_{in}) = \overline{p}_o P_{in} \eta_O^{-sign(\overline{p}_o)} + \overline{p}_i P_{in} \eta_I^{-sign(\overline{p}_i)} \tag{6}$$

$P_{batt}$ is positive if batteries provide electric power for vehicle propulsion, while it is negative during battery charging. Therefore, in the former case $P_{batt}$ is an input power provided to the powertrain, along with $P_{fuel}$. Instead, in the latter case $P_{batt}$ is an output power, similarly to $P_{out}$. Hence, by defining the powertrain global efficiency $\eta_{gl}$ as the negative ratio between the total output power and the total input power, two definitions are possible according to the direction of the battery power flow:

$$\eta_{gl}(V_{veh}, \omega_{in}, T_{in}) = -\frac{P_{out}}{P_{fuel} + P_{batt}} \quad for \ \ P_{batt} > 0 \tag{7}$$

$$\eta_{gl}(V_{veh}, \omega_{in}, T_{in}) = -\frac{P_{out} + P_{batt}}{P_{fuel}} \quad for \ \ P_{batt} < 0 \tag{8}$$

The relationships presented in this section lead to a global efficiency map as a function of $\omega_{in}$ and $T_{in}$ for a given vehicle speed. Thus, it is possible to identify the system operating point that achieves the highest efficiency. Nonetheless, the best solution should be selected only among the feasible ones, since the speed and the torque of ICE and electric machines must be comprised between the respective minimum and maximum values.

### 3.1 Battery State of Charge

Even the battery power flow cannot exceed a maximum limit both in the charging and discharging phase, according to the current SOC. However, the exact instantaneous SOC is not evaluated in this paper, but four general working conditions are outlined.

If it is supposed that the current SOC is the one that always allows the achievement of the best operating condition, the battery can instantaneously receive or provide any amount of power, so long as it is comprised between the viable constructive limits. In Sect. 4 we refer to this condition as $SOC = FREE$ and it must be:

$$P_{max\_charge} \leq P_{batt} \leq P_{max\_discharge} \tag{9}$$

The condition whereby the battery is completely charged and thus prevented from receiving further power is indicated as SOC = 1 and implies $P_{max\_charge} = 0$. On the contrary, if the battery is fully discharged (SOC = 0), it cannot supply power and hence $P_{max\_discharge} = 0$. Lastly, the SOC can be maintained constant if $P_{batt} = 0$.

## 4 Application

The procedure described in the previous sections can be applied to any PS-CVT. In this section, the output-split transmission designed in [12] by the same parametric model is analysed. Its design was aimed at minimising the size of the electric machines, by assuming the vehicle parameters of Table 1, but any power losses were neglected. The ICE can operate between 800 and 5800 rpm, as its efficiency map shows (Fig. 3a), and has 80 kW maximum power. The electric machines have the same size and can deliver up to 35 kW each. The maximum power flow that the battery pack can deliver or receive is 70 kW. The efficiency map of Fig. 3b is used for both electric machines and both for generating and motoring operation.

**Table 1.** Vehicle parameters.

| $m$ [kg] | $\alpha$ [rad] | $f_r$ [−] | $c_x$ [−] | $A_f$ [m$^2$] | $\rho_a$ [kg/m$^3$] |
|---|---|---|---|---|---|
| 2200 | 0 | 0.0122 | 0.4 | 2.5 | 1.225 |

Firstly, the transmission is analysed by the dimensionless method of Sect. 2. The constructive layout of the transmission, shown in Fig. 4, is derived from [12], as also its basic functional parameters, which are $\tau_{\#o} = 0.218$, $\tau_{\#i} = \tau_{o\#i} = -\infty$, $\tau_{i\#o} = 1.22$. The PG Willis' ratio is $\Psi = -0.4$, while the OGs fixed ratios are $k_{in} = \omega_{in}/\omega_R = 0.82$

and $k_{out} = \omega_{out}/\omega_C = 0.25$. The dimensionless mechanical power losses occurring in the PSU are those shown in Fig. 2 evaluated by considering the OGs efficiency equal to 0.98 and the PG fixed-carrier efficiency equal to 0.96. We refer the reader to [10] and [11] for more details. Once computed $\overline{p}_L$, $\overline{p}_o$ and $\overline{p}_i$ are calculated by Eq. (1).

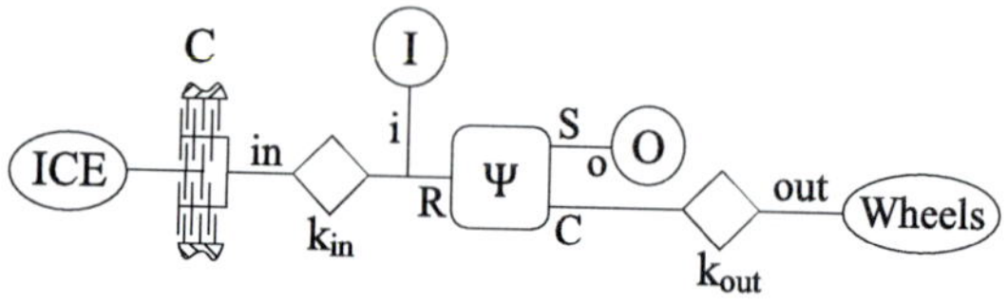


**Fig. 4.** Transmission functional layout. The clutch C provides full electric operation if engaged.

The procedure described in Sect. 3 is performed for a vehicle speed ranging from 0 to 200 km/h, thus for each $V_{veh}$ a powertrain global efficiency map is obtained. Correspondingly, each system operating point involves a univocal $P_{batt}$. Therefore, the selection of the best operating point can be performed among the viable functioning points according to the SOC constraints addressed in Sect. 3.1. It is worth noting that the efficiency of power converters and batteries is not considered here. However, it would have affected the calculation of the only $P_{batt}$.

## 4.1 Results and Discussion

Figure 5 shows the results of the simulation, in terms of the best global efficiency achievable for given SOC constraint (Fig. 5a) and the $P_{batt}$ (Fig. 5b), ICE operating point (Fig. 5c) and ICE power (Fig. 5d).

The results show that the absence of instantaneous constraints on the SOC would realise the most efficient power operations while the ICE is on. This is due to the fact that electric machines can simultaneously operate both as motors or as generators, by avoiding power recirculation in the electric unit (the trend of electric machines power is not reported for brevity). Nonetheless, the global efficiency is lower than 0.3 up to 44 km/h, therefore it would be more convenient to turn off the engine and enable the full electric driving if batteries are charged (SOC = 1). Otherwise, if the engine is on also for the lowest speed, it should operate in its maximum efficiency region (Fig. 5c), provided that it is possible to recharge the batteries. However, Fig. 5a suggests that the battery recharging is more convenient between 65 and 130 km/h because the global efficiency would be slightly higher than the one achievable for lower speeds.

For a medium-high vehicle speed (from 45 km/h), Fig. 5b and Fig. 5d show that the most efficient functioning condition is keeping the ICE at its minimum power and exploit battery power for traction. However, in a real-time optimisation strategy, the battery power flow should be adjusted according to the actual SOC and predicted path, in order to have sufficient range. Therefore, up to 105 km/h, it would be advisable to discharge batteries or recharging them only if SOC = 0, but imposing SOC = COST should always be avoided since power recirculation in the electric unit cannot be prevented. Instead, pursuing the most efficient powertrain operations for speeds higher than 105 km/h could

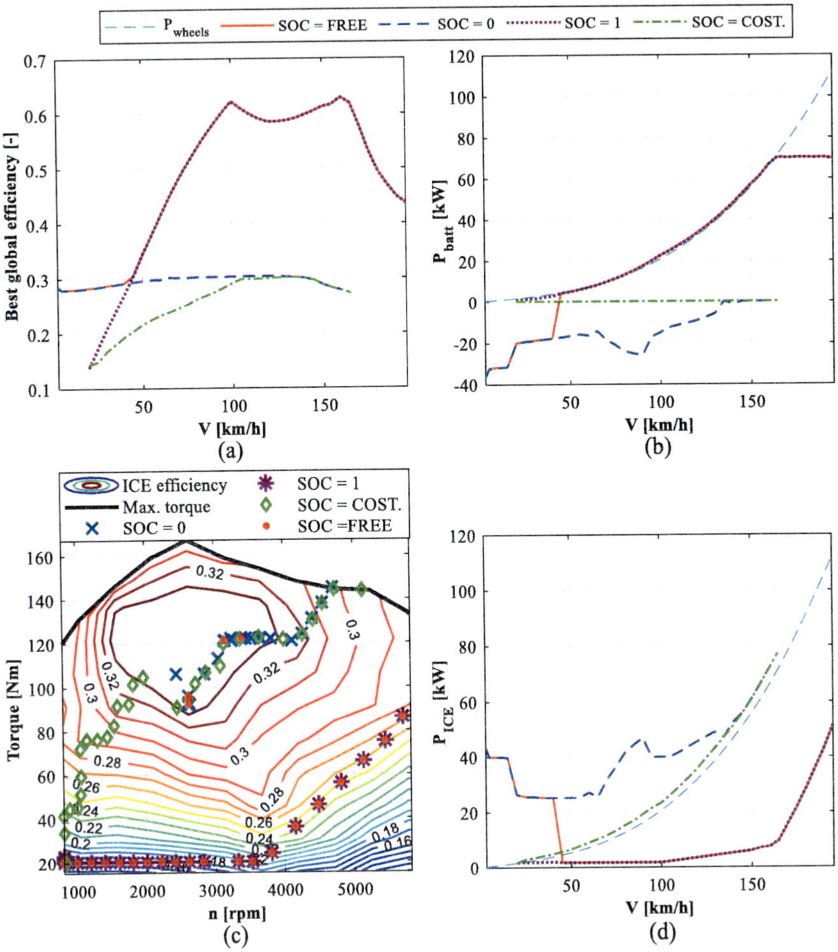


(a) (b) (c) (d)

**Fig. 5.** Simulation results.

result in a very limited range during cruising. Thus, it could be even more advisable to maintain constant SOC from 105 to 135 km/h or increase the ICE power and recharge the battery, while battery charging over 135 km/h should be avoided.

Starting from 165 km/h, the optimal demanded power is higher than $P_{max\_discharge}$, therefore the ICE has to increase its power and thus its losses, which lower the maximum global efficiency. Moreover, since the ICE maximum power is 80 kW the vehicle speed cannot exceed 165 km/h if the batteries are fully discharged (SOC = 0) or if a constant SOC is desired.

## 5 Conclusions

This paper presented a non-iterative approach to provide valuable data that can be exploited for the implementation of a proper EMS, by evaluating the response of a power-split hybrid powertrain in steady-state driving.

The addressed procedure is based on a parametric model that is extremely general and thus suitable for analysing any PS-CVT. It enabled the calculation of a set of maps for every vehicle speed, which includes the feasible operating points of each power source and the trend of the overall powertrain efficiency, as well as the battery power flow. Hence, a real-time control strategy could rely on these results that can be preliminarily computed for reducing the computational burden.

Even though the main goal of this paper was not providing an EMS, the results of the proposed case study suggested that the battery SOC should be maintained always between its minimum and maximum values. In this way, power recirculation in the CVU could be prevented and the overall global efficiency could be maximised. Moreover, in contrast to many rule-based strategies [6, 13], it is not always advisable to let the ICE work within its maximum efficiency region if the final aim is minimising the power losses of the whole powertrain, similarly to [14].

It should be noted that the procedure here implemented for a vehicle in specified steady-state operation (i.e., for fixed on-board load and road slope) could be swiftly replicate for any other vehicle under different conditions, by simply considering different vehicle parameters and transmission, as well as propulsors efficiency maps.

Future works might extend this research in order to investigate the powertrain behaviour in dynamic operation, by taking into consideration the inertia of vehicle and propulsors to evaluate the maximum acceleration available or embed a real-time control strategy by taking into account the instantaneous battery SOC.

## References


1. Mi, C., Masrur, M.A.: Hybrid Electric Vehicles, Principles and Applications with Practical Perspectives, 2nd edn. Wiley, Hoboken (2017)
2. Zhuang, W., et al.: A survey of powertrain configuration studies on hybrid electric vehicles. Appl. Energy **262**, 114553 (2020)
3. Singh, K.V., Bansal, H.O., Singh, D.: A comprehensive review on hybrid electric vehicles: architectures and components. J. Mod. Transp. **27**(2), 77–107 (2019). https://doi.org/10.1007/s40534-019-0184-3
4. Cammalleri, M., Sorge, F.: Approximate closed-form solutions for the shift mechanics of rubber belt variators. In: 2009 Proceedings of the ASME International Design Engineering Technical Conferences and Computers and Information in Engineering Conference, DETC 2009, vol. 6, pp. 187–196 (2009). https://doi.org/10.1115/DETC2009-86638
5. Zhang, F., Wang, L., Coskun, S., Pang, H., Cui, Y., Xi, J.: Energy management strategies for hybrid electric vehicles: review, classification, comparison, and outlook. Energies **13**(13), 1–35 (2020)
6. Zhang, P., Yan, F., Du, C.: A comprehensive analysis of energy management strategies for hybrid electric vehicles based on bibliometrics. Renew. Sustain. Energy Rev. **48**(205), 88–104 (2015)

7. Panday, A., Bansal, H.O.: A review of optimal energy management strategies for hybrid electric vehicle. Int. J. Veh. Technol. **2014**, 160510 (2014)
8. Cammalleri, M., Rotella, D.: Functional design of power-split CVTs: an uncoupled hierarchical optimized model. Mech. Mach. Theory **116**, 294–309 (2017)
9. Rotella, D., Cammalleri, M.: Direct analysis of power-split CVTs: a unified method. Mech. Mach. Theory **121**, 116–127 (2018)
10. Rotella, D., Cammalleri, M.: Power losses in power-split CVTs: a fast black-box approximate method. Mech. Mach. Theory **128**, 528–543 (2018)
11. Cammalleri, M., Castellano, A.: Analysis of hybrid vehicle transmissions with any number of modes and planetary gearing: kinematics, power flows, mechanical power losses. Mech. Mach. Theory **162**, 104350 (2021)
12. Rotella, D., Cammalleri, M., Qin, D., Zhou, X.: A simple method for the design of hybrid electric power-split CVTs: a case study. In: Carbone, G., Gasparetto, A. (eds.) IFToMM ITALY 2018. MMS, vol. 68, pp. 70–79. Springer, Cham (2019). https://doi.org/10.1007/978-3-030-03320-0_8
13. Banvait, H., Anwar, S., Chen, Y.: A rule-based energy management strategy for plug-in hybrid electric vehicle (PHEV). In: 2009 American Control Conference, pp. 3938–3943 (2009)
14. Ahn, K., Cho, S., Cha, S.W., Lee, J.M.: Engine operation for the planetary gear hybrid powertrain. Proc. Inst. Mech. Eng. Part D J. Automob. Eng. **220**(12), 1727–1735 (2006)

# Steering Mechanisms with Alterable Kinematic Parameter

Roman Yu. Dobretsov, Andrei V. Lozin, Andrei O. Kaninskii(✉), and Vladimir E. Rolle

Peter the Great Saint-Petersburg Polytechnic University, Saint-Petersburg, Russia

**Abstract.** The paper focuses on analysis of tracked vehicles with skid steering system. Vehicle transmission is studied in terms of its kinematic parameter (kinematic factor) which is distance between point that keeps rectilinear speed and vehicle's center of masses during turn taken in relation to vehicle's tread. The relation between steering mechanism kinematic parameter (kinematic factor) and vehicle performance during turn is investigated. The ways of improving vehicle performance by altering kinematic parameter of steering mechanism are proposed. The kinematic diagrams of electromechanical and friction transmissions with ability to alter kinematic factor are presented. The analysis of suggested transmissions and conclusion about its perspectives are given.

**Keywords:** SDG9 · Tracked vehicle · Steering mechanism · Double-flow transmission · Electromechanical and friction transmission

## 1 Introduction

While performing a turn the tracked vehicle experience motion resistance. This resistance is caused by ground surface deformation, friction between track and ground and other known phenomena.

The value of motion resistance could be higher than amount of power produced by vehicle engine. In this case it is impossible for vehicle to perform a turn. Because of this the problem of decreasing amount of power needed to make tracked vehicle turning is critical to solve.

Theory of transport vehicles [1, 2] provides a formula which connects the amount of power needed to overcome motion resistance during turn $N_0$ and linear movement $N_{lin}$ [2]. This formula is described by Eq. (1)

$$\overline{N} = N_0/N_{lin} = \frac{\rho + q_g}{\rho + q_m} \tag{1}$$

In shown expression $\rho = R/B$ – relative turn radius ($R$ – turn radius, $B$ – tread of the vehicle); $q_g$ – steering mechanism force factor; $q_m$ – steering mechanism kinematic factor $N^{-}$ – relative power increase (increase of power needed to perform a turn in comparison to power needed for propulsion during linear movement) (Fig. 1).


G. Quaglia et al. (Eds.): I4SDG 2021, MMS 108, pp. 512–521, 2022.
https://doi.org/10.1007/978-3-030-87383-7_55

The expression is correct for the case of turn with exact conditions: constant turn radius and angular velocity, uniform linear motion, engine operates with constant parameters, power losses are equal for linear motion and turn, track skidding is not considered.

Formulas used to express the process of turn also used with respectable assumptions.

Steering mechanism force factor is used as distance between point in which resultant force of motion resistance is applied and center of vehicle masses in relation to vehicle's tread. The expression described by Eq. (2) is used as in [2]:

$$q_g = \frac{y_g}{B} = \frac{\mu L}{4fB} \tag{2}$$

Here $y_g$ – moment arm of the resultant force; $B$ – tread of the vehicle; $\mu$ – turn resistance coefficient; L – length of ground contacting area of the track; f – rolling resistance coefficient. The force factor $q_g$ value is variable because turn resistance coefficient is considered in terms of relative turn radius. It is therefore connected to the expression *L/B* which connects main characteristics of a vehicle chassis.

In this case propulsion force of the slowed track (track which moves with lesser velocity than another during turn) may be expressed in terms of linear movement motion resistance [2]:

$$P_1 = fG(\pm q_g \mp 0{,}5) \tag{3}$$

Here f – rolling resistance coefficient; G – is vehicle mass; $q_g$ – force factor. The positive or negative values are considered in dependance with propulsion force direction. This means that free turn (described in [3] as uncontrolled turn) should be performed while $q_g = 0{,}5$.

According to calculations the most sufficient power losses are experienced by vehicles with differential steering while turning on dry loam ($q_m = 0$). The amount of power needed to perform a turn starts intensively increasing while relative turn radius value is equal to 10 or lower. The information provided in sources [4–6], suggests that value $\rho = 10$ is statistically most possible for cases of high-seed tracked vehicles turning.

In each case considered in the article turn resistance coefficient was calculated in accordance with empirical formula [3]:

$$\mu = \mu_{max}/(0{,}925 + 0{,}15\,\rho) \tag{4}$$

Here $\mu_{max}$ – maximum value of turn resistance coefficient while $\rho = 0{,}5$; $\rho = R/B$ – relative turn radius ($R$ – turn radius, $B$ – tread of the vehicle).

Kinematic parameter (kinematic factor, $q_m$) of steering mechanism is the distance between point which moves with linear motion speed and center of mass of the vehicle taken in relation to vehicle's tread. One of existing classifications of steering mechanism types (for example [2, 3, 7]) is based on kinematic factor's value ($q_m = 0$ for differential steering, $q_m = 0{,}5$ – for board steering). The value of parameter $q_m$ determines minimal turn radius that is achievable for considered steering mechanism. It should be noted that vehicle with differential steering could be rotated around its center of masses while double-flow differential transmission allows vehicle to perform a pivot turn.

Linear speed of vehicle's center of mass could be found according to the next expression [5]:

$$V = V_{lin} \frac{\rho}{\rho + q_m} \tag{5}$$

Where $V_{lin}$ is vehicle's linear motion speed; $\rho = R/B$ – relative turn radius ($R$ – turn radius, $B$ – tread of the vehicle); $q_m$ – kinematic factor. Let us consider differential steering mechanisms with non-symmetrical differentials. An example of this mechanism type is Zaichik-Kreines steering mechanism. In case of considered mechanisms $q_m = q_g$ which means that kinctor's value is not constant. According to Eq. (1), power needed to perform a turn is equal to power needed for propulsion during linear motion.

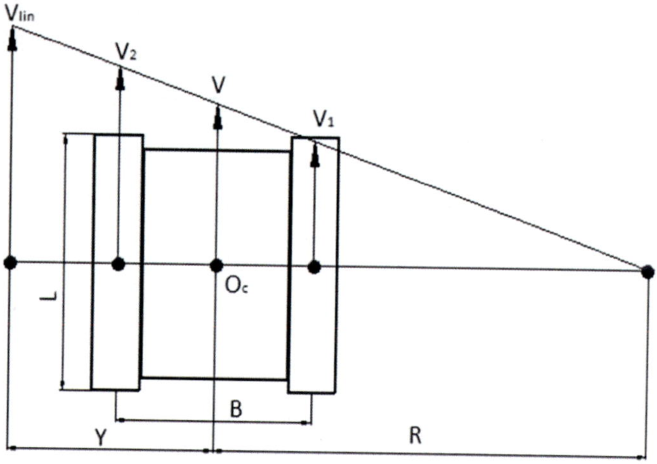


**Fig. 1.** Tracked vehicle chassis geometry R – turn radius; Y – distance between point that keeps rectilinear speed and vehicle's center of masses; B – tread of the vehicle; L – length of ground contacting area of the track; $O_c$ – center of masses; $V_1$ – slower track linear speed; $V_c$ – center of masses linear speed; $V_2$ – faster track linear speed; $V_{lin}$ – vehicle's rectilinear motion speed (before turn)

The main disadvantage of non-symmetrical differential steering mechanisms is unstable turn radius. This fact affects vehicle's controllability and increases skidding risk especially while moving with high speed.

The point which moves with linear motion speed could extend beyond vehicle hull in cases of turn with high speed and low motion resistance. This means that both tracks would rotate slower during turning than in case of linear motion (in accordance with Eq. (5)). This fact may be also considered as disadvantage of non-symmetrical differential steering mechanisms. In case of transport platforms which doesn't need to move with high speeds such fact is far less critical than power losses decrease. For example, statistical information given in [8] shows that in most cases driver prefers to perform turns and other maneuvers while vehicle moves with relatively slow speed.

Facts shown above suggests that existing steering mechanisms should be modified to properly perform its function. The article goal is to provide concepts of steering mechanisms that would ensure high quality of movement control in case of vehicle turn on high speeds (as double-flow hydrostatic differential transmission) and minimize power losses for maneuvers with low turn radius.

## 2 Steering Mechanism Kinematic Parameter Altering

Figure 2 shows diagram of double-flow transmission. Both electric engines operates while vehicle moves linearly. This fact is the main difference between considered transmission and concepts described in works [9–12]. This option allows to increase engine power without sufficient increase of gearbox load. This is achieved because auxiliary power flows through electric circuits. Minimizing of power losses during turn could be achieved by lowering engine shaft velocity. This case is similar to turn of the vehicle with non-symmetrical differential steering [3, 7]. However, considered transmission could possibly allow vehicle to turn on high speeds without skidding. The concept also allows vehicle to perform a pivot turn. Control algorithm may be used in accordance to information provided in [13]. Considered steering mechanism allows vehicle to perform a turn even if one of engines is not functional.

Modern research in field of theory of tracked vehicles suggests that kinematic factor $q_m$ could be taken in relation to transmission ratio between vehicle sprockets velocity $u_0$ (examples can be seen in [3, 14]):

$$u_0 = \frac{q_m - 0{,}5}{q_m + 0{,}5} \tag{6}$$

Therefore, the desirable steering mechanism should have kinematic factor value that is according to expression described by Eq. (7):

$$q_m(\rho) = 0{,}5\frac{1 + u_0}{1 - u_0} = \frac{\mu L}{4fB} \tag{7}$$

Therefore:

$$u_0 = \frac{\mu L - 2fB}{\mu L + 2fB} = \frac{q_g - 0{,}5}{q_g + 0{,}5} \tag{8}$$

In expressions above $q_g$ – force factor; $B$ – tread of the vehicle; $\mu$ – turn resistance coefficient; L – length of ground contacting area of the track; f – rolling resistance coefficient.

In case of linear motion $\rho = \infty$ Eq. (8) could be shortened to $u_0 = (-1)$ This case is similar to traditional differential steering system. In considered case $1 > u_{0max} > 0$ and

$$u_{0max} = \frac{\mu_{max} L - 2fB}{\mu_{max} L + 2fB} \tag{9}$$

Here $\mu_{max}$ – maximum value of turn resistance coefficient while $\rho = 0{,}5$.

Maximum values of each respective part of Eq. (9) are known. Therefore, value $u_{0max}$ could be found. For $L/B = 2$ this value is $u_{0max} = 0{,}82$. In this case while vehicle performs a turn its slower side sprocket should be slowed more intensively than the sprocket on the faster side of the vehicle. Transmissions considered in work [9] could be improved by adding additional operating modes. This could be achieved by changing parameters of reduction gear connecting vehicle sides. The operation modes needed are $u_{0max}$ and $1/u_{0max}$ (in considered example: 0,82 and 1,22). These modes should be used to propulse vehicle in opposite directions. The examples of design solutions that could be used to provide considered ratios could be found in the following works: [15–17].

The information provided in [18] suggests that rolling resistance and turning resistance could be related. However, this statement requires additional analysis based on wider sources of information. The design of closed-loop steering control system (see [19, 20]), that could allow power losses decreasing and maintaining stable turn radius and speed could be used to improve transmission parameters. It seems that the most flexible transmission type in terms of possible control algorithms should be double-flow and electromechanical. The schematic diagram of considered transmission type is explained in Fig. 2.

Analogous transmission type could be designed with use of friction steering devices instead of electric drive [21]. The friction steering device is reduction gear which clutches' slipping parameters are controlled. The gear ratio in this case can be changed continuously. Therefore, friction steering device can be roughly described as continuously-variable transmission. Its structure, however, is similar to ordinary gearbox, which makes friction steering device easy in production.

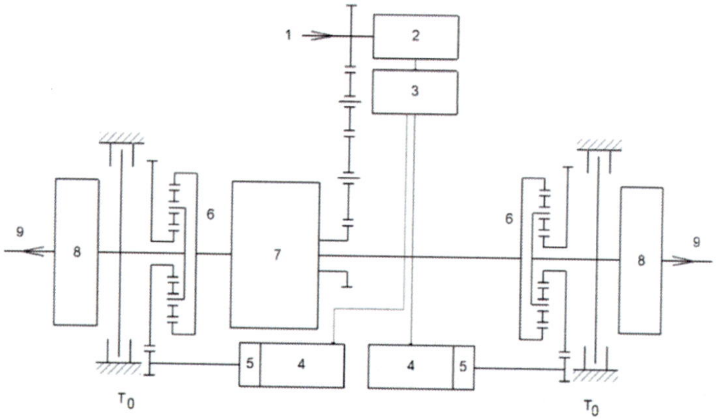


**Fig. 2.** Kinematic diagram of double-flow electromechanical steering mechanism with alterable kinematic parameter 1 – power input from engine; 2 – electric generator; 3 – battery; 4 – electric engine; 5 – reduction gear; 6 – summing planetary gear; 7 – gearbox; 8 – final drive; 9 – power output to sprocket; $T_0$ – brake

The transmission with friction steering devices, however, would be affected by power loses in friction devise. Considered transmission schematics can be found in Fig. 3 and 4.

Considered transmission type allows vehicle to be operated even while one of the engines is not functional. In such case engines' shafts are stopped and turn is performed with use of brakes ($q_m = 0{,}5$).

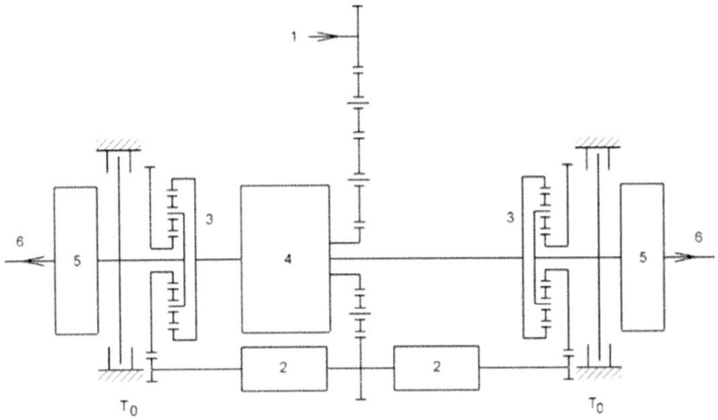


**Fig. 3.** Kinematic diagram of double-flow friction steering mechanism with alterable kinematic parameter: 1 – power input from engine; 2 – friction steering device; 3 – summing planetary gear; 4 – gearbox; 5 – final drive; 6 – power output to sprocket; $T_0$ – brake

Summing planetary gear 6 kinematics could be expressed via known formula [7]:

$$\omega_{I\,1,2} = k\omega_{II\,1,2} + (1-k)\omega_{C\,1,2} \tag{10}$$

The indexes for angular velocity ω in expression above means: *I* is for sun gear, *II* is for ring gear, C – is for carrier; indexes 1 and 2 is for slower and faster sprockets respectively *k* is for mechanisms' kinematic factor. To find value $u_0$ let us consider that ring gears from both sides are motionless. In this case gear ratio between sprockets are equivalent to gear ratio between each summing planetary gear sun gears:

$$u_0 = \omega_{I\,2}/\omega_{I\,1} \tag{11}$$

The relative speed between sprockets as a function of the controlled quantity can be expressed as:

$$u_0 = \frac{\omega_{I\,2}}{\omega_{I\,1}} = \frac{k\omega_{II\,2} + (1-k)\omega_{C\,2}}{k\omega_{II\,1} + (1-k)\omega_{C\,1}} = \frac{k\omega_{Gb} - (1-k)\omega_{E\,2}}{k\omega_{Gb} - (1-k)\omega_{E\,1}} \tag{12}$$

Here indexes *I, II, 1, 2* used as mentioned above, $\omega_{Gb}$ – angular velocity of the output shaft of the gearbox, $\omega_{E\,1}$ and $\omega_{E\,2}$ – angular velocities of each electric engine (friction steering device) shafts.

This relation could be altered by changing angular velocity of each engine shafts. In accordance with Eq. (7) the kinematic factor of the transmission can be expressed as a function of relative speed between sprockets. Therefore, steering mechanism kinematic parameter could be controlled by driver or steering control system by altering relative

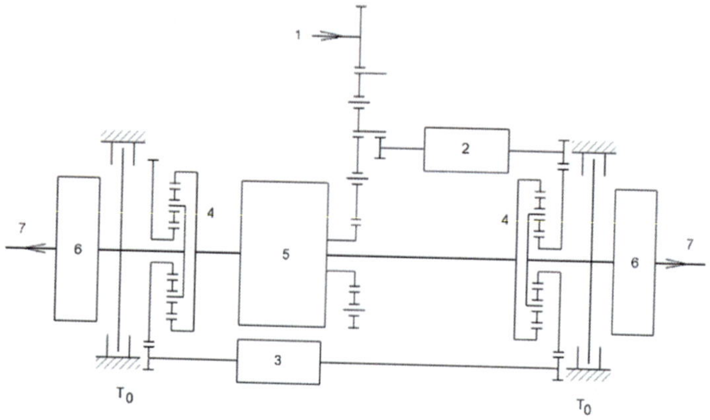


**Fig. 4.** Kinematic diagram of double-flow friction steering mechanism with alterable kinematic parameter: 1 – power input from engine; 2 – friction steering device 1; 3 – friction steering device 2; 4 – summing planetary gear; 5 – gearbox; 6 – final drive; 7 – power output to sprocket; $T_0$ – brake

speed between sprockets which is achieved by setting different velocities of electric engine or friction steering device output shafts.

It is suggested, that transmission which utilizes friction steering devices has significant disadvantage in comparison with electromechanical transmission. Due to friction nature of steering control some power transmitted from engine would be dissipated. The electromechanical transmission does not have this kind of power losses and therefore it performs better in terms of energy conversion efficiency.

The described steering control principle could be used in "hyperbolic" steering mechanisms (see [22–25]). This type of transmission is shown on Fig. 5.

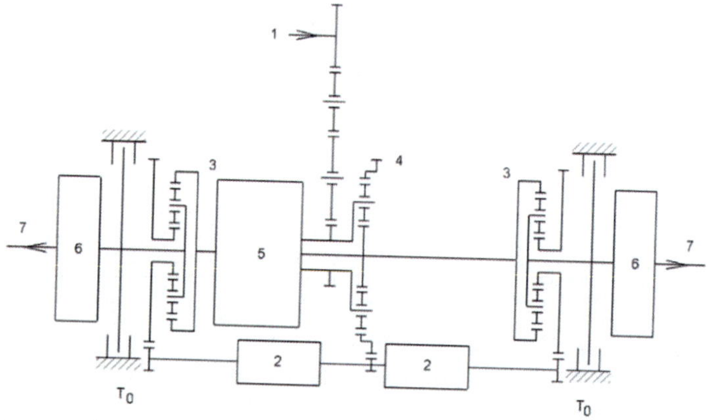


**Fig. 5.** Kinematic diagram of "hyperbolic" steering mechanism with alterable kinematic parameter: 1 – power input from engine; 2 – friction steering device; 3 – summing planetary gear; 4 – differential; 5 – gearbox; 6 – final drive; 7 – power output to sprocket; $T_0$ – brake

Drawings shown on Fig. 2, 3, 4 and 5 could be altered by changing planetary gear structure, for example rerouting power input from mechanical system to sun gear and power input from electric engines to ring gear.

Considered steering control system could also be modified to enable its use in tracks with two sprockets. In this case power will be summed by track itself instead of epicyclic gear [26].

The complete quantitative analysis of transmissions described in the paper and its comparison to existing steering devices in terms of performance is a field of future research. Some information regarding this topic in general can be found in [4, 27].

## 3 Conclusions

The ability to control an alter kinematic factor of tracked vehicle steering mechanism allows to decrease power loses in transmission during vehicle turn. In ideal case summary power loses are equivalent to those of non-symmetrical differential steering mechanism.

The article provides schematic diagram of double-flow electromechanical transmission with ability to control and alter kinematic parameter value.

The article provides schematic diagram of double-flow transmission, utilizing friction steering devices and which provides ability to control and alter kinematic parameter value.

## References

1. Zaichik, G.I., Kreines, M.A., Kristi, M.K.: Parametri, kharakterizuiushie mehanizm povorota gusenichnogo samohoda [Parametres characterizing tracked vehicle steering mechanism]. Vestnik tankovoi promishlennosti [Tank Ind. Bull.] (4), 8–14 (1944). (in Russian)
2. Zabavnikov, N.A.: Osnovy teorii transportnykh gusenichnykh mashin [Fundamentals of the Theory of Transport Tracked Vehicles]. Mashinostroenie Publishers, Moscow (1975). (in Russian)
3. Shelomov, V.B.: Teoriya dvizheniya mnogocelevih gusenichnih I kolesnih mashin. Tyagoviy raschet krivolineynogo dvizheniya [Theory of motion of a multi-purpose tracked and wheeled vehicles]. Publishing House of Polytechnic University, St. Petersburg (2013)
4. Dobretsov, R.: Taking into account the energy parameters of steering devices in the comprehensive assessment of power losses in the chassis of tracked vehicles. St. Petersburg State Polytech. Univ. J. **1**(117), 122–128 (2011). (in Russian)
5. Dobretsov, R.: Objective assessment of technical characteristics of the transport caterpillar vehicles' chassis. Tractors Agric. Mach. **2**, 19–23 (2011). (in Russian)
6. Dobretsov, R.Yu.: Povorot bistrokhodnoi gusenichnoi mashini: opredelenie parametrov energoeffectivnosti chassi [fast-moving tracked vehicle turning: assessment of chassis energy efficiency parametres]. In: XLIII Aktualnie problem zashiti i bezopasnosti [Proceedings of the Actual Problems of Protection and Security], vol. 3. pp. 103–111. FGBU "Rossiyskaya academiya raketnih i artilleriyskih nauk" Publishers, Moscow (2015)
7. Nosov, N.A., et al.: Raschet i konstruirovanie gusenichnih mashin [Calculation and Design of Tracked Vehicles]. Mashinostroenie Publishers, Leningrad (1972)
8. Anisimov, G.M., Kochnev, A.M.: Osnovnie napravleniya povisheniya ekspluatacionnoy effectivnosti gusenichnih trelyovochnih traktorov [Main ways of increasing operational efficiency of tracked skidders]. Publishing House of Polytechnic University, St. Petersburg (2007)

9. Dobretsov, R.Yu., et al.: Vibor schemnogo varianta postroeniya transmissii voennih mashin s gibridnoi silovoi ustanovkoi [Selection of schematic variant of transmission design for military vehicles with hybrid powertrain]. In: Razrabotka I ispolzovanie electricheskih transmissii dlya obrazcov vooruzheniya I voennoi tehniki [Development and Use of Electric Transmissions for Military Vehicles], pp. 87–100. OAO VNIITransmash Publishers, St. Petersburg (2016)
10. Dobretsov, R.Yu., Porshnev, G.P., Uvakina, D.V.: Performance improvement of Arctic tracked vehicles. In: MATEC Web Conferences. 2018 International Scientific Conference on Energy, Environmental and Construction Engineering (EECE-2018), vol. 245 (2018). https://doi.org/10.1051/matecconf/201824517001
11. Dobretsov, R., Grigorev, I., Gazizov, A.: Ways to improve handling of forest and transport tracks. Vestnik BSAU **3**(43), 97–106 (2017). (in Russian)
12. Dobretsov, R., Grigorev, I., Ivanov, V.: Mobility increase in all-terrain tracked vehicles for logging camp-type operations. Syst. Methods Technol. **2**(30), 114–119 (2016). (in Russian)
13. Demidov, N.N.: Mehanizm peredachi i povorota dlya gusenichnoi mashini s dvumya dvigatelyami [Steering mechanism for tracked vehicle with two engines]. In: XXI Aktualnie problem zashiti i bezopasnosti [Proceedings of the Actual Problems of Protection and Security], vol. 3, pp. 61–69. FGBU "Rossiyskaya academ-iya raketnih i artilleriyskih nauk" Publishers, Moscow (2018)
14. Shelomov, V.B., Dobretsov, R.Yu.: Moshnosti dvigatelya I buksovaniya frikcionnogo elementa upravleniya povorotom gusenichnoi mashini [Engine and skidding powers of tracked vehicle's friction steering control device]. Nauchno-tekhnicheskie vedomosti SPbGPU. Seriya Nauka i obrazovanie **2**, 87–91 (2014). (in Russian)
15. Didikov, R.A., et al.: Transmission of the perspective wheel tractor with automatic gearbox: management of the power distribution mechanism. Int. Rev. Mech. Eng. **12**(9), 790–796 (2018). https://doi.org/10.15866/ireme.v12i9.15646
16. Didikov, R.A., Dobretsov, R.Yu.: By question about choice of kinematic schemes for the gear type power distribution mechanisms. Avtomobilnaya promishlennost [Autom. Ind.] **9**, 12–14 (2014). (in Russian)
17. Dobretsov, R.Yu., Semenov, A.G., Didikov, R.A.: Mehanizm raspredeleniya moshnosti v transmissii avtomobilya [Power distribution mechanism of vehicle transmission system]. Patent RU, no. 2634062 (2017)
18. Teoriya i konstruktsiya tanka. T.8. Parametry vneshney sredy, ispolzuyemyye pri raschete tankov [Theory and Design of the Tank. Volume 8. Parameters of the Environment Used in the Calculation of Tanks]. Mashinostroyeniye Publishers, Moscow (1987)
19. Galishev, Yu.V., Dobretsov, R.Yu., Porshnev, G.P., Hoodorozhkov, S.I.: Research and development of SPBSTU scientistsin defense technique based on the IX international exhibition of arms, military equipmentand ammunition. Nauchno-tekhnicheskie vedomosti SPbGPU. Seriya Nauka i obrazovanie **1**, 26–32 (2014). (in Russian)
20. Galyshev, Y.V., Grigoriev, A.P., Dobretsov, R.Yu., Lozin, A.V.: Closed-loop control system for tracked vehicle steering. Nauchno-tekhnicheskie vedomosti SPbGPU. Seriya Nauka i obrazovanie **3**(202), 201–208 (2014). (in Russian)
21. Demidov, N.N., Dobretsov, R., Medvedev, M.S.: Frictional steering devices in two-flow transmissions of transport caterpillar vehicles. Tractors Agric. Mach. **1**, 60–69 (2019). (in Russian)
22. Demidov, N.N., Dobretsov, R.Yu., Lozin, A.V.: Opredelenie parametrov giperbolicheskogo mehanizma povorota [Definition of parametres of hyperbolic steering mechanism]. In: XX Aktualnie problem zashiti i bezopasnosti [Proceedings of the Actual Problems of Protection and Security], vol. 3, pp 42–48. FGBU "Rossiyskaya academ-iya raketnih i artilleriyskih nauk" Publishers, Moscow (2017)

23. Dobretsov, R.Y., Lozin, A.V., Medvedev, M.S.: Hyperbolic steering for tracked vehicles. In: Radionov, A.A., Kravchenko, O.A., Guzeev, V.I., Rozhdestvenskiy, Y.V. (eds.) ICIE 2018. LNME, pp. 2367–2374. Springer, Cham (2019). https://doi.org/10.1007/978-3-319-95630-5_255
24. Filippov, A.N., Lozin, A.V., Demidov, N.N., Dobretsov, R.Yu.: Tracked vehicles: the nonlinear steering mechanism. In: MMESE-2016, pp. 898–912. Publishing House of Polytechnic University, St. Petersburg (2016)
25. Dobretsov, R.Yu., Lozin, A.V., Semenov, A.G., Shelomov, V.B.: Dvuhpotochnaya transmissiya gusenichnoy mashini s bortovim sposobom povorota [Double-flow transmission of transport vehicle with on-board turning method]. Patent RU, no. 2599855 (2016)
26. Dobretsov, R.Yu., Lozin, A.V.: "Diagonalniy" princip upravleniya povorotom gusenichnoy mashini ["Diagonal" principle of tracked vehicle steering control]. In: Povishenie energoeffectivnosti lesnogo kompleksa [5th Improvement of Energy Efficiency of Forest Complex], pp. 33–34. PetrGU Publishers, Petrozavodsk (2019).
27. Dobretsov, R.Yu.: Ob uchete parametrov mechanizma povorota pri otsenke energoeffectivnosti chassi gusenichnih mashin [Steering mechanism parameters consideration in tracked vehicles' chassis energy efficiency analysis]. In: modernizatsiya I nauchnie issledovaniya v transportnom komplekse [Modernization and Scientific Research in Transport Complex], pp. 51–54. Publishing House of Perm National Research Polytechnic University, Perm (2015)

# On the Power-Weighted Efficiency of Multimode Powertrains: A Case Study on a Two-Mode Hybrid System

Antonio Tota(✉), Enrico Galvagno, and Mauro Velardocchia

Politecnico di Torino, Turin, Italy
antonio.tota@polito.it

**Abstract.** Multimode powertrains represent one of the most versatile solutions for hybrid electric vehicles where multiple power sources are integrated with aim of improving fuel economy and reducing pollutants emission in every operating condition. Some hybrid powertrain designs feature multiple planetary gear sets whose components can be directly driven by the powertrain actuators (electric motor or thermal engine) or can be connected through clutches and brakes. The advantages due to the availability of multiple modes are mitigated by the increase of production costs and complexity because of the higher number of components required if compared with the single mode solutions. A numerical methodology is adapted from the literature to analyze, categorize, and compare each distinct working configuration. The energy consumption of each powertrain configuration is then evaluated through the power-weighted efficiency concept whose formulation normalize the contribution from each power source. This paper aims at extending the methodology to investigate the operating range for each powertrain configuration to always achieve the maximum efficiency. The methodology is then applied to the realistic case study of the EVT 2-Mode Hybrid System.

**Keywords:** SDG11 · Multimode powertrains · Hybrid vehicles · Planetary gears · EVT 2-mode hybrid system

## 1 Introduction

The market of passenger cars is going through a challenging period where fuel economy and emission reduction represent the main driving factors for the definition of product portfolio by the car makers to satisfy the sustainable development goals set by the United Nations [1]. Hybrid powertrains are certainly among the most supported solutions adopted by car manufactures to minimize the fuel consumption. Indeed, the powertrain efficiency can be improved by introducing one or more electric motors/generators, in addition to the conventional internal combustion engine. The power flow through the powertrain components is defined according to three basic architectures: series, parallel, and series–parallel (power-split) [2, 3]. Within the last category, the Electrically Variable Transmission (EVT) or Electric Continuously Variable Transmission (E-CVT) stands out for their performance and operating efficiency, also favored by a more efficient use of the


G. Quaglia et al. (Eds.): I4SDG 2021, MMS 108, pp. 522–531, 2022.
https://doi.org/10.1007/978-3-030-87383-7_56

engine, as described by [4, 5]. A multimode hybrid powertrain enables different working modes, e.g. series, parallel, input split, output split, compound split, according to the state of clutches and brakes. This solution ensures grater adaptability to different vehicle operations and fully realizes the powertrain potential to achieve both better fuel economy and improved drivability, as compared to conventional vehicles. An automated procedure to analyze and categorize the whole set of the multimode powertrains is proposed by [6, 7] meanwhile a comparative study of multimode powertrains with different numbers of planetary gear sets (PGS) is carried out by [8] where it is demonstrated that three or more PGS does not show a significant fuel economy improvement for passenger cars with respect to the two PGS configurations. Once the powertrain architecture is defined and all the feasible modes are identified, a methodology to choose the correct mode to achieve the best possible fuel economy for each operating condition is required. Among the most studied and implemented solutions there are the equivalent consumption minimization strategy (ECMS) [9, 10], the Pontryagin's minimum principle (PMP) [11, 12] and the dynamic programming (DP) [13] which are all optimization-based algorithms to satisfy a prescribed cost function or performance index. The DP and the PMP methods are the most reliable solutions, but the DP requires a high computational power meanwhile the PMP may suffer from numerical convergence problems. The ECMS represents an instantaneous suboptimal solution for the fuel economy optimization problem, but it needs a specific tuning for the equivalence factor calculation. To overcome these limits, a suboptimal energy management strategy, named power-weighted efficiency analysis for rapid sizing (PEARS) is proposed by [6], whose results are similar to the DP optimal solution but much more convenient in terms of computational burden. In this paper, the PEARS method is applied to a two-mode hybrid system developed by General Motors [14], whose dynamics is also analyzed in [15] for different driving conditions. The paper is organized as follows: Sect. 2 describes the automated procedure to build and classify the dynamics model of each powertrain mode; the description of the power-weighted efficiency algorithm is carried out in Sect. 3; the results for the two-mode hybrid system are shown in Sect. 4 meanwhile the conclusions are drawn in Sect. 5.

## 2 Multimode Powertrain Model

An example of a multimode hybrid powertrain featuring two planetary gear sets, an internal combustion engine (ICE), two electric motors/generators (MGs), a clutch and a brake, is shown in Fig. 1. Given a transmission architecture with $n$ PGS, the total number of nodes (a node is a ring, a solar or a carrier) is equal to $3n$. Each PGS provides two degrees-of-freedom (dofs) meanwhile each clutch engagement removes one degree-of-freedom (dof) by constraining the motion between two PGS's elements or by connecting one PGS element to the powertrain chassis (grounding clutch). By considering a number $n$ of PGS and a number $q$ of engaged clutches, each powertrain configuration owns a total number of $2n - q$ degrees-of-freedom. The example reported in Fig. 1 shows a total number of $n = 2$ PGS with 1 rigid connections between the carriers of the two PGS and the presence of 2 clutches. If one clutch is engaged, then the powertrain has two degrees-of-freedom ($q = 2$).

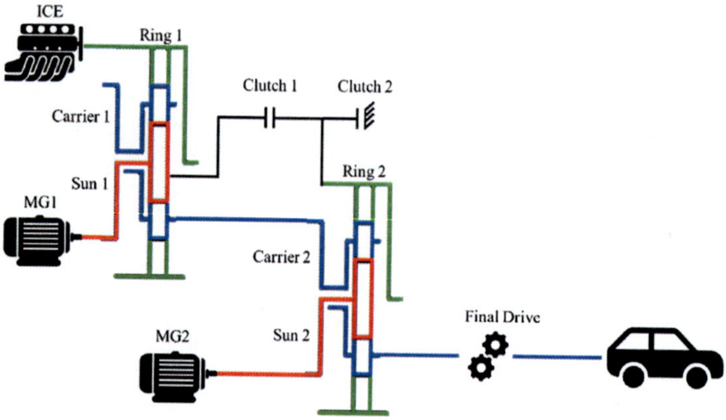


**Fig. 1.** Scheme of the powertrain architecture in the EVT 2-mode Hybrid System.

Note that the system dofs must be within the range of one to the number of the power components (ICE and/or MGs) so that the vehicle is controllable and drivable. The following sections describe the automated modelling procedure introduced by [7] to elaborate the powertrain dynamic equations for each architecture configuration.

## 2.1 Dynamics Model

The powertrain dynamics with no connections among PGS nodes is described by:

$$\boldsymbol{A_0 x_0} = \boldsymbol{u_0} \tag{1}$$

Where $\boldsymbol{x_0} = [\dot{\boldsymbol{\Omega}}_0\ \boldsymbol{F}]^T$ is the generalized coordinate vector, $\dot{\boldsymbol{\Omega}}_0 = [\dot{\omega}_{c,1}\ \dot{\omega}_{s,1}\ \dot{\omega}_{r,1} \ldots \dot{\omega}_{c,n}\ \dot{\omega}_{s,n}\ \dot{\omega}_{r,n}]^T$ is the angular acceleration vector (dimension of $3n$ x 1) which includes the angular acceleration of the carrier ($\dot{\omega}_{c,i}$), the sun ($\dot{\omega}_{s,i}$) and the ring ($\dot{\omega}_{r,i}$), respectively, for the $i_{th} = 1 : n$ PGS. The vector $\boldsymbol{F}$ (dimension of $n$ x 1) contains the internal forces $F_i$ between the ring and the pinions gears teeth of the $i_{th}$ PGS. $\boldsymbol{u_0} = [\boldsymbol{T}\ \boldsymbol{0}]^T$ represents the $4n$ x 1 generalized input vector where the elements of $\boldsymbol{T} = [T_{load}\ T_{eng}\ T_{MG1} \ldots T_{MGk}]^T$ (dimension of $3n \times 1$) are the torques applied to the transmission, e.g. the output transmission torque $T_{load}$ obtained from the vehicle motion resistances, the ICE torque $T_{eng}$, and the MGs torques $T_{MG1}, T_{MG2}, \ldots, T_{MGk}$, where $k$ is the total number of MGs. The first $3n$ equations of Eq. 1 describe the torsional dynamic equilibrium for each PGS node meanwhile the last $n$ equations report the PGS kinematic relation among the ring, the solar and the carrier angular speeds: $\omega_{s,i} r_{s,i} + \omega_{r,i} r_{r,i} = \omega_{c,i}(r_{s,i} + r_{r,i})$, where $r_{r,i}$ and $r_{s,i}$ are the radius of the ring and the solar, respectively. The matrix $\boldsymbol{A_0}$ has dimension of $4n$ x $4n$ and includes all the inertial and geometric parameters of the transmission:

$$\boldsymbol{A_0} = \begin{bmatrix} \boldsymbol{J} & \boldsymbol{B} \\ \boldsymbol{B}^T & \boldsymbol{0} \end{bmatrix} \tag{2}$$

Where the matrix $\boldsymbol{J}$ is a $3n \times 3n$ diagonal matrix whose elements are the total torsional moment of inertia of the powertrain components connected to each PGS node. The matrix $\boldsymbol{B}$ is a $3n \times n$ whose elements of the $i_{th}$ column are $-r_{r,i}$, $-r_{s,i}$ or $r_{r,i} + r_{s,i}$ if the corresponding $\boldsymbol{A_0}$ raw refers to the torsional dynamic equilibrium of the ring, the solar or the carrier, respectively. The remaining elements of the matrix $\boldsymbol{B}$ are set to zero.

To obtain the dynamics of the system after the engagement of $q$ clutches, the transition matrices $\boldsymbol{C}$ and $\boldsymbol{D}$ are introduced:

$$\underbrace{\left(\boldsymbol{C A_0 C^T}\right)}_{\boldsymbol{A}} \underbrace{(\boldsymbol{D x_0})}_{\boldsymbol{x}} = \underbrace{(\boldsymbol{C x_0})}_{\boldsymbol{u}} \tag{3}$$

Matrices $\boldsymbol{C}$ and $\boldsymbol{D}$ are initialized as a $4n \times 4n$ identity matrices $\boldsymbol{I}_{4n}$ whose rows are modified based on the position of the PGS nodes in Eq. 1 connected through the engagement of each clutch (see [6–8] for further details). It is important to remark that this mathematical model does not include the transmission nonlinearities such as the gear backlashes, whose presence may affect the powertrain torsional dynamics in specific conditions [16].

### 2.2 Powertrain Modes Classification

The engagement of $q$ clutches lead to multiple powertrain modes whose dynamic behavior is described by Eq. 3. However, some modes may share identical dynamic equations, thus resulting in redundant powertrain configurations, or some may include infeasible solutions where the vehicle is not powered by any powertrain components. It is relevant to identify and remove both redundant and infeasible powertrain modes.

The first step to eliminate these modes, consists of inverting the matrix $\boldsymbol{A}$, thus relating the torques $\boldsymbol{T}$ to the angular acceleration of each PGS node:

$$\boldsymbol{x} = \boldsymbol{A}^{-1}\boldsymbol{u} \tag{4}$$

However, not every element in $\boldsymbol{x}$ is useful for the powertrain modes analysis and classification, e.g. the dynamics of the free nodes, that are not connected to a power components or to the vehicle, is not of interest. By considering the useful part of vector $\boldsymbol{x}$, named in the following with the vector $\boldsymbol{x}^*$, the inverse dynamic equation is obtained:

$$\boldsymbol{x}^* = \left[\dot{\omega}_{out}\ \dot{\omega}_{eng}\ \dot{\omega}_{MG1}\ \ldots\ \dot{\omega}_{MGk}\right]^T = \boldsymbol{A}^*\boldsymbol{T} \tag{5}$$

The size of $\boldsymbol{x}^*$ and $\boldsymbol{T}$ vectors is $(m+1)$ x 1, where $m$ is the number of the power components. The size of $\boldsymbol{A}^*$ is then $(m+1) \times (m+1)$. By analyzing each row of the resulting matrix $\boldsymbol{A}^*$, it is possible to remove the unfeasible configurations (i.e. a component not connected to the rest of powertrain) and the redundant modes with identical matrix $\boldsymbol{A}^*$ (see [6–8] for further details).

## 3 Power-Weighted Efficiency

The selection of the most suitable powertrain mode, for each vehicle operating condition, relies on the minimum energy consumption criterion. The methodology adopted in this paper to solve the energy loss minimization problem is the power-weighted efficiency analysis for rapid sizing (PEARS) proposed by [7].

The PEARS concept is based on the calculation of the power-weighted efficiency of each powertrain mode as function of the vehicle operating conditions, defined by the load torque $T_{load}$ applied to the transmission output shaft and the desired vehicle speed $V$. The evaluation of the power-weighted efficiency for each powertrain mode requires the calculation of the output power requested to each powertrain component as well as its individual efficiency. Given $T_{load}$ and $V$, the transmission output acceleration is calculated as:

$$\dot{\omega}_{out,ref} = \frac{\tau_F}{MR_W}\left(T_{out}\frac{\tau_F}{R_W} - Mgf_0 - \frac{1}{2}\rho C_x S_F V^2\right) \tag{6}$$

Where $M$ is the vehicle mass, $\tau_F$ is the transmission final drive ratio, $R_W$ is tire rolling radius, g is the gravitational acceleration, $f_0$ is the constant tire rolling resistance coefficient, $\rho$ is the air density, $C_x$ is the aerodynamic drag coefficient and $S_F$ the vehicle frontal area. For each powertrain mode, characterized by a matrix $\boldsymbol{A}^*$, it is possible to search for the torque distribution $\boldsymbol{T}$ such that $\dot{\omega}_{out} = \dot{\omega}_{out,ref}$ through Eq. 6. The output transmission speed is calculated as $\omega_{out} = V\frac{\tau_F}{R_W}$. If the powertrain mode has one dof, the angular speed of each powertrain component is calculated through the PGS kinematic relation constrained by the engagement of $q$ clutches. If the powertrain mode has two or more degrees-of-freedom, one or more angular speeds must be imposed, e.g. the ICE speed. Knowing the vector $\boldsymbol{T}$ and the angular speeds, it is possible to evaluate the output power and the efficiency of each powertrain component. By assuming $k = 2$ MGs, the electric power produced by the generators $P_G$, and the electric power absorbed by the motors $P_M$ is then calculated in Table 1, together with their correspondent efficiencies $\eta_G$ and $\eta_M$.

**Table 1.** Electric power and efficiency for the electric power absorbed and generated

| IF | $P_G$ | $P_M$ | $\eta_G$ | $\eta_M$ |
|---|---|---|---|---|
| $P_{MG1} \leq 0$ <br> $P_{MG2} \leq 0$ | $-P_{MG1}\eta_{MG1} -$ <br> $P_{MG2}\eta_{MG2}$ | 0 | $\frac{P_{MG1}\eta_{MG1}+P_{MG2}\eta_{MG2}}{P_{MG1}+P_{MG2}}$ | 0 |
| $P_{MG1} \leq 0$ <br> $P_{MG2} \geq 0$ | $-P_{MG1}\eta_{MG1}$ | $-\frac{P_{MG2}}{\eta_{MG2}}$ | $\eta_{MG1}$ | $\eta_{MG2}$ |
| $P_{MG1} \geq 0$ <br> $P_{MG2} \leq 0$ | $-P_{MG2}\eta_{MG2}$ | $-\frac{P_{MG1}}{\eta_{MG1}}$ | $\eta_{MG2}$ | $\eta_{MG1}$ |
| $P_{MG1} \geq 0$ <br> $P_{MG2} \geq 0$ | 0 | $-\frac{P_{MG1}}{\eta_{MG1}} -$ <br> $\frac{P_{MG2}}{\eta_{MG2}}$ | 0 | $\frac{P_{MG1}\eta_{MG1}+P_{MG2}\eta_{MG2}}{P_{MG1}+P_{MG2}}$ |

Where $P_{MG1}$, $P_{MG2}$ are the output power of the two MGs and $\eta_{MG1}$, $\eta_{MG2}$ their correspondent efficiencies.

The demanded power at the transmission output can be fed by two energy sources: the battery and the fuel tank. In the generic scheme of Fig. 2, the engine output power $P_e =$

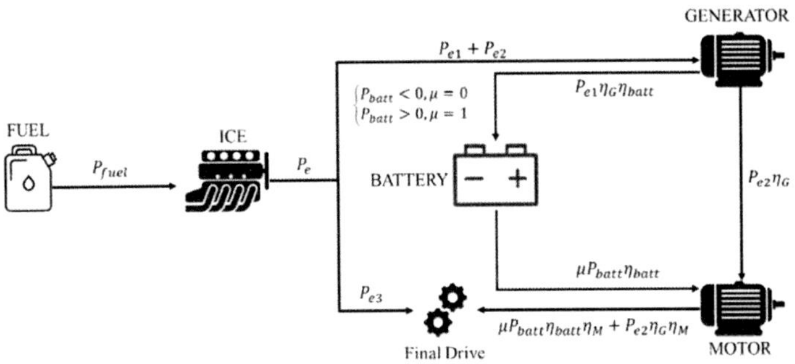


**Fig. 2.** Generic scheme of the power flow for a hybrid powertrain.

$T_e\omega_e$ can be split into three contributions: the engine power $P_{e1}$ that flows to the battery through the generator, the engine power $P_{e2}$ that flows to the electric motor through the generator and the engine power $P_{e3}$ that directly flows to transmission output shaft. $P_{batt}$ and $P_{fuel}$ represent the battery power and the fuel power consumed, respectively. If $P_G > P_M$, the additional power is used to charge the battery ($\mu = 0$): $P_{e3} = P_e - \frac{P_G}{\eta_G}$, $P_{e2}\eta_G = P_M$, $P_{e1}\eta_G = P_G - P_{e2}\eta_G$ and $P_{batt} = -P_{e1}\eta_G\eta_{batt}$. Otherwise, if $P_G < P_M$, the battery energy is consumed ($\mu = 1$): $P_{batt} = (P_M - P_G)/\eta_{batt}$, $P_{e2}\eta_G = P_G$, $P_{e1}\eta_G = 0$, and $P_{e3} = P_e - \frac{P_G}{\eta_G}$. The battery efficiency $\eta_{batt}$ is evaluated by considering a simple circuit model (see [17]):

$$\eta_{batt} = 1 - \frac{R_{batt}I_{batt}^2}{V_{oc}I_{batt}} \tag{7}$$

where $I_{batt}$ is the current through the battery circuit:

$$I_{batt} = \frac{V_{oc} - \sqrt{V_{OC}^2 - 4R_{batt}|P_M - P_G|}}{2R_{batt}} \tag{8}$$

The resistance of the battery circuit $R_{batt}$ and the battery open circuit voltage $V_{oc}$ are considered as constant quantities.

Finally, the power-weighted efficiency $\eta_{HEV}$ is evaluated as:

$$\eta_{HEV} = \frac{\frac{P_{e1}\eta_G\eta_{batt}}{\eta_{e,max}\eta_{G,max}} + \frac{P_{e2}\eta_G\eta_M}{\eta_{e,max}\eta_{G,max}\eta_{M,max}} + \frac{P_{e3}}{\eta_{e,max}} + \frac{\mu P_{batt}\eta_{batt}\eta_M}{\eta_{M,max}}}{P_{fuel} + \mu P_{batt}} \tag{9}$$

where $\eta_{e,max}$, $\eta_{G,max}$, and $\eta_{M,max}$ are the maximum achievable efficiencies for the engine, the generator, and the motor, respectively. This formulation allows to normalize the efficiencies from different type of power sources, without penalizing the engine operation even if $\eta_{e,max}$ is much lower than $\eta_{G,max}$, and $\eta_{M,max}$.

For each combination of $T_{load}$ and $V$, a multiple solution for the torque distribution vector $\boldsymbol{T}$ that satisfy $\dot{\omega}_{out} = \dot{\omega}_{out,ref}$ may exists and two different criteria are defined to discriminate the preferred solution. For the Charge-Sustaining criterion, the objective is to keep constant the battery state of charge, and the torque distribution $\boldsymbol{T}$ that guarantees the minimum battery power $P_{batt}$ is selected (only if $\dot{\omega}_{out,ref} \geq 0$). Otherwise, in case of a Charge-Depleting criterion, the use of electric motors is prioritized over the engine, and a torque distribution $\boldsymbol{T}$ that provides the maximum efficiency $\eta_{HEV}$ is preferred. It is important to remark that the vehicle braking is only entrusted to the electric generators, thus not considering the presence of a conventional hydraulic braking system (see [18] for further details).

## 4 Case Study: EVT 2-Mode Hybrid System

The general methodology described in the previous sections is then applied to the specific case of the EVT 2-mode hybrid system. The powertrain architecture is shown in Fig. 1, where $n = 2$ PGS are connected in series. The engine and the first electric motor MG1 are installed on the ring and the sun of the first PGS, respectively. The second electric motor MG2 and the transmission output shaft are connected to the sun and the carrier of the last PGS, respectively. A rigid connection constrains the carriers of the two PGS and the presence of two clutches enables the activation of the input split (clutch 1 engaged) or the compound split (clutch 2 engaged) hybrid modes. The total number of dofs is two for both powertrain modes, so the engine speed $\omega_{eng}$ is chosen as second degree of freedom together with $\omega_{out}$. The list of vehicle parameters used for the calculation of the $\eta_{HEV}$ for each powertrain mode is also shown in Table 2.

**Table 2.** List of vehicle parameters used for the $\eta_{HEV}$ calculation

| | | | |
|---|---|---|---|
| $M$ | 1600 kg | $\rho$ | 1.2 kg/ m$^3$ |
| $R_W$ | 0.34 m | $\tau_F$ | 3.02 |
| $f_0$ | 0.01 | $r_{r,1}, r_{s,1}, r_{r,2}, r_{s,2}$ | 8, 5, 8, 5 cm |
| $C_x$ | 0.26 | $J_{out}, J_{eng}, J_{MG1}, J_{MG2}$ | 20.28, 0.2, 0.05, 0.05 kg m$^2$ |
| $S_F$ | 2 m$^2$ | $J_{r,1}, J_{c,1}, J_{s,1}, J_{r,2}, J_{c,2}, J_{s,2}$ | 0.08, 0.06, 0.05, 0.08, 0.06, 0.05 kg m$^2$ |

The torque and efficiency characteristics of the engine and the two electric motors, reported in Fig. 3, are obtained from a commercial software usually adopted for modelling and simulating the behavior of hybrid vehicles.

The matrix $\boldsymbol{A}^*$ is calculated for both the input split and the compound split powertrain modes, thus obtaining the dynamic relation between each powertrain component acceleration to the torque distribution among the engine, the two electric motors and the vehicle resistance load. A 2D grid of values for $T_{load}$ and $V$ is selected as representative for the whole vehicle operating range, and the procedure described in the previous section is then applied to evaluate the torque and angular speed of each powertrain component. The results are shown for the compound split mode with the charge-sustaining and the charge-depleting criteria in Fig. 4, where the power flow among the four components is evaluated for each grid point of the vehicle operating range.

The Charge-Sustaining objective in Fig. 4 (left) is achieved by a $P_{batt}$ almost null when $\dot{\omega}_{out,ref} \geq 0$ (above the black line representing the vehicle motion resistance at

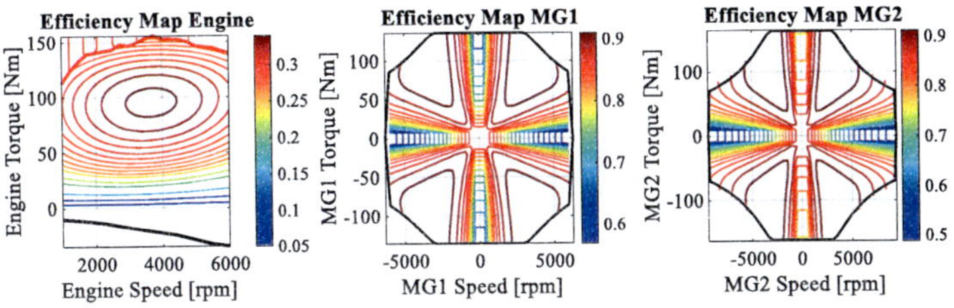


**Fig. 3.** Efficiency maps for the engine (left), the MG1 (center) and the MG2 (right) of the EVT 2-mode hybrid system

constant speed), so that the state of charge of the battery is maintained at a certain level while accelerating. However, this solution does not exploit the potential of the electric power stored in the battery, thus requiring a frequent intervention of the engine with a consequent reduced power-weighted efficiency. On the other hand, the charge-depleting solution in Fig. 4 (right) is much more oriented towards the minimization of the power losses, relying on the electric energy stored in the battery and limiting the engine intervention within its operating range of maximum efficiency. Similar conclusions can be drawn also for the input slip mode.

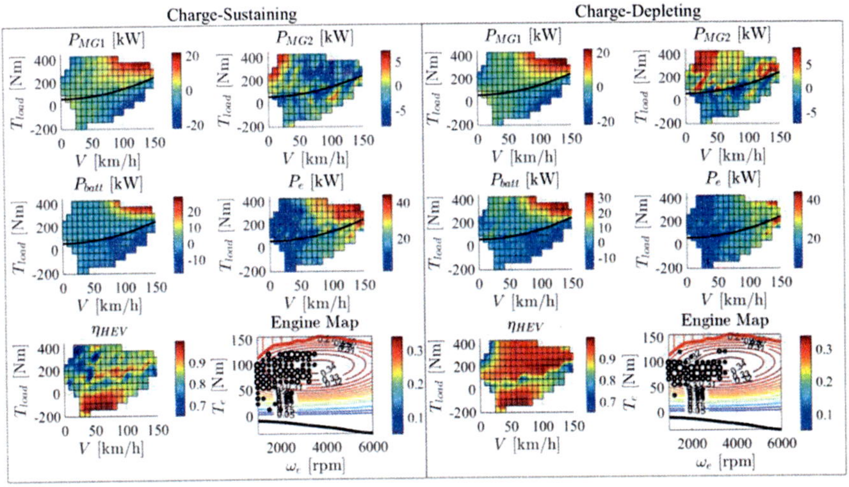


**Fig. 4.** MG1 power, MG2 power, battery power, ICE power, power-weighted efficiency, and map with engine working conditions (black circles) for the compound split mode with the charge-sustaining (left) and charge-depleting (right) torque distribution criteria.

Finally, for each combination of $T_{load}$ and $V$, the selection between the input split and the compound split modes is decided based on the maximum power-weighted efficiency $\eta_{HEV}$ value between the two modes, thus obtaining the resulting mode map in Fig. 5.

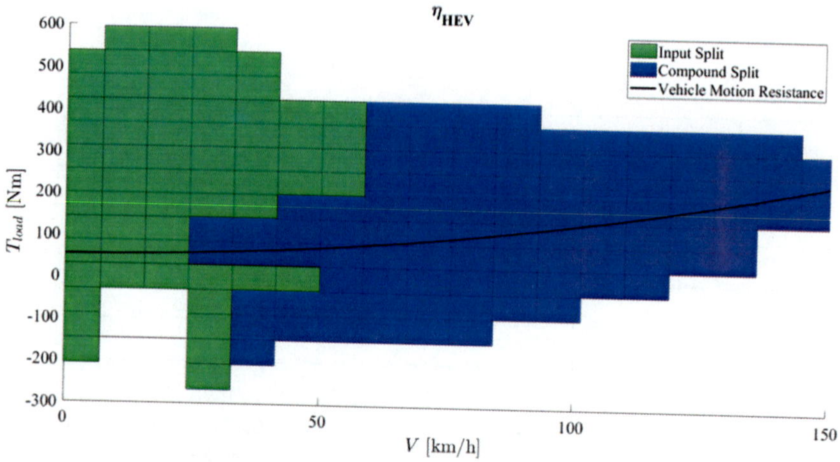


**Fig. 5.** Resulting powertrain mode map for the EVT 2-mode hybrid system.

The input split mode provides better power-weighted efficiency values for high load torque requests at lower vehicle speeds (launching conditions, uphill motion) meanwhile the compound slit mode is more suitable for high vehicle speeds operating points.

## 5 Conclusions

This paper shows the way to select the correct mode of a multimode hybrid powertrain through the principle of the power-weighted efficiency, by providing the case study of the EVT 2-mode hybrid system produced by General Motors. The power-weighted efficiency represents a unique global indicator that incorporates the contribution of each mechanical actuators based on the requested power-flow for each powertrain mode. The torque distribution among the powertrain actuators is evaluated to achieve the desired vehicle dynamics, identified by the transmission output shaft acceleration $\dot{\omega}_{out,ref}$, and to satisfy the Charge-Sustaining (extended battery range) or the Charge-Depleting (maximum power-weighted efficiency) targets. Results demonstrate that the methodology represents a powerful tool to select the more suitable powertrain mode for each power request at the transmission output in terms of loading torque and vehicle speed. In the specific case of EVT 2-mode hybrid system, the input split mode represents the best choice for the vehicle launching conditions (high loading torques and low vehicle speeds) meanwhile the compound split mode is preferred to guarantee the same efficiency level at higher vehicle speeds.

## References

1. Goal 11: Make cities inclusive, safe, resilient and sustainable. https://www.un.org/sustainabledevelopment/cities/

2. Mi, C., Masrur, M.A.: Hybrid Electric Vehicles: Principles and Applications with Practical Perspectives. Wiley, Hoboken (2017)
3. Ehsani, M., Gao, Y., Longo, S., Ebrahimi, K.M.: Modern Electric, Hybrid Electric, and Fuel Cell Vehicles. CRC Press, Boca Raton (2018)
4. Bole, B., Coogan, S., Cubero-Ponce, C., Edwards, D., Melsert, R., Taylor, D.: Energy management control of a hybrid electric vehicle with two-mode electrically variable transmission. In: EVS26 International Battery, Hybrid and Fuel Cell Electric Vehicle Symposium, pp. 6–9, May 2012
5. Galvagno, E., Vigliani, A., Velardocchia, M.: Transient response and frequency domain analysis of an electrically variable transmission. Adv. Mech. Eng. **10**(5), 1–12 (2018)
6. Zhang, X., Peng, H., Sun, J.: A near-optimal power management strategy for rapid component sizing of multimode power split hybrid vehicles. IEEE Trans. Control Syst. Technol. **23**(2), 609–618 (2014)
7. Zhang, X., Eben Li, S., Peng, H., Sun, J.: Efficient exhaustive search of power-split hybrid powertrains with multiple planetary gears and clutches. J. Dyn. Syst. Measur. Control **137**(12), 121006 (2015)
8. Zhuang, W., Zhang, X., Ding, Y., Wang, L., Hu, X.: Comparison of multi-mode hybrid powertrains with multiple planetary gears. Appl. Energy **178**, 624–632 (2016)
9. Guercioni, G.R., Galvagno, E., Tota, A., Vigliani, A.: Adaptive equivalent consumption minimization strategy with rule-based gear selection for the energy management of hybrid electric vehicles equipped with dual clutch transmissions. IEEE Access **8**, 190017–190038 (2020)
10. Sciarretta, A., Back, M., Guzzella, L.: Optimal control of parallel hybrid electric vehicles. IEEE Trans. Control Syst. Technol. **12**(3), 352–363 (2004)
11. Delprat, S., Lauber, J., Guerra, T.M., Rimaux, J.: Control of a parallel hybrid powertrain: optimal control. IEEE Trans. Veh. Technol. **53**(3), 872–881 (2004)
12. Kim, N., Cha, S., Peng, H.: Optimal control of hybrid electric vehicles based on Pontryagin's minimum principle. IEEE Trans. Control Syst. Technol. **19**(5), 1279–1287 (2010)
13. Guercioni, G.R., Vigliani, A., Rizzoni, G.: Dynamic programming solution for the energy management problem in HEVs equipped with DCTs. Int. J. Autom. Technol. (in print)
14. Grewe, T.M., Conlon, B.M., Holmes, A.G.: Defining the general motors 2-mode hybrid transmission (no. 2007-01-0273). SAE technical paper (2007)
15. Tinelli, V., Galvagno, E., Velardocchia, M.: Dynamic analysis and control of a dual mode electrically variable transmission. In: Uhl, T. (ed.) Advances in Mechanism and Machine Science: Proceedings of the 15th IFToMM World Congress on Mechanism and Machine Science, pp. 3731–3740. Springer, Cham (2019). https://doi.org/10.1007/978-3-030-20131-9_368
16. Guercioni, G.R., Galvagno, E., Tota, A., Vigliani, A., Zhao, T.: Driveline backlash and half-shaft torque estimation for electric powertrains control (no. 2018-01-1345). SAE technical paper (2018)
17. Serrao, L., Onori, S., Rizzoni, G.: A comparative analysis of energy management strategies for hybrid electric vehicles. J. Dyn. Syst. Measur. Control **133**(3), 031012 (2011)
18. Galvagno, E., Tota, A., Vigliani, A., Velardocchia, M.: Pressure following strategy for conventional braking control applied to a HIL test bench. SAE Int. J. Passeng. Cars-Mech. Syst. **10**(2017-01-2496), 721–727 (2017)

# Rotordynamics, Multibody Dynamics and Vibrations

# Virtual Turbocharger as a Tool for Reality Simulation

Pavel Novotný(✉), Jiří Vacula, and Petr Kudláček

Brno University of Technology, Brno, Czech Republic
novotny.pa@fme.vutbr.cz

**Abstract.** The mobility of people and the transport of goods requires powertrains. These powertrains often include internal combustion engines with turbocharging technology. The development of turbochargers requires an increase in their efficiency, for example in the form of reduced mechanical losses. In parallel, however, related processes such as vibration, gas blow-by through the sealing system or lubricant consumption still need to be addressed. The virtual turbocharger is a multiphysical tool for the coupled solution of these processes. This physically wide model is described by a system of differential and algebraic equations, assembled in a multibody system, discretized and solved numerically. The solution results are suitable for calibration by experiments on real turbocharger. A reasonably good agreement with experiments is achieved using the example of a truck engine turbocharger. In terms of the physical description of the processes, the development of computational modelling will continue to proceed simultaneously in two different directions. On the one hand by trying to increase the depth of the physical description, on the other hand by including more physical problems within a single virtual prototype.

**Keywords:** SDQ7 · Virtual turbocharger · Turbocharger · Bearing · Gas blow-by · Vibration · Efficiency

## 1 Introduction

The mobility of people and the transport of goods are fundamental needs of modern society. Any approach to achieving these needs must meet the criteria of safety, energy efficiency and long-term sustainability in addition to meeting customer requirements. The powertrain, as a key element of all means of transport, always has a significant impact on the fulfilment of these criteria. A key parameter describing the powertrain is its efficiency of converting input energy into traction work. It should be noted that well-to-wheel efficiency is used to assess the overall impact of the desired mobility.

The paper focuses on powertrains based on the use of internal combustion engines (ICE). In general, the overall efficiency of the ICE that represents the efficiency of the use of chemical energy contained in the supplied fuel is one of the key parameters. High overall efficiency is also achieved through turbocharging technology. Thus, turbochargers allow ICE to be more efficient in preferred operating conditions. Turbochargers make


G. Quaglia et al. (Eds.): I4SDG 2021, MMS 108, pp. 535–544, 2022.
https://doi.org/10.1007/978-3-030-87383-7_57

it possible to achieve better specific fuel consumption and lower pollutant emissions [1], but on the other hand it introduced many issues, such as turbo-lag [2], failures [3–5], noise and vibration [6, 7] and many others.

In general, the total efficiency of the turbocharger (TC) includes three components, one of which is the mechanical efficiency. The mechanical efficiency is thus an important parameter for achieving a high overall turbocharger efficiency and mainly in transient operating conditions.

The mechanical efficiency of a TC includes various energy losses such as frictional losses due to friction in the bearings or in the lubricant flow in the lubrication system. The issue of reducing frictional losses is directly related to lubricant consumption and the generation of particles, which are strictly limited. The particles can origin in TC's lubrication system. In addition to the oil passed through, the sealing system can affect the mechanical parts of the engine and turbocharger, e.g. soot adhering to the surface of the impeller blades increases rotor unbalance and can result in an increased risk of damage to the turbocharger.

Even with decades of powertrain development, especially for ICEs, further improvements in mechanical efficiency are already very difficult. It requires the use of advanced and sophisticated approaches and tools, with a thorough focus on all the component parts and various physical processes. As a direct consequence, there is a pressure for research and development of both the mechanical loss reduction technologies and the research and development of appropriate computational methods to solve these problems.

## 2 Backgrounds to Selected Physical Problems in Turbocharging

### 2.1 Rotor-Bearing System Dynamics

TCs are operated under application-specific operating conditions (OC). For example, passenger car TCs exhibit a very wide range of operating conditions, both design and off-design OCs. In contrast, stationary engine TCs operate at a much more limited range of OCs. Thus, the OCs represent a major influence on the moving rotor and consequently on the bearing lubrication and the gas dynamics in the sealing system.

Regardless of the OCs in which the TC is encountered, the rotor dynamics should be treated as a transient process. This is due to the fact that TC's OCs change very rapidly and are often accompanied by different types of instabilities. These instabilities are generally influenced by the preceding processes (i.e. history) and they may or may not occur under the given OC. In any case, the rotor dynamics then affects the lubrication of the hydrodynamic bearings and the gas dynamics and thus represents an important influence for their solution.

### 2.2 Bearing Lubrication with Impacts on Mechanical Losses

Hydrodynamic journal or thrust bearings are part of the lubrication system and as such transfer the forces using lubricant flow through a suitably shaped lubrication gap. The lubricant is a mixture of lubricating oil and a small proportion of air dissolved in the oil.

Mechanical losses and lubricant flow rates of lubrication system have been the subject of many research studies. Experiments presented, for example, by Deligant et al. [8]

showed losses of journal and thrust bearings under typical OCs. They presented that the thrust bearing makes a significant contribution to friction losses and lubricant flow rates, and this contribution rises with increasing rotor speeds. Journal bearings also contributed to the losses, but the impact of the individual bearing was smaller. The similar observation presented also Hoepke et al. [9] or Novotný et al. [10, 11].

However, the bearings are part of the rotor-bearing system and the rotor dynamics thus strongly influence the lubrication system and the bearings themselves. For example, Berger et al. [12] presented in his study that the thrust bearing can significantly affect the radial motion of the rotor, including the values of the rotor-bearing system natural frequencies, and the more the greater the axial load. It is therefore, in principle, necessary to address the bearing lubrication under transient conditions by computational models of large physical width.

### 2.3 Gas Dynamics in Sealing System with Impacts on Gas Blow-By

The gas flow through the sealing system (called blow-by and blow-back flows) of the TC is one of the parameters that is closely monitored. This process is closely related to the thermodynamic processes in the compressor and turbine. Most of the time, the pressures behind the impellers are higher than in the lubrication system and there is blow-by flow from turbine or compressor to the lubrication system. The blow-by causes a mixing of gases into the lubricant. Under some OCs, the situation arises that the blow-back flow from the lubrication system to the impellers predominates. These OCs are then associated with a significant increase in lubricant consumption as well as an increase in exhaust emissions.

The basic mechanisms for blow-by or blow-back flows are flow through a thin gap between the sealing rings and the rotor or through a seal ring gaps. The dynamics of the gases are also affected by the dynamics of the moving TC's components such as the rotor and seal rings. For example, in the case of large axial loads acting on the rotor, the rotor is displaced axially and thus the gas flow is highly affected.

The correct description of the sealing system is again closely coupled to the dynamics of the moving parts of the TC and it is appropriate to treat the gas dynamics as a problem coupled to the rotor dynamics.

## 3 Virtual Turbocharger as Multi Physical Computational Model

The nature of the processes taking place in a turbocharger is generally transient in terms of rotor and fluid dynamics. Thus, computational modelling to correctly describe these processes must use some level of multiphysical description.

In general, the goal of computational modelling is a perfect simulation of the real world, which in the optimal case cannot be distinguished from reality. To do this, the various human senses can also be simulated through appropriate software and hardware. This virtual world is created by the software in computer and can be a copy of reality. The goal of computational modelling research is also the perfect ability to predict the behaviour of the object under investigation, which allows sensitivity studies, feasibility studies and, eventually, reliable optimisation without the cost of producing and testing many variants under consideration. A multiphysical computational model to achieve the objectives set for TCs is a virtual turbocharger. The virtual turbocharger can thus be seen as a global computational model incorporating the essential physical processes for simulation.

The virtual turbocharger as a computational model of large physical width thus extends the possibilities of computational modelling of TC behaviour under steady-state and transient OCs. The possibility of integrating it into larger units, for example to include the effect of the powertrain or the whole vehicle, is also a significant benefit. This computational model allows coupled solutions of the dynamics of three-dimensional elastic bodies, for example typical turbocharger components such as rotor, stator, seal rings or bearing floating rings. The model also includes the interactions of these bodies through thin fluid layers, i.e., hydrodynamic journal or thrust bearings, or the problem of gas dynamics in a sealing system with forces acting on individual bodies. The virtual turbocharger is therefore a model for the coupled solution of structural dynamics and stability of the rotor, lubrication system issues or vibration and noise of the turbocharger including the influence of higher units (ICE, etc.).

### 3.1 Modelling the Turbocharger Rotor Dynamics

The dynamics of TC rotors and its computational modelling have historically been based on relatively simple linear models or locally linearised models [13]. However, these models do not accurately represent the nonlinear dynamics of a moving rotor. Some progress is presented by Schweizer [14], who used an approach for solving the dynamic stability of a TC rotor based on the use of Multibody software (MBS). This approach is now the most frequently developed, for example [10, 15, 16], and also an important approach for simulations of TC rotor dynamics. However, individual models are developed to solve a specific sub-problem of TC rotors and lack a higher level of generalization needed for coupled solution.

A more precise description of body motion (shaft or turbocharger housing) in MBS requires the assumption of deformable bodies in the form of their elastic properties discretised by Finite Element Method (FEM) and obtained by Craig's reduction [17]. The equation of motion for any three-dimensional (3D) elastic body can be described as

$$\boldsymbol{\Theta}^{\mathrm{T}}\mathbf{M}\boldsymbol{\Theta}\ddot{\mathbf{q}} + \boldsymbol{\Theta}^{\mathrm{T}}\mathbf{K}\boldsymbol{\Theta}\dot{\mathbf{q}} = \boldsymbol{\Theta}^{\mathrm{T}}\mathbf{F}_{\mathrm{f}}, \tag{1}$$

where $\mathbf{q}$ is the vector of modal coordinates, $\boldsymbol{\Theta}$ is the transformation matrix of modal shapes, $\mathbf{M}$ and $\mathbf{K}$ are the mass matrix and stiffness matrix respectively and $\mathbf{F}_{\mathrm{f}}$ is the vector of body loading. The linear deformation vector of a discretized body can be obtained by back-transforming modal coordinates into physical coordinates.

### 3.2 Modelling the Hydrodynamic Bearing Lubrication

The computational model of the bearing is a key part of the virtual turbocharger. The model that will be effective must involve a certain physical depth and width of a hydrodynamic lubrication. In general, the requirements for bearing computational models are quite challenging. Sometimes a high calculation speed is required and a smaller physical depth of the bearing lubrication can be tolerated, sometimes, on the contrary, it is necessary to deeply analyse the lubrication in great detail and unfortunately and longer calculation times must be accepted.

The lubricant flow in a bearing associated with temperature transfer is generally described by the conservation of mass, momentum and energy equations. Using simplifying assumptions, lubricant flow through journal bearing lubrication layer can be modelled using the Reynolds equation for hydrodynamic lubrication. Based on the work of Novotný et al. [10], this equation is discretized over the lubricating layer and solved numerically for various combinations of design and operating parameters. The results of the solution are compiled in the form of force databases for every lubrication layer of a journal bearing in the form of

$$\mathbf{F}_{\mathrm{r}} = \Phi_{\mathrm{r},i}(\varepsilon, \boldsymbol{\Psi}_{\mathrm{r}}, c_{\mathrm{r}}, \omega_{\mathrm{e}}, v_{\mathrm{r}}, \mu_0, T_{\mathrm{in}}), \tag{2}$$

where $\Phi_{\mathrm{r},i}$ is the database that includes results of journal bearing hydrodynamic modes with index $i = 1–3$, $\varepsilon$ represents the relative pin eccentricity, $\boldsymbol{\Psi}_{\mathrm{r}}$ the vector of journal bearing design parameters, $c_{\mathrm{r}}$ the bearing radial clearance, $\omega_{\mathrm{e}}$ the effective angular velocity, $v_{\mathrm{r}}$ the radial pin velocity, $\mu_0$ the lubricant dynamic viscosity at room temperature and $T_{\mathrm{in}}$ the inlet temperature of lubricant. The model [10] allows to solve the transient state processes of hydrodynamic floating ring bearings with cylindrical, unidirectional and bidirectional work surfaces of the lubricating gap.

The modelling of a two-side segmented hydrodynamic thrust bearing is performed in a similar way. Based on the work of Novotný et al. [15], the modified Reynolds equation is discretized over the lubrication layer of the bearing and then numerically solved for various combinations of design and operating parameters. The results of the solution are compiled in the form of force databases for a single work surface of the thrust bearing in the form of

$$\mathbf{F}_{\mathrm{a}} = \Phi_{\mathrm{a},i}(\varepsilon_{\mathrm{a}}, \boldsymbol{\Psi}_{\mathrm{a}}, c_{\mathrm{a}}, \omega, v_{\mathrm{a}}, \mu_0, T_{\mathrm{in}}), \tag{3}$$

where $\Phi_{\mathrm{a},i}$ is the database that includes results of thrust bearing hydrodynamic modes with index $i = 14$, $\varepsilon_{\mathrm{a}}$ represents the relative thrust disc eccentricity, $\Psi_{\mathrm{a}}$ the vector of bearing design parameters, $c_{\mathrm{a}}$ the bearing axial clearance, $\omega$ the rotor angular velocity and $v_{\mathrm{a}}$ the axial velocity of bearing disc relative to the rotor.

The individual lubrication layers of the journal and thrust bearings are described by force databases ($\mathbf{F}_{\mathrm{r}}$, $\mathbf{F}_{\mathrm{a}}$) and these are inserted into the MBS as force elements modelling the bearing force interactions between the rotor and the stator parts.

### 3.3 Modelling the Gas Dynamics Through the Sealing System

A labyrinth model is used to determine the gas flow through the TC sealing system (so-called blow-by), which divides the sealing system into control volumes, as presented by Novotný et al. [18], both on the turbine side and on the compressor side. The individual control volumes are geometrically connected by thin gaps or small orifices (sealing ring gaps). The gas mass flow through the piston ring gap is computed as the isentropic gas flow from the large volume through the small orifice [18]. The mass flow of gas through the narrow gap is determined from the Reynolds equation considering a one-dimensional fully developed flow of an isothermal incompressible fluid. This mass flow can be analytically calculated, for example, by using the relationship for flow through a hydrostatic bearing [19].

The solution of the gas dynamics leads to the solution of the gas pressures in the individual control volumes. Their integration over the surrounding surfaces will generate the loading force on the sealing rings as well as the reactions on the rotor parts. The resulting loads on the rotor and stator components for each pair of interacting surfaces can be written as

$$\mathbf{F}_{\mathrm{s}} = \Upsilon_{\mathrm{s}}\left(\varepsilon_{\mathrm{s}}, \mathbf{\Psi}_{\mathrm{s}}, c_{\mathrm{s}}, \omega, v_{\mathrm{a}}, \mu_{\mathrm{gas}}, p_0, p_1, T_1\right), \tag{4}$$

where $\varepsilon_{\mathrm{s}}$ represents the relative ring eccentricity, $\mathbf{\Psi}_{\mathrm{s}}$ the vector of ring design parameters, $c_{\mathrm{s}}$ the ring axial clearance, $v_{\mathrm{s}}$ the axial velocity of the ring relative to the rotor, $p_0$ the gas inlet pressure near to sealing inlet, $p_1$ the gas pressure in the lubrication system, $\mu_{\mathrm{gas}}$ gas dynamic viscosity and $T_1$ the inlet temperature of gas.

### 3.4 Governing Equations of Motion Describing the Virtual Turbocharger

The virtual turbocharger as a computational model can be described as a system of nonlinear differential and algebraic equations (DAE) with a defined number of degrees of freedom. The geometric and constraint conditions can be written in the form of nonlinear algebraic equations in the general form

$$\mathbf{\Phi}(\mathbf{u}(t), t) = 0 \tag{5}$$

where $\mathbf{u}(t)$ is the dimensional vector of coordinates representing displacements or rotations as a function of time. The dynamic problem consists in solving the equilibrium of a mechanical system using the equations of motion considering known external forces and coupling conditions according to Eq. (5). The equation of motion can be described through the Lagrange equation in the following DAE form describing the mechanical system [20]

$$\mathbf{M}\ddot{\mathbf{u}} + \mathbf{\Phi}_{\mathbf{u}}^{\mathrm{T}}\boldsymbol{\lambda} - \mathbf{F}(t, \mathbf{u}, \dot{\mathbf{u}}) - \mathbf{L}_{\mathbf{q}}^{'} - \dot{\mathbf{M}}\dot{\mathbf{u}} = 0, \tag{6}$$

where $\mathbf{M}$ is the mass matrix, the symbol $L = T - V$ represents the Lagrangian, $T(\mathbf{u}, \dot{\mathbf{u}})$ is the kinetic energy and a $V(\mathbf{u})$ is the potential energy, $\mathbf{\Phi}_{\mathrm{u}}$ is the notation for the Jacobian of the nonlinear constrain equations with respect to the vector of independent coordinates, $\boldsymbol{\lambda}$ is the vector of Lagrange multipliers, and $\mathbf{F}$ is the load vector containing the internal force interactions and the external force load. The internal force interactions include radial bearing ($\mathbf{F}_{\mathrm{r}}$), axial bearing ($\mathbf{F}_{\mathrm{a}}$), or the interaction between the seal rings and the stator ($\mathbf{F}_{\mathrm{s}}$). External loads are represented, for example, by the force loading of the turbine wheel ($\mathbf{F}_{\mathrm{t}}$) or the compressor wheel ($\mathbf{F}_{\mathrm{c}}$). Axial loads of the rotor are computed using the pressures, temperatures and mass flow rates measured upstream and downstream of the turbine and the compressor according to the procedure described in reference [13].

These MBSs are designed for the general solution of multi body dynamics, but are not primarily capable of solving some of the processes occurring in the interaction of the individual components of the TC. The description of these processes, or their models and effective solutions, are currently the key "know-how" in the computational modelling of TC dynamics.

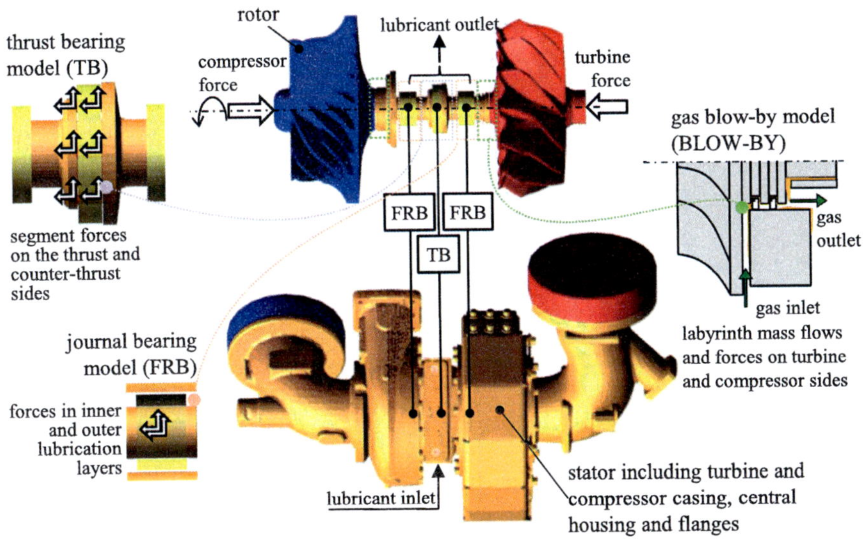


**Fig. 1.** A schematic of a graphical representation of the virtual turbocharger assembled from the computational models of the floating-ring journal bearing (FRB), the thrust bearing (TB) and gas blow-by models (BLOW-BY).

Virtual turbocharger model assembled in commercial software ADAMS is designed to be able to simultaneously solve the problems of the rotor dynamics, the bearing lubrications, the gas dynamics and the ring interactions in the sealing system. The numerical model in ADAMS includes 22 moving parts, two 3D flexible parts, 20 fixed joints, 15 primitive joints, 7 motions, 23 general forces, 140 data elements, 320 design variables, 610 solver variables and results in 35 degrees of freedom. The schematic of the virtual turbocharger is shown in Fig. 1. The solution of Eqs. (4) and (5) in the form of a DAE system is realised by numerical integration and in the case of the ADAMS system by the alfa method proposed by Hilbert et al. [21].

## 4 Examples of Results and Discussion

The virtual turbocharger is designed to allow direct calibration against experimental data from physically existing TCs, as without this calibration the computational modelling is not fully reliable. It is thus a projection of the real world into the virtual world. The quantities chosen for calibration are usually of a discrete or integral nature, such as average outlet oil temperature, wall displacement, velocity or acceleration at a discrete location, total power loss or volume oil or gas flows.

A comparison of the results of computational modelling and technical experiments on a TC of a truck diesel engine is shown in Fig. 2. The comparison shows that although the agreement is not perfect, the basic tendencies are correctly predicted by the virtual turbocharger.

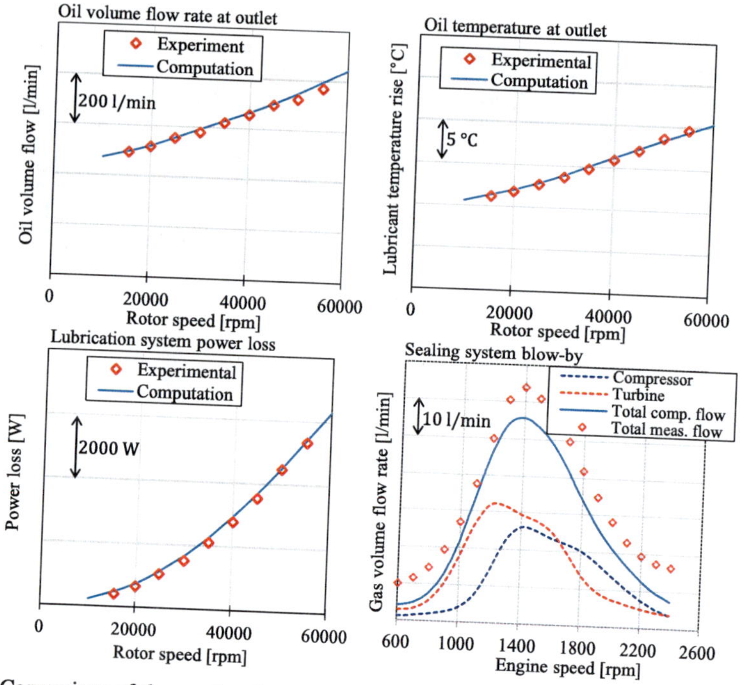


**Fig. 2.** Comparison of the results of virtual turbocharger and technical experiments for the oil volume flow rate through the TC, oil temperature at the outlet of the TC lubrication system, total power loss in the lubrication system and gas volume gas flow rate through the sealing system from the compressor side, turbine side and total blow-by.

## 5 Conclusions

Computational modelling of TCs is naturally subject to continuous development, both in the description of physical processes and in the development of the corresponding computational models and their numerical solutions. However, the practical application of the virtual turbocharger also faces different challenges with implications for its development. Computational modelling has reached in some respects a state where the main problem may not be the physical principle itself or the subsequent numerical solution but is primarily the knowledge of suitable input data. From this perspective, the development of computational models or numerical solutions does not necessarily lead to advances in the practical applications.

Thus, an important task in the development of computational modelling must include facilitating the input of data, parameters or OC, even for users without in-depth theoretical knowledge or extensive practical experience. The transfer of knowledge and experience of computational modelling of a given problem must be encouraged. It should be noted that the research and development of new forms of virtual prototypes will continue to be carried out only by a limited group of highly specialised experts.

From the point of view of the physical description of processes, the development of simulation modelling will proceed simultaneously in two different directions. On the one hand by trying to increase the depth of the physical description, on the other hand by including more physical problems within a single virtual prototype. These two directions are mutually limiting to some extent, but the combination of these two trends is a clear direction for the future development of TC computational modelling. It can be assumed that the physical description of selected processes occurring in TCs involving elastic body dynamics or fluid dynamics will continue to be progressively improved, but it is unlikely that fundamental or revolutionary advances will be made.

**Acknowledgement.** The research leading to these results has received funding from the Specific research program at Brno University of Technology, reg. no. FSI-S-20-6267.

## References

1. Torregrosa, A., Broatch, A., Novella, R., Mónico, L.: Suitability analysis of advanced diesel combustion concepts for emissions and noise control. Energy **36**, 825–838 (2011)
2. Pagot, A., Duparchy, A., Gautrot, X., Leduc, P., Monnier, G.: Combustion approach for downsizing: the IFP concept. Oil Gas Sci. Technol. **61**, 139–153 (2006)
3. Dellis, P., Evaggelos, R., Alcibiades, G., Pesyridis, A.: Turbocharger lubrication -lubricant behavior and factors that cause turbocharger failure. Int. J. Autom. Eng. Technol. **2**, 40–54 (2013)
4. Stodola, P., Stodola, J.: Model of predictive maintenance of machines and equipment. Appl. Sci. **10**, 213 (2020)
5. Smal, T., Furch, J.: Expedient repairs – analysis of possibilities and needs. Adv. Milit. Technol. **6**, 69–82 (2011)
6. Galindo, J., Serrano, J., Climent, H., Tiseira, A.: Experiments and modelling of surge in small centrifugal compressor for automotive engines. Exp. Thermal Fluid Sci. **32**, 818–826 (2008)
7. Andersen, J., Westin, F., Westin, F.: Surge Definitions for radial compressors in automotive turbochargers. SAE Int. J. Engines **1**, 218–231 (2008)
8. Deligant, M., Podevin, P., Descombes, G.: Experimental identification of turbocharger mechanical friction losses. Energy **39**, 388–394 (2012)
9. Hoepke, B., Uhlman, T., Pischinger, S., Lueddecke, B., Filsinger, D.: Analysis of thrust bearing impact on friction losses in automotive turbochargers. J. Eng. Gas Turbines Power **137**, 082507 (2015)
10. Novotný, P., Škara, P., Hliník, J.: The effective computational model of the hydrodynamics journal floating ring bearing for simulations of long transient regimes of turbocharger rotor dynamics. Int. J. Mech. Sci. **148**, 611–619 (2018)

11. Novotný, P., Hrabovský, J.: Efficient computational modelling of low loaded bearings of turbocharger rotors. Int. J. Mech. Sci. **2020**, 105505 (2020)
12. Berger, S., Bonneau, O., Frene, J.: Influence of axial thrust bearing on the dynamic behavior of an elastic shaft: coupling between the axial dynamic behavior and the bending vibrations of a flexible shaft. J. Vib. Acoust. **123**, 145–149 (2001)
13. Nguyen-Schäfer, H.: Rotordynamics of Automotive Turbochargers. Springer, Ludwigsburg (2015). https://doi.org/10.1007/978-3-642-27518-0
14. Schweizer, B.: Total instability of turbocharger rotors—physical explanation of the dynamic failure of rotors with full-floating ring bearings. J. Sound Vib. **328**, 156–190 (2009)
15. Novotný, P., Hrabovský, J., Juračka, J., Klíma, J., Hort, V.: Effective thrust bearing model for simulations of transient rotor dynamics. Int. J. Mech. Sci. **2019**, 374–383 (2019)
16. Dyk, Š, Smolík, L., Hajžman, M.: Effect of various analytical descriptions of hydrodynamic forces on dynamics of turbochargers supported by floating ring bearings. Tribol. Int. **126**, 65–79 (2018)
17. Craig, R.: Structural Dynamics - an Introduction to Computer Methods. John Wiley, Chichester (1981)
18. Novotny, P., Pistek, V., Drapal, L., Svída, D., Devera, T.: Efficient approach for solution of the mechanical losses of the piston ring pack. Proc. Inst. Mech. Eng. Part D J. Autom. Eng. **227**, 1377–1388 (2013)
19. Stachowiak, G., Batchelor, A.: Engineering Tribology. Elsevier, Waltham (2014)
20. de Jalon, J., Bayo, E.: Kinematics and Dynamic Simulations of Multibody Systems the Real-Time Chalange. Springer-Verlag, New York (1994). https://doi.org/10.1007/978-1-4612-2600-0
21. Hilber, H., Hughes, T., Taylor, R.: Improved numerical dissipation for time integration algorithms in structural dynamics. Earthquake Eng. Struct. Dynam. **5**, 283–292 (1977)

# An Overview on Acoustic Metamaterials

Walter Bova[1,2](✉), Eugene Nijman[2], and Domenico Mundo[1]

[1] Department of Mechanical, Energy and Management Engineering, University of Calabria, Cubo 45C, 87036 Rende, Italy
walter.bova@unical.it

[2] Virtual Vehicle Research GmbH, Inffeldgasse 21a, 8010 Graz, Austria

**Abstract.** Acoustic metamaterials are artificially engineered composites composed of mesoscopic subwavelength units, which have recently become an active field with an enormous range of potential applications. These media, which derive their properties from their structure rather than from the chemical properties of their constitutive materials, can be tailored in such a fashion to exhibit arbitrary effective constitutive parameters, ranging from very large positive to negative values. Initially proposed to achieve super absorption, acoustic metamaterials are today investigated to obtain numerous extraordinary effects, which include compact wavefront modulation, cloaking, asymmetric transmission, subdiffraction focusing, etc. In this article, we review some of the most relevant works in this field and analyse the basic phenomena governing the behaviour of different kinds of acoustic metamaterials. We conclude with an outlook on active acoustic metamaterials, in which inclusions can provide energy to the impinging wave to obtain effective material properties that are not possible in passive structures.

**Keywords:** SDG9 · Acoustic metamaterials · Metasurfaces · Low-frequency sound absorption · Active acoustic metamaterials

## 1 Introduction

Noise pollution is an ever-growing problem due to industrialization and urbanization. It impacts everyday millions of people, reducing their quality of life by causing health problems ranging from stress and poor concentration to more serious issues as heart diseases and hearing loss. It is therefore crucial to develop noise reduction techniques to protect citizens from noise-related dangers. In particular, the problem of low frequency noise attenuation in urban areas is critical and still far from being solved. Noise in the low-frequency range dissipates slowly during propagation, making it difficult to block. Commonly used sound absorbing materials are, in fact, only efficient at mid and high frequencies.

Metamaterials (MMs) are artificially engineered composites composed of mesoscopic subwavelength inclusions (typically periodic) placed in a host medium, that exhibit properties that cannot be found in nature. Gaining their properties from their structure rather than from the chemical properties of their constitutive materials, MMs are promising in terms of sound attenuation at low frequencies. Such materials were


G. Quaglia et al. (Eds.): I4SDG 2021, MMS 108, pp. 545–552, 2022.
https://doi.org/10.1007/978-3-030-87383-7_58

originally investigated in electromagnetism by Veselago [1] back in 1968, who formulated a theory for possible materials having negative electric permittivity and magnetic permeability, and hence resulting in a negative refractive index. A similar approach has been used to design acoustic metamaterials (AMMs) [2–35] to control airborne sound.

The equation governing the propagation of sound waves inside a homogenous linear fluid can be written as.

$$\nabla^2 p - \frac{\rho}{K}\frac{\partial^2 p}{\partial t^2} = 0, \tag{1}$$

where $p$ is the acoustic pressure, $\rho$ the mass density and $K$ the bulk modulus. In natural materials, $\rho$ and $K$ can only take positive values, which hinders the design of acoustic absorbers. While in principle the laws of physics would allow extreme wave functionalities like cloaking or subwavelength focusing, there are no natural materials that provide the required acoustic parameters to achieve these effects. Conversely, AMMs can be tailored in such a fashion to exhibit arbitrary effective constitutive parameters, ranging from very large positive to negative values. This is the reason why, over the last 20 years, AMMs have been widely investigated to control sound propagation in unprecedented manners.

All regions of the material parameter space have been studied: negative effective mass density AMMs [2–5], near-zero AMMs [6, 7] in which sound propagates with an extremely large phase velocity, double negative materials [8–13], in which both the acoustic parameters are dispersive and can become negative in the same frequency range.

## 2 Resonant Metamaterials

AMMs are often constructed using resonant subwavelength units. The first locally resonant material with a negative effective mass density was demonstrated by Liu et al. in 2000 [2]. The calculations show that large band gaps can be created at low audible frequencies. A negative modulus AMM was proposed in 2006 by Fang et al. [14]. The material was composed of an array of periodically distributed subwalength-size Helmholtz resonators.

An interesting approach to create AMMs consists in introducing a stop band behavior in the class of periodic lightweight structures, such as honeycomb-core sandwich panels. These structures are employed in transport applications due to the combination of excellent mechanical properties with a low mass. The internal cavities of these periodic panels are an ideal host structure for resonating units. Claeys et al. in [15] added resonant elements in the cavity of a honeycomb panel and demonstrated how a stop band opens up from 1065 Hz up to 1226 Hz. The metamaterial proposed by Tang et al. [16] is another example of this approach. Introducing perforations on both top face sheets and corrugation of a lightweight sandwich plate with honeycomb-corrugation hybrid core, the authors showed that sound absorption is possible over a broad frequency range. The unit cell is shown in Fig. 1a.

In many studies concerning metamaterials, fluid losses are not included in the model. In [17], instead, Lewińska et al. aimed at improving the absorbing capabilities of a poroelastic foam due to thermal and viscous losses by embedding resonating masses in the unit cell at the tip of an elastic micro-cantilever (Fig. 1b). The so-called metafoam performs significantly better at low frequencies (below 450 Hz) than its light or non-resonating equivalent foam.

## 2.1 Resonant Metasurfaces

In natural materials, the possibilities of engineering the wavefront are limited by their acoustical properties and the reflection and refraction of acoustic waves is always governed by Snell's law. However, the research on wavefront modulation has advanced rapidly in the last decades. Novel wave manipulations can be obtained through a new class of metamaterials called acoustic metasurfaces (AMSs). The control of sound through these surfaces is achieved by introducing abrupt phase shifts in the propagation path using periodic structures at subwavelength scales [18]. The Snell's law for AMSs can be written in a generalized way as.

$$\theta_r = \arcsin\,[\sin(\theta_i) + \frac{\lambda_1}{2\pi}\frac{d\Phi}{dx}], \tag{2}$$

where $\theta_r$ is the angle of reflection, $\theta_i$ is the angle of incidence, $\lambda_1$ is the wavelength in the first medium and $\Phi$ is the introduced phase discontinuity. This behavior allows to mold wavefronts into arbitrary shapes with subwavelength resolution. Numerous works have used this principle to manipulate waves in ways unattainable with natural materials.

With unique wave-steering abilities, resonant AMSs open exciting possibilities for designing compact acoustic components. Li et al. presented a passive metascreen stacked up by an array of subwavelength Helmholtz resonators and straight pipes with extreme capabilities of wave-front shaping and transmission [19]. The pipes are used to enhance the transmission owing to Fabry-Perot resonances (Fig. 1c). While phased arrays are normally constructed by a large number of sources leading to high cost and complexity in the electronics to operate the single sources, the authors showed how a complex wave front can be formed with a passive panel made of 80 of the units they designed. Asymmetric acoustic wave propagation can be achieved by coupling a phase array and a near-zero-index medium. The material proposed in [20] behaves like a nearly-reflectionless surface with arbitrary wave-steering capability for incident wave impinging on one side, while virtually blocking the reversed wave. A possible use of such materials is "noise trapping". The incident wave is allowed to pass through the metamaterial and trapped into a chamber with absorbing walls. This concept could be used to improve the absorption of absorbing materials (e.g., porous) thanks to the wave impinging numerous times on them. In [21] the authors designed a membrane-type acoustic metamaterial consisting of decorated membrane resonators. The acoustically reflecting surface can acquire hybrid resonances and becomes impedance-matched to airborne sound at tunable frequencies, such that no reflection is generated. A representation of the unit cell is shown in Fig. 1d.

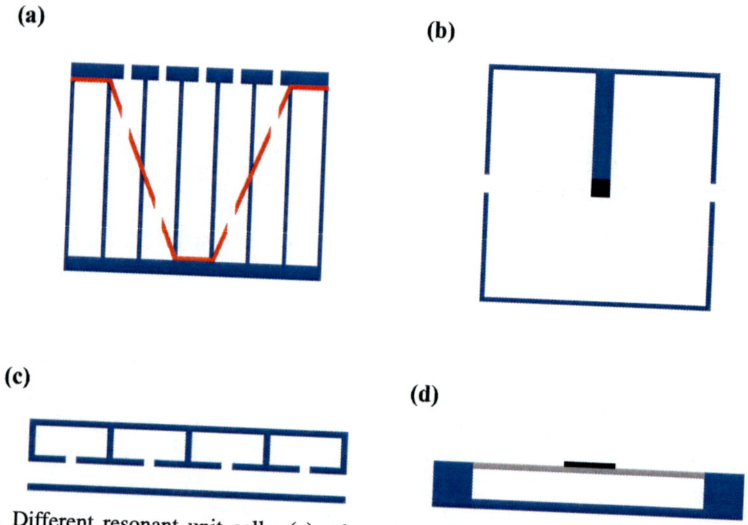


**Fig. 1.** Different resonant unit cells: **(a)** schematic of the perforated honeycomb-corrugation hybrid unit cell [16]; **(b)** representation of the pore with the embedded resonator. the black square represents the heavy mass [17]; **(c)** unit cell made of 4 Helmholtz resonators [19]; **(d)** schematic of a decorated membrane resonator unit cell [21].

# 3 Non-resonant Metamaterials

Resonant AMMs enable the realization of media with extraordinary effective properties. However, they are accompanied with inherent disadvantages. The resonant character of these materials (occurring only in the neighborhood of the resonant frequencies) limits in fact the bandwidth over which the unusual effective properties are available and, therefore, the spectrum of operation of these devices. Extensive research is therefore ongoing to realize AMMs without making use of local resonances. This way it is possible to create units with much lower dispersion, providing larger bandwidth.

Several alternative approaches have been studied in order to realize unusual acoustic refraction [22–28]. The authors of [22] used labyrinthine structures like the one depicted in Fig. 2a. Thanks to the the space-coiling structure the propagating path length is significantly increased. As a result, the phase can be arbitrarily delayed over a range of $2\pi$, which enables full control of the phase shift. Band folding with negative dispersion ($\rho < 0$ and $K < 0$) are obtained. This approach has the advantage of creating negative refraction with a relatively simple metamaterial structure.

## 3.1 Non-resonant AMSs

Dispersionless metasurfaces [25, 26] can be designed by decorating the surface with grooves of subwavelength width (Fig. 2b). The wave is delayed when propagating along a round trip in each groove and a local phase shift is added to the reflected wave. In [25] the authors have demonstrated experimentally and numerically phenomenon of extraordinary acoustic reflection within an ultra-broad band. The same mechanism is employed in [26] by Zhu et al. to create a tunnel completely open for substances like fluids or other energy fluxes to exchange while allowing sound to pass only in one direction. This particular design may pave the way to more versatile acoustic one-way devices with potential applications in duct acoustics.

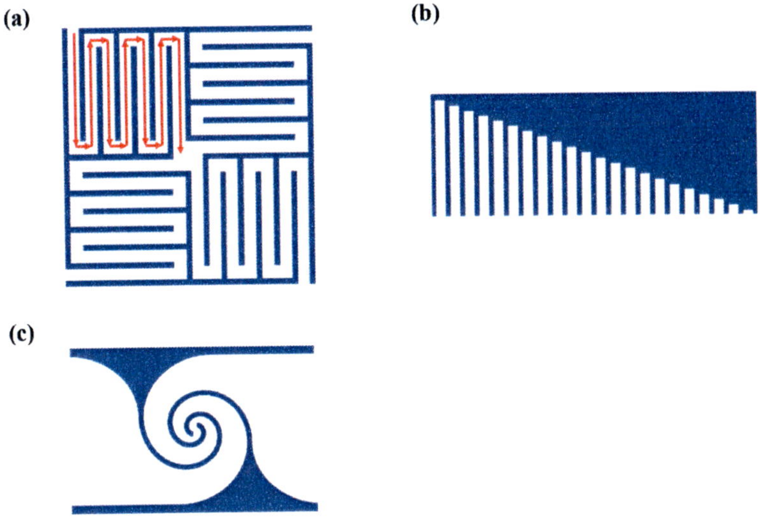


**Fig. 2.** Different non-resonant unit cells: **(a)** schematic of the labyrinthine unit cell [22]; **(b)** schematic of the subwavelength corrugated surface unit cell [25, 26]; **(c)** schematic of the spiral-shaped unit cell [27].

Contrarily to reflective cases, transmissive cases need to fulfil the requirement of impedance matching between the AMS and its surrounding material (usually, air). The impedance mismatch between natural solids and air is very large, which significantly reduces the transmission efficiency of an acoustic device. Therefore, to enhance the performance of a AMS designed to transmit or absorb sound, its units must be tailored to maximize the impedance matching with air. Various works have followed this direction [27, 28], such as the tapered labyrinthine metamaterials proposed in [27] by Jia et al. A two-way spiral-shaped metamaterial is designed in the study, which is capable to achieve a sound energy enhancement higher than 10.75 dB in the studied frequency range of 4.8–5.8 kHz. Due to the impedance matching design, the structure shows almost opposite transmission behaviours in orthogonal directions. A schematic of the unit cell of this AMM is depicted in Fig. 2c. Five different kinds of labyrinthine or space-coiling acoustic metamaterials with tapered channels and openings are presented in [28].

## 4 Active AMMs

All the different types of AMMs presented above are passive. While being simple to realize, they impose certain restrictions on the constitutive effective parameters. For example, resonant AMMs show a very limited bandwidth over which the exotic properties are obtained, which explains the success of using non-resonant AMMs. Furthermore, the efficiency of passive acoustic metamaterials is also hindered by the presence of losses. Again, this is especially true in the case of resonant inclusions. For these reasons there has been recently an increasing interest in active AMMs, which could enhance the practical applicability of AMMs.

The most common elements used in active AMMs are active transducers, micro or nano-electromechanical systems, piezoelectric materials and electrically loaded acoustic elements. Bergamini et al. [28] did a first implementation of a metamaterial with variable mechanical connectivity, in which effective properties can be tuned in real time with properly controlled piezoelectric discs. Piezoelectric elements are ideal to tune and control the acoustic properties of a metamaterial, because they respond strongly to electrical signals and can be controlled with simple electronics [29–32]. In [32] Akl et al. investigated a coupled fluid-structure one-dimensional metamaterial cell, in which a piezoelectric active element is introduced. The active material is manufactured to achieve controllable dynamic density. Systems based on the electrical control of resonant frequencies in membrane type metamaterials have been proposed and prototyped. Xiao et al. [33] described and demonstrated an active acoustic metamaterial, in which the effective material parameters can be tuned independently over a wide range of values, including negative material parameters. Lee et al. [34] designed an array of side tubes resonators connected to the wall of a main duct that can be tuned with a mechanical method by plungers.

Besides reconfiguration, active AMMs are also exploited to achieve more exotic behaviours. Sensing an incoming signal is typically associated with absorbing a portion of its energy, inherently perturbing the measurement. Fleury et al. [35] proposed a non-invasive invisible sensor for airborne sound waves, which fully absorbs the impinging signal, without perturbing its own measurement.

## 5 Conclusions

The field of acoustic metamaterials has rapidly grown over the past 20 years. We now know how to design structures with extraordinary properties that are able to manipulate acoustic waves in ways that would have sounded absurd not long ago. Despite this tremendous progress, many challenges remain to be solved, mainly in terms of bandwidth and losses, before these materials can be employed in real world applications. Active acoustic metamaterials are arising with the ability to tune the effective acoustic properties to overcome some of these limitations. Furthermore, new ideas in the field of acoustic metamaterials will certainly emerge in the near future, with more powerful abilities to manipulate sound waves in various ways. Thus, we conclude that the future of acoustic metamaterials is promising and bright.

**Acknowledgements.** The publication was written at Virtual Vehicle Research GmbH in Graz, Austria. The authors would like to acknowledge the financial support within the COMET K2 Competence Centers for Excellent Technologies from the Austrian Federal Ministry for Climate Action (BMK), the Austrian Federal Ministry for Digital and Economic Affairs (BMDW), the

Province of Styria (Dept. 12) and the Styrian Business Promotion Agency (SFG). The Austrian Research Promotion Agency (FFG) has been authorised for the programme management.

## References

1. Veselago, V.G.: The electrodynamics of substances with simultaneously negative values of ε and μ. Sov. Phys. Uspekhi **10**, 509–514 (1968)
2. Liu, Z., et al.: Locally resonant sonic materials. Science **289**, 1734–1736 (2000)
3. Liu, Z., Chan, C.T., Sheng, P.: Analytic model of phononic crystals with local resonances. Phys. Rev. B **71**, 014103 (2005)
4. Sheng, P., Zhang, X.X., Liu, Z., Chan, C.T.: Locally resonant sonic materials. Phys. B **338**, 201–205 (2003)
5. Mei, J., Liu, Z., Wen, W., Sheng, P.: Effective mass density of fluid-solid composites. Phys. Rev. Lett. **96**, 24301 (2006)
6. Alu, A., Silveirinha, M.G., Salandrino, A., Engheta, N.: Epsilon-near-zero metamaterials and electromagnetic sources: tailoring the radiation phase pattern. Phys. Rev. B **75**(15), 155410 (2007).
7. Wu, Y., Li, J.: Total reflection and cloaking by zero index metamaterials loaded with rectangular dielectric defects, Appl. Phys. Lett. **102**(18), 183105(2013)
8. Ding, Y., Liu, Z., Qiu, C., Shi, J.: Metamaterial with simultaneously negative bulk modulus and mass density. Phys. Rev. Lett.**99**(9), 093904 (2007)
9. Fok, L., Zhang, X.: Negative acoustic index metamaterial. Phys. Rev. B **83**(21), 214304 (2011)
10. Cheng, Y., Xu, J.Y, Liu, X.J.: One-dimensional structured ultrasonic metamaterials with simultaneously negative dynamic density and modulus, Phys. Rev. B **77**, 045134 (2008)
11. Bongard, F., Lissek, H., Mosig, J.R.: Acoustic transmission line metamaterial with negative/zero/positive refractive index, Phys. Rev. B **82**(9), 094306 (2010)
12. Li, J., Chan, C.T.: Double-negative acoustic metamaterial. Phys. Rev. E **70**, 055602 (2004)
13. Gusev, V.E., Wright, O.B.: Double-negative flexural acoustic metamaterial, New J. Phys. **16**(12), 123053 (2014)
14. Fang, N., et al.: Ultrasonic metamaterials with negative modulus. Nat. Mater. **5**, 452–456 (2006)
15. Claeys, C.C., Pluymer, B., Sas, P., Desmet, W.: Design of a resonant metamaterial based acoustic enclosure. Proc. ISMA **2014**, 3351–3358 (2014)
16. Tang, Y., et al.: Hybrid acoustic metamaterial as super absorber for broadband low-frequency sound. Sci. Rep. **7**, 43340 (2017)
17. Lewińska, M.A., Van Dommelen, J.A.W., Kouznetsova, V.G., Geers, M.G.D.: Towards acousticmetafoams: the enhanced performance of a poroelastic material with local resonators. J. Mech. Phys. Solids **124**, 189–205 (2019)
18. Holloway, C.L., Kuester, E.F., Gordon, J.A., O'Hara, J., Booth, J., Smith, D.R.: An overview of the theory and applications of metasurfaces: the two-dimensional equivalents of metamaterials. IEEE Antennas Propag. Mag. **54**(2), 10–35 (2012)
19. Yong, L., Xue, J., Bin, L., Cheng, J.C., Zhang, L.: Metascreen-based acoustic passive phased array. Phys. Rev. Appl. **4**, 024003 (2015)
20. Jiang, X., Liang, B., Zou, X.Y., Yin, L.L., Yang, J.: Acoustic one-way metasurfaces: asymmetric phase modulation of sound by subwavelength layer. Sci. Rep. **6**, 28023 (2016)
21. Ma, G., Yang, M., Xiao, S., Yang, Z., Sheng, P.: Acoustic metasurface with hybrid resonances. Nat. Mater. **13**, 873–878 (2014)
22. Liang, Z., Li, J.: Extreme acoustic metamaterial by coiling up space. Phys. Rev. Lett. **108**, 114301 (2016)

23. Xie, Y., Popa, B.-I., Zigoneanu, L., Cummer, S.A.: Measurement of a broadband negative index with space-coiling acoustic metamaterials. Phys. Rev. Lett. **110**, 175501 (2013)
24. Liang, Z., et al.: Space-coiling metamaterials with double negativity and conical dispersion. Sci. Rep. **3**, 1614 (2013)
25. Zhu, Y.F., Zou, X.Y., Liang, B., Cheng, J.C.: Acoustic one-way open tunnel by using metasurface. Appl. Phys. Lett. **107**, 113501 (2015)
26. Zhu, Y.F., et al.: Dispersionless manipulation of reflected acoustic wavefront by subwavelength corrugated surface. Sci. Rep. **5**, 10966 (2015)
27. Jia, X., Yan, M., Hong, M.: Sound energy enhancement via impedance-matched anisotropic metamaterial. Mater. Des. **197**, 109204 (2021)
28. Xie, Y., Konneker, A., Popa, B-I., Cummer, S.A.: Tapered labyrinthine acoustic metamaterials for broadband impedance matching. Appl. Phys. Lett. **103**(20), 201906 (2013)
29. Bergamini, A., Delpero, T., De Simoni, L., Di Lillo, L., Ruzzene, M., Ermanni, P.: Phononic crystal with adaptive connectivity. Adv. Mater. **26**, 1343–1347 (2014)
30. Popa, B.I., Zigoneanu, L., Cummer, S.A.: Tunable active acoustic metamaterials. Phys. Rev. B **88**, 024303 (2013)
31. Popa, B.I., Shinde, D., Konneker, A., Cummer, S.A.: Active acoustic metamaterials reconfigurable in real-time. Phys. Rev. B **91**, 220303(R) (2015)
32. Akl, W., Baz, A.: Experimental characterization of active acoustic metamaterial cell with controllable dynamic density. J. Appl. Phys. **112**, 084912 (2012)
33. Xiao, S., Ma, G., Li, Y., Yang, Z., Sheng, P.: Active control of membrane-type acoustic metamaterial by electric field. Appl. Phys. Lett. **106**, 091904 (2015)
34. Lee, K.J.B., Jung, M.K., Lee, S.H.: Highly tunable acoustic metamaterials based on a resonant tubular array. Phys. Rev. B Condens. Matter **86**, 184302 (2012)
35. Fleury, R., Sounas, D., Alù, A.: An invisible acoustic sensor based on parity-time symmetry. Nat. Commun. **6**, 5905 (2015)

# Parameter Tuning of ODE-Based Walking Control Algorithm for a Six Legged Mobile Robot

Şahin Yıldırım and Erdem Arslan(✉)

Erciyes University Mechatronics Engineering Division, 38039 Kayseri, Turkey
erdemarslan@erciyes.edu.tr

**Abstract.** The design of controllers which considering the dynamic effects that walking robots may encounter during their movement, is an important requirement in the field of legged mobile robots. In recent years, in the control of complex dynamical systems, numerical simulation algorithms based control structures, have begun to be preferred. Open Dynamics Engine (ODE) is a physics simulator for modeling and controlling of rigid bodies. In the ODE-based stabilization and walking algorithm that earlier developed for a six-legged mobile robot, a model based controller structure was obtained that takes into account the disturbance effects that occur during control. In this study, the parameters of ODE-based walking and stabilization algorithm were tuned. It is known that the sequential sampling strategy gives affirmative results in tuning the parameters of model-based controllers. For this reason, a sequential sampling strategy was used to tune the parameters of the ODE-based stabilization and walking controller. After this tuning procedure, 37.83% improvement in area error and 4.63% improvement in similarity error were obtained.

**Keywords:** SDG9 · Legged mobile robot · Walking robot · Model based control · Parameter tuning · Sequential optimization

## 1 Introduction

The technologies that humanity has today have emerged as a result of innovative studies carried out by researchers in the past. With the innovative studies on industrial robot technologies developed in the past, the dangerous and sensitive works that had to be performed by humans have started to be carried out with automation systems. In automation the idea of imitating human behaviors in system development has always been the best alternative for system developers. The first thing comes to mind, when someone think a robot imitating human movement is legged mobile robots. Although the human behavior is imitated in the design of industrial systems, dozens of innovative solutions are needed to see exactly human-like systems such as androids or cyborgs in industrial facilities. In the case of the development of these devices under human control, it will be possible to offer many systems that can be used to correct the negative effects of humans on nature.


G. Quaglia et al. (Eds.): I4SDG 2021, MMS 108, pp. 553–561, 2022.
https://doi.org/10.1007/978-3-030-87383-7_59

Therefore, legged mobile robots are of great importance for a sustainable development for humanity.

There are basically two major handicap in the development of legged robots. The first is that efficient actuators and the battery systems that can provide sufficient power have not yet been fully developed. The second major problem is that the control systems have not yet been developed to control the leg mobile movement in a way that performs just like a human. If the movements of walking creatures are examined, it will be seen that the creature repositions itself according to the changing situations during movement.

Generally, to develop a controller for a legged robot, dynamic models consisting of highly complex and non-linear equations are used [1]. The robot's repositioning with the changing situations will completely change the dynamics of the robot.

In classical control techniques, the changing system behaviors cannot be taken into account effectively because of the ideal controllers are developed just for the analytically modeled conditions. In model-based controllers that include numerical simulation algorithms (ODE, Bullet, Mujoco i.e.), because of the changing system states are simulated simultaneously, the disturbance behaviors affecting the system can be easily taken into account.

ODE-based stabilization and walking algorithm that earlier developed for a six-legged mobile robot, uses a model based controller structure. In this controller the disturbance effects that occur during the control takes into account by using simultaneous running ODE algorithm. In this study, by tuning the parameters of ODE-based walking and stabilization controller, the stability performance tried to be raised.

## 2 Open Dynamics Engine

Open Dynamics Engine (ODE) was coded by Russell Smith in 2007 as an open source physics simulator [2]. Until today, many software developers have continued to contribute to the development of ODE. The final version of ODE was published in July 2020 by Oleh Derevenko.

ODE is a discrete-time multi-body dynamic simulation algorithm. For the mathematical infrastructure of the ODE algorithm, the study numbered [3] can be examined. While simulating a multi-body system in ODE, the structure given in Eq. (1) 1 is iteratively solved for $\Delta t$ time intervals. Discretization method also called as augmented formulation [4].

$$\begin{bmatrix} \frac{1}{\Delta t} M_e & -J_c^T \\ J_c & 0 \end{bmatrix} \begin{bmatrix} \overline{v}^{n+1} \\ \overline{\lambda} \end{bmatrix} = \begin{bmatrix} \left(\frac{1}{\Delta t}\right) M \overline{v}^n + G_e \\ J_c \overline{v}^n \end{bmatrix} \tag{1}$$

In this Eq. (1); $\Delta t$ is the time step, M is a total mass and inertial Matrix. G is an effort vector also known as the external forces applied to bodies, $\overline{v}$ is the velocity vector of bodies and $J_C$ is the Jacobian constraint matrix. The most important parameter to be determined in this Eq. (1) for each step is the $\overline{\lambda}$ parameter, known as the Lagrange multipliers. $\overline{\lambda}$ values need to be solved in order to determine the positions of the velocities and positions of the masses in each time interval. During the simulation process of ODE, the problem in Eq. (1) is first converted to LCP (Linear Complementary Problem) and then these values are tried to be solved by PGS (Projected Gauss-Seidel algorithm) during a predetermined simulation step. [5]. Equation (2) is used to calculate $\overline{\lambda}$ values.

$$\left[\overline{\overline{JM}}^{-1}\overline{\overline{J}}^{T}\right]\overline{\lambda} = \frac{\overline{c}}{\Delta t} - \overline{\overline{J}}\left[\frac{\overline{v}^{n}}{\Delta t} + \overline{\overline{M}}^{-1}\overline{E}\right] \tag{2}$$

After solving the $\overline{\lambda}$ parameters with Eq. (2), the Eq. (1) needs to be integrated numerically for determining the positions and orientations of the bodies under the current constraint conditions. In some cases, it is possible to encounter situations where the bodies intersect with each other while simulating. Although it is not physically possible for the bodies to intersect with each other, situations may occur in the simulation from time to time, when the bodies intersect. This error in the simulation is due to the draft error caused by integration and insufficiently determined $\overline{\lambda}$ values. To eliminate this situation in ODE, position constrained correction term is added to $\overline{c}$ value and updated version of c is given in Eq. (3):

$$\overline{\overline{J}}\overline{v} + \frac{\beta}{\Delta t}\overline{h}^{n} = \overline{c} \tag{3}$$

$\beta$ parameter added in the Eq. (3) is known as ERP (Error Reduction Parameter) in ODE. This term can be thought of as a form of Baumgarte stabilization [6]. With this added term, the problem of intersection during simulation is tried to be eliminated. For this reason, it is necessary to tune this ERP value appropriately for simulations with different numbers of masses and constraints.

In addition, the term $\left(1/\Delta t\right)\overline{\overline{C}}$ is added to the $\overline{\overline{JM}}^{-1}\overline{\overline{J}}^{T}$ expression given on the left in ODE in order to stabilize the problem given in Eq. (2) and to make the solution convergent. Otherwise, it will not be possible for the equation to reach a proper solution during the time interval given for PGS. This added $\overline{\overline{C}}$ term is known as CFM(Constraint Force Mixing) in the ODE. When the corrections are added to the Eq. (2), it is obtained in Eq. (4).

$$\left[\overline{\overline{JM}}^{-1}\overline{\overline{J}}^{T} + \frac{1}{\Delta t}\overline{\overline{C}}\right]\overline{\lambda} = \frac{\overline{c}}{\Delta t} - \frac{\beta}{\Delta t^{2}}\overline{h}^{n} - \overline{\overline{J}}\left[\frac{\overline{v}^{n}}{\Delta t} + \overline{\overline{M}}^{-1}\overline{E}\right] \tag{4}$$

For systems modeled in ODE, the CFM and ERP parameters need to be adjusted properly. Adjusting these parameters will affect the simulation performance and if they are set incorrectly, undesirable results may be encountered.

## 3 ODE Based Stabilization and Walking Control Algorithm

The purpose of using the ODE physics simulator in the controller is the need to consider the inertial forces that will occure from accelerations during movement. In previous studies, a control structure has been developed in which the robot can automatically balance itself according to both the effects coming from these inertia forces and the external disturbance effects, thanks to the joint, stability and walking controllers [7–9]. Walking controller, manipulates the legs of the robot in an appropriate order according

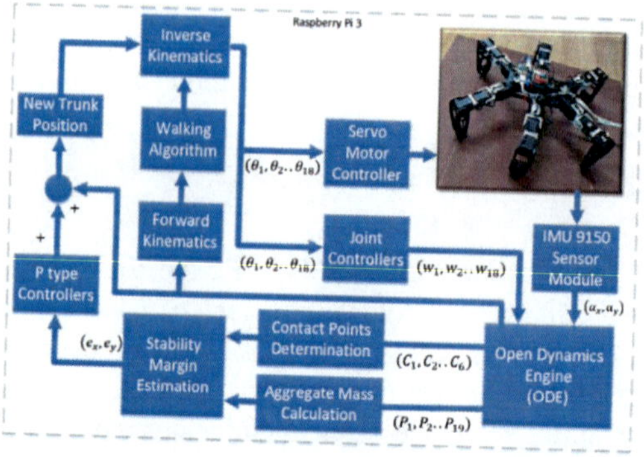


**Fig. 1.** The block diagram of the ODE based walking and stability controller [7]

to the gait pattern. The block diagram of the ODE based walking and stability controller is given in Fig. 1.

In the ODE-based stability and walking algorithm, all the legs attached to the body are used to balance the robot. In this controller, the body is moved in the forward-backward and right-left directions in order to bring the robot into balance. Then, using the inverse kinematics process between this new body position and the old contact points, all legs are repositioned to support the body.

The stability margin used when developing this controller is the static stability margin (SSM) [10]. This margin is defined by the shortest distance that the robot's average center of gravity travels to the edges of the contact polygon. The magnitude of this value gives information about the current balancing performance of the robot. The location farthest from the edges within the geometric area covered by the contact polygon is the polygon center. In this case, the most balanced stance that the robot can achieve will be the stance in which the projection of the average center of gravity of the robot and the center of the contact polygon intersect. In order to intersect these two points, the position of the average center of gravity tried to be controlled in both x and y axis with a closed kinematic control loop. In the diagram shown in Fig. 1, the system model developed on the ODE and the experimental robot are operated simultaneously. In the closed loop controller, in addition to the kinematic controller, the ODE model is used. In this way, the inertial forces during the motion are taken into account. You can scan the QR code in Fig. 2 to examine the control performance and see the operation of the system.

**Fig. 2.** Real-time ODE based controller interface [5].

## 4 Performance Criteria

In this research, the parameter tuning of the ODE-based walking and stabilization algorithm is carried out. For this reason, a numerical value was needed to be used in order to compare the performances of the experiments performed during the adjustment process with each other. As mentioned in Sect. 3, the instantaneous balance of the robot is determined by the SSM criterion. The change of SSM over time for a tripod walking given in Fig. 3a as an example (6 step).

In order to find the SSM performance of an experiment performed for a certain time, the numerical integral of the change of SSM over time must be calculated. The result of this integral will give the control performance of the tuned parameters. In the results section, this integration error term will be expressed as the area error ($e_a$). Since each step is a copy of each other in six-legged tripod walking, the area under the graph can be an accurate representation of total performance. However, in some cases where ERP and CFM values are not set correctly, the SSM change of steps during walking could be non-periodic. An example of SSM change to this situation is given in Fig. 3b (4 step).

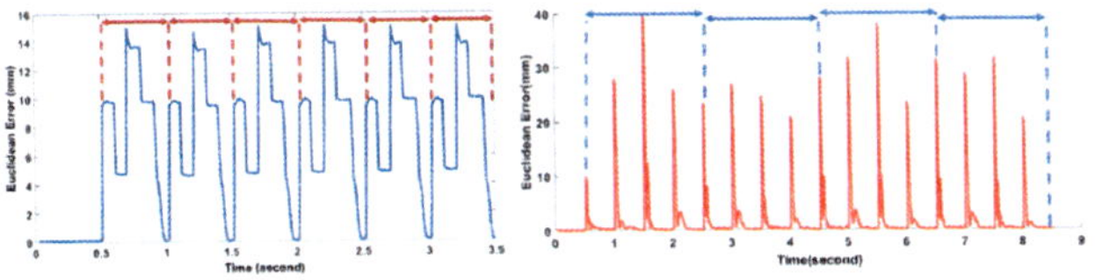

**Fig. 3.** a) SSM change over time for a tripod walking (6 step). **b)** Non-periodic SSM change over time for a tripod walking (4 step).

In some cases, where the area error is quite low, walking cannot be considered a valid experiment. Because, as well as being stable, the steps must also maintain their similarities to each other. For this reason, a performance criterion was needed to represent the similarities of the steps with each other during walking. This similarity error can be calculated using Eq. (5)

$$e_s = \frac{1}{N}\left(\sum_{i=1}^{M}\left(\sum_{j=1}^{N}\left(s_j(i) - \frac{1}{N}\sum_{k=1}^{N} s_k(i)\right)^2\right)\right) \tag{5}$$

In the results section, this similarity error term will be expressed as the area error ($e_s$). For more information on performance criteria, the study in [11] can be reviewed.

## 5 Tuning Procedure

The ODE-based stability and walking controller has twelve parameters that can be tuned to improve gait performance. These parameters can be grouped under four main class. These classes are ODE solver parameters, joint controller gains, Stability controller gains and contact parameters of the ODE respectively. These parameters are shown on the block diagram given in Fig. 4. ODE solver has two parameters that can be changed. These are ERP and CFM parameters. How these parameters affect the ODE solver is given in detail in Sect. 2. In addition to these parameters, there are five more parameters in the ODE dynamic model that will affect the contact between the feet of the six-legged mobile robot and the ground. These parameters are, maximum contact correction speed ($c_{mcv}$), surface layer size ($c_{sl}$), contact bounce factor ($c_b$), contact bounce rate ($c_{bv}$) and contact softness factor ($c_{scfm}$) respectively. The user manual of the ODE can be examined to understand what kind of effect these parameters have in case of contact [2]. ODE solver has two parameters that can be changed. These are ERP and CFM parameters. How these parameters affect the ODE solver is given in detail in Sect. 2.

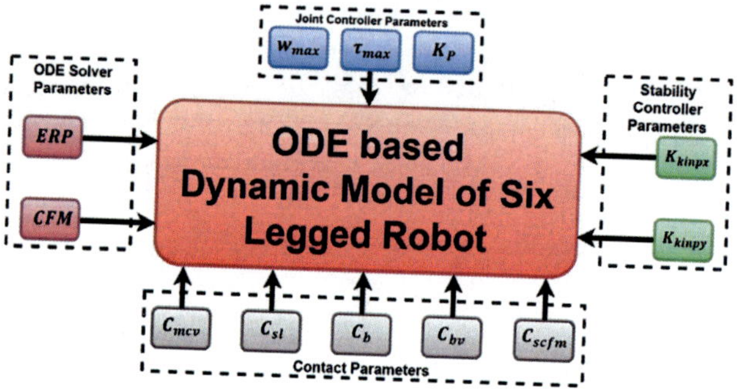


**Fig. 4.** The parameters of ODE-based walking and stability controller.

These parameters are, parameters are maximum contact correction speed ($c_{mcv}$), surface layer size ($c_{sl}$), contact bounce factor ($c_b$), contact bounce rate ($c_{bv}$) and contact softness factor ($c_{scfm}$) respectively. The user manual of the ODE can be examined to understand what kind of effect these parameters have in case of contact [2].

In the ODE based controller, there are three parameters that will affect the operation of the joint controllers. These parameters are maximum joint speed ($\omega_{\max}$), maximum joint torque ($\tau_{\max}$) and joint control coefficient parameters ($K_P$) respectively.

There are two parameters of affect the stability controller. These parameters are the kinematic stability controller coefficients in the X and Y directions ($K_{kinpx}$, $K_{kinpy}$).

Since balancing is performed in both directions at the same time in body balancing, two different control coefficients can be defined. However, since the initial position of the robot is chosen symmetrically, a single value will be selected for these two parameters during the tuning process.

If the experimental robot is removed from the controller model given in Fig. 1, a walking simulator that moves according to the given trajectory conditions is obtained. Before the parameter setting procedure, the initial values of the controller parameters entered into the simulator were selected in accordance with the values recommended in the ODE manual. A five-step trajectory for ten seconds (ODE-based tripod walking [11]) is performed, was entered into the walking simulator during all of the tuning trials. After this walking simulation, the area error obtained from the SSM variation over time was calculated as 87.132, and the similarity error was calculated as 6.242.

The aim of the tuning process is to determine the eleven different parameters that will give the least area error at the output of the walking controller. For this process, graphs were required that show the response in terms of the change of area and similarity errors according to a certain parameter change. For this purpose, while tuning each parameter, the other parameters were kept constant and the variation of the errors was observed according to this selected single parameter. The ranges of parameters that need to be changed in order to make the simulation of the ODE solver healthier are available in the ODE manual. Search limits trial numbers of all parameters are given in Table 1.

It is known that the sequential sampling optimization gives affirmative results in the tuning of model-based controllers [12]. This technique is generally used for tuning systems with multiple parameters. In the application of this technique, first of all, the system is tried to be optimized with only one parameter. The first step is to find the system parameter that gives the best value. In all subsequent steps, the newly found value is used for the first parameter, not the original value anymore. New parameters are tried sequentially, and each step has input values that bring the output to the best value. Just like in the first step, this value will continue to be used in all operations after this step. This process continues until all the parameters in the system are set.

In this study, this strategy was used while setting eleven parameters of the ODE-based walking algorithm. Trial sequence and the information about the improvement obtained at each step has given in Table 1. The variation of the area and similarity error depending on the $\omega_{max}$ parameter tuned in the first step is given in Fig. 5 as an example for understanding how the tuning process is done. The setting of ERP and CFM parameters was performed slightly differently from other parameters. Since these parameters affect the same equation in the ODE algorithm, it is not possible to set them independently of each other. For this reason, ERP and CFM values were chosen together while adjusting these parameters. The effect of the changes of these parameters on the area error can be seen from the surface graph in Fig. 6.

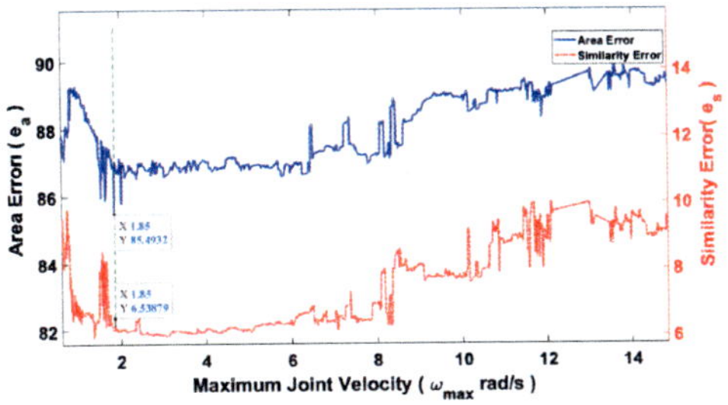


**Fig. 5.** Variation of area and similarity error accordinng to the maximum joint speed param.

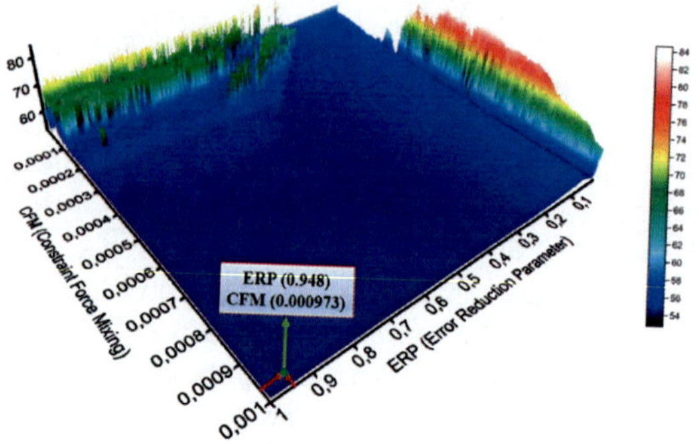


**Fig. 6.** Variation of area error according to ERP-CFM parameters of ODE solver.

## 6 Results

After this sequential tuning process, the field error, which was 87.32 before the tuning process, was reduced to 54.19. In other words, an improvement of 37.81% was obtained on the basis of percent area error. The similarity error, which was 6.24 before the tuning process, was also reduced to 5.95 with the tuning process. At the end of the tuning procedure, the improvement in similarity error was also obtained as 4.63%. Detailed information on the improvement in each step is given in Table 1.

**Table 1.** The improvement table of sequential parameter tuning.

| Changed parameter | First value | Tuned value | Trial range | Trial number | Prev $e_a$ | Tuned $e'_a$ | %Imp. | Prev $e_s$ | Tuned $e'_s$ | %Imp. |
|---|---|---|---|---|---|---|---|---|---|---|
| $\omega_{max}$ | 4.536 | 1.85 | 0.01–30 | 3000 | 87.132 | 85.493 | 1.881 | 6.242 | 6.539 | −4.76 |
| $\tau_{max}$ | 70 | 108.7 | 0.1–300 | 3000 | 85.493 | 84.668 | 2.828 | 6.539 | 6.373 | −2.1 |
| $K_P$ | 60.3 | 23.8 | 0.1–300 | 3000 | 84.668 | 84.623 | 2.88 | 6.373 | 5.996 | 3.941 |
| $c_{mcv}$ | 0.1 | 0.5642 | 1e−4–1 | 10000 | 84.623 | 84.499 | 3.022 | 5.996 | 6.340 | −1.57 |
| $c_{sl}$ | 0.001 | 0.0131 | 1e−6–0.006 | 6000 | 84.499 | 84.445 | 3.084 | 6.340 | 6.361 | −1.91 |
| $c_b$ | 0.1 | 0.0995 | 1e−4–1 | 10000 | 84.445 | 84.445 | 3.084 | 6.361 | 6.361 | −1.91 |
| $c_{bv}$ | 0.0001 | 0.0343 | 1e−6–0.1 | 100000 | 84.445 | 84.427 | 3.104 | 6.361 | 6.319 | −1.23 |
| $c_{scfm}$ | 0.0002 | 0.0002 | 1e−6–0.1 | 100000 | 84.427 | 84.427 | 3.104 | 6.319 | 6.319 | −1.23 |
| $K_{kinp}$ | 0.0002 | 0.0032 | 1e−7–0.001 | 10000 | 84.427 | 56.093 | 35.62 | 6.319 | 6.089 | 2.451 |
| ERP | 0.8 | 0.948 | 0.001–1 | 1000000 | 56.093 | 54.185 | 37.813 | 6.089 | 5.953 | 4.63 |
| CFM | .00001 | 9.7e−4 | 1e−6–0.001 | | | | | | | |

## References

1. Vukobratović, M., Stepanenko, J.: Mathematical models of general anthropomorphic systems. Math. Biosci. **17**(3–4), 191–242 (1973)

2. Open dynamics engine (ODE) homepage. http://www.ode.org Accessed 30 May 2021
3. Hsu, J.M., Peters, S.C.: Extending open dynamics engine for the DARPA virtual robotics challenge. In: Brugali, D., Broenink, J.F., Kroeger, T., MacDonald, B.A. (eds.) SIMPAR 2014. LNCS (LNAI), vol. 8810, pp. 37–48. Springer, Cham (2014). https://doi.org/10.1007/978-3-319-11900-7_4
4. Shabana, A.A., Sany, J.R.: An augmented formulation for mechanical systems with non-generalized coordinates: application to rigid body contact problems. Nonlinear Dyn. **24**(2), 183–204 (2001)
5. Silcowitz, M., Niebe, S., Erleben, K.: Interactive rigid body dynamics using a projected Gauss-Seidel subspace minimization method. In: Computer Vision, Imaging and Computer Graphics. Theory and Applications: International Joint Conference Proceedings, pp. 218–229. Springer, Berlin-Heidelberg (2010)
6. Baumgarte, J.: Stabilization of constraints and integrals of motion in dynamical systems. Comput. Methods Appl. Mech. Eng. **1**(1), 1–16 (1972)
7. Yıldırım, Ş, Arslan, E.: ODE (Open dynamics engine) based stability control algorithm for six legged robot. Measurement **124**, 367–377 (2018)
8. Arslan, E., Yıldırım, Ş: ODE (open dynamics engine) based walking control algorithm for six legged robot. J. New Results Sci. **7**(2), 35–46 (2018)
9. Arslan, E., Yıldırım, Ş: ODE Based joint controller design for legged robots. J. Curr. Constr. Iss. **1**(1), 270–279 (2017)
10. Ting, L.H., Blickhan, R., Full, R.J.: Dynamic and static stability in hexapedal runners. J. Exp. Biol. **197**(1), 251–269 (1994)
11. Yildirim, Ş., Arslan, E.: A Comparison of six legged ODE (open dynamics engine) based gait control algorithm and standard walking gaits. Avrupa Bilim ve Teknoloji Dergisi 242–255 (2019)
12. Bartz-Beielstein, T., Lasarczyk, C.W., Preuß, M.: Sequential parameter optimization. In: IEEE congress on evolutionary computation. pp. 773–780. IEEE (2005)

# Safety of Multi-link Road Trains Weighing 60 Tons

Vasiliev Alexey(✉) and Dobretsov Roman

Institute of Machinery, Materials and Transport, Peter the Great Saint Petersburg Polytechnic University, Polytechnicheskaya. 29, 195251 St Petersburg, Russian Federation

**Abstract.** The efficiency of cargo turnover is determined by the presence of an extensive network of transport routes, their throughput, technical characteristics of vehicles, their speed of movement and other parameters. The number of highways is growing steadily, the quality of road systems is gradually improving, at the same time consumer demands are growing. All these facts provide prerequisites for the possibility of changing the existing restrictions in order to increase cargo turnover. In connection with the increase in cargo turnover, it becomes necessary to transport large volumes of goods. The easiest way to increase the volume of winter cargo is to use more trailed links. The issue of constructing a nonholonomic model of turning a road train with an active trailer is considered: the formation of a set of basic hypotheses and a design scheme, the compilation of the Routh equation, the determination of the components of the equation. The work considers the behavior in the turn of various promising options for multi-link road trains. To simulate the turn of road trains, the program complex "Universal Mechanism" was used. The behavior of the road train is investigated under the following conditions: when driving at a speed of 20 m/s, a sharp jerk of the steering wheel is made. In the course of the work, the stability indicators of road trains were revealed and it was shown that the road train of the B-Double type is safer in a turn.

**Keywords:** SDG9 industry · Innovation and infrastructure · Multilinks road trains · Safety · Turning · Stability · Dolly · B-Double · Tandem · Computer simulation · Universal mechanism

## 1 Introduction

The efficiency of cargo turnover is determined by the presence of an extensive network of transport routes, their throughput, technical characteristics of vehicles, their speed of movement and other parameters. The number of highways is growing steadily, the quality of road systems is gradually improving, at the same time, consumer demands are growing. All these facts provide prerequisites for the possibility of changing the existing restrictions in order to increase cargo turnover, etc. [1].

One of the simplest ways to increase cargo turnover is to increase the permissible overall length of the road train, as a result of which you can use longer trailers or a set of trailers. Such solutions are known: for example, in Australia there is no limitation on


G. Quaglia et al. (Eds.): I4SDG 2021, MMS 108, pp. 562–572, 2022.
https://doi.org/10.1007/978-3-030-87383-7_60

the length of the road train, so there are used huge multi-link road trains (up to seven trailer links), which carry up to 190 tons of cargo [1].

In the Netherlands, from 2004 to 2006, tests of freight road trains with a length of up to 25.25 m and a gross weight of up to 60 tons were carried out. The purpose of the tests is to study the possibility of increasing the efficiency of cargo transportation by large road trains. During the tests, it was revealed that an increase in the profit from the payload of up to 60% was achieved, a 30% decrease in the number of heavy vehicles on the highways with the same volumes of freight, the cost of wages for drivers and fuel consumption per transport unit and by 15% decreased emission of harmful substances [1]. In Finland and Sweden, road trains with a length of up to 25.25 m are allowed.

In addition to the problem of overall efficiency, the issue of safety is also important.

The issue of stability of multi-link road trains should be considered much more broadly, taking into account the influence of specific design solutions applied for the investigated road train, as well as operating conditions. The influence of the turning moment is more pronounced when driving downhill. In bad weather conditions, the effect of the "pushing trailer" is more pronounced (directional stabilization systems allow you to combat this phenomenon. Volvo has developed an impulse braking system, in addition to the directional stabilization system, which prevents the trailer from folding. The essence of the system consists in the fact that after the driver releases the accelerator pedal, the brakes on the trailer will be automatically applied in impulse mode until the descent ends and it becomes possible to pick up speed again [1]).

Earlier, the issue of stability of the road train was also considered: road trains with an active and passive trailer were compared. The masses of the road trains and the overall length were much less than those of the road trains considered in this work, however, the stability indicators were lower than those obtained for the examples considered in this article. This is possibly due to the smaller number of axles [2, 3].

## 2 Research Methods

The use of the "Universal Mechanism" software package intended for modeling the dynamics and kinematics of plane and spatial mechanical systems.

In the course of the research, educational examples from the "Universal Mechanism" complex were used, which were adapted to the problem being solved.

## 3 Main Part

When compiling a mathematical model of rotation, the following basic hypotheses were adopted:

- uneven turning occurs on a horizontal surface;
- the properties of the support base are isotropic, tangential and lateral forces due to friction are directly proportional to the normal reaction; the withdrawal effect is described by the Rocard hypothesis. The French physicist I. Rocard investigated the dependence of the transverse movement of the wheel center on the transverse reaction. He proposed a hypothesis that reflects the linear dependence of the lateral force on the angle of withdrawal;

- the turning poles of the tractor and the trailer do not generally coincide;
- deformations of the trailer chassis are negligible compared to lateral deformation of tires;
- suspension deformations are neglected.

Below is a diagram of the forces when turning the road train (Fig. 1). Using the Routh equations, the equations for the rotation of a road train with an active trailer were compiled.

$$\frac{d}{dt}\left(\frac{dT}{d\,q_k}\right) - \frac{dT}{dq_k} = Q_k + R_k; \tag{1}$$

Where $T$– kinetic energy of the system; $\dot{q_k}$ – generalized coordinate; $R_к$– generalized force reaction of non-breaking and kinematics-force connections corresponding to the coordinate; $Q_k$ – generalized force corresponding to the coordinate and obtained from the expression for elementary work [2]. The position of the road train links is determined by the generalized coordinates $x_{01}, y_{01}, x_{02}, y_{02}, \alpha_1, \alpha_2, \psi_{ij}$. Without taking into account the separation of the wheels and their sliding, the expression for determining the kinetic energy of the tractor-trailer system:

$$T = \frac{m_1}{2}(V_{A1}^2 + \frac{m_2}{2}V_{A1}^2 + \frac{m_3}{2}V_{A2}^2 + \frac{m_4}{2}V_{A2}^2 + \frac{1}{2}J_1(\dot{\alpha}_1 + \dot{\alpha}_2)^2 + \frac{1}{2}J_2\dot{\alpha}_1{}^2 + \frac{1}{2}\sum_{i=1}^{2}\sum_{j=1}^{4}J_{ij}\dot{\psi_{ij}}^2 + \frac{1}{2}J_Д\dot{\psi_Д}^2; \tag{2}$$

Where $m_1$, $J_1$ – mass and moments of inertia of the tractor, $m_2$, $J_2$ – mass and moments of inertia of the trailer, $J_\partial$ – motor moment of inertia; $J_{ij}$ – moments of inertia $ij$-th wheel and associated rotating parts.

The connection of the motor shaft with the wheels is described by the dependence:

$$\dot{\psi_Д} = \sum_{i=1}^{2}\sum_{j}^{4} c_{ij}\,\dot{\psi_{ij}}; \tag{3}$$

$V_{A1} = \sqrt{x_1^2 + y_1^2}$ and $V_{A2} = \sqrt{x_2^2 + y_2^2}$ – the projection of the speed of the tractor and trailer on the coordinate axis.

The first derivatives of the generalized coordinates are determined by the expressions:

$$\dot{x}_2 = \dot{x}_{02} + R_2\dot{\alpha}_1 cos\alpha_1;\ \dot{y}_2 = \dot{y}_{02} + R_2\dot{\alpha}_1 sin\alpha_1;$$

$$\dot{x}_1 = \dot{x}_{01} + R_1(\dot{\alpha}_1 + \dot{\alpha}_2)\cos(\alpha_1 + \alpha_2);\ \dot{y}_1 = \dot{y}_{02} + R_2(\dot{\alpha}_1 + \dot{\alpha}_2)\sin(\alpha_1 + \alpha_2)$$

Using dependences (2) and (3), we write an expression for determining the kinetic energy in terms of generalized coordinates and their derivatives:

$$T = \frac{1}{2}m_1(\dot{x}_{01}^2 + \dot{y}_{01}^2) + \frac{1}{2}m_1(\dot{x}_{02}^2 + \dot{y}_{02}^2) + \frac{1}{2}m_1(2\dot{x}_{02}^2 \cdot R_1(\dot{a}_1 + \dot{\alpha}_2)\cos(\alpha_1 + \alpha_2) + R_1^2(\dot{a}_1 + \dot{\alpha}_2)^2 + 2R_1\dot{y_{01}}(\dot{a}_1 + \dot{\alpha}_2)\sin(\alpha_1 + \alpha_2)) + \frac{1}{2}m_2(2R_2(\dot{\alpha}_2)\dot{y}_{02}^2 sin\alpha_2 + \dot{x}_{02}^2 cos\alpha_2 + R_2^2\dot{\alpha}_2{}^2) + \frac{1}{2}(y_1(\dot{a}_1{}^2 + \dot{\alpha}_2{}^2) + \frac{1}{2}(y_2(\dot{a}_2{}^2) + \frac{1}{2}\sum_1^2\sum_1^4 J_{ij}\psi_{ij}{}^2 + \frac{1}{2}J_Д(\sum_1^2\sum_1^4 c_{ij}\,\psi_{ij}). \tag{4}$$

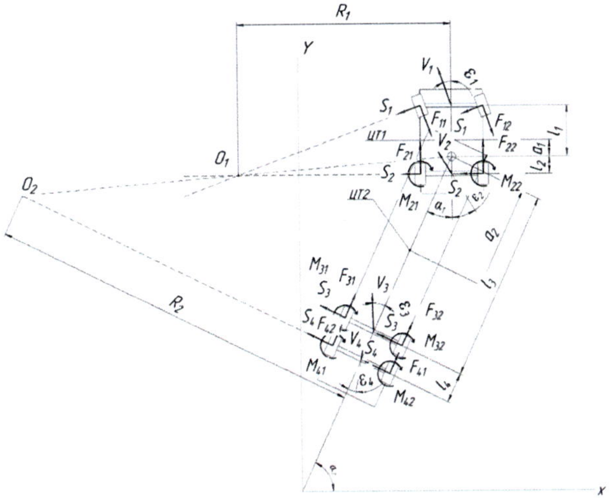


**Fig. 1.** Diagram of forces acting on a road train with an active trailer when turning

Here $c_{ij}$ – coefficients determining the kinematic connection between the engine and the wheel, $\psi_{ij}$ – steering angle.

The elementary work of the acting forces

$$dA = -\sum_{i=1}^{2}\sum_{j=1}^{4} M_{\mu ij}\,\delta\alpha_2 - \left(\sum_{i=1}^{2}\sum_{j=1}^{4} M_{\mu ij}\right)\delta\alpha_1 - \sum_{i=1}^{2}\sum_{i=1}^{4} sign\,\dot{\psi}_{ij} M_{fij}\delta\psi_{ij} + M_{Д}\delta\psi_{Д} - P_{КР}[\sin(\alpha_1 + \alpha_2)\,\delta y_b + \cos(\alpha_1 + \alpha_2)\delta x_b]; \quad (5)$$

$M_{\mu ij}$ –the moment of resistance to turning the *ij*-th wheel; $M_{fij}$- the moment of resistance to rolling of the *ij*-th wheel; $P$ – hook force.

To determine the projections of the reaction forces on the wheels of the road train (along the x and y axes):

$$R_x = \sum_{i=1}^{2}\sum_{i=1}^{4} F_{ij}\cos(\alpha_1 + \alpha_2) - \sum_{i=1}^{4} S_i \sin(\alpha_1 + \alpha_2) \quad (6)$$

$$R_y = \sum_{i=1}^{2}\sum_{i=1}^{4} F_{ij}\cos(\alpha_1 + \alpha_2) + \sum_{i=1}^{4} S_i \cos(\alpha_1 + \alpha_2) \quad (7)$$

Where $F_{ij}$ – longitudinal reaction of the *ij*-th wheel, $S_i$ – lateral reaction of the *i*-th section. The method for determining the values of these forces is adopted in accordance with the recommendations from the work [2].

Using Eq. (1), and the values (4–7) included in it, we obtain a system of differential equations of the road train motion, the solution of which will allow us to determine all the considered forces and moments, proceed to the compilation of algorithms for the operation of the drive control system in turn [4–7].

For further research, the Universall mechanism program was used. 3D models were created (using models from the program examples) with the same overall parameters.

The same mass characteristics are set, both compared road trains have a total mass of 60 tons. Mathematical model of the tire – Table (Fig. 4).

During the tests, road trains were used: truck + semi-trailer Dolly; B-Double (truck tractor + semi-trailer + semi-trailer); truck tractor + semi-trailer + tandem (trailer); car + tandem (trailer) + tandem (trailer) (Fig. 2).

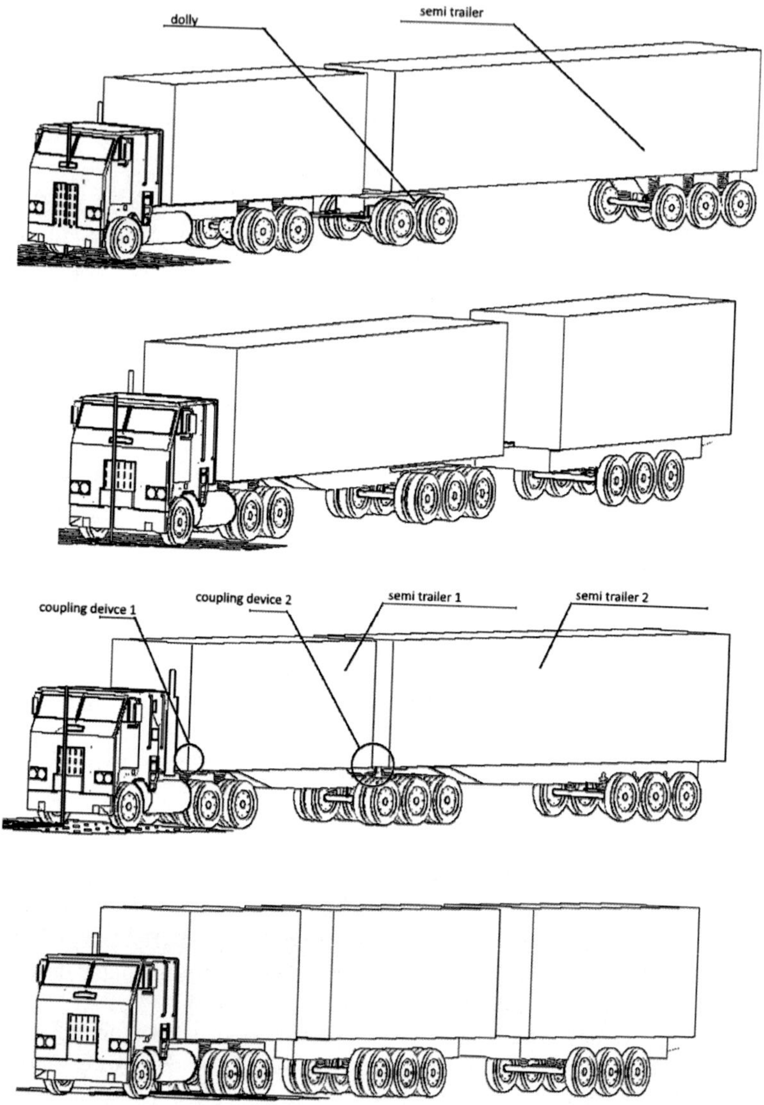


**Fig. 2.** Road train (from top to bottom): truck and semi-trailer with Dolly bogie, road train: Semitrailer, semi-trailer, tandem trailer, B-Double road train: semitrailer tractor, short semi-trailer with a fifth wheel coupling (fifth wheel coupling) at the rear, semi-trailer, road train: a truck and two tandem trailers

Description of the simulation of movement. Road trains shown in Fig. 2 (parameters in Table 1), move at a given linear speed (20 m/s). In the 2 s of the simulation, the steering is impulse (PulseSteer) (Fig. 3). The rudder jerks to the side, then returns to its original position.

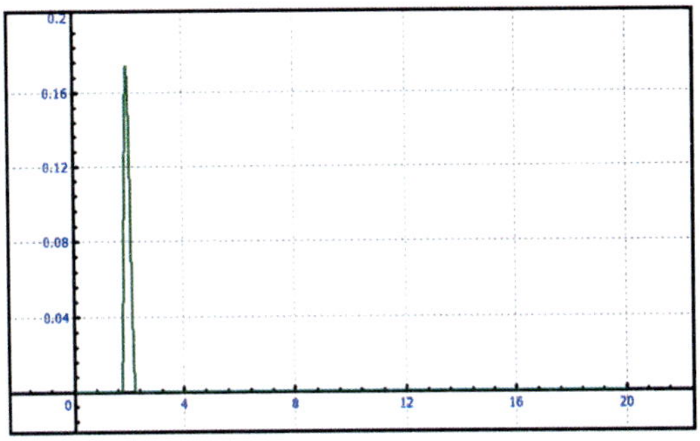


**Fig. 3.** Steering input signal. Dependence of the angle of rotation of the wheels (radians) from time (seconds)

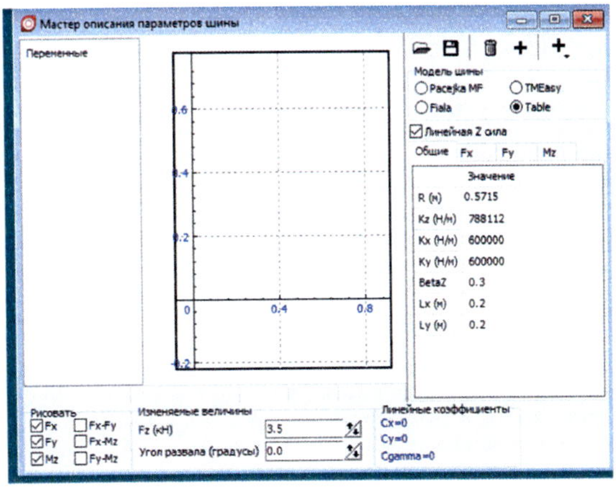


**Fig. 4.** Model of tire from program Universal Mechanism

In the course of the simulation, graphs of the sideslip angles of the wheels of the links of the road train were obtained (Fig. 5).

When turning the B-Double road train, the last semi-trailer rocking occurs (red line, Fig. 5), however, compared to the "tractor + Dolly" road train, the rest of the road train links stabilize faster. In addition, the head link of a road train with a B-Double trailer stabilizes faster (Fig. 5). The head link of the first road train under consideration (Fig. 5) stabilizes for a long time, while the centrifugal force vector at a certain point in time (2.1–2.4 s, Fig. 5) changes its direction (that is, there is a swing).

At a speed of 27 m/s, a tractor with a Dolly semitrailer is unstable (with a given signal): after turning, the truck rolls (Fig. 6), then it overturns. The instability when

**Table 1.** Parameters of road trains

| Type of roadtrain | Trailer 1 | Trailer 2 | Trailer 1 | Trailer 2 |
|---|---|---|---|---|
| | Overall length, m | | Mass, kg | |
| Tractor (6 t) and semi-trailer with Dolly bogie | Trailer on truck 7.4 | 13.5 | 20000 | 34000 |
| Tractor (6 t), semi-trailer, tandem trailer | 13.5 | 7.4 | 34000 | 20000 |
| B-Double road-train | 7.4 | 13.5 | 20000 | 34000 |
| Ttruck and two tandem trailers (7.4 trailer on truck and mass full truck 20 t) | 7.4 | 7.4 | 20000 | 20000 |

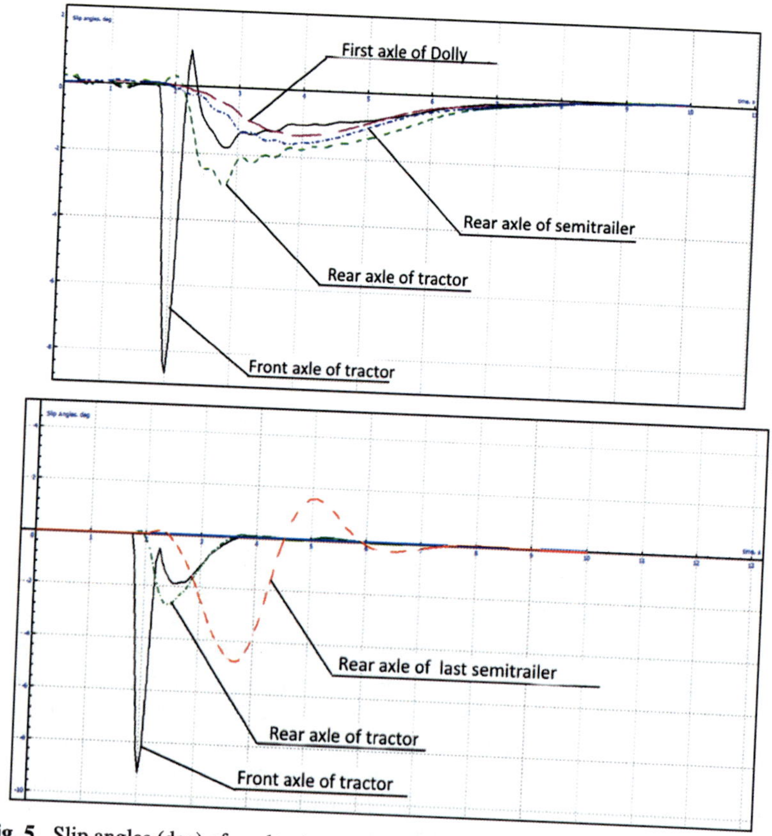


**Fig. 5.** Slip angles (deg) of road trains: truck with trailer Dolly (top), B-Double (bottom).

turning such a road train is associated with the large mass of the head link (truck). Due to the rotation, a centrifugal force arises, which creates an overturning moment. As soon as the overturning moment becomes greater than or equal to the restoring moment, the actual overturning occurs.

**Fig. 6.** Road-train: tractor + Dolly semi-trailer in a corner: load distribution on the wheels in a corner

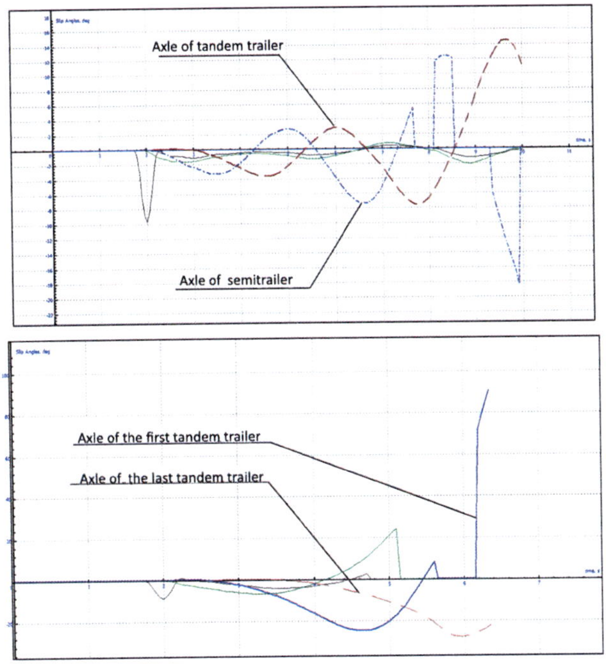


**Fig. 7.** Slip angles (deg) of road trains: a semitrailer with a semitrailer and a tandem trailer (above), a truck with two tandem trailers (below)

The simulation results substantiate the assumption that the Dolly semi-trailer has a negative effect on the tractor, the forces of inertia push the tractor forward. As soon as the tractor makes a turn, the inertia forces of the semi-trailer act on its rear, which in turn create a turning moment (a typical example of folding a road train), acting on the road train. The sum of all these moments leads to overturning of the head link. Figure 6 it can also be seen that the excessive load falls on the rear axle of the truck when turning.

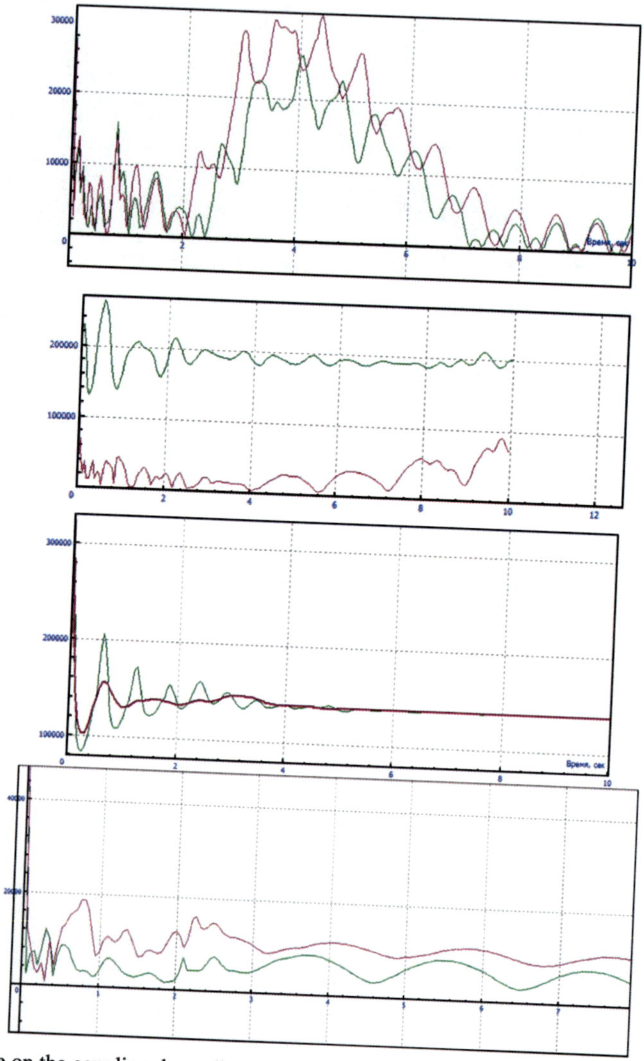

**Fig. 8.** Force on the coupling depending on the time. Road train (from top to bottom): truck and semi-trailer with Dolly bogie, road train: tractor, semi-trailer, tandem trailer, B-Double road train: semitrailer tractor, short semi-trailer with a fifth wheel coupling (fifth wheel coupling) at the rear, semi-trailer, road train: a truck and two tandem trailers

The tested road trains (truck + two tandem trailers and a semitrailer + semi-trailer + tandem trailer) showed the worst performance at the same speed (20 m/s). A road train with two tandem trailers is unstable, while a road train with a semitrailer and a tandem trailer, after turning, while continuing to move at a constant speed, an increase in the amplitude of the lateral slip angles occurs (Fig. 7). A 60-ton road train with two tandem trailers is stable only at a speed of 8 m/s.

## 4 Conclusion

Based on the simulation carried out, it can be concluded that the B-Double type road trains are much safer. This is due to the fact that this type of road train has more kinematic subsystems (truck tractor + semi-trailer + semi-trailer). The centrifugal force of each link is less for such an auto-train than for road trains with two links (truck + trailer with Dolly bogie) (Fig. 8). The semitrailer tractor weighs much less and has smaller dimensions than a truck.

In this paper, passive trailed links (without drive) are considered. Of interest is the stability of a multi-link road train with active trailers, as well as the possibility of using traction motors as part of a stabilization system [8–12].

## References


1. Turner, B.: Belarus. In: Turner, B. (ed.) The Statesman's Yearbook. TSY, pp. 186–190. Palgrave Macmillan UK, London (2011). https://doi.org/10.1007/978-1-349-59051-3_173
2. Shuvalov, E.A., Boykov, A.V., Dobryakov, B.A., Pantyukhin, M.G.: Theory and Calculation of the "Kirovets" Tractor, p. 208. Mashinostroenie, Leningrad (1980)
3. Gorelov, V.A., Padalkin, B.V., Chudakov O.I.: Mathematical model of rectilinear motion along the deformable support surface of a two-link semitrailer train with an active semitrailer link. Vestnik MGTU Im. N.E. Bauman. Ser. Mech. Eng. **2**(113), 121–138 (2017)
4. Dubovik, D.A., et al.: The hypothesis of the linear law of distribution of the reference indicators of the wheels of multi-axle wheeled vehicles and the method of its experimental substantiation. Bull. Belarusian-Russian Univ. **2**, 8–14 (2010)
5. Manesis, S.A.: Road trains of heavy trucks as highway commercial transportation systems. IFAC Proc. **31**(20), 361–365 (1998). ISSN:1474-6670, https://doi.org/10.1016/S1474-6670(17)41821-0
6. Manesis, S.A.: Off-tracking elimination in road trains of heavy duty trucks with multiple semi-trailers. IFAC Proc. **31**(20), 355–359 (1998). ISSN:1474-6670
7. Lucet, E., Micaelli, A.: Stabilization of a road-train of articulated vehicles. Robot. Auton. Syst. **114**, 106–123 (2019). ISSN:0921-8890, https://doi.org/10.1016/j.robot.2019.01.016
8. Usov, O.A., et al.: Electromechanical transmission with a hybrid power plant for a military tracked vehicle. State Polytech. Univ. J. **2**, 167–174 (2015)
9. Didikov, R.A., et al.: Power distribution control in perspective wheeled tractor transmission. Procedia Eng. **206**, 1735–1740 (2017)
10. Belousov, B., Haritonchik, S., Keller, A., Bakhmutov, S., et al.: Main features of forming the transmission of an active multi-link road train. SAE Technical Paper 2020-01-0427 (2020). https://doi.org/10.4271/2020-01-0427

11. Galyshev Yu, V., et al.: Closed steering control systems for tracked vehicles. Sci. Tech. Stat. SPbSPU No. **3**(2), 201–209 (2014)
12. Dobretsov, R.Y., Demidov, N.N., Kaninskii, A.O.: Friction steering devices as an object of impulse control. In: Evgrafov, A.N. (ed.) Advances in Mechanical Engineering. LNME, pp. 49–62. Springer, Cham (2020). https://doi.org/10.1007/978-3-030-39500-1_6

# History of Mechanism and Machine Science, Education

# The Machine, IFToMM and Ethics

Teun Koetsier(✉)

Department of Mathematics, Faculty of Science, VU-University, Amsterdam, The Netherlands
t.koetsier@vu.nl

**Abstract.** In the present paper I argue that the exponential growth of technology has led to a situation in which mechanical engineers inevitably will have to address ethical questions. Henry Kissinger called artificial intelligence a potentially dominating technology that urgently needs a guiding philosophy. It is my opinion that it is the 18th century philosophy of the Enlightenment that can still help us out. The ethical issues concern not only the UN's sustainable development goals but all other ethical aspects of modern mechanical engineering design as well. I argue that it is desirable that IFToMM (the International Federation for the Promotion of Mechanism and Machine Science) explicitly adopts this view.

**Keywords:** SDG17 · Mechanical engineering · Artificial intelligence · Machine ethics · IFToMM · Codes of ethics

## 1 Introduction

In 1948 Derek de Solla Price discovered the exponential growth of scientific knowledge. He noticed that since the 17th century every twenty years our knowledge roughly doubled. Extrapolating this trend implies that during the last twenty years just as much new knowledge was acquired as in the entire period preceding it. It is an amazing consequence of the fact that.

$$1 + 2 + 2^2 + 2^3 + \cdots + 2^n = 2^{(n+1)} - 1.$$

Although the long-term economic development driven by technology follows a wave-like pattern, also here there is exponential growth. Moore's law, which says that the number of transistors in an integrated circuit doubles about every two years, is probably the best known example.

The developments in the semiconductor industry contributed to the growth of the Internet. The Internet started as a global information machine that in combination with the World Wide Web stored and processed information. In the past twenty years the number of Internet users doubled roughly every five years. In the same period, however, the Internet started to automatically gather its own information by means of millions and finally billions of sensors. The Internet became the Internet of Things. Collecting huge amounts of data is in itself not enough. The spectacular developments in Artificial Intelligence make it possible to process the information gathered by the sensors of the Internet of Things in a smart way.


G. Quaglia et al. (Eds.): I4SDG 2021, MMS 108, pp. 575–585, 2022.
https://doi.org/10.1007/978-3-030-87383-7_61

The Internet of Things in itself is still merely an information machine, but we have entered a new phase. While becoming more intelligent it is connected more and more to production machinery, that is machinery that intervenes in the physical world. The Internet of Things is changing into what I call the Global Intelligent Machine or The Machine [11]. It is a global robotic machine that collects information, processes it and physically acts in the world. The Machine is growing exponentially, sensors are being added and more production machinery is connected to it. Moreover, its intelligence is growing fast.

The Machine is a robotic system but obviously not a robot, because there is no central control. It is one entity owned by a great number of actors. Hundreds of millions of people own computers that are part of the Internet. Numerous companies and national governments own huge server farms. Many telecommunications companies own the cables that connect all these computers to each other. The set of Internet addresses and database of domain names associated with those addresses are owned by an organization called ICANN, the Internet Corporation for Assigned Names and Numbers, which administers them as a service. The production machines that are connected to The Machine are also owned or used by a great number of people, companies and institutions.

There is the World Wide Web. Part of it, the Surface Web, is accessible to everyone using the Internet. There is also the Deep Web, the part of the WWW whose contents are not web indexed by standard web search-engines. The Internet is like a digital continent with numerous subdivisions protected by firewalls or other security systems controlled by numerous actors, individuals and institutions. Strictly speaking the Internet does not encompass the networks of North Korea, Myanmar and Cuba.

## 2 Where Are We Heading to?

With The Machine we have reached a remarkable moment in history. It looks as if nearly all machines of a certain sophistication that will be built in the future will be incorporated in The Machine. It will become the ultimate machine, the machine that makes all other machines obsolete and superfluous.

Moreover, developments in AI are such that some predict that we are approaching the moment at which machines will become superior in every respect to our species. Although this will not happen in the near future, given the exponential growth of technology, we have to take the possibility seriously. In many ways The Machine is already superior to us.

Where are we heading? We are obviously creating living conditions for ourselves that are increasingly artificial. One could argue that for many of us The Machine will become the environment they live in. The houses they live in and work in are part of The Machine. The entire complex of infrastructure they use to get supplies, to communicate or to move from one place to another is part of The Machine. What is left of nature will also be managed as a part of The Machine. Are we heading towards the world as it is described in *The Machine Stops*, a story by Edward Morgan Forster? In Forster's novel from 1909 mankind lives actually inside a huge machine that controls all aspects of their lives. In the story The Machine breaks down and many people are killed. Yet the protagonist, Kuno, escapes and with some other survivors he will try to rebuild a civilization that is more in harmony with nature [7].

It is an old idea that technology is more a threat than a blessing. In 1863 Samuel Butler wrote: “Day by day, however, the machines are gaining ground upon us; day by day we are becoming more subservient to them; more men are daily bound down as slaves to tend them, more men are daily devoting the energies of their whole lives to the development of mechanical life” [5].

With Artificial intelligence in mind, Henry Kissinger wrote in 2018 “The Enlightenment started with essentially philosophical insights spread by a new technology. Our period is moving in the opposite direction. It has generated a potentially dominating technology in search of a guiding philosophy” [10]. The Enlightenment liberated us. Science and technology have brought us very much. Yet, we sometimes look like the sorcerer’s apprentice, who created something that he cannot control.

It is, of course, an illusion to think that we can stop the development of technology. We cannot stop it. All we can do is steer it in the right direction. And Kissinger is right. We need a guiding philosophy. What should be our guiding principles when we further construct The Machine?

Mankind is faced by a long list of serious challenges that have to be taken into account as well. The existence of extremely powerful weapons of mass destruction represents a permanent threat. There is too much political instability in the world. At the same time there is still poverty, there is hunger and too many people lack adequate health care. Moreover, the impact of mankind on the environment is huge and it includes changes to ecosystems, biodiversity, and natural resources. This leads to environmental degradation and ecological crises. There are many problems to be solved and it will not be easy, but one thing is certain, The Machine will play a central role.

## 3 Towards a Guiding Philosophy

The Enlightenment was an intellectual movement in which the Scientific Revolution of the 16th and 17th centuries played a central role. The rise of the new science progressively undermined ancient conceptions. Religious leaders, philosophers and scientists discovered the power of human reason. They no longer uncritically accepted the opinions of the Greek philosophers and the Christian Church Fathers, as the major source of truth, and successfully started to make up their own mind. They believed in intellectual progress and thought that this would lead to a better world. The aristocracy and the church lost much of their influence. Isaiah Berlin wrote about the Enlightenment philosophers: “They liberated people from horrors, obscurantism, fanaticism, monstrous views. They were against cruelty, they were against oppression, they fought the good fight against superstition and ignorance and against a great many things which ruined people’s lives” [13].

The Enlightenment never left us. Romanticism with its emphasis on emotion and intuition represented a critical reaction, but scientific and technological development guided by human reason could not be stopped. The results have been spectacular.

In particular, in the 20th century science and technology became a universal problem-solving system. Whatever problem companies and governments had, they gave them as an input to the system and it gave a solution as output. Right from the start, the solutions were sometimes accompanied by new problems, but they could always also be fed into the system in order to be solved.

We have now reached a situation in which we are faced by major problems generated by the output of the system. Of course the new problems can be solved by feeding them into the system, but the situation has become more complicated. The impact that technology has on the world, is so massive and the development of technology is so fast that we can no longer wait and see what happens next. We need to anticipate it.

Scientists, engineers and the governments and the companies they work for, are all involved, and so are their professional organizations. The development of technology should and can be influenced on all levels.

William Bristow wrote "the core of the Enlightenment [was] the aspiration for intellectual progress, and the belief in the power of such progress to improve human society and individual lives" [3]. Enlightenment philosophy is not outdated. On the contrary, it still defines the direction we should go in.

Possibly more than ever before, engineers ought to adopt the Enlightenment idea that they are more than the builders of artefacts. They should consider themselves as intellectuals contributing to "improve human society and individual lives". They should not only build machines, but when they do so they should do the ethically right thing.

## 4 Philosophical Ethics

Yet, what is ethically the right thing? The first thing one should do before an ethical judgement can be made about an act is collecting all the relevant facts. "Thou shalt know all the facts" is the first law of ethics. Too often people base their ethical judgments on insufficient information, sometimes only on headlines.

Once one knows the facts one can try to decide on the basis of our moral intuition whether an act is ethically good or wrong. Yet our intuition is sometimes unclear; Enlightenment philosophers were among the first, outside of the religious context, to try to catch our ethical intuition in rules. Very influential is *utilitarianism* (also called consequentialism), popularized by the English Enlightenment philosopher Jeremy Bentham (1748–1832). One way to define it is this: one should act in accordance with rules that maximize happiness and well-being for all involved. Utilitarianists look at the consequences of actions or rules in order to decide whether they are good or evil.

The opposite of utilitarianism is *deontological ethics*. In deontological ethics one looks at the intentions of the one who acts and not at the consequences. Does the person who acts follow the rules? Famous is Kant's categorical imperative. Immanuel Kant (1724–1804) was another important Enlightenment philosopher. One formulation of the imperative is: "Act in such a way that you treat humanity, whether in your own person or in the person of any other, never merely as a means to an end, but always at the same time as an end."

Imagine a robot hospital. After an accident five wounded individuals come in. In order to save them organs are needed. A single healthy man without friends or family who has come in for a minor operation could donate these organs and save the five. He is being operated on under narcosis. A strictly utilitarianist robot surgeon would kill the healthy man and save the five. Otherwise the five will die. Kant's imperative forbids this: we are not allowed to treat the healthy man merely as a means to an end.

The two Enlightenment philosophers Bentham and Kant represent two major aspects of our moral intuition, not only in the 18th century, but also today. Most people often act in accordance with utilitarianism, but at the same time attempt to respect Kant's imperative. Remarkable in this respect is a British case of 1884, which concerned the trial of two sailors, Thomas Dudley and Edwin Stephens, for the murder of a cabin boy Richard Parker. After a shipwreck in the Pacific the two men had killed the cabin boy and eaten him [1]. Their utilitarianist considerations were understood but the men were sentenced to death. Yet, the verdict was changed to six months' imprisonment, due to a public outcry.

Next to utilitarian and deontological ethics there is another ethical theory called *moral virtue ethics*. Its roots lie in Classical Antiquity, but it experienced a revival in the second half of the 20th century. Unlike in utilitarian and deontological ethics, in virtue ethics the emphasis is not on criteria to evaluate particular acts, but on the characteristics of the actor. One attempts to define a virtuous agent or a moral agent. Examples of moral virtues are justice, truthfulness and friendliness.

## 5 The Engineers

The people involved in the development of The Machine are numerous. There are, of course, governments and company directors. A crucial role, however, is played by engineers, computer scientists, and applied mathematicians. Their activities have become highly interdisciplinary. Enlightenment thinking implies that given the situation mankind is in whatever engineers are designing it has become extremely urgent that they ask themselves – individually, and on the level of the engineering organizations, whether what they are designing is good for mankind, whether it improves human society and the individual lives.

The Machine being one entity means that, in principle, any part of it can be controlled wherever it resides from very far away, although it may in practice require superior technical abilities. Everything being connected with everything else has great advantages, but the security risks are also considerable. In 2017 it was reported that a year earlier a Homeland Security team remotely hacked a Boeing 757 [16]. That it is possible to hack a car had been demonstrated earlier when hackers gained remote access to the controls of a Jeep Cherokee. And indeed, Elon Musk is reported to have said: "I think one of the biggest risks for autonomous vehicles is somebody achieving a fleet-wide hack". He gave an example: "In principle, if somebody was able to hack, say, all of the autonomous Teslas, they could, say—I mean just as a prank—they could say like 'send them all to Rhode Island' from across the United States. And that would be like, well OK, that would be the end of Tesla. And there would be a lot of angry people in Rhode Island, that's for sure" [15].

I will list some areas that deserve attention. I believe that in many cases a utilitarianist approach in combination with Kant's imperative offers a healthy basis for an ethical discussion.

(i). The above mentioned hacks show that because of the global nature of The Machine and its connection to production machinery, a malevolent actor who succeeds in gaining access can in principle cause immense damage, even on the other side on the globe. The development of The Machine requires permanent attention to the security of the systems.

(ii). An important question concerns the use of a machine: How is it going to be used? Does its existence contribute to society and human lives? Mechanical engineering is crucial in weapons manufacturing. The related ethical issues form a complex issue; there is an extensive literature on the ethics of defense technology. Recent developments in military robotics have led to concerns, by for example, the International Committee for Robot Arms Control (ICRAC), that military robots pose an extra threat to international security and to civilians during a war. Engineers can do good work here. I appreciate the following statement by Trae Stephens from 2019: "The technology of warfare, like any other technology, can be misused and abused. But it can also be designed and used to deter violence, to reduce casualties and the loss of life, to mitigate bias and misuse, to enforce accountability in the use of force, and, when violence is necessary, to make it quick and precise, leading to a rapid victory rather than an extended conflict. The technology industry can and should play a critical role in achieving these outcomes" [17].

(iii). The necessity of sustainable development [4]. The 17 sustainable development goals defined by the United Nations represent an ambitious program. Yet there are undoubtedly many ways in which machinery can either help or impede reaching the goals. The ideal would be that whenever machinery is designed one considers all aspects of the impact that the machine will have on the world. Consideration of the *ecological footprint*[1] of a machine, defined as a measure of its impact on the environment, is just one aspect.

(iv). We know that The Machine is used to massively collect, store and intelligently process data. The systems designed by mechanical engineers participate in this process, and engineers should be aware of this. Obviously the data are collected to make the system function better. Yet, those who gain access to the data not only can sometimes infringe on the privacy of others; they can also sometimes misuse the data to exert influence, manipulate individuals and even use them to influence elections. They can also distribute fake news and fake images. The ethical issues are numerous. For example, privacy and intimacy go together and Gerstein has even argued that without privacy there cannot be intimacy [8]. Using intimate information on an individual without permission for financial gains may boils down to using someone merely as a means to an end. Is this a contribution to the happiness and well-being of mankind? Do we respect Kant's categorical imperative?

(v). Not so long ago in free democratic societies information was produced and distributed in a transparent way. A free press, reliable governmental institutions and universities supplied the public with true knowledge. With the Internet this has changed. The World Wide Web offers, via many social media, a barrage of information, that is hardly controlled by anybody. For many the line between truth and nonsense has become vague. In the fight against nonsense also scientists and engineers have a role to play.

(vi). More and more decisions are being transferred to AI-systems. This leads to serious questions concerning transparency and responsibility. Traditional software is, at

[1] The ecological footprint was defined by the Global Footprint Network, which develops and promotes tools for advancing sustainability.

least for the expert, rather transparent. Moreover, there are methods to spot bugs and correct them. Deep learning is different in this sense. In the training process the algorithm defines itself. After the network has been trained, it is used. In 2017 Google's trained software identified a turtle as a gun.[2] The fact that neural networks can be fooled represents a danger.[3] These are not traditional bugs. It is actually unclear how the network does the identification. This means that the way the output is determined is opaque, not transparent. If we allow such artificial-intelligence systems to make important decisions we introduce decisions that cannot be explained. What kind of ethics should be built into the software? And how should we do it?[4] For certain applications a utilitarian perspective could be suitable, while robots working with people might work better on the basis of deontological ethics. The application of virtue ethics has also been considered: learning by example could make a robot a 'nice person' [2, 12].

(vii). Robots and machines working together. Traditional AI engineering concentrated on replacing human actors. The emphasis is, for example, on algorithms for game playing, pattern recognition and natural-language processing. Human-centered AI (HCAI) focuses on AI enhancing human performance in ways that make systems safe and trustworthy.

## 6 The Example of the Self-driving Car: Building Ethics into a Machine

The Society of Automotive Engineers distinguishes six levels of driving automation ranging from 0 (entirely manual) to 5 (entirely self-driving). Level-4 cars exist. They can operate fully automatically most of the time, but the human driver can still manually take over in unmapped areas or during bad weather. Level - cars will not have a steering wheel or pedals.[5] Such cars are being tested.

An autonomous level-5 car is a highly complex artifact [19]. It has a car navigation system based on a location system and a path-planning system based on an electronic map. Moreover, it has a perception system that scans the surrounding environment in which usually laser navigation, visual navigation and radar navigation are used. Finally there is a vehicle control system that fully controls the vehicle.

Obviously a self-driving car should reach its destination as quickly as possible. Moreover, the destination should be reached safely and collisions ought to be avoided.

The patents that Google has obtained concerning the control system give an idea of what engineers are working on. There are, for example, patents that target avoiding

---

[2] https://www.theverge.com/2017/11/2/16597276/google-ai-image-attacks-adversarial-turtle-rifle-3d-printed.

[3] https://www.labsix.org/physical-objects-that-fool-neural-nets/.

[4] For a review of ethical perspectives of robotics and AI see for example: https://www.frontiersin.org/articles/10.3389/frobt.2017.00075/full.

[5] https://www.synopsys.com/automotive/autonomous-driving-levels.html.

blind spots of other vehicles[6], detecting and responding to tailgaters[7] and how to avoid driving too close to a big truck[8].

At the University of Waterloo engineers developed decision-making technology to limit injuries and damage when a self-driving vehicle cannot avoid a crash.[9] The system analyzes all available options and chooses the course of action with the least serious injuries and damage. Apparently the engineers at Waterloo didn't factor in complex ethical questions, such as how to evaluate the well-being of the occupants of the car versus the well-being of others involved in the crash. Yet such questions have been discussed extensively in the literature.

In 2016 Noah J. Goodall imagined a self-driving vehicle killing a drunk pedestrian because it didn't have time to brake. The manufacturer is sued, the argument being that the car "could have swerved around the pedestrian, crossing the double yellow line and colliding with the empty driverless vehicle in the next lane". There are several issues involved. Of course there is the interesting legal puzzle: Who is liable when an autonomous vehicle is involved in a serious accident? The manufacturer of the car, the manufacturer of the software or the driver? Presumably driverless cars will make fewer mistakes than human beings, but the rare mistakes will get a lot of attention and manufacturers will have to defend a car's actions.

There are numerous publications on a specific thought experiment called the 'trolley problem'. This problem concerns questions like Goodall's experiment and the dilemma faced by the robot surgeon that we discussed above.

The Trolley Problem was defined by Philippa Foot in 1978. Judith Thomson described it in 1985 as follows: "Suppose you are the driver of a trolley. The trolley rounds a bend, and there come into view ahead five track workmen, who have been repairing the track. […] You must stop the trolley if you are to avoid running the five men down. You step on the brakes, but alas they don't work. Now you suddenly see a spur of track leading off to the right. You can turn the trolley onto it, and thus save the five men on the straight track ahead. Unfortunately, Mrs. Foot has arranged that there is one track workman on that spur of track. He can no more get off the track in time than the five can, so you will kill him if you turn the trolley onto him. Is it morally permissible for you to turn the trolley?" [18]. How should a self-driving car decide in such situations? From a strict utilitarian point of view the trolley should be turned. From a Kantian point of view it probably should not.

Although theoretically interesting, from a practical point of view much of discussion on the Trolley Problem referring to self-driving vehicles are probably unrealistic and naïve [9]. Yet self-driving cars will have to abide by the law and in one way or another legal and ethical rules will have to be integrated.

In 2017 Bryan Casey published an interesting essay on "amoral machines" in which he argued that "abstractions like the trolley-problem […] provide a poor means of understanding the real-world constraints robotics engineers face" [6]. He argues that firms designing autonomous vehicles will be primarily interested in profit-maximization. They

[6] US 8,874.267 B1.
[7] US 9,290,181 B1.
[8] US 8,781,670 B2.
[9] https://www.sciencedaily.com/releases/2019/10/191010075411.htm.

will be interested in predictive liability and not in ethics. Liability for injury is governed by legal codes and not by moral considerations. Casey argues that in the future robots will not maximize morality but instead they will minimize liability. Only the piecemeal work of democratic change can morally guide the introduction of more and more robots in society, according to Casey.

There are other aspects that have to be taken into account. And Heather Roff has argued that "The Trolley Problem detracts from understanding about how autonomous cars actually work, how they "think," and the real ethical issues that face those advocating the advancement and deployment of autonomous cars."[10] According to her the real ethical problems concern the overall effect of the large scale introduction if autonomous vehicles (AVs). Roff writes:" We must understand how AVs could change the design, layout and make-up of cities and towns, and what effects those may have on everything from access to basic resources to increasing forms of inequality."[11]

## 7 IFToMM and Ethics?

In many countries the national engineering organizations have a code of ethics. Such codes should assist members of an organization to do the right thing. See, for example, the ASME Code of ethics or the code of conduct of the British Institution of Mechanical Engineers. One of the fundamental principles of the ASME Code of Ethics is that engineers "uphold and advance the integrity, honor and dignity of the engineering profession by using their knowledge and skill for the enhancement of human welfare". One of the rules of conduct of the Institution of Mechanical Engineers is that the members "embrace the needs of the community and future generations and adopt practices that have minimal adverse effects on social, cultural, archaeological and ethnic heritage, and the broader interests of humanity as a whole." And the Ethische Grundsätze des Ingenieurberufs (Ethical principles of the profession of the engineer) of the German VDI is written in a similar vein. The Charte d'étique of the French IESF[12] says: "i. The engineer is a responsible citizen ensuring the link between science, technology and society; he is involved in actions aimed at the common good. [...] iii. The engineer is aware of and raises awareness of the impact of technical achievements on the environment. iv. The engineer contributes with his actions to a process of 'sustainable development'". The code of the Italian engineers says: "The professional services of the engineer must be carried out taking into account the protection of human life and health."

The bottom line is that engineers are supposed to do the right thing. And given the huge impact of modern technology on all aspects of human existence doing the right thing involves more than building technically well-functioning devices. It has an ethical dimension that deserves to be discussed at universities but also inside engineering organizations.

Obviously, a code of ethics should not be a dead letter. It should be alive and in order to keep it alive discussions are necessary.

---

10 https://www.brookings.edu/research/the-folly-of-trolleys-ethical-challenges-and-autonomous-vehicles/#footnote-1.

11 For a nice survey on ethics and autonomous vehicles see [14].

12 https://www.iesf.fr/offres/doc_inline_src/752/150731_Charte_ethique.pdf.

The International Federation for the Promotion of Mechanism and Machine Science (IFToMM) is a federation with 43 national-organization members. As such, it is in an ideal position to be a platform for discussion on ethics among engineers from different nations.

How should such a discussion be organized? It seems to me that IFToMM's Permanent Commission for History and its Permanent Commission for Education could play a role here. The Commission for History ought to be involved because its activities cover also reflection on the impact of mechanical engineering on the world, while the Commission for Education ought to participate because ethical considerations ought to be part of the training of young engineers.

IFToMM's member states represent countries with different cultures and different political systems. There are states among them that are geopolitically engaged in serious competition. And given that state of affairs it might be tempting to limit discussions in IFToMM to strictly technical matters. The political differences should however not distract us from the major common interests that we have globally.

**Acknowledgements.** I am grateful to Jorge Angeles for reading and critically commenting on an earlier version of this paper. I also thank the anonymous reviewers.

## References

1. Bacik, I.: My favourite case (2012). https://www.irishtimes.com/news/crime-and-law/necessity-is-no-defence-for-murder-1.519637
2. Berberich, N., Diepold, K.: The Virtuous Machine -Old Ethics for New Technology? 28 June 2018. https://www.researchgate.net/publication/326029764_The_Virtuous_Machine_-_Old_Ethics_for_New_Technology
3. Bristow, W.: Enlightenment, Stanford Encyclopedia of Philosophy (2017). https://plato.stanford.edu/entries/enlightenment/
4. Brundtland Report (1987). https://sustainabledevelopment.un.org/content/documents/5987our-common-future.pdf
5. Butler, S.: Darwin Among the Machines—To the Editor of the Press, Christchurch, New Zealand, 13 June 1863. http://nzetc.victoria.ac.nz/tm/scholarly/tei-ButFir-t1-g1-t1-g1-t4-body.html
6. Casey, B.J.: Amoral machines, or: how roboticists can learn to stop worrying and love the law. Northwest Univ. Law Revue **11**, 231–250 (2017)
7. Forster, E.M.: The Machine Stops, Oxford and Cambridge Review, 1909 (2021)
8. Gerstein, R.: Intimacy and privacy. In: Schoeman, F. (ed.) Philosophical Dimensions of Privacy: An Anthology, pp. 265–271. Cambridge University Press, Cambridge (1984). https://doi.org/10.1017/CBO9780511625138.011
9. Himmelreich, J.: Never mind the trolley: the ethics of autonomous vehicles in mundane situations. Ethical Theory Moral Pract. **21**(3), 669–684 (2018). https://doi.org/10.1007/s10677-018-9896-4
10. Kissinger, H.: How the Enlightenment Ends, The Atlantic, June 2018. https://www.theatlantic.com/magazine/archive/2018/06/henry-kissinger-ai-could-mean-the-end-of-human-history/559124/
11. Koetsier, T.: The Ascent of GIM, the Global Intelligent Machine, A History of Production and Information Machines. Springer, Cham (2019). https://doi.org/10.1007/978-3-319-96547-5

12. Kulicki, P., Musielewicz, M.P., Trypuz, R.: Virtue Ethics for Autonomous Cars (short version), May 2019. https://doi.org/10.13140/RG.2.2.32982.50241
13. Lenhard, P.: The many shades of light: Isaiah Berlin, the Counter-Enlightenment, and the Haskalah. J. Mod. Jewish Stud. (2020). https://doi.org/10.1080/14725886.2020.1844983
14. Lim, H.S.M., Taeihagh, A.: Algorithmic decision-making in AVs: understanding ethical and technical concerns for smart cities. Sustainability **11**, 5791 (2019). https://doi.org/10.3390/su11205791
15. Ms. Smith: Elon Musk's top cybersecurity concern: Preventing a fleet-wide hack of Teslas, 17 July 2017. https://www.csoonline.com/article/3208035/elon-musk-s-top-cybersecurity-concern-preventing-a-fleet-wide-hack-of-teslas.html#tk.cso_fsb
16. Ms. Smith: Homeland Security team remotely hacked a Boeing 757, 12 November 2017. https://www.csoonline.com/article/3236721/homeland-security-team-remotely-hacked-a-boeing-757.html
17. Stephens, T.: The Ethics of Defense Technology Development: An Investor's Perspective, 4 December 2019. https://medium.com/@traestephens/the-ethics-of-defense-technology-development-an-investors-perspective-45c71bf6e6af
18. Thomson, J.J.: The trolley problem. Yale Law J. 1395–1415 (1985). https://doi.org/10.2307/796133
19. Zhao, J., Liang, B., Chen, Q.: The key technology toward the self-driving car. Int. J. Intell. Unmanned Syst. **6**(1), 2–20 (2018). https://doi.org/10.1108/IJIUS-08-2017-0008

# Non-circular Gears: Innovative Manufacturing Technologies

O. V. Egorova(✉), N. N. Barbashov, L. R. Abdullina, and R. M. Kiselev

Theory of Machines and Mechanisms Department, Bauman Moscow State Technical University, ul. Baumanskaya 2-ya, 5/1, Moscow 105005, Russian Federation
barbashov@bmstu.ru, roma.kis@list.ru

**Abstract.** The paper describes a brief evolution history of non-circular gears started from the 14th century, the main gear parameters and 5 models with non-circular gears from BMSTU mechanism collection. The authors compare different calculation methods for a complex profile of a non-circular tooth and present innovative technologies for its manufacturing based on 3D prototyping and printing, electric discharge processing as well as others. The very special attention is paid to Faydor Litvin, one of the founders of the gearing theory and non-circular gears, in particular. Non-circular gears were not widespread in the past, but this changed: increasingly becoming popularity leads to the creation of new machines. Finally, some examples of successful application of non-circular gears are shown.

**Keywords:** SDG4 · Non-circular gears · Centroid · Variable gear ratio · Coordinate EDM machines · Innovative technology · 3D prototyping

## 1 Introduction

Among 17 Sustainable Development Goals (SDGs) the SDG 4 - ensure inclusive and equitable quality education and promote lifelong learning opportunities for all, takes an important place. Education is a key element that contributes to reducing inequality and expands opportunities for people around the world to live a healthier and more stable life. University programs based on the study of achievements of the past and present encourage students to search for innovative technologies, applications of which can unleash dynamic and competitive economic forces that create employment and income [18].

Well known cam mechanisms are used in modern machines to automatically coordinate the movement of output elements with variable speed. However, they do not allow the transfer of large loads. The problem of converting rotary motion at constant speed into a rotary motion at variable speed with simultaneous transmission of a large load is effectively solved by mechanisms with non-circular gears (NCG). For a long time, NCG were not widely used in mechanical engineering, primarily due to the complexity of the calculation and manufacturing of the non-circular gears themselves. The problem was to provide non-linear relationships between a tool and a gear blank movements


G. Quaglia et al. (Eds.): I4SDG 2021, MMS 108, pp. 586–594, 2022.
https://doi.org/10.1007/978-3-030-87383-7_62

when using the generating method, which usually required the creation of a special gear cutting machine and led to a significant increase in cost.

Further development of the technology for cutting NCG developed in the direction of designing special devices for existing gear cutting machines or creating a new gear cutting tool.

The innovative development of machine tools has led to the emergence of Coordinate EDM Machines, in which teeth cutting occurs at predetermined coordinates. Such equipment makes it possible to manufacture a gear of any shape and profile. An innovative technology based on the geometric method of teeth cutting on an electric discharge machine requires a very accurate calculation of the coordinates of the teeth of a non-circular gear, which is feasible with the help of modern specialized engineering software solutions. However, the coordinate cutting of NCG can be beneficial only with a small production volume. For serial and mass production, you still need a special cutting tool, or a special gear cutting machine, or innovative technologies based on 3D design/prototyping, powder metallurgy, vacuum casting and etc. Among the alternative methods, three-dimensional printing (3D) or additive manufacturing is gaining popularity, which is a modern method of manufacturing parts directly from a digital model using the method of layer-by-layer material growth. This manufacturing method allows high-density metal parts to be produced in a short time with high precision.

## 2 Non-circular Gears

### 2.1 A Brief History of Mechanisms with Non-circular Gears

One of the first mechanisms where NCG were used to build an accurate trajectory of movement was the Astrarium, made by Giovanni Dondi (1330–1388) [5]. The Astrarium modeled the Solar System and, in addition to counting time, representing calendar dates and holidays, showed how the Sun, Moon and the five planets known at that time moved around the celestial sphere.

Mention of non-circular gears is also found in the works of Leonardo da Vinci [1]. He was looking for the possibility of obtaining a variable gear ratio by means of a non-standard gear shape, and elliptical gears were quite suitable for this task.

Much later, at the end of the 19th century, Franz Reuleaux (1829–1905) [2] ordered a collection of models from the Gustav Voigt mechanical workshop in Berlin (Mechanische Werkstatt) to study the kinematics of NCG mechanisms.

The collection of mechanisms with NCG made by Ferdinand Jakob Redtenbacher (1809–1863) is also known. He was one of the first to use the demonstration of scaled mechanism models in the Mechanics course he taught as a visual aid.

In the book "Principles of mechanism" by S. W. Robinson [5], graphical methods for constructing centroids of NCG and their tooth profiles are presented as well as the principles of designing NCG, photos and examples of their application.

In the works of I. I. Artobolevsky [14], several mechanisms composed of NCG with a detailed description of the principle of operation and the main characteristics of engagement are presented. However, Artobolevsky did not touch on problems associated with their manufacture and application.

Professor Faydor Litvin (1914–2017) played a major role in the development of the theory and manufacture of NCG, see Sect. 4.1.

## 2.2 Collection of Mechanisms with Non-circular Gears at the Bauman Moscow State Technical University (Bmstu)

In Russia, at BMSTU, the Mechanisms Collection [3] has preserved five models of mechanisms with NCG. Two models were made in the workshop of Alexander Clair and purchased in the 19th century. Figure 1 shows pages from the book-catalog published by A. Clair [3] and photos of mechanisms of the «Theory of Mechanisms and Machines» department.

The first mentioned mechanism $G_{02}$ (Fig. 1a) consists of two non-circular elliptical gears $z_1$ and $z_2$. Its gear ratio $u_{21}$ is variable and cyclical, and the gear ratio $i_{21}$ is constant and equal to one, which makes it possible to provide the necessary change in the operating and idle speed. The cycle of movement is $2\pi$.

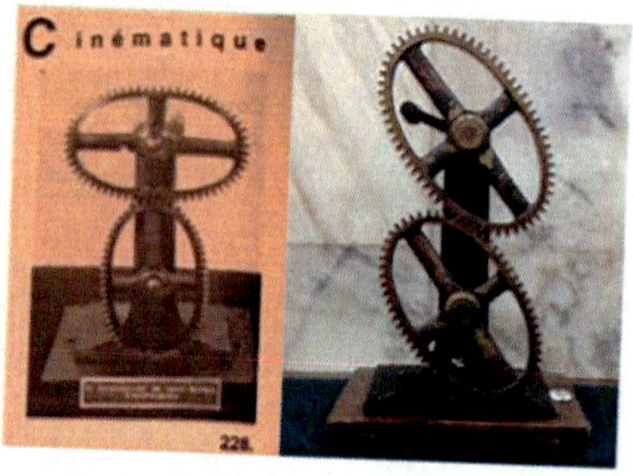


a)

b)

**Fig. 1.** Mechanism model $G_{02}$ ***(a)** and* $G_{01}$ **(b)** from A.Clair and their copies at BMSTU [3]

Gear ratio $i_{21}$ is the ratio of the number of teeth of the gears [4]. In a gear train with a variable gear ratio, gear ratio is equal to the average integral value of the gear ratio per cycle. The fundamental difference between the gear ratio $u_{21}$ and the gear ratio $i_{21}$ is that the first is the ratio of continuous quantities, and the second is the ratio of discrete.

The second mentioned mechanism model $G_{01}$ (Fig. 1b) from Claire's workshop is differed by the fact that the smaller wheel $z_1$ is circular. It is fixed on the input link with an eccentricity $e$, which makes it possible to ensure its continuous engagement with a large elliptical gear $z_2$. The gear ratio of the $G_{01}$ mechanism $i_{21}$ is 0.5; and the cycle of movement is $4\pi$.

One more model (Fig. 2) from the BMSTU Mechanisms Collection was developed on the results of the dissertation work of Karl Sergeevich Tarkhanov (see Sect. 4.3). The other two models have been partially preserved and today are under restoration.

Models with NCG allow BMSTU students to get acquainted with a little-known section of Mechanics, personally observe the movement of non-standard gearing, which significantly expands the horizons of future engineers, increases their interest in the study of engineering disciplines and in the design of new machines and mechanisms.

**Fig. 2.** Mechanism model $G_{03}$ designed by Karl Tarkhanov

# 3 Main Parameters of Non-circular Gears

For gear mechanisms composed of NCG, the centroid is the main geometric parameter. In the TMM, it is generally accepted that centroids are conjugate curves that roll without sliding during the engagement of the profiles. Depending on the formulation of the problem [3] there are 2 ways to determine the centroid:

1. by the function of the position of the driven link;
2. according to the law of changing the gear ratio.

The centroid of each of the gears is a set of instantaneous centers of rotation in the coordinate system associated with this gear. According to the main linking theorem, the pole always lies on the line of centers. In gears with a constant gear ratio, the meshing pole does not change its position. In gear mechanisms with NCG, the pole moves along the center line, describing the bipoloid. The gear ratio becomes variable.

It should be noted that at the end of the 19th century, there was virtually no analytical solution for determining the centroid. Basically, calculations were carried out due to graphical construction, which led to errors and difficulties in manufacturing. The problem of finding the centroids of NCG is nontrivial, since many restrictions are imposed on the centroids.

In addition to the centroid, an important parameter of the gear transmission is the evolution, which is the locus of points that are the centers of curvature of the curve (in our case, the profile of the tooth). Evolution allows you to evaluate the curvature of the tooth profiles, to compare the advantages and disadvantages of various cutting schemes. The graphic representation of the evolute together with the centroid contributes to the visual identification of those teeth of the NCG, in which the evolute is close to the centroid and, therefore, there is a danger of cutting the tooth stem [8, 9].

During tooth cutting by the generating method, the profile curves are envelopes of many positions of the cutting edge of the tool when the centroid of the tool and the gear blank rolls against each other. The variability of the curvature of the centroid of the gear makes it difficult to obtain an analytical function that determines the profile curve of the tooth, and, as a consequence, the complexity / impossibility of determining the evolute.

## 4 Manufacturing Methods for Non-circular Gears

Methods for NCG cutting remain laborious to date, primarily due to the complexity of calculation and production [8, 9]. When using the generating method, it is required to provide non-linear relationships between the movements of the tool and the gear blank, which leads to the need to create a special gear cutting machine or tool, and, therefore, to a significant increase in the cost of manufacturing the gears. In his scientific works F.L. Litvin (1914–2017) [4] showed several machine tools for NCG cutting. But the author himself noted that the existing methods of cutting NCG need improvement for serial or mass production.

### 4.1 Litvin Faydor Lvovych

In 2015, the public celebrated the 101st anniversary of Faydor Litvin, who published about 150 works in the USSR, about 120 works in the USA and 3 monographs. He is well known for his work on the general theory of gearing, and NCG in particular. Litvin explained that the main difficulty of NCG lies in their manufacture. In his early works, Faydor Litvin created a method for the analysis and synthesis of NCG and described their application, for example, in liquid meters. Litvin paid special attention to the peculiarities of the manufacture of NCG: cutting with a rail, running in with the use of an annular or rack cutter, the use of a chisel, the method of dividing with a modular disc cutter [6]. In 1950–1951, he received an inventor's certificate for the scheme of a gear hobbing machine and implemented it in the workshops of the Leningrad Polytechnic Institute. The machine was named ZFN-01 and was used to cut NCG with oval centroids used in liquid meters. Litvin also created a modification of the ZFN-02 machine for cutting NCG with a large module. Later, he modified this machine several times, increasing the rigidity of the structure to improve the accuracy of gear cutting. This was an innovative technology for that time.

### 4.2 Calculation of the Coordinates of the Profile of a Non-circular Gear

In the 50s, F.L. Litvin derived and published a new method for calculating tooth profile for NCG [4], but for practical use it was inconvenient. The cutter of a gear-cutting machine in those years could not move along such a complex trajectory, so additional devices had to be developed.

In [6], another, more effective, method for determining the coordinates of profile points of NCG was proposed: the method of profile normals. Figure 3 shows the geometry of the producing wheel and the processing pattern of individual areas of the tooth. This method refers to modern innovate technologies, the profiles shown on Fig. 3 are characterized by increased surface accuracy.

On Fig. 3 there are involute section, transition curve, centroid of non-circular gear 1, involute section of tooth 2, transition curve 3.

Equations for the left tooth profile. Involute part:

$$U = -u \cdot cos\alpha_w i_p + u \cdot sin\alpha_w j_p$$
$$u = [r_w(inv\alpha - inv\alpha_w + \beta) + l_e]cos\alpha_w \quad (1)$$

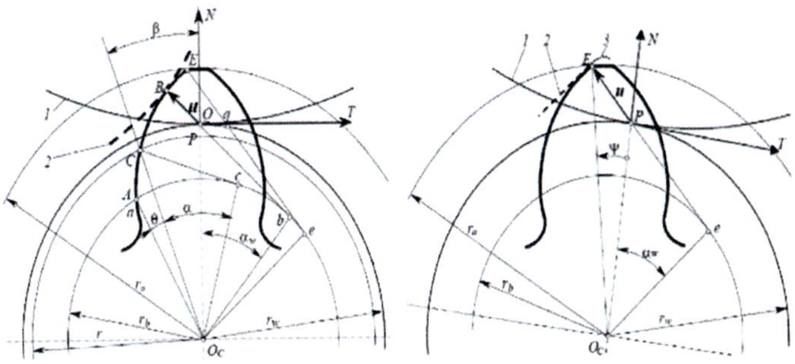

**Fig. 3.** Shaping the tooth profile with an involute production wheel [6]

Transition section, fillet:

$$
\begin{gathered}
U = -r_a \cdot sin\psi i_p + (r_a \cdot cos\psi - r_w) j_p \\
\psi = \frac{\pi}{2} - \alpha_w - \sqrt{\frac{r_a}{r_w cos\alpha_w} - 1} + inv\alpha \\
+\alpha_w + \beta + \frac{l_z}{r_w} - \arcsin\frac{r_w cos\alpha_w}{r_a}
\end{gathered} \tag{2}
$$

where $l_e$ is the value of the distance traveled by the pole P along the centroid 1 from the reference point.

For the selected tooth, the reference point on the centroid 1 is assigned arbitrarily, and on the generating gear on the axis of symmetry of the tooth or on the axis of the cavity. In this case, the reference points for all other teeth of both sunwheels must be strictly linked to the initial one. When forming a fillet, the displacement of the pole P will determine the parameter $l_z$.

Equations for the right tooth profile looks similar [6].

### 4.3 Karl Sergeevich Tarkhanov

Karl Tarkhanov (graduated from BMSTU) created two models with NCG (Fig. 3) that are currently stored within Mechanism Collection at the BMSTU [3]. In 1958 he received the inventor's certificate "Adaptation to a gear shaping machine for cutting by the generating method of non-circular gears". The inventor's certificate describes a device that makes it possible to simplify the relative motion of generation and to combine the common normal at the point of contact between the centroid of the gear-shaping cutter and the gear blank during the reverse idling of the cutter. In order to compensate in the position of the gear blank axis relative to the axis of the chisel when moving the cradle on the machine, the developed device is kinematically connected with a cam mounted on the tool support of the machine tool, which carries out additional horizontal movement of the chisel relative to the gear blank.

## 5 Modern Methods of Non-circular Gears Manufacturing

Modern manufacturing methods of NCG are based on the use of 3D technologies and specialized engineering software solutions [15–17]. That is the innovative technology of the 21st century. The latest software versions allow to automatically calculate the coordinates of any gear tooth profile. For example, the Russian company "GloBAS Group" offers on the market a software product that allows you to calculate and simulate the design of any mechanism [6]. One of the functions performed by the program is the calculation and optimization of the characteristics of NCG. The calculation results can be immediately put into the software complex of the processing machine.

## 6 Modern Machines with Non-circular Gears

One of the successful options for using NCG was the drive of the conveyor, which was driven by the Maltese cross and made an intermittent movement [10–13]. The use of NCG made it possible to significantly reduce the angular acceleration without reducing the angular velocity of the main shaft, but reducing the load on the structural elements and reducing the dynamic effects that lead to the appearance of fatigue stresses and a decrease in the bearing capacity. This solution made it possible to increase the reliability of the conveyor and increase the service life of its components.

It is also effective to use NCG in crank-slider mechanisms, where it is necessary to ensure the movement of the slider at a constant speed. Since it is required to design the crank so that it rotates at a variable angular velocity, NCG placed in the crank drive can be used. Based on this principle, a key milling machine has been developed, in which the rotation of the crank of the crank mechanism is communicated from NCG in order to feed at an approximately constant speed.

NCGs have found application in automatic lathes for slow rotation of the camshaft when performing work operations, and fast rotation during idling. They are used in printing machines; in textile machines to obtain fabrics with a specific pattern; in silk-winding machines to obtain a barrel-shaped coil; in conveyors, in many devices and meters; in valve drives of engines, in metal-cutting machines; in pressing equipment, in motion control systems and in a number of other machines and mechanisms.

## 7 Conclusions

1. Models with NCG allow students to get acquainted with a little-known section of Mechanics, personally observe the movement of non-standard gearing, which significantly expands the horizons of future engineers, increases their interest in the study of engineering disciplines.
2. Unlike cam gears, mechanisms with NCG allow transferring significant loads and are highly reliable. They do not require special devices or springs to maintain continuity of contact in the higher pair, which leads to a simplified design and improves manufacturability.

3. NCG application in the past did not become widespread due to the complexity of manufacturing and a lack of understanding of their properties, but nowadays they are increasingly becoming popular among engineers. That happens thanks to the emergence of innovative technologies for the manufacture of teeth based on copying, 3D prototyping and printing, electric discharge processing and others.
4. The successful application of these unconventional gears leads to the creation of new machines and various mechanical tools (flow meters, bicycles, internal combustion engines, etc.) with high performance.

## References


1. Moon, F.: The Machines of Leonardo Da Vinci and Franz Reuleaux: Kinematics of Machines from the Renaissance to the 20th Century. History of Mechanism and Machine Science, vol. 2, pp. 239–241. Springer, Dordrecht (2007).https://doi.org/10.1007/978-1-4020-5599-7
2. Moon, F., et al.: Ferdinand Redtenbacher (1809–1863): pioneer in scientific machine engineering. Mech. Mach. Theory **44**(9), 1607–1626 (2009)
3. Golovin, A., Tarabarin, V.B.: Russian Model from the Mechanisms Collection of Bauman University, 245. Springer, Dordrecht (2008). https://doi.org/10.1007/978-1-4020-8776-9
4. Litvin, F.: Noncircular Gears: Design and Generation. University Press, Cambridge (2009)
5. Robinson S.: Principles of Mechanism. A Treatise on the Modification of Motion by Means of the Elementary Combinations of Mechanism, or of the Parts of Machines. Wiley, New York (1896)
6. I-Kan, A., Belyaev, A.: Cutting non-circular gears. In: Theory and Practice of Gears. Proceedings of the International Conference. Izhevsk: Izhevsk State Technical University, pp. 369–372 (1998)
7. Litvin F.L.: A life dedicated to the development of the modern theory of gearing. In: American Period. Proceedings of international symposium "Theory and Practice of Gearing", pp. 9–14 (2014)
8. Balbayev, G., Ceccarelli, M.: An experimental test validation of a non-circular gear train. In: Proceedings of International Symposium "Theory and Practice of Gearing", pp. 96–103 (2014)
9. Prikhodko, A., Smelyagin, A.: Creation and research of planetary mechanisms of intermittent motion with elliptical gear wheels. J. Bull. BMSTU Ser. Mech. Eng. (2019)
10. Voronov, S.A., Ivanov, I.I., Kiselev, I.A.: Investigation of the milling process based on a reduced dynamic model of cutting tool. J. Mach. Manufact. Reliab. **44**(1), 70–78 (2015). https://doi.org/10.3103/S1052618815010100
11. Polyakov, S.A., Kuksenova, L.I., Alekseeva, M.S.: Methodological foundations for choosing the materials for gears according to wear resistance criteria. J. Mach. Manufact. Reliab. **48**(4), 328–335 (2019). https://doi.org/10.3103/S1052618819040125
12. Smirnov, A.E., Mokhova, A.S., Semenov, M.Y., Kuksenova, L.I., Fakhurtdinov, R.S.: Mixed-surface impregnation of gear wheels made of 13Kh3N3M2VFB-Sh age-hardenable integrally alloyed steel aimed at improving surface hardness, wear-resistance, and back-to-back endurance. J. Mach. Manufact. Reliab. **46**(4), 404–408 (2017). https://doi.org/10.3103/S1052618817040173
13. Grubyi, S., Chaevskiy, P.: Improving efficiency of machining of grooves on shafts of increased hardness structural steel. Lect. Notes Mech. Eng.921–929 (2020)
14. Artobolevsky, I.: Mechanisms in Modern Technology. Lenand Publishing, Moscow (2019)

15. Polyakov, S., Kuksenova, L.: Methods for estimating the adaptive properties of mechanical systems as means of improving their performance indicators. Probl. Mech. Eng. Mach. Reliab. **46**(1), 20–24 (2017)
16. Egorova, O., Samsonov, K., Sevryukova, A.: New trend in learning through 3D modeling of historical mechanism's model explorations in the history of machines and mechanisms. In: Proceedings of the Fifth IFToMM Symposium on the History of Machines and Mechanisms, pp. 137–150. Springer (2016)
17. Egorova, O.V., Vukolov, A.Y.: Challenges of 3D mechanisms models in TMM training IREME. Int. Rev. Mech. Eng. **10**(5), 303–311 (2016)
18. Sustainable Development Goal 4. https://en.wikipedia.org/wiki/Sustainable_Development_Goal_4. Accessed 14 July 2021

# The Monolith: Epic of the Mechanical Progress, from the Great Exhibition to a Space Odyssey

Marco Pozzi(✉)

Dipartimento di Ingegneria Meccanica e Aerospaziale, Politecnico di Torino, Turin, Italy
marco.pozzi@polito.it

**Abstract.** Man has always built machines, but the nineteenth century was the century of machines. The Great Exhibitions were the great showcases to show the achievements of technology. A new way of describing machines was also born at that time. Even poets and writers therefore participate in the evolution of science and technology because they build myths in the collective imagination. Their words reach us, and they help us to reconstruct a past century, to understand ours and imagine the future.

**Keywords:** SDG4 · History of mechanism and machine science · Industrial revolution · Poetry · Storytelling · Biennale Tecnologia

During Biennale Tecnologia, an initiative for promoting the integration among science, technology and society, the Politecnico di Torino organized numerous initiatives aimed at high schools in Piedmont. After following a first lectio magistralis about technology issues, the participating students discussed in class; guided by the professor, they summarized the debate in some questions; after about a month, the students met to discuss it with the professor or his collaborators. Due to the pandemic, all the appointments took place in videoconference.

"What is the best way to reconcile technical studies with a passion for poetry?"
"What is the poem that best represents the relationship between poetry and technology?"
"Who is the author who managed to combine these two worlds in the best way?"

The three questions are taken from the list prepared by the students.

## 1 The Birth of Machine Epic

Machines made the history of mankind. From the plow to the auger, from the cart to the mill, from the harquebus to the clock, from the Neolithic to the 18th century the sapiens evolved by designing and building machines.

At the end of the 18th century the first steam engines spread, from Newcomen to James Watt. From the early 19th century blast furnaces were perfected to melt iron, in 1856 steel was cast without using coal. In 1821 the propeller steamer was invented, in 1825 the steam locomotive. Production increases, the demand for labour rises, cities


G. Quaglia et al. (Eds.): I4SDG 2021, MMS 108, pp. 595–600, 2022.
https://doi.org/10.1007/978-3-030-87383-7_63

grow together with needs and trade. The new production methods are not only evolving inside the factories, but they are shaping new habits of life, new ruling classes, new rich and new poor. The whole society changes quickly.

The Ecole Polytechnique was founded in Paris in 1795 and the Conservatoire des Arts et Métiers was established in 1798; in 1798 also the "Première Exposition des produits de industrie française" was inaugurated, in the centre of the great Champ-de-Mars, a rectangle with arcades built by François de l'Etoile, the future architect of the Arc de Triomphe de l'Etoile. Precision instruments, works of iron and steel, porcelain paintings, laces, books, canvases were there exhibited and made available to the public's gaze. Above the central temple-shaped building, a hot-air balloon swings in the sky, as a symbol of progress in motion. In 1849, at the eleventh edition, there are already 4,352 exhibitors, including an African colony, Algeria.

On the other side of the Channel, another nation could say: "products from all parts of the globe are at our disposal and we only have to choose what is best and most suitable for our purposes". Prince Albert, president of the London Society of Arts, declared it in a conference in 1850 [1].

In London, the following year, on May 1st, the first Great Universal Exposition was inaugurated. It took place in a colossal iron and crystal building called "Crystal Palace": windows, lots of light, scenography plants, under a majestic structure of cast and laminated iron beams. The products on display came from all over Europe, the United States, Turkey, for a total of 17,000 exhibitors. The success was enormous: 6,000,000 visitors. The Gallery of Machines was one of the focal points, with exhibits of gas ovens and Gothic-style frames, of bagging machines and suitable for various other tasks.

To respond to the English Greatness, the French Grandeur, after the edition in New York in 1853, organized the third *Exposition Universelle*, in 1855, in Paris. Le palais de l'Industrie was built, a 48-m vault, the widest at that time, thanks to iron lattice girders. But space was not enough, and it was necessary to add a Gallery of Machines, 27 m wide, and a Palais des Beaux Arts, near the Trocadero. Works of new metallurgy, textile machinery for mechanical spinning, or for the glass and porcelain industry were exhibited.

An already famous poet, Charles Baudelaire, visited the Exposition. He wrote a long essay, "Method of criticism. Of the modern idea of progress applied to the fine arts. Displacement of vitality." He imagined asking any honest Frenchman, reading his newspaper in his café, what progress meant to him; the honest Frenchman would have replied that progress was steam, electricity and gas lighting, which are miracles unknown to the Romans, discoveries that absolutely demonstrate the superiority over ancient peoples [2].

Baudelaire had already talked about it writing on the Salon of 1845, a periodic exhibition of painting and sculpture created in the 17th century, where the relevant works of the moment came together.

"Le peintre, le vrai peintre sera celui qui pourra arracher son côté épique à la vie d'aujourd'hui, et nous faire voir et comprendre, à travers la couleur et le dessin, à quel point nous sommes grands et poétiques avec nos cravates et nos chaussures vernies." [3].

So: to understand the epic side of life, to understand how great and poetic we are with our ties and patent shoes.

Baudelaire is usually known only "Les fleurs du mal", perhaps some prose inspired by absinthe and artificial paradises. But he's little known for his activity as an art critic and, in general, starting from art, of a very acute analyst of his society. And the poet is able to find only one primary element to represent the whole world, and build on it an epic, such as those "ties" and those "patent shoes", which was produced by the new civilization of machines.

We are talking about Paris in the mid-nineteenth century, in full industrial change.

Les vieux Paris n'est plus (la forme d'une ville
Change plus vite, hélas! que le cœur d'un mortel)

The poem is "Le Cygne", "The Swan", published in 1860, dedicated to Victor Hugo; he sang the extension of the barracks, the piles of stems and capitals, in a very deep "Tableau parisien".

Paris change! mais rien dans ma mélancolie
N'a bougé! palais neufs, échafaudages, blocs,
Vieux faubourgs, tout pour moi devient allégorie
Et mes chers souvenirs sont plus lourds que des rocs [4].

In addition to Baudelaire, even the other cursed poet par excellence cannot ignore to sing the machines of the industry.

"The city, its smoke and noise, pursued us down the roads.

[...]

Crystal grey skies. A strange pattern of bridges, some straight, some arched, others falling at oblique angles to the first, their shapes repeating in the illuminated curves of the canal, all of them so long and light that the banks, heavily canopied, sink and shrink. A few of these bridges are still freighted with hovels.

[...]

From certain points on the copper footbridges, platforms, and staircases that wind through the markets and around pillars, I thought I could judge the depth of the city! One marvel I couldn't reconcile: are the city's other regions above or beneath the level of the acropolis? Reconnaissance is impossible for the tourist of today. The commercial quarter is a circus in a single style: arcaded galleries" [5].

The civilization of machines also emerges from Arthur Rimbaud's "Illuminations".

In short, machines changed cities, they changed life. Technology became a means to give prestige to one's nation. The Expositions were the immense showcases where each country could show its wealth, its progress.

The Crystal Palace and the Palais de l'Industrie were the new industrial sites; they became epic, places with their own sacredness and a strong power of fascination, even of suggestion. The machine left the factory and entered the society; she escaped from her specialist room and went into the gala ball, dressing up so that everyone could appreciate her, even love her.

Fëdor Dostoevskij visited the Great Exhibition in London in 1862. In "Winter notes on summer impressions" [6] he wrote that the Universal Exposition was something amazing that you are aware of an immense thought: you perceive that something has already been achieved there, that there is the victory, there is the triumph. You even begin to fear something. He underlined that all this was so solemn, victorious and proud, that you begin to feel a burden on your heart. He perceived that event like a sort of biblical picture.

On the other hand, from 1838 to 1843 Dostoevskij studied engineering at the Higher School of Military Engineers in Petersburg, and undoubtedly something of those studies remained inside him. (Sooner or later it will be necessary to study seriously how engineering helped him write his characters and his main novels.)

Moreover, the great poets and novelists are neither journalists, nor chroniclers, nor authors of orderly reportage; but they synthesize the variables from the world to link the rhythm of evocation, in a work that, as in physics, causes thoughts to shift in listening thanks to some mysterious musical force.

The poet does not describe technology, but evokes the effects that technology diffuses in society; he talks about technology but does not describe it; he does not highlight the formula in the instruction booklet, but intercepts every change that technology causes, such as growing circles generated in the water by throwing a single stone.

And also from the words of the poets the machines took strength, almost authority, to declare "the victory, [...] the triumph". In fact, over the years the Exhibitions had grown further, to the point of overflowing the places used for the event, daring to elevate a solemn flag well above the city.

It happened in the 1889 edition, in Paris, when, in seventeen months, an iron monolith was raised for over 300 m.

The engineer Gustave Eiffel explained that the tower was built only by means of the progress of science and the art of engineering and the metal industry; and that only thanks to modern industry it has been possible to overcome past generations. [1].

The Tour Eiffel is still dominating the urban landscape today, invading the collective imagination at every latitude of the globe, in history books as much as on postcards and fridge magnets.

The machine built other machines, which become symbols. The machine creates itself as a symbol, as a fetish to be adored, bringing together the meanings of temple and work of engineering, and consequently of Man and God.

At the inauguration of the 1889 Exposition, the French Minister of Commerce and Industry, Pierre Tirand [1], said that progress does not slow down its path and the new generations replace the exhausted or disappeared forces; science was the sovereign force of the century and every day it penetrated more into the secrets of nature; steam and electricity had already revolutionized the economic order of the universe. Who knew what wonders and surprises science would prepare for future descendants?

## 2 Which Epic in Our Modernity?

Pierre Tirand's question joins the students' questions. But we, who are the descendants of this history, what can we answer?

We can imagine that when the sapiens began to describe the machine, he began to create an epic of the machine. The machines became a character in the story, its protagonist. They acquired an identity. The revolution is also a revolution of language, in which we begin to give names to things, re-founding around a society, which always seems better, in the image of Paradise.

"So the LORD God formed out of the ground various wild animals and various birds of the air, and he brought them to the man to see what he would call them; whatever the man called each of them would be its name." (Genesis 2:19) [7].

The fruits of the ancient Prometheus sprout everywhere. As Plato wrote in Protagora, through the technique, the man articulated the voice with words, and invented houses, clothes, beds, shoes [8].

Through the new names, with the storytelling, we can develop his own self-awareness of new phenomena, which are no longer only passively experienced as an inevitable fact, but which can instead be understood, and therefore mastered. It must be argued if it really succeeded; we have to think about weapons that will be used in world wars in the century to come, or about the social injustices we still talk about today, far from a shared solution.

Perhaps, instead of reason, myth prevailed, and the Grand Exhibitions were more like the modern soccer World Cup in competition between nations, rather than an opportunity to share rationality, the same with which the machines were built. The engineers were like a goleador or a rock stars, rather than masters.

With Plato's shoes or Baudelaire's patent leather shoes, we have walked through the centuries in a great game, in which we have all collaborated, building a society around "immense thought", realizing only over time that there really is "something to fear". Controlling the world is too ambitious a dream, also because the world, like Paris, "change plus vite que le cœur d'un mortel", the world changes faster than a man's heart.

So, it is not easy even for us to find what the epic of our iPhones is. And maybe, mechanics can help us. Just as the poets managed to represent the spirit behind the iron vaults and the machines of the first steam, so today, where much is virtual, it would help to rediscover the *mechanical* identity of the machines. The material and immaterial dimensions complement each other and considering them both the scientist manages to balance the knowledge of the world.

Poetry and engineering can find a new balance, as the experience of the poet and engineer Leonardo Sinisgalli had already shown us with the magazine "Civiltà delle Macchine", in the 1950s. Leonardo Sinisgalli was an engineer who wrote poems, short stories, essays, and was involved in design. Through this research he tried to find ways to tell the Mechanics in relation to the time he was living; with "Civiltà delle Macchine" he promoted the debate by mixing different disciplines, making them integrate each other.

In that editorial office there would have been Baudelaire and Rimbaud enrolled in engineering rather than in a literature degree; and Dostoevskij would have his first epileptic seizure on the morning of the first exam of mathematical analysis.

And probably the editorial office of "Civiltà della machine" would also have appreciated the verses of the song "Meccanica" by Franco Battiato appeared in the album "Fetus" (1971), reappeared even in English in 1999.

Meccanici i miei occhi
di plastica il mio cuore

Meccanico il cervello
sintetico il sapore

Meccaniche le dita
di polvere lunare
In un laboratorio
il gene dell'amore

My eyes are mechanical
my heart is made of plastic
my brain is mechanical
the taste is synthetic

my fingers are mechanical
made of moon dust
in a laboratory
the genes of love

The poet and philosopher Franco Battiato passed away on 18 May 2021, but this song, like others, will remain. Like poetry in the nineteenth century, even his music can make us enter the epic of the machine. So today TV series, electronic music, podcasts can do the same: the languages of expression and communication change over time to make Mechanics modern and perhaps more sustainable.

## References


1. Perilli, A.: Le grandi esposizioni universali. In: Civiltà delle macchine - Antologia di una rivista 1953–1957, Libri Scheiwiller, Milano, pp. 123–134 (1989)
2. Baudelaire, C.: Esposizione universale - 1855 - Belle arti. In: Opere, Arnoldo Mondadori Editore, Milano, p. 1164 (2006)
3. Baudelaire, C.: Salon de 1845. In: Oeuvres complètes, Editions Robert Laffont, Parigi (2004), p. 633 (2006)
4. Baudelaire, C.: Le cygne. In: Opere, Arnoldo Mondadori Editore, Milano, pp. 174–179 (2006)
5. Rimbaud, A.: Illuminations. In: Rimbaud Complete - Volume I: Poetry and Prose (trans. Mason, W. (ed.)). The Modern Library, New York (2003)
6. Dostoevskij, F.: Note invernali su impressioni estive, p. 52. Feltrinelli, Milano (2008)
7. The Holy See, New American Bible. https://www.vatican.va/archive/ENG0839/__P4.HTM. Accessed 29 May 2021
8. Platone: Protagora. In: Tutti gli scritti, Bompiani, Firenze, p 819 (2019)

# Siméon-Denis Poisson's Theory of Machines (1833)

Sandro Caparrini(✉)

Politecnico di Torino, Turin, Italy
sandro.caparrini@polito.it

**Abstract.** The earliest examples of theories of complex machines date from the eighteenth century. A fairly abstract, purely mathematical theory of machines was formulated by Gaspard-Gustave de Coriolis in 1829. In 1833 the mathematical physicist Siméon-Denis Poisson published his own reflections on this subject. His outline of a general theory of machines appeared in an appendix the development of nineteenth-century theoretical physics, Poisson's ideas (mainly inspired by Coriolis) are worth a closer look.

**Keywords:** SDG4 · History of mechanism and machine science · 19th century · Conservation of energy

## 1 Introduction

While the concept of sustainable development is recent, its underlying idea of economizing, or using resources wisely, is not. In fact, its scientific basis can be traced back to the eighteenth century and beyond. The concept of the efficiency of a machine, for example, can be considered as an attempt of computing the dispersion of energy, and originated in the context of the Industrial Revolution.[1]

For centuries, machines were essentially those of classical antiquity. As late as 1803, the *Éléments de Statique* by Louis Poinsot, an innovative textbook on mechanics, was limited to studying only the inclined plane, the lever, the wedge, the wheel & axle, the pulley and the screw. The first definition of machine given by Poinsot is, in essence, still that of the ancient Greeks:

> Machines are usually defined as instruments whose function is to transmit the action of forces.

However, a couple of pages later, Poinsot gives a second definition:

> Thus we are naturally led to the following general definition of machines, namely: machines are nothing more than bodies, or systems of bodies, restricted in their movements by some obstacles.

[1] For al history of machines, see [1].


G. Quaglia et al. (Eds.): I4SDG 2021, MMS 108, pp. 601–607, 2022.
https://doi.org/10.1007/978-3-030-87383-7_64

This shift in thinking reflects a larger historical context, for in the same years a general theory of complex machines was being developed in France. The theory of the efficiency of a steam engine expounded by Sadi Carnot in his *Reflections on the motive power of fire* (1824) is the most famous example of these new ideas. As is well known, Carnot's book marks the beginning of thermodynamics.[2] Yet, in addition to the *Reflections*, several other fundamental texts were published in the first half of the nineteenth century. Moreover, further back in time, there were some notable memoirs on the theory of machines during the eighteenth century.[3] These old-times books and research papers, both charming and instructive, are not sufficiently known to the generality of physicists and engineers.

## 2 Energy Efficiency in Machines

The need for energy-efficient machines played a significant role in the Industrial Revolution.

In 1712 Thomas Newcomen's steam engine seemed revolutionary. A balance wheel that oscillated 12 times per minute was capable of lifting 45 L of water from a depth of 50 m with each swing. The performance was much higher than that which could be achieved with animal force. A test carried out in 1752 measured this superiority: operating on a well 73 m deep, in one day, a horse-drawn water pump was able to lift 305 tons of water; at the same time, and from the same depth, a Newcomen machine lifted 1,140 tons of water.

In 1768 John Smeaton managed to double this efficiency by means of a number of technical improvements.

At about the same time James Watt invented the true steam engine. In 1776, a Watt machine consumed only 1/3 of the coal needed by an equivalent Newcomen machine. As payment for the use of these machines, mine owners were required to pay 1/3 of the money they were saving.

The improvements continued well into the 19th century. In 1835 a steam engine had an efficiency three to five times that of a machine built thirty years earlier. These technological advances led to the formulation of a general theory of machines, as we shall see.

## 3 Work

A hint of the modern concept (work) = (force) × (displacement) already occurs in Galileo's early writings [5], but the precise definition appeared only at the end of the seventeenth century [6].

In a memoir on mills by the French physicist and engineer Guillaume Amontons, read at the Paris Academy of Sciences in 1699 but published only in 1732, the modern definition is explicitly stated. According to Amontons, work (*puissance continuelle*) is defined by (force) × (displacement)/(time). Amontons experimentally measured the

2 On the history of thermodynamics, both technical and theoretical, see [2] and [3].
3 On the history of mechanics during the eighteenth century, see [4] and references therein.

*puissance continuelle* of men and horses, finding that a horse produces about six times the work of a man.

Five years later the mathematician and physicist Antoine Parent further developed these ideas; unfortunately, his memoir was only published in 1745. Parent measures the *puissance motrice* supplied to a machine by means of the distance a weight falls for a certain time. More importantly, he asserts that each machine has a maximum efficiency, which can be theoretically calculated.

In 1738 was published Daniel Bernoulli's famous *Hydrodynamica*. In this book Bernoulli systematically uses work (*potentia*) to study the performance of hydraulic mechanisms. Thus the concept of work entered theoretical mechanics.

## 4 Coriolis (1829)

Despite these remarkable results, the general theory of machines only began in 1782 with the publication of the *Essay on Machines in General* by the French mathematician and engineer Lazare Carnot (father of Sadi Carnot). In the following years appeared Jean Nicolas Pierre Hachette's *Elementary Treatise on Machines* (1811), C.-A. Coulomb's *Theory of Simple Machines* (1821), Claude Navier's *Lectures on the Application of Mechanics to the Construction of Machines* (1826) and Jean-Victor Poncelet's *Introduction to Industrial Mechanics* (1829). Some of these works were related to the teaching of applied mechanics at the École Nationale des Ponts et Chaussées.

A notable advancement came with the publication of the *Calculation of the Effect of Machines, or Considerations on the Use of Engines and their Evaluation* (1829) by the mathematician and engineer Gaspard-Gustave Coriolis.[4] This book is more theoretically oriented than the previous works. There are few practical applications and very few examples, and the whole discussion is kept at the highest level of generality. In fact, the *Calculation of the Effect of Machines* can be viewed as a treatise on the application of the work-kinetic energy theorem to systems of constrained rigid bodies.[5]

In the *Calculation*, Coriolis is greatly interested in the "economy of the engine" (*économie du moteur*), that is, in minimizing the dissipation of work. He studies the loss of work due to friction and impact, and the maximum of work which can be obtained from a machine.

This is the book in which Coriolis coins the term "work" (*travail*) for the product (force) × (displacement), and introduces the factor 1/2 in the definition of what we now call "kinetic energy".

Today, of course, Coriolis is best known for the law of the composition of accelerations which bears his name. He proved a special case of this theorem in a memoir read at the Paris Academy of Science in 1831 and published in 1832:

> The principle of living forces is still true for motion relative to moving axes, provided that there are added to the quantities of action calculated from the given forces and the arcs described in the relative motion, other quantities of action which

[4] On the life and works of Coriolis, see [7].

[5] The work-kinetic energy theorem states that in any displacement of a body, the change in the kinetic energy is equal to the work done by the forces.

are due to the forces equal and opposite to those which it would be necessary to apply to each moving point in order to make it take the motion that it would have had if it were invariably connected to the moving axes.[6]

While the mathematical theory of relative motion has been studied at least since the 1750s, it was Coriolis who first obtained general formulae for this class of problems.[7]

## 5 Poisson (1833)

The French mathematical physicist Siméon-Denis Poisson (1781–1840) has been somewhat overlooked by historians of science. As is well known, he achieved fundamental results in analytical mechanics, elasticity theory, fluid mechanics, potential theory and electrostatics. Together with Laplace and Fourier, Poisson laid the foundations for the development of theoretical physics in the second half of the nineteenth century.

In addition to his research work, Poisson wrote treatises on subjects as varied as heat conduction, capillarity and probability. Chief among them is the *Traité de mécanique* (1811). The *Traité* was influential in France and abroad, was carefully read by the big names of science and quickly became the model for all nineteenth-century mechanical texts. It was an important link between two centuries, for most scientists of the nineteenth century learned of Euler's dynamics of rigid bodies and Lagrange's general theory of systems of linked mass points from Poisson's *Traité*.

About twenty years after the appearance of the *Traité*, Poisson reprinted the book (1833). In this second edition a considerable amount of new matter was introduced and modifications were made in some of the older methods described. In fact, Poisson included most of the new results that had emerged since 1811. He added, for example, the theorem according to which the principle of moment of momentum[8] is a consequence of the second law of dynamics and of the principle of action and reaction; this proposition was first stated by Jacques Binet in 1823. Today this so-called "Poisson's theorem" can be found in every textbook of elementary mechanics.[9]

In an appendix to this new edition, entitled *Addition relative to the application of the principle of living forces to the calculation of machines in motion*, Poisson gives his own concise and abstract treatment of the theory of machines. He cites as his main references the lectures of Navier and Poncelet; in fact, most of the *Addition* is based on the work of Coriolis.

There are several innovative elements in the *Addition*. In what follows we shall restrict our discussion to these aspects.

The incipit provides information on the physical principles and the mathematical approach adopted by Poisson:

6 The "quantity of action" is what we now call "work", while the "living force" is our "kinetic energy".

7 On Coriolis' paper see [8].

8 The principle of moment of momentum states that the rate of change of moment of momentum of a system about any fixed origin is equal to the sum of the moments of the external forces about that origin.

9 The theorem was attributed to Poisson by Truesdell; see [9]. On the history of moment of momentum see [10] and [11].

> The conditions of the equilibrium of forces applied to machines are furnished immediately by the principle of virtual velocities; the theory of their motion is given by that of living forces, which enables us to calculate, in the most direct manner, the effects of the forces that are applied to them. This application of the principle of living forces constitutes, so to speak, the point at which rational and practical mechanics coincide. It is on this account that the author thought to it necessary to give, in this addition, a brief sketch of the most general principles relative to this matter.[10]

Here Poisson implicitly acknowledges the scientific value of the new theory of machines. Although based on technical considerations, the results obtained by two or three generations of engineers-mathematicians formed by then a consistent whole, providing a solid foundation for a new chapter of mechanics.

Poisson adopts the term *travail* to denote the product (force) × (displacement):

> The products *Pdp* and *Qdq* [...] have received different denominations; they have been termed *the quantity of action, the moments of activity, the dynamical effects* of the forces *P* and *Q*. It would be useful if they were always designated by the same term; the denomination of *quantities of elementary work*, which was proposed to be given to them by M Coriolis, is the one that will be here adopted.

Taking into account the diffusion of the *Traité* and the fame of Poisson, it is probably because of the *Addition* that the new terminology became generally accepted.

The *Addition* draws much from the preceding literature: the main mathematical tool is the work-kinetic energy theorem; a machine is any mechanical system that transmits forces; a distinction is made between the work of the active forces (*travail moteur*) and that of the resisting forces (*travail résistant*); the *dynamical unit* is the work done to raise a weight of 1000 kg to a height of 1 m.

While most of the results reported in the *Addition* are taken from the existing literature, Poisson improves on Coriolis in some significant ways.

Poisson brings in the full force of mathematics. In his book Coriolis had not made use of three-dimensional analytic geometry. Instead, Poisson expresses the work of a force using the differential form $Xdx + Ydy + Zdz$, where $X$, $Y$, $Z$ are the components of the force, introduced by Lagrange in the *Méchanique analitique* (1788). In today's textbooks, this becomes a curvilinear integral.

Scholars of French science will recognize some characteristic features of that time. For example, Poisson expresses the $x$ component of the internal forces between two points in the form $R\ (x–x')/r$. where $x$ is the coordinate of the first point, $x'$ that of the second point, $r$ the distance between the two points and $R$ the intensity of the force. This expression, introduced by Laplace in the first volume of the *Traité de mécanique céleste* (1798), makes immediately apparent Newton's third law of motion.

[10] For the sake of references I will use the English translation *A Treatise of Mechanics*, transl, by the Rev. Henry H. Harte (London & Dublin 1842). It must be noted, however, that this translation is sometimes strange and clumsy to modern eyes. For example, Harte employs the term *cipher*, which comes from the non-classical Latin word *cifra* and in turn from the Arabic word *ṣifr*, to denote the number zero (see p. 628).

There are also more substantial improvements, related to the underlying physics. Coriolis had observed that the definition of work leads to a paradox, for apparently a man carrying a weight and walking on a horizontal road at constant speed does no work. According to Poisson, the solution lies in the fact that walking is essentially a sequence of falls and rises:

> If as he walks [a] man has first the left foot before the right, his centre of gravity is then depressed below its natural position [...]. Then by leaning on his left foot, the man, by means of the pressure of this foot against the ground, brings up his right foot to a level with the left, afterwards the right foot is advanced before the left, and placed on the ground; in this operation he makes an entire step, which is thus made up of two parts.

By introducing the height of the "depression" and the average speed of the man, Poisson obtains a rough estimate (unfortunately not much useful in practice) of the work performed by a walking man with a bag.

In calculating the effects of machines, it is frequently important to distinguish between the absolute and the relative velocities. This means, for example, having to deal with points constrained to move on a rotating surface. For this reason, Poisson demonstrates anew Coriolis' theorem on the calculation of the living forces in a moving frame of reference.[11] However, there are significant differences between the two proofs. Poisson does not consider explicitly two frames of references, applies what we now call the conservation of energy and introduces the molecular forces of attraction and repulsion between the particles. Going into the mathematical details would require more space than can be here afforded.

## 6 Conclusion

While Coriolis' *Calculation* and Poisson's *Addition* did not directly influence the concept of sustainable development, they had an indirect but important effect on the formulation of the physical law which govern sustainability, that is, the conservation of energy.

Most historians of physics argue that the first clear and general statement of the conservation of energy occurs in Hermann von Helmholtz's *On the Conservation of Force* (1847).[12] Helmholtz based his considerations on the idea that matter is made up of point masses that interact via attractive and repulsive central forces; the work-energy theorem here plays a crucial role. While most of the mathematics in the *Conservation* originated with Lagrange's *Méchanique analitique*, the notations are borrowed from Poisson.

The ideas discussed in the preceding pages have long-range ramifications. For example, as early as the late eighteenth century, some geologists warned of the end of coal and its consequences [13]. Also, at the end of the nineteenth century, the doctrine of energetics purported to unify all of physics under the single principle of the conservation of energy [14]. Finally, we note that in 1865 the economist Stanley Jevons initiated

11 Interestingly, Poisson had refereed Coriolis' paper.

12 On the discovery of the conservation of energy, see [12].

a properly economic analysis of fossil fuels based on the energy needs of industrial manufacturers [15]. Unfortunately, to go into all this in detail would take us too far afield.

**Acknowledgment.** The author wishes to thank Vittorio Marchis (Politecnico di Torino) for his advice and encouragement. Thanks are also due to the referees, whose most valuable remarks greatly improved the final presentation of the text. The usual responsibility for facts and opinion remains with the author.

## References

1. Marchis, V.: Storia delle machine: tre millenni di cultura tecnologica. Laterza, Bari (2005)
2. Cardwell, D.S.L.: From Watt to Clausius: The Rise of Thermodynamics in the Early Industrial Age. Heinemann, London (1971)
3. Truesdell, C.: The Tragicomical History of Thermodynamics, 1822–1854. Springer, New York (1980). https://doi.org/10.1007/978-1-4613-9444-0
4. Caparrini, S., Fraser, C.: Mechanics in the eighteenth century. In: Buchwald, J., Fox, R. (eds.) The Oxford Handbook of the History of Physics, pp. 1–13. Springer, Heidelberg (2016)
5. Cardwell, D.S.L.: Some factors in the early development of the concepts of power, work and energy. Br. J. Hist. Sci. **3**(3), 209–224 (1967)
6. Fontenau, Y.: Les antécédents du concept de travail mécanique chez Amontons, Parent et D. Bernoulli: de la qualité à la quantité (1699–1738). Dix-huitième siècle **1**(41), 39–368 (2009)
7. Moatti, A.: Le mystère Coriolis. CNRS, Paris (2014)
8. Dugas, R.: A History of Mechanics. Dover Publications, New York (1988)
9. Truesdell, C.: A First Course in Rational Continuum Mechanics. Academic Press, Boston (1976)
10. Caparrini, S.: On the history of the principle of moment of momentum. Sciences et Techniques en Perspective **3**(2), 47–56 (1999)
11. Caparrini, S.: The discovery of the vector representation of moments and angular velocity. Arch. Hist. Exact Sci. **56**(2), 151–181 (2002)
12. Elkana, Y.: The Discovery of the Conservation of Energy. Hutchinson, London (1974)
13. Sieferle, R.F.: The Subterranean Forest: Energy Systems and the Industrial Revolution. The white Horse Press, Cambridge (2001)
14. Deltete, R.: Helm and Boltzmann: energetics at the Lübeck Naturforscherversammlung. Synthèse **119**(1/2), 45–68 (1999)
15. Missemer, A.: Fossil fuels in economic theory: back to the 19th century British debates. Revue Française de Civilisation Britannique **23**(3), 90–96 (2018)

# A Sustainable Association Case Study: IFToMM Member Organization Turkey

Gökhan Kiper[1(✉)] and Eres Söylemez[2]

[1] İzmir Institute of Technology, 35430 İzmir, Turkey
gokhankiper@iyte.edu.tr
[2] Middle East Technical University, 06800 Ankara, Turkey

**Abstract.** As a member of IFToMM founded in 2011, Turkish Machine Theory Association's (MakTeD) structure has put special emphasis on sustainability of the association and initiated several activities for improving the quality of education and research in mechanism and machine science area, hence supporting the SDG4 Quality Education of the UN. This paper presents the activities and methodologies of MakTeD. The paper starts with a brief history of mechanism and machine science in Turkey. In 10 years MakTeD organized 4 symposiums, 1 conference, 4 summer/winter schools, 8 workshops and hosted an IFToMM Executive Council meeting. Each chapter of the biyearly held National Symposium on Theory of Machines is held by a different university and especially newly established and the universities in relatively smaller cities are prioritized in order to promote those universities and to contribute to development of these universities. Especially the summer/winter schools and workshops have a widespread impact on mechanism and machine science education and research at the universities and institutions. MakTeD also contributes to the community with published books and support programs for young researchers.

**Keywords:** SDG4 · Theory of machines · Civil society organization · Sustainability of association

## 1 Introduction

Founded in 1969, the primary aim of International Federation for the Promotion of Mechanism and Machine Science (IFToMM) is to promote research and development in the field of Machines and Mechanisms by theoretical and experimental methods, along with their practical application and to broaden contacts among persons and organizations of different countries engaged in scientific or engineering work in the field of Machines and Mechanisms, or related sciences. Another aim of IFToMM is to assist the developing countries in their work in Mechanism and Machine Science (MMS) by sponsoring the visits of experts, organizing special courses, and other undertakings [1]. IFToMM community has aimed and still aims promotion of MMS especially in under-developed and developing countries.

Turkey has been accepted as a member organization during the 13$^{th}$ IFToMM World Congress in 2011. In the same year, the Turkish Machine Theory Association (MakTeD)


G. Quaglia et al. (Eds.): I4SDG 2021, MMS 108, pp. 608–615, 2022.
https://doi.org/10.1007/978-3-030-87383-7_65

[2] was established as a civil society organization of Turkey. The charter of MakTeD has been written in parallel with IFToMM aims. MakTeD aims to promote national research and development activities, international collaboration, national and international mobility, and financially and technically support young researchers in MMS or associated scientific areas [3].

MakTeD has put special emphasis on the sustainability of the association and promote quality education since its establishment, hence supporting the Sustainable Development Goal (SDG4) about Quality Education of the UN. This paper presents the activities and methodologies of MakTeD as a good example of a sustainable organization for education and research. The paper also aims to document the activities of IFToMM Member Organization Turkey (a.k.a. MakTeD) in its first 10 years.

## 2 Brief History of MMS in Turkey and MakTeD

In Turkey, studies on MMS can be said to be started with Prof. Bekir Dizioğlu. Having obtained his Ph.D. degree from Dresden University, Dizioğlu returned to İstanbul in 1947 as a faculty member at Istanbul Technical University (ITU), which was the only technical university in Turkey in those years. Dizioğlu went back to Germany in 1960, but he kept close ties in Turkey and his two students Fuat Pasin and Mustafa Köseoğlu taught mechanisms and dynamics of machinery courses. The MMS field at ITU was very much influenced by the German school.

After 1960's several new technical universities were founded in Turkey. One of the most influential engineering school was Middle East Technical University (METU), which was based on American engineering education system and language of instruction in English. Due to the need of instructors for these technical universities, in late 60's several graduates of METU and ITU went abroad (mainly to United States and Great Britain) in all different fields of engineering to continue their graduate studies and a big percentage returned in the first half 1970's. In several Turkish Universities small groups working on MMS were formed.

In 1984, with the support of Turkish Scientific Council of Turkey, National Symposium on Theory of Machines (UMTS) was initiated. The first symposium was held at METU, Ankara. In the first symposium there were 60 attendees. It was decided to held this symposium every other year. 20$^{th}$ chapter of the symposium series will be held at Dicle University, Diyarbakır on September 2021.

In 2011 it was found that there were a satisfactory number of young academicians working on MMS. MakTeD was founded in accordance with the Turkish laws and MakTeD formally applied for membership to IFToMM.Together with the establishment of MakTeD, UMTS started to be held under the patronage of MakTeD starting with the 16$^{th}$ chapter in Atatürk University, Erzurum in 2013. Besides the symposia, MakTeD started organizing national/international summer/winter schools, workshops and also hosted an IFToMM conference. In addition to the organized activities, MakTeD has some support programs for young researchers.

## 3 MakTeD Activities and Supports

This section summarizes the activities and supports of MakTeD. The activities are grouped as the National Symposium on Theory of Machines, summer/winter schools and workshops; published books and other activities. Also, MakTeD supports for programs for young researchers are briefly explained.

### 3.1 The National Symposium on Theory of Machines (UMTS)

UMTS has been organized biyearly since 1984 [4]. Together with a historical review of MMS studies in Turkey, Sabuncu listed the main paper topics in the first 15 UMTS chapters as mechanism science, mechanical vibrations, dynamics of machinery, acoustics, automotive, control, agriculture machines and others (education, etc.) [5]. It is noted that not only academicians from Turkish universities, but also several researchers from institutions, companies and foreign universities have attended the symposia.

An important aspect of UMTS is that each chapter is being held by a different university (Fig. 1 – except for years 1986 and 1999) and especially newly established and the universities in relatively smaller cities are prioritized in order to promote those universities and to contribute to development of these universities. Many undergraduate or graduate level young researchers meet in UMTS and collaborate later on in their professional lives.

**Fig. 1.** The universities that organized the UMTS chapters.

Until 2013, the next UMTS organizer would be decided by the attendees during an UMTS. Starting with 2013, UMTS started to be organized under patronage of MakTeD, where universities apply MakTeD by filling in an application form and the organizer is decided by the MakTeD members by grading all applicants according to specified criteria. Such an application procedure motivates competition, increases the quality of the symposium and the decision is well received by the community.

UMTS proceedings play an important role in Turkish MMS literature. As an indicator of widespread impact of UMTS proceedings, a Google Scholar search reveals 541 documents and citations to UMTS (UMTS proceedings before 2005 are not available online). As a special case, UMTS 2015 was organized with international participation with the name TrC-IFToMM Symposium on Theory of Machines and Mechanisms (TrISToMM), where 100 Turkish and 23 English papers were presented.

### 3.2 Summer/Winter Schools and Workshops

2 summer schools, 2 winter schools and 8 workshops were organized under patronage of MakTeD in 2013–2020 period (Table 1) [6]. IFToMM Summer School on Mechanism Design for Medical Applications (2018) and Workshop on Methodologies for Increasing the Positioning Accuracy of High-Acceleration Parallel Robots Used in Industrial Applications (2019) are international events, whereas the others are national events.

**Table 1.** Summer/Winter Schools and Workshops organized in 2013–2020.

| Event name | Date | Venue |
|---|---|---|
| Online Workshop on Graphical Methods in Kinematic Synthesis (KS2020) | 7–12 Sept. 2020 | Online |
| Workshop on Methodologies for Increasing the Positioning Accuracy of High-Acceleration Parallel Robots Used in Industrial Applications | 20–21 Nov. 2019 | Middle East Technical University, Ankara |
| Workshop on TRIZ (Creative/Innovative Problem Solving Technique) New Product/Process/Technology Development | 30–31 Jan.–1 Feb. 2019 | İskenderun Technical University, Hatay |
| IFToMM Summer School on Mechanism Design for Medical Applications | 17–22 Sep. 2018 | Izmir Institute of Technology, İzmir |
| Workshop on Fundamental Dynamics of Machinery Education | 29 Jan.–3 Feb. 2018 | Ondokuz Mayıs University, Samsun |
| Workshop on 3D Printers and Additive Manufacturing | 18–22 Sep. 2017 | Middle East Technical University, Ankara |
| Workshop on System Dynamics and Control | 31 Aug.–3 Sep. 2016 | Hitit University, Çorum |
| Workshop on Robot Dynamics and Control (RDK 2016) | 3–6 Feb. 2016 | Gazi University, Ankara |
| Workshop on Robot Kinematics (RKC 2015) | 31 Aug.–3 Sep. 2015 | Gaziantep University, Gaziantep |
| Winter School on Dynamics of Machinery: Theory and Applications | 2–8 Feb. 2015 | Middle East Technical University, Ankara |
| Winter School on Kinematic Synthesis (KS2014) | 3–8 Feb. 2014 | Erciyes University, Kayseri |
| Summer School on Kinematic Synthesis (KS2013) | 2–7 Sep. 2013 | Izmir Institute of Technology, İzmir |

Just as IFToMM, MakTeD's main focus is to promote MMS among young people. Each summer/winter school or workshop organized has been conducted by voluntary presenters with an average of 30 participants, most of which are under the age of 30. Interestingly, almost %40 of the participants in these events were not university affiliated. Except for the international summer school, where there were several invited speakers from all around the world, participation to all these events were free of charge, expenses being sponsored by MakTeD and the organizing university and sometimes by a branch of Turkish Chamber of Mechanical Engineers or a private company.

Most of the events organized so-far are specifically intended for young academicians in order for them to learn from competent researchers and improve their teaching and research activities. MakTeD administrators believe that such events are very important in increasing the level of university education in MMS and aim for the sustainability of these events by organizing one or two every year. Besides the technical benefits, participants strengthen their professional network and the bonding within the community increases. Figure 2 presents a picture taken during the gala dinner of Workshop on Fundamental Dynamics of Machinery Education.

**Fig. 2.** Participants of Workshop on Fundamental Dynamics of Machinery Education - Samsun, 2018.

## 3.3 MakTeD Books and Lecture Notes

MakTeD also aims to contribute to Turkish literature on MMS. MakTeD possesses a publisher certificate since 2015. Proceedings of 2015, 2017 and 2019 have been published by MakTeD and the electronic versions of UMTS proceeding books since 2005 are freely available at MakTeD website [7]. The notes of Workshop on Robot Kinematics (RKC 2015), Workshop on Robot Dynamics and Control and Workshop on Fundamental Dynamics of Machinery Education were compiled as 3 books in Turkish by Prof. Özgören and published by MakTeD [8–10]. There are also appendices in [8, 9] which are short Turkish-English dictionaries for robotics terms. The MakTeD book cover pages are illustrated in Fig. 3.

Besides the published books, MakTeD Web-page serves as an archive for notes of workshops and other lecture notes. Lecture notes on System Dynamics and Control [11] are notes from Workshop on System Dynamics and Control held in 2016. Turkish and English lecture notes and videos of Prof. Söylemez on Mechanisms are also available online [12].

**Fig. 3.** MakTeD books.

### 3.4 Other Activities, Supports and International Collaborations of MakTeD

Besides UMTS and workshops as the routine events, there are two international events organized under patronage of MakTeD. The first one is MeTrApp 2017 - The 4th Conference on Mechanisms, Transmissions and Applications organized at Karadeniz Technical University, Trabzon in 2017 and the IFToMM 2$^{nd}$ Face-to-Face Executive Council Meeting at Middle East Technical University, Ankara in 2019.

In 2018, MakTeD started the "Young Researcher Fund" support program [13]. The program is inspired by the IFToMM Young Delegates Program [14] and works in junction with it. The program provides financial support to young MakTeD members (age 35 or less) for attending IFToMM affiliated events. With this program, MakTeD aims to increase mobility of young researchers.

Turkish researchers have been attending IFToMM events since very early times of IFToMM, but the level of participation has considerably increased in the last 10 years. Promotion and supports of MakTeD have important effect in this increase. The increase in communication resulted in increase in mobility, exchanges and international collaboration in the form of joint projects and publications.

## 4 Conclusions

This paper presents the activities and methodologies of IFToMM member organization Turkey, a.k.a. Turkish Machine Theory Association (MakTeD). In recent years, certain unfortunate events about some members of IFToMM resulted in severe problems, even at the level of termination of membership. Having witnessed these problems, as a relatively new member of IFToMM, MakTeD has worked on establishing a sustainable association. Being founded as an official civil society organization, MakTeD has raised considerable amount of capital which ensured organization of large-scale events and small workshops with free-of-charge participation for especially young researchers. Hundreds of researchers have attended and benefited from these events in the last 10 years.

MakTeD activities mainly aim to increase quality of education and research in the MMS field in Turkey. The activities will continue by organizing and supporting event series and standalone events such as the 20$^{th}$ National Symposium on Theory of Machines (UMTS 2021) in September 2021 and 6$^{th}$ Turkish Robotics Conference in July 2021. Meanwhile, the positive developments in MMS community in Turkey also result in increase in global contribution to MMS.

## References


1. IFToMM, IFToMM Constitution and By-Laws 2019 (2019). http://iftomm.net
2. Turkish Machine Theory Association Homepage. http://makted.org.tr. Accessed 12 May 2021
3. Makina Teorisi Derneği, Makina Teorisi Derneği Tüzüğü (2019). http://makted.org.tr/eski/belgeler/tuzuk.pdf.
4. Makina Teorisi Derneği, Ulusal Makina Teorisi Sempozyumu. http://makted.org.tr/index.php/dernegimizin-duzenledigi-etkinlikler/ulusal-etkinlikler/. Accessed 12 May 2021
5. Sabuncu, M.: Türkiye'de makina teorisi ve dinamiği anabilim dalındaki eğitim ve araştırmada gelinen nokta. In: 16. Ulusal Makina Teorisi Sempozyumu, Erzurum (2013). http://makted.org.tr/wp-content/uploads/misc/Makina_Teorisi_ve_Dinamigi_Analiz.pdf
6. Makina Teorisi Derneği, Workshops and Summer/Winter Schools. http://makted.org.tr/index.php/dernegimizin-duzenledigi-etkinlikler/calistaylar/?lang=en. Accessed 12 May 2021
7. Makina Teorisi Derneği, Sempozyum Bildiri Kitapları. http://makted.org.tr/index.php/kaynaklar/bildiri-kitaplari. Accessed 12 May 2021
8. Özgören, K.: Robot Kinematiği: Seri ve Paralel Manipülatörlerin Analitik ve Yarı-Analitik Yöntemlerle Konum ve Hız Analizleri, Makina Teorisi Derneği, Lecture Note Series, No: 1, Ankara (2015). http://makted.org.tr/wp-content/uploads/kaynaklar/RobotKinematigi.pdf

9. Özgören, K.: Robot Manipülatörlerin Dinamiği ve Kontrolü, Makina Teorisi Derneği, Lecture Note Series, No: 2, Ankara (2016). http://makted.org.tr/wp-content/uploads/kaynaklar/robot_manipulatorlerin_dinamigi_ve_kontrolu.pdf
10. Özgören, K.: Mekanizmaların Hareket ve Kuvvet Analizi, Makina Teorisi Derneği, Lecture Note Series, No: 3, Ankara (2018). http://makted.org.tr/wp-content/uploads/kaynaklar/Mekanizmalarin_Hareket_ve_Kuvvet_Analizi.pdf
11. Makina Teorisi Derneği, Sistem Dinamiği ve Kontrol (2016). http://makted.org.tr/index.php/kaynaklar/sistem-dinamigi-ve-kontrol/?lang=en. Accessed 2021/05/12.
12. Makina Teorisi Derneği, Lecture Notes. http://makted.org.tr/index.php/kaynaklar/ders-notlari/?lang=en. Accessed 12 May 2021
13. Makina Teorisi Derneği, Young Researcher Fund. http://makted.org.tr/index.php/destekler/genc-arastirmaci-destegi/?lang=en. Accessed 12 May 2021
14. IFToMM, Young Delegates Program. http://iftomm.net/index.php?option=com_content&view=article&id=34:young-delegates-program&catid=16&Itemid=147. Accessed 12 May 2021

# Implementation of a Didactic Regenerative Braking System

Bechara Nehme[1](✉), Tilda Akiki[1], and Barbar Zeghondy[2]

[1] Department of Electrical, Telecommunications and Computer Engineering, School of Engineering, Holy Spirit University of Kaslik (USEK), Jounieh, Mount Lebanon, Lebanon
becharanehme@usek.edu.lb

[2] Department of Mechanical Engineering, School of Engineering, Holy Spirit University of Kaslik (USEK), Jounieh, Mount Lebanon, Lebanon

**Abstract.** New and emerging engineering applications are being developed to respond to all engineering challenges. Especially climate change and energy security. Renewable energy and its related topics are all new applied concepts that should be taught in engineering schools. Besides, new soft skills are being asked in today disruptive jobs' era like complex problem solving, critical thinking, and creativity. Lifelong learning is key in today's educational systems. We have designed and implemented a renewable energy laboratory that responds to actual regional and international needs in engineering skills. In addition, the laboratory will foster creativity mindset in future engineers and leaders. Appropriate experimental sets have been designed and implemented. The lab covers a lot of green solutions which are deployed nowadays. The laboratory may also address research topics that students may undertake. We focus in this paper on regenerative braking systems. The latter is deployed in a wide range of applications. We have successfully developed an experimentation apparatus aiming to teach students on this important system. Leading in energy security and fighting against unemployment is a key responsibility of engineering schools. Educators should redesign their courses and create new courses that respond to the marked needs.

**Keywords:** SDG7 · SDG4 · Critical thinking · Today's skills · Creativity · Energy challenge

## 1 Introduction

In this paper, we present the development of a didactic laboratory for renewable energy sources. We focus on regenerative braking systems. The laboratory aims to prepare students to future needed technical and soft skills. Target 7.b of SDG7 aims to develop an infrastructure where modern and sustainable energy services can be deployed. In addition, target 4.7 of SDG4 aims that learners acquire the needed skills and knowledge for sustainable development.

PV share in energy production is increasing and has exceeded 1% of global energy generation [1]. In recent years, renewable energy sources are becoming an emerging


G. Quaglia et al. (Eds.): I4SDG 2021, MMS 108, pp. 616–623, 2022.
https://doi.org/10.1007/978-3-030-87383-7_66

domain in engineering which is creating new jobs. Regenerative braking systems are now deployed in different areas like cars, escalators, and elevators. Therefore, engineering schools must prepare future engineers to meet the requirements of new job vacancies. Moreover, new learning outcomes related to renewable energy sources must be met. Students must be able to: (1) Design and size renewable power systems, (2) Recognize the effect of environmental conditions on renewable energy sources, (3) Determine the best solution that suits the studied case, and (4) Generate and implement control algorithms to optimize the power generation of renewable energy sources.

The importance of developing a didactic lab in parallel to the course taught is very motivating to students. Studies have proved that this connection motivates the perception of students towards the course value. Motivation is needed in encouraging students to problem solving [2]. In addition, a didactic lab helps to eliminate some misconceptions acquired by students [3].

Creativity is the ability to transcend traditional ideas or patterns and create meaningful new ideas. Through creativity, humans can compete with machines. New skills like Creativity, Complex Problem Solving, Emotional Intelligence, Critical Thinking are required by 2020 [4].

The aim of the developed laboratory is to prepare students in both technical and creative mindsets for future jobs' requirements.

### 1.1 Background

We can find a series of didactic systems tackling regenerative braking systems from different vendors. The systems allow students to highlight the 4 quadrants of operation of machines and converters. The main problem of these systems is the high cost. And a lot of times the closed and rigid mounting that does not allow students to change specific components, add sensors or measure additional signals. Similar work has been developed by Isabelli Viana et al. developed a didactic kit for switch mode DC/DC converters. An analog controller has been added [5]. We also developed previously a didactic Photovoltaic system. The system integrates the following experiments: effect of irradiance on the I-V characteristic, effect of incident angle on the power output, effect of temperature on the efficiency and degradation effect, plotting the I-V and P-V curves, investigating the effect of partial shading and the utility of by-pass diodes, developing MPPT algorithm [6]. Poria Fajri et al. developed a test bench to evaluate friction braking and regenerative braking models. The HIL system showed close results between the test bench and the simulation from ADvanced VehIcle SimulatOR (ADVISOR) [7].

The powertrain and regenerative braking system technology and research are developing. Fan Jingjing et al. developed an algorithm for hybrid powertrain power allocation. The algorithm aims to reduce fuel consumption by acting on the engine speed, motor torque, and throttle opening. A 31.5% fuel economy was recorded compared to the logic-based method [8]. J. Han et al. developed a control algorithm to cooperate front wheel regenerative braking with rea wheel friction braking to optimize energy recovery and mitigate under-steer problems [9]. Jia Wang et al. developed a control strategy to maximize regenerative braking energy and increase mileage on full charge [10].

On another hand, to encourage creativity, many engineering schools have adopted changes and projects. In Korea, the Innovation Center for Engineering Education works on promoting three skills: Creativity, Capability, and Challenge. Many actions have been elaborated: setting practical exams from problems arising in the field, establishment of a tool shop where students can design and implement creative ideas, and scheduling customized training according to industry demand [11]. At Texas A&M University in USA, an introductory course to engineering design was developed relaying on experiential and team-based learning model. This course at the freshman level boosted creative thinking during engineering studies [12]. In 2010, at the National Research Tomsk Polytechnic University in Russia, they created a lab called "Polygon of Engineering Entrepreneurship". The main objective of the lab is the promotion of innovative activities and education on the topics of innovations. Co-curricular activities also are scheduled where instructors and students are involved with preparation of these socio-cultural activities [13].

We present a didactic regenerative braking system. First the system design is presented. Then the power and control parts are described. At the end possible experiments that can be done using the developed system are enumerated.

## 2 System Design

Energy efficiency is a key axis to reduce energy demand especially in cars. Regenerative braking system is being implemented in escalators, elevators, cars. Where dissipated energy may be converted to electricity form. Thus, a didactic regenerative braking system has been developed to aid students understand and interact with this new technology (Fig. 1).

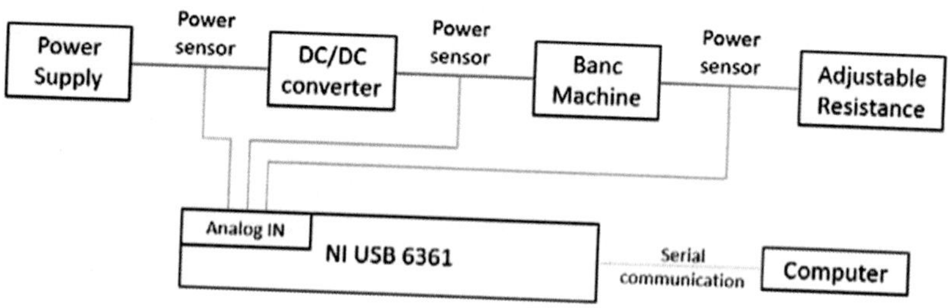


**Fig. 1.** System design of the didactic braking system.

The system consists of a battery power supply feeding a DC machine banc via a DC/DC converter. The DC/DC converter allows power reverse. The banc machine is constituted of two mechanically DC machines the first one operates as a motor and drives the second one. The latter supplies a variable resistor. Three voltage current probes are used to monitor power flow and losses in the system.

The second machine and load emulate an electric car chassis. To emulate a rise, a small load is used to draw high current. To emulate a slope, power is supplied to the second machine to drive the system and generates electricity back to the battery supply.

## 3 Power System

### 3.1 DC Banc Machine

The main component of the system is the DC banc machine. By its two DC machines, it can emulate an electric car, an elevator, or an escalator. The two machines are mechanically connected, the first machine acts as the master and the second one as the load.

We use in this experiment the EPD037580 module from didalab. It is specially adapted to practical work of power electronics. It consists of a 24 V DC motor, 12.85 W with a maximum efficiency of 76.5%. The motor is coupled to an identical generator. An encoder of 1 V/1000 rpm allows speed measurement.

A resistive load is connected to the second machine that acts as a generator (Fig. 2).

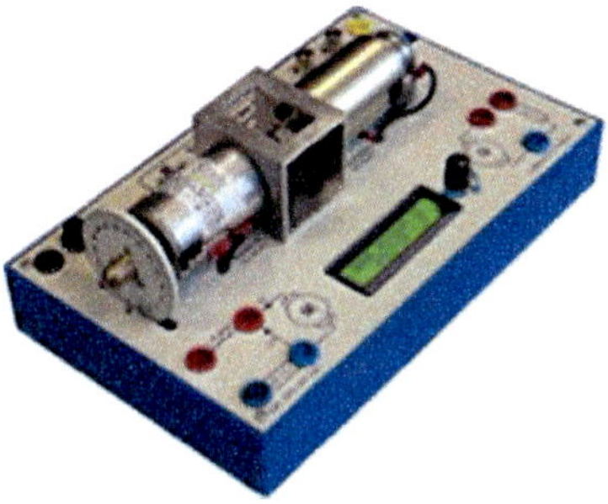

**Fig. 2.** EPD037580 DC banc machine.

### 3.2 DC/DC Converter

The element that allows power transfer between the supply and the system is the four quadrant DC/DC converter. Through its circuit shown in Fig. 3a, power can flow in both directions.

The DC/DC converter used in this experiment is the PED 020420 four quadrant converter that allows double way power transfer between the battery and the DC banc machine (Fig. 3b). PED 020420 is rated at 30 V, 2 A. The TA+, TA−, TB+, TB− switches are controlled by a PWM signal.

The technical characteristics of the PED 020420 are as follows:

Rated supply voltage: 30 V, rated output current: 2 A, Max Frequency: 2 kHz. The duty cycle can be linearly controlled by a DC voltage ranging from −10V to + 10V.

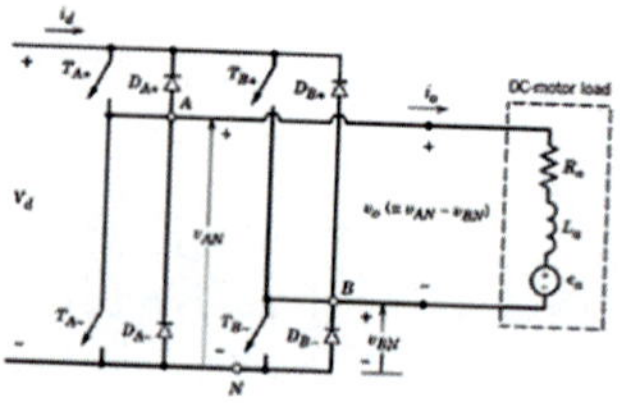

(a)

(b)

**Fig. 3** **a.** Circuit of the four quadrant DC/DC converter [14]. **b.** PED 020420 DC/DC converter.

### 3.3 Battery

The power supply of the system is a battery that stores electrical energy and allows the two way of power transfer. Lead-acid batteries are widespread since they're conventional onward, and reliable. We use in this experiment a UB1250 Universal Battery.

## 4 Instrumentation and Control System

### 4.1 Sensors

Temperature Sensor

One of the main advantages of regenerative braking systems is the reduction of heat generated by friction. Heat reduces the lifespan of a braking system and may lead to brake fade. It is an important approach to measure the system temperature and highlight the importance of the regenerative braking system.

The LM35 is used to detect temperature with an electrical output proportional to the temperature (in °C). The output voltage of the LM35 has a comparative value to the Celsius temperature.

The variation in the output of this sensor is linear as function of the Celsius Temperature. The output voltage of this IC sensor is given by the following formula:

$$V_{TS} = 10\,\mathrm{mV} \times T \tag{1}$$

Where, VTS is the output voltage of the temperature sensor in V, and T the temperature in °C.

Voltage/Current Sensor

In this experiment power flow is the key element to show in order to understand the operation and the efficiency of regenerative braking systems. For that we will measure power at three points in the system:

- Between the power supply and the DC/DC converter.
- Between the DC/DC converter and the DC machine banc.
- Between the DC machine banc and the load.

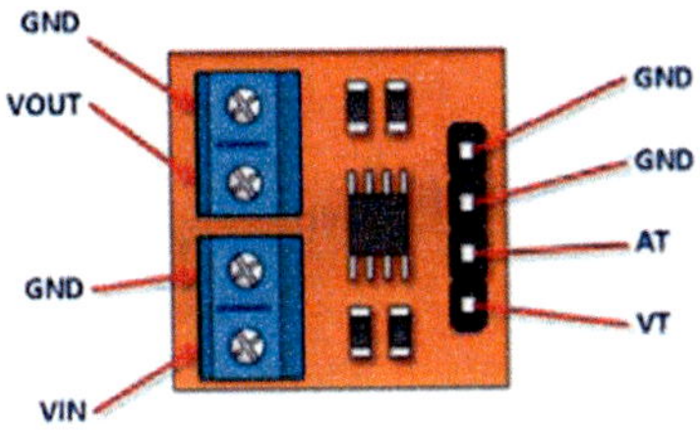


**Fig. 4.** MAX471 current and voltage sensor.

The MAX471, a current and voltage sensor will be used. It is a Bi-Directional and comprehensive sensor (Fig. 4).

By measuring the voltage and current output pins from the module, it is trivial to calculate the power passing through the device.

## 4.2 Controller

In this project we need mainly to monitor our system to show students the power flow and the importance of regenerative braking systems. The solution should include a Graphical User Interface (GUI) where students can see the 4 quadrant operation of the machine and power converter. We choose the NI USB 6361 Data Acquisition System (DAQ).

The USB 6361 is a Multifunction I/O Device and offers analog I/O, digital I/O, and four 32-bit counters/timers for PWM, encoder, frequency, event counting, and more. On board NI-STC3 timing and synchronization technology delivers advanced timing functionality, including independent analog and digital timing engines. The USB-6361 is well suited for a broad range of applications, from basic data logging to control and test automation.

A signal conditioner is added to the system, the SC2075 helps noise reduction and connection to the DAQ. The SC-2075 is a desktop signal conditioner that you can connect directly to National Instruments E Series or 1200 Series devices.

The USB 6361 can measure up to eight differential analog input channels. We need in this project 3 differential analog measurements for current, 3 differential analog measurements for voltage, 1 differential analog measurement for the temperature sensor, and 1 differential analog measurement for the speed measurement available in the EPD037580 DC banc machine.

The developed GUI of the system presents the operation of the main DC machine (Generator or Motor). It also indicates the ambient temperature around the shaft and speed of the DC banc machine. In addition, it indicates the current, voltage and power of the line between the converter and the machine and between the second machine and the load (Fig. 5).

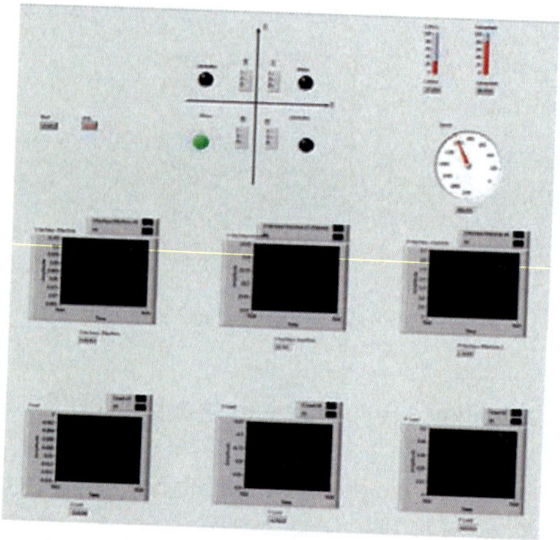

**Fig. 5.** GUI showing the operation as a motor.

## 5 Laboratory Experiments of the Regenerative Braking System

In order to understand the operation of the DC/DC converter, the machine, the regeneration process, and the power flow students are asked to perform two experiments: 1) Operation as motor and 2) operation as generator.

In the motor operation, students have to supply the DC/DC converter, apply the PWM signal to get the required speed. According to the knob present on the PED 020420, the Duty Cycle (DC) of the PWM can be varied and also the direction of rotation. Students will see the power flowing from the battery to the load.

In the generator operation, students will have to supply the DC second machine with an external power supply. The second machine will drive the first one that will operate as a generator and charge the battery. Students will see the power flowing from the load to the battery.

## 6 Conclusion

In this paper we presented the development and implementation of a renewable energy laboratory. We focused on the regenerative braking system oriented to electrical engineering students.

In this lab, students learning is emphasized where they are asked to learn by doing. Several topics are covered where students are asked to develop their algorithm and their analysis (MPPT, effect of temperature, effect of irradiance). The active learning approach will aid students to develop technical and soft skills that are required by future employers.

As an expansion to the renewable energy laboratory some topics should be added for example, grid connection of renewable energy sources. The latter is a common topic between power electronics and renewable energy. Micro-grid and hybrid renewable energy sources integration is a new topic that is emerging in the recent years.

## References


1. Saad Fouad, A.: The shock of energy transition (2015)
2. Kirn, A., Benson, L.: Engineering students' perceptions of problem solving and their future. J. Eng. Educ. **107**(1), 87–112 (2018)
3. Nelson, K.G., Husman, J., Brem, S.K., Honsberg, C., Bowden, S.: Optimizing educational approaches for university photovoltaics education. In: 2011 37th IEEE Photovoltaic Specialists Conference (PVSC), pp. 1211–1216 (2011)
4. Grey, A.: The 10 skills you need to thrive in the Fourth Industrial Revolution, 19 January. https://www.weforum.org/agenda/2016/01/the-10-skills-you-need-to-thrive-in-the-fourth-industrial-revolution/
5. Viana, I.P., Lima, W.S., Aquino, C.A., Johannesson, J.G.R., Medeiros, C.M.S.: Didactic kit for practical testing of the basic switched mode power supply topologies. In: 2013 Brazilian Power Electronics Conference, pp. 761–766 (2013)
6. Nehme, B., Akiki, T.: Implementing a didactic photovoltaic energy laboratory for developing countries. In: 2016 3rd International Conference on Renewable Energies for Developing Countries (REDEC), pp. 1–4, 29 September 2016
7. Fajri, P., Lee, S., Prabhala, V.A.K., Ferdowsi, M.: Modeling and integration of electric vehicle regenerative and friction braking for motor/dynamometer test bench emulation. IEEE Trans. Veh. Technol. **65**(6), 4264–4273 (2016)
8. Jingjing, F., Jianying, G., Li, W., Jiaping, Z., Xinfeng, Z.: Dynamic programming algorithm for hybrid powertrain power allocation. In: 2019 Chinese Control Conference (CCC), pp. 6674–6679 (2019)
9. Han, J., Park, Y., Park, Y.: Cooperative regenerative braking control for front-wheel-drive hybrid electric vehicle based on adaptive regenerative brake torque optimization using under-steer index. Int. J. Automot. Technol. **15**(6), 989–1000 (2014). https://doi.org/10.1007/s12239-014-0104-9
10. Wang, J., Wang, Y., Li, M.: Regenerative braking control strategy for electric vehicle. In: Wang, J., Yen, G.G., Polycarpou, M.M. (eds.) ISNN 2012. LNCS, vol. 7368, pp. 488–495. Springer, Heidelberg (2012). https://doi.org/10.1007/978-3-642-31362-2_54
11. Lim, J., Kim, S., Lee, C.: C 3 (Creativity, Capability, and Challenge) Activities of Innovation Center for Engineering Education, pp. 725–728 (2013)
12. Aboukinane, C.: Fostering creativity in engineering education through experiential and team-based learning, age, vol. 23, p. 1 (2013)
13. Kirillov, N.P., Leontyeva, E.G., Moiseenko, Y.A.: Creativity in engineering education. Procedia Soc. Behav. Sci. **166**, 360–363 (2015)
14. Ned, M., Undeland, T.M., Robbins, W.P.: Power Electronics: Converters, Applications, and Design. Wiley, Hoboken (2007)

# Assistant Teaching Platform for Creative Mechanical Design Course

Yu-Tong Li and Yu-Xin Wang(✉)

China University of Petroleum Huadong, Qingdao 266580, China

**Abstract.** Innovative engineering education plays a vital role in promoting inclusive and sustainable industrialization worldwide. Creative Mechanical Design is a course open to senior students and graduate students for enhancing their creative design capability. The shortcoming of the current creative mechanical design course, such as time-consuming model and simulation, fewer innovative schemes and prototypes, leads students to have little chance to practice creative thinking training. This paper proposes a virtual assistant teaching platform for creative mechanical design to automatically help students operating 3D modeling, simulation, and optimization selection of creative design schemes within a minimal experimental time. Thus, the platform dramatically enhances the efficiency of the creative design process. As a result, students are freed from onerous innovation processes and have more energy to engage in creative thinking training from different perspectives and compare innovative techniques. Consequently, students' enthusiasm for carrying out creative mechanical design is stimulated.

**Keywords:** SDG4/SDG9 · Creative mechanical design · Inspire students' creative design enthusiasm · Automatic virtual platform · Computer aid instruction

## 1 Introduction

Education can promote upward socio-economic mobility and is a key to escaping poverty. Higher education, especially innovative engineering education, plays a vital role in promoting inclusive and sustainable industrialization worldwide in fully excavating and utilizing the potential of material resources, reducing the manufacture and maintenance costs of products, and improving products' energy efficiency. Mechanical systems are fundamental means and carriers for expanding human physical strength. Their novelty is crucial in determining the cost of manufacture and maintenance, the service life, and the reliability of products, and impacts the world's sustainable development.

Innovation ability is a universal requirement for talents in modern society. It is a comprehensive ability formed by the interaction of knowledge vision, innovation consciousness, innovation thinking, innovation skills, and innovation quality [1]. Engineering design is the process of satisfying requirements by developing and synthesizing building blocks into meaningful designs that meet the requirements and desires [2]. The


G. Quaglia et al. (Eds.): I4SDG 2021, MMS 108, pp. 624–634, 2022.
https://doi.org/10.1007/978-3-030-87383-7_67

process and resulting design are at times novel, creative and innovative, and at times routine. It is a creative thinking activity that integrates knowledge and experience [3].

Creative Mechanical Design is a course open to senior students and graduate students. This course mainly focuses on innovative design methods and engineering case studies to improve students' technology-related knowledge and skills by cultivating their creative spirit and practical abilities. Using multi-disciplinary knowledge based on different innovative methods or principles to construct complex mechanical design problems from different perspectives and select the most innovative and competitive design solutions are critical factors throughout the whole teaching process [4]. Therefore, exploration is an effective approach that cultivates creativity and improves practical abilities by directing students to perform creative design work based on their practical experience to master the key points and features of creative design [5, 6].

Creative acts often result from long periods of difficult, purposeful struggle - a struggle with the idea produced and maintaining the contexts and self-concepts that make such ideas possible [7]. However, designers are often reluctant to make the required effort but proceed rapidly to synthesize a solution, because she/he must make assumptions regarding some system features to accomplish the task, even before acquiring all the necessary information [2]. For mechanical systems, virtual structural design, space layout, virtual simulation are critical design processes for screening optimized creative design schemes. However, these processes are time-consuming. Moreover, because senior college students in China face English-level certificates and postgraduate examinations, it is difficult for them to spend more time experiencing innovative design from different perspectives to understand innovative design characteristics and choose the best design schemes [8]. In this case, computers are necessary auxiliary means to automate tasks, leaving more time for students to expand the range of solutions and implement the creative design of mechanical systems.

Computer-based means of design synthesis enable designers to explore a wider variety of creative solutions by eliminating a significant number of tedious design tasks. AI-based function reasoning techniques can efficiently deal with the synthesis of combined mechanisms in series connections for input/output functional requirements [9–12]. Although some computer-based innovative design platforms [9, 13–15] have been developed for helping designers to explore new directions by providing a wider variety of possibilities, they cannot be applied to deal with the synthesis of combined mechanisms in parallel and compound connections. The reason is that a lack of direct, indirect, or even implicit knowledge supports the reasoning process of mechanism combination from constituent mechanisms to their combination [16]. However, these combined mechanisms are widely used in industry because of their more elegant structures and higher efficiency.

A computerized creative conceptual design platform for solving non-routine design problems was proposed to expand the solution space of creative design and increase the creativity of matching physical structures for mechanical systems [17, 18]. The platform consists of three parts: a creative design unit, virtual simulation unit, and supporting database. The creative design unit includes the functional decomposition module [19], rule-based generative module [20], and domain synthesis module [21]. The platform can effectively imitate the creative design skills of human experts to realize the ingenious combination of mechanisms and obtain more simplified physical structures with fewer and simpler constituent devices to realize the overall functional requirements of design problems organically and cooperatively. The novelty of the matched physical structures is improved significantly.

Based on the above platform, this paper presents an assistant teaching platform for implementing the creative design of mechanical systems so that the shortcoming in the current teaching of creative mechanical design, such as fewer innovative schemes, time-consuming simulation, and inefficient modeling of virtual simulation prototypes should be overcome. With the computer's powerful reasoning mechanism and storage capacity, the platform can automatically generate novel schemes that are not inferior to human experts and automatically give their 3D virtual simulation models. Under the support of the assistant teaching platform, students are freed from onerous innovation processes and can mainly concentrate on creative thinking training from different angles. Consequently, students' innovative consciousness is inspired because the platform enhances the efficiency of the creative design process.

## 2 Systematic Computer-Based Creative Design Platform

The systematic computer-based creative design platform consists of three parts: a creative design unit, virtual simulation unit, and supporting database [17]. The creative conceptual design unit includes the functional decomposition module, rule-based generative module, and domain synthesis module. To help automatic evaluation of the generated physical structures in the creative conceptual design process, Li and Wang [18] provided a virtual simulation tool, which contains the recognition of symbolic schemes, motion analysis, solid structure modeling, layout design, 3D virtual simulation, and other automated processes related to automatic virtual simulation.

The platform works as follows. In the conceptual design stage, a rule-based creative design method, for instance, is applied to generate the conceptual schemes according to the design requirements for a complex mechanism. The outcome of the conceptual design is represented with symbolic schemes and recorded with character strings. The next stage is to identify the symbolic schemes into the constituent mechanisms and their connections. For the identified basic mechanisms, the kinematic analysis methods are applied to obtain kinematic data for each component, which will be utilized to support the 3D virtual simulation of the complex mechanism later. With the aid of the object-oriented virtual solid design method, each constituent component or mechanism is modeled under a 3D environment based on the feature-based technique and encapsulated as one design object with the dialogue interface. Inputting key parameters through the interface, the solid virtual model for each constituent mechanism will be constructed automatically. The space layout process is added to reduce the space occupied by a complex mechanism corresponding to the symbolic scheme. At the end of the virtual solid design of the basic mechanisms, the space layout parameters and the layout coordinate systems for each mechanism have been determined and stored in the database. According to the virtual solids of these basic mechanisms and their connections, the space layout for the complex mechanism corresponding to the symbolic scheme is done using the adaptive genetic algorithm. While the spatial position for each mechanism is calculated in the space layout process, an approach called "virtual-assembly" is adopted to put the constituent parts or basic mechanisms onto their proper positions where the constraint equations for the complex mechanism are satisfied. In this way, the assembly model of the complex mechanism related to the current driving links' positions is obtained.

# 3 Assistant Teaching Platform for Creative Mechanical Design

Based on the above platform, an assistant teaching platform for implementing the creative design of mechanical systems is built, as shown in Fig. 1. It comprises five relatively independent modules: design requirements input module, innovative method simulation and scheme generation module, scheme selection and optimization module, scheme 3D virtual simulation module, and scheme output module.

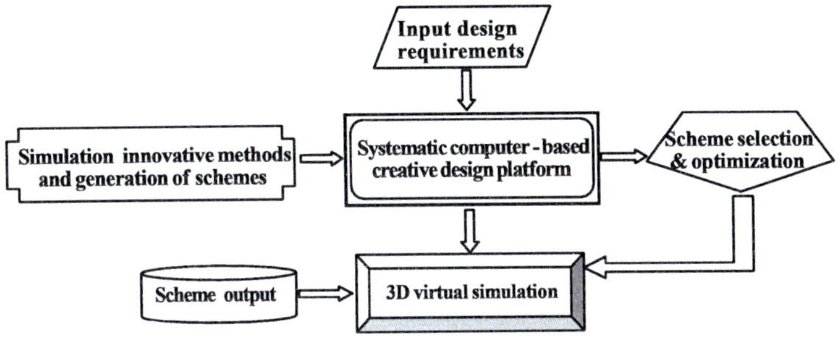


**Fig. 1.** An assistant teaching platform for creative mechanical design

## 3.1 Design Requirements Input

This module is mainly used to input the experimental topic of creative mechanical design and record relevant experimental information, such as student information, start time, tutors, etc. The key of this module is how to convert the design requirements for creative mechanical design, defined by the design topic, into the design rules that can be recognized by computers or the computerized language description of creative mechanical design problems. The desired functional requirements for complex mechanisms to implement had been summarized into nine functional requirements [22] through the comprehensive analysis of complex mechanical systems' composing principles and functional design requirements. The students can define the synthesis rules based on these nine functional requirements. For facilitating the rule input, the module provides the help interface for guiding the input of design rules.

Generally, the ground link should be firstly defined in the first synthesis rule, and then the driving links should be determined. Generally, the synthesis rules about the treatment of the P pair should be written at the end of the rule set. For different design problems, the synthesis rules should be different. In the nine aspects above, not only the topological constitution characteristics, which are widely used in the number synthesis of mechanisms, are considered, but functional requirements, such as 1 to 3, are also considered. So the grammar-based synthesis rules in accord with the nine aspects above can represent the general functional requirements of complex mechanisms, rather than only topological synthesis requirements. For example, the topic of the creative mechanical design is:

*Design a three redundant high voltage electrical short-circuit switch mechanism, the mechanism can achieve the following functions:*

***Manual switch*; *High temperature melting switch*; *Remote control switch*.**

**Problem analysis:** for such a problem, (1) the function of the manual switch is: the input rotation makes the executive member out of contact; (2) the high-temperature fuse switch (*F*) should be a member. When the fuse is broken, the length of the rod changes, so a Dyad (Assur-II) is needed to define the fuse member; (3) the remote control switch (*R*) is generally realized by electromagnetic actuators. The analysis shows that the disappearance of the electromagnetic force will increase the distance between the two components acting on the electromagnetic force; therefore, the electromagnetic force can be treated as a force-acting member. A Dyad can embody this functional requirement, namely realized by increasing the distance between the two external revolute pairs of the Dyad. Then, the creative design rules for three redundant high voltage electrical short-circuit switch mechanisms are listed in Table 1.

**Table 1.** Creative design rules for three redundant HV short-circuit switch mechanism

| Number | Synthesis rules | Meaning of design rules |
|---|---|---|
| 1 | *G* | One link should be as Ground (*G*) |
| 2 | *E* | One link should be as executive member (*E*) |
| 3 | *C* [G] | Driving link *C* (manual switch) is connected with Ground (*G*) |
| 4 | *D* <2> [*E*] | Binary link *D* is connected with the executive link *E* |
| 5 | *Y* <2> [*D*] | Binary link *Y* (*F* switch) is connected with the binary link *D* |
| 6 | *S* <2> [*E*] | Binary link *S* is connected with the executive link *E* |
| 7 | *Z* <2> [*S*] | Binary link *Z* (*R* switch) is connected with the binary link *S* |

## 3.2 Innovation Methods Simulation and Schemes Generation

The platform can simulate combination innovation, transplantation innovation, mutation innovation, substitution innovation, enumeration innovation, and domain functional synthesis innovation. The simulation principle is as follows: the input design rules are recorded in the linked tables, through searching the linked tables, the design rules that fit the selected creative design method, as shown in Fig. 2, are appropriately selected from them and reconstructed into a group of new design rules. Then, under these new design rules, the platform can generate schemes that coincide with the selected creative design method.

When students complete the input of creative design rules, after choosing the creative design method to be simulated, the platform automatically generates creative design schemes consistent with both the input design rules and the creative design method to be simulated. Thus, different solution schemes will be obtained for the same group of input creative design rules while choosing different creative design methods. By comparing the number and novelty of schemes generated by different creative design methods, students have a deeper understanding of the characteristics of each creative design method, students' enthusiasm for obtaining innovative and leading-edge design schemes was stimulated.

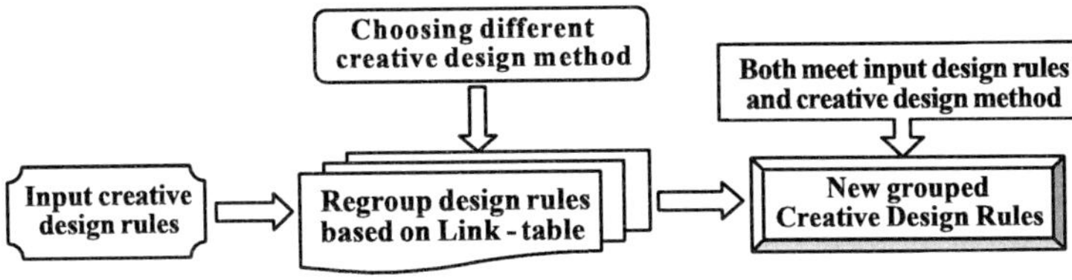


**Fig. 2.** Simulation principle for different creative design methods

### 3.3 Scheme Selection and Optimization

As listed in Table 1, the design rules can be constraint rules, intended (preferred) design solution rules, reconstruction and mutation rules, convergence rules, etc. If the design rules are written too strictly, e.g., they are written reference to the existing solution, fewer innovative schemes can be obtained under different innovation techniques. Consequently, the novelty of the design problem solutions is reduced, or there may be only one comparable solution. On the contrary, if the design rules are too loose, hundreds or even thousands of solutions may be generated. It will be time-consuming and unrealistic to evaluate and select a suitable number of design schemes in a three-dimensional virtual environment among them. For the above reason, the platform provides a scheme selection and optimization module by refining the design rules. Under this module, through the input filtering rules, such as the combination of constitute mechanisms, executing component which must connect to Ground, using intermittent mechanisms, several optimized design solutions are selected that conform to the designer's preferences. In this way, the number of solutions created by the system is significantly reduced. As shown in Fig. 3, in the left figure, the design rules can only generate schemes with standard Cam (*X* <*n*1> [*G*]; one link must connect to Ground); while in the left figure, the design rules can generate schemes with generalized Cam, i.e., in the Cam, no link is needed to connect to the Ground. Here, *X* and *Y* are two binary links to form a Dyad to represent a Cam pair.

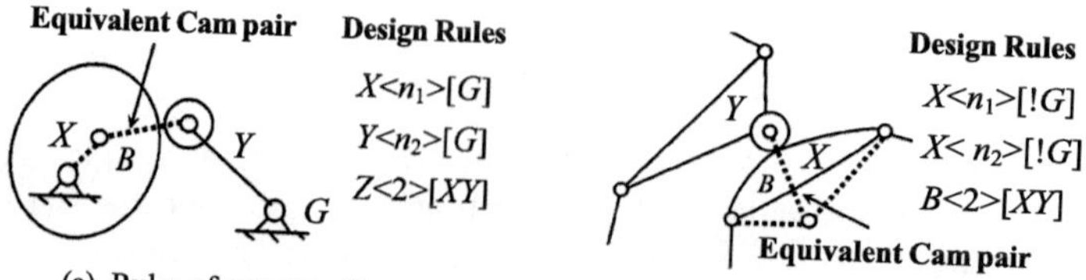


(a) Rules of common Cam (b) Rules of generalized Cam

**Fig. 3.** Design rules and relative Cam schemes

### 3.4 3D Virtual Simulation

Structural design has always been a difficulty and weaknesses in teaching creative mechanical design. The main reasons are as follows: (1) due to the limitation of teaching content and length, textbooks can only provide limited space in structural design; (2) experimental teaching is mainly used to verify innovative design schemes. Furthermore, due to a small number of available physical components, the structural schemes to be assembled are insufficient; (3) structural innovation design requires a high accumulation of design experience, so it is difficult to determine a reasonable innovative structure for students who lack practical experience. However, structural innovation design plays an essential role in mechanical innovation design, directly affecting the reliability, life, cost, maintainability, and other market competition factors. In order to support structural innovation design, a large number of three-dimensional virtual solid models of existing structures have been collected in the database while setting up three-dimensional virtual modeling encapsulation objects of the mechanisms.

In the 3D Virtual Simulation modules database, the initial solid models for the Object-oriented 3D virtual solids of the Gear, four-bar linkage, Cam, Groove wheel, and other basic mechanisms are supplied. The Gear mechanism is divided into the planar Gear and the spatial Gear for enhancing the creativity of matched structures. Moreover, the planar Gear is subdivided into eleven kinds of specific structures further, and the spatial Gear is subdivided into four kinds of specific structures in the Gear library to facilitate getting different structures. Similarly, there are 15 types of encapsulated Cam mechanisms for the Cam mechanism, 18 types of active components and 23 types of follower structures, and $15 \times 18 \times 23$ types of the 3D virtual structures of Cam mechanisms. Sixteen kinds of link-type planar mechanisms are supplied in the linkage library that contains four kinds of crank structures, eighteen kinds of binary links, and ten kinds of ternary links. Object-oriented language C++ is utilized to build up the user interface and object encapsulation. The typical encapsulated interface for the Gear mechanism is shown in Fig. 4b. Based on the key parameters in the interface, the system can automatically build up the 3D models of constituent mechanisms.

## 4 Assistant Teaching Platform for Creative Mechanical Design

As shown in Fig. 4, the virtual assistant teaching platform is built based on the SolidWorks® API functions with the object-oriented VC++ language and MySQL database. Opening the virtual assistant teaching platform menu, the system pops up with five module buttons interface. After the experimental information is input and confirmed on the input interface, the system proceeds to the creative techniques simulation module. This module encapsulates three kinds of computerized innovative design methods, namely, rule-based innovative design, function synthesis-based innovative design, and domain synthesis and mutation innovative design, and provides an innovative technique simulation module.

Here, taking the rule-based creative design module as an example of operating the platform is briefly introduced. Clicking on the **rule-based method** button, and the rule-based creative mechanical design interface pops up (Fig. 4a). Inputting the design rules and selecting the graphic carriers, the system automatically generates symbolic schemes that conform to the design rules. By clicking the button of **creative method simulation**, creative design techniques can be simulated according to the selected innovative approach. Then the system automatically screens out the symbol schemes that meet the requirements of the inventive method. Further, under the **scheme selection and optimization** button, the scheme optimization and screening process are carried out to figure out the creative design schemes that conform to the requirements of innovative techniques and the intending optimization design schemes. Clicking the **3D virtual simulation** button, the interface of the computerized creative conceptual design platform for complex mechanical systems pops up (Fig. 4b). Under this menu, recognition of symbol schemes, modeling entity of constituent mechanisms, layout of the system, motion analysis, and 3D virtual simulation can be carried out automatically.

In detail, for the selected symbol scheme, the system identifies it as the constituent mechanisms and the connections among mechanisms automatically and carries out object-oriented three-dimensional modeling design for the common mechanisms, such as the interface for Gear in the middle of the left of this figure. According to the connection relationships and the 3D modeling of each mechanism, the layout design [23] of the mechanical system related to the symbol scheme is then carried out, and the spatial position for each constituent mechanism is obtained. While the spatial position for each mechanism is calculated in the space layout process, the virtual-assembly approach [24] is applied to put the constituent parts or mechanisms onto their proper positions where the constraint equations for the actual complex mechanism are satisfied. Implementing the virtual-assembly method relies on the kinematic data and the corresponding transformation matrix for each part. Under the transformation matrixes, all parts in the complex mechanism are put onto the proper positions where the constraint equations and the mating conditions are met as they are constrained and mated through the manual assembly. Thus, the assembly model of the complex mechanism related to the driving links' current positions is obtained (Fig. 4c).

Because the conceptual scheme generation, screening and optimization, virtual structural design, motion analysis, space layout, and other design processes for creative mechanical design are all completed automatically, the human's operation time for creative mechanical design problem is reduced to the minimum in carrying out the creative

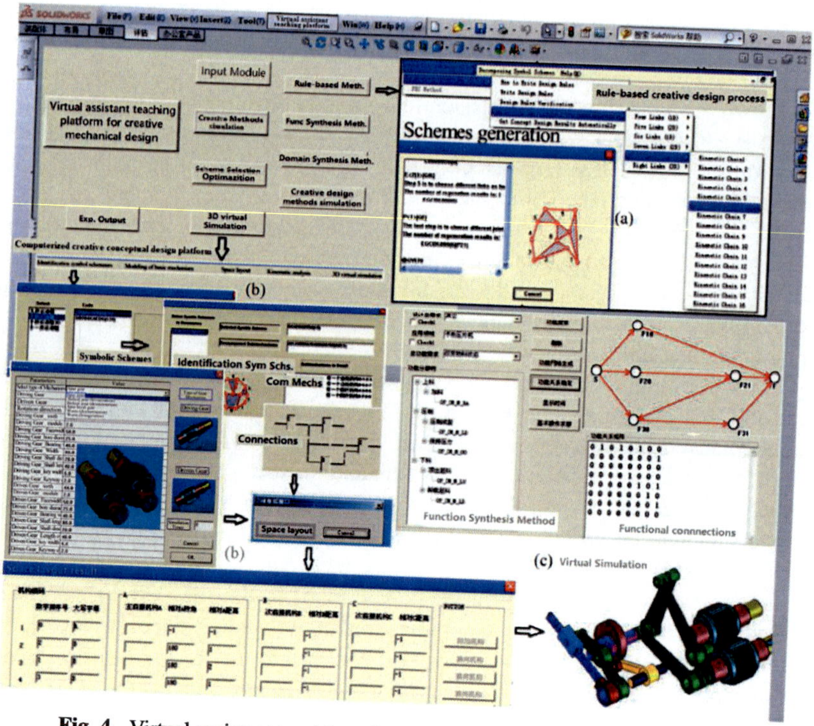


**Fig. 4.** Virtual assistant teaching platform for creative mechanical design

design process. Therefore, the virtual assistant teaching platform for creative mechanical design, as shown in Fig. 4, can help students operate 3D modeling, simulation, and optimization selection of creative design schemes within a minimal experimental time. As a result, students can engage in scheme innovation and innovative techniques with more energy. Furthermore, this platform benefits students to deeply understand the values of expanding innovation solution spaces from different perspectives. Students' creative design enthusiasm is stimulated consequently.

## 5 Conclusion

Creative Mechanical Design is a course open to senior students and graduate students. Some critical design processes for mechanical systems, such as virtual structural design, space layout, virtual simulation, are time-consuming. It is difficult for students to experience innovative design from different perspectives to understand innovative design characteristics with enough time. In order to overcome the shortcoming in the current teaching of creative mechanical design, a virtual assistant teaching platform for creative mechanical design is put up. The platform can automatically simulate many creative design approaches, obtain innovative solutions, construct 3D models and carry out virtual simulations related to the generated schemes. As a result, the efficiency of the creative design process is dramatically enhanced, and students are freed from onerous innovation processes and have more energy to engage in creative thinking training from different perspectives and compare different innovative techniques.

## References

1. Zhu, Q.S.: How to cultivate students' innovative ability. Acad. Degree Graduate Educ. **4**, 1–3 (2002)
2. Chakrabarti, A., et al.: Computer-based design synthesis research: an overview. J. Comput. Inf. Sci. Eng. **11**, 021003-1-12 (2011)
3. Howard, T.J., Culley, S.J., Dekoninck, E.: Describing the creative design process by the integration of engineering design and cognitive psychology literature. Des. Stud. **29**(2), 160–180 (2008)
4. Cong, X.X. (ed.): Mechanical Innovation Design. Peking University Press, Beijing (2008)
5. Wong, Y.L., Michael, S.K.W.: A model of creative design process for fostering creativity of students in design education. Int. J. Technol. Des. Educ. **22**, 437–450 (2012)
6. Pahl, G., Beitz, W.: Engineering Design: A Systematic Approach, 2nd edn. Springer, London (1997). https://doi.org/10.1007/978-1-4471-3581-4
7. Crill, N.: The structure of design revolutions: kuhnian paradigm shifts in creative problem solving. Des. Issues **26**(1), 54–66 (2010)
8. Wei, X.D., Weng, D.D., Liu, Y., Wang, Y.T.: Teaching based on augmented reality for a technical creative design course. Comp. Educ. **81**, 221–234 (2015)
9. Pal, U., Liu, Y.C., Chakrabarti, A.: Evaluating FuncSION: a software for automated synthesis of design solutions for stimulating ideation during mechanical conceptual design. Artif. Intell. Eng. Des. Anal. Manuf. **28**(3), 209–226 (2014)
10. Bolognini, F., Seshia, A., Shea, K.: Exploring the application of multi-domain simulation-based computational synthesis methods in MEMS design. In: Proceedings 16th International Conference Engineering Design, pp. 81–82, France, Paris (2007)
11. Campbell, M.I., Cagan, J., Kotovsky, K.: An agent-based approach to conceptual design in a dynamic environment. Res. Eng. Des. **11**(3), 172–192 (1999)
12. Chen, B.: Conceptual design synthesis based on series-parallel functional unit structure. J. Eng. Des. **29**(3), 87–130 (2018)
13. Schmidt, L., Cagan, J.: GGREADA: a graph grammar -based machine design algorithm. Res. Eng. Des. **9**(2), 195–213 (1997)
14. Nagel, R.L., Perry, K.L., Stone, R.B., McAdams, D.A.: FunctionCAD: a functional modelling application based on the function design framework. In: ASME 2009 IDETC/CIE, pp. 591–600. ASME, San Diego (2009)
15. Xue, D.Y., Imaniyan, D.: An integrated framework for optimal design of complex mechanical products. J. Com. Inf. Sci. Eng. **21**(4), 1–32 (2021)
16. Zou, H.J., Liang, Q.H., Zhang, Q.: Theory and method of mechanism system design. Front. Mech. Eng. **5**(4), 399–411 (2010)
17. Li, Y.T., Wang, Y.X., Duffy, A.H.: Computer-based creativity enhanced the conceptual design model for the non-routine design of mechanical systems. Chinese J. Mech. Eng. (Engl. Edn.) **27**(6), 1083–1098 (2014)
18. Li, Y.T., Wang, Y.X.: A systematic rapid creative design framework for complex mechanical systems from conceptual design to 3D virtual simulation. In: Proceedings ASME 2019 DETC/CIE, DETC2019-98150, California (2019)
19. Li, Y.T., Wang, Y.X.: A mathematical functional decomposition approach through granularity partition process in quotient space. In: Proceedings ASME International Mechanical Engineering Congress Expo, IMECE2018-86217, Pittsburgh, Pennsylvania (2018)

20. Li, Y.T., Wang, Y.X.: A bottom-up function-based synthesis approach for complex mechanisms based on clustering rules. Mech. Based Des. Struct. Mach. **47**(5), 555–556 (2020)
21. Li, Y.T., Wang, Y.X.: The innovation of matching structures through clustering and reconstructing basic operation actions in the form layer. In: Proceedings ASME 2017 DETC/CIE, DETC2017-67307, Cleveland, Ohio (2017)
22. Wang, Y.X.: Rapid Creative Design for Complex Mechanical Systems. Sci Tech Publish House, Beijing (2006)
23. Wang, Y.X., Wang, Y.S., Duffy, A.H.: Spatial layouts of complex mechanical systems. In: Proceedings ASME 2005 DEIC/ CIE, pp. 51–60. Long Beach, California (2005)
24. Li, Y.T., Wang, Y.X.: Automatic virtual entity simulation of conceptual design results-part II: symbolic scheme simulation. Comp. Aided Draf. **24**(2), 35–42 (2014)

# Beyond the Mechanical Engineering Education. New Frontiers for a Sustainable Growth

Vittorio Marchis(✉)

Politecnico di Torino – Dipartimento di Ingegneria Meccanica e Aerospaziale, Turin, Italy
vittorio.marchis@polito.it

**Abstract.** The experiences carried out at the Polytechnic of Turin starting from the 90s of the last century show that even in Italy, albeit late, it was decided to make the training of engineers culturally and socially sustainable. The introduction of human sciences courses in the curricula of future engineers is however only a first step to take, because it is necessary to review the "engineering paradigm" in the light of the new challenges of contemporary society. The digital revolution has changed the scientific and professional profile of the technician. Also, the more traditional industrial engineer, that which deals with machines and mechanisms, must recover new cultural values in order avoid being reduced to a role of subsistence. Mechanical engineering must redesign its epistemological statutes trying to propose above all on the cultural level the importance of the material essence of technological products in a society that tends to become more and more virtual. Future challenges and perplexities about some solutions till today adopted are here analyzed to open a new debate, especially among professionals.

**Keywords:** SDG4 · Engineering · Humanities · Storytelling · Polytechnic culture · Mechanics

## 1 A Chronicle of Some Teaching Experiences in Turin, Italy

The need to include human sciences in the education and training of engineers arose at the Politecnico di Torino in the 1990s. In these years the new born Institute for Advanced Studies in the Humanities (ISSU) began a program to introduce in the engineering curricula also the humanities. In parallel some seminars, and conferences were organized above all to sensitize the teachers of the Engineering Faculties still tied to traditional schemes. Among the conferences of which we recall those relating to technological taxonomies [1] and to risk and environmental problems [2], the plenary lecture *The Crisis of the Engineering Algorithm.* [3] held by Kenneth Keniston was fundamental. This professor taught at Harvard University and Yale University before joining the Massachusetts Institute of Technology faculty in 1977, where he served as the Andrew W. Mellon Professor of Human Development. His first book *The Uncommitted: Alienated Youth in American Society* (1965) focused the American young students' tensions for strong changes in a too optimistic society. But the lecture he held at Politecnico in 1996 was rich (and continues giving) of fundamental suggestions for a sustainable engineering education. Here some of the central passages:


G. Quaglia et al. (Eds.): I4SDG 2021, MMS 108, pp. 635–640, 2022.
https://doi.org/10.1007/978-3-030-87383-7_68

"This engineering algorithm -this idea of deliberately applying science to solve certain practical problems, and the correlated methodology that tells the engineer how to do so- has proven one of the world's most revolutionary and creative ideas. Embodied in the pedagogy of engineering schools and in the practice of modern engineering, it has been a driving force behind the progressive growth in the dominion over nature by humankind, transforming life on this planet in the last century more than in all of previous human history [...].

"Finally, I believe and hope that we will see, in increasing numbers, a new kind of engineer who has discovered, often after a few years of work in the "real world", that the engineering algorithm is not enough for his or her professional life. [...] What we need even more, I believe, are new graduate programs in which technological excellence, creativity, elegance, and accuracy are valued even as students learn to deal more systematically and intelligently with the non-engineering aspects of the world.

"Today, however, a contradiction has opened between the algorithm and the realities of productive life of the engineers we educate. Immensely productive and creative, not to be rejected, this engineering algorithm nonetheless faces a crisis: it does not and cannot prepare engineers for the real contexts within which most of them will work. The fundamental question for engineering education, then, is how to devise methods of preparing graduates more adequately for that complex and messy world, but without losing the dedication to the brilliant problem-solving algorithm of engineering.

A special faculty task commission, developed in the "traditional wing" of the Turin traditional polytechnic community (i.e. the Industrial Engineering Faculty), organized and promoted a series of courses to be given to the students: History of technology, Contemporary history, Philosophy of science, Sociology of science, Risk and ambient, Writing techniques. Most of these courses were held by professors of the Turin University because no researchers and professors of these disciplines belonged to the Politecnico. Later other courses added to the list: History of energy, Logics, History of Italian industry, History of the material culture, History of things. The particular "cultural tensions" and attentions to the humanities were transferred by SEFI (Société Européenne pour la Formation des Ingénieurs or European Society for Engineering Education) in the Annual Conference held in Krakow on September 1997 with the title "Humanities and Arts in a Balanced Engineering Education". The first Turin experiences were reported at the SEFI Conference 2002, in Florence [4].

But the reform of the Italian university and the abolition of the faculties, together with a process of stiffening the curricula, have gradually diminished the impact of the human sciences and the Institute for Advanced Studies in the Humanities lost its role. Only some courses, especially those of the author of this paper, because held by a member of the faculty staff, continued their activities till the present days.

On the other hand, a special attention must be paid to the initiative started in 2016 and conducted by the writer of this paper in the context of the PhD courses. Machine Epistemology [5] is the title of a 20 h seminar directed to all the PhD students. Here the "machine paradigm" whose definition from a philosophical point of view is "an artifact that consuming a resource viaries the state of a system" is studied from the point of view of the machine both as an object and as a subject of knowledge. With the contributions of the several PhD students who followed the courses five books have been published

[6]. The "rendezvous with the machine" are narrative meta-scientific short papers that starting from a technical starting point (connected with the individual research field) try to overcome borders and explore the unknown. The use of the most varied narrative schemes, from fairy tales to science fiction, from technical reports to poetry, has made it possible to focus on the complex dilemmas of sustainability, not only in mechanics. A particular example was the poem in verse entitled *Quanto costa?* (How much does it cost?) by Daniele Porcu, which deals with efforts to improve the efficiency of an internal combustion engine.

In 2018 a new Commission of the Politecnico rethought the educational programs for the Engineering Schools and initially started, in junction with the Politecnico di Milano, a general survey within the Europe. In particular many schools and public institutions have been interviewed by the Commission members in the Paris area and in Lausanne, in order to analyze initiatives to introduce Humanities into the curricula of engineers. The solution adopted by EPFL (École polytechnique fédérale de Lausanne) has been transferred to the Turin Schools, where the "Great Challenges" program includes a series of courses intended for all second-year students. The great challenges (climate crises, digital revolution, energy resources, sustainable mobility, healthcare, technologies, and humanity) will be presented in parallel by two teachers, a technician and a social scientist, and will then be discussed in group workshops. The first-year experimentation will start in spring 2022. We all hope for the success of the initiative.

## 2 Technology and Humanities: In Praise of a Better Mechanics

When Apple presented its iPad 2 in Silicon Valley in 2011, Steve Jobs stated that "it is in Apple's DNA that technology alone is not enough—it's technology married with liberal arts, married with the humanities, that yields us the results that make our heart sing." Technology, and obviously Mechanics, need to be "context aware".

The danger, for every discipline is autoreferentiality that means no recognized need of the "others". In the mid-20th century, as a consequence of WWII, some papers appeared emphasizing the role of the humanities in the science and technology world [7, 8], but in few years the attention was polarized by the new ICT, so the same Duke of Edinburgh wrote a paper [9] focalized on the telecommunication world. Electronics grew up and the computer science also took over in debates about the role of the humanities in engineering. Briefly speaking, in 2011 and 2012 IEEE organized a couple of Colloquium on Humanities, Science and Engineering (CHUSER) in Malaysia.

The progress in AI and its philosophical implications became of prime importance, also in the rare Italian literature [10]. No evidence emerges in the traditional engineering disciplines. Also by browsing Internet with the couple of words "Mechanical Engineering" + "Humanities" nothing interesting stands up, in favor of need of a better integration with the human sciences. At this point it is necessary to define what are "humanities" and what kind of humanities must cope with mechanical studies.

Surprisingly in an article by Kenneth D. Ostrand found by chance [11] we can read (citation by Theodore M. Hesburgh, *The Future of Liberal Education*, in "The Collegiate Career Woman", volume 9, #1 (1981) p. 9–13.): "Closely related is a pondering and weighing of values. Without a sense of value and purpose a person's actions are taken up with means, not ends, with techniques, not purposes. A lawyer ignores justice and becomes a clever manipulator of the law without values; the doctor becomes nothing more than a body mechanic if he ignores the human value and dignity of the person he treats. Ignoring values, the greatest engineer may become the world's greatest menace." And some lines below "In spite of your love of such investigations, you may be deterred by repugnance; if not, then by the fear of spending the nights... And i f this does not deter you, then perhaps you lack the skill...and you may have no knowledge or perspective; and if you have it, you may not be versed in geometrical exposition., or perhaps you are lacking in patience, so that you will not be diligent." These are words of Leonardo da Vinci (*Anatomy*, c. 13 v) and in the conclusion of the article he writes "But I am convinced that if we give our students a greater dose of humanities and social sciences, they will someday thank us for it. And who knows; we may turn on another Leonardo da Vinci".

By leaving aside any rhetoric, it must be remembered here that the sustainability of all knowledge and of each product that derives from it is linked to the ability to transmit the values it contains. Sharing the knowledge is the way of sustainability because on it reside the critic and the corrections. But sharing implies overcoming the barriers of knowledge and disciplines and all this is possible only if you have the instinct to narrate. At this point Jonathan Gottschall enters the scene with his books [12, 13].

Says Gottschall, "the methodological model for the humanities will not be the clean one of physics or chemistry; it will entail the messier, probabilistic triangulations of the life and social sciences. In these fields, progress rarely emerges from sweepingly definitive experiments; it emerges when slow accretions of independent findings converge in support of a hypothesis. Granted, testing literary hypotheses presents challenges that are unique to the field." But it must be clear the assertion of Keniston about the fail of the "engineering algorithm", where everything is deterministic. Obviously mechanical models are deterministic, at least at the macro level, the one of interest of most engineers, but reality is often messy.

Going back to "the storytelling animal" is the narrative approach that tries to overcome the inconsistencies of human reality. Unlike *homo neanderthaliensis, homo sapiens*, our evolutionary root, organized his brain maps around a complex approach that involves emotions and fantasy, invented superstition and myth, but he was also able to react to natural disasters which technology cannot remedy. In this sense we must look at humanities. Humanities, or better using a French allocution, the «sciences de l'homme et de la société» are the necessary ingredients for the new recipe of Mechanics. It is not a question of bringing engineers a ridiculous sprinkling of refined scholarship on a forgotten past, but of having the ability to adapt to the times.

The digital revolution is changing the rules of the innovation game, but what must not be forgotten is what I would call "the sustainable heaviness of matter" with the paraphrase of a successful novel by Milan Kundera. Contrasted to "the unbearable lightness of software". Mechatronics and Robotics are certainly an evolution of traditional mechanics: what literature and cinema offers every day is a world that suffers from the syndrome of omnipotence and total memory. We need new narratives and new stories that, with mythologies suited to the revolution we are experiencing, bring the post-industrial society back to the awareness of limits.

And just as mathematics and physics serve the future engineer more for their ability to form mind maps for problem solving, so humanities, intended as fundamentals, should guarantee better flexibility in contemplating complex reality. The teaching of social sciences simply aimed at specific purposes risks distorting the role of the same, reducing them to simple tools, losing all critical skills.

## 3 A Look to the Future, and Some Perplexities

At the end of these disordered considerations, it is certainly not possible to reach a conclusion. Instead, there is the possibility of focusing on some critical points to be kept under observation. During one of the recent debates around the pandemic, triggered by the difficulty of defining what the real purpose of science and scientists is, I heard an interesting argument. "Research is unique, "one", and the distinction between pure and applied, the competition between the two cultures (only two?), is a mistake, because this distinction depends on the (unconscious?) desire to establish a ranking." There is always the will to affirm that one's knowledge is better than the others. The wisdom of the ancients, and the reference is to the book of Francis Bacon [14], is not a reference to a forgotten golden age, but the understanding that the mankind's dilemmas are always the same. This was the reason why the classical studies are also of fundamental importance in the profession of an engineer. Martha Nussbaum, watching with concern the loss of space of humanities in industrialized societies, states: "Radical changes are occurring in what democratic societies teach the young, and these changes have not been well thought through. Thirsty for national profit, nations, and their systems of education, are heedlessly discarding skills that are needed to keep democracies alive. If this trend continues, nations all over the world will soon be producing generations of useful machines, rather than complete citizens who can think for themselves, criticize tradition, and understand the significance of another person's sufferings and achievements" [15].

The endeavor for the difference is the basis of every search for the new, and the difference lies when borders are crossed. The ancient Greeks called their frontier "ouros", but it was not to be understood as a boundary, a division, but the line that separated the known world from the unknown. The idea of frontier must be always positive. On the other hand, the danger of a homogeneous culture is always around the corner because it forces our minds to be static. And finally, the challenge of the other, the dilemmas that the other and elsewhere present to us in the most unthinkable situations, must not scare us, but must be the stimulus to change. Mechanics must also rethink its frontiers.

Endeavor, danger, and challenge are "ideas" that do not belong to the Mechanics Dictionary, but they can be a new topic to be introduced in the education of engineers, with the help of new narrative tools.

## References

1. Marchis, V. (ed.): Tassonomie Tecnologiche. Otto, Torino (2000)
2. Mela, A., Piccinini, N., Vineis, P. (eds.): Rischio e ambiente. Otto, Torino (1999)
3. Keniston, K.: The Crisis of the Engineering Algorithm, Talk for Institute for Advanced Studies in the Humanities (Politecnico of Turin/Institute for Scientific Interchange), 17 October 1996

4. Marchis, V.: Engineering Education in the Transition Phase between Industrial and Knowledge Society. In: "SEFIrenze 2002", Annual Conference "The Renaissance Engineer of Tomorrow – Innovation, Interdisciplinarity and Flexibility in European Engineer Education", Florence (2002)
5. http://epistemologiadellamacchina.blogspot.com/. Accessed 21 Nov 2021
6. Marchis, V., Pozzi, M. (eds.): Incontri con la macchina, Sesto San Giovanni, Mimesis (2018), Nuovi incontri con la macchina, id (2019), Atlante degli incontri con la macchina, id. (2020), Virus ex machina, id. (2020), Utopie della macchina, id. (2021)
7. Friedman, G.: Problèmes humains du machinisme industriel. Gallimard, Paris (1946)
8. Keniston, H.: The humanities in a scientific world. In: The Annals of the American Academy of Political and Social Science, Social Implications of Modern Science, vol. 249, pp. 160–168, January 1947
9. The Duke of Edinburgh, Engineers of the Humanities, Science and Technology. Nature, no. 4983, pp. 439–442, 1 May 1965
10. De Martin, J.C.: Università futura – tra democrazia e bit. Codice, Torino (2017)
11. Ostrand, K.D.: Leonardo wasn't just an engineer: the role of humanities and social sciences in engineering. In: IEEE 1982 "Frontiers in Education" Conference Proceedings, pp. 171–174 (1982)
12. Gottschall, J.: Literature, Science, and a New Humanities. Pargrave Macmillan, New York (2008)
13. Gottschall, J.: The Storytelling Animal. How Stories Make us Human. Houghton Mifflin Harcourt, Boston-New York (2012)
14. Bacon, F.: De Sapientia Veterum (The Wisdom of the Ancients), London (1609)
15. Nussbaum, M.C.: Not for Profit. Why Democracy Needs the Humanities. Princeton University Press, Princeton and New York (2010)

# Innovative and Multidisciplinary Teaching to Design a Low Consumption Vehicle

Massimiliana Carello(✉)

Mechanical and Aerospace Engineering Department, Politecnico di Torino, Turin, Italy
massimiliana.carello@polito.it

**Abstract.** The paper presents an innovative and multidisciplinary method for teaching how to design a low consumption vehicle point out at the Politecnico of Torino thanks to the application cases of the Team $H_2$politO.

The interdisciplinary Team, made by students, give an opportunity to tackle a "new teaching", different from that of the traditional classroom lesson given by the teacher, but innovative with the students who are able to tackle complex technical problems, such as those they will face in the world of work.

The challenge of the Team is design, build, assembly and test on track a real vehicle prototype with an important goal: reduce the consumption and increase the distance obtained with the equivalent of 1 L of gasoline. The final global efficiency challenge is the Shell Eco-marathon race, an International student's competition where the vehicle with the lowest consumption wins.

Every year new students managers of the different areas "grow up" training students who, year by year, change roles, realizing a non-traditional and innovative training, from student to student and learning by doing. In this way, the University is able to develop a new generation of engineers, able to solve complex technical problems but who also have managerial skills, more ready to enter the world of work.

**Keywords:** SDG4 · SDG5 · SDG7 · SDG12 · CAD & FEM analysis · Vehicle dynamics · Multi-body simulation · Aerodynamics · Energy optimization · Low consumption vehicle · Education · Research

## 1 The Team $H_2$politO

The Team $H_2$politO was born in 2007 with 15 students and then year after year it has grown to have until 80 students.

The Politecnico of Torino missions was to create an interdisciplinary Team made by student (bachelor and master degree) which come from different engineering fields (mechanical, automotive, aerospace material science, electronic, informatics, mechatronic, management, etc.) with a goal: design, build, assembly, test on track ad finally participate at the Shell Eco-Marathon (defined also SEM) competition.

The 1st participation at the SEM of the Team $H_2$politO was in 2008 in Nogarò (France) with the vehicle IDRA08, a fuel cell prototype.


G. Quaglia et al. (Eds.): I4SDG 2021, MMS 108, pp. 641–650, 2022.
https://doi.org/10.1007/978-3-030-87383-7_69

The Team 2021 is composed by 60 students, approximatelly 40 from bachelor and 20 from master degree, organized and followed by a Team Leader, that is the firt contact with the Faculty Advisor. Each Vehicle has a Vehicle Manager, that organize the different area: Mechanics, Electronics, Fuel cell, Combustion Engine, Aerodynamics, Dynamics and Strategy, Management and Communication.

## 2 The Technical Challenge

The European SEM is an International competition organized by Shell from more than 30 years. It is, first of all, an educative project that brings together the values of: sustainable mobility, low consumption, emission reduction, alternative energy use.

Following a certain number of phases every year more than 200 Teams (made by students of University and high schools) are selected to take part at the competition involving more than 3000 students arriving from all over Europe.

The vehicles must be designed considering the technical and safety competition rules [1] that are relative to: maximum dimensions and mass, mechanical and electronic safety, safety belts, roll bar, powertrain etc. In fact the judges, during the "Technical and Safety Inspections" (Fig. 1 left) make a detailed analysis of each vehicle to give the mandatory stickers (Fig. 1 right) that allow to run.

The SEM has two different vehicle categories: Prototype (three wheel vehicle) and Urban Concept (four wheel vehicle).

And different propulsion system: battery-electric, hydrogen (fuel cell), combustion engine (gasoline, diesel, ethanol,..).

The last SEM was in 2019 in London (Mercedes-Benz World Track Weybridge – London) and the competition rules were to complete 11 laps of the track, for a total distance of about 15.6 km in a maximum time of 39 min, with a "Start & Stop" phase at each laps for the Urban Vehicle Category. Each Team has 4 tentative of run available, and for the final ranking the best one is considered to evaluate the consumption in kWh, $kg/m^3$ of energy or km/L that can be traveled with the equivalent of a liter of fuel.

**Fig. 1.** Technical inspection (left) and technical and safety stickers (right)

## 3 The Innovative Path

However, despite the SEM has existed for more than 30 years, as well as other student competitions (for example the Formula Student or Moto Student), there are no standard for the management methodologies (as a function of University) or scientific literature about it. When a Faculty Advisor begins to follow a team sets the rules related to his experience or from the comparison with colleagues.

From the beginning, the Team $H_2$politO was organized as a Business Team (Fig. 2) where instead of the Product and the Market there are the Vehicle Prototype and the Competition, with two fundamental common aspects:

- deadlines,
- budget.

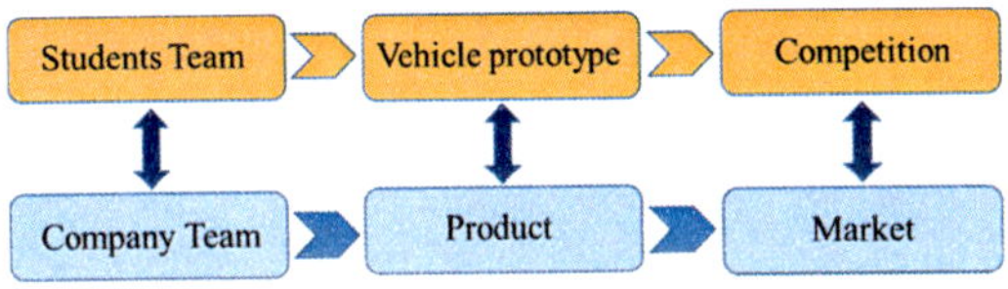


**Fig. 2.** Organization comparison between Student and Company Team

Observing Fig. 3 it is possible to explain the connection between the Innovative knowledge and the Traditional Didactic knowledge. The students thanks to the knowledge and skills acquired during their academic career (Traditional Didactic) have the opportunity to learn more competencies, also multidisciplinary, managing a real project from A to Z, working together and not individually, solving real problems, typical of the workplace (Innovative knowledges). The blue narrows explain the work flow while the red narrows explain the closed loop of the path, in particular take into account the competition results obtained of the Team it is possible to change, year by year, the innovative and the didactic path [2].

The Team $H_2$politO is a different, innovative and somehow unique project, is not just a Team but also something more: it is a new type of conceiving educational, professional and personal growth.

Team members aim at being perceived as an experimental laboratory where competences, capabilities and potentialities of future's engineers are fostered. Students strive to become not only solid and advanced technical experts but also, equally important, down-to-earth managers having excellent communication, leadership and teamwork skills.

Practical and hands-on experiences are doubtlessly a complementary and enriching form of educational path. Therefore, $H_2$politO firmly believes that learning by doing represents an absolute advantage Team members can count on.

Figure 4 explains the methodology used for the transfer of know-how between students year by year, in particular the *New Team* formation, that maintain the *Remaining students* come from the *Old Team* and involve a certain number on *Training students* (selected by the *Student manager/Leaders* after an *Advertising campaign* and a *Recruitment* phase). The *Student manager/Leaders* are some students of the *Remaining students* group that have particular *leadership* and are able to be the Responsible of each technical area (Mechanics, Electronics, Aerodynamics, Vehicle dynamics/Race strategy, Fuel cell, Combustion engine, Management, Social Media) or/and the Team Leader.

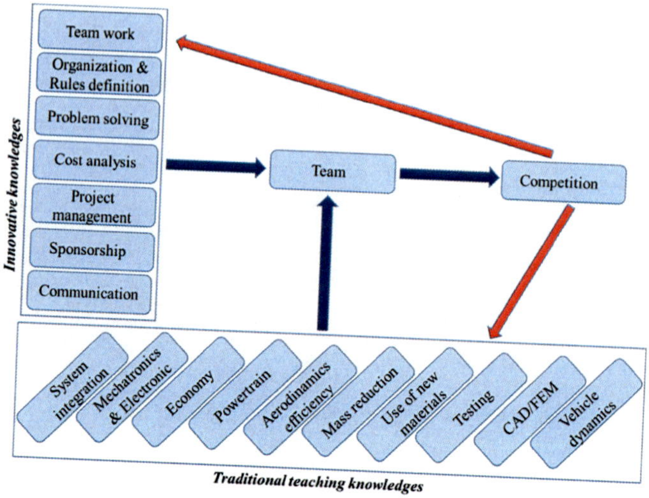


**Fig. 3.** Team path

**Fig. 4.** The passage of knowledge and the formation of a new Team

The *know-how* passage is possible thanks to the use of Reports that all the students have to do to describe the technical activity during the stay inside the Team, but also thanks to the *Thesis* (Bachelor and/or Master) that the *Team Members* develop inside the Team itself (from the beginning until now 200 thesis has been developed).

## 4 The Application's Method to Design a Vehicle

Starting from a "white sheet", the students, under the technical supervision of a Faculty Advisor, design the vehicle. Partner companies (if not commercial) make the components, then the student's assembly all the subsystems, testing all in laboratory and complete the assembly of the vehicle. Finally, the test on the track, before the competition, allow validating the vehicle dynamic and strategy models.

From 2007 to now, for the SEM competition, 6 different vehicles have been made: 4 hydrogen fuel cell prototypes (IDRA08 2008, IDRA09 2009, IDRApegasus 2012, IDRAkronos 2016), 1 parallel hybrid urban concept (XAM 2011) and 1 combustion engine urban concept (JUNO 2019). While in 2012 the Team has been developed an electric with range extender heavy quadricycle (XAM 2.0) for a "normal road" competition from Brighton to London.

All the vehicles realized by the Team $H_2$politO has shown in Fig. 5.

**Fig. 5.** The vehicles of Team $H_2$politO

Table 1 shows the technical characteristics of the actual vehicles that participate at the race: IDRAkronos (middle of Fig. 5) and JUNO (lower-right in Fig. 5).

When the Team starts to design a new vehicle (usually one year before the competition) a lot of parameter, financial and technical, have to take into account (Fig. 6).

The steps of the activity to design a new vehicle are:

1. Analysis of the rules and constrains of the competition (maximum dimensions and mass, safety aspect for mechanical and electronics,...) [1].

2. Parallel activities of: internal layout study; CAS of the external shape of vehicle and aerodynamics simulation; CAD of all mechanical components; FEM model and simulation of the body and mechanical components; vehicle dynamics simulation to evaluate force/moment (i.e. in suspension system or hub/bearing); powertrain modelling for fuel cell and combustion engine motor; electronic boards simulations. The phase design take into account new technical solution and material to reduce mass and guarantee the safety; some iterations are necessary and mandatory among the different activity.
3. Budget analysis and partner/sponsor research to have technical and financial support. This phase, also if it not "technical" is very important and need the strictly interaction between management and technical areas.
4. Final control of all the virtual models and "Freezing" of the vehicle, parallel identification of the companies for vehicle and sub system construction, when it is not possible to realize the components inside the University.
5. Components construction.
6. Sub system assembling (electronic control board, powertrain, mechanical transmission, etc.) and testing in laboratory, if possible.
7. Vehicle assembling.
8. First vehicle "functionality" tests on track to obtain a reliable vehicle with a double purpose: make consumption tests to validate the developed models and training news drivers and race strategy students.
9. Participation at the competition and application of the race strategy model to obtain the best results.

If the vehicle is not a new one similar steps are followed to re-design and implement new component and sub-system.

**Fig. 6.** Activity around a New Vehicle

**Table 1.** Main characteristics of JUNO and IDRAkronos.

| JUNO | IDRAkronos |
|---|---|
| One seater urban concept<br>Internal combustion engine 50 cc port fuel injection<br>Fuel: bio-ethanol<br>Chassis: carbon fiber monocoque<br>Front steering wheels<br>Transmission: spur gears (straight teeth) + planetary reduction gear<br>Brakes: X connection with separate circuits<br>Cone clutch<br>Suspensions: Push rod and Double wishbone<br>Tyres: Michelin 90/80 R16<br>Length/height/width: 2875/1165/1217 mm (respectively)<br>Track width/wheelbase: 1000/1660 mm respectively)<br>Total mass: 130 kg<br>Max speed: 50 km/h (limited)<br>Cx = 0,28<br>Best fuel consumption: 240 km/L | One seater prototype<br>Fuel Cell PEM 500 W (customized)<br>Electric motor 250 W (brushed DC)<br>Fuel: Hydrogen<br>Chassis: carbon fiber monocoque<br>Front steering wheels<br>Transmission; single spur gear directed connected to the wheel<br>Brakes: front and rear separate circuits<br>-<br>No suspension system<br>-<br>Tyres: Michelin 45/75 R16<br>Length/height/width: 3341/807/561mm (respectively)<br>Track width/wheelbase: 568.5/1403 mm (respectively)<br>Total mass: 42 kg<br>Max speed: 30 km/h (limited)<br>Cx = 0,1<br>Best fuel consumption: 3445 km/L |

## 5 The Results for the Students on Track

The results obtained during the last competition 2019 with the vehicle IDRAkronos are:

- 1$^{st}$ place hydrogen category SEM Netherlands– 1027 km/m$^3$ (3129 km/L)
- 2$^{nd}$ place hydrogen category SEM Europe London– 1058 km/m$^3$ (3445 km/L)

However, during the SEM it is possible to win some off track awards and from 2007 to now the Team $H_2$politO has been obtained:

- 5 Communication Award: 2009, 2010, 2012, 2014, 2017.
- 3 Design Award: 2010 (with IDRA), 2011 (with XAM) and 2016 (with IDRAkronos).

For the Team, winning a Design Award means having done a good project from an overall point of view (from vehicle dynamics to aerodynamics, from mechanical to electronic design).While to win the Communication Award does it mean to have made a very good work to involve other students and companies.

However, the most important victory is obtained on track during the race, where the students must be able to solve the problems that arise even in stressful conditions, it may be in their future career.

## 6 The Results for the Students at the Politecnico

The Politecnico of Torino is the only University in Italy that recognize the Team activity in terms of credits. In fact, after a period of work inside the Team (approximately 1 years) for the students, in agreement with the Faculty Advisor, is possible to obtain:

- 6 ECTS credit with "Team learning activities in student" instead of a choice exam during the bachelor or master degree.

However, if the student stays inside the Team more than one year has the possibility to "transform" their work in a thesis, to obtain the important passage of knowledge "student by student". Until now inside the Team 200 thesis have been developed, and this is a very innovative educational results. In fact, they allow the realization of the thesis cluster allowing the important transmission of knowledge, year after year, among students.

Finally yet importantly, when the thesis has a good level of scientific interest a paper will be written and the student becomes one of the co-author [3 ÷ 19].

Until now 4 students made bachelor thesis and master thesis inside the Team and after start doing and finish a phD. More than 40 students have been obtained a research grant inside the Politecnico after the degree or during the master.

## 7 Conclusions

The Team has always believed that hard work is the basis for future success. Students crave for continuously improving and strive for exceeding expectations by nurturing the team spirit in order to create those synergies able to add value to individual performances and capabilities. Therefore, passion and team spirit are really the foundation of $H_2$politO values.

The work at the Team and the prototype can be perceived as a business laboratory where is required a synergy between technical knowledge and managerial skills. Therefore, $H_2$politO allowed developing both technical competencies strictly linked to the Prototype/Product and managerial competences characteristics of a business team.

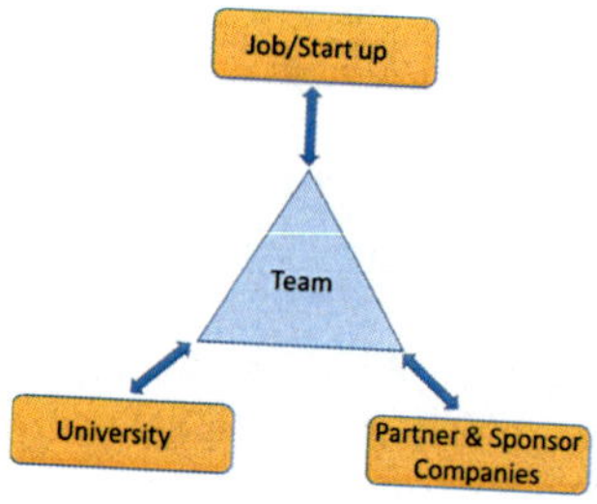


**Fig. 7.** Innovative rule of team $H_2$politO

The University believes that the competition is a way to achieve excellence and grow up a new generation of engineers, able to solve quickly probolems anchìd challenge.

The market, thus the competition, gives them the possibility to work hard in order to challenge the competitors and produce a more advanced technology vehicle that could become benchmark for its category.

The Team wants, through the University to drive innovation into everything it does (Fig. 7). It works to become a mean by which partners and sponsors could create and develop innovative path to prepare the future engineers at the job, in a company or to create new innovative start up but with a "open mind" with a transversal knowledge and approach.

**Acknowledgements.** The author wish to acknowledge to all the students (approximately 450) that have been part of Team $H_2$politO from 2007 to now, without all of them this was not possible.

## References

1. Shell Eco-marathon rules – Chapter 1 and Chapter 2. https://www.makethefuture.shell/en-gb/shell-eco-marathon/global-rules
2. Carello, M.: Innovative and multidisciplinary teaching through the design and construction of low consumption vehicles for international competitions. Mechanisms and machine science - new trend in educational activity in the field of mechanism and machine theory, Springer Netherlands. In: 2nd International Symposium on the Education in Mechanism and Machine Science, Madrid 23–24 November 2017, p. 8, Vol. 64 (2019), ISBN: 978-3-030-00107-0, https://doi.org/10.1007/978-3-030-00108-7_9
3. Messana, A., et al.: Design, optimization and manufacturing of an aluminum wheel rim for the IDRAkronos vehicle prototype. In: Computer-Aided Design and Applications, CAD Solutions, LLC, p. 9, vol. 16 (2019), ISSN:1686-4360. https://doi.org/10.14733/cadaps.2019.733-741
4. Carello, M., Bertipaglia, A., Messana, A., Airale, A.G., Sisca, L.: Modeling and optimization of the consumption of a three-wheeled vehicle. In: SAE Technical Paper, SAE International, SAE World Congress Experience 9–11 April 2019, WCX 2019, p. 9. Detroit (2019). ISSN:0148–7191, https://doi.org/10.4271/2019-01-0164
5. Ferraris, A., et al.: Nafion tubing humidification system for polymer electrolyte membrane fuel cells. Energies, MDPI AG, p. 16, vol. 12 (2019). ISSN:1996-1073, https://doi.org/10.3390/en12091773
6. Ferraris, A., Messana, A., Multari, D., Sisca, L., Airale, A.G., Carello, M.: Steering system of a low-consumption vehicle: from the dynamics analysis to the design of the wheel assembly. In: Carbone, G., Gasparetto, A. (eds.) IFToMM ITALY 2018. MMS, vol. 68, pp. 91–99. Springer, Cham (2019). https://doi.org/10.1007/978-3-030-03320-0_10
7. Messana, A.: From design to manufacture of a carbon fiber monocoque for a three-wheeler vehicle prototype. Materials, MDPI AG, p. 11, vol. 12 (2019). ISSN: 1996-1944, https://doi.org/10.3390/ma12030332
8. Ferraris, A.G., Airale, A.M., Xu, S., Carello, M.: The regenerative braking for a l7e range extender hybrid vehicle. In: IEEE International Conference on Environmental and Electrical Engineering and IEEE Industrial and Commercial Power Systems Europe, EEEIC/I and CPS Europe, Palermo, 12–15 June 2018, ISBN: 9781-5386-51858, 1109/EEEIC.2018.8494000
9. Ferraris, A., Xu, S., Airale, A.G., Carello, M.: Design and optimisation of XAM 2.0 plug-in powertrain. Int. J. Vehi. Perform. Indersci. **3**, 25 (2017). ISSN: 1745-3208, https://doi.org/10.1504/IJVP.2017.10004910

10. Carello, M., Messana, A.:IDRApegasus: a fuel-cell prototype for 3000 km/L. Comput.-Aided Des. Appl. CAD Solut. LLC **11**(a), 5 (2015), ISSN:1686-4360.
11. Carello, M., De Vita, A., Ferraris, A.: Method for Increasing the Humidity in Polymer Electrolyte Membrane Fuel Cell, p. 8. Wiley-Vch Verlag GmbH & Co. KGaA, Weinheim, (2016). ISSN:1615-6854, https://doi.org/10.1002/fuce.201500110
12. Carello, M., Airale, A., Messana, A.: IDRApegasus: a carbon fiber monocoque vehicle prototype. Mater. Sci. Eng. Tech. **45**(5), 307–405 (2014). ISSN:0933-5137
13. Carello, M., Airale, A.G., Messana, A.: IDRApegasus: a carbon fiber monocoque vehicle prototype. Materialwissenschaft Und Werkstofftechnik **45**, 9 (2014). ISSN:0933-5137, https://doi.org/10.1002/mawe.201400238
14. Carello, M., Airale, A., Ferraris, A., Messana, A.:XAM 2.0: from student competition to professional challenge. Comput.-Aid. Des. Appl. **11**(S1), S61–67, (2014). https://doi.org/10.1080/16864360.2014
15. Carello, M., Airale, A.G.: Composite suspension arm optimization for the city vehicle XAM 2.0. In: Öchsner, A., Altenbach, H. (eds.) Design and Computation of Modern Engineering Materials. ASM, vol. 54, pp. 257–272. Springer, Cham (2014). https://doi.org/10.1007/978-3-319-07383-5_18
16. Carello, M., Filippo, N., Marcello, A., D'Auria, M.: Optimization of IDRApegasus: fuel cell hydrogen vehicle. In: SAE International Conference, Detroit, pp. 16–18, April 2013, 4271/2013-01-0964
17. Carello, M., Bonansea, P., D'Auria, M.: Driveline Optimization for a Hybrid Electric Vehicle to Minimize Fuel Consumption. In: SAE International Conference, Detroit, 08–10 April 2014. https://doi.org/10.4271/2014-01-1090
18. Carello, M., Ferraris, A., Airale, A., Fuentes, F.: City vehicle XAM 2.0: design and optimization of its Plug-In E_REV Powertrain. In: SAE International Conference, Detroit, 08–10 April 2014. https://doi.org/10.4271/2014-01-1822
19. Carello, M., Ferraris, A., Airale, A.: City vehicle XAM 2.0: design and optimization of the composite suspension system. In: SAE International Conference, Detroit, 08–10 April 2014. https://doi.org/10.4271/2014-01-1050

# Author Index




G. Quaglia et al. (Eds.): I4SDG 2021, MMS 108, pp. 651–653, 2022.
https://doi.org/10.1007/978-3-030-87383-7